Heat Capacities and Entropies of Organic Compounds in the Condensed Phase

Journal of
Physical and Chemical Reference Data

David R. Lide, Jr., Editor

The Journal of Physical and Chemical Reference Data (ISSN 0047-2689) is published quarterly by the American Chemical Society (1155 16th St., N. W., Washington, DC 20036) and the American Institute of Physics (335 E. 45th St., New York, NY 10017) for the National Bureau of Standards. Second-class postage paid at Washington, DC and additional mailing offices. POSTMASTER: Send address changes to Membership and Subscription Services, P. O. Box 3337, Columbus, Ohio 43210.

The objective of the Journal is to provide critically evaluated physical and chemical property data, fully documented as to the original sources and the criteria used for evaluation. Critical reviews of measurement techniques, whose aim is to assess the accuracy of available data in a given technical area, are also included. The Journal is not intended as a publication outlet for original experimental measurements such as are normally reported in the primary research literature, nor for review articles of a descriptive or primarily theoretical nature.

Supplements to the Journal are published at irregular intervals and are not included in subscriptions to the Journal. They contain compilations which are too lengthy for a journal format.

The Editor welcomes appropriate manuscripts for consideration by the Editorial Board. Potential contributors who are interested in preparing a compilation are invited to submit an outline of the nature and scope of the proposed compilation, with criteria for evaluation of the data and other pertinent factors, to:

David R. Lide, Jr., Editor
J. Phys. Chem. Ref. Data
National Bureau of Standards
Washington, D.C. 20234

One source of contributions to the Journal is The National Standard Reference Data System (NSRDS), which was established in 1963 as a means of coordinating on a national scale the production and dissemination of critically evaluated reference data in the physical sciences. Under the Standard Reference Data Act (Public Law 90-396) the National Bureau of Standards of the U.S. Department of Commerce has the primary responsibility in the Federal Government for providing reliable scientific and technical reference data. The Office of Standard Reference Data of NBS coordinates a complex of data evaluation centers, located in university, industrial, and other Government laboratories as well as within the National Bureau of Standards, which are engaged in the compilation and critical properties retrieved from the world scientific literature. The participants in this NBS-sponsored program, together with similar groups under private or other Government support which are pursuing the same ends, comprise the National Standard Reference Data System.

The primary focus of the NSRDS is on well-defined physical and chemical properties of well-characterized materials or systems. An effort is made to assess the accuracy of data reported in the primary research literature and to prepare compilations of critically evaluated data which will serve as reliable and convenient reference sources for the scientific and technical community.

Information for Contributors

Manuscripts submitted for publication must be prepared in accordance with *Instructions for Preparation of Manuscripts for the Journal of Physical and Chemical Reference Data,* available on request from the Editor.

New and renewal subscriptions should be sent with payment to the Office of the Controller at the American Chemical Society, 1155 Sixteenth Street, N.W., Washington, DC 20036. **Address changes**, with at least six weeks advance notice, should be sent to Manager, Membership and Subscription Services, American Chemical Society, P.O. Box 3337, Columbus, OH 43210. Changes of address must include both old and new addresses and ZIP codes and, if possible, the address label from the mailing wrapper of a recent issue. Claims for missing numbers will not be allowed: if loss was due to failure of the change-of-address notice to be received in the time specified; if claim is dated (a) North America: more than 90 days beyond issue date, (b) all other foreign: more than one year beyond issue date.

Members of AIP member and affiliate societies requesting member subscription rates should direct subscriptions, renewals, and address changes to American Institute of Physics, Dept. S/F, 335 E. 45th St., NY 10017.

Subscription Prices (1984) *(not including supplements)*	U.S.A.	Foreign (surface mail)	Optional air freight Europe Mideast N. Africa	Asia and Oceania
Members (of ACS or of AIP member of affiliated society)	$ 48.00	$ 55.00	$ 68.00	$ 77.00
Regular rate	$190.00	$197.00	$210.00	$219.00

Rates above do not apply to nonmember subscribers in Japan, who must enter subscription orders with Maruzen Company Ltd., 3-10 Nihonbashi 2-chome, Chuo-ku, Tokyo 103, Japan. Tel: (03) 272-7211.

Back numbers are available at a cost of $55 per single copy and $220 per volume.

Orders for reprints, supplements, and back numbers should be addressed to the American Chemical Society, 1155 Sixteenth Street, N. W., Washington, DC 20036. Prices for reprints and supplements are listed at the end of this issue.

Microfilm subscriptions of the *Journal of Physical and Chemical Reference Data* are available on 16 mm and 35 mm. This journal also appears in Sec. I of *Current Physics Microform* (CPM) along with 26 other journals published by the American Institute of Physics and its member societies. *A Microfilm Catalog* is available on request.

Journal of
**Physical and
Chemical
Reference Data**

Volume 13, 1984
Supplement No. 1

Heat Capacities and Entropies of Organic Compounds in the Condensed Phase

Eugene S. Domalski, William H. Evans, and Elizabeth D. Hearing

*Chemical Thermodynamics Data Center,
Chemical Thermodynamics Division,
Center for Chemical Physics,
National Bureau of Standards, Washington, DC 20234*

Published by the **American Chemical Society** and the **American Institute of Physics** for the **National Bureau of Standards**

Library of Congress Catalog Card Number 84-70909

International Standard Book Number
0-88318-447-8

American Institute of Physics, Inc.

335 East 45th Street

New York, New York 10017

Printed in the United States of America

Foreword

The *Journal of Physical and Chemical Reference Data* is published jointly by the American Institute of Physics and the American Chemical Society for the National Bureau of Standards. Its objective is to provide critically evaluated physical and chemical property data, fully documented as to the original sources and the criteria used for evaluation. One of the principal sources of material for the journal is the National Standard Reference Data System (NSRDS), a program coordinated by NBS for the purpose of promoting the compilation and critical evaluation of property data.

The regular issues of the *Journal of Physical and Chemical Reference Data* are published quarterly and contain compilations and critical data reviews of moderate length. Longer monographs, volumes of collected tables, and other material unsuited to a periodical format are published separately as *Supplements* to the *Journal*. This monograph, "Heat Capacities and Entropies of Organic Compounds in the Condensed Phase" by Eugene S. Domalski, William H. Evans, and Elizabeth D. Hearing is presented as a Supplement No. 1 to Volume 13 of the *Journal of Physical and Chemical Reference Data*.

David R. Lide, Jr., Editor
Journal of Physical and Chemical Reference Data

Heat Capacities and Entropies of Organic Compounds in the Condensed Phase

Eugene S. Domalski, William H. Evans,[a] and Elizabeth D. Hearing[b]

Chemical Thermodynamics Data Center, Chemical Thermodynamics Division, Center for Chemical Physics, National Bureau of Standards, Washington, DC 20234

Heat capacities and entropies have been compiled for approximately 1400 organic compounds in the liquid and solid phases. Values for the enthalpies and entropies of phase transitions—solid state, fusion, and vaporization—which were encountered as part of this evaluation and tabulation are included. An attempt was made to include articles which were published from about 1925 through most of 1982. Some earlier papers have been included which report data for compounds not otherwise studied. Over 800 references have been examined and evaluated. The data given for each compound in the tabulation are: empirical formula, physical state, reference code, compound name(s), heat capacity, entropy, and, where available, phase-transition data, Wiswesser Line Notation for the compound, formula weight, and a rating which indicates the estimated overall quality of the reported data.

Key words: condensed phase; entropy; evaluated data; heat capacity; organic compounds; phase transitions; WLN.

[a] 14100 Canterbury Lane, Rockville, MD 20853.
[b] 2447 Regina Drive, Clarksburg, MD 20871.

Contents

1. Introduction

This paper provides heat capacity and entropy data on approximately 1400 organic compounds in the liquid and solid phases. Data on the enthalpies and entropies of phase transitions which have been determined from calorimetric measurements are also included. Over 800 references have been examined and evaluated.

The need for such a compilation has been voiced for many years by scientists and engineers in a variety of disciplines. Some of these disciplines are: chemical process engineering, thermodynamic data evaluation, cryogenics, chemical hazard assessment, thermochemistry, resource conservation and recovery, solid state physics, combustion engineering, and hazardous waste disposal.

The information provided in this paper supplements two other NBS publications. The first is NBS Report 10487, Thermodynamic Data for Industrial Incinerators[1]. The contents of this report were incorporated into an American Society of Mechanical Engineers (ASME) publication entitled: "Combustion Fundamentals for Waste Incineration" and have been available from ASME since 1974[2]. The compilative effort contained in NBS Report 10487 was sponsored by the ASME Research Committee on Industrial and Municipal Wastes, and the report lists data on the enthalpies of formation at 298 K for over 1000 organic compounds in the gaseous, liquid, and solid phases.

The second publication is NBSIR 78–1479, Thermodynamic Data for Waste Incineration[3]. The general orientation of this report was to make thermodynamic data available on chemical mixtures, polymers, composite materials, solid wastes, and materials not easily identifiable by a single stoichiometric formula. The orientation of this report was directed at incinerator engineers involved in the disposal of waste materials. A total of 331 materials has been compiled with specific heat capacity and gross heat of combustion being the major property entries. This publication has also been available from ASME under the same title since 1979[4] and was sponsored by the ASME Research Committee on Industrial and Municipal Wastes.

2. Applicability of the Compiled Data

The data reported here have a wide range of applicability. Heat capacity data are needed in the adjustment of reaction enthalpies and entropies to various temperatures of interest in the condensed phase region and in the calculation of heat balances for physical processes. One of the more useful applications of entropy data is through combination with enthalpy data to obtain the corresponding Gibbs energies, ΔG°, and equilibrium constants, K. Information on Gibbs energies allows one to calculate whether a particular reaction will occur spontaneously under specified conditions of temperature, pressure, activities and/or concentrations of reactants and products. At constant temperature and pressure, the standard Gibbs energy change is defined by:

$$\Delta G^\circ = \Delta H^\circ - T\Delta S^\circ ,$$

where ΔH° and ΔS° are the standard enthalpy and standard entropy changes for the reaction under consideration at constant temperature, T. The superscript degree (°) indicates that the substance is in its standard state. For a pure liquid or solid, the standard state is the substance in the condensed phase under a pressure of one atmosphere.

For a process in equilibrium at a given temperature and pressure, the Gibbs energy change is zero. For a process which occurs spontaneously at a given temperature and pressure, the standard Gibbs energy change is negative, i.e., ΔG° is less than zero. The following equation relates the standard Gibbs energy change to the equilibrium constant, i.e.,

$$\Delta G^\circ = -RT\ln K ,$$

where R is the gas constant, T is the absolute temperature, and $\ln K$ is the natural logarithm of the equilibrium constant. This equation provides a means of determining whether a particular process or reaction is thermodynamically possible (spontaneous) under a specified set of conditions.

In general, both the heat capacity and entropy data serve as a reference data-base for estimating and correlating thermodynamic properties for organic compounds for which no experimental data are available.

3. Scope of the Search

References containing heat capacity data for organic compounds were located by a search of the files of the Chemical Thermodynamics Data Center at the National Bureau of Standards; additional references were located through the Bulletin of Chemical Thermodynamics[5], the International Critical Tables[6], the Landolt-Börnstein Physikalisch-Chemische Tabellen[7], and volume 6, and its supplement, of Thermophysical Properties of Matter[8,9].

The original papers were examined to obtain the data given here, to determine what corrections should be applied, and to evaluate qualitatively the reported measurements.

An attempt has been made to include all significant papers published from about 1925 through most of 1982; in addition, a few earlier papers which report data for compounds not otherwise studied have been included. References 6 and 7 above contain most of the literature prior to 1925 and can be consulted by the reader interested in the historical development of measurements and data on the heat capacities of organic compounds.

The aim of the search was to obtain heat capacity and entropy data for organic compounds in the condensed phase at "room temperature"; however the temperature range included is about 200–450 K. This extended range was chosen so that the user would have, where possible, values for temperatures of interest, even if "room temperature" was not included. This would also allow the reader to extrapolate to temperatures outside of the reported range.

In addition to the heat capacity and entropy data, values of the enthalpy and entropy of phase transitions - solid state, fusion, vaporization, and sublimation - obtained from calorimetric measurements are included. No specific search for these phase transition properties was made; the data reported were obtained as a by-product of the heat capacity search .

Corrections for relative atomic mass (atomic weight), temperature scale, and energy units have been made, where appropriate. Values reported at "298 K" with the ice point taken as 273.1, 273.15, or 273.16 K are assumed to be at 298.15 K; the correction for this small change is much less than the precision and accuracy of the data. Correction for the energy units in most measurements made since about 1930 is likewise within the uncertainty of the data and has usually been omitted. Older data are of a lower precision, so that the correction is not needed. Transition temperatures are, in general, those reported by the investigator. It was felt that the effort necessary to convert each investigator's temperature scale to the present scale was not warranted. Thus, the reported values may have a systematic error of up to 0.1 K. Fortunately, the modern high-precision measurements are usually based on the current 1968 International Practical Temperature Scale.

4. Arrangement of the Data

The table of heat capacities, entropies, and phase transition properties given here contains the data entries for the various compounds. The entries in the table are arranged in the order of the empirical formulae of the compounds; isomers are further separated by their Wiswesser Line Notation.

Under a given chemical species the data from the pertinent papers are included. The data from each paper form a separate entry, complete with identification of the source. When there are several entries for a compound, they are arranged chronologically by year.

For each entry the data given are: compound empirical formula, physical state, reference code, compound name(s), followed by values for the heat capacity, entropy, and, where available, phase-transition data. The entry is completed by the formula weight, the Wiswesser Line Notation for the compound, and a graduated indication of the quality of the data.

The formula given is the empirical formula for the compound; water of hydration is shown as $\cdot(n)H_2O$. The elements are arranged in the order C, H(D, T), followed by the other elements in alphabetical order of their chemical symbols.

One or more names are given for each compound. No attempt has been made to adhere to a rigorously systematic nomenclature. Common names and systematic names are used; alternate names have been given freely. All names used appear in the Compound Name-Formula Index in section 9, which should assist the reader who is aware of the compound name but not its empirical formula.

The bibliography is given in Section 10. The reference code is of the form XXAAA/BBB N where XX are the last two digits of the year of publication of the paper, AAA the first three letters of the last name of the first author, and BBB those of the last name of the second author (if present). Authors after the first two are disregarded. N is a digit from 2 to 9 used to indicate a second, third,...... paper with the same year and author codes. Thus, 60BRO/SMI 2 refers to a paper by Brown and Smith appearing in 1960, the second one with authors BRO... and SMI... 45PIT is a 1945 paper by Pitzer. The full citation appears in the bibliography, arranged according to the reference codes. For years before 1900 the reference codes are in bold face type.

Where the authors have given a table of smoothed values for the heat capacity, the value at 298 K (interpolated if necessary) or the value nearest to that temperature is given. If the experimental measurements are represented only by a smoothing equation, this is used to calculate the value given. If only the unsmoothed experimental results are given by the authors, one of these is given, with the appropriate temperature. Such a selection is accompanied by a remark.

The third-law entropy is given at 298 K or at the temperature closest to this temperature. The value is that obtained by the authors; we have not reintegrated the heat capacity data to reevaluate the entropy.

Phases are indicated by g, liq, c,I, c,II, etc. In general, no attempt has been made to specify the crystalline form of the solid phases; c,I, is used for the form stable at the normal melting point. For each phase transition, the appropriate process, i.e., c/liq, the temperature in kelvins, the enthalpy and entropy change for the isothermal process, and, where appropriate, the pressure, are given. The entropy change ΔS is taken as $\Delta H/T$ unless indicated otherwise.

Energy values are given in both calories and joules; the calorie is defined as 4.184 J. As most of the data are reported in calories, the value in joules is obtained by multiplying by 4.184; usually the same number of decimal places is retained. Occasionally, rounding errors occur and the conversion from calories to joules (or vice versa) is not exact. Pressures are given in kilopascals; one standard atmosphere is taken as 101.325 kPa.

The Wiswesser Line Notation is used to represent the structure of the compound[10].

The formula weight is based upon the 1979 IUPAC Table of Atomic Weights[11]. When the formula weight differs from that originally used by the authors, appropriate corrections to the values have been made.

An indication of our general evaluation of the data reported is given as A (high quality), B (good), C (average), D (low quality). This rating is based upon the method used, the details of the measurements as reported, the number of measurements, purity of the sample, calibrations, and corrections applied to the data; it is intended as a guide to those data we feel are more reliable. In addition, the number of significant figures given for the numerical values indicates roughly the quality of the data. In general, papers which are rated as being of high quality provide a detailed description of the cryostat used, the experimental procedure, the purity and characterization of the sample, calibration results, both raw and smoothed data for the temperature range over which measurements were made, and comment on the precision or accuracy of their data. An absence of numerical or descriptive information or poor agreement with a detailed and accurate study can lead to a low rating.

All of the names used to identify the compounds are included in the Compound Name–Formula Index with the appropriate empirical formulae. Prefixes, such as tert-, ortho-, α-, 1,2- (but not Iso) are disregarded in the alphabetization of the names.

The sequencing of the compounds is based on the empirical formula. The formulae are sorted alphabetically by the first atomic symbol, then by the number of atoms of this element present (the Hill Indexing System[12]). In this compilation C, carbon, is always the first element. This arranged list of formulae is then sorted by the second atomic symbol (H, hydrogen, if present) and then by the number of atoms of this element. The sorting proceeds alphabetically thereafter for each element present. The following list illustrates this scheme:

C
CCl_4
$CHCl_3$
CH_4
CO_2
$C_2HCl_3O_2$
$C_2H_2Cl_2$
C_2H_3Cl
C_2H_4
$C_2H_4O_2$
C_2H_6
C_2H_6O

Isomeric compounds are further sorted by their Wiswesser Line Notation:

$C_4H_{10}O$	2O2	$C_2H_5OC_2H_5$
	3O1	$C_3H_7OCH_3$
	Q1Y1&1	$(CH_3)_2CHCH_2OH$
	QX1&1&1	$(CH_3)_3COH$
	QY2&1	$C_2H_5CH(CH_3)OH$

5. Definitions

Heat Capacity. The heat capacity may be defined as the ratio of the energy supplied to a system to the temperature change in the system. The heat capacity may be an average value over a temperature range or the limiting ratio over an infinitesmal temperature change. If the system is maintained at constant volume, the heat capacity, C_v, is the ratio of the change in internal energy with temperature,

$$C_v = (\partial U/\partial T)_v .$$

If the system is maintained at constant pressure, the heat capacity, C_p, is the ratio of the change in enthalpy with temperature,

$$C_p = (\partial H/\partial T)_p .$$

The heat capacity reported in this compilation is the value at constant pressure and is that corresponding to a gram formula weight (or mole) of a specified substance. In simple terms, it is sometimes described as the amount of energy needed to change a gram-formula weight of substance by one kelvin.

Experimentally, the heat capacity is obtained by measuring the enthalpy change over a small temperature change and is referred to the midpoint of the temperature range: $C_p = \Delta H/(T_2-T_1)$, $(T_2-T_1)/2$. The measurements are usually made with the sample under its own vapor pressure; the correction to the standard pressure, 101.325 kPa (1 atm) is usually negligible for solids and liquids below their boiling point. In some papers, data are reported for mean temperatures over a large temperature range; these are noted as "mean temperature" and the temperature given refers to the midpoint of the experiments nearest to 25 °C.

Entropy. The entropy, S, of a system is a measure of its randomness and is related to the number of possible available states of energy for a collection of atoms or molecules. The number of available energy states and thus the amount of randomness increases with temperature and the entropy of a substance at constant volume or constant pressure will be greater at a high temperature than at a low temperature. If the number of available states for a group of atoms or molecules is greater in one configuration than in another, the entropy will be greater.

Entropy and heat capacity are related mathematically by the following expressions:

at constant volume,

$$dS_v = (C_v/T)dT, \text{ and}$$

at constant pressure,

$$dS_p = (C_p/T)dT.$$

The (calorimetric) entropy is obtained by integration of the measured values of C_p/T from the lowest temperature of measurement to the reported temperature. Various extrapolation methods have been used to extrapolate from the lowest experimental temperature to zero kelvin. Appropriate values of the entropies of phase changes are added. The entropy at zero kelvin is taken as zero for the stable crystalline state, with the addition of residual (zero point) entropy, not removed by the extrapolation, due to random ordering, optical isomerization, multiple electronic ground states for the molecule, etc. Thus, $S^\circ_T = S^\circ_0$ (zero point) + $\int_0^{T_1} (C_p/T)\mathrm{d}T$ (extrapolation) + $\int_{T_1}^{T_2} (C_p/T)\mathrm{d}T + \Delta H_{T_2}/T_2$ (phase change) + $\int_{T_2}^{T_3} (C_p/T)\mathrm{d}T + \Delta H_{T_3}/T_3$ (phase change) + $\int_{T_3}^{T} (C_p/T)\mathrm{d}T$.

Phase Transitions. A process by which a substance undergoes a change of physical state, i.e., solid-solid, solid-liquid, solid-gas, or liquid-gas, is known as a phase transition or phase change. The phase change is accompanied by a transfer of energy (commonly referred to as latent heat) and a change in volume while both temperature and pressure remain constant. For a phase change which is carried out reversibly (i.e., under equilibrium conditions) at a constant temperature and pressure, the total Gibbs energy for a given mass of substance remains unchanged. If there is an enthalpy (or heat) change, then it follows that there will also be an entropy change for the process, because: $\Delta H - T\Delta S = 0$, or $\Delta S = \Delta H/T$. It should be noted that these equations are applicable only for the temperature and pressure at which the two given phases are in equilibrium.

For phase changes - solid-solid transition, fusion, sublimation, vaporization - encountered in the table in section 8, ΔH refers to the isothermal enthalpy change at the transition temperature. Corrections can be applied for premelting effects to reduce the observed measurements to the isothermal state. The pressure, unless specified, is that of the substance at the transition temperature; the correction to standard pressure is usually negligible at ordinary pressures for solid transitions and fusions. The entropy is taken as $\Delta H/T$ and refers to the phases at the equilibrium pressure; this restriction is significant only for gases. Exceptions are indicated in the table in section 8.

Some investigators have reported the measurement of anomalous phase changes in which the volume and entropy are continuous, but the heat capacity is discontinuous. During such phase changes no latent heat is present and the shape of the curve of the heat capacity plotted as a function of temperature often resembles the Greek letter lambda at the transition point and is called a "lambda transition". In order to differentiate these anomalous transitions from ordinary phase changes, it has become customary to identify ordinary phase changes as phase changes of the first order and atypical phase changes as those of the second order. The discontinuity which occurs in a first order transition is a commonly observed phenomenon, however, the discontinuity associated with a second order transition has been more difficult to identify and/or interpret. Sometimes the discontinuous nature of the heat capacity is questioned in a second order transition because experimental measurements show a peak or hump at the transition temperature rather than an unambiguous discontinuity.

A phase change which is accompanied by changes in entropy and volume and whose first-order derivatives of the Gibbs energy with respect to temperature and pressure change discontinuously is known as a phase change of the first order,

$$S = -(\partial G/\partial T)_P\,,$$

$$V = (\partial G/\partial P)_T\,.$$

A phase change which is accompanied by changes in the heat capacity, volume expansivity, and isothermal compressibility and whose second-order derivatives of the Gibbs energy with respect to temperature and pressure change discontinuously is known as a phase change of the second order,

$$C_p/T = (\partial S/\partial T)_P = -(\partial^2 G/\partial T^2)_P\,,$$

$$\kappa V = -(\partial V/\partial P)_T = -(\partial^2 G/\partial P^2)_T\,,$$

$$\beta V = (\partial V/\partial T)_P = (\partial^2 G/\partial T \partial P)_{T,P}\,,$$

where κ is the molar isothermal compressibility and β is the molar volume expansivity. The relationship between these parameters is given below by Ehrenfest's equations[13]. For one mole of substance:

$$\mathrm{d}P/\mathrm{d}T = [C_p(\mathrm{f}) - C_p(\mathrm{i})]/[TV(\beta(\mathrm{f}) - \beta(\mathrm{i}))]\,,$$

$$\mathrm{d}P/\mathrm{d}T = [\beta(\mathrm{f}) - \beta(\mathrm{i})]/[\kappa(\mathrm{f}) - \kappa(\mathrm{i})]\,,$$

where i and f represent the initial and final states of the phase change.

6. Acknowledgments

The authors thank Elton Caviness, Joyce J. Grimes, Ruby N. Boyd, and Jennifer C. Colbert for their efforts in the early stages of this work in the NBS Chemical Thermodynamics Data Center. We also express our thanks to William J. Wiswesser for proof-reading the individual Wiswesser Line Notations for all of the organic compounds for which data have been tabulated, to Nancy W. Young for inputting the compiled data into the computer for machine manipulation and tabulation purposes, and to Carla G. Messina and Constance L. Seymour for their help in merging computer files, editing, and introducing data-format modifications. Financial support for this effort is acknowledged from the NBS Office of Standard Reference Data.

7. References for the Introductory Discussion

[1]G. T. Armstrong and E. S. Domalski, *Thermodynamic Data for Industrial Incinerators,* NBS Report 10487 (May 1972).

[2]ASME Research Committee on Industrial and Municipal Wastes, *Combustion Fundamentals for Waste Incineration,* 212 pp., American Society of Mechanical Engineers, New York, NY (1974).

[3]E. S. Domalski, W. H. Evans, and T. L. Jobe, Jr., *Thermodynamic Data for Waste Incineration,* NBSIR 78-1479 (August 1978).

[4]ASME Research Committee on Industrial and Municipal Wastes, *Thermodynamic Data for Waste Incineration,* 160 pp., American Society of Mechanical Engineers, New York, NY (1979).

[5]R. D. Freeman, Editor, *Bulletin of Chemical Thermodynamics,* Nos. 20–23, (1977–1980); E. F. Westrum, Jr., Editor, *Bulletin of Thermodynamics and Thermochemistry,* Nos. 6–19 (1963–1976).

[6]E. W. Washburn, Editor-in-Chief, *International Critical Tables of Numerical Data Physics, Chemistry and Technology,* **5**, McGraw–Hill Book Co., New York, NY (1929).

[7]Landolt-Börnstein, *Zahlenwerte und Funktionen aus Physik, Chemie, Astronomie, Geophysik und Technik, Sechste Auflage,* unter vorbereitender Mitwirkung von J. D'Ans, A. Eucken, G. Joos, W. A. Roth, herausgegeben von J. Bartels, H. Borchers, P. Ten Bruggencate, H. Hausen, K.H. Hellwege, Kl. Schäfer, E. Schmidt, II. Band, Eigenschaften der Materie in Ihren Aggregatzuständen, 4. Teil, Kalorische Zustandsgrössen, herausgegeben von Kl. Schäfer, and E. Lax, Springer-Verlag, Berlin-Gottingen-Heidelberg, (1961).

[8]Y. S. Touloukian and T. Makita, *Specific Heat of Nonmetallic Liquids and Gases,* **6**, in *Thermophysical Properties of Matter,* Plenum Press, New York, NY (1970).

[9]Y. S. Touloukian and T. Makita, *Specific Heat of Nonmetallic Liquids and Gases* in *Thermophysical Properties of Matter,* Supplement to **6**, Plenum Press, New York, NY (1976).

[10]E. G. Smith and P. A. Baker, Editors, *The Wiswesser Line-Formula Notation,* 3rd Edition, Chemical Information Management, Inc. Cherry Hill, NJ (1975).

[11]N. E. Holden, Pure App. Chem. **52**, 2349–2384 (1980).

[12]E. A. Hill, J. Amer. Chem. Soc. **22**, 478–494 (1900).

[13]P. Ehrenfest, Proc. Amsterdam Acad. **36**, 153–157 (1933).

8. Table of Heat Capacities, Entropies, and Phase Transition Properties

C (c) 34JAC/PAR
Graphite; Carbon, graphite
Heat Capacity 293.5 K, C_p = 2.031 cal·mol^{-1}·K^{-1}
8.498 J·mol^{-1}·K^{-1}
Temperature range 93–294 K. Value is unsmoothed experimental datum.
Entropy 298.1 K, S = 1.36 cal·mol^{-1}·K^{-1}
5.69 J·mol^{-1}·K^{-1}
Extrapolation below 90 K, 0.182 cal·mol^{-1}·K^{-1}.
Molecular Weight 12.0110
Wiswesser Line Notation C
Evaluation A(C_p),B(S)
Acheson No. 38 graphite

C (c) 38PIT
Diamond; Carbon, diamond
Heat Capacity 287.96 K, C_p = 1.355 cal·mol^{-1}·K^{-1}
5.669 J·mol^{-1}·K^{-1}
Temperature range 70–288 K. Value is unsmoothed experimental datum.
Entropy 298.15 K, S = 0.585 cal·mol^{-1}·K^{-1}
2.448 J·mol^{-1}·K^{-1}
Extrapolation below 70.8 K, 0.0077 cal·mol^{-1}·K^{-1}.
Molecular Weight 12.0110
Wiswesser Line Notation C
Evaluation A(C_p),B(S)

C (c) 53DES/TYL
Graphite, Acheson; Carbon, graphite
Heat Capacity 298.15 K, C_p = 2.038 cal·mol^{-1}·K^{-1}
8.527 J·mol^{-1}·K^{-1}
Temperature range 13–300 K
Entropy 298.15 K, S = 1.372 cal·mol^{-1}·K^{-1}
5.740 J·mol^{-1}·K^{-1}
Molecular Weight 12.0110
Wiswesser Line Notation C
Evaluation A

C (c) 55DES
Graphite; Carbon, graphite
Heat Capacity 298.15 K, C_p = 1.874 cal·mol^{-1}·K^{-1}
7.841 J·mol^{-1}·K^{-1}
Temperature range 17–300 K
Entropy 298.15 K, S = 1.2895 cal·mol^{-1}·K^{-1}
5.3953 J·mol^{-1}·K^{-1}
Molecular Weight 12.0110
Wiswesser Line Notation C
Evaluation A

C (c) 57DES/TYL
Graphite, Acheson, irradiated; Carbon, irradiated graphite
Heat Capacity 298.15 K, C_p = 2.136 cal·mol^{-1}·K^{-1}
8.937 J·mol^{-1}·K^{-1}
Temperature range 13–300 K
Entropy 298.15 K, S = 1.492 cal·mol^{-1}·K^{-1}
6.243 J·mol^{-1}·K^{-1}
Molecular Weight 12.0110
Wiswesser Line Notation C
Evaluation A
Stored energy of about 475 cal g^{-1}.

C (c) 58DES/MOR
Diamond; Carbon, diamond
Heat Capacity 277.68 K, C_p = 1.2686 cal·mol^{-1}·K^{-1}
5.3078 J·mol^{-1}·K^{-1}
Temperature range 13–277 K. Value is unsmoothed experimental datum.
Entropy 100.00 K, S = 0.0172 cal·mol^{-1}·K^{-1}
0.0720 J·mol^{-1}·K^{-1}
Agreement with DeSorbo (1953) data above 100 K.
Molecular Weight 12.0110
Wiswesser Line Notation C
Evaluation A

C (c) 62VIC
Diamond; Carbon, diamond
Heat Capacity 298.15 K, C_p = 1.462 cal·mol^{-1}·K^{-1}
6.117 J·mol^{-1}·K^{-1}
Temperature range 273–1073 K
Molecular Weight 12.0110
Wiswesser Line Notation C
Evaluation A

C (c) 65MCD
Graphite; Carbon, graphite
Heat Capacity 300 K, C_p = 2.05 cal·mol^{-1}·K^{-1}
8.58 J·mol^{-1}·K^{-1}
Temperature range 298–1723 K
Molecular Weight 12.0110
Wiswesser Line Notation C
Evaluation A
Special Spectroscopic Electrode Grade SPK.

C (c) 70LUT/VOL
Graphite; Carbon, graphite
Heat Capacity 298.15 K, C_p = 2.146 cal·mol^{-1}·K^{-1}
8.979 J·mol^{-1}·K^{-1}
Temperature range 57–320 K
Entropy 298.15 K, S = 1.419 cal·mol^{-1}·K^{-1}
5.937 J·mol^{-1}·K^{-1}
T^2 extrapolation below 50 K
Molecular Weight 12.0110
Wiswesser Line Notation C
Evaluation B
Prepared by electrode technique from petroleum coke and coal tar pitch. Heat treated at 3000 °C.

C (c) 70LUT/VOL
Graphite; Carbon, graphite
Heat Capacity 298.15 K, C_p = 2.025 cal·mol^{-1}·K^{-1}
8.473 J·mol^{-1}·K^{-1}
Temperature range 52–315 K
Entropy 298.15 K, S = 1.349 cal·mol^{-1}·K^{-1}
5.644 J·mol^{-1}·K^{-1}
T^2 extrapolation below 50 K
Molecular Weight 12.0110
Wiswesser Line Notation C
Evaluation B
Sample with density 2.0 g·cm^{-3} prepared from petroleum coke and coal tar pitch by heat treatment under pressure at 2400 °C. Density of 2.1 g·cm^{-3} obtained by same process with addition of metal catalysts. Both heat treated at 3000 °C.

Table of Heat Capacities, Entropies, and Phase Transition Properties — Continued

C (c) 70LUT/VOL
Graphite, pyrolytic; Carbon, pyrolytic graphite
Heat Capacity 298.15 K, C_p = 1.925 cal·mol^{-1}·K^{-1}
8.054 J·mol^{-1}·K^{-1}
Temperature range 51–311 K
Entropy 298.15 K, S = 1.298 cal·mol^{-1}·K^{-1}
5.431 J·mol^{-1}·K^{-1}
T^2 extrapolation below 90 K
Molecular Weight 12.0110
Wiswesser Line Notation C
Evaluation B
Prepared by deposition from methane on hot graphite surface (2100 °C). Heat treated at 3000 °C.

C (c) 70LUT/VOL
Graphite, natural Taiguinski; Carbon, natural graphite
Heat Capacity 298.15 K, C_p = 1.925 cal·mol^{-1}·K^{-1}
8.054 J·mol^{-1}·K^{-1}
Temperature range 51–311 K
Entropy 298.15 K, S = 1.298 cal·mol^{-1}·K^{-1}
5.431 J·mol^{-1}·K^{-1}
T^2 extrapolation below 50 K
Molecular Weight 12.0110
Wiswesser Line Notation C
Evaluation B
Values are taken as those of pyrolytic graphite.

C (c) 70LUT/VOL
Carbon, baked; Baked carbon
Heat Capacity 298.15 K, C_p = 2.211 cal·mol^{-1}·K^{-1}
9.251 J·mol^{-1}·K^{-1}
Temperature range 52–302 K
Entropy 298.15 K, S = 1.482 cal·mol^{-1}·K^{-1}
6.201 J·mol^{-1}·K^{-1}
T^2 extrapolation below 50 K
Molecular Weight 12.0110
Wiswesser Line Notation C
Evaluation B

C (amorp) 70TAK/WES
Carbon, glassy; Glassy carbon
Heat Capacity 298.15 K, C_p = 2.055 cal·mol^{-1}·K^{-1}
8.598 J·mol^{-1}·K^{-1}
Temperature range 5–350 K
Entropy 298.15 K, S = 1.406 cal·mol^{-1}·K^{-1}
5.883 J·mol^{-1}·K^{-1}
Values actually S–S_0; there may be a residual entropy.
Molecular Weight 12.0110
Wiswesser Line Notation C
Evaluation A

C (c) 72SHE/BEL
Graphite; Carbon, graphite
Heat Capacity 298 K, C_p = 1.94 cal·mol^{-1}·K^{-1}
8.12 J·mol^{-1}·K^{-1}
Temperature range 273–3650 K, C_p calculated from equation applicable to the temperature range 273–1000 K.
Molecular Weight 12.0110
Wiswesser Line Notation C
Evaluation B

C (c) 80TAY/GRO
Graphite; Carbon, GPCO graphite
Heat Capacity 300 K, C_p = 2.015 cal·mol^{-1}·K^{-1}
8.43 J·mol^{-1}·K^{-1}
Temperature range 300–2400 K
Molecular Weight 12.0110
Wiswesser Line Notation
Evaluation A

$CBrCl_3$ (liq) 59BEN/THO
Bromotrichloromethane
Heat Capacity 298 K, C_p = 35.7 cal·mol^{-1}·K^{-1}
149.4 J·mol^{-1}·K^{-1}
Mean value 25 to 50 °C
Molecular Weight 198.2740
Wiswesser Line Notation GXGGE
Evaluation C

CBr_4 (c) 39FRE/HIL
Tetrabromomethane; Carbon tetrabromide
Heat Capacity 298 K, C_p = 35.0 cal·mol^{-1}·K^{-1}
148.4 J·mol^{-1}·K^{-1}
Temperature range 298–423 K
Phase Changes
c,II/c,I 320.0 K, ΔH = 1420 cal·mol^{-1}
5941 J·mol^{-1}
ΔS = 4.44 cal·mol^{-1}·K^{-1}
18.57 J·mol^{-1}·K^{-1}
c,I/liq 363.2 K, ΔH = 945 cal·mol^{-1}
3954 J·mol^{-1}
ΔS = 2.60 cal·mol^{-1}·K^{-1}
10.89 J·mol^{-1}·K^{-1}
Molecular Weight 331.6270
Wiswesser Line Notation EXEEE
Evaluation B

CBr_4 (c) 56MAR/STA
Tetrabromomethane; Carbon tetrabromide
Heat Capacity 300.6 K, C_p = 30.75 cal·mol^{-1}·K^{-1}
128.66 J·mol^{-1}·K^{-1}
Temperature range 22 to 84 °C
Phase Changes
c,II/c,I 320.0 K, ΔH = 1594 cal·mol^{-1}
6669 J·mol^{-1}
ΔS = 4.98 cal·mol^{-1}·K^{-1}
20.84 J·mol^{-1}·K^{-1}
Molecular Weight 331.6270
Wiswesser Line Notation EXEEE
Evaluation B

CBr_4 (liq) 48KUR
Tetrabromomethane; Carbon tetrabromide
Heat Capacity 373 K, C_p = 38.8 cal·mol^{-1}·K^{-1}
162.3 J·mol^{-1}·K^{-1}
Temperature range 96 to 182 °C, mean C_p, three temperatures.
Molecular Weight 331.6270
Wiswesser Line Notation EXEEE
Evaluation D

CCl_2F_2 (liq) 31BUF/FLE
Dichlorodifluoromethane; Freon 12
Heat Capacity 290 K, C_p = 30.3 cal·mol^{-1}·K^{-1}
126.8 J·mol^{-1}·K^{-1}
Heat capacity measured at 230 K and 290 K using two different methods. C_p at 230 K is 25.4 cal·mol^{-1}·K^{-1}.
Molecular Weight 120.9138
Wiswesser Line Notation GXGFF
Evaluation C

Table of Heat Capacities, Entropies, and Phase Transition Properties — Continued

CCl_2O (liq) 48GIA/JON
Carbonyl chloride; Phosgene
Heat Capacity 280 K, C_p = 24.09 cal·mol⁻¹·K⁻¹
100.79 J·mol⁻¹·K⁻¹
Temperature range 15–280 K
Entropy 280.71 K, S = 47.35 cal·mol⁻¹·K⁻¹
198.11 J·mol⁻¹·K⁻¹

A value of S_0 of 1.63 cal·mol⁻¹·K⁻¹ has been added to the calorimetric value of S_{280}–S_0.

Phase Changes

c/liq	145.37 K,	ΔH =	1371 cal·mol⁻¹
			5736 J·mol⁻¹
		ΔS =	9.43 cal·mol⁻¹·K⁻¹
			39.46 J·mol⁻¹·K⁻¹
liq/g	280.71 K,	ΔH =	5832 cal·mol⁻¹
			24401 J·mol⁻¹
		ΔS =	20.78 cal·mol⁻¹·K⁻¹
			86.93 J·mol⁻¹·K⁻¹
		P =	101.325 kPa

Molecular Weight 98.9164
Wiswesser Line Notation GVG
Evaluation A

CCl_2O (liq) 60GIA/OTT
Carbonyl chloride; Phosgene
Heat Capacity 160.55 K, C_p = 24.62 cal·mol⁻¹·K⁻¹
102.59 J·mol⁻¹·K⁻¹
Temperature range 13–160 K. Value is unsmoothed experimental datum.
Entropy 280.76 K, S = 46.08 cal·mol⁻¹·K⁻¹
192.80 J·mol⁻¹·K⁻¹
Data for liquid from 48GIA/JON

Phase Changes

c,III/liq	139.19 K,	ΔH =	1131 cal·mol⁻¹
			4732 J·mol⁻¹
		ΔS =	8.26 cal·mol⁻¹·K⁻¹
			34.00 J·mol⁻¹·K⁻¹
c,II/liq	142.09 K,	ΔH =	1335 cal·mol⁻¹
			5586 J·mol⁻¹
		ΔS =	9.40 cal·mol⁻¹·K⁻¹
			39.31 J·mol⁻¹·K⁻¹
c,I/liq	145.37 K,	ΔH =	1373 cal·mol⁻¹
			5745 J·mol⁻¹
		ΔS =	9.44 cal·mol⁻¹·K⁻¹
			39.52 J·mol⁻¹·K⁻¹

Molecular Weight 98.9164
Wiswesser Line Notation GVG
Evaluation A

CCl_3F (liq) 40BEN/MCH
Fluorotrichloromethane; Freon 11
Heat Capacity 298.15 K, C_p = 30.28 cal·mol⁻¹·K⁻¹
126.7 J·mol⁻¹·K⁻¹
Temperature range 261–347 K, Data calculated from equation.
Molecular Weight 137.3684
Wiswesser Line Notation GYGGF
Evaluation B

CCl_3F (liq) 41OSB/GAR
Fluorotrichloromethane; Freon 11
Heat Capacity 298.15 K, C_p = 29.05 cal·mol⁻¹·K⁻¹
121.55 J·mol⁻¹·K⁻¹
Temperature range 15–290 K. Value for saturated liquid.
Entropy 298.15 K, S = 53.92 cal·mol⁻¹·K⁻¹
225.60 J·mol⁻¹·K⁻¹
Value for saturated liquid.

Phase Changes

c/liq	162.68 K,	ΔH =	1647.6 cal·mol⁻¹
			6893.6 J·mol⁻¹
		ΔS =	10.13 cal·mol⁻¹·K⁻¹
			42.38 J·mol⁻¹·K⁻¹
liq/g	290.40 K,	ΔH =	6025 cal·mol⁻¹
			25209 J·mol⁻¹
		ΔS =	20.75 cal·mol⁻¹·K⁻¹
			86.81 J·mol⁻¹·K⁻¹
		P =	80.33 kPa

Molecular Weight 137.3684
Wiswesser Line Notation GXGGF
Evaluation A

CCl_4 (c) 71ATA/CHI
Tetrachloromethane; Carbon tetrachloride
Heat Capacity 46 K, C_p = 10.57 cal·mol⁻¹·K⁻¹
44.22 J·mol⁻¹·K⁻¹
Temperature range 3–46 K
Molecular Weight 153.8230
Wiswesser Line Notation GXGGG
Evaluation A

CCl_4 (liq) 22LAT
Tetrachloromethane; Carbon tetrachloride
Heat Capacity 290 K, C_p = 32.0 cal·mol⁻¹·K⁻¹
133.9 J·mol⁻¹·K⁻¹
Temperature range 39.1–290 K
Entropy 298 K, S = 49.1 cal·mol⁻¹·K⁻¹
205.4 J·mol⁻¹·K⁻¹

Phase Changes

c,II/c,I	224.6 K,	ΔH =	1100 cal·mol⁻¹
			4600 J·mol⁻¹
		ΔS =	4.9 cal·mol⁻¹·K⁻¹
			20.5 J·mol⁻¹·K⁻¹
c,I/liq	249 K,	ΔH =	644 cal·mol⁻¹
			2694 J·mol⁻¹
		ΔS =	2.6 cal·mol⁻¹·K⁻¹
			10.8 J·mol⁻¹·K⁻¹

Molecular Weight 153.8230
Wiswesser Line Notation GXGGG
Evaluation B

CCl_4 (liq) 24WIL/DAN
Tetrachloromethane; Carbon tetrachloride
Heat Capacity 303 K, C_p = 30.8 cal·mol⁻¹·K⁻¹
128.9 J·mol⁻¹·K⁻¹
Temperature range 303–330 K, equation only.
Molecular Weight 153.8230
Wiswesser Line Notation GXGGG
Evaluation C

CCl_4 (liq) 25WIL/DAN
Tetrachloromethane; Carbon tetrachloride
Heat Capacity 293.2 K, C_p = 30.6 cal·mol⁻¹·K⁻¹
128.0 J·mol⁻¹·K⁻¹
Temperature range 20 to 50 °C
Molecular Weight 153.8230
Wiswesser Line Notation GXGGG
Evaluation B

Table of Heat Capacities, Entropies, and Phase Transition Properties — Continued

CCl_4 (liq) 32RIC/WAL
Tetrachloromethane; Carbon tetrachloride
Heat Capacity 298.1 K, C_p = 31.2 cal·mol^{-1}·K^{-1}
130.5 J·mol^{-1}·K^{-1}
Temperature range 293–323 K
Molecular Weight 153.8230
Wiswesser Line Notation GXGGG
Evaluation C

CCl_4 (liq) 33 KOL/UDO
Tetrachloromethane; Carbon tetrachloride
Heat Capacity 288.3 K, C_p = 30.2 cal·mol^{-1}·K^{-1}
126.4 J·mol^{-1}·K^{-1}
One temperature
Molecular Weight 153.8230
Wiswesser Line Notation GXGGG
Evaluation C

CCl_4 (liq) 34KOL/UDO 2
Tetrachloromethane; Carbon tetrachloride
Heat Capacity 288.3 K, C_p = 30.2 cal·mol^{-1}·K^{-1}
126.4 J·mol^{-1}·K^{-1}
One temperature
Molecular Weight 153.8230
Wiswesser Line Notation GXGGG
Evaluation C

CCl_4 (liq) 37STU
Tetrachloromethane; Carbon tetrachloride
Heat Capacity 298.1 K, C_p = 31.70 cal·mol^{-1}·K^{-1}
132.63 J·mol^{-1}·K^{-1}
Temperature range 90–320 K
Entropy 298.1 K, S = 52.4 cal·mol^{-1}·K^{-1}
219.2 J·mol^{-1}·K^{-1}
Extrapolation below 91 K; 17.76 cal·mol^{-1}·K^{-1}.
Phase Changes
c,II/c,I 225.63 K, ΔH = 1100 cal·mol^{-1}
4602 J·mol^{-1}
ΔS = 4.88 cal·mol^{-1}·K^{-1}
26.40 J·mol^{-1}·K^{-1}
c,I/liq 250.37 K, ΔH = 581 cal·mol^{-1}
2431 J·mol^{-1}
ΔS = 2.32 cal·mol^{-1}·K^{-1}
9.71 J·mol^{-1}·K^{-1}
Molecular Weight 153.8230
Wiswesser Line Notation GXGGG
Evaluation B(C_p),C(S)

CCl_4 (liq) 37VOL
Tetrachloromethane; Carbon tetrachloride
Heat Capacity 298 K, C_p = 31.8 cal·mol^{-1}·K^{-1}
133.1 J·mol^{-1}·K^{-1}
One temperature
Molecular Weight 153.8230
Wiswesser Line Notation GXGGG
Evaluation B

CCl_4 (liq) 39PHI
Tetrachloromethane; Carbon tetrachloride
Heat Capacity 301.2 K, C_p = 31.8 cal·mol^{-1}·K^{-1}
133.1 J·mol^{-1}·K^{-1}
One temperature
Molecular Weight 153.8230
Wiswesser Line Notation GXGGG
Evaluation C

CCl_4 (liq) 41ZHD
Tetrachloromethane; Carbon tetrachloride
Heat Capacity 298.1 K, C_p = 31.6 cal·mol^{-1}·K^{-1}
132.2 J·mol^{-1}·K^{-1}
Temperature range 5 to 46 °C
Molecular Weight 153.8230
Wiswesser Line Notation GXGGG
Evaluation C

CCl_4 (liq) 44HIC/HOO
Tetrachloromethane; Carbon tetrachloride
Heat Capacity 298.15 K, C_p = 31.47 cal·mol^{-1}·K^{-1}
131.67 J·mol^{-1}·K^{-1}
Temperature range 15–300 K
Entropy 298.15 K, S = 51.25 cal·mol^{-1}·K^{-1}
214.39 J·mol^{-1}·K^{-1}
Phase Changes
c,II/c,I 225.35 K, ΔH = 1095 cal·mol^{-1}
4582 J·mol^{-1}
ΔS = 4.6 cal·mol^{-1}·K^{-1}
20.33 J·mol^{-1}·K^{-1}
c,I/liq 250.3 K, ΔH = 601 cal·mol^{-1}
2515 J·mol^{-1}
ΔS = 2.40 cal·mol^{-1}·K^{-1}
10.05 J·mol^{-1}·K^{-1}
Molecular Weight 153.8230
Wiswesser Line Notation GXGGG
Evaluation A

CCl_4 (liq) 48KUR
Tetrachloromethane; Carbon tetrachloride
Heat Capacity 298 K, C_p = 30.8 cal·mol^{-1}·K^{-1}
128.8 J·mol^{-1}·K^{-1}
Temperature range –20 to 72 °C, mean C_p, four temperatures.
Molecular Weight 153.8230
Wiswesser Line Notation GXGGG
Evaluation D

CCl_4 (liq) 55STA/TUP
Tetrachloromethane; Carbon tetrachloride
Heat Capacity 298 K, C_p = 31.69 cal·mol^{-1}·K^{-1}
132.59 J·mol^{-1}·K^{-1}
Temperature range 295–339 K
Molecular Weight 153.8230
Wiswesser Line Notation GXGGG
Evaluation B

CCl_4 (liq) 57HAR/MOE
Tetrachloromethane; Carbon tetrachloride
Heat Capacity 300 K, C_p = 31.29 cal·mol^{-1}·K^{-1}
130.9 J·mol^{-1}·K^{-1}
Temperature range 243–303 K
Molecular Weight 153.8230
Wiswesser Line Notation GXGGG
Evaluation B

CCl_4 (liq) 67RAS/GAN
Tetrachloromethane; Carbon tetrachloride
Heat Capacity 293 K, C_p = 31.4 cal·mol^{-1}·K^{-1}
131.5 J·mol^{-1}·K^{-1}
Temperature range 293–333 K
Molecular Weight 153.8230
Wiswesser Line Notation GXGGG
Evaluation C

Table of Heat Capacities, Entropies, and Phase Transition Properties — Continued

CCl_4 (liq) 71DES/BHA
Tetrachloromethane; Carbon tetrachloride
Heat Capacity 298 K, C_p = 31.3 cal·mol^{-1}·K^{-1}
131.0 J·mol^{-1}·K^{-1}
Temperature range 298–318 K
Molecular Weight 153.8230
Wiswesser Line Notation GXGGG
Evaluation B

CCl_4 (liq) 72ARE/MIL
Tetrachloromethane; Carbon tetrachloride
Heat Capacity 256.10 K, C_p = 31.5 cal·mol^{-1}·K^{-1}
131.8 J·mol^{-1}·K^{-1}
Temperature range 243–256 K. Value is unsmoothed experimental datum.
Phase Changes
c,II/liq 245.70 K, ΔH = 442 cal·mol^{-1}
1848 J·mol^{-1}
ΔS = 1.80 cal·mol^{-1}·K^{-1}
7.52 J·mol^{-1}·K^{-1}
c,I/liq 250.28 K, ΔH = 611 cal·mol^{-1}
2558 J·mol^{-1}
ΔS = 2.44 cal·mol^{-1}·K^{-1}
10.22 J·mol^{-1}·K^{-1}
stable phase
Molecular Weight 153.8230
Wiswesser Line Notation GXGGG
Evaluation A

CCl_4 (liq) 73SUB/RAS
Tetrachloromethane; Carbon tetrachloride
Heat Capacity 298.15 K, C_p = 31.3 cal·mol^{-1}·K^{-1}
130.8 J·mol^{-1}·K^{-1}
Temperature range 298–323 K
Molecular Weight 153.8230
Wiswesser Line Notation GXGGG
Evaluation B

CCl_4 (liq) 75GRO/BEN
Tetrachloromethane; Carbon tetrachloride
Heat Capacity 298.15 K, C_p = 31.52 cal·mol^{-1}·K^{-1}
131.9 J·mol^{-1}·K^{-1}
One temperature
Molecular Weight 153.8230
Wiswesser Line Notation GXGGG
Evaluation B

CCl_4 (liq) 76FOR/BEN
Tetrachloromethane; Carbon tetrachloride
Heat Capacity 298.15 K, C_p = 31.40 cal·mol^{-1}·K^{-1}
131.36 J·mol^{-1}·K^{-1}
One temperature
Molecular Weight 153.8230
Wiswesser Line Notation GXGGG
Evaluation B

CCl_4 (liq) 79WIL/FAR
Tetrachloromethane; Carbon tetrachloride
Heat Capacity 298.15 K, C_p = 31.40 cal·mol^{-1}·K^{-1}
131.36 J·mol^{-1}·K^{-1}
One temperature.
Molecular Weight 153.8230
Wiswesser Line Notation GXGGG
Evaluation B

CCl_4 (liq) 79GRO/HAM
Tetrachloromethane; Carbon tetrachloride
Heat Capacity 298.15 K, C_p = 31.5 cal·mol^{-1}·K^{-1}
131.6 J·mol^{-1}·K^{-1}
One temperature
Molecular Weight 153.8230
Wiswesser Line Notation GXGGG
Evaluation B

CCl_4 (liq) 82TAN
Tetrachloromethane; Carbon tetrachloride
Heat Capacity 298.15 K, C_p = 31.39 cal·mol^{-1}·K^{-1}
131.34 J·mol^{-1}·K^{-1}
Temperature range 293.15, 298.15, 303.15 K, Data at three temperatures.
Molecular Weight 153.8230
Wiswesser Line Notation GXGGG
Evaluation A

CF_4 (liq) 69SMI/PAC 2
Tetrafluoromethane; Carbon tetrafluoride; Freon 14
Heat Capacity 145 K, C_p = 19.14 cal·mol^{-1}·K^{-1}
80.08 J·mol^{-1}·K^{-1}
Temperature range 12–145 K. See also 69SMI/PAC.
Entropy 145.12 K, S = 34.41 cal·mol^{-1}·K^{-1}
143.97 J·mol^{-1}·K^{-1}
Phase Changes
c,II/c,I 76.27 K, ΔH = 408.5 cal·mol^{-1}
1709.2 J·mol^{-1}
ΔS = 5.36 cal·mol^{-1}·K^{-1}
22.41 J·mol^{-1}·K^{-1}
c,I/liq 89.56 K, ΔH = 170.2 cal·mol^{-1}
712.1 J·mol^{-1}
ΔS = 1.90 cal·mol^{-1}·K^{-1}
7.95 J·mol^{-1}·K^{-1}
liq/g 145.12 K, ΔH = 2823.6 cal·mol^{-1}
11814 J·mol^{-1}
ΔS = 19.46 cal·mol^{-1}·K^{-1}
81.41 J·mol^{-1}·K^{-1}
P = 101.325 kPa
Molecular Weight 88.0046
Wiswesser Line Notation FXFFF
Evaluation A

$CHBr_3$ (liq) 32TRE
Tribromomethane; Bromoform
Heat Capacity 298 K, C_p = 32.3 cal·mol^{-1}·K^{-1}
135.1 J·mol^{-1}·K^{-1}
One temperature
Molecular Weight 252.7309
Wiswesser Line Notation EYEE
Evaluation B

$CHBr_3$ (liq) 48KUR
Tribromomethane; Bromoform
Heat Capacity 298 K, C_p = 31.2 cal·mol^{-1}·K^{-1}
130.5 J·mol^{-1}·K^{-1}
Temperature range 9 to 147 °C, mean C_p, four temperatures.
Molecular Weight 252.7309
Wiswesser Line Notation EYEE
Evaluation D

Table of Heat Capacities, Entropies, and Phase Transition Properties — Continued

$CHClF_2$ (liq) 40BEN/MCH
Chlorodifluoromethane; Freon 22
Heat Capacity 298.15 K, C_p = 27.27 $cal{\cdot}mol^{-1}{\cdot}K^{-1}$
114.10 $J{\cdot}mol^{-1}{\cdot}K^{-1}$
Temperature range 256–328 K, Data calculated from equation.
Molecular Weight 86.4687
Wiswesser Line Notation GYFF
Evaluation B

$CHClF_2$ (liq) 57NEI/WHI
Chlorodifluoromethane; Freon 22
Heat Capacity 232.50 K, C_p = 22.23 $cal{\cdot}mol^{-1}{\cdot}K^{-1}$
93.01 $J{\cdot}mol^{-1}{\cdot}K^{-1}$
Temperature range 16–230 K
Entropy 232.50 K, S = 43.00 $cal{\cdot}mol^{-1}{\cdot}K^{-1}$
179.91 $J{\cdot}mol^{-1}{\cdot}K^{-1}$

Phase Changes
c,II/c,I 59 K, ΔH = 16 $cal{\cdot}mol^{-1}$
67 $J{\cdot}mol^{-1}$
ΔS = 0.27 $cal{\cdot}mol^{-1}{\cdot}K^{-1}$
1.14 $J{\cdot}mol^{-1}{\cdot}K^{-1}$
λ-type transition
c,I/liq 115.73 K, ΔH = 985.5 $cal{\cdot}mol^{-1}$
4123.3 $J{\cdot}mol^{-1}$
ΔS = 8.52 $cal{\cdot}mol^{-1}{\cdot}K^{-1}$
35.63 $J{\cdot}mol^{-1}{\cdot}K^{-1}$
liq/g 232.50 K, ΔH = 4832 $cal{\cdot}mol^{-1}$
20217 $J{\cdot}mol^{-1}$
ΔS = 20.78 $cal{\cdot}mol^{-1}{\cdot}K^{-1}$
86.95 $J{\cdot}mol^{-1}{\cdot}K^{-1}$
P = 101.325 kPa
Molecular Weight 86.4687
Wiswesser Line Notation GYFF
Evaluation A

$CHCl_2F$ (liq) 40BEN/MCH
Dichlorofluoromethane; Freon 21
Heat Capacity 298.15 K, C_p = 26.92 $cal{\cdot}mol^{-1}{\cdot}K^{-1}$
112.6 $J{\cdot}mol^{-1}{\cdot}K^{-1}$
112.6 $J{\cdot}mol^{-1}{\cdot}K^{-1}$
Temperature range 261–338 K, Data calculated from equation.
Molecular Weight 102.9233
Wiswesser Line Notation GYGF
Evaluation B

$CHCl_3$ (liq) 24WIL/DAN
Trichloromethane; Chloroform
Heat Capacity 303 K, C_p = 27.9 $cal{\cdot}mol^{-1}{\cdot}K^{-1}$
116.7 $J{\cdot}mol^{-1}{\cdot}K^{-1}$
Temperature range 295–315 K, equation only.
Molecular Weight 119.3779
Wiswesser Line Notation GYGG
Evaluation C

$CHCl_3$ (liq) 25WIL/DAN
Trichloromethane; Chloroform
Heat Capacity 293.2 K, C_p = 27.6 $cal{\cdot}mol^{-1}{\cdot}K^{-1}$
115.5 $J{\cdot}mol^{-1}{\cdot}K^{-1}$
Temperature range 20 to 50 °C
Molecular Weight 119.3779
Wiswesser Line Notation GYGG
Evaluation B

$CHCl_3$ (liq) 32RIC/WAL
Trichloromethane; Chloroform
Heat Capacity 298.1 K, C_p = 27.0 $cal{\cdot}mol^{-1}{\cdot}K^{-1}$
113.0 $J{\cdot}mol^{-1}{\cdot}K^{-1}$
Temperature range 293–323 K
Molecular Weight 119.3779
Wiswesser Line Notation GYGG
Evaluation C

$CHCl_3$ (liq) 39PHI
Trichloromethane; Chloroform
Heat Capacity 303.6 K, C_p = 33.4 $cal{\cdot}mol^{-1}{\cdot}K^{-1}$
139.7 $J{\cdot}mol^{-1}{\cdot}K^{-1}$
One temperature
Molecular Weight 119.3779
Wiswesser Line Notation GYGG
Evaluation C

$CHCl_3$ (liq) 48KUR
Trichloromethane; Chloroform
Heat Capacity 298 K, C_p = 28.0 $cal{\cdot}mol^{-1}{\cdot}K^{-1}$
117.1 $J{\cdot}mol^{-1}{\cdot}K^{-1}$
Temperature range −52 to 51 °C, mean C_p, four temperatures.
Molecular Weight 119.3779
Wiswesser Line Notation GYGG
Evaluation D

$CHCl_3$ (liq) 55STA/TUP
Trichloromethane; Chloroform
Heat Capacity 298 K, C_p = 27.29 $cal{\cdot}mol^{-1}{\cdot}K^{-1}$
114.18 $J{\cdot}mol^{-1}{\cdot}K^{-1}$
Temperature range 284–329 K
Molecular Weight 119.3779
Wiswesser Line Notation GYGG
Evaluation B

$CHCl_3$ (liq) 57HAR/MOE
Trichloromethane; Chloroform
Heat Capacity 300 K, C_p = 27.11 $cal{\cdot}mol^{-1}{\cdot}K^{-1}$
113.4 $J{\cdot}mol^{-1}{\cdot}K^{-1}$
Temperature range 243–303 K
Molecular Weight 119.3779
Wiswesser Line Notation GYGG
Evaluation B

$CHCl_3$ (liq) 67RAS/GAN
Trichloromethane; Chloroform
Heat Capacity 293 K, C_p = 27.8 $cal{\cdot}mol^{-1}{\cdot}K^{-1}$
116.2 $J{\cdot}mol^{-1}{\cdot}K^{-1}$
Temperature range 293–333 K
Molecular Weight 119.3779
Wiswesser Line Notation GYGG
Evaluation C

CHF_3 (liq) 62VAL/BRO
Trifluoromethane; Fluoroform; Freon 23
Heat Capacity 190.97 K, C_p = 20.66 $cal{\cdot}mol^{-1}{\cdot}K^{-1}$
86.44 $J{\cdot}mol^{-1}{\cdot}K^{-1}$
Temperature range 15–190.97 K
Entropy 190.97 K, S = 36.10 $cal{\cdot}mol^{-1}{\cdot}K^{-1}$
151.04 $J{\cdot}mol^{-1}{\cdot}K^{-1}$

Table of Heat Capacities, Entropies, and Phase Transition Properties — Continued

Phase Changes
c/liq 117.97 K, ΔH = 970 cal·mol^{-1}
4058 J·mol^{-1}
ΔS = 8.22 cal·mol^{-1}·K^{-1}
34.40 J·mol^{-1}·K^{-1}
liq/g 190.97 K, ΔH = 3994 cal·mol^{-1}
16711 J·mol^{-1}
ΔS = 20.91 cal·mol^{-1}·K^{-1}
87.50 J·mol^{-1}·K^{-1}
P = 101.325 kPa
Molecular Weight 70.0141
Wiswesser Line Notation FYFF
Evaluation A

$CHLiO_2$ (c) 75FER/SAN
Lithium formate
Phase Changes
c,II/c,I 496 K, ΔH = 430 cal·mol^{-1}
1800 J·mol^{-1}
ΔS = 0.87 cal·mol^{-1}·K^{-1}
3.6 J·mol^{-1}·K^{-1}
c,I/liq 546 K, ΔH = 3870 cal·mol^{-1}
16190 J·mol^{-1}
ΔS = 7.1 cal·mol^{-1}·K^{-1}
29.7 J·mol^{-1}·K^{-1}
Molecular Weight 51.9587
Wiswesser Line Notation VHO .LI
Evaluation C

CHN (liq) 39GIA/RUE
Hydrogen cyanide
Heat Capacity 300 K, C_p = 16.97 cal·mol^{-1}·K^{-1}
71.00 J·mol^{-1}·K^{-1}
Temperature range 15–300 K
Entropy 298.86 K, S = 27.01 cal·mol^{-1}·K^{-1}
113.01 J·mol^{-1}·K^{-1}
Phase Changes
near 170 K, ΔH = 3.8 cal·mol^{-1}
15.9 J·mol^{-1}
ΔS = 0.022 cal·mol^{-1}·K^{-1}
0.092 J·mol^{-1}·K^{-1}
Second order transition
c/liq 259.90 K, ΔH = 2009 cal·mol^{-1}
8406 J·mol^{-1}
ΔS = 7.73 cal·mol^{-1}·K^{-1}
32.34 J·mol^{-1}·K^{-1}
liq/g 298.85 K, ΔH = 6027 cal·mol^{-1}
25217 J·mol^{-1}
ΔS = 20.17 cal·mol^{-1}·K^{-1}
84.38 J·mol^{-1}·K^{-1}
P = 101.325 kPa
Molecular Weight 27.0256
Wiswesser Line Notation NCH
Evaluation A

$CHNaO_2$ (c) 60WES/CHA
Sodium formate
Heat Capacity 298.15 K, C_p = 19.76 cal·mol^{-1}·K^{-1}
87.68 J·mol^{-1}·K^{-1}
Temperature range 5–350 K
Entropy 298.15 K, S = 24.80 cal·mol^{-1}·K^{-1}
103.76 J·mol^{-1}·K^{-1}
Molecular Weight 68.0075
Wiswesser Line Notation VHO .NA
Evaluation A

$CHNaO_2$ (c) 75FER/SAN
Sodium formate
Heat Capacity 340 K, C_p = 21.1 cal·mol^{-1}·K^{-1}
88.3 J·mol^{-1}·K^{-1}
Temperature range 340–560 K
Molecular Weight 68.0075
Wiswesser Line Notation VHO .NA
Evaluation B

CHO_2Rb (c) 75FER/SAN
Rubidium formate
Phase Changes
c,II/c,I 368 K, ΔH = 60 cal·mol^{-1}
250 J·mol^{-1}
ΔS = 0.16 cal·mol^{-1}·K^{-1}
0.7 J·mol^{-1}·K^{-1}
Molecular Weight 130.4855
Wiswesser Line Notation VHO .RB
Evaluation C

$(CH_2)_n$ (c) 62DAI/EVA 3
Polyethylene; Marlex 50 polymer
Heat Capacity 298.15 K, C_p = 6.283 cal·mol^{-1}·K^{-1}
26.29 J·mol^{-1}·K^{-1}
Temperature range 20–310 K, Data for "Marlex 50" (low pressure) polythene.
Entropy 298.15 K, S = 5.907 cal·mol^{-1}·K^{-1}
24.72 J·mol^{-1}·K^{-1}
Molecular Weight 14.0268
Wiswesser Line Notation /*1*/
Evaluation A

$(CH_2)_n$ (c) 62DAI/EVA 3
Polyethylene; Rigidex 50 polymer
Heat Capacity 298.15 K, C_p = 6.219 cal·mol^{-1}·K^{-1}
26.02 J·mol^{-1}·K^{-1}
Temperature range 20–310 K, Data for "Rigidex 50" (low pressure) polythene.
Entropy 298.15 K, S = 5.863 cal·mol^{-1}·K^{-1}
24.53 J·mol^{-1}·K^{-1}
Molecular Weight 14.0268
Wiswesser Line Notation /*1*/
Evaluation A

$(CH_2)_n$ (c) 62DAI/EVA 3
Polyethylene; W.N.C. 18 polymer
Heat Capacity 298.15 K, C_p = 7.761 cal·mol^{-1}·K^{-1}
32.47 J·mol^{-1}·K^{-1}
Temperature range 90–300 K, Data for "W.N.C. 18" (high pressure) polythene.
Entropy 298.15 K, S = 6.22 cal·mol^{-1}·K^{-1}
26.02 J·mol^{-1}·K^{-1}
Data extrapolated from 90 K.
Molecular Weight 14.0268
Wiswesser Line Notation /*1*/
Evaluation A

$(CH_2)_n$ (c) 65WUN
Polyethylene; Marlex 50 polymer
Heat Capacity 298.15 K, C_p = 5.23 cal·mol^{-1}·K^{-1}
21.88 J·mol^{-1}·K^{-1}
Temperature range 180–410 K. Values per gram formula weight.
Molecular Weight 14.0268
Wiswesser Line Notation /*1*/
Evaluation A
99% Crystallinity, extrapolated to 100%. Number–average molecular weight, 9800, weight–average = 130000.

Table of Heat Capacities, Entropies, and Phase Transition Properties — Continued

$(CH_2)_n$ (c) 73CHA/BES
Polyethylene, branched
Heat Capacity 298.15 K, C_p = 7.22 $cal \cdot mol^{-1} \cdot K^{-1}$
30.20 $J \cdot mol^{-1} \cdot K^{-1}$
Temperature range 2–360 K. Values per unit formula weight.
Entropy 298.15 K, S = 6.19 $cal \cdot mol^{-1} \cdot K^{-1}$
25.91 $J \cdot mol^{-1} \cdot K^{-1}$
Values are S–S_0
Molecular Weight 14.0268
Wiswesser Line Notation /*1*/
Evaluation A
Branched polyethylene, SRM 1476. Density 0.9247 $g \cdot cm^{-3}$ at 23 °C.

$(CH_2)_n$ (c) 73CHA/BES
Polyethylene, branched, annealed
Heat Capacity 298.15 K, C_p = 7.23 $cal \cdot mol^{-1} \cdot K^{-1}$
30.26 $J \cdot mol^{-1} \cdot K^{-1}$
Temperature range 2–360 K
Entropy 298.15 K, S = 6.15 $cal \cdot mol^{-1} \cdot K^{-1}$
25.74 $J \cdot mol^{-1} \cdot K^{-1}$
Values are S–S_0.
Molecular Weight 14.0268
Wiswesser Line Notation /*1*/
Evaluation A
Branched polyethylene, SRM1476, annealed. Density 0.9272 $g \cdot cm^{-3}$ at 23 °C.

$(CH_2)_n$ (c) 73CHA/BES
Polyethylene, linear
Heat Capacity 298.15 K, C_p = 6.14 $cal \cdot mol^{-1} \cdot K^{-1}$
25.68 $J \cdot mol^{-1} \cdot K^{-1}$
Temperature range 2–360 K. Values per unit formula weight.
Entropy 298.15 K, S = 5.86 $cal \cdot mol^{-1} \cdot K^{-1}$
24.52 $J \cdot mol^{-1} \cdot K^{-1}$
Values are S–S_0.
Molecular Weight 14.0268
Wiswesser Line Notation /*1*/
Evaluation A
Linear polyethlene, SRM 1475. Density, 0.95 $g \cdot cm^{-3}$ at 23 °C.

$(CH_2)_n$ (c) 74CHA
Polyethylene, linear high density
Heat Capacity 298.15 K, C_p = 5.40 $cal \cdot mol^{-1} \cdot K^{-1}$
22.60 $J \cdot mol^{-1} \cdot K^{-1}$
Temperature range 5–360 K. Value per monomer unit.
Entropy 298.15 K, S = 5.50 $cal \cdot mol^{-1} \cdot K^{-1}$
23.02 $J \cdot mol^{-1} \cdot K^{-1}$
Value per monomer unit. S–S_0.
Molecular Weight 14.0268
Wiswesser Line Notation /*1*/
Evaluation A
Extrapolated to 100% crystallinity, from data on other samples.

$(CH_2)_n$ (c) 74CHA
Polyethylene, linear high density
Heat Capacity 298.15 K, C_p = 5.49 $cal \cdot mol^{-1} \cdot K^{-1}$
22.98 $J \cdot mol^{-1} \cdot K^{-1}$
Temperature range 5–360 K. Value per monomer unit.
Entropy 298.15 K, S = 5.55 $cal \cdot mol^{-1} \cdot K^{-1}$
23.22 $J \cdot mol^{-1} \cdot K^{-1}$
Value per monomer unit. S–S_0.
Molecular Weight 14.0268
Wiswesser Line Notation /*1*/
Evaluation A
Density 0.993 $g \cdot cm^{-3}$ at 23 °C. Pressure crystallized.

$(CH_2)_n$ (c) 74CHA
Polyethylene, linear high density
Heat Capacity 298.15 K, C_p = 5.73 $cal \cdot mol^{-1} \cdot K^{-1}$
23.96 $J \cdot mol^{-1} \cdot K^{-1}$
Temperature range 5–360 K. Value per monomer unit.
Entropy 298.15 K, S = 5.65 $cal \cdot mol^{-1} \cdot K^{-1}$
23.64 $J \cdot mol^{-1} \cdot K^{-1}$
Value per monomer unit. S–S_0.
Molecular Weight 14.0268
Wiswesser Line Notation /*1*/
Evaluation A
Density 0.981 $g \cdot cm^{-3}$ at 23 °C. Slow–melt crystallized.

$(CH_2)_n$ (c) 75CHA/WES
Polyethylene, branched, DYNH CT–1660
Heat Capacity 298.15 K, C_p = 7.70 $cal \cdot mol^{-1} \cdot K^{-1}$
32.20 $J \cdot mol^{-1} \cdot K^{-1}$
Temperature range 5–350 K. Values per CH_2 unit.
Entropy 298.15 K, S = 6.30 $cal \cdot mol^{-1} \cdot K^{-1}$
26.37 $J \cdot mol^{-1} \cdot K^{-1}$
Does not include zero–point entropy.
Molecular Weight 14.0268
Wiswesser Line Notation /*1*/
Evaluation A
Branched polyethylene, density 0.91$g \cdot cm^{-3}$

$(CH_2)_n$ (c) 75CHA/WES
Polyethylene, linear, Marlex 50 polymer
Heat Capacity 298.15 K, C_p = 5.92 $cal \cdot mol^{-1} \cdot K^{-1}$
24.79 $J \cdot mol^{-1} \cdot K^{-1}$
Temperature range 5–350 K. Values per CH_2 unit.
Entropy 298.15 K, S = 5.75 $cal \cdot mol^{-1} \cdot K^{-1}$
24.04 $J \cdot mol^{-1} \cdot K^{-1}$
Does not include zero–point entropy.
Molecular Weight 14.0268
Wiswesser Line Notation /*1*/
Evaluation A
Linear polyethylene, density 0.973 $g \cdot cm^{-3}$

$(CH_2)_n$ (c) 76CHA
Polyethylene, linear high molecular weight
Heat Capacity 298.15 K, C_p = 5.97 $cal \cdot mol^{-1} \cdot K^{-1}$
24.97 $J \cdot mol^{-1} \cdot K^{-1}$
Temperature range 5–380 K. Value per monomer unit.
Entropy 298.15 K, S = 5.27 $cal \cdot mol^{-1} \cdot K^{-1}$
22.04 $J \cdot mol^{-1} \cdot K^{-1}$
Value per monomer unit. S–S_0.
Molecular Weight 14.0268
Wiswesser Line Notation /*1*/
Evaluation A
Produced by Ziegler–type vapor polymerization. Molecular weight 2.7–3.0 × 10^6. Approximately 45% crystalline. Data for quenched sample.

Table of Heat Capacities, Entropies, and Phase Transition Properties — Continued

CH_2Br_2 (liq) 48KUR
Dibromomethane; Methylene bromide
Heat Capacity 298 K, C_p = 30.4 cal·mol^{-1}·K^{-1}
127.2 J·mol^{-1}·K^{-1}
Temperature range −22 to 98 °C, mean C_p, four temperatures.
Molecular Weight 173.8348
Wiswesser Line Notation E1E
Evaluation D

CH_2Br_2 (liq) 57HAR/MOE
Dibromomethane; Methylene bromide
Heat Capacity 300 K, C_p = 25.17 cal·mol^{-1}·K^{-1}
105.3 J·mol^{-1}·K^{-1}
Temperature range 243–303 K
Molecular Weight 173.8348
Wiswesser Line Notation E1E
Evaluation B

CH_2Cl_2 (liq) 37PER
Dichloromethane; Methylene chloride
Heat Capacity 292.5 K, C_p = 24.1 cal·mol^{-1}·K^{-1}
100.8 J·mol^{-1}·K^{-1}
Temperature range −58 to 19 °C. Value is unsmoothed experimental datum.
Molecular Weight 84.9328
Wiswesser Line Notation G1G
Evaluation B

CH_2Cl_2 (liq) 37PER 2
Dichloromethane; Methylene chloride
Heat Capacity 292.5 K, C_p = 24.0 cal·mol^{-1}·K^{-1}
100.5 J·mol^{-1}·K^{-1}
Temperature range −58 to 19 °C. Value is unsmoothed experimental datum.
Molecular Weight 84.9328
Wiswesser Line Notation G1G
Evaluation B

CH_2Cl_2 (liq) 40RIE
Dichloromethane; Methylene chloride
Heat Capacity 298.1 K, C_p = 23.9 cal·mol^{-1}·K^{-1}
100.0 J·mol^{-1}·K^{-1}
Temperature range −47 to 41 °C
Molecular Weight 84.9328
Wiswesser Line Notation G1G
Evaluation A

CH_2Cl_2 (liq) 41RIE
Dichloromethane; Methylene chloride
Heat Capacity 298 K, C_p = 23.9 cal·mol^{-1}·K^{-1}
100.0 J·mol^{-1}·K^{-1}
Temperature range −47 to 41 °C
Molecular Weight 84.9328
Wiswesser Line Notation G1G
Evaluation A

CH_2Cl_2 (liq) 48KUR
Dichloromethane; Methylene chloride
Heat Capacity 298 K, C_p = 30.9 cal·mol^{-1}·K^{-1}
129.3 J·mol^{-1}·K^{-1}
Temperature range −76 to 41 °C, mean C_p, four temperatures.
Molecular Weight 84.9328
Wiswesser Line Notation G1G
Evaluation D

CH_2I_2 (liq) 48KUR
Diiodomethane; Methylene iodide
Heat Capacity 298 K, C_p = 32.0 cal·mol^{-1}·K^{-1}
133.9 J·mol^{-1}·K^{-1}
Temperature range 12 to 164 °C, mean C_p, three temperatures.
Molecular Weight 267.8358
Wiswesser Line Notation I1I
Evaluation D

$(CH_2O)_n$ (c) 59DAI/IVI
Polyoxymethylene
Heat Capacity 300 K, C_p = 8.1 cal·mol^{-1}·K^{-1}
33.9 J·mol^{-1}·K^{-1}
Temperature range 300–333 K, mean value. Value per monomer unit.
Molecular Weight 30.0262
Wiswesser Line Notation /*O1*/
Evaluation B

$(CH_2O)_n$ (c) 62DAI/EVA
Polyoxymethylene
Heat Capacity 300 K, C_p = 10.23 cal·mol^{-1}·K^{-1}
42.79 J·mol^{-1}·K^{-1}
Temperature range 20–300 K, Data given for Delrin. Data also given for trioxan copolymer where C_p(300 K) = 9.83 cal·mol^{-1}, 41.11 J·mol^{-1}·K^{-1}.
Entropy 300 K, S = 10.67 cal·mol^{-1}·K^{-1}
44.65 J·mol^{-1}·K^{-1}
Temperature range 20–300 K, Data given for Delrin. Data also given for trioxan copolymer where S(300 K) = 10.33 cal·mol^{-1}·K^{-1}, 43.21 J·mol^{-1}·K^{-1}.
Molecular Weight 30.0262
Wiswesser Line Notation /*O1*/
Evaluation A

CH_2O_2 (liq) **81REI**
Methanoic acid; Formic acid
Heat Capacity 298 K, C_p = 22.8 cal·mol^{-1}·K^{-1}
95.4 J·mol^{-1}·K^{-1}
Temperature range 291–385 K
Molecular Weight 46.0256
Wiswesser Line Notation VHQ
Evaluation D

CH_2O_2 (liq) 20GIB/LAT
Methanoic acid; Formic acid
Heat Capacity 291.5 K, C_p = 23.5 cal·mol^{-1}·K^{-1}
98.3 J·mol^{-1}·K^{-1}
Temperature range 71–292 K. Value is unsmoothed experimental datum.
Entropy 298 K, S = 34.2 cal·mol^{-1}·K^{-1}
143.1 J·mol^{-1}·K^{-1}
Used Berthelot's value, 2420 cal·mol^{-1} for ΔH fusion. Extrapolation below 70 K, no details.
Molecular Weight 46.0256
Wiswesser Line Notation VHQ
Evaluation B(C_p),C(S)

CH_2O_2 (liq) 29PAR/KEL
Methanoic acid; Formic acid
Entropy 298.1 K, S = 30.7 cal·mol^{-1}·K^{-1}
128.4 J·mol^{-1}·K^{-1}
Extrapolation below 90 K, 7.1 cal·mol^{-1}·K^{-1}. Revision of previous data.
Molecular Weight 46.0256
Wiswesser Line Notation VHQ
Evaluation C

Table of Heat Capacities, Entropies, and Phase Transition Properties — Continued

CH_2O_2 (liq) 34RAD/JUL
Methanoic acid; Formic acid
Heat Capacity 290 K, C_p = 23.9 cal·mol^{-1}·K^{-1}
100.0 J·mol^{-1}·K^{-1}
One temperature
Molecular Weight 46.0256
Wiswesser Line Notation VHQ
Evaluation C

CH_2O_2 (liq) 41STO/FIS
Methanoic acid; Formic acid
Heat Capacity 298.15 K, C_p = 23.67 cal·mol^{-1}·K^{-1}
99.04 J·mol^{-1}·K^{-1}
Temperature range 15–300 K
Entropy 298.15 K, S = 31.51 cal·mol^{-1}·K^{-1}
131.84 J·mol^{-1}·K^{-1}
Includes 0.69 cal·mol^{-1}·K^{-1} for zero–point entropy.
Phase Changes
c/liq 281.40 K, ΔH = 3031 cal·mol^{-1}
12678 J·mol^{-1}
ΔS = 10.77 cal·mol^{-1}·K^{-1}
45.05 J·mol^{-1}·K^{-1}
Molecular Weight 46.0256
Wiswesser Line Notation VHQ
Evaluation A

CH_2S_3 (liq) 63GAT/KRE
Dihydrosulfide carbon sulfide; Trithiocarbonic acid
Heat Capacity 273 K, C_p = 35.5 cal·mol^{-1}·K^{-1}
146.5 J·mol^{-1}·K^{-1}
Temperature range –95 to 20 °C
Entropy 298 K, S = 52 cal·mol^{-1}·K^{-1}
218 J·mol^{-1}·K^{-1}
Extrapolation below –95 °C. Estimated uncertainty ±6 cal·mol^{-1}·K^{-1}
Phase Changes
c/liq 246.3 K, ΔH = 2010 cal·mol^{-1}
8410 J·mol^{-1}
ΔS = 8.2 cal·mol^{-1}·K^{-1}
34.1 J·mol^{-1}·K^{-1}
Molecular Weight 110.2068
Wiswesser Line Notation SUYSHSH
Evaluation B(C_p),D(S)

CH_3DO (liq) 49STA/GUP
Methanol–d$_1$; Methyl alcohol–d$_1$
Heat Capacity 270 K, C_p = 19.04 cal·mol^{-1}·K^{-1}
79.66 J·mol^{-1}·K^{-1}
Temperature range 90–270 K
Phase Changes
c,II/c,I 161.1 K, ΔH = 155.8 cal·mol^{-1}
651.9 J·mol^{-1}
ΔS = 0.97 cal·mol^{-1}·K^{-1}
4.05 J·mol^{-1}·K^{-1}
c,I/liq 173.5 K, ΔH = 726 cal·mol^{-1}
3038 J·mol^{-1}
ΔS = 4.18 cal·mol^{-1}·K^{-1}
17.51 J·mol^{-1}·K^{-1}
Molecular Weight 33.0482
Wiswesser Line Notation Q1 &1H–2
Evaluation B

CH_3Br (liq) 38EGA/KEM
Bromomethane; Methyl bromide
Heat Capacity 280 K, C_p = 18.84 cal·mol^{-1}·K^{-1}
78.83 J·mol^{-1}·K^{-1}
Temperature range 15–280 K
Entropy 276.71 K, S = 37.08 cal·mol^{-1}·K^{-1}
155.14 J·mol^{-1}·K^{-1}
Phase Changes
c,II/c,I 173.78 K, ΔH = 113 cal·mol^{-1}
473 J·mol^{-1}
ΔS = 0.65 cal·mol^{-1}·K^{-1}
2.72 J·mol^{-1}·K^{-1}
c,I/liq 179.47 K, ΔH = 1429 cal·mol^{-1}
5979 J·mol^{-1}
ΔS = 7.96 cal·mol^{-1}·K^{-1}
33.3 J·mol^{-1}·K^{-1}
liq/g 276.71 K, ΔH = 5715 cal·mol^{-1}
23912 J·mol^{-1}
ΔS = 20.65 cal·mol^{-1}·K^{-1}
86.42 J·mol^{-1}·K^{-1}
P = 101.325 kPa
Molecular Weight 94.9387
Wiswesser Line Notation E1
Evaluation A

CH_3Br (liq) 48KUR
Bromomethane; Methyl bromide
Heat Capacity 283 K, C_p = 27.4 cal·mol^{-1}·K^{-1}
114.6 J·mol^{-1}·K^{-1}
Temperature range –67 to 9 °C, mean C_p, three temperatures.
Molecular Weight 94.9387
Wiswesser Line Notation E1
Evaluation D

CH_3Cl (liq) 24SHO
Chloromethane; Methyl chloride
Heat Capacity 298 K, C_p = 19.4 cal·mol^{-1}·K^{-1}
81.2 J·mol^{-1}·K^{-1}
Temperature range –30 to 40 °C
Molecular Weight 50.4877
Wiswesser Line Notation G1
Evaluation C

CH_3Cl (liq) 40AWB/GRI
Chloromethane; Methyl chloride
Heat Capacity 293.15 K, C_p = 19.3 cal·mol^{-1}·K^{-1}
80.8 J·mol^{-1}·K^{-1}
Temperature range 243–303 K, C_p reported at 20 °C = 0.382 cal·g^{-1}·K^{-1} and at 30 °C = 0.390 cal·g^{-1}·K^{-1}.
Molecular Weight 50.4877
Wiswesser Line Notation G1
Evaluation B

CH_3Cl (liq) 40MES/AST
Chloromethane; Methyl chloride
Heat Capacity 249.67 K, C_p = 18.07 cal·mol^{-1}·K^{-1}
75.60 J·mol^{-1}·K^{-1}
Temperature range 12–249.67 K. Value is unsmoothed experimental datum.
Entropy 248.94 K, S = 33.48 cal·mol^{-1}·K^{-1}
140.08 J·mol^{-1}·K^{-1}

Table of Heat Capacities, Entropies, and Phase Transition Properties — Continued

Phase Changes
c/liq 175.44 K, ΔH = 1537 cal·mol^{-1}
6431 J·mol^{-1}
ΔS = 8.76 cal·mol^{-1}·K^{-1}
36.66 J·mol^{-1}·K^{-1}
liq/g 248.94 K, ΔH = 5147 cal·mol^{-1}
21535 J·mol^{-1}
ΔS = 20.68 cal·mol^{-1}·K^{-1}
86.51 J·mol^{-1}·K^{-1}
P = 101.325 kPa
Molecular Weight 50.4877
Wiswesser Line Notation G1
Evaluation A
Correction in 40MES/AST 2

CH_3ClFOP (liq) 64FUR/REI
Methylphosphonyl chlorofluoride
Heat Capacity 298.15 K, C_p = 36.61 cal·mol^{-1}·K^{-1}
153.17 J·mol^{-1}·K^{-1}
Temperature range 15–335 K
Entropy 298.15 K, S = 51.72 cal·mol^{-1}·K^{-1}
216.40 J·mol^{-1}·K^{-1}
Phase Changes
c/liq 250.70 K, ΔH = 2833 cal·mol^{-1}
11853 J·mol^{-1}
ΔS = 11.30 cal·mol^{-1}·K^{-1}
47.28 J·mol^{-1}·K^{-1}
Molecular Weight 116.4593
Wiswesser Line Notation OPGF1
Evaluation A

CH_3Cl_2OP (c) 64FUR/REI
Methylphosphonyl dichloride
Heat Capacity 298.15 K, C_p = 31.34 cal·mol^{-1}·K^{-1}
131.12 J·mol^{-1}·K^{-1}
Temperature range 15–335 K
Entropy 298.15 K, S = 39.40 cal·mol^{-1}·K^{-1}
164.84 J·mol^{-1}·K^{-1}
Phase Changes
c/liq 306.14 K, ΔH = 4320 cal·mol^{-1}
18076 J·mol^{-1}
ΔS = 14.11 cal·mol^{-1}·K^{-1}
59.04 J·mol^{-1}·K^{-1}
Molecular Weight 132.9139
Wiswesser Line Notation OPGG1
Evaluation A

CH_3F_2OP (liq) 64FUR/REI
Methylphosphonyl difluoride
Heat Capacity 298.15 K, C_p = 34.69 cal·mol^{-1}·K^{-1}
145.14 J·mol^{-1}·K^{-1}
Temperature range 15–335 K
Entropy 298.15 K, S = 49.79 cal·mol^{-1}·K^{-1}
208.34 J·mol^{-1}·K^{-1}
Phase Changes
c/liq 236.34 K, ΔH = 2839 cal·mol^{-1}
11878 J·mol^{-1}
ΔS = 12.01 cal·mol^{-1}·K^{-1}
50.26 J·mol^{-1}·K^{-1}
Molecular Weight 100.0047
Wiswesser Line Notation OPFF1
Evaluation A

CH_3I (liq) 48KUR
Iodomethane; Methyl iodide
Heat Capacity 298 K, C_p = 35.4 cal·mol^{-1}·K^{-1}
148.1 J·mol^{-1}·K^{-1}
Temperature range –56 to 35 °C, mean C_p, five temperatures.
Molecular Weight 141.9392
Wiswesser Line Notation I1
Evaluation D

CH_3I (liq) 57HAR/MOE
Iodomethane; Methyl iodide
Heat Capacity 300 K, C_p = 19.76 cal·mol^{-1}·K^{-1}
82.68 J·mol^{-1}·K^{-1}
Temperature range 243–303 K
Molecular Weight 141.9392
Wiswesser Line Notation I1
Evaluation B

CH_3I (liq) 62LOW/MOE
Iodomethane; Methyl iodide
Heat Capacity 298.2 K, C_p = 19.78 cal·mol^{-1}·K^{-1}
82.76 J·mol^{-1}·K^{-1}
Temperature range 293–308 K
Molecular Weight 141.9392
Wiswesser Line Notation I1
Evaluation A

CH_3NO (liq) 07WAL
Methanamide; Formamide
Heat Capacity 292 K, C_p = 25 cal·mol^{-1}·K^{-1}
105 J·mol^{-1}·K^{-1}
One temperature
Molecular Weight 45.0408
Wiswesser Line Notation ZVH
Evaluation D

CH_3NO (liq) 65SOM/COO
Methanamide; Formamide
Heat Capacity 298 K, C_p = 25.7 cal·mol^{-1}·K^{-1}
107.6 J·mol^{-1}·K^{-1}
One temperature
Phase Changes
c/liq 275.72 K, ΔH = 1907 cal·mol^{-1}
7980 J·mol^{-1}
ΔS = 6.92 cal·mol^{-1}·K^{-1}
28.94 J·mol^{-1}·K^{-1}
Molecular Weight 45.0408
Wiswesser Line Notation ZVH
Evaluation B

CH_3NO (liq) 67RAS/GAN
Methanamide; Formamide
Heat Capacity 293 K, C_p = 25.1 cal·mol^{-1}·K^{-1}
105.2 J·mol^{-1}·K^{-1}
Temperature range 293–373 K
Molecular Weight 45.0408
Wiswesser Line Notation ZVH
Evaluation C

CH_3NO (liq) 74VIS/SOM
Methanamide; Formamide
Heat Capacity 298.15 K, C_p = 25.84 cal·mol^{-1}·K^{-1}
108.11 J·mol^{-1}·K^{-1}
One temperature
Molecular Weight 45.0408
Wiswesser Line Notation ZVH
Evaluation A

Table of Heat Capacities, Entropies, and Phase Transition Properties — Continued

CH_3NO (liq) 77VOR/PRI
Methanamide; Formamide
Heat Capacity 298.15 K, C_p = 25.84 cal·mol⁻¹·K⁻¹
108.11 J·mol⁻¹·K⁻¹
One temperature
Molecular Weight 45.0408
Wiswesser Line Notation ZVH
Evaluation A

CH_3NO_2 (liq) 07WAL
Nitromethane
Heat Capacity 289 K, C_p = 25 cal·mol⁻¹·K⁻¹
105 J·mol⁻¹·K⁻¹
One temperature
Molecular Weight 61.0402
Wiswesser Line Notation WN1
Evaluation D

CH_3NO_2 (liq) 25WIL
Nitromethane
Heat Capacity 298 K, C_p = 24 cal·mol⁻¹·K⁻¹
100 J·mol⁻¹·K⁻¹
Temperature range 288–343 K. Equation only.
Molecular Weight 61.0402
Wiswesser Line Notation WN1
Evaluation C

CH_3NO_2 (liq) 47JON/GIA
Nitromethane
Heat Capacity 298.15 K, C_p = 25.33 cal·mol⁻¹·K⁻¹
105.98 J·mol⁻¹·K⁻¹
Temperature range 15–300 K
Entropy 298.15 K, S = 41.05 cal·mol⁻¹·K⁻¹
171.75 J·mol⁻¹·K⁻¹
Phase Changes
c/liq 244.77 K, ΔH = 2319 cal·mol⁻¹
9703 J·mol⁻¹
ΔS = 9.47 cal·mol⁻¹·K⁻¹
39.64 J·mol⁻¹·K⁻¹
liq/g 298.15 K, ΔH = 9147 cal·mol⁻¹
38271 J·mol⁻¹
ΔS = 30.68 cal·mol⁻¹·K⁻¹
128.36 J·mol⁻¹·K⁻¹
P = 4.89 kPa
Molecular Weight 61.0402
Wiswesser Line Notation WN1
Evaluation A

CH_3NO_2 (liq) 50HOU/MAS
Nitromethane
Heat Capacity 313 K, C_p = 26.0 cal·mol⁻¹·K⁻¹
108.8 J·mol⁻¹·K⁻¹
Temperature range 313–363 K
Molecular Weight 61.0402
Wiswesser Line Notation WN1
Evaluation B

CH_3NO_2 (liq) 69BER/WES
Nitromethane
Heat Capacity 308 K, C_p = 25.39 cal·mol⁻¹·K⁻¹
106.22 J·mol⁻¹·K⁻¹
Temperature range 308–473 K
Molecular Weight 61.0402
Wiswesser Line Notation WN1
Evaluation B

CH_3NO_3 (liq) 53GRA/SMI
Methyl nitrate
Heat Capacity 298.2 K, C_p = 37.57 cal·mol⁻¹·K⁻¹
157.19 J·mol⁻¹·K⁻¹
Temperature range 13–295 K
Entropy 298.2 K, S = 51.86 cal·mol⁻¹·K⁻¹
216.98 J·mol⁻¹·K⁻¹
Phase Changes
c/liq 190.2 K, ΔH = 1970 cal·mol⁻¹
8242 J·mol⁻¹
ΔS = 10.36 cal·mol⁻¹·K⁻¹
43.33 J·mol⁻¹·K⁻¹
Molecular Weight 77.0396
Wiswesser Line Notation WNO1
Evaluation A

CH_3NaO (c) 57GRE/WES
Sodium methoxide
Heat Capacity 298.15 K, C_p = 16.60 cal·mol⁻¹·K⁻¹
69.45 J·mol⁻¹·K⁻¹
Temperature range 5–340 K
Entropy 298.15 K, S = 26.43 cal·mol⁻¹·K⁻¹
110.58 J·mol⁻¹·K⁻¹
Phase Changes
Anomolous region near 34 K with excess enthalpy of 11.5 cal·mol⁻¹, excess entropy of 0.43 cal·mol⁻¹·K⁻¹.
Molecular Weight 54.0239
Wiswesser Line Notation O1 .NA
Evaluation A

CH_4N_2O (c) 20GIB/LAT
Urea
Heat Capacity 298.0 K, C_p = 27.6 cal·mol⁻¹·K⁻¹
115.5 J·mol⁻¹·K⁻¹
Temperature range 86–300 K. Value is unsmoothed experimental datum.
Entropy 298 K, S = 41 cal·mol⁻¹·K⁻¹
172 J·mol⁻¹·K⁻¹
Extrapolation below 86 K, no details.
Molecular Weight 60.0554
Wiswesser Line Notation ZVZ
Evaluation B(C_p),C(S)

CH_4N_2O (c) 33PAR/HUF
Urea
Heat Capacity 298.0 K, C_p = 22.38 cal·mol⁻¹·K⁻¹
93.64 J·mol⁻¹·K⁻¹
Temperature range 93–298 K. Value is unsmoothed experimental datum.
Entropy 298.1 K, S = 25.2 cal·mol⁻¹·K⁻¹
105.4 J·mol⁻¹·K⁻¹
Extrapolation below 90 K, 7.93 cal·mol⁻¹·K⁻¹.
Molecular Weight 60.0554
Wiswesser Line Notation ZVZ
Evaluation B(C_p),C(S)

CH_4N_2O (c) 40CAM/CAM
Urea
Heat Capacity 293 K, C_p = 16.4 cal·mol⁻¹·K⁻¹
68.6 J·mol⁻¹·K⁻¹
One temperature
Molecular Weight 60.0554
Wiswesser Line Notation ZVZ
Evaluation C

Table of Heat Capacities, Entropies, and Phase Transition Properties — Continued

CH_4N_2O (c) 46RUE/HUF
Urea
Heat Capacity 298.15 K, C_p = 22.26 cal·mol^{-1}·K^{-1}
93.14 J·mol^{-1}·K^{-1}
Temperature range 19–318 K
Entropy 298.15 K, S = 25.00 cal·mol^{-1}·K^{-1}
104.60 J·mol^{-1}·K^{-1}
Molecular Weight 60.0554
Wiswesser Line Notation ZVZ
Evaluation A

CH_4N_2O (c) 66SAS/YOK
Urea
Heat Capacity 298.15 K, C_p = 21.5 cal·mol^{-1}·K^{-1}
90.0 J·mol^{-1}·K^{-1}
Temperature range 90–298 K
Molecular Weight 60.0554
Wiswesser Line Notation ZVZ
Evaluation A

CH_4N_2S (c) 70VAN/WES
Ammonium thiocyanate
Heat Capacity 298.15 K, C_p = 29.98 cal·mol^{-1}·K^{-1}
125.4 J·mol^{-1}·K^{-1}
Temperature range 5–340 K
Entropy 298.15 K, S = 33.52 cal·mol^{-1}·K^{-1}
140.2 J·mol^{-1}·K^{-1}
Molecular Weight 76.1160
Wiswesser Line Notation NCSH &ZH
Evaluation A

$CH_4N_4O_2$ (c) 73KRI/LIC
Nitroguanidine
Heat Capacity 298 K, C_p = 30.9 cal·mol^{-1}·K^{-1}
129.3 J·mol^{-1}·K^{-1}
Temperature range 200–460 K. Equation only.
Molecular Weight 104.0682
Wiswesser Line Notation WNMYZUM
Evaluation C

CH_4O (c) 25MAA/WAL
Methanol; Methyl alcohol
Heat Capacity 173 K, C_p = 25 cal·mol^{-1}·K^{-1}
105 J·mol^{-1}·K^{-1}
Temperature range 93–173 K
Phase Changes
c/liq 176 K, ΔH = 525 cal·mol^{-1}
2196 J·mol^{-1}
ΔS = 3.0 cal·mol^{-1}·K^{-1}
12.5 J·mol^{-1}·K^{-1}
Molecular Weight 32.0420
Wiswesser Line Notation Q1
Evaluation C

CH_4O (c) 37AHL/BLA
Methanol; Methyl alcohol
Heat Capacity 20.5 K, C_p = 1.29 cal·mol^{-1}·K^{-1}
5.40 J·mol^{-1}·K^{-1}
Temperature range 5–28 K
Entropy 16.25 K, S = 0.267 cal·mol^{-1}·K^{-1}
1.117 J·mol^{-1}·K^{-1}
Molecular Weight 32.0420
Wiswesser Line Notation Q1
Evaluation A

CH_4O (liq) 81REI
Methanol; Methyl alcohol
Heat Capacity 298 K, C_p = 19.9 cal·mol^{-1}·K^{-1}
83.3 J·mol^{-1}·K^{-1}
Temperature range 288–335 K
Molecular Weight 32.0420
Wiswesser Line Notation Q1
Evaluation D

CH_4O (liq) 07WAL
Methanol; Methyl alcohol
Heat Capacity 291 K, C_p = 19 cal·mol^{-1}·K^{-1}
79 J·mol^{-1}·K^{-1}
One temperature
Molecular Weight 32.0420
Wiswesser Line Notation Q1
Evaluation D

CH_4O (liq) 25PAR
Methanol; Methyl alcohol
Heat Capacity 290.1 K, C_p = 19.1 cal·mol^{-1}·K^{-1}
79.9 J·mol^{-1}·K^{-1}
Temperature range 89–290 K. Value is unsmoothed experimental datum.
Entropy 298.1 K, S = 32.6 cal·mol^{-1}·K^{-1}
136.4 J·mol^{-1}·K^{-1}
Extrapolation below 90 K, 9.74 cal·mol^{-1}·K^{-1}.
Phase Changes
c,II/c,I 161.1 K, ΔH = 141 cal·mol^{-1}
590 J·mol^{-1}
ΔS = 0.88 cal·mol^{-1}·K^{-1}
3.66 J·mol^{-1}·K^{-1}
c,I/liq 175.3 K, ΔH = 759 cal·mol^{-1}
3176 J·mol^{-1}
ΔS = 4.33 cal·mol^{-1}·K^{-1}
18.12 J·mol^{-1}·K^{-1}
Molecular Weight 32.0420
Wiswesser Line Notation Q1
Evaluation B(C_p),C(S)

CH_4O (liq) 29KEL 5
Methanol; Methyl alcohol
Heat Capacity 292.0 K, C_p = 19.1 cal·mol^{-1}·K^{-1}
79.9 J·mol^{-1}·K^{-1}
Temperature range 16–293 K. Value is unsmoothed experimental datum.
Entropy 298.15 K, S = 30.3 cal·mol^{-1}·K^{-1}
126.8 J·mol^{-1}·K^{-1}
Phase Changes
c,II/c,I 157.4 K, ΔH = 154.3 cal·mol^{-1}
645.6 J·mol^{-1}
ΔS = 0.98 cal·mol^{-1}·K^{-1}
4.10 J·mol^{-1}·K^{-1}
c,I/liq 175.2 K, ΔH = 757.0 cal·mol^{-1}
3167 J·mol^{-1}
ΔS = 4.32 cal·mol^{-1}·K^{-1}
18.08 J·mol^{-1}·K^{-1}
Molecular Weight 32.0420
Wiswesser Line Notation Q1
Evaluation B

Table of Heat Capacities, Entropies, and Phase Transition Properties — Continued

CH_4O (liq) 29PAR/KEL
Methanol; Methyl alcohol
Entropy 298.1 K, S = 31.0 cal·mol⁻¹·K⁻¹ 129.7 J·mol⁻¹·K⁻¹
Extrapolation below 90 K, 8.2 cal·mol⁻¹·K⁻¹. Revision of previous data.
Molecular Weight 32.0420
Wiswesser Line Notation Q1
Evaluation C

CH_4O (liq) 29MIT/HAR
Methanol; Methyl alcohol
Heat Capacity 270 K, C_p = 18.7 cal·mol⁻¹·K⁻¹ 78.2 J·mol⁻¹·K⁻¹
Temperature range 190–270 K
Molecular Weight 32.0420
Wiswesser Line Notation Q1
Evaluation B

CH_4O (liq) 31FIO/GIN
Methanol; Methyl alcohol
Heat Capacity 313.15 K, C_p = 19.97 cal·mol⁻¹·K⁻¹ 83.56 J·mol⁻¹·K⁻¹
Temperature range 40 to 110 °C
Molecular Weight 32.0420
Wiswesser Line Notation Q1
Evaluation A

CH_4O (liq) 39PHI
Methanol; Methyl alcohol
Heat Capacity 300.8 K, C_p = 20.7 cal·mol⁻¹·K⁻¹ 86.6 J·mol⁻¹·K⁻¹
One temperature
Molecular Weight 32.0420
Wiswesser Line Notation Q1
Evaluation C

CH_4O (liq) 49STA/GUP
Methanol; Methyl alcohol
Heat Capacity 270 K, C_p = 18.11 cal·mol⁻¹·K⁻¹ 75.77 J·mol⁻¹·K⁻¹
Temperature range 90–270 K
Phase Changes

c,II/c,I	157.8 K,	ΔH =	170 cal·mol⁻¹ 711 J·mol⁻¹
		ΔS =	1.08 cal·mol⁻¹·K⁻¹ 4.51 J·mol⁻¹·K⁻¹
c,I/liq	175.4 K,	ΔH =	755 cal·mol⁻¹ 3159 J·mol⁻¹
		ΔS =	4.30 cal·mol⁻¹·K⁻¹ 18.01 J·mol⁻¹·K⁻¹

Molecular Weight 32.0420
Wiswesser Line Notation Q1
Evaluation B

CH_4O (liq) 50HOU/MAS
Methanol; Methyl alcohol
Heat Capacity 323 K, C_p = 20.6 cal·mol⁻¹·K⁻¹ 86.2 J·mol⁻¹·K⁻¹
Temperature range 323–353 K
Molecular Weight 32.0420
Wiswesser Line Notation Q1
Evaluation B

CH_4O (liq) 60SWI/ZIE
Methanol; Methyl alcohol
Heat Capacity 311 K, C_p = 19.3 cal·mol⁻¹·K⁻¹ 80.8 J·mol⁻¹·K⁻¹
Mean value 21 to 56 °C
Molecular Weight 32.0420
Wiswesser Line Notation Q1
Evaluation C

CH_4O (liq) 62KAT
Methanol; Methyl alcohol
Heat Capacity 298.2 K, C_p = 20.5 cal·mol⁻¹·K⁻¹ 85.8 J·mol⁻¹·K⁻¹
Temperature range 10 to 60 °C
Molecular Weight 32.0420
Wiswesser Line Notation Q1
Evaluation B

CH_4O (liq) 70PAZ/PAZ
Methanol; Methyl alcohol
Heat Capacity 313.2 K, C_p = 20.5 cal·mol⁻¹·K⁻¹ 85.8 J·mol⁻¹·K⁻¹
One temperature
Molecular Weight 32.0420
Wiswesser Line Notation Q1
Evaluation B

CH_4O (liq) 71CAR/WES
Methanol; Methyl alcohol
Heat Capacity 298.15 K, C_p = 19.39 cal·mol⁻¹·K⁻¹ 81.13 J·mol⁻¹·K⁻¹
Temperature range 5–332 K
Entropy 298.15 K, S = 30.40 cal·mol⁻¹·K⁻¹ 127.19 J·mol⁻¹·K⁻¹
Phase Changes

c,II/c,I	157.34 K,	ΔH =	152.0 cal·mol⁻¹ 636.0 J·mol⁻¹
		ΔS =	0.97 cal·mol⁻¹·K⁻¹ 4.04 J·mol⁻¹·K⁻¹
c,I/liq	175.59 K,	ΔH =	768.5 cal·mol⁻¹ 3215.4 J·mol⁻¹
		ΔS =	4.38 cal·mol⁻¹·K⁻¹ 18.31 J·mol⁻¹·K⁻¹

Molecular Weight 32.0420
Wiswesser Line Notation Q1
Evaluation A

CH_4O (liq) 71DES/BHA
Methanol; Methyl alcohol
Heat Capacity 298 K, C_p = 20.0 cal·mol⁻¹·K⁻¹ 83.7 J·mol⁻¹·K⁻¹
Temperature range 298–318 K
Molecular Weight 32.0420
Wiswesser Line Notation Q1
Evaluation B

CH_4O_3 (liq) 81REI
Orthoformic acid
Heat Capacity 298 K, C_p = 37.2 cal·mol⁻¹·K⁻¹ 155.6 J·mol⁻¹·K⁻¹
Temperature range 293–406 K
Molecular Weight 64.0408
Wiswesser Line Notation QYQQ
Evaluation D

Table of Heat Capacities, Entropies, and Phase Transition Properties — Continued

CH_4S (liq) 42RUS/OSB

Methanethiol; Methyl mercaptan

Heat Capacity 280 K, C_p = 21.28 cal·mol⁻¹·K⁻¹ = 89.04 J·mol⁻¹·K⁻¹

Temperature range 15–280 K

Entropy 279.12 K, S = 39.01 cal·mol⁻¹·K⁻¹ = 163.22 J·mol⁻¹·K⁻¹

Phase Changes

c,II/c,I	137.6 K,	ΔH =	52.5 cal·mol⁻¹
			219.7 J·mol⁻¹
		ΔS =	0.38 cal·mol⁻¹·K⁻¹
			1.60 J·mol⁻¹·K⁻¹
c,I/liq	150.16 K,	ΔH =	1411 cal·mol⁻¹
			5904 J·mol⁻¹
		ΔS =	9.40 cal·mol⁻¹·K⁻¹
			39.32 J·mol⁻¹·K⁻¹
liq/g	279.12 K,	ΔH =	5872 cal·mol⁻¹
			24568 J·mol⁻¹
		ΔS =	21.04 cal·mol⁻¹·K⁻¹
			88.02 J·mol⁻¹·K⁻¹
		P =	101.325 kPa

Molecular Weight 48.1026

Wiswesser Line Notation SH1

Evaluation A

CH_5N (liq) 37AST/SIL

Aminomethane; Methylamine

Heat Capacity 259.28 K, C_p = 24.33 cal·mol⁻¹·K⁻¹ = 101.80 J·mol⁻¹·K⁻¹

Temperature range 14–259 K. Value is unsmoothed experimental datum.

Entropy 298.15 K, S = 35.9 cal·mol⁻¹·K⁻¹ = 150.2 J·mol⁻¹·K⁻¹

For superheated liquid, using extrapolated heat capacities.

Phase Changes

c/liq	179.70 K,	ΔH =	1466 cal·mol⁻¹
			6134 J·mol⁻¹
		ΔS =	8.16 cal·mol⁻¹·K⁻¹
			34.13 J·mol⁻¹·K⁻¹
liq/g	266.84 K,	ΔH =	6169 cal·mol⁻¹
			25811 J·mol⁻¹
		ΔS =	23.129 cal·mol⁻¹·K⁻¹
			96.73 J·mol⁻¹·K⁻¹

Molecular Weight 31.0572

Wiswesser Line Notation Z1

Evaluation A

$CH_6AlNO_8S_2 \cdot 12H_2O$ (c) 68ASH/STE

Methyl ammonium alum

Heat Capacity 300 K, C_p = 188.2 cal·mol⁻¹·K⁻¹ = 787.4 J·mol⁻¹·K⁻¹

Temperature range 5–300 K

Entropy 300 K, S = 182.6 cal·mol⁻¹·K⁻¹ = 764.2 J·mol⁻¹·K⁻¹

Phase Changes

c,II/c,I	176.18 K,	ΔS =	2.32 cal·mol⁻¹·K⁻¹
			9.70 J·mol⁻¹·K⁻¹

90 K Anomaly: Schottky type anomaly between 65 and 120 K, maximum at 90 K, entropy estimated to be between 6 and 9 J·mol⁻¹·K⁻¹.

176 K Anomaly: Due to free rotation of methylammonium group (CH_3NH_3); observed entropy change at 176.18 K is 9.70 J·mol⁻¹·K⁻¹.

Molecular Weight 467.3442

Wiswesser Line Notation AL Z1& S-O4*2 QH-12-

Evaluation A

CH_6ClN (c) 46AST/ZIE

Methylammonium chloride

Heat Capacity 298.15 K, C_p = 21.73 cal·mol⁻¹·K⁻¹ = 90.92 J·mol⁻¹·K⁻¹

Temperature range 12–298 K

Entropy 298 K, S = 33.11 cal·mol⁻¹·K⁻¹ = 138.53 J·mol⁻¹·K⁻¹

Using metastable c,II below 220 K gives S = 33.15 cal·mol⁻¹·K⁻¹.

Phase Changes

c,III/c,II	220.4 K,	ΔH =	425 cal·mol⁻¹
			1778 J·mol⁻¹
		ΔS =	1.93 cal·mol⁻¹·K⁻¹
			8.07 J·mol⁻¹·K⁻¹
c,II/c,I	264.5 K,	ΔH =	674 cal·mol⁻¹
			2820 J·mol⁻¹
		ΔS =	2.55 cal·mol⁻¹·K⁻¹
			10.66 J·mol⁻¹·K⁻¹

Molecular Weight 67.5181

Wiswesser Line Notation Z1 &GH

Evaluation A

CH_6ClN_3O (c) 41SAT/SOG 4

Semicarbazide hydrochloride

Heat Capacity 323 K, C_p = 34.3 cal·mol⁻¹·K⁻¹ = 143.5 J·mol⁻¹·K⁻¹

Temperature range 0 to 100 °C. Mean value.

Molecular Weight 111.5309

Wiswesser Line Notation ZVMZ &GH

Evaluation C

Same data in 40SAT/SOG 5.

CH_6N_2 (liq) 51AST/FIN

Methylhydrazine

Heat Capacity 298.15 K, C_p = 32.25 cal·mol⁻¹·K⁻¹ = 134.93 J·mol⁻¹·K⁻¹

Temperature range 15–298 K

Entropy 298.15 K, S = 39.66 cal·mol⁻¹·K⁻¹ = 165.94 J·mol⁻¹·K⁻¹

Phase Changes

c/liq	220.79 K,	ΔH =	2490 cal·mol⁻¹
			10418 J·mol⁻¹
		ΔS =	11.28 cal·mol⁻¹·K⁻¹
			47.19 J·mol⁻¹·K⁻¹
liq/g	298.15 K,	ΔH =	9648 cal·mol⁻¹
			40367 J·mol⁻¹
		ΔS =	32.36 cal·mol⁻¹·K⁻¹
			135.39 J·mol⁻¹·K⁻¹
		P =	6.62 kPa

Molecular Weight 46.0718

Wiswesser Line Notation ZM1

Evaluation A

$CH_6N_2O_2$ (c) 28CLU/HAR

Ammonium carbamate

Heat Capacity 295.5 K, C_p = 31.31 cal·mol⁻¹·K⁻¹ = 131.00 J·mol⁻¹·K⁻¹

Temperature range 13–296 K

Molecular Weight 78.0706

Wiswesser Line Notation ZVQ &ZH

Evaluation B

Table of Heat Capacities, Entropies, and Phase Transition Properties — Continued

CKNS (c) 70VAN/WES
Potassium thiocyanate
Heat Capacity 298.15 K, C_p = 21.16 $cal{\cdot}mol^{-1}{\cdot}K^{-1}$
88.53 $J{\cdot}mol^{-1}{\cdot}K^{-1}$
Temperature range 5–340 K
Entropy 298.15 K, S = 29.70 $cal{\cdot}mol^{-1}{\cdot}K^{-1}$
124.3 $J{\cdot}mol^{-1}{\cdot}K^{-1}$
Molecular Weight 97.1760
Wiswesser Line Notation K SCN
Evaluation A

COS (liq) 37KEM/GIA
Carbonyl sulfide
Heat Capacity 220 K, C_p = 17.03 $cal{\cdot}mol^{-1}{\cdot}K^{-1}$
71.25 $J{\cdot}mol^{-1}{\cdot}K^{-1}$
Temperature range 20–220 K
Entropy 222.91 K, S = 32.58 $cal{\cdot}mol^{-1}{\cdot}K^{-1}$
136.31 $J{\cdot}mol^{-1}{\cdot}K^{-1}$
Phase Changes
c/liq 134.33 K, ΔH = 1130 $cal{\cdot}mol^{-1}$
4728 $J{\cdot}mol^{-1}$
ΔS = 8.41 $cal{\cdot}mol^{-1}{\cdot}K^{-1}$
35.20 $J{\cdot}mol^{-1}{\cdot}K^{-1}$
liq/g 222.91 K, ΔH = 4423 $cal{\cdot}mol^{-1}$
18506 $J{\cdot}mol^{-1}$
ΔS = 19.84 $cal{\cdot}mol^{-1}{\cdot}K^{-1}$
83.02 $J{\cdot}mol^{-1}{\cdot}K^{-1}$
P = 101.325 kPa
Molecular Weight 60.0704
Wiswesser Line Notation SCO
Evaluation A

CS_2 (liq) 37BRO/MAN
Carbon disulfide
Heat Capacity 297.43 K, C_p = 18.17 $cal{\cdot}mol^{-1}{\cdot}K^{-1}$
76.02 $J{\cdot}mol^{-1}{\cdot}K^{-1}$
Temperature range 15–297 K. Value is unsmoothed experimental datum.
Entropy 298.15 K, S = 36.10 $cal{\cdot}mol^{-1}{\cdot}K^{-1}$
151.0 $J{\cdot}mol^{-1}{\cdot}K^{-1}$
Phase Changes
c/liq 161.11 K, ΔH = 1049 $cal{\cdot}mol^{-1}$
4389 $J{\cdot}mol^{-1}$
ΔS = 6.51 $cal{\cdot}mol^{-1}{\cdot}K^{-1}$
27.24 $J{\cdot}mol^{-1}{\cdot}K^{-1}$
Molecular Weight 76.1310
Wiswesser Line Notation SCS
Evaluation A

CS_2 (liq) 39MAZ 3
Carbon disulfide
Heat Capacity 293 K, C_p = 18.6 $cal{\cdot}mol^{-1}{\cdot}K^{-1}$
77.8 $J{\cdot}mol^{-1}{\cdot}K^{-1}$
Temperature range −100 to 20 °C
Molecular Weight 76.1310
Wiswesser Line Notation SCS
Evaluation C

CS_2 (liq) 39PHI
Carbon disulfide
Heat Capacity 301.2 K, C_p = 18.2 $cal{\cdot}mol^{-1}{\cdot}K^{-1}$
76.1 $J{\cdot}mol^{-1}{\cdot}K^{-1}$
One temperature
Molecular Weight 76.1310
Wiswesser Line Notation SCS
Evaluation C

CS_2 (liq) 45ZHD
Carbon disulfide
Heat Capacity 294.81 K, C_p = 17.90 $cal{\cdot}mol^{-1}{\cdot}K^{-1}$
74.89 $J{\cdot}mol^{-1}{\cdot}K^{-1}$
Temperature range 7 to 31 °C. Value is unsmoothed experimental datum.
Molecular Weight 76.1310
Wiswesser Line Notation SCS
Evaluation B

CS_2 (liq) 55STA/TUP
Carbon disulfide
Heat Capacity 298 K, C_p = 18.88 $cal{\cdot}mol^{-1}{\cdot}K^{-1}$
78.99 $J{\cdot}mol^{-1}{\cdot}K^{-1}$
Temperature range 286–317 K
Molecular Weight 76.1310
Wiswesser Line Notation SCS
Evaluation B

CSe_2 (liq) 66GAT/DRA
Carbon diselenide
Heat Capacity 298 K, C_p = 21.2 $cal{\cdot}mol^{-1}{\cdot}K^{-1}$
88.7 $J{\cdot}mol^{-1}{\cdot}K^{-1}$
Temperature range −190 to 50 °C
Entropy 298 K, S = 39.5 $cal{\cdot}mol^{-1}{\cdot}K^{-1}$
165.3 $J{\cdot}mol^{-1}{\cdot}K^{-1}$
Phase Changes
c/liq 229.5 K, ΔH = 1520 $cal{\cdot}mol^{-1}$
6360 $J{\cdot}mol^{-1}$
ΔS = 6.6 $cal{\cdot}mol^{-1}{\cdot}K^{-1}$
27.7 $J{\cdot}mol^{-1}{\cdot}K^{-1}$
Molecular Weight 169.9310
Wiswesser Line Notation –SE–C–SE–
Evaluation B

$C_2Br_2F_4$ (liq) 82KOS/ZHO
1,2–Dibromotetrafluoroethane
Heat Capacity 298.15 K, C_p = 40.82 $cal{\cdot}mol^{-1}{\cdot}K^{-1}$
170.8 $J{\cdot}mol^{-1}{\cdot}K^{-1}$
Temperature range 8–300 K
Entropy 298.15 K, S = 71.56 $cal{\cdot}mol^{-1}{\cdot}K^{-1}$
299.4 $J{\cdot}mol^{-1}{\cdot}K^{-1}$
Phase Changes
c,I/liq 162.83 K, ΔH = 1681.8 $cal{\cdot}mol^{-1}$
7036.7 $J{\cdot}mol^{-1}$
ΔS = 10.33 $cal{\cdot}mol^{-1}{\cdot}K^{-1}$
43.22 $J{\cdot}mol^{-1}{\cdot}K^{-1}$
Molecular Weight 259.8236
Wiswesser Line Notation FXFFEXFFE
Evaluation A

C_2ClF_3 (liq) 51OLI/GRI
Chlorotrifluoroethene; Chlorotrifluoroethylene; Trifluorochloroethene; Trifluorochloroethylene
Heat Capacity 244.80 K, C_p = 29.26 $cal{\cdot}mol^{-1}{\cdot}K^{-1}$
122.42 $J{\cdot}mol^{-1}{\cdot}K^{-1}$
Temperature range 16–245 K
Entropy 244.80 K, S = 52.74 $cal{\cdot}mol^{-1}{\cdot}K^{-1}$
220.66 $J{\cdot}mol^{-1}{\cdot}K^{-1}$
Phase Changes
c/liq 115.0 K, ΔH = 1327.1 $cal{\cdot}mol^{-1}$
5552.6 $J{\cdot}mol^{-1}$
ΔS = 11.54 $cal{\cdot}mol^{-1}{\cdot}K^{-1}$
48.28 $J{\cdot}mol^{-1}{\cdot}K^{-1}$
Molecular Weight 116.4702
Wiswesser Line Notation GYFUYFF
Evaluation A

Table of Heat Capacities, Entropies, and Phase Transition Properties — Continued

$(C_2ClF_3)_n$ (c) 52HOF
Polytrifluorochloroethylene; Polytrifluorovinyl chloride
Heat Capacity 298 K, C_p = 24.3 cal·mol^{-1}·K^{-1}
101.7 J·mol^{-1}·K^{-1}
Temperature range 0 to 241 °C. Values given for air–quenched and slow–cooled samples. Values per monomer unit slow–cooled samples. Unsmoothed experimental datum.
Molecular Weight 116.4702
Wiswesser Line Notation /*XGFXFF*/
Evaluation B

$(C_2ClF_3)_n$ (liq) 57YAR/KAY
Polytrifluorochloroethylene; Polytrifluorovinyl chloride
Heat Capacity 298 K, C_p = 27.8 cal·mol^{-1}·K^{-1}
116.3 J·mol^{-1}·K^{-1}
Temperature range 298–373 K. Equation only. Values per monomer unit.
Molecular Weight 116.4702
Wiswesser Line Notation /*XGFXFF*/
Evaluation B

C_2ClF_5 (liq) 55AST/WIL
Pentafluorochloroethane
Heat Capacity 234.04 K, C_p = 34.72 cal·mol^{-1}·K^{-1}
145.27 J·mol^{-1}·K^{-1}
Temperature range 15–234 K. Value is unsmoothed experimental datum.
Entropy 234.04 K, S = 59.28 cal·mol^{-1}·K^{-1}
248.03 J·mol^{-1}·K^{-1}

Phase Changes

c,II/c,I	80.24 K,	ΔH =	628 cal·mol^{-1}
			2628 J·mol^{-1}
		ΔS =	7.83 cal·mol^{-1}·K^{-1}
			32.75 J·mol^{-1}·K^{-1}
c,I/liq	173.71 K,	ΔH =	449 cal·mol^{-1}
			1879 J·mol^{-1}
		ΔS =	2.58 cal·mol^{-1}·K^{-1}
			10.82 J·mol^{-1}·K^{-1}
liq/g	234.04 K,	ΔH =	4639 cal·mol^{-1}
			19410 J·mol^{-1}
		ΔS =	19.82 cal·mol^{-1}·K^{-1}
			82.93 J·mol^{-1}·K^{-1}
		P =	101.325 kPa

Molecular Weight 154.4670
Wiswesser Line Notation GXFFXFFF
Evaluation A

$C_2Cl_2F_4$ (liq) 37PER
1,2–Dichloro–1,1,2,2–tetrafluoroethane; Freon 114
Heat Capacity 293.3 K, C_p = 40.6 cal·mol^{-1}·K^{-1}
169.9 J·mol^{-1}·K^{-1}
Temperature range −188 to 20 °C. Value is unsmoothed experimental datum.
Molecular Weight 170.9216
Wiswesser Line Notation GXFFXGFF
Evaluation C
Probably a mixture of isomers.

$C_2Cl_2F_4$ (liq) 37PER 2
1,2–Dichloro–1,1,2,2–tetrafluoroethane; Freon 114
Heat Capacity 293.3 K, C_p = 40.6 cal·mol^{-1}·K^{-1}
169.9 J·mol^{-1}·K^{-1}
Temperature range −188 to 20 °C. Value is unsmoothed experimental datum.
Molecular Weight 170.9216
Wiswesser Line Notation GXFFXGFF
Evaluation C
Probably a mixture of isomers. Shows peaks in heat capacity at −180 °C, −140 °C, and −100 °C, which may be glassy transitions.

$C_2Cl_2F_4$ (liq) 81KOL/KOS
1,2–Dichloro–1,1,2,2–tetrafluoroethane; Freon 114
Heat Capacity 298.15 K, C_p = 39.2 cal·mol^{-1}·K^{-1}
164.0 J·mol^{-1}·K^{-1}
Temperature range 8–300 K
Entropy 298.15 K, S = 67.4 cal·mol^{-1}·K^{-1}
282.0 J·mol^{-1}·K^{-1}

Phase Changes

c,III/c,II	109.3 K,	ΔH =	289.7 cal·mol^{-1}
			1212 J·mol^{-1}
		ΔS =	2.65 cal·mol^{-1}·K^{-1}
			11.09 J·mol^{-1}·K^{-1}
c,II/c,I	134.6 K,	ΔH =	628.1 cal·mol^{-1}
			2628 J·mol^{-1}
		ΔS =	4.67 cal·mol^{-1}·K^{-1}
			19.5 J·mol^{-1}·K^{-1}
c,I/liq	180.62 K,	ΔH =	361 cal·mol^{-1}
			1510 J·mol^{-1}
		ΔS =	2.00 cal·mol^{-1}·K^{-1}
			8.36 J·mol^{-1}·K^{-1}

Molecular Weight 170.9216
Wiswesser Line Notation GXFFXGFF
Evaluation A

$C_2Cl_3F_3$ (liq) 38RIE
1,1,2–Trichloro–1,2,2–trifluoroethane; Freon 113
Heat Capacity 298.2 K, C_p = 42.3 cal·mol^{-1}·K^{-1}
177.0 J·mol^{-1}·K^{-1}
Temperature range −30 to 61 °C
Molecular Weight 187.3762
Wiswesser Line Notation GXGFXGFF
Evaluation B

$C_2Cl_3F_3$ (liq) 39RIE
1,1,2–Trichloro–1,2,2–trifluoroethane; Freon 113
Heat Capacity 298.2 K, C_p = 42.3 cal·mol^{-1}·K^{-1}
177.0 J·mol^{-1}·K^{-1}
Temperature range −30 to 61 °C
Molecular Weight 187.3762
Wiswesser Line Notation GXGFXGFF
Evaluation B

$C_2Cl_3F_3$ (liq) 40BEN/MCH
1,1,2–Trichloro–1,2,2–trifluoroethane; Freon 113
Heat capacity 298.15 K, C_p = 42.97 cal·mol^{-1}·K^{-1}
179.8 J·mol^{-1}·K^{-1}
Temperature range 243–336 K, Data calculated from equation.
Molecular Weight 187.3762
Wiswesser Line Notation GXGFXGFF
Evaluation B

Table of Heat Capacities, Entropies, and Phase Transition Properties — Continued

$C_2Cl_3F_3$ (liq) 81KOL/KOS
1,1,2–Trichloro–1,2,2–trifluoroethane; Freon 113
Heat Capacity 298.15 K, C_p = 41.3 cal·mol^{-1}·K^{-1}
172.8 J·mol^{-1}·K^{-1}
Temperature range 8–300 K
Entropy 298.15 K, S = 69.2 cal·mol^{-1}·K^{-1}
289.5 J·mol^{-1}·K^{-1}
Phase Changes
c,II/c,I 82.5 K, ΔH = 198.6 cal·mol^{-1}
830.9 J·mol^{-1}
ΔS = 2.41 cal·mol^{-1}·K^{-1}
10.1 J·mol^{-1}·K^{-1}
c,I/liq 236.92 K, ΔH = 589.6 cal·mol^{-1}
2467 J·mol^{-1}
ΔS = 2.49 cal·mol^{-1}·K^{-1}
10.4 J·mol^{-1}·K^{-1}
Molecular Weight 187.3762
Wiswesser Line Notation GXGFXGFF
Evaluation A

C_2Cl_4 (liq) 81REI
Tetrachloroethene; Tetrachloroethylene
Heat Capacity 298 K, C_p = 35.1 cal·mol^{-1}·K^{-1}
146.9 J·mol^{-1}·K^{-1}
Temperature range 291–410 K
Molecular Weight 165.8340
Wiswesser Line Notation GYGUYGG
Evaluation D

C_2Cl_4 (liq) 48KUR
Tetrachloroethene; Tetrachloroethylene
Heat Capacity 298 K, C_p = 33.4 cal·mol^{-1}·K^{-1}
139.7 J·mol^{-1}·K^{-1}
Temperature range 16 to 119 °C, mean C_p, two temperatures.
Molecular Weight 165.8340
Wiswesser Line Notation GYGUYGG
Evaluation D

C_2Cl_4 (liq) 82GRO/ING
Tetrachloroethene; Tetrachloroethylene
Heat Capacity 298.15 K, C_p = 35.01 cal·mol^{-1}·K^{-1}
146.5 J·mol^{-1}·K^{-1}
Temperature range 298.15 K, One data point given.
Molecular Weight 165.8340
Wiswesser Line Notation GYGUYGG
Evaluation A

$C_2Cl_4F_2$ (liq) 57YAR/KAY
1,1,2,2–Tetrachloro–1,2–difluoroethane
Heat Capacity 298 K, C_p = 41.5 cal·mol^{-1}·K^{-1}
173.6 J·mol^{-1}·K^{-1}
Temperature range 298–373 K. Equation only.
Molecular Weight 203.8308
Wiswesser Line Notation GXGFXGGF
Evaluation B

$C_2Cl_4F_2$ (c) 78KOS/KOL
1,1,2,2–Tetrachloro–1,2–difluoroethane
Heat Capacity 298.15 K, C_p = 41.95 cal·mol^{-1}·K^{-1}
175.52 J·mol^{-1}·K^{-1}
Temperature range 12–310 K. Complete data deposited at VINITI, No. 3512–77., 20 Aug 1977.
Entropy 298.15 K, S = 65.45 cal·mol^{-1}·K^{-1}
273.84 J·mol^{-1}·K^{-1}
Low entropy of fusion indicates possible zero point entropy (disorder).
Phase Changes
c/liq 299.70 K, ΔH = 883.5 cal·mol^{-1}
3696.6 J·mol^{-1}
ΔS = 2.95 cal·mol^{-1}·K^{-1}
12.33 J·mol^{-1}·K^{-1}
Molecular Weight 203.8308
Wiswesser Line Notation GXGFXGFF
Evaluation B

$C_2Cl_4F_2$ (liq) 78KIS/SUG
1,1,2,2–Tetrachloro–1,2–difluoroethane
Heat Capacity 300 K, C_p = 42.68 cal·mol^{-1}·K^{-1}
178.59 J·mol^{-1}·K^{-1}
Temperature range 13–310 K
Entropy 300 K, S = 67.74 cal·mol^{-1}·K^{-1}
283.43 J·mol^{-1}·K^{-1}
Phase Changes
c/liq 297.91 K, ΔH = 876 cal·mol^{-1}
3666 J·mol^{-1}
ΔS = 2.94 cal·mol·K^{-1}
12.31 J·mol^{-1}·K^{-1}
Molecular Weight 203.8308
Wiswesser Line Notation GXGFXGGF
Evaluation A

C_2Cl_6 (c) 75RAK/GUT
Hexachloroethane
Heat Capacity 298.15 K, C_p = 47.38 cal·mol^{-1}·K^{-1}
198.24 J·mol^{-1}·K^{-1}
Temperature range 13.7–360 K
Entropy 298.15 K, S = 56.72 cal·mol^{-1}·K^{-1}
237.3 J·mol^{-1}·K^{-1}
Entropy estimated at 13 K to be 0.74 cal·mol^{-1}·K^{-1}.
Phase Changes
c,III/c,II 317.4 K, ΔH = 632 cal·mol^{-1}
2644 J·mol^{-1}
ΔS = 1.99 cal·mol^{-1}·K^{-1}
8.33 J·mol^{-1}·K^{-1}
c,II/c,I 344.6 K, ΔH = 1653 cal·mol^{-1}
6916 J·mol^{-1}
ΔS = 4.80 cal·mol^{-1}·K^{-1}
20.1 J·mol^{-1}·K^{-1}
Molecular Weight 236.7400
Wiswesser Line Notation GXGGXGGG
Evaluation A

C_2F_3N (liq) 61PAC/BOB
Trifluoroacetonitrile; Trifluoromethyl cyanide
Heat Capacity 205 K, C_p = 28.23 cal·mol^{-1}·K^{-1}
118.11 J·mol^{-1}·K^{-1}
Temperature range 15–205 K
Entropy 205.47 K, S = 44.08 cal·mol^{-1}·K^{-1}
184.43 J·mol^{-1}·K^{-1}
Phase Changes
c/liq 128.73 K, ΔH = 1188 cal·mol^{-1}
4971 J·mol^{-1}
ΔS = 9.23 cal·mol^{-1}·K^{-1}
38.62 J·mol^{-1}·K^{-1}
liq/g 205.47 K, ΔH = 4262 cal·mol^{-1}
17832 J·mol^{-1}
ΔS = 20.74 cal·mol^{-1}·K^{-1}
86.79 J·mol^{-1}·K^{-1}
P = 101.325 kPa
Molecular Weight 95.0239
Wiswesser Line Notation NCXFFF
Evaluation A

Table of Heat Capacities, Entropies, and Phase Transition Properties — Continued

C_2F_4 (liq) 53FUR/MCC
Tetrafluoroethene; Tetrafluoroethylene
Heat Capacity 200 K, C_p = 26.95 cal·mol⁻¹·K⁻¹
112.77 J·mol⁻¹·K⁻¹
Temperature range 6–210 K
Entropy 197.53 K, S = 44.03 cal·mol⁻¹·K⁻¹
184.23 J·mol⁻¹·K⁻¹
Phase Changes
c/liq 142.00 K, ΔH = 1843.8 cal·mol⁻¹
7714.5 J·mol⁻¹
ΔS = 12.98 cal·mol⁻¹·K⁻¹
54.33 J·mol⁻¹·K⁻¹
liq/g 197.53 K, ΔH = 4020 cal·mol⁻¹
16821 J·mol⁻¹
ΔS = 20.35 cal·mol⁻¹·K⁻¹
85.16 J·mol⁻¹·K⁻¹
P = 101.325 kPa
Molecular Weight 100.0156
Wiswesser Line Notation FYFUYFF
Evaluation A

$(C_2F_4)_n$ (c) 52FUR/MCC
Polytetrafluoroethylene, annealed; Teflon, annealed
Heat Capacity 280 K, C_p = 23.33 cal·mol⁻¹·K⁻¹
97.6 J·mol⁻¹·K⁻¹
Temperature range 15–370 K. Values per monomer unit.
Phase Changes
Transition region 280–310 K
Molecular Weight 100.0156
Wiswesser Line Notation /*X*FF/
Evaluation A

$(C_2F_4)_n$ (c) 52FUR/MCC
Polytetrafluoroethylene, molded; Teflon, molded
Heat Capacity 280 K, C_p = 24.19 cal·mol⁻¹·K⁻¹
101.2 J·mol⁻¹·K⁻¹
Temperature range 15–370 K. Values per monomer unit.
Phase Changes
Transition region 280–310 K
Molecular Weight 100.0156
Wiswesser Line Notation /*X*FF/
Evaluation A

$(C_2F_4)_n$ (c) 52FUR/MCC
Polytetrafluoroethylene, powder; Teflon, powder
Heat Capacity 280 K, C_p = 24.76 cal·mol⁻¹·K⁻¹
103.6 J·mol⁻¹·K⁻¹
Temperature range 15–370 K. Values per monomer unit.
Phase Changes
Transition region 280–310 K
Molecular Weight 100.0156
Wiswesser Line Notation /*X*FF/
Evaluation A

$(C_2F_4)_n$ (c) 55MAR/DOL
Polytetrafluoroethylene, drawn; Teflon, drawn
Heat Capacity 313 K, C_p = 23.73 cal·mol⁻¹·K⁻¹
99.29 J·mol⁻¹·K⁻¹
Temperature range 250–393 K. Value calc. from equation, valid above 313 K, C_p = per monomer unit.
Molecular Weight 100.0156
Wiswesser Line Notation /*X*FF/
Evaluation B

$(C_2F_4)_n$ (c) 55MAR/DOL
Polytetrafluoroethylene, powder; Teflon, powder
Heat Capacity 313 K, C_p = 23.27 cal·mol⁻¹·K⁻¹
97.36 J·mol⁻¹·K⁻¹
Temperature range 250–393 K. Value calc. from equation, valid above 313 K, C_p = per monomer unit.
Molecular Weight 100.0156
Wiswesser Line Notation /*X*FF/
Evaluation B

$(C_2F_4)_n$ (c) 65DOU/HAR
Polytetrafluoroethylene, powder; Teflon, powder
Heat Capacity 350 K, C_p = 24.3 cal·mol⁻¹·K⁻¹
101.6 J·mol⁻¹·K⁻¹
Temperature range 350–573 K. Values per monomer unit.
Molecular Weight 100.0156
Wiswesser Line Notation /*X*FF/
Evaluation B

$(C_2F_4)_n$ (c) 65DOU/HAR
Polytetrafluoroethylene, quenched; Teflon, quenched
Heat Capacity 350 K, C_p = 25.4 cal·mol⁻¹·K⁻¹
106.4 J·mol⁻¹·K⁻¹
Temperature range 350–725 K. Values per monomer unit.
Molecular Weight 100.0156
Wiswesser Line Notation /*X*FF/
Evaluation B

C_2F_4O (liq) 72PAC/HOD
Trifluoroacetyl fluoride
Heat Capacity 214.0 K, C_p = 33.21 cal·mol⁻¹·K⁻¹
138.95 J·mol⁻¹·K⁻¹
Temperature range 14–214 K
Entropy 214.1 K, S = 51.74 cal·mol⁻¹·K⁻¹
216.48 J·mol⁻¹·K⁻¹
Phase Changes
c/liq 113.69 K, ΔH = 1164 cal·mol⁻¹
4870 J·mol⁻¹
ΔS = 10.24 cal·mol⁻¹·K⁻¹
42.84 J·mol⁻¹·K⁻¹
liq/g 214.10 K, ΔH = 4605 cal·mol⁻¹
19207 J·mol⁻¹
ΔS = 21.51 cal·mol⁻¹·K⁻¹
89.99 J·mol⁻¹·K⁻¹
P = 101.325 kPa
Molecular Weight 116.0150
Wiswesser Line Notation FXFFVF
Evaluation A

C_2F_6 (liq) 48PAC/AST
Hexafluoroethane
Heat Capacity 195 K, C_p = 31.38 cal·mol⁻¹·K⁻¹
131.3 J·mol⁻¹·K⁻¹
Temperature range 12–195 K
Entropy 194.87 K, S = 49.88 cal·mol⁻¹·K⁻¹
250.54 J·mol⁻¹·K⁻¹

Table of Heat Capacities, Entropies, and Phase Transition Properties — Continued

Phase Changes
c,II/c,I 103.98 K, ΔH = 893 cal·mol^{-1}
3736 J·mol^{-1}
ΔS = 8.59 cal·mol^{-1}·K^{-1}
35.93 J·mol^{-1}·K^{-1}
c,I/liq 173.10 K, ΔH = 642 cal·mol^{-1}
2686 J·mol^{-1}
ΔS = 3.71 cal·mol^{-1}·K^{-1}
15.52 J·mol^{-1}·K^{-1}
liq/g 194.87 K, ΔH = 3860 cal·mol^{-1}
16150 J·mol^{-1}
ΔS = 19.81 cal·mol^{-1}·K^{-1}
82.88 J·mol^{-1}·K^{-1}
P = 101.325 kPa
Molecular Weight 138.0124
Wiswesser Line Notation FXFFXFFF
Evaluation A

$C_2HD_3Br_2$ (liq) 49WUY/JUN
1,2–Dibromoethane–d$_3$
Heat Capacity 310 K, C_p = 34.69 cal·mol^{-1}·K^{-1}
145.14 J·mol^{-1}·K^{-1}
One temperature
Molecular Weight 190.8802
Wiswesser Line Notation E2E &2–1/2–1/2–1/H–2 3
Evaluation C

C_2HCl_3 (liq) 33TRE/WAT
Trichloroethene; Trichloroethylene
Heat Capacity 298 K, C_p = 29.8 cal·mol^{-1}·K^{-1}
124.7 J·mol^{-1}·K^{-1}
One temperature
Molecular Weight 131.3889
Wiswesser Line Notation GYGU1G
Evaluation B

C_2HCl_3 (liq) 48KUR
Trichloroethene; Trichloroethylene
Heat Capacity 298 K, C_p = 28.7 cal·mol^{-1}·K^{-1}
120.1 J·mol^{-1}·K^{-1}
Temperature range 12 to 80 °C, mean C_p, three temperatures.
Molecular Weight 131.3889
Wiswesser Line Notation GYGU1G
Evaluation D

C_2HCl_3O (liq) 81REI
2,2,2–Trichloroethanal; Chloral; α,α,α–Trichloroacetaldehyde
Heat Capacity 298 K, C_p = 36.2 cal·mol^{-1}·K^{-1}
151.5 J·mol^{-1}·K^{-1}
Temperature range 294–383 K
Molecular Weight 147.3883
Wiswesser Line Notation VHXGGG
Evaluation D

C_2HCl_5 (liq) 48KUR
Pentachloroethane
Heat Capacity 298 K, C_p = 46.9 cal·mol^{-1}·K^{-1}
196.2 J·mol^{-1}·K^{-1}
Temperature range 16 to 154 °C, mean C_p, three temperatures.
Molecular Weight 202.2949
Wiswesser Line Notation GYGXGGG
Evaluation D

$C_2H_2D_2Br_2$ (liq) 49WUY/JUN
1,2–Dibromoethane–1,2–d$_2$
Heat Capacity 310 K, C_p = 33.97 cal·mol^{-1}·K^{-1}
142.13 J·mol^{-1}·K^{-1}
One temperature
Molecular Weight 189.8740
Wiswesser Line Notation E2E &2–1/2–2/H–2 2
Evaluation C

$C_2H_2D_2Br_2$ (liq) 49WUY/JUN
1,2–Dibromoethane–1,1–d$_2$
Heat Capacity 310 K, C_p = 34.33 cal·mol^{-1}·K^{-1}
143.64 J·mol^{-1}·K^{-1}
One temperature
Molecular Weight 189.8740
Wiswesser Line Notation E2E &2–1/H–2 2
Evaluation C

$C_2H_2Br_2O_2$ (c) 61GLA/TIM
Dibromoethanoic acid; Dibromoacetic acid
Heat Capacity 301.37 K, C_p = 29.80 cal·mol^{-1}·K^{-1}
124.68 J·mol^{-1}·K^{-1}
Temperature range –180 to 37 °C. Value is unsmoothed experimental datum.
Molecular Weight 217.8446
Wiswesser Line Notation QVYEE
Evaluation B

$C_2H_2Br_4$ (liq) 48KUR
1,1,2,2–Tetrabromoethane
Heat Capacity 298 K, C_p = 39.6 cal·mol^{-1}·K^{-1}
165.7 J·mol^{-1}·K^{-1}
Temperature range 15 to 132 °C, mean C_p, three temperatures.
Molecular Weight 345.6538
Wiswesser Line Notation EYEYEE
Evaluation D

$C_2H_2Cl_2$ (liq) 48KUR
1,2–Dichloroethene; 1,2–Dichloroethylene
Heat Capacity 298 K, C_p = 26.8 cal·mol^{-1}·K^{-1}
112.1 J·mol^{-1}·K^{-1}
Temperature range –31 to 54 °C, mean C_p, three temperatures.
Molecular Weight 96.9438
Wiswesser Line Notation G1U1G
Evaluation D

$C_2H_2Cl_2$ (liq) 34MEH 2
cis–1,2–Dichloroethene; cis–1,2–Dichloroethylene
Heat Capacity 288 K, C_p = 27.2 cal·mol^{-1}·K^{-1}
113.8 J·mol^{-1}·K^{-1}
One temperature
Molecular Weight 96.9438
Wiswesser Line Notation G1U1G –C
Evaluation C

$C_2H_2Cl_2$ (liq) 34MEH 2
trans–1,2–Dichloroethene; trans–1,2–Dichloroethylene
Heat Capacity 288 K, C_p = 27.0 cal·mol^{-1}·K^{-1}
113.0 J·mol^{-1}·K^{-1}
One temperature
Molecular Weight 96.9438
Wiswesser Line Notation G1U1G –T
Evaluation C

Table of Heat Capacities, Entropies, and Phase Transition Properties — Continued

$C_2H_2Cl_2$ (liq) 59HIL/MCD
1,1–Dichloroethene; Vinylidene chloride
Heat Capacity 298.15 K, C_p = 26.60 cal·mol⁻¹·K⁻¹ / 111.29 J·mol⁻¹·K⁻¹
Temperature range 13–290 K
Entropy 298.15 K, S = 48.17 cal·mol⁻¹·K⁻¹ / 201.54 J·mol⁻¹·K⁻¹
Phase Changes
c/liq 150.59 K, ΔH = 1557 cal·mol⁻¹ / 6514 J·mol⁻¹
ΔS = 10.34 cal·mol⁻¹·K⁻¹ / 43.26 J·mol⁻¹·K⁻¹
liq/g 298.15 K, ΔH = 6328 cal·mol⁻¹ / 26476 J·mol⁻¹
ΔS = 21.22 cal·mol⁻¹·K⁻¹ / 88.80 J·mol⁻¹·K⁻¹
P = 80.03 kPa
Molecular Weight 96.9438
Wiswesser Line Notation GYGU1
Evaluation A

$C_2H_2Cl_2O_2$ (c) 61GLA/TIM
Dichloroethanoic acid; Dichloroacetic acid
Heat Capacity 280.31 K, C_p = 43.57 cal·mol⁻¹·K⁻¹ / 182.30 J·mol⁻¹·K⁻¹
Temperature range –180 to 7 °C. Value is unsmoothed experimental datum.
Phase Changes
c/liq 286.5 K, ΔH = 2950 cal·mol⁻¹ / 12340 J·mol⁻¹
ΔS = 10.3 cal·mol⁻¹·K⁻¹ / 43.1 J·mol⁻¹·K⁻¹
Molecular Weight 128.9426
Wiswesser Line Notation QVYGG
Evaluation B

$C_2H_2Cl_2O_2$ (liq) 02LOU
Dichloroethanoic acid; Dichloroacetic acid
Heat Capacity 380 K, C_p = 45 cal·mol⁻¹·K⁻¹ / 188 J·mol⁻¹·K⁻¹
Temperature range 22 to 196 °C; mean value given.
Molecular Weight 128.9426
Wiswesser Line Notation QVYGG
Evaluation D

$C_2H_2Cl_4$ (liq) 48KUR
1,1,2,2–Tetrachloroethane
Heat Capacity 298 K, C_p = 39.5 cal·mol⁻¹·K⁻¹ / 165.3 J·mol⁻¹·K⁻¹
Temperature range 15 to 145 °C, mean C_p, four temperatures.
Molecular Weight 167.8498
Wiswesser Line Notation GYGYGG
Evaluation D

$C_2H_2O_4$ (c) 29PAR/KEL
Ethanedioic acid; Oxalic acid
Entropy 298.1 K, S = 28.7 cal·mol⁻¹·K⁻¹ / 120.1 J·mol⁻¹·K⁻¹
Extrapolation below 90 K, 9.2 cal·mol⁻¹·K⁻¹. Revision of previous data.
Molecular Weight 90.0354
Wiswesser Line Notation QVVQ
Evaluation C

$C_2H_2O_4$ (c) 39SAT/SOG
Ethanedioic acid; Oxalic acid
Heat Capacity 323 K, C_p = 28.2 cal·mol⁻¹·K⁻¹ / 118.0 J·mol⁻¹·K⁻¹
Temperature range 0 to 100 °C. Mean value given.
Molecular Weight 90.0354
Wiswesser Line Notation QVVQ
Evaluation C

$C_2H_2O_4$ (c) 64DAV
Ethanedioic acid; Oxalic acid
Heat Capacity 340 K, C_p = 35 cal·mol⁻¹·K⁻¹ / 146 J·mol⁻¹·K⁻¹
Temperature range 298–373 K. Mean value. Temperature range uncertain.
Molecular Weight 90.0354
Wiswesser Line Notation QVVQ
Evaluation D

$C_2H_2O_4$ (c) 82LUF/REE
Ethanedioic acid; Oxalic acid
Heat Capacity 298.15 K, C_p = 25.31 cal·mol⁻¹·K⁻¹ / 105.9 J·mol⁻¹·K⁻¹
Temperature range 5–320 K
Entropy 298.15 K, S = 27.62 cal·mol⁻¹·K⁻¹ / 115.6 J·mol⁻¹·K⁻¹
Molecular Weight 90.0354
Wiswesser Line Notation QVVQ
Evaluation A

$C_2H_3DBr_2$ (liq) 49WUY/JUN
1,2–Dibromoethane–d_1
Heat Capacity 310 K, C_p = 33.05 cal·mol⁻¹·K⁻¹ / 138.28 J·mol⁻¹·K⁻¹
One temperature
Molecular Weight 188.8678
Wiswesser Line Notation E2E &2/H–2
Evaluation C

C_2H_3Br (liq) 34MEH 2
Bromoethene; Vinyl bromide
Heat Capacity 288 K, C_p = 25.7 cal·mol⁻¹·K⁻¹ / 107.5 J·mol⁻¹·K⁻¹
One temperature
Molecular Weight 106.9497
Wiswesser Line Notation E1U1
Evaluation C

C_2H_3Br (liq) 34MEH
Bromoethene; Vinyl bromide
Heat Capacity 288 K, C_p = 25.8 cal·mol⁻¹·K⁻¹ / 107.9 J·mol⁻¹·K⁻¹
One measurement
Molecular Weight 106.9497
Wiswesser Line Notation E1U1
Evaluation C

$(C_2H_3Cl)_n$ (gls) 77CHA
Polyvinyl chloride
Heat Capacity 298.15 K, C_p = 14.11 cal·mol⁻¹·K⁻¹ / 59.03 J·mol⁻¹·K⁻¹
Temperature range 6–380 K. Value per monomer unit.
Entropy 298.15 K, S = 17.11 cal·mol⁻¹·K⁻¹ / 71.60 J·mol⁻¹·K⁻¹
Per monomer unit. Value is S–S_0.

Table of Heat Capacities, Entropies, and Phase Transition Properties — Continued

Molecular Weight 62.4987
Wiswesser Line Notation /*YG1*/
Evaluation A
One sample of suspension–polymerized PVC, two samples of bulk–polymerized PVC. Data represent annealed bulk material.

$C_2H_3ClF_2$ (liq) 37PER
1,1–Difluoro–1–chloroethane
Heat Capacity 291.6 K, C_p = 31.2 cal·mol^{-1}·K^{-1}
130.5 J·mol^{-1}·K^{-1}
Temperature range –180 to 18 °C. Value is unsmoothed experimental datum.
Phase Changes
c/liq 142.4 K, ΔH = 642 cal·mol^{-1}
2686 J·mol^{-1}
ΔS = 4.51 cal·mol^{-1}·K^{-1}
18.86 J·mol^{-1}·K^{-1}
Molecular Weight 100.4955
Wiswesser Line Notation GXFF1
Evaluation B

$C_2H_3ClF_2$ (liq) 37PER 2
1,1–Difluoro–1–chloroethane
Heat Capacity 291.6 K, C_p = 31.2 cal·mol^{-1}·K^{-1}
130.5 J·mol^{-1}·K^{-1}
Temperature range –180 to 18 °C. Value is unsmoothed experimental datum for saturated liquid.
Phase Changes
c/liq 142.4 K, ΔH = 642 cal·mol^{-1}
2686 J·mol^{-1}
ΔS = 4.51 cal·mol^{-1}·K^{-1}
18.86 J·mol^{-1}·K^{-1}
Molecular Weight 100.4955
Wiswesser Line Notation GXFF1
Evaluation B
Sample probably contains other isomers in small amounts.

$C_2H_3ClF_2$ (liq) 42RIE
1,1–Difluoro–1–chloroethane
Heat Capacity 294.9 K, C_p = 31.2 cal·mol^{-1}·K^{-1}
130.5 J·mol^{-1}·K^{-1}
Temperature range –61 to 21 °C. Value is unsmoothed experimental datum for saturated liquid.
Molecular Weight 100.4955
Wiswesser Line Notation GXFF1
Evaluation A

$C_2H_3ClF_2$ (liq) 41RIE 2
1,1–Difluoro–1–chloroethane
Heat Capacity 294.8 K, C_p = 31.2 cal·mol^{-1}·K^{-1}
130.5 J·mol^{-1}·K^{-1}
Temperature range –61 to 22 °C. Value is unsmoothed experimental datum.
Molecular Weight 100.4955
Wiswesser Line Notation GXFF1
Evaluation A

C_2H_3ClO (liq) 81REI
Acetyl chloride
Heat Capacity 298 K, C_p = 28.0 cal·mol^{-1}·K^{-1}
117.2 J·mol^{-1}·K^{-1}
Temperature range 289–343 K
Molecular Weight 78.4981
Wiswesser Line Notation GV1
Evaluation D

$C_2H_3ClO_2$ (liq) 50URA/SID
Chloroacetic acid
Heat Capacity 321.05 K, C_p = 43.0 cal·mol^{-1}·K^{-1}
179.9 J·mol^{-1}·K^{-1}
Temperature range 47 to 65 °C. Value is unsmoothed experimental datum. Maxima at 50.7, 56.9, and 61.2 °C.
Molecular Weight 94.4975
Wiswesser Line Notation QV1G
Evaluation B

$C_2H_3Cl_3$ (liq) 44RUB/LEV
1,1,1–Trichloroethane; Methyl chloroform
Heat Capacity 299.59 K, C_p = 34.49 cal·mol^{-1}·K^{-1}
144.31 J·mol^{-1}·K^{-1}
Temperature range 12–300 K. Value is unsmoothed experimental datum.
Entropy 298.15 K, S = 54.37 cal·mol^{-1}·K^{-1}
227.48 J·mol^{-1}·K^{-1}
Phase Changes
c,II/c,I 224.20 K, ΔH = 1786 cal·mol^{-1}
7473 J·mol^{-1}
ΔS = 7.97 cal·mol^{-1}·K^{-1}
33.33 J·mol^{-1}·K^{-1}
c,I/liq 240.2 K, ΔH = 450 cal·mol^{-1}
1880 J·mol^{-1}
ΔS = 1.9 cal·mol^{-1}·K^{-1}
7.8 J·mol^{-1}·K^{-1}
ΔH estimated because of errors in C_p above 225 K. Not used in calculation of entropy.
liq/g 286.53 K, ΔH = 7962 cal·mol^{-1}
33313 J·mol^{-1}
ΔS = 27.79 cal·mol^{-1}·K^{-1}
116.26 J·mol^{-1}·K^{-1}
P = 10.26 kPa
Molecular Weight 133.4047
Wiswesser Line Notation GXGG1
Evaluation A

$C_2H_3Cl_3$ (liq) 50CRO/SMY
1,1,1–Trichloroethane; Methyl chloroform
Heat Capacity 257.2 K, C_p = 33.2 cal·mol^{-1}·K^{-1}
138.9 J·mol^{-1}·K^{-1}
Temperature range 117–260 K. Value is unsmoothed experimental datum.
Phase Changes
c,III/c,II 205 K, ΔH = 50 cal·mol^{-1}
210 J·mol^{-1}
ΔS = 0.2 cal·mol^{-1}·K^{-1}
1.0 J·mol^{-1}·K^{-1}
c,II/c,I 223.6 K, ΔH = 1780 cal·mol^{-1}
7450 J·mol^{-1}
ΔS = 8.0 cal·mol^{-1}·K^{-1}
33.3 J·mol^{-1}·K^{-1}
c,I/liq 240.1 K, ΔH = 450 cal·mol^{-1}
1880 J·mol^{-1}
ΔS = 1.9 cal·mol^{-1}·K^{-1}
7.8 J·mol^{-1}·K^{-1}
Molecular Weight 133.4047
Wiswesser Line Notation GXGG1
Evaluation C

Table of Heat Capacities, Entropies, and Phase Transition Properties — Continued

$C_2H_3Cl_3$ (liq) 73AND/COU

1,1,1–Trichloroethane; Methyl chloroform

Heat Capacity 298.15 K, C_p = 34.51 cal·mol^{-1}·K^{-1}
144.4 J·mol^{-1}·K^{-1}

Temperature range 10–310 K

Entropy 298.15 K, S = 54.18 cal·mol^{-1}·K^{-1}
226.7 J·mol^{-1}·K^{-1}

Phase Changes

c,II/c,I 224.80 K, ΔH = 1790 cal·mol^{-1}
7490 J·mol^{-1}
ΔS = 7.96 cal·mol^{-1}·K^{-1}
33.32 J·mol^{-1}·K^{-1}

c,I/liq 243.13 K, ΔH = 562 cal·mol^{-1}
2350 J·mol^{-1}
ΔS = 2.31 cal·mol^{-1}·K^{-1}
9.67 J·mol^{-1}·K^{-1}

Molecular Weight 133.4047
Wiswesser Line Notation GXGG1
Evaluation A

$C_2H_3Cl_3$ (c) 50CRO/SMY

1,1,2–Trichloroethane

Heat Capacity 251.7 K, C_p = 34.2 cal·mol^{-1}·K^{-1}
143.1 J·mol^{-1}·K^{-1}

Temperature range 117–252 K. Value is unsmoothed experimental datum.

Phase Changes

c/liq 237.1 K, ΔH = 2720 cal·mol^{-1}
11380 J·mol^{-1}
ΔS = 11.5 cal·mol^{-1}·K^{-1}
48.0 J·mol^{-1}·K^{-1}

Molecular Weight 133.4047
Wiswesser Line Notation GYG1G
Evaluation C

$C_2H_3F_3$ (liq) 44RUS/GOL

1,1,1–Trifluoroethane; Freon 143

Heat Capacity 220 K, C_p = 26.21 cal·mol^{-1}·K^{-1}
109.66 J·mol^{-1}·K^{-1}

Temperature range 15–226 K

Entropy 225.85 K, S = 43.56 cal·mol^{-1}·K^{-1}
224.10 J·mol^{-1}·K^{-1}

Phase Changes

c,II/c,I 156.35 K, ΔH = 71 cal·mol^{-1}
297 J·mol^{-1}
ΔS = 0.45 cal·mol^{-1}·K^{-1}
1.90 J·mol^{-1}·K^{-1}

c,I/liq 161.82 K, ΔH = 1480 cal·mol^{-1}
6192 J·mol^{-1}
ΔS = 9.15 cal·mol^{-1}·K^{-1}
38.26 J·mol^{-1}·K^{-1}

liq/g 224.40 K, ΔH = 4583 cal·mol^{-1}
19175 J·mol^{-1}
ΔS = 20.42 cal·mol^{-1}·K^{-1}
85.45 J·mol^{-1}·K^{-1}
P = 94.54 kPa

Molecular Weight 84.0409
Wiswesser Line Notation FXFF1
Evaluation A

$C_2H_3KO_2$ (c) 75FER/SAN

Potassium acetate

Phase Changes

c,II/c,I 423 K, ΔH = 100 cal·mol^{-1}
420 J·mol^{-1}
ΔS = 0.24 cal·mol^{-1}·K^{-1}
1.0 J·mol^{-1}·K^{-1}

c,I/liq 578.7 K, ΔH = 3640 cal·mol^{-1}
15230 J·mol^{-1}
ΔS = 6.3 cal·mol^{-1}·K^{-1}
26.3 J·mol^{-1}·K^{-1}

Molecular Weight 98.1428
Wiswesser Line Notation OV1 .KA
Evaluation C

$C_2H_3LiO_2$ (c) 75FER/SAN

Lithium acetate

Phase Changes

c/liq 557 K, ΔH = 2840 cal·mol^{-1}
11880 J·mol^{-1}
ΔS = 5.1 cal·mol^{-1}·K^{-1}
21.3 J·mol^{-1}·K^{-1}

Molecular Weight 65.9855
Wiswesser Line Notation OV1 .LI
Evaluation C

C_2H_3N (liq) 07WAL

Acetonitrile; Methyl cyanide

Heat Capacity 291 K, C_p = 21 cal·mol^{-1}·K^{-1}
88 J·mol^{-1}·K^{-1}

One temperature

Molecular Weight 41.0524
Wiswesser Line Notation NC1
Evaluation D

C_2H_3N (liq) 65PUT/MCE

Acetonitrile; Methyl cyanide

Heat Capacity 298.15 K, C_p = 21.86 cal·mol^{-1}·K^{-1}
91.46 J·mol^{-1}·K^{-1}

Temperature range 20–300 K

Entropy 298.15 K, S = 35.76 cal·mol^{-1}·K^{-1}
149.62 J·mol^{-1}·K^{-1}

Phase Changes

c,II/c,I 216.9 K, ΔH = 214.6 cal·mol^{-1}
897.9 J·mol^{-1}
ΔS = 0.99 cal·mol^{-1}·K^{-1}
4.14 J·mol^{-1}·K^{-1}

c,I/liq 229.32 K, ΔH = 1952 cal·mol^{-1}
8167 J·mol^{-1}
ΔS = 8.51 cal·mol^{-1}·K^{-1}
35.61 J·mol^{-1}·K^{-1}

liq/g 298.15 K, ΔH = 7941 cal·mol^{-1}
33225 J·mol^{-1}
ΔS = 26.63 cal·mol^{-1}·K^{-1}
111.44 J·mol^{-1}·K^{-1}
P = 11.83 kPa

Molecular Weight 41.0524
Wiswesser Line Notation NC1
Evaluation A

C_2H_3N (liq) 71HAL/BAL

Acetonitrile; Methyl cyanide

Heat Capacity 297 K, C_p = 19.7 cal·mol^{-1}·K^{-1}
82.5 J·mol^{-1}·K^{-1}

One temperature

Molecular Weight 41.0524
Wiswesser Line Notation NC1
Evaluation C

Table of Heat Capacities, Entropies, and Phase Transition Properties — Continued

C_2H_3N (liq) 79VIS/SOM
Acetonitrile; Methyl cyanide
Heat Capacity 298.15 K, C_p = 21.9 cal·mol^{-1}·K^{-1}
91.7 J·mol^{-1}·K^{-1}
One temperature
Molecular Weight 41.0524
Wiswesser Line Notation NC1
Evaluation B

$(C_2H_3NO)_n$ (c) 81FIN/KUM
Polyglycine I
Heat Capacity 298.15 K, C_p = 24.51 cal·mol^{-1}·K^{-1}
102.5 J·mol^{-1}·K^{-1}
Temperature range 150–375 K, Polyglycine I (β–sheet). Data given graphically. C_p calculated from equation.
Molecular Weight 75.0670
Wiswesser Line Notation /*MV1*/
Evaluation B

$(C_2H_3NO)_n$ (c) 81FIN/KUM
Polyglycine II
Heat Capacity 298.15 K, C_p = 22.34 cal·mol^{-1}·K^{-1}
93.46 J·mol^{-1}·K^{-1}
Temperature range 150–375 K, Polyglycine II (3_1helix). Data given graphically. C_p calculated from equation.
Molecular Weight 75.0670
Wiswesser Line Notation /*MV1*/
Evaluation B

$C_2H_3NaO_2$ (c) 55STR
Sodium acetate
Heat Capacity 291.18 K, C_p = 21.05 cal·mol^{-1}·K^{-1}
88.07 J·mol^{-1}·K^{-1}
Temperature range 13–292 K. Value is unsmoothed experimental datum.
Entropy 298.15 K, S = 29.42 cal·mol^{-1}·K^{-1}
123.09 J·mol^{-1}·K^{-1}
Molecular Weight 82.0343
Wiswesser Line Notation OV1 .NA
Evaluation A

$C_2H_3NaO_2$ (c,IV) 75FER/SAN
Sodium acetate
Heat Capacity 340 K, C_p = 26.7 cal·mol^{-1}·K^{-1}
111.7 J·mol^{-1}·K^{-1}
Temperature range 340–610 K
Phase Changes
c,IV/c,III 414 K, ΔH = 70 cal·mol^{-1}
290 J·mol^{-1}
ΔS = 0.17 cal·mol^{-1}·K^{-1}
0.7 J·mol^{-1}·K^{-1}
c,III/c,II 465 K, ΔH = 100 cal·mol^{-1}
420 J·mol^{-1}
ΔS = 0.2 cal·mol^{-1}·K^{-1}
0.8 J·mol^{-1}·K^{-1}
c,II/c,I 527 K, ΔH = 40 cal·mol^{-1}
170 J·mol^{-1}
ΔS = 0.08 cal·mol^{-1}·K^{-1}
0.3 J·mol^{-1}·K^{-1}
c,I/liq 601.3 K, ΔH = 4290 cal·mol^{-1}
17950 J·mol^{-1}
ΔS = 7.1 cal·mol^{-1}·K^{-1}
29.9 J·mol^{-1}·K^{-1}
Molecular Weight 82.0343
Wiswesser Line Notation OV1 .NA
Evaluation B

$C_2H_3NaO_2{\cdot}3H_2O$ (c) 32STU
Sodium acetate trihydrate
Heat Capacity 325 K, C_p = 77 cal·mol^{-1}·K^{-1}
322 J·mol^{-1}·K^{-1}
Temperatures above and below melting point.
Phase Changes
c/liq 331.7 K, ΔH = 4840 cal·mol^{-1}
20250 J·mol^{-1}
ΔS = 15 cal·mol^{-1}·K^{-1}
61 J·mol^{-1}·K^{-1}
Molecular Weight 136.0799
Wiswesser Line Notation OV1 .NA &QH 3
Evaluation D

$C_2H_4D_2O_2$ (liq) 62RAB/NIK
1,2–Dihydroxyethane–d_2; 1,2–Ethanediol–d_2; Ethylene glycol–d_2
Heat Capacity 298 K, C_p = 37.27 cal·mol^{-1}·K^{-1}
155.94 J·mol^{-1}·K^{-1}
Temperature range 10 to 55 °C
Molecular Weight 64.0806
Wiswesser Line Notation Q2Q &1/3/H–2 2
Evaluation B

$C_2H_4D_2O_2$ (liq) 67NIK/RAB 2
1,2–Dihydroxyethane–d_2; 1,2–Ethanediol–d_2; Ethylene glycol–d_2
Heat Capacity 298.15 K, C_p = 37.80 cal·mol^{-1}·K^{-1}
158.16 J·mol^{-1}·K^{-1}
Temperature range 80–300 K
Phase Changes
c/liq 258.8. K, ΔH = 2330 cal·mol^{-1}
9749 J·mol^{-1}
ΔS = 9.00 cal·mol^{-1}·K^{-1}
37.67 J·mol^{-1}·K^{-1}
Molecular Weight 64.0806
Wiswesser Line Notation Q2Q &1/3/H–2 2
Evaluation B

C_2H_4BrCl (liq) 39RAI
1–Bromo–2–chloroethane
Heat Capacity 300 K, C_p = 31.1 cal·mol^{-1}·K^{-1}
130.1 J·mol^{-1}·K^{-1}
Temperature range 90–320 K. Data graphically only. Value read from graph.
Phase Changes
c,II/c,I 182 K, ΔH = 740 cal·mol^{-1}
3100 J·mol^{-1}
ΔS = 4.1 cal·mol^{-1}·K^{-1}
17.0 J·mol^{-1}·K^{-1}
Lambda–type transition. Temperature is maximum in specific heat curve.
c,I/liq 256.4 K, ΔH = 2300 cal·mol^{-1}
9625 J·mol^{-1}
ΔS = 9.0 cal·mol^{-1}·K^{-1}
37.5 J·mol^{-1}·K^{-1}
Molecular Weight 143.4106
Wiswesser Line Notation G2E
Evaluation C

Table of Heat Capacities, Entropies, and Phase Transition Properties — Continued

$C_2H_4Br_2$ (liq) 39RAI
1,2-Dibromoethane; Ethylene dibromide
Heat Capacity 300 K, C_p = 31.8 cal·mol^{-1}·K^{-1}
129.5 J·mol^{-1}·K^{-1}
Temperature range 90–320 K. Data graphically only. Value read from graph.
Phase Changes
c,II/c,I 250.6 K, ΔH = 450 cal·mol^{-1}
1880 J·mol^{-1}
ΔS = 1.8 cal·mol^{-1}·K^{-1}
7.5 J·mol^{-1}·K^{-1}
c,I/liq 283.1 K, ΔH = 2590 cal·mol^{-1}
10835 J·mol^{-1}
ΔS = 9.1 cal·mol^{-1}·K^{-1}
38.3 J·mol^{-1}·K^{-1}
Molecular Weight 187.8616
Wiswesser Line Notation E2E
Evaluation C

$C_2H_4Br_2$ (liq) 40PIT 2
1,2-Dibromoethane; Ethylene dibromide
Heat Capacity 298.15 K, C_p = 32.5 cal·mol^{-1}·K^{-1}
136.0 J·mol^{-1}·K^{-1}
Temperature range 15–300 K
Entropy 298.15 K, S = 53.37 cal·mol^{-1}·K^{-1}
223.30 J·mol^{-1}·K^{-1}
Phase Changes
c,II/c,I 249.54 K, ΔH = 463.8 cal·mol^{-1}
1940.5 J·mol^{-1}
ΔS = 1.86 cal·mol^{-1}·K^{-1}
7.78 J·mol^{-1}·K^{-1}
c/I/liq 283.0 K, ΔH = 2615.8 cal·mol^{-1}
10944.5 J·mol^{-1}
ΔS = 9.24 cal·mol^{-1}·K^{-1}
38.67 J·mol^{-1}·K^{-1}
Molecular Weight 187.8616
Wiswesser Line Notation E2E
Evaluation A

$C_2H_4Br_2$ (liq) 48KUR
1,2-Dibromoethane; Ethylene dibromide
Heat Capacity 298 K, C_p = 32.2 cal·mol^{-1}·K^{-1}
134.7 J·mol^{-1}·K^{-1}
Temperature range 16 to 127 °C, mean C_p, four temperatures.
Molecular Weight 187.8616
Wiswesser Line Notation E2E
Evaluation D

$C_2H_4Br_2$ (liq) 49WUY/JUN
1,2-Dibromoethane; Ethylene dibromide
Heat Capacity 310 K, C_p = 32.30 cal·mol^{-1}·K^{-1}
135.14 J·mol^{-1}·K^{-1}
One temperature
Molecular Weight 187.8616
Wiswesser Line Notation E2E
Evaluation C

$C_2H_4Br_2$ (liq) 69WIL/SCH
1,2-Dibromoethane; Ethylene dibromide
Heat Capacity 298.15 K, C_p = 32.3 cal·mol^{-1}·K^{-1}
135.0 J·mol^{-1}·K^{-1}
Temperature range 20, 30, 40C
Molecular Weight 187.8616
Wiswesser Line Notation E2E
Evaluation B

$C_2H_4Cl_2$ (liq) 81REI
1,2-Dichloroethane; Ethylene dichloride
Heat Capacity 298 K, C_p = 29.2 cal·mol^{-1}·K^{-1}
122.2 J·mol^{-1}·K^{-1}
Temperature range 290–364 K
Molecular Weight 98.9596
Wiswesser Line Notation G2G
Evaluation D

$C_2H_4Cl_2$ (liq) 39RAI
1,2-Dichloroethane; Ethylene dichloride
Heat Capacity 300 K, C_p = 31.3 cal·mol^{-1}·K^{-1}
131.0 J·mol^{-1}·K^{-1}
Temperature range 90–320 K. Data graphically only. Value read from graph.
Phase Changes
c,II/c,I 175 K, ΔH = 680 cal·mol^{-1}
2845 J·mol^{-1}
ΔS = 3.9 cal·mol^{-1}·K^{-1}
16.2 J·mol^{-1}·K^{-1}
Lambda–type transition. Temperature is maximum in specific heat curve.
c,I/liq 237.6 K, ΔH = 2090 cal·mol^{-1}
8745 J·mol^{-1}
ΔS = 8.80 cal·mol^{-1}·K^{-1}
36.8 J·mol^{-1}·K^{-1}
Molecular Weight 98.9596
Wiswesser Line Notation G2G
Evaluation C

$C_2H_4Cl_2$ (liq) 40PIT 2
1,2-Dichloroethane; Ethylene dichloride
Heat Capacity 298.15 K, C_p = 30.8 cal·mol^{-1}·K^{-1}
128.9 J·mol^{-1}·K^{-1}
Temperature range 15–308 K
Entropy 298.15 K, S = 49.84 cal·mol^{-1}·K^{-1}
208.53 J·mol^{-1}·K^{-1}
Phase Changes
Anomalous region at 175–180 K, probably a lambda–type transition. No transition heat or temperature given.
c/liq 237.2 K, ΔH = 2112.0 cal·mol^{-1}
8836.6 J·mol^{-1}
ΔS = 8.90 cal·mol^{-1}·K^{-1}
37.25 J·mol^{-1}·K^{-1}
Molecular Weight 98.9596
Wiswesser Line Notation G2G
Evaluation A

$C_2H_4Cl_2$ (liq) 48KUR
1,2-Dichloroethane; Ethylene dichloride
Heat Capacity 298 K, C_p = 29.4 cal·mol^{-1}·K^{-1}
123.0 J·mol^{-1}·K^{-1}
Temperature range –25 to 86 °C, mean C_p, four temperatures.
Molecular Weight 98.9596
Wiswesser Line Notation G2G
Evaluation D

$C_2H_4Cl_2$ (liq) 51SIE/CRU
1,2-Dichloroethane; Ethylene dichloride
Heat Capacity 293 K, C_p = 29.7 cal·mol^{-1}·K^{-1}
124.3 J·mol^{-1}·K^{-1}
One temperature
Molecular Weight 98.9596
Wiswesser Line Notation G2G
Evaluation B

Table of Heat Capacities, Entropies, and Phase Transition Properties — Continued

$C_2H_4Cl_2$ (liq) 55STA/TUP
1,2–Dichloroethane; Ethylene dichloride
Heat Capacity 298 K, C_p = 31.00 cal·mol^{-1}·K^{-1}
129.70 J·mol^{-1}·K^{-1}
Temperature range 284–348 K
Molecular Weight 98.9596
Wiswesser Line Notation G2G
Evaluation B

$C_2H_4Cl_2$ (liq) 55RUI
1,2–Dichloroethane; Ethylene dichloride
Heat Capacity 298.15 K, C_p = 30.96 cal·mol^{-1}·K^{-1}
129.54 J·mol^{-1}·K^{-1}
Temperature range 7–50 °C
Molecular Weight 98.9596
Wiswesser Line Notation G2G
Evaluation B

$C_2H_4Cl_2$ (liq) 67RAS/GAN
1,2–Dichloroethane; Ethylene dichloride
Heat Capacity 293 K, C_p = 30.9 cal·mol^{-1}·K^{-1}
129.2 J·mol^{-1}·K^{-1}
Temperature range 293–353 K
Molecular Weight 98.9596
Wiswesser Line Notation G2G
Evaluation C

$C_2H_4Cl_2$ (liq) 69WIL/SCH
1,2–Dichloroethane; Ethylene dichloride
Heat Capacity 298.15 K, C_p = 30.7 cal·mol^{-1}·K^{-1}
128.6 J·mol^{-1}·K^{-1}
Temperature range 20, 30, 40 °C
Molecular Weight 98.9596
Wiswesser Line Notation G2G
Evaluation B

$C_2H_4Cl_2$ (liq) 79WIL/FAR
1,2–Dichloroethane; Ethylene dichloride
Heat Capacity 298.15 K, C_p = 30.83 cal·mol^{-1}·K^{-1}
128.99 J·mol^{-1}·K^{-1}
One temperature.
Molecular Weight 98.9596
Wiswesser Line Notation G2G
Evaluation B

$C_2H_4Cl_2$ (liq) 79WIL/GRO
1,2–Dichloroethane; Ethylene dichloride
Heat Capacity 298.15 K, C_p = 30.8 cal·mol^{-1}·K^{-1}
129.0 J·mol^{-1}·K^{-1}
One temperature
Molecular Weight 98.9596
Wiswesser Line Notation G2G
Evaluation B

$C_2H_4Cl_2$ (liq) 81REI
1,1–Dichloroethane; Ethylidene chloride
Heat Capacity 298 K, C_p = 28.9 cal·mol^{-1}·K^{-1}
120.9 J·mol^{-1}·K^{-1}
Temperature range 287–344 K
Molecular Weight 98.9596
Wiswesser Line Notation GYG1
Evaluation D

$C_2H_4Cl_2$ (liq) 48KUR
1,1–Dichloroethane; Ethylidene chloride
Heat Capacity 298 K, C_p = 30.4 cal·mol^{-1}·K^{-1}
127.2 J·mol^{-1}·K^{-1}
Temperature range –51 to 55 °C, mean C_p, four temperatures.
Molecular Weight 98.9596
Wiswesser Line Notation GYG1
Evaluation D

$C_2H_4Cl_2$ (liq) 56LI/PIT
1,1–Dichloroethane; Ethylidene chloride
Heat Capacity 298.15 K, C_p = 30.18 cal·mol^{-1}·K^{-1}
126.27 J·mol^{-1}·K^{-1}
Temperature range 14–294 K
Entropy 298.15 K, S = 50.61 cal·mol^{-1}·K^{-1}
211.75 J·mol^{-1}·K^{-1}
Phase Changes
c/liq 176.18 K, ΔH = 1881 cal·mol^{-1}
7870 J·mol^{-1}
ΔS = 1068 cal·mol^{-1}·K^{-1}
44.67 J·mol^{-1}·K^{-1}
liq/g 293 K, ΔH = 7409 cal·mol^{-1}
31000 J·mol^{-1}
ΔS = 25.29 cal·mol^{-1}·K^{-1}
105.81 J·mol^{-1}·K^{-1}
P = 30.32 kPa
Molecular Weight 98.9596
Wiswesser Line Notation GYG1
Evaluation A

$C_2H_4N_2O_2$ (c) 65EGA/WAK
Oxamide
Heat Capacity 298.15 K, C_p = 27.22 cal·mol^{-1}·K^{-1}
113.89 J·mol^{-1}·K^{-1}
Temperature range 10–310 K
Entropy 298.15 K, S = 28.23 cal·mol^{-1}·K^{-1}
118.11 J·mol^{-1}·K^{-1}
Molecular Weight 88.0658
Wiswesser Line Notation ZVVZ
Evaluation A
Triclinic form

$C_2H_4N_4$ (c) 52STE/BER
Dicyandiamide
Heat Capacity 294.63 K, C_p = 28.14 cal·mol^{-1}·K^{-1}
117.74 J·mol^{-1}·K^{-1}
Temperature range 15–300 K. Value is unsmoothed experimental datum.
Entropy 298.15 K, S = 30.90 cal·mol^{-1}·K^{-1}
129.29 J·mol^{-1}·K^{-1}
Molecular Weight 84.0804
Wiswesser Line Notation NCMYZUM
Evaluation A

$C_2H_4N_4$ (c) 64DAV
Dicyandiamide
Heat Capacity 340 K, C_p = 34 cal·mol^{-1}·K^{-1}
142 J·mol^{-1}·K^{-1}
Temperature range 298–373 K. Mean value. Temperature range uncertain.
Molecular Weight 84.0804
Wiswesser Line Notation NCMYZUM
Evaluation D

Table of Heat Capacities, Entropies, and Phase Transition Properties — Continued

C_2H_4O (liq) 49GIA/GOR
Oxirane; Ethylene oxide
Heat Capacity 285 K, C_p = 20.77 cal·mol^{-1}·K^{-1}
86.90 J·mol^{-1}·K^{-1}
Temperature range 15–283 K
Entropy 283.60 K, S = 35.72 cal·mol^{-1}·K^{-1}
149.45 J·mol^{-1}·K^{-1}
Phase Changes
c/liq 160.65 K, ΔH = 1236.4 cal·mol^{-1}
5173.1 J·mol^{-1}
ΔS = 7.70 cal·mol^{-1}·K^{-1}
32.20 J·mol^{-1}·K^{-1}
liq/g 283.66 K, ΔH = 6101 cal·mol^{-1}
25527 J·mol^{-1}
ΔS = 21.51 cal·mol^{-1}·K^{-1}
89.99 J·mol^{-1}·K^{-1}
P = 101.325 kPa
Molecular Weight 44.0530
Wiswesser Line Notation T3OTJ
Evaluation A

C_2H_4O (liq) 47CON/ELV
Ethanal; Acetaldehyde
Heat Capacity 273 K, C_p = 23.00 cal·mol^{-1}·K^{-1}
96.21 J·mol^{-1}·K^{-1}
One temperature
Molecular Weight 44.0530
Wiswesser Line Notation VH1
Evaluation B

$C_2H_4O_2$ (liq) 81REI
Ethanoic acid; Acetic acid
Heat Capacity 298 K, C_p = 29.5 cal·mol^{-1}·K^{-1}
123.5 J·mol^{-1}·K^{-1}
Temperature range 292–358 K
Molecular Weight 60.0524
Wiswesser Line Notation QV1
Evaluation D

$C_2H_4O_2$ (liq) 25PAR/KEL
Ethanoic acid; Acetic acid
Heat Capacity 294.7 K, C_p = 29.5 cal·mol^{-1}·K^{-1}
123.4 J·mol^{-1}·K^{-1}
Temperature range 87–295 K. Value is unsmoothed experimental datum.
Entropy 298.1 K, S = 46.3 cal·mol^{-1}·K^{-1}
193.7 J·mol^{-1}·K^{-1}
Extrapolation below 90 K. 18.36 cal·mol^{-1}·K^{-1}.
Phase Changes
c/liq 289.8 K, ΔH = 2803 cal·mol^{-1}
11728 J·mol^{-1}
ΔS = 9.67 cal·mol^{-1}·K^{-1}
40.47 J·mol^{-1}·K^{-1}
Molecular Weight 60.0524
Wiswesser Line Notation QV1
Evaluation B(C_p),C(S)

$C_2H_4O_2$ (liq) 29PAR/KEL
Ethanoic acid; Acetic acid
Entropy 298.1 K, S = 38.2 cal·mol^{-1}·K^{-1}
159.8 J·mol^{-1}·K^{-1}
Extrapolation below 90 K, 10.2 cal·mol^{-1}·K^{-1}. Revision of previous data.
Molecular Weight 60.0524
Wiswesser Line Notation QV1
Evaluation C

$C_2H_4O_2$ (liq) 32NEU
Ethanoic acid; Acetic acid
Heat Capacity 297.1 K, C_p = 29.0 cal·mol^{-1}·K^{-1}
121.3 J·mol^{-1}·K^{-1}
Temperature range 23.9–80.5 °C. Value is unsmoothed experimental datum.
Molecular Weight 60.0524
Wiswesser Line Notation QV1
Evaluation c

$C_2H_4O_2$ (liq) 34RAD/JUL
Ethanoic acid; Acetic acid
Heat Capacity 298 K, C_p = 28.8 cal·mol^{-1}·K^{-1}
120.5 J·mol^{-1}·K^{-1}
One temperature
Molecular Weight 60.0524
Wiswesser Line Notation QV1
Evaluation C

$C_2H_4O_2$ (liq) 58SWI/ZIE
Ethanoic acid; Acetic acid
Heat Capacity 332 K, C_p = 33.4 cal·mol^{-1}·K^{-1}
139.7 J·mol^{-1}·K^{-1}
Mean value 22 to 96 °C
Molecular Weight 60.0524
Wiswesser Line Notation QV1
Evaluation C

$C_2H_4O_2$ (liq) 82MAR/AND
Ethanoic acid ; Acetic acid
Heat Capacity 298.15 K, C_p = 29.42 cal·mol^{-1}·K^{-1}
123.1 J·mol^{-1}·K^{-1}
Temperature range 13–450 K, Data also given by equation.
Entropy 298.15 K, S = 37.76 cal·mol^{-1}·K^{-1}
158.0 J·mol^{-1}·K^{-1}
Phase Changes
c,I/liq 298.69 K, ΔH = 2801.1 cal·mol^{-1}
11720 J·mol^{-1}
ΔS = 9.67 cal·mol^{-1}·K^{-1}
40.5 J·mol^{-1}·K^{-1}
Molecular Weight 60.0524
Wiswesser Line Notation QV1
Evaluation A

$C_2H_4O_2$ (liq) 34MEH 2
Methyl methanoate; Methyl formate
Heat Capacity 288 K, C_p = 29.0 cal·mol^{-1}·K^{-1}
121.3 J·mol^{-1}·K^{-1}
One temperature
Molecular Weight 60.0524
Wiswesser Line Notation VHO1
Evaluation C

$C_2H_4O_2$ (liq) 71HAL/BAL
Methyl methanoate; Methyl formate
Heat Capacity 297 K, C_p = 22.8 cal·mol^{-1}·K^{-1}
95.5 J·mol^{-1}·K^{-1}
One temperature
Molecular Weight 60.0524
Wiswesser Line Notation VHO1
Evaluation C

Table of Heat Capacities, Entropies, and Phase Transition Properties — Continued

$C_2H_4O_2$ (liq) 79FUC
Methyl methanoate; Methyl formate
Heat Capacity 298.15 K, C_p = 28.6 cal·mol^{-1}·K^{-1}
119.7 J·mol^{-1}·K^{-1}
One temperature
Molecular Weight 60.0524
Wiswesser Line Notation VHO1
Evaluation B

C_2H_5DO (liq) 62RAB/NIK
Ethanol–d$_1$; Ethyl alcohol–d$_1$
Heat Capacity 298 K, C_p = 27.78 cal·mol^{-1}·K^{-1}
116.23 J·mol^{-1}·K^{-1}
Temperature range 15 to 55 °C
Molecular Weight 47.0750
Wiswesser Line Notation Q2 &1/H–2
Evaluation B

C_2H_5DO (liq) 67NIK/RAB
Ethanol–d$_1$; Ethyl alcohol–d$_1$
Heat Capacity 250 K, C_p = 24.40 cal·mol^{-1}·K^{-1}
102.09 J·mol^{-1}·K^{-1}
Temperature range 80–250 K
Phase Changes
c,II/c,I 113.3 K, ΔH = 800 cal·mol^{-1}
3347 J·mol^{-1}
ΔS = 7.06 cal·mol^{-1}·K^{-1}
29.54 J·mol^{-1}·K^{-1}
c,I/liq 156.9 K, ΔH = 1030 cal·mol^{-1}
4310 J·mol^{-1}
ΔS = 6.56 cal·mol^{-1}·K^{-1}
27.47 J·mol^{-1}·K^{-1}
Molecular Weight 47.0750
Wiswesser Line Notation Q2 &1/H–2
Evaluation A

C_2H_5Br (liq) 48KUR
Bromoethane; Ethyl bromide
Heat Capacity 298 K, C_p = 24.1 cal·mol^{-1}·K^{-1}
100.8 J·mol^{-1}·K^{-1}
Temperature range –50 to 37 °C, mean C_p, five temperatures.
Molecular Weight 108.9655
Wiswesser Line Notation E2
Evaluation D

C_2H_5Cl (liq) 40RIE
Chloroethane; Ethyl chloride
Heat Capacity 298.1 K, C_p = 26.0 cal·mol^{-1}·K^{-1}
108.8 J·mol^{-1}·K^{-1}
Temperature range –48 to 46 °C
Molecular Weight 64.5145
Wiswesser Line Notation G2
Evaluation A

C_2H_5Cl (liq) 41RIE
Chloroethane; Ethyl chloride
Heat Capacity 298 K, C_p = 26.0 cal·mol^{-1}·K^{-1}
108.8 J·mol^{-1}·K^{-1}
Temperature range –48 to 45 °C
Molecular Weight 64.5145
Wiswesser Line Notation G2
Evaluation A

C_2H_5Cl (liq) 48GOR/GIA
Chloroethane; Ethyl chloride
Heat Capacity 290 K, C_p = 24.69 cal·mol^{-1}·K^{-1}
103.3 J·mol^{-1}·K^{-1}
Temperature range 13–287 K
Entropy 285.42 K, S = 44.52 cal·mol^{-1}·K^{-1}
186.27 J·mol^{-1}·K^{-1}
Phase Changes
c/liq 134.82 K, ΔH = 1064 cal·mol^{-1}
4452 J·mol^{-1}
ΔS = 7.893 cal·mol^{-1}·K^{-1}
33.02 J·mol^{-1}·K^{-1}
liq/g 285.42 K, ΔH = 5892 cal·mol^{-1}
24652 J·mol^{-1}
ΔS = 20.64 cal·mol^{-1}·K^{-1}
86.37 J·mol^{-1}·K^{-1}
P = 101.325 kPa
Molecular Weight 64.5145
Wiswesser Line Notation G2
Evaluation A

C_2H_5Cl (liq) 48KUR
Chloroethane; Ethyl chloride
Heat Capacity 288 K, C_p = 26.1 cal·mol^{-1}·K^{-1}
109.6 J·mol^{-1}·K^{-1}
Temperature range –67 to 15 °C, mean C_p, three temperatures.
Molecular Weight 64.5145
Wiswesser Line Notation G2
Evaluation D

C_2H_5I (liq) 48KUR
Iodoethane; Ethyl iodide
Heat Capacity 298 K, C_p = 27.5 cal·mol^{-1}·K^{-1}
115.1 J·mol^{-1}·K^{-1}
Temperature range –37 to 70 °C, mean C_p, three temperatures.
Molecular Weight 155.9660
Wiswesser Line Notation I2
Evaluation D

C_2H_5NO (liq) 74VIS/SOM
N–Methylmethanamide; N–Methylformamide
Heat Capacity 298.15 K, C_p = 30.14 cal·mol^{-1}·K^{-1}
126.1 J·mol^{-1}·K^{-1}
One temperature
Molecular Weight 59.0676
Wiswesser Line Notation VHM1
Evaluation A

C_2H_5NO (liq) 79VIS/SOM
N–Methylmethanamide; N–Methylformamide
Heat Capacity 298.15 K, C_p = 29.9 cal·mol^{-1}·K^{-1}
125.2 J·mol^{-1}·K^{-1}
One temperature
Molecular Weight 59.0676
Wiswesser Line Notation VHM1
Evaluation B

C_2H_5NO (c) 40CAM/CAM
Ethanamide; Acetamide
Heat Capacity 293 K, C_p = 15.9 cal·mol^{-1}·K^{-1}
66.5 J·mol^{-1}·K^{-1}
One temperature
Molecular Weight 59.0676
Wiswesser Line Notation ZV1
Evaluation C

Table of Heat Capacities, Entropies, and Phase Transition Properties — Continued

$C_2H_5NO_2$ (liq) 66LIU/ZIE
Nitroethane
Heat Capacity 298.15 K, C_p = 32.08 cal·mol⁻¹·K⁻¹ / 134.22 J·mol⁻¹·K⁻¹
Temperature range 80–300 K
Phase Changes
c/liq 183.69 K, ΔH = 2355 cal·mol⁻¹ / 9853 J·mol⁻¹
ΔS = 12.82 cal·mol⁻¹·K⁻¹ / 53.64 J·mol⁻¹·K⁻¹
Molecular Weight 75.0670
Wiswesser Line Notation WN2
Evaluation A

$C_2H_5NO_2$ (c) 33PAR/HUF
Aminoethanoic acid; Glycine
Heat Capacity 299.5 K, C_p = 24.02 cal·mol⁻¹·K⁻¹ / 100.50 J·mol⁻¹·K⁻¹
Temperature range 93–300 K. Value is unsmoothed experimental datum.
Entropy 298.1 K, S = 26.1 cal·mol⁻¹·K⁻¹ / 109.2 J·mol⁻¹·K⁻¹
Extrapolation below 90 K, 7.55 cal·mol⁻¹·K⁻¹.
Molecular Weight 75.0670
Wiswesser Line Notation Z1VQ
Evaluation B(C_p),C(S)

$C_2H_5NO_2$ (c) 60HUT/COL
Aminoethanoic acid; Glycine
Heat Capacity 298.15 K, C_p = 23.71 cal·mol⁻¹·K⁻¹ / 99.20 J·mol⁻¹·K⁻¹
Temperature range 11–305 K
Entropy 298.15 K, S = 24.74 cal·mol⁻¹·K⁻¹ / 103.51 J·mol⁻¹·K⁻¹
Molecular Weight 75.0670
Wiswesser Line Notation Z1VQ
Evaluation A

$C_2H_5NO_2$ (c) 75SPI/WAD
Aminoethanoic acid; Glycine
Heat Capacity 298.15 K, C_p = 23.7 cal·mol⁻¹·K⁻¹ / 99.3 J·mol⁻¹·K⁻¹
One temperature
Molecular Weight 75.0670
Wiswesser Line Notation Z1VQ
Evaluation B

$C_2H_5NO_3$ (liq) 54GRA/SMI
Ethyl nitrate
Heat Capacity 298 K, C_p = 40.7 cal·mol⁻¹·K⁻¹ / 170.3 J·mol⁻¹·K⁻¹
Temperature range 21–293 K
Entropy 298 K, S = 59.08 cal·mol⁻¹·K⁻¹ / 247.2 J·mol⁻¹·K⁻¹
Phase Changes
c/liq 178.6 K, ΔH = 2038 cal·mol⁻¹ / 8527 J·mol⁻¹
ΔS = 11.41 cal·mol⁻¹·K⁻¹ / 47.74 J·mol⁻¹·K⁻¹
Molecular Weight 91.0664
Wiswesser Line Notation WNO2
Evaluation B

$C_2H_5NO_4$ (c) 39SAT/SOG
Ammonium acid oxalate
Heat Capacity 323 K, C_p = 36.5 cal·mol⁻¹·K⁻¹ / 152.7 J·mol⁻¹·K⁻¹
Temperature range 0 to 100 °C. Mean value.
Molecular Weight 107.0658
Wiswesser Line Notation QVVQ &ZH
Evaluation C

$C_2H_5N_3O_2$ (c) 82LUF/REE
Biuret
Heat Capacity 298.15 K, C_p = 31.39 cal·mol⁻¹·K⁻¹ / 131.3 J·mol⁻¹·K⁻¹
Temperature range 5–320 K
Entropy 298.15 K, S = 34.91 cal·mol⁻¹·K⁻¹ / 146.1 J·mol⁻¹·K⁻¹
Molecular Weight 103.0894
Wiswesser Line Notation ZVMVZ
Evaluation A

C_2H_6 (liq) 30WIE/HUB
Ethane
Heat Capacity 200 K, C_p = 17.80 cal·mol⁻¹·K⁻¹ / 74.48 J·mol⁻¹·K⁻¹
Temperature range 67–305.2 K, Heat capacity of saturated liquid given to 295 K is 32.54 cal·mol⁻¹·K⁻¹, 136.1 J·mol⁻¹·K⁻¹.
Phase Changes
c,I/liq 89.50 K, ΔH = 667.5 cal·mol⁻¹ / 2793 J·mol⁻¹
ΔS = 7.46 cal·mol⁻¹·K⁻¹ / 31.2 J·mol⁻¹·K⁻¹
liq/g 184.46 K
Molecular Weight 30.0694
Wiswesser Line Notation 2H
Evaluation A

C_2H_6 (liq) 37WIT/KEM
Ethane
Heat Capacity 180 K, C_p = 17.26 cal·mol⁻¹·K⁻¹ / 72.22 J·mol⁻¹·K⁻¹
Temperature range 15–185 K
Entropy 184.1 K, S = 30.29 cal·mol⁻¹·K⁻¹ / 126.7 J·mol⁻¹·K⁻¹
Entropy from 0–15 K calculated using a Debye function.
Phase Changes
c,I/liq 89.87 K, ΔH = 682.9 cal·mol⁻¹ / 2857 J·mol⁻¹
ΔS = 7.60 cal·mol⁻¹·K⁻¹ / 31.8 J·mol⁻¹·K⁻¹
liq/g 184.1 K, ΔH = 3514 cal·mol⁻¹ / 14703 J·mol⁻¹
ΔS = 19.09 cal·mol⁻¹·K⁻¹ / 79.87 J·mol⁻¹·K⁻¹
Molecular Weight 30.0694
Wiswesser Line Notation 2H
Evaluation A

C_2H_6 (liq) 76ROD
Ethane
Heat Capacity 100.32 K, C_p = 16.36 cal·mol⁻¹·K⁻¹ / 68.44 J·mol⁻¹·K⁻¹
Temperature range 93–301 K (saturation line), 91–330 K, pressures from 0 to 33MPa.
Molecular Weight 30.0694
Wiswesser Line Notation 2H
Evaluation A

Table of Heat Capacities, Entropies, and Phase Transition Properties — Continued

C_2H_6 (liq) 76ROD 2
Ethane
Heat Capacity 94 K, C_p = 16.4 cal·mol^{-1}·K^{-1}
68.5 J·mol^{-1}·K^{-1}
From data 90.3–94 K. Average value over range.
Phase Changes
c,II/c,I 89.77 K, ΔH = 582.6 cal·mol^{-1}
2437.5 J·mol^{-1}
ΔS = 6.49 cal·mol^{-1}·K^{-1}
27.15 J·mol^{-1}·K^{-1}
Molecular Weight 30.0694
Wiswesser Line Notation 2H
Evaluation B

C_2H_6Cd (liq) 56LI
Dimethyl cadmium; Cadmium dimethyl
Heat Capacity 298.15 K, C_p = 31.55 cal·mol^{-1}·K^{-1}
132.01 J·mol^{-1}·K^{-1}
Temperature range 15–300 K
Entropy 298.15 K, S = 48.25 cal·mol^{-1}·K^{-1}
201.88 J·mol^{-1}·K^{-1}
Phase Changes
c,II/c,I 254.35 K, ΔH = 363.5 cal·mol^{-1}
1520.9 J·mol^{-1}
ΔS = 1.43 cal·mol^{-1}·K^{-1}
5.98 J·mol^{-1}·K^{-1}
c,I/liq 270.48 K, ΔH = 1873 cal·mol^{-1}
7837 J·mol^{-1}
ΔS = 6.92 cal·mol^{-1}·K^{-1}
28.97 J·mol^{-1}·K^{-1}
liq/g 291.5 K, ΔH = 9153 cal·mol^{-1}
38296 J·mol^{-1}
ΔS = 31.4 cal·mol^{-1}·K^{-1}
131.4 J·mol^{-1}·K^{-1}
P = 3.033 kPa
Molecular Weight 142.4794
Wiswesser Line Notation 1–CD–1
Evaluation A

$C_2H_6N_4O_4$ (c) 73KRI/LIC
Ethylenedinitramine
Heat Capacity 298 K, C_p = 41.9 cal·mol^{-1}·K^{-1}
175.3 J·mol^{-1}·K^{-1}
Temperature range 200–448 K. Equation only.
Molecular Weight 150.0938
Wiswesser Line Notation WNM2MNW
Evaluation C

C_2H_6O (liq) 41KEN/SAG
2–Oxapropane; Dimethyl ether; Methoxymethane
Heat Capacity 240 K, C_p = 24.45 cal·mol^{-1}·K^{-1}
102.30 J·mol^{-1}·K^{-1}
Temperature range 14–240 K
Entropy 200 K, S = 35.03 cal·mol^{-1}·K^{-1}
146.57 J·mol^{-1}·K^{-1}
Phase Changes
c/liq 131.66 K, ΔH = 1179.8 cal·mol^{-1}
4936.3 J·mol^{-1}
ΔS = 8.96 cal·mol^{-1}·K^{-1}
37.49 J·mol^{-1}·K^{-1}
liq/g 248.34 K, ΔH = 5141 cal·mol^{-1}
21510 J·mol^{-1}
ΔS = 20.70 cal·mol^{-1}·K^{-1}
86.61 J·mol^{-1}·K^{-1}
P = 101.325 kPa
Molecular Weight 46.0688
Wiswesser Line Notation 1O1
Evaluation A

C_2H_6O (liq) 81REI
Ethanol; Ethyl alcohol
Heat Capacity 298 K, C_p = 26.8 cal·mol^{-1}·K^{-1}
112.1 J·mol^{-1}·K^{-1}
Temperature range 288–346 K
Molecular Weight 46.0688
Wiswesser Line Notation Q2
Evaluation D

C_2H_6O (liq) 07WAL
Ethanol; Ethyl alcohol
Heat Capacity 293 K, C_p = 26 cal·mol^{-1}·K^{-1}
109 J·mol^{-1}·K^{-1}
One temperature
Molecular Weight 46.0688
Wiswesser Line Notation Q2
Evaluation D

C_2H_6O (liq) 24WIL/DAN
Ethanol; Ethyl alcohol
Heat Capacity 303 K, C_p = 27.5 cal·mol^{-1}·K^{-1}
115.1 J·mol^{-1}·K^{-1}
Temperature range 303–333 K, Equation only.
Molecular Weight 46.0688
Wiswesser Line Notation Q2
Evaluation C

C_2H_6O (liq) 25PAR
Ethanol; Ethyl alcohol
Heat Capacity 298.0 K, C_p = 27.1 cal·mol^{-1}·K^{-1}
113.4 J·mol^{-1}·K^{-1}
Temperature range 87–298 K. Value is unsmoothed experimental datum.
Entropy 298.1 K, S = 42.3 cal·mol^{-1}·K^{-1}
177.0 J·mol^{-1}·K^{-1}
Extrapolation below 90 K, 13.19cal·mol^{-1}·K^{-1}.
Phase Changes
c/liq 158.7 K, ΔH = 1186 cal·mol^{-1}
4962 J·mol^{-1}
ΔS = 7.47 cal·mol^{-1}·K^{-1}
21.22 J·mol^{-1}·K^{-1}
Molecular Weight 46.0688
Wiswesser Line Notation Q2
Evaluation B(C_p),C(S)

C_2H_6O (liq) 29KEL 2
Ethanol; Ethyl alcohol
Heat Capacity 294.31 K, C_p = 26.26 cal·mol^{-1}·K^{-1}
109.87 J·mol^{-1}·K^{-1}
Temperature range 16–298 K. Value is unsmoothed experimental datum.
Entropy 298.15 K, S = 38.4 cal·mol^{-1}·K^{-1}
160.7 J·mol^{-1}·K^{-1}
Phase Changes
c/liq 158.5 K, ΔH = 1200 cal·mol^{-1}
5021 J·mol^{-1}
ΔS = 7.57 cal·mol^{-1}·K^{-1}
31.68 J·mol^{-1}·K^{-1}
Molecular Weight 46.0688
Wiswesser Line Notation Q2
Evaluation B

Table of Heat Capacities, Entropies, and Phase Transition Properties — Continued

C_2H_6O (liq) 29MIT/HAR
Ethanol; Ethyl alcohol
Heat Capacity 270 K, C_p = 25.4 cal·mol⁻¹·K⁻¹ ; 106.3 J·mol⁻¹·K⁻¹
Temperature range 190–270 K
Molecular Weight 46.0688
Wiswesser Line Notation Q2
Evaluation B

C_2H_6O (liq) 29PAR/KEL
Ethanol; Ethyl alcohol
Entropy 298.1 K, S = 38.4 $cal{\cdot}mol^{-1}{\cdot}K^{-1}$; 160.7 $J{\cdot}mol^{-1}{\cdot}K^{-1}$
Extrapolation below 90 K, 9.3 $cal{\cdot}mol^{-1}{\cdot}K^{-1}$. Revision of previous data.
Molecular Weight 46.0688
Wiswesser Line Notation Q2
Evaluation C

C_2H_6O (liq) 31FIO/GIN
Ethanol; Ethyl alcohol
Heat Capacity 313.15 K, C_p = 28.37 $cal{\cdot}mol^{-1}{\cdot}K^{-1}$; 118.72 $J{\cdot}mol^{-1}{\cdot}K^{-1}$
Temperature range 40 to 110 °C
Molecular Weight 46.0688
Wiswesser Line Notation Q2
Evaluation A

C_2H_6O (liq) 36ERN/WAT
Ethanol; Ethyl alcohol
Heat Capacity 298 K, C_p = 24.7 $cal{\cdot}mol^{-1}{\cdot}K^{-1}$; 103.3 $J{\cdot}mol^{-1}{\cdot}K^{-1}$
One temperature
Molecular Weight 46.0688
Wiswesser Line Notation Q2
Evaluation C

C_2H_6O (liq) 39BYK
Ethanol; Ethyl alcohol
Heat Capacity 298 K, C_p = 26.7 $cal{\cdot}mol^{-1}{\cdot}K^{-1}$; 111.7 $J{\cdot}mol^{-1}{\cdot}K^{-1}$
One temperature
Molecular Weight 46.0688
Wiswesser Line Notation Q2
Evaluation A

C_2H_6O (liq) 60SWI/ZIE
Ethanol; Ethyl alcohol
Heat Capacity 316 K, C_p = 28.4 $cal{\cdot}mol^{-1}{\cdot}K^{-1}$; 118.8 $J{\cdot}mol^{-1}{\cdot}K^{-1}$
Mean value 21 to 66 °C
Molecular Weight 46.0688
Wiswesser Line Notation Q2
Evaluation C

C_2H_6O (liq) 61GRE
Ethanol; Ethyl alcohol
Heat Capacity 298.15 K, C_p = 26.76 $cal{\cdot}mol^{-1}{\cdot}K^{-1}$; 111.96 $J{\cdot}mol^{-1}{\cdot}K^{-1}$
Temperature range 16–350 K
Entropy 298.15 K, S = 38.53 $cal{\cdot}mol^{-1}{\cdot}K^{-1}$; 161.21 $J{\cdot}mol^{-1}{\cdot}K^{-1}$
Molecular Weight 46.0688
Wiswesser Line Notation Q2
Evaluation A
Reevaluation of 29KEL2 and 31FIO/GIN

C_2H_6O (liq) 62RAB/NIK
Ethanol; Ethyl alcohol
Heat Capacity 298 K, C_p = 26.83 $cal{\cdot}mol^{-1}{\cdot}K^{-1}$; 112.26 $J{\cdot}mol^{-1}{\cdot}K^{-1}$
Temperature range 15 to 55 °C
Molecular Weight 46.0688
Wiswesser Line Notation Q2
Evaluation B

C_2H_6O (liq) 67NIK/RAB
Ethanol; Ethyl alcohol
Heat Capacity 250 K, C_p = 23.31 $cal{\cdot}mol^{-1}{\cdot}K^{-1}$; 97.53 $J{\cdot}mol^{-1}{\cdot}K^{-1}$
Temperature range 80–250 K
Phase Changes
c,II/c,I 111.4 K, ΔH = 750 $cal{\cdot}mol^{-1}$; 3138 $J{\cdot}mol^{-1}$
ΔS = 6.73 $cal{\cdot}mol^{-1}{\cdot}K^{-1}$; 28.17 $J{\cdot}mol^{-1}{\cdot}K^{-1}$
c,I/liq 158.8 K, ΔH = 1110 $cal{\cdot}mol^{-1}$; 4644 $J{\cdot}mol^{-1}$
ΔS = 6.99 $cal{\cdot}mol^{-1}{\cdot}K^{-1}$; 29.24 $J{\cdot}mol^{-1}{\cdot}K^{-1}$
Molecular Weight 46.0688
Wiswesser Line Notation Q2
Evaluation A

C_2H_6O (liq) 70PAZ/PAZ
Ethanol; Ethyl alcohol
Heat Capacity 313.2 K, C_p = 28.3 $cal{\cdot}mol^{-1}{\cdot}K^{-1}$; 118.4 $J{\cdot}mol^{-1}{\cdot}K^{-1}$
One temperature
Molecular Weight 46.0688
Wiswesser Line Notation Q2
Evaluation B

C_2H_6O (liq) 76FOR/BEN
Ethanol; Ethyl alcohol
Heat Capacity 298.15 K, C_p = 26.85 $cal{\cdot}mol^{-1}{\cdot}K^{-1}$; 112.33 $J{\cdot}mol^{-1}{\cdot}K^{-1}$
One temperature
Molecular Weight 46.0688
Wiswesser Line Notation Q2
Evaluation B

C_2H_6O (liq) 77HAI/SUG
Ethanol; Ethyl alcohol
Heat Capacity 298.15 K, C_p = 26.9 $cal{\cdot}mol^{-1}{\cdot}K^{-1}$; 112.5 $J{\cdot}mol^{-1}{\cdot}K^{-1}$
Temperature range 14–300 K. Also glass, supercooled liquid, metastable crystal.
Entropy 298.15 K, S = 38.21 $cal{\cdot}mol^{-1}{\cdot}K^{-1}$; 159.86 $J{\cdot}mol^{-1}{\cdot}K^{-1}$
Phase Changes
c,II/liq 127.5 K, ΔH = 158 $cal{\cdot}mol^{-1}$; 659 $J{\cdot}mol^{-1}$
ΔS = 1.24 $cal{\cdot}mol^{-1}{\cdot}K^{-1}$; 5.19 $J{\cdot}mol^{-1}{\cdot}K^{-1}$
c,I/liq 159.00 K, ΔH = 1179 $cal{\cdot}mol^{-1}$; 4931 $J{\cdot}mol^{-1}$
ΔS = 7.42 $cal{\cdot}mol^{-1}{\cdot}K^{-1}$; 31.01 $J{\cdot}mol^{-1}{\cdot}K^{-1}$
Molecular Weight 46.0688
Wiswesser Line Notation Q2
Evaluation A

Table of Heat Capacities, Entropies, and Phase Transition Properties — Continued

C_2H_6O (liq) 79BRO/ZIE
Ethanol; Ethyl alcohol
Heat Capacity 298.15 K, C_p = 26.80 cal·mol^{-1}·K^{-1}
112.15 J·mol^{-1}·K^{-1}
Temperature range 159–306 K. Results as equation only.
Molecular Weight 46.0688
Wiswesser Line Notation Q2
Evaluation B

C_2H_6OS (liq) 60KEN/LIN
Dimethyl sulfoxide
Heat Capacity 298.15 K, C_p = 35.6 cal·mol^{-1}·K^{-1}
149.0 J·mol^{-1}·K^{-1}
Temperature range 298, 343 K
Molecular Weight 78.1288
Wiswesser Line Notation OS1&1
Evaluation B

C_2H_6OS (liq) 70CLE/WES
Dimethyl sulfoxide
Heat Capacity 298.15 K, C_p = 36.61 cal·mol^{-1}·K^{-1}
153.18 J·mol^{-1}·K^{-1}
Temperature range 5–350 K
Entropy 298.15 K, S = 45.12 cal·mol^{-1}·K^{-1}
188.78 J·mol^{-1}·K^{-1}
Phase Changes
c/liq 291.67 K, ΔH = 3434 cal·mol^{-1}
14368 J·mol^{-1}
ΔS = 11.77 cal·mol^{-1}·K^{-1}
49.26 J·mol^{-1}·K^{-1}
Molecular Weight 78.1288
Wiswesser Line Notation OS1&1
Evaluation A

C_2H_6OS (liq) 79VIS/SOM
Dimethyl sulfoxide
Heat Capacity 298.15 K, C_p = 37.3 cal·mol^{-1}·K^{-1}
155.9 J·mol^{-1}·K^{-1}
One temperature
Molecular Weight 78.1288
Wiswesser Line Notation OS1&1
Evaluation B

$C_2H_6O_2$ (liq) 25PAR/KEL
1,2–Dihydroxyethane; 1,2–Ethanediol; Ethylene glycol
Heat Capacity 293.0 K, C_p = 35.7 cal·mol^{-1}·K^{-1}
149.4 J·mol^{-1}·K^{-1}
Temperature range 88–293 K. Value is unsmoothed experimental datum.
Entropy 298.1 K, S = 42.9 cal·mol^{-1}·K^{-1}
179.5 J·mol^{-1}·K^{-1}
Extrapolation below 90 K, 11.46 cal·mol^{-1}·K^{-1}.
Phase Changes
c/liq 260.8 K, ΔH = 2778 cal·mol^{-1}
11623 J·mol^{-1}
ΔS = 10.65 cal·mol^{-1}·K^{-1}
44.57 J·mol^{-1}·K^{-1}
Molecular Weight 62.0682
Wiswesser Line Notation Q2Q
Evaluation B(C_p),C(S)

$C_2H_6O_2$ (liq) 29PAR/KEL
1,2–Dihydroxyethane; 1,2–Ethanediol; Ethylene glycol
Entropy 298.1 K, S = 39.9 cal·mol^{-1}·K^{-1}
166.9 J·mol^{-1}·K^{-1}
Extrapolation below 90 K, 8.5 cal·mol^{-1}·K^{-1}. Revision of previous data.
Molecular Weight 62.0682
Wiswesser Line Notation Q2Q
Evaluation C

$C_2H_6O_2$ (liq) 32NEI/KUR
1,2–Dihydroxyethane; 1,2–Ethanediol; Ethylene glycol
Heat Capacity 293.4 K, C_p = 34.8 cal·mol^{-1}·K^{-1}
145.6 J·mol^{-1}·K^{-1}
Temperature range 20.2 to 78.4 °C. Value is unsmoothed experimental datum.
Molecular Weight 62.0682
Wiswesser Line Notation Q2Q
Evaluation C

$C_2H_6O_2$ (liq) 62RAB/NIK
1,2–Dihydroxyethane; 1,2–Ethanediol; Ethylene glycol
Heat Capacity 298 K, C_p = 35.58 cal·mol^{-1}·K^{-1}
148.87 J·mol^{-1}·K^{-1}
Temperature range 10 to 55 °C
Molecular Weight 62.0682
Wiswesser Line Notation Q2Q
Evaluation B

$C_2H_6O_2$ (liq) 65TUN/MIS
1,2–Dihydroxyethane; 1,2–Ethanediol; Ethylene glycol
Heat Capacity 298 K, C_p = 35.2 cal·mol^{-1}·K^{-1}
147.3 J·mol^{-1}·K^{-1}
One temperature
Molecular Weight 62.0682
Wiswesser Line Notation Q2Q
Evaluation B

$C_2H_6O_2$ (liq) 67NIK/RAB 2
1,2–Dihydroxyethane; 1,2–Ethanediol; Ethylene glycol
Heat Capacity 298.15 K, C_p = 35.93 cal·mol^{-1}·K^{-1}
150.33 J·mol^{-1}·K^{-1}
Temperature range 80–300 K
Phase Changes
c/liq 260.6 K, ΔH = 2380 cal·mol^{-1}
9958 J·mol^{-1}
ΔS = 9.13 cal·mol^{-1}·K^{-1}
38.21 J·mol^{-1}·K^{-1}
Molecular Weight 62.0682
Wiswesser Line Notation Q2Q
Evaluation B

$C_2H_6O_2$ (liq) 70PAZ/PAZ
1,2–Dihydroxyethane; 1,2–Ethanediol; Ethylene glycol
Heat Capacity 301.2 K, C_p = 36.0 cal·mol^{-1}·K^{-1}
150.6 J·mol^{-1}·K^{-1}
Temperature range 28, 40 °C
Molecular Weight 62.0682
Wiswesser Line Notation Q2Q
Evaluation B

Table of Heat Capacities, Entropies, and Phase Transition Properties — Continued

$C_2H_6O_2$ (liq) 72KAW/OTA
1,2–Dihydroxyethane; 1,2–Ethanediol; Ethylene glycol
Heat Capacity 303 K, C_p = 34.7 cal·mol^{-1}·K^{-1}
145.2 J·mol^{-1}·K^{-1}

One temperature
Molecular Weight 62.0682
Wiswesser Line Notation Q2Q
Evaluation B

$C_2H_6O_2$ (liq) 79STE/TAM
1,2–Dihydroxyethane; 1,2–Ethanediol; Ethylene glycol
Heat Capacity 298 K, C_p = 35.7 cal·mol^{-1}·K^{-1}
149.3 J·mol^{-1}·K^{-1}

Temperature range 273–493 K
Molecular Weight 62.0682
Wiswesser Line Notation Q2Q
Evaluation B

$C_2H_6O_2S$ (c) 70CLE/WES
Dimethyl sulfone
Heat Capacity 298.15 K, C_p = 29.96 cal·mol^{-1}·K^{-1}
125.35 J·mol^{-1}·K^{-1}

Temperature range 5–410 K
Entropy 298.15 K, S = 34.77 cal·mol^{-1}·K^{-1}
145.48 J·mol^{-1}·K^{-1}

Phase Changes
c/liq 382.01 K, ΔH = 4374 cal·mol^{-1}
18301 J·mol^{-1}
ΔS = 11.45 cal·mol^{-1}·K^{-1}
47.91 J·mol^{-1}·K^{-1}

Molecular Weight 94.1282
Wiswesser Line Notation WS1&1
Evaluation A

C_2H_6S (liq) 42OSB/DOE
Dimethylsulfide; 2–Thiapropane
Heat Capacity 298.15 K, C_p = 28.23 cal·mol^{-1}·K^{-1}
118.11 J·mol^{-1}·K^{-1}

Temperature range 11–287 K
Entropy 298.15 K, S = 46.94 cal·mol^{-1}·K^{-1}
196.40 J·mol^{-1}·K^{-1}

Phase Changes
c/liq 174.85 K, ΔH = 1908.4 cal·mol^{-1}
7984.7 J·mol^{-1}
ΔS = 10.91 cal·mol^{-1}·K^{-1}
45.67 J·mol^{-1}·K^{-1}
liq/g 291.06 K, ΔH = 6688 cal·mol^{-1}
27983 J·mol^{-1}
ΔS = 22.98 cal·mol^{-1}·K^{-1}
96.14 J·mol^{-1}·K^{-1}
P = 35.40 kPa

Molecular Weight 62.1294
Wiswesser Line Notation 1S1
Evaluation A

C_2H_6S (liq) 52MCC/SCO
Ethanethiol; Ethyl mercaptan
Heat Capacity 299.05 K, C_p = 28.20 cal·mol^{-1}·K^{-1}
117.99 J·mol^{-1}·K^{-1}

Temperature range 14–315 K. Unsmoothed experimental datum.
Entropy 298.15 K, S = 49.48 cal·mol^{-1}·K^{-1}
207.02 J·mol^{-1}·K^{-1}

Phase Changes
c/liq 195.26 K, ΔH = 1189 cal·mol^{-1}
4975 J·mol^{-1}
ΔS = 6.09 cal·mol^{-1}·K^{-1}
25.48 J·mol^{-1}·K^{-1}

Molecular Weight 62.1294
Wiswesser Line Notation SH2
Evaluation A

$C_2H_6S_2$ (liq) 50SCO/FIN
2,3–Dithiabutane; Dimethyl disulfide
Heat Capacity 298.15 K, C_p = 34.92 cal·mol^{-1}·K^{-1}
146.11 J·mol^{-1}·K^{-1}

Temperature range 13–352 K
Entropy 298.15 K, S = 56.26 cal·mol^{-1}·K^{-1}
235.39 J·mol^{-1}·K^{-1}

Phase Changes
c/liq 188.44 K, ΔH = 2197.1 cal·mol^{-1}
9192.7 J·mol^{-1}
ΔS = 11.66 cal·mol^{-1}·K^{-1}
48.78 J·mol^{-1}·K^{-1}

Molecular Weight 94.1894
Wiswesser Line Notation 1SS1
Evaluation A

C_2H_7N (liq) 39AST/EID
Dimethylamine
Heat Capacity 280.44 K, C_p = 32.69 cal·mol^{-1}·K^{-1}
136.77 J·mol^{-1}·K^{-1}

Temperature range 14–280 K. Value for saturated liquid.
Entropy 280.03 K, S = 41.55 cal·mol^{-1}·K^{-1}
173.85 J·mol^{-1}·K^{-1}

Saturated liquid at boiling point.
Phase Changes
c/liq 180.97 K, ΔH = 1420 cal·mol^{-1}
5941 J·mol^{-1}
ΔS = 7.85 cal·mol^{-1}·K^{-1}
32.83 J·mol^{-1}·K^{-1}
liq/g 280.03 K, ΔH = 6330 cal·mol^{-1}
26485 J·mol^{-1}
ΔS = 22.60 cal·mol^{-1}·K^{-1}
94.58 J·mol^{-1}·K^{-1}
P = 101.325 kPa

Molecular Weight 45.0840
Wiswesser Line Notation 1M1
Evaluation A

$C_2H_7NO_3S$ (c) 40HUF/FOX
2–Aminoethanesulfonic acid; Taurine
Heat Capacity 300.3 K, C_p = 33.59 cal·mol^{-1}·K^{-1}
140.54 J·mol^{-1}·K^{-1}

Temperature range 90–298 K. Value is unsmoothed experimental datum.
Entropy 298.15 K, S = 36.8 cal·mol^{-1}·K^{-1}
154.0 J·mol^{-1}·K^{-1}

Extrapolation below 90 K, 10.56 cal·mol^{-1}·K^{-1}.
Molecular Weight 125.1422
Wiswesser Line Notation Z2SWQ
Evaluation B(C_p),C(S)

Table of Heat Capacities, Entropies, and Phase Transition Properties — Continued

$C_2H_7N_3O_3$ (c) 40HUF/ELL
Guanidine carbonate
Heat Capacity 298.1 K, C_p = 61.87 cal·mol^{-1}·K^{-1}
258.86 J·mol^{-1}·K^{-1}
Temperature range 86–298 K. Value is unsmoothed experimental datum.
Entropy 298.1 K, S = 70.59 cal·mol^{-1}·K^{-1}
295.35 J·mol^{-1}·K^{-1}
Extrapolation below 90 K, 22.31 cal·mol^{-1}·K^{-1}.
Molecular Weight 121.0956
Wiswesser Line Notation ZYZUM 2 &QVQ
Evaluation A(C_p),C(S)

$C_2H_8N_2$ (liq) 50HOU/MAS
1,2–Diaminoethane; Ethylenediamine
Heat Capacity 313 K, C_p = 42.7 cal·mol^{-1}·K^{-1}
178.7 J·mol^{-1}·K^{-1}
Temperature range 313–333 K
Molecular Weight 60.0986
Wiswesser Line Notation Z2Z
Evaluation B

$C_2H_8N_2$ (liq) 75MES/FIN
1,2–Diaminoethane; Ethylenediamine
Heat Capacity 298.15 K, C_p = 41.25 cal·mol^{-1}·K^{-1}
172.59 J·mol^{-1}·K^{-1}
Temperature range 11–335 K
Entropy 298.15 K, S = 48.38 cal·mol^{-1}·K^{-1}
202.42 J·mol^{-1}·K^{-1}
Phase Changes
c,II/c,I 189.0 K, ΔH = 116.5 cal·mol^{-1}
487.4 J·mol^{-1}
ΔS = 0.62 cal·mol^{-1}·K^{-1}
2.58 J·mol^{-1}·K^{-1}
c,I/liq 284.29 K, ΔH = 5397.5 cal·mol^{-1}
22583.1 J·mol^{-1}
ΔS = 18.99 cal·mol^{-1}·K^{-1}
79.44 J·mol^{-1}·K^{-1}
Molecular Weight 60.0986
Wiswesser Line Notation Z2Z
Evaluation A

$C_2H_8N_2$ (liq) 53AST/WOO
N,N–Dimethylhydrazine
Heat Capacity 298.15 K, C_p = 39.21 cal·mol^{-1}·K^{-1}
164.05 J·mol^{-1}·K^{-1}
Temperature range 13–300 K
Entropy 298.15 K, S = 47.86 cal·mol^{-1}·K^{-1}
200.25 J·mol^{-1}·K^{-1}
Phase Changes
c/liq 215.95 K, ΔH = 2407.4 cal·mol^{-1}
10072.6 J·mol^{-1}
ΔS = 11.15 cal·mol^{-1}·K^{-1}
46.64 J·mol^{-1}·K^{-1}
liq/g 298.15 K, ΔH = 8366 cal·mol^{-1}
35003 J·mol^{-1}
ΔS = 28.1 cal·mol^{-1}·K^{-1}
117.4 J·mol^{-1}·K^{-1}
P = 20.90 kPa
Molecular Weight 60.0986
Wiswesser Line Notation ZN1&1
Evaluation A

$C_2H_8N_2O_4$ (c) 39SAT/SOG
Ammonium oxalate
Heat Capacity 323 K, C_p = 50.6 cal·mol^{-1}·K^{-1}
211.7 J·mol^{-1}·K^{-1}
Temperature range 0 to 100 °C. Mean value.
Molecular Weight 124.0962
Wiswesser Line Notation QVVQ &ZH 2
Evaluation C

$C_2H_{12}CdCl_4N_2$ (c) 81RAH/CLA
Tetrachlorobis–(methylammonium) cadmium II
Heat Capacity 298.15 K, C_p = 62.6 cal·mol^{-1}·K^{-1}
261.9 J·mol^{-1}·K^{-1}
Temperature range 2.3–301 K
Entropy 298.15 K, S = 98.2 cal·mol^{-1}·K^{-1}
410.9 J·mol^{-1}·K^{-1}
Phase Changes
c,IV/c,III 164.2 K, ΔH = 418 cal·mol^{-1}
1749 J·mol^{-1}
ΔS = 2.58 cal·mol^{-1}·K^{-1}
10.8 J·mol^{-1}·K^{-1}
c,III/c,II 282 K, ΔH = 15.9 cal·mol^{-1}
66.5 J·mol^{-1}
ΔS = 0.06 cal·mol^{-1}·K^{-1}
0.25 J·mol^{-1}·K^{-1}
c,II/c,I 484 K, ΔH = no data given.
Molecular Weight 304.3455
Wiswesser Line Notation CD Z1&2 G4
Evaluation B

$C_2H_{12}Cl_4MnN_2$ (c) 82WHI/GRA
Tetrachlorobis–(methylammonium) manganese II
Heat Capacity 298.15 K, C_p = 63.0 cal·mol^{-1}·K^{-1}
263.6 J·mol^{-1}·K^{-1}
Temperature range 10–300 K
Entropy 298.15 K, S = 96.3 cal·mol^{-1}·K^{-1}
402.9 J·mol^{-1}·K^{-1}
Phase Changes
c,III/c,II 94.37 K, ΔH = 174 cal·mol^{-1}
728 J·mol^{-1}
ΔS = 1.85 cal·mol^{-1}·K^{-1}
7.74 J·mol^{-1}·K^{-1}
c,II/c,I 257.02 K, ΔH = 28 cal·mol^{-1}
117 J·mol^{-1}
ΔS = 0.108 cal·mol^{-1}·K^{-1}
0.452 J·mol^{-1}·K^{-1}
Molecular Weight 260.8802
Wiswesser Line Notation MN Z1&2 G4
Evaluation A

$C_2D_4Br_2$ (liq) 49WUY/JUN
1,2–Dibromoethane–d_4
Heat Capacity 310 K, C_p = 35.64 cal·mol^{-1}·K^{-1}
149.12 J·mol^{-1}·K^{-1}
One temperature
Molecular Weight 191.8864
Wiswesser Line Notation E2E &2/H–2 4
Evaluation C

Table of Heat Capacities, Entropies, and Phase Transition Properties — Continued

C_2N_2 (liq) 39RUE/GIA
Cyanogen
Heat Capacity 255 K, C_p = 25.27 cal·mol^{-1}·K^{-1}
105.73 J·mol^{-1}·K^{-1}
Temperature range 15–252 K
Entropy 252.0 K, S = 33.18 cal·mol^{-1}·K^{-1}
138.83 J·mol^{-1}·K^{-1}
Phase Changes
c/liq 245.32 K, ΔH = 1938 cal·mol^{-1}
8109 J·mol^{-1}
ΔS = 7.90 cal·mol^{-1}·K^{-1}
33.05 J·mol^{-1}·K^{-1}
liq/g 252.0 K, ΔH = 5576 cal·mol^{-1}
23330 J·mol^{-1}
ΔS = 22.13 cal·mol^{-1}·K^{-1}
92.58 J·mol^{-1}·K^{-1}
P = 101.325 kPa
Molecular Weight 52.0354
Wiswesser Line Notation NCCN
Evaluation A

$C_2N_6O_{12}$ (c,I) 70KRI/LIC
Hexanitroethane
Heat Capacity 291 K, C_p = 78 cal·mol^{-1}·K^{-1}
327 J·mol^{-1}·K^{-1}
Temperature range 190–350 K
Phase Changes
c,II/c,I 291 K, ΔH = 2964 cal·mol^{-1}
12400 J·mol^{-1}
ΔS = 10.2 cal·mol^{-1}·K^{-1}
42.6 J·mol^{-1}·K^{-1}
Temperature range 289–292 K; 291 K assumed.
Molecular Weight 300.0550
Wiswesser Line Notation WNXNWNWXNWNWNW
Evaluation C

$C_2Na_2O_4$ (c) 37CHE/CHE
Sodium oxalate
Heat Capacity 281 K, C_p = 31 cal·mol^{-1}·K^{-1}
130 J·mol^{-1}·K^{-1}
Temperature range 273–373 K. Mean values, three temperatures.
Molecular Weight 133.9992
Wiswesser Line Notation OVVO .NA 2
Evaluation C

C_2O_4Pb (c) 60KAP/STR
Lead(II) oxalate
Heat Capacity 299.37 K, C_p = 24.54 cal·mol^{-1}·K^{-1}
102.7 J·mol^{-1}·K^{-1}
Temperature range 66–300 K, Value is unsmoothed experimental datum.
Entropy 298.16 K, S = 34.9 cal·mol^{-1}·K^{-1}
146.0 J·mol^{-1}·K^{-1}
Extrapolation below 66 K, 6.7 cal·mol^{-1}·K^{-1}.
Molecular Weight 295.2196
Wiswesser Line Notation OVVO .PB
Evaluation B

$C_{2.2}H_{6.5}N_2O$ (c) 72GAN/PAR
Urea–1–dodecene adduct; 1–Dodecene–urea adduct
Heat Capacity 298.15 K, C_p = 30.3 cal·mol^{-1}·K^{-1}
126.6 J·mol^{-1}·K^{-1}
Temperature range 12–300 K. Values for one mole of urea in adduct.
Entropy 298.15 K, S = 33.8 cal·mol^{-1}·K^{-1}
141.3 J·mol^{-1}·K^{-1}
Does not include possible zero-point entropy.
Phase Changes
Anomalous region 225–235 K, ΔH = 2 J·mol^{-1}(urea), ΔS = 0.010 J·mol^{-1}·K^{-1}.
c,II/c,I 82.0 K, ΔH = 102 cal·mol^{-1}
426 J·mol^{-1}
ΔS = 1.2 cal·mol^{-1}·K^{-1}
5.2 J·mol^{-1}·K^{-1}
Molecular Weight 77.8975
Wiswesser Line Notation 11U1 &ZVZ
Evaluation B

$C_{2.2}H_{6.6}N_2O$ (c) 69COP/PAR
Urea–n–undecane adduct; n–Undecane–urea adduct
Heat Capacity 298.15 K, C_p = 30.9 cal·mol^{-1}·K^{-1}
129.2 J·mol^{-1}·K^{-1}
Temperature range 12–300 K. Values for one mole urea in adduct.
Entropy 298.15 K, S = 33.0 cal·mol^{-1}·K^{-1}
138.0 J·mol^{-1}·K^{-1}
Does not include possible zero-point entropy.
Phase Changes
Anomolous region 225–235 K, with ΔH = 14 J·mol^{-1}(urea) and ΔS = 0.061 J·mol^{-1}·K^{-1}.
c,II/c,I 122.4 K, ΔH = 34.6 cal·mol^{-1}
144.6 J·mol^{-1}
ΔS = 0.28 cal·mol^{-1}·K^{-1}
1.18 J·mol^{-1}·K^{-1}
Molecular Weight 77.2496
Wiswesser Line Notation ZVZ &11H
Evaluation B

$C_{2.3}H_{6.7}N_2O$ (c) 69COP/PAR
Urea–1–hexadecene adduct; 1–Hexadecene–urea adduct
Heat Capacity 298.15 K, C_p = 31.0 cal·mol^{-1}·K^{-1}
129.6 J·mol^{-1}·K^{-1}
Temperature range 12–300 K. Values for one mole urea in adduct.
Entropy 298.15 K, S = 33.7 cal·mol^{-1}·K^{-1}
141.2 J·mol^{-1}·K^{-1}
Does not include possible zero-point entropy.
Phase Changes
Anomolous region 225–235 K, with ΔH = 23 J·mol^{-1}(urea) and ΔS = 0.101 J·mol^{-1}·K^{-1}.
c,II/c,I 141.7 K, ΔH = 33.5 cal·mol^{-1}
140.2 J·mol^{-1}
ΔS = 0.24 cal·mol^{-1}·K^{-1}
0.99 J·mol^{-1}·K^{-1}
Molecular Weight 79.3563
Wiswesser Line Notation 15U1 &ZVZ
Evaluation B

Table of Heat Capacities, Entropies, and Phase Transition Properties — Continued

$C_{23}H_{67}N_2O$ (c) 69COP/PAR

Urea-1-decene adduct; 1-Decene-urea adduct

Heat Capacity 298.15 K, C_p = 32.0 cal·mol⁻¹·K⁻¹ / 133.7 J·mol⁻¹·K⁻¹

Temperature range 12–300 K. Values for one mole urea in adduct.

Entropy 298.15 K, S = 34.8 cal·mol⁻¹·K⁻¹ / 145.4 J·mol⁻¹·K⁻¹

Does not include possible zero point entropy.

Phase Changes

Anomolous region 225–235 K, with ΔH = 10 J·mol⁻¹(urea) and ΔS = 0.045 J·mol⁻¹·K⁻¹.

Molecular Weight 79.2721

Wiswesser Line Notation ZVZ &9U1

Evaluation B

$C_{24}H_{68}N_2O$ (c) 72GAN/PAR

1-Octadecene-urea adduct; Urea-1-octadecene adduct

Heat Capacity 298.15 K, C_p = 30.6 cal·mol⁻¹·K⁻¹ / 128.1 J·mol⁻¹·K⁻¹

Temperature range 12–300 K. Value for one mole of urea in adduct.

Entropy 298.15 K, S = 33.9 cal·mol⁻¹·K⁻¹ / 142.0 J·mol⁻¹·K⁻¹

Does not include possible zero-point entropy.

Phase Changes

Anomalous region 225-235 K, ΔH = 12 J·mol⁻¹(urea) and ΔS = 0.049 J·mol⁻¹·K⁻¹.

c,II/c,I	155.9 K,	ΔH =	460 cal·mol⁻¹
			1926 J·mol⁻¹
		ΔS =	3.0 cal·mol⁻¹·K⁻¹
			12.4 J·mol⁻¹·K⁻¹

Molecular Weight 79.8893

Wiswesser Line Notation 17U1 &ZVZ

Evaluation B

$C_{24}H_{68}N_2O$ (c) 69COP/PAR

Urea-1-eicosene adduct; 1-Eicosene-urea adduct

Heat Capacity 298.15 K, C_p = 30.9 cal·mol⁻¹·K⁻¹ / 129.2 J·mol⁻¹·K⁻¹

Temperature range 12–300 K. Values for one mole urea in adduct.

Entropy 298.15 K, S = 33.4 cal·mol⁻¹·K⁻¹ / 139.6 J·mol⁻¹·K⁻¹

Does not include possible zero-point entropy.

Phase Changes

Anomolous region 225–235 K, with ΔH = 1 J·mol⁻¹(urea) and ΔS = 0.002 J·mol⁻¹·K⁻¹.

c,II/c,I	153.9 K,	ΔH =	23.9 cal·mol⁻¹
			100.0 J·mol⁻¹
		ΔS =	0.16 cal·mol⁻¹·K⁻¹
			0.65 J·mol⁻¹·K⁻¹

Molecular Weight 79.6929

Wiswesser Line Notation 19U1 &ZVZ

Evaluation B

C_3F_6O (liq) 67PLA/PAC

Hexafluoropropanone; Hexafluoroacetone

Heat Capacity 245 K, C_p = 43.33 cal·mol⁻¹·K⁻¹ / 181.29 J·mol⁻¹·K⁻¹

Temperature range 12–244 K

Entropy 245.87 K, S = 68.50 cal·mol⁻¹·K⁻¹ / 286.60 J·mol⁻¹·K⁻¹

Phase Changes

c/liq	147.70 K,	ΔH =	2003.5 cal·mol⁻¹
			8382.6 J·mol⁻¹
		ΔS =	13.58 cal·mol⁻¹·K⁻¹
			56.75 J·mol⁻¹·K⁻¹
liq/g	245.87 K,	ΔH =	5166 cal·mol⁻¹
			21615 J·mol⁻¹
		ΔS =	21.01 cal·mol⁻¹·K⁻¹
			87.91 J·mol⁻¹·K⁻¹
		P =	101.325 kPa

Molecular Weight 166.0228

Wiswesser Line Notation FXFFVXFFF

Evaluation A

C_3F_8 (liq) 67PAC/PLA

Perfluoropropane; Octafluoropropane

Heat Capacity 235 K, C_p = 43.35 cal·mol⁻¹·K⁻¹ / 181.38 J·mol⁻¹·K⁻¹

Temperature range 14–236 K

Entropy 236.42 K, S = 68.65 cal·mol⁻¹·K⁻¹ / 287.23 J·mol⁻¹·K⁻¹

Phase Changes

c,II/c,I	99.39 K,	ΔH =	849.8 cal·mol⁻¹
			3555.6 J·mol⁻¹
		ΔS =	8.55 cal·mol⁻¹·K⁻¹
			35.77 J·mol⁻¹·K⁻¹
c,I/liq	125.45 K,	ΔH =	114.1 cal·mol⁻¹
			477.4 J·mol⁻¹
		ΔS =	0.91 cal·mol⁻¹·K⁻¹
			3.81 J·mol⁻¹·K⁻¹
liq/g	236.42 K,	ΔH =	4723.0 cal·mol⁻¹
			19761 J·mol⁻¹
		ΔS =	19.98 cal·mol⁻¹·K⁻¹
			83.58 J·mol⁻¹·K⁻¹
		P =	101.325 kPa

Molecular Weight 188.0202

Wiswesser Line Notation FXFFXFFXFFF

Evaluation A

$C_3H_2ClF_5$ (liq) 74VOR/KOL

3-Chloro-1,1,1,3,3-pentafluoropropane

Heat Capacity 298.15 K, C_p = 46.96 cal·mol⁻¹·K⁻¹ / 196.48 J·mol⁻¹·K⁻¹

Temperature range 12–300 K. Data in paper deposited at VINITI, No. 6783–73, 25 Sept. 1973.

Entropy 298.15 K, S = 74.48 cal·mol⁻¹·K⁻¹ / 403.67 J·mol⁻¹·K⁻¹

Phase Changes

c/liq	165.42 K,	ΔH =	2503 cal·mol⁻¹
			10473 J·mol⁻¹
		ΔS =	15.13 cal·mol⁻¹·K⁻¹
			63.31 J·mol⁻¹·K⁻¹

Molecular Weight 168.4938

Wiswesser Line Notation GXFF1XFFF

Evaluation A

Table of Heat Capacities, Entropies, and Phase Transition Properties — Continued

$C_3H_2Cl_3F_3$ (liq) 71KOL/VOR
1,1,1-Trichloro-3,3,3-trifluoropropane
Heat Capacity 298.15 K, C_p = 47.78 cal·mol^{-1}·K^{-1}
199.91 J·mol^{-1}·K^{-1}
Temperature range 12–300 K. Data given in paper deposited at VINITI, No. 1760–70, 21 May 1970.
Entropy 298.15 K, S = 74.43 cal·mol^{-1}·K^{-1}
311.42 J·mol^{-1}·K^{-1}
Phase Changes
c/liq 232.69 K, ΔH = 3362 cal·mol^{-1}
14067 J·mol^{-1}
ΔS = 14.45 cal·mol^{-1}·K^{-1}
60.45 J·mol^{-1}·K^{-1}
Molecular Weight 201.4030
Wiswesser Line Notation GXGG1XFFF
Evaluation A

$C_3H_2N_2$ (c,I) 68WES/WUL
Malononitrile; Dicyanomethane
Heat Capacity 298.15 K, C_p = 26.36 cal·mol^{-1}·K^{-1}
110.29 J·mol^{-1}·K^{-1}
Temperature range 5–320 K
Entropy 298.15 K, S = 31.30 cal·mol^{-1}·K^{-1}
130.96 J·mol^{-1}·K^{-1}
Entropy as calculated from data on undercooled, c,I from 5 K and from data on c,II, c,II/c,I transition, and c,I is the same.
Phase Changes
Transition between stable c,II and metastable c,I between 255–270 K; maximum in C_p at 260.3 with ΔH = 302 cal·mol^{-1} and ΔS = 1.15 cal·mol^{-1}·K^{-1}. Metastable c,I can be undercooled to 5 K.
c,I/liq 304.99 K, ΔH = 2580 cal·mol^{-1}
10795 J·mol^{-1}
ΔS = 8.46 cal·mol^{-1}·K^{-1}
35.39 J·mol^{-1}·K^{-1}
Molecular Weight 66.0622
Wiswesser Line Notation NC1CN
Evaluation A

$C_3H_3Cl_2F_3$ (liq) 72KOL/VOR
1,1,1-Trifluoro-3,3-dichloropropane
Heat Capacity 298.15 K, C_p = 45.72 cal·mol^{-1}·K^{-1}
191.29 J·mol^{-1}·K^{-1}
Temperature range 12–300 K
Entropy 298.15 K, S = 70.52 cal·mol^{-1}·K^{-1}
295.06 J·mol^{-1}·K^{-1}
Phase Changes
c,III/c,II 156.4 K, ΔH = −215.2 cal·mol^{-1}
−900.4 J·mol^{-1}
Metastable crystal transition. Non-equilibrium.
c,II/c,I 167.7 K, ΔH = 48.0 cal·mol^{-1}
200.8 J·mol^{-1}
ΔS = 0.29 cal·mol^{-1}·K^{-1}
1.20 J·mol^{-1}·K^{-1}
c,I/liq 182.16 K, ΔH = 2422 cal·mol^{-1}
10134 J·mol^{-1}
ΔS = 13.30 cal·mol^{-1}·K^{-1}
55.63 J·mol^{-1}·K^{-1}
Molecular Weight 166.9579
Wiswesser Line Notation GYG1XFFF
Evaluation A

C_3H_3N (liq) 45DAV/WIE
Acrylonitrile; Cyanoethene; Vinyl cyanide
Heat Capacity 298 K, C_p = 27 cal·mol^{-1}·K^{-1}
113 J·mol^{-1}·K^{-1}
One temperature
Molecular Weight 53.0634
Wiswesser Line Notation NC1U1
Evaluation D

C_3H_3N (liq) 71HAL/BAL
Acrylonitrile; Cyanoethene; Vinyl cyanide
Heat Capacity 297 K, C_p = 25.5 cal·mol^{-1}·K^{-1}
106.7 J·mol^{-1}·K^{-1}
One temperature
Molecular Weight 53.0634
Wiswesser Line Notation NC1U1
Evaluation C

C_3H_3N (liq) 72FIN/MES
Acrylonitrile; Cyanoethene; Vinyl cyanide
Heat Capacity 298.15 K, C_p = 26.00 cal·mol^{-1}·K^{-1}
108.78 J·mol^{-1}·K^{-1}
Temperature range 12–350 K
Entropy 298.15 K, S = 42.76 cal·mol^{-1}·K^{-1}
178.91 J·mol^{-1}·K^{-1}
Phase Changes
c,II/c,I 162.5 K, ΔH = 284.0 cal·mol^{-1}
1188.3 J·mol^{-1}
ΔS = 1.75 cal·mol^{-1}·K^{-1}
7.31 J·mol^{-1}·K^{-1}
c,I/liq 189.63 K, ΔH = 1489.0 cal·mol^{-1}
6230.0 J·mol^{-1}
ΔS = 7.85 cal·mol^{-1}·K^{-1}
32.85 J·mol^{-1}·K^{-1}
Molecular Weight 53.0634
Wiswesser Line Notation NC1U1
Evaluation A

C_3H_3NS (liq) 68GOU/WES 2
Thiazole
Heat Capacity 298.15 K, C_p = 28.92 cal·mol^{-1}·K^{-1}
121.00 J·mol^{-1}·K^{-1}
Temperature range 5–340 K, Glass transition 145–175 K.
Entropy 298.15 K, S = 40.62 cal·mol^{-1}·K^{-1}
169.95 J·mol^{-1}·K^{-1}
Phase Changes
c/liq 239.48 K, ΔH = 2292 cal·mol^{-1}
9590 J·mol^{-1}
ΔS = 9.57 cal·mol^{-1}·K^{-1}
40.04 J·mol^{-1}·K^{-1}
Molecular Weight 85.1234
Wiswesser Line Notation T5N CSJ
Evaluation A

C_3H_3NS (liq) 69SOU/GOU 2
Thiazole
Heat Capacity 298.15 K, C_p = 28.92 cal·mol^{-1}·K^{-1}
121.00 J·mol^{-1}·K^{-1}
Temperature range 4–350 K
Entropy 298.15 K, S = 40.62 cal·mol^{-1}·K^{-1}
169.95 J·mol^{-1}·K^{-1}
Phase Changes
Anomalous region 145–175 K
c/liq 239.53 K, ΔH = 2292.0 cal·mol^{-1}
9539.7 J·mol^{-1}
ΔS = 9.57 cal·mol^{-1}·K^{-1}
40.04 J·mol^{-1}·K^{-1}
Molecular Weight 85.1234
Wiswesser Line Notation T5N CSJ
Evaluation A

Table of Heat Capacities, Entropies, and Phase Transition Properties — Continued

$C_3H_3N_3$ (c) 79BRI/MIL
s–Triazine
Heat Capacity 298.15 K, C_p = 22.8 cal·mol^{-1}·K^{-1}
95.6 J·mol^{-1}·K^{-1}
Temperature range 160–382 K
Phase Changes
Transition between 130 & 177 K with ΔH = 75 J·mol^{-1}.
c,I/liq 353.9 K, ΔH = 3486 cal·mol^{-1}
14584 J·mol^{-1}
ΔS = 9.85 cal·mol^{-1}·K^{-1}
41.21 J·mol^{-1}·K^{-1}
Molecular Weight 81.0768
Wiswesser Line Notation T6N CN ENJ
Evaluation B

$C_3H_4ClF_3$ (liq) 72KOL/VOR
1,1,1–Trifluoro–3–chloropropane
Heat Capacity 298.15 K, C_p = 40.89 cal·mol^{-1}·K^{-1}
171.08 J·mol^{-1}·K^{-1}
Temperature range 12–300 K
Entropy 298.15 K, S = 64.93 cal·mol^{-1}·K^{-1}
271.67 J·mol^{-1}·K^{-1}
Phase Changes
c,III/c,II 116.0 K, ΔH = –605.4 cal·mol^{-1}
–2533.0 J·mol^{-1}
Metastable crystal transition. Non–equilibrium.
c,II/c,I 169.8 K, ΔH = 1073 cal·mol^{-1}
4489 J·mol^{-1}
ΔS = 6.32 cal·mol^{-1}·K^{-1}
26.44 J·mol^{-1}·K^{-1}
c,I/liq 179.40 K, ΔH = 1270 cal·mol^{-1}
5314 J·mol^{-1}
ΔS = 7.08 cal·mol^{-1}·K^{-1}
29.62 J·mol^{-1}·K^{-1}
Molecular Weight 132.5128
Wiswesser Line Notation G2XFFF
Evaluation A

$C_3H_4ClF_3$ (liq) 74KOL/VOR
1,1,1–Trifluoro–3–chloropropane
Heat Capacity 298.15 K, C_p = 40.89 cal·mol^{-1}·K^{-1}
171.08 J·mol^{-1}·K^{-1}
Temperature range 12–300 K
Entropy 298.15 K, S = 64.93 cal·mol^{-1}·K^{-1}
271.67 J·mol^{-1}·K^{-1}
Phase Changes
c,III/c,II 116 K, ΔH = –605 cal·mol^{-1}
–2531 J·mol^{-1}
Maximum temperature of metastable phase. ΔH obtained from total heat 115–137 K, calculated to 116 K. Not reversible.
c,II/c,I 169.8 K, ΔH = 1073 cal·mol^{-1}
4489 J·mol^{-1}
ΔS = 6.32 cal·mol^{-1}·K^{-1}
26.44 J·mol^{-1}·K^{-1}
c,I/liq 179.32 K, ΔH = 1207 cal·mol^{-1}
5050 J·mol^{-1}
ΔS = 6.73 cal·mol^{-1}·K^{-1}
28.16 J·mol^{-1}·K^{-1}
Molecular Weight 132.5128
Wiswesser Line Notation G2XFFF
Evaluation A

$C_3H_4Cl_4$ (liq) 74KOL/VOR
1,1,1,3–Tetrachloropropane
Heat Capacity 298.15 K, C_p = 46.94 cal·mol^{-1}·K^{-1}
196.40 J·mol^{-1}·K^{-1}
Temperature range 12–300 K
Entropy 298.15 K, S = 67.95 cal·mol^{-1}·K^{-1}
284.30 J·mol^{-1}·K^{-1}
Phase Changes
c,II/c,I 219.9 K, ΔH = 527 cal·mol^{-1}
2205 J·mol^{-1}
ΔS = 2.40 cal·mol^{-1}·K^{-1}
10.03 J·mol^{-1}·K^{-1}
c,I/liq 237.74 K, ΔH = 2507 cal·mol^{-1}
10489 J·mol^{-1}
ΔS = 10.55 cal·mol^{-1}·K^{-1}
44.12 J·mol^{-1}·K^{-1}
Molecular Weight 181.8766
Wiswesser Line Notation GXGG2G
Evaluation A

$C_3H_4O_3$ (c) 73VAS/KOR
Ethylene carbonate
Heat Capacity 298.15 K, C_p = 28.07 cal·mol^{-1}·K^{-1}
117.44 J·mol^{-1}·K^{-1}
Temperature range 52–310 K. Full data deposited in VINITI, No. 326–73, 21 June 1973.
Entropy 298.15 K, S = 31.68 cal·mol^{-1}·K^{-1}
132.54 J·mol^{-1}·K^{-1}
Extrapolation below 52 K
Phase Changes
c/liq 309.49 K, ΔH = 3178 cal·mol^{-1}
13295 J·mol^{-1}
ΔS = 10.27 cal·mol^{-1}·K^{-1}
42.96 J·mol^{-1}·K^{-1}
Molecular Weight 88.0628
Wiswesser Line Notation T5OVOTJ
Evaluation B

$C_3H_4O_3$ (liq) 58PEP
Ethylene carbonate
Heat Capacity 323.15 K, C_p = 32.0 cal·mol^{-1}·K^{-1}
133.9 J·mol^{-1}·K^{-1}
One temperature
Molecular Weight 88.0628
Wiswesser Line Notation T5OVOTJ
Evaluation C

Table of Heat Capacities, Entropies, and Phase Transition Properties — Continued

$C_3H_5D_3O_3$ (liq) 62RAB/NIK
1,2,3-Trihydroxypropane-d_3;
1,2,3-Propanetriol-d_3; Glycerol-d_3
Heat Capacity 298 K, C_p = 55.21 cal·mol⁻¹·K⁻¹ / 231.00 J·mol⁻¹·K⁻¹
Temperature range 10 to 55 °C
Molecular Weight 95.1130
Wiswesser Line Notation Q1YQ1Q &1/4/6/H-2 3
Evaluation B

$C_3H_5Br_3$ (liq) 48KUR
1,2,3-Tribromopropane
Heat Capacity 298 K, C_p = 39.8 cal·mol⁻¹·K⁻¹ / 166.5 J·mol⁻¹·K⁻¹
Temperature range 17 to 218 °C, mean C_p, six temperatures.
Molecular Weight 280.7845
Wiswesser Line Notation E1YE1E
Evaluation D

C_3H_5Cl (liq) 81REI
3-Chloropropene-1
Heat Capacity 298 K, C_p = 29.9 cal·mol⁻¹·K⁻¹ / 125.1 J·mol⁻¹·K⁻¹
Temperature range 289–334 K
Molecular Weight 76.5255
Wiswesser Line Notation G2U1
Evaluation D

C_3H_5ClO (liq) 81REI
Propanoyl chloride; Propionyl chloride
Heat Capacity 298 K, C_p = 35.2 cal·mol⁻¹·K⁻¹ / 147.3 J·mol⁻¹·K⁻¹
Temperature range 291–365 K
Molecular Weight 92.5249
Wiswesser Line Notation GV2
Evaluation D

$C_3H_5Cl_3$ (liq) 41NEL/NEW
1,2,3-Trichloropropane
Heat Capacity 298 K, C_p = 43.9 cal·mol⁻¹·K⁻¹ / 183.6 J·mol⁻¹·K⁻¹
Temperature range 0 to 60 °C. Equation only.
Molecular Weight 147.4315
Wiswesser Line Notation G1YG1G
Evaluation B

$C_3H_5Cl_3$ (liq) 48KUR
1,2,3-Trichloropropane
Heat Capacity 298 K, C_p = 41.3 cal·mol⁻¹·K⁻¹ / 172.8 J·mol⁻¹·K⁻¹
Temperature range 17 to 155 °C, mean C_p, three temperatures.
Molecular Weight 147.4315
Wiswesser Line Notation G1YG1G
Evaluation D

$C_3H_5CsO_2$ (c) 75FER/SAN
Cesium propionate
Phase Changes

Transition	T	Property	Value
c,III/c,II	314 K,	ΔH =	320 cal·mol⁻¹
			1340 J·mol⁻¹
		ΔS =	1.02 cal·mol⁻¹·K⁻¹
			4.3 J·mol⁻¹·K⁻¹
c,II/c,I	419 K,	ΔH =	460 cal·mol⁻¹
			1925 J·mol⁻¹
		ΔS =	1.1 cal·mol⁻¹·K⁻¹
			4.6 J·mol⁻¹·K⁻¹
c,I/liq	580 K,	ΔH =	2800 cal·mol⁻¹
			11715 J·mol⁻¹
		ΔS =	4.8 cal·mol⁻¹·K⁻¹
			20.2 J·mol⁻¹·K⁻¹

Molecular Weight 205.9767
Wiswesser Line Notation OV2 .CS
Evaluation C

$C_3H_5KO_2$ (c) 75FER/SAN
Potassium propionate
Phase Changes

Transition	T	Property	Value
c,III/c,II	258 K,	ΔH =	80 cal·mol⁻¹
			330 J·mol⁻¹
		ΔS =	0.3 cal·mol⁻¹·K⁻¹
			1.3 J·mol⁻¹·K⁻¹
c,II/c,I	352.5 K,	ΔH =	410 cal·mol⁻¹
			1715 J·mol⁻¹
		ΔS =	1.2 cal·mol⁻¹·K⁻¹
			4.9 J·mol⁻¹·K⁻¹
c,I/liq	638.3 K,	ΔH =	4810 cal·mol⁻¹
			20125 J·mol⁻¹
		ΔS =	7.5 cal·mol⁻¹·K⁻¹
			31.5 J·mol⁻¹·K⁻¹

Molecular Weight 112.1696
Wiswesser Line Notation OV2 .KA
Evaluation C

$C_3H_5LiO_2$ (c) 75FER/SAN
Lithium propionate
Phase Changes

Transition	T	Property	Value
c,II/c,I	533 K,	ΔH =	800 cal·mol⁻¹
			3350 J·mol⁻¹
		ΔS =	1.5 cal·mol⁻¹·K⁻¹
			6.3 J·mol⁻¹·K⁻¹
c,I/liq	606.5 K,	ΔH =	3790 cal·mol⁻¹
			15860 J·mol⁻¹
		ΔS =	6.2 cal·mol⁻¹·K⁻¹
			26.1 J·mol⁻¹·K⁻¹

Also metastable fusion at 584.6 K, ΔH = 4260 cal·mol⁻¹.
Molecular Weight 80.0123
Wiswesser Line Notation OV2 .LI
Evaluation C

C_3H_5N (liq) 07WAL
Propionitrile; Ethyl cyanide; Cyanoethane
Heat Capacity 290 K, C_p = 28 cal·mol⁻¹·K⁻¹ / 117 J·mol⁻¹·K⁻¹
One temperature
Molecular Weight 55.0792
Wiswesser Line Notation NC2
Evaluation D

C_3H_5N (liq) 62WEB/KIL
Propionitrile; Ethyl cyanide; Cyanoethane
Heat Capacity 298.15 K, C_p = 28.56 cal·mol⁻¹·K⁻¹ / 119.50 J·mol⁻¹·K⁻¹
Temperature range 15–300 K
Entropy 298.15 K, S = 45.25 cal·mol⁻¹·K⁻¹ / 189.33 J·mol⁻¹·K⁻¹

Table of Heat Capacities, Entropies, and Phase Transition Properties — Continued

Phase Changes
c,II/c,I 176.96 K, ΔH = 407.9 cal·mol^{-1}
1706.7 J·mol^{-1}
ΔS = 2.31 cal·mol^{-1}·K^{-1}
9.64 J·mol^{-1}·K^{-1}
c,I/liq 180.37 K, ΔH = 1202.2 cal·mol^{-1}
5030.0 J·mol^{-1}
ΔS = 6.67 cal·mol^{-1}·K^{-1}
27.89 J·mol^{-1}·K^{-1}
liq/g 298.15 K, ΔH = 8632 cal·mol^{-1}
36116 J·mol^{-1}
ΔS = 28.9 cal·mol^{-1}·K^{-1}
121.1 J·mol^{-1}·K^{-1}
P = 6.29 kPa
Molecular Weight 55.0792
Wiswesser Line Notation NC2
Evaluation A

C_3H_5N (liq) 71HAL/BAL
Propionitrile; Ethyl cyanide; Cyanoethane
Heat Capacity 297 K, C_p = 27.8 cal·mol^{-1}·K^{-1}
112.9 J·mol^{-1}·K^{-1}
One temperature
Molecular Weight 55.0792
Wiswesser Line Notation NC2
Evaluation C

$(C_3H_5NO)_n$ (c) 75DAU/DEL
Poly-L-alanine, α-helix
Heat Capacity
Temperature range 1–300 K. C_p data given graphically.
Entropy 273 K, S = 24.3 cal·mol^{-1}·K^{-1}
101.7 J·mol^{-1}·K^{-1}
Molecular Weight 89.0938
Wiswesser Line Notation /*MV2*/
Evaluation B

$(C_3H_5NO)_n$ (c) 75DAU/DEL
Poly-L-alanine, β-sheet
Heat Capacity
Temperature range 1–300 K. C_p data given graphically.
Entropy 273 K, S = 27.9 cal·mol^{-1}·K^{-1}
116.7 J·mol^{-1}·K^{-1}
Molecular Weight 89.0938
Wiswesser Line Notation /*MV2*/
Evaluation B

C_3H_5NS (liq) 36KUR/VOS
Ethyl isothiocyanate
Heat Capacity 290 K, C_p = 25.4 cal·mol^{-1}·K^{-1}
106.3 J·mol^{-1}·K^{-1}
One temperature
Molecular Weight 87.1392
Wiswesser Line Notation SCN2
Evaluation D

$C_3H_5NaO_2$ (c,III) 75FER/SAN
Sodium propionate
Heat Capacity 340 K, C_p = 35.9 cal·mol^{-1}·K^{-1}
150.2 J·mol^{-1}·K^{-1}
Temperature range 340–570 K
Phase Changes
c,III/c,I 482 K, ΔH = 1760 cal·mol^{-1}
7360 J·mol^{-1}
ΔS = 3.65 cal·mol^{-1}·K^{-1}
15.3 J·mol^{-1}·K^{-1}
Taken as sum of data for transitions at 470 and 494 K at average temperature.

c,I/liq 562.4 K, ΔH = 3200 cal·mol^{-1}
13390 J·mol^{-1}
ΔS = 5.7 cal·mol^{-1}·K^{-1}
23.8 J·mol^{-1}·K^{-1}
Molecular Weight 96.0611
Wiswesser Line Notation OV2 .NA
Evaluation B

$C_3H_5O_2Rb$ (c) 75FER/SAN
Rubidium propionate
Phase Changes
c,III/c,II 317 K, ΔH = 360 cal·mol^{-1}
1510 J·mol^{-1}
ΔS = 1.1 cal·mol^{-1}·K^{-1}
4.8 J·mol^{-1}·K^{-1}
c,II/c,I 564.3 K, ΔH = 710 cal·mol^{-1}
2970 J·mol^{-1}
ΔS = 1.3 cal·mol^{-1}·K^{-1}
5.3 J·mol^{-1}·K^{-1}
c,I/liq 623.1 K, ΔH = 3480 cal·mol^{-1}
14560 J·mol^{-1}
ΔS = 5.6 cal·mol^{-1}·K^{-1}
23.4 J·mol^{-1}·K^{-1}
Molecular Weight 158.5391
Wiswesser Line Notation OV2 .RB
Evaluation C

C_3H_6 (liq) 31HUF/PAR
Propene; Propylene
Heat Capacity 210.3 K, C_p = 21.5 cal·mol^{-1}·K^{-1}
90.0 J·mol^{-1}·K^{-1}
Temperature range 69–210 K. Value is unsmoothed experimental datum.
Phase Changes
c/liq 88.2 K, ΔH = 701 cal·mol^{-1}
2933 J·mol^{-1}
ΔS = 7.9 cal·mol^{-1}·K^{-1}
33.3 J·mol^{-1}·K^{-1}
Molecular Weight 42.0804
Wiswesser Line Notation 2U1
Evaluation B

C_3H_6 (liq) 46RUE/POW
Cyclopropane
Heat Capacity 240 K, C_p = 19.44 cal·mol^{-1}·K^{-1}
81.34 J·mol^{-1}·K^{-1}
Temperature range 14–240 K
Entropy 240.34 K, S = 34.09 cal·mol^{-1}·K^{-1}
142.63 J·mol^{-1}·K^{-1}
Phase Changes
c/liq 145.57 K, ΔH = 1301 cal·mol^{-1}
5443 J·mol^{-1}
ΔS = 8.94 cal·mol^{-1}·K^{-1}
37.39 J·mol^{-1}·K^{-1}
liq/g 240.34 K, ΔH = 4793 cal·mol^{-1}
20054 J·mol^{-1}
ΔS = 19.94 cal·mol^{-1}·K^{-1}
83.44 J·mol^{-1}·K^{-1}
P = 101.325 kPa
Molecular Weight 42.0804
Wiswesser Line Notation L3TJ
Evaluation A

Table of Heat Capacities, Entropies, and Phase Transition Properties — Continued

$(C_3H_6)_n$ (gls) 62DAI/EVA 4
Polypropylene, atactic
Heat Capacity 298.15 K, C_p = 21.41 cal·mol^{-1}·K^{-1}
89.59 J·mol^{-1}·K^{-1}
Temperature range 20–310 K
Entropy 298.15 K, S = 18.88 cal·mol^{-1}·K^{-1}
78.99 J·mol^{-1}·K^{-1}
When extrapolated to 100% crystallinity, the entropy is 15.0 cal·mol^{-1}·K^{-1}.
Phase Changes
c,I/gls 249 K
Molecular Weight 42.0804
Wiswesser Line Notation /*Y1&1*/
Evaluation A

$(C_3H_6)_n$ (gls) 62DAI/EVA 4
Polypropylene, isotactic
Heat Capacity 298.15 K, C_p = 18.45 cal·mol^{-1}·K^{-1}
77.18 J·mol^{-1}·K^{-1}
Temperature range 20–310 K
Entropy 298.15 K, S = 17.22 cal·mol^{-1}·K^{-1}
72.05 J·mol^{-1}·K^{-1}
When extrapolated to 100% crystallinity, the entropy is 14.8 cal·mol^{-1}·K^{-1}.
Phase Changes
c,I/gls 260 K
Molecular Weight 42.0804
Wiswesser Line Notation /*Y1&1*/
Evaluation A

$(C_3H_6)_n$ (c) 68GEE/MEL
Polypropylene, syndiotactic
Heat Capacity 298.15 K, C_p = 16.6 cal·mol^{-1}·K^{-1}
69.6 J·mol^{-1}·K^{-1}
Temperature range 180–460 K. Values per C_3H_6 unit
Entropy 298.15 K, S = 18.08 cal·mol^{-1}·K^{-1}
75.66 J·mol^{-1}·K^{-1}
Extrapolation below 180 K, 47.0 J·mol^{-1}·K^{-1}. Values per C_3H_6 unit.
Molecular Weight 42.0804
Wiswesser Line Notation /*Y1&1*/
Evaluation B(C_p),C(S)
Glass transition at 270 K. Results corrected to 100% crystalline from 75%.

$C_3H_6Br_2$ (liq) 48KUR
1,3–Dibromopropane
Heat Capacity 298 K, C_p = 38.0 cal·mol^{-1}·K^{-1}
159.0 J·mol^{-1}·K^{-1}
Temperature range 16 to 160 °C, mean C_p, three temperatures.
Molecular Weight 201.8884
Wiswesser Line Notation E3E
Evaluation D

$C_3H_6Br_2$ (c) 50CRO/SMY
1,3–Dibromopropane
Heat Capacity 245.7 K, C_p = 37.3 cal·mol^{-1}·K^{-1}
156.1 J·mol^{-1}·K^{-1}
Temperature range 117–246 K. Value is unsmoothed experimental datum.
Phase Changes
c/liq 238.6 K, ΔH = 3500 cal·mol^{-1}
14640 J·mol^{-1}
ΔS = 14.7 cal·mol^{-1}·K^{-1}
61.4 J·mol^{-1}·K^{-1}
Molecular Weight 201.8884
Wiswesser Line Notation E3E
Evaluation C

$C_3H_6Br_2$ (liq) 48KUR
1,2–Dibromopropane
Heat Capacity 298 K, C_p = 41.3 cal·mol^{-1}·K^{-1}
172.8 J·mol^{-1}·K^{-1}
Temperature range 10 to 133 °C, mean C_p, three temperatures.
Molecular Weight 201.8884
Wiswesser Line Notation EY1&1E
Evaluation D

$C_3H_6ClNO_2$ (liq) 50CRO/SMY
2–Chloro–2–nitropropane
Phase Changes
c,II/c,I 213.8 K, ΔH = 2280 cal·mol^{-1}
9540 J·mol^{-1}
ΔS = 10.7 cal·mol^{-1}·K^{-1}
44.6 J·mol^{-1}·K^{-1}
c,I/liq 251.6 K, ΔH = 320 cal·mol^{-1}
1340 J·mol^{-1}
ΔS = 1.3 cal·mol^{-1}·K^{-1}
5.3 J·mol^{-1}·K^{-1}
Molecular Weight 123.5389
Wiswesser Line Notation WNXG1&1
Evaluation C

$C_3H_6Cl_2$ (liq) 72MIL
2,2–Dichloropropane
Heat Capacity 270 K, C_p = 36.1 cal·mol^{-1}·K^{-1}
151.4 J·mol^{-1}·K^{-1}
Temperature range 137–267 K
Phase Changes
c,II/c,I 188 K, ΔH = 1429 cal·mol^{-1}
5979 J·mol^{-1}
ΔS = 7.6 cal·mol^{-1}·K^{-1}
31.8 J·mol^{-1}·K^{-1}
Temperature not certain; given explicitly as 184.8 K and in Table as 188 K.
c,I/liq 239.25 K, ΔH = 560 cal·mol^{-1}
2341 J·mol^{-1}
ΔS = 2.3 cal·mol^{-1}·K^{-1}
9.8 J·mol^{-1}·K^{-1}
Molecular Weight 112.9864
Wiswesser Line Notation GXG1&1
Evaluation B

$C_3H_6Cl_2$ (liq) 48KUR
1,2–Dichloropropane
Heat Capacity 298 K, C_p = 36.9 cal·mol^{-1}·K^{-1}
154.4 J·mol^{-1}·K^{-1}
Temperature range 11 to 156 °C, mean C_p, three temperatures.
Molecular Weight 112.9864
Wiswesser Line Notation GY1&1G
Evaluation D
Data may be poor. Highest reported temperature too far above listed boiling point.

$C_3H_6F_4N_2$ (liq) 70REE/SEE
1,2–Bis(difluoramino)propane
Heat Capacity 298.15 K, C_p = 51 cal·mol^{-1}·K^{-1}
213 J·mol^{-1}·K^{-1}
One temperature
Molecular Weight 146.0874
Wiswesser Line Notation FNFY1&1NFF
Evaluation B

Table of Heat Capacities, Entropies, and Phase Transition Properties — Continued

$C_3H_6N_2O_4$ (c,I) 58BIL/NOL
2,2–Dinitropropane
Heat Capacity 298 K, C_p = 49.3 cal·mol^{-1}·K^{-1}
206.3 J·mol^{-1}·K^{-1}
From PVT data –5 to 25 °C. Mean value.
Phase Changes
c,II/c,I 268 K, ΔH = 2920 cal·mol^{-1}
12220 J·mol^{-1}
ΔS = 10.9 cal·mol^{-1}·K^{-1}
45.6 J·mol^{-1}·K^{-1}
Calculated from Clausius–Clapeyron equation, from dp/dT and density measurements.
Molecular Weight 134.0914
Wiswesser Line Notation WNX1&1&NW
Evaluation C

$C_3H_6N_2O_4$ (c,I) 78GOD/RAC
2,2–Dinitropropane
Heat Capacity 298 K, C_p = 51 cal·mol^{-1}·K^{-1}
213 J·mol^{-1}·K^{-1}
Temperature range 100–347 K. Data graphically only, value estimated.
Phase Changes
c,III/c,II 259.7 K, ΔH = 447 cal·mol^{-1}
1870 J·mol^{-1}
ΔS = 1.72 cal·mol^{-1}·K^{-1}
7.20 J·mol^{-1}·K^{-1}
c,II/c,I 267.7 K, ΔH = 2695 cal·mol^{-1}
11275 J·mol^{-1}
ΔS = 10.07 cal·mol^{-1}·K^{-1}
42.12 J·mol^{-1}·K^{-1}
c,I/liq 324.5 K, ΔH = 630 cal·mol^{-1}
2635 J·mol^{-1}
ΔS = 1.94 cal·mol^{-1}·K^{-1}
8.12 J·mol^{-1}·K^{-1}
Molecular Weight 134.0914
Wiswesser Line Notation WNX1&1&NW
Evaluation D(C_p),B(Phase changes)

$C_3H_6N_2S$ (c) 82LEB/NOV
2–Imino–4–thiazolidone
Heat Capacity 277.8 K, C_p = 25.3 cal·mol^{-1}·K^{-1}
106.0 J·mol^{-1}·K^{-1}
Temperature range 66–300 K
Molecular Weight 102.1538
Wiswesser Line Notation T5MYSTJ BUM
Evaluation B

$C_3H_6N_6$ (c) 52STE/BER
Melamine
Heat Capacity 299.95 K, C_p = 37.31 cal·mol^{-1}·K^{-1}
156.11 J·mol^{-1}·K^{-1}
Temperature range 15–300 K. Value is unsmoothed experimental datum.
Entropy 298.15 K, S = 35.63 cal·mol^{-1}·K^{-1}
149.08 J·mol^{-1}·K^{-1}
Molecular Weight 126.1206
Wiswesser Line Notation T6N CN ENJ BZ DZ FZ
Evaluation A

$C_3H_6N_6O_6$ (c) 73KRI/LIC
1,3,5–Trinitro–1,3,5–triazacyclohexane; Hexogen
Heat Capacity 298 K, C_p = 59.5 cal·mol^{-1}·K^{-1}
248.9 J·mol^{-1}·K^{-1}
Temperature range 200–475 K. Equation only.
Molecular Weight 222.1170
Wiswesser Line Notation T6N CN ENTJ ANW CNW ENW
Evaluation C

C_3H_6O (c) 25MAA/WAL
Propanone; Acetone; Dimethyl ketone
Heat Capacity 173 K, C_p = 23 cal·mol^{-1}·K^{-1}
96 J·mol^{-1}·K^{-1}
Temperature range 93–173 K
Phase Changes
c/liq 178.5 K, ΔH = 1140 cal·mol^{-1}
4770 J·mol^{-1}
ΔS = 6.4 cal·mol^{-1}·K^{-1}
26.7 J·mol^{-1}·K^{-1}
Molecular Weight 58.0798
Wiswesser Line Notation 1V1
Evaluation C

C_3H_6O (liq) 81REI
Propanone; Acetone; Dimethyl ketone
Heat Capacity 298 K, C_p = 32.0 cal·mol^{-1}·K^{-1}
133.9 J·mol^{-1}·K^{-1}
Temperature range 289–352 K
Molecular Weight 58.0798
Wiswesser Line Notation 1V1
Evaluation D

C_3H_6O (liq) 07WAL
Propanone; Acetone; Dimethyl ketone
Heat Capacity 291 K, C_p = 30 cal·mol^{-1}·K^{-1}
126 J·mol^{-1}·K^{-1}
One temperature
Molecular Weight 58.0798
Wiswesser Line Notation 1V1
Evaluation D

C_3H_6O (liq) 16BRA
Propanone; Acetone; Dimethyl ketone
Heat Capacity 283 K, C_p = 29.0 cal·mol^{-1}·K^{-1}
121.3 J·mol^{-1}·K^{-1}
Mean value, 0 to 20 °C
Molecular Weight 58.0798
Wiswesser Line Notation 1V1
Evaluation C

C_3H_6O (liq) 25PAR/KEL
Propanone; Acetone; Dimethyl ketone
Heat Capacity 289.4 K, C_p = 29.8 cal·mol^{-1}·K^{-1}
124.7 J·mol^{-1}·K^{-1}
Temperature range 70–290 K. Value is unsmoothed experimental datum.
Entropy 298.1 K, S = 52.0 cal·mol^{-1}·K^{-1}
217.6 J·mol^{-1}·K^{-1}
Extrapolation below 90 K, 17.12 cal·mol^{-1}·K^{-1}.
Phase Changes
c/liq 177.6 K, ΔH = 1360 cal·mol^{-1}
5690 J·mol^{-1}
ΔS = 7.66 cal·mol^{-1}·K^{-1}
32.03 J·mol^{-1}·K^{-1}
Molecular Weight 58.0798
Wiswesser Line Notation 1V1
Evaluation B(C_p),C(S)

C_3H_6O (liq) 25WIL/DAN
Propanone; Acetone; Dimethyl ketone
Heat Capacity 293.2 K, C_p = 30.1 cal·mol^{-1}·K^{-1}
125.9 J·mol^{-1}·K^{-1}
Temperature range 20 to 40 °C
Molecular Weight 58.0798
Wiswesser Line Notation 1V1
Evaluation B

Table of Heat Capacities, Entropies, and Phase Transition Properties — Continued

C_3H_6O (liq) 28PAR/KEL
Propanone; Acetone; Dimethyl ketone
Heat Capacity 298.4 K, C_p = 29.6 cal·mol⁻¹·K⁻¹ / 123.8 J·mol⁻¹·K⁻¹
Temperature range 70–289 K. Value is unsmoothed experimental datum.
Entropy 298.1 K, S = 52.7 cal·mol⁻¹·K⁻¹ / 220.5 J·mol⁻¹·K⁻¹
Extrapolation below 70 K, 14.35 cal·mol⁻¹·K⁻¹.
Phase Changes
c/liq 177.6 K, ΔH = 1360 cal·mol⁻¹ / 5690 J·mol⁻¹; ΔS = 7.66 cal·mol⁻¹·K⁻¹ / 32.0 J·mol⁻¹·K⁻¹
Molecular Weight 58.0798
Wiswesser Line Notation 1V1
Evaluation B(C_p); C(S)

C_3H_6O (liq) 29KEL 3
Acetone; Propanone; Dimethyl ketone
Heat Capacity 296.99 K, C_p = 29.80 cal·mol⁻¹·K⁻¹ / 124.68 J·mol⁻¹·K⁻¹
Temperature range 16–298 K. Value is unsmoothed experimental datum.
Entropy 298.15 K, S = 47.9 cal·mol⁻¹·K⁻¹ / 200.4 J·mol⁻¹·K⁻¹
Phase Changes
Hump in C_p curve at about 126 K, with excess entropy of about 0.1 cal·mol⁻¹·K⁻¹.
c/liq 176.62 K, ΔH = 1366 cal·mol⁻¹ / 5715 J·mol⁻¹; ΔS = 7.73 cal·mol⁻¹·K⁻¹ / 32.36 J·mol⁻¹·K⁻¹
Molecular Weight 58.0798
Wiswesser Line Notation 1V1
Evaluation A

C_3H_6O (liq) 29MIT/HAR
Propanone; Acetone; Dimethyl ketone
Heat Capacity 260 K, C_p = 29.7 cal·mol⁻¹·K⁻¹ / 124.3 J·mol⁻¹·K⁻¹
Temperature range 200–260 K
Molecular Weight 58.0798
Wiswesser Line Notation 1V1
Evaluation B

C_3H_6O (liq) 29PAR/KEL
Propanone; Acetone; Dimethyl ketone
Entropy 298.1 K, S = 47.8 cal·mol⁻¹·K⁻¹ / 200.0 J·mol⁻¹·K⁻¹
Extrapolation below 90 K, 12.9 cal·mol⁻¹·K⁻¹. Revision of previous data.
Molecular Weight 58.0798
Wiswesser Line Notation 1V1
Evaluation C

C_3H_6O (liq) 32TRE
Propanone; Acetone; Dimethyl ketone
Heat Capacity 298 K, C_p = 29.8 cal·mol⁻¹·K⁻¹ / 124.7 J·mol⁻¹·K⁻¹
One temperature
Molecular Weight 58.0798
Wiswesser Line Notation 1V1
Evaluation B

C_3H_6O (liq) 33TRE/WAT
Propanone; Acetone; Dimethyl ketone
Heat Capacity 298 K, C_p = 29.8 cal·mol⁻¹·K⁻¹ / 124.7 J·mol⁻¹·K⁻¹
One temperature
Molecular Weight 58.0798
Wiswesser Line Notation 1V1
Evaluation B

C_3H_6O (liq) 39PHI
Propanone; Acetone; Dimethyl ketone
Heat Capacity 302.4 K, C_p = 30.7 cal·mol⁻¹·K⁻¹ / 128.4 J·mol⁻¹·K⁻¹
One temperature
Molecular Weight 58.0798
Wiswesser Line Notation 1V1
Evaluation C

C_3H_6O (liq) 55STA/TUP
Propanone; Acetone; Dimethyl ketone
Heat Capacity 298 K, C_p = 30.65 cal·mol⁻¹·K⁻¹ / 128.24 J·mol⁻¹·K⁻¹
Temperature range 288–323 K
Molecular Weight 58.0798
Wiswesser Line Notation 1V1
Evaluation B

C_3H_6O (liq) 62LOW/MOE
Propanone; Acetone; Dimethyl ketone
Heat Capacity 298.2 K, C_p = 30.01 cal·mol⁻¹·K⁻¹ / 125.56 J·mol⁻¹·K⁻¹
Temperature range 253–308 K
Molecular Weight 58.0798
Wiswesser Line Notation 1V1
Evaluation A

C_3H_6O (liq) 67RAS/GAN
Propanone; Acetone; Dimethyl ketone
Heat Capacity 293 K, C_p = 30.2 cal·mol⁻¹·K⁻¹ / 126.3 J·mol⁻¹·K⁻¹
Temperature range 293–333 K
Molecular Weight 58.0798
Wiswesser Line Notation 1V1
Evaluation C

C_3H_6O (liq) 71DES/BHA
Propanone; Acetone; Dimethyl ketone
Heat Capacity 298 K, C_p = 31.0 cal·mol⁻¹·K⁻¹ / 129.7 J·mol⁻¹·K⁻¹
Temperature range 298–318 K
Molecular Weight 58.0798
Wiswesser Line Notation 1V1
Evaluation B

C_3H_6O (liq) 79SAL/PEA
Propanone; Acetone; Dimethyl ketone
Heat Capacity 298.15 K, C_p = 30.1 cal·mol⁻¹·K⁻¹ / 125.9 J·mol⁻¹·K⁻¹
One temperature
Molecular Weight 58.0798
Wiswesser Line Notation 1V1
Evaluation B

Table of Heat Capacities, Entropies, and Phase Transition Properties — Continued

C_3H_6O (liq) 64OET

Propylene oxide; 2-Methyloxirane

Heat Capacity 298.15 K, C_p = 28.77 cal·mol^{-1}·K^{-1}
120.37 J·mol^{-1}·K^{-1}

Temperature range 11–300 K

Entropy 298.15 K, S = 46.91 cal·mol^{-1}·K^{-1}
196.27 J·mol^{-1}·K^{-1}

Phase Changes

c/liq 161.22 K, ΔH = 1561.4 cal·mol^{-1}
6532.9 J·mol^{-1}
ΔS = 9.68 cal·mol^{-1}·K^{-1}
40.52 J·mol^{-1}·K^{-1}

Molecular Weight 58.0798
Wiswesser Line Notation T3OTJ B1
Evaluation A

C_3H_6O (liq) 66BEA/CLE

Propylene oxide; 2-Methyloxirane

Heat Capacity 298.15 K, C_p = 29.9 cal·mol^{-1}·K^{-1}
125.1 J·mol^{-1}·K^{-1}

Temperature range 90–300 K

Entropy 298.15 K, S = 46.5 cal·mol^{-1}·K^{-1}
194.6 J·mol^{-1}·K^{-1}

Extrapolation below 90 K, 116 cal·mol^{-1}·K^{-1}

Phase Changes

c/liq 161.25 K, ΔH = 1570 cal·mol^{-1}
6569 J·mol^{-1}
ΔS = 9.74 cal·mol^{-1}·K^{-1}
40.74 J·mol^{-1}·K^{-1}

Molecular Weight 58.0798
Wiswesser Line Notation T3OTJ B1
Evaluation A(C_p),C(S)

C_3H_6O (liq) 81REI

3-Propen-1-ol; Allyl alcohol

Heat Capacity 298 K, C_p = 33.2 cal·mol^{-1}·K^{-1}
138.9 J·mol^{-1}·K^{-1}

Temperature range 291–369 K

Molecular Weight 58.0798
Wiswesser Line Notation Q1U2
Evaluation D

C_3H_6O (liq) 76CON/GIN

Oxetane; Trimethylene oxide

Heat Capacity 298 K, C_p = 23.8 cal·mol^{-1}·K^{-1}
99.6 J·mol^{-1}·K^{-1}

One temperature

Molecular Weight 58.0798
Wiswesser Line Notation T4OTJ
Evaluation B

C_3H_6O (liq) 81REI

Propanal; Propaldehyde

Heat Capacity 298 K, C_p = 32.2 cal·mol^{-1}·K^{-1}
134.7 J·mol^{-1}·K^{-1}

Temperature range 288–328 K

Molecular Weight 58.0798
Wiswesser Line Notation VH2
Evaluation D

$(C_3H_6O)_n$ (gls) 70YOS/SAK

Poly(oxacyclobutane)

Heat Capacity 302.13 K, C_p = 30.1 cal·mol^{-1}·K^{-1}
125.9 J·mol^{-1}·K^{-1}

Temperature range 90–320 K, C_p at 302.13 K is unsmoothed experimental datum. Data also given graphically. C_p reported as 2.170 J·g^{-1}·K^{-1} at 302.13 K for annealed sample. C_p given for quenched sample from 95–205 K.

Phase Changes

c,I/gls 195 K,
c,I/liq 305 K, ΔH = 2256 cal·mol^{-1}
9439 J·mol^{-1}
ΔS = 7.40 cal·mol^{-1}·K^{-1}
30.96 J·mol^{-1}·K^{-1}

Melting region extended from 200 to 300 K

Molecular Weight 58.0798
Wiswesser Line Notation /*T4OTJ*/
Evaluation B

$C_3H_6O_2$ (liq) 71HAL/BAL

Methyl ethanoate; Methyl acetate

Heat Capacity 297 K, C_p = 29.6 cal·mol^{-1}·K^{-1}
123.7 J·mol^{-1}·K^{-1}

One temperature

Molecular Weight 74.0792
Wiswesser Line Notation 1VO1
Evaluation C

$C_3H_6O_2$ (liq) 02LOU

Propanoic acid; Propionic acid

Heat Capacity 350 K, C_p = 41 cal·mol^{-1}·K^{-1}
172 J·mol^{-1}·K^{-1}

Mean value 20 to 136°C

Molecular Weight 74.0792
Wiswesser Line Notation QV2
Evaluation D

$C_3H_6O_2$ (liq) 34RAD/JUL

Propanoic acid; Propionic acid

Heat Capacity 289 K, C_p = 38.1 cal·mol^{-1}·K^{-1}
159.4 J·mol^{-1}·K^{-1}

One temperature

Molecular Weight 74.0792
Wiswesser Line Notation QV2
Evaluation C

$C_3H_6O_2$ (liq) 71KON/WAD

Propanoic acid; Propionic acid

Heat Capacity 298.15 K, C_p = 36.1 cal·mol^{-1}·K^{-1}
151 J·mol^{-1}·K^{-1}

One temperature

Molecular Weight 74.0792
Wiswesser Line Notation QV2
Evaluation B

$C_3H_6O_2$ (liq) 82MAR/AND

Propanoic acid; Propionic acid

Heat Capacity 298.15 K, C_p = 36.52 cal·mol^{-1}·K^{-1}
152.8 J·mol^{-1}·K^{-1}

Temperature range 13–450 K, Data also given by equation.

Entropy 298.15 K, S = 45.65 cal·mol^{-1}·K^{-1}
191.0 J·mol^{-1}·K^{-1}

Table of Heat Capacities, Entropies, and Phase Transition Properties — Continued

Phase Changes
c,I/liq 252.65 K, ΔH = 2547.8 cal·mol^{-1}
10660 J·mol^{-1}
ΔS = 10.08 cal·mol^{-1}·K^{-1}
42.19 J·mol^{-1}·K^{-1}
Molecular Weight 74.0792
Wiswesser Line Notation QV2
Evaluation A

$C_3H_6O_2$ (liq) 76CON/GIN
1,3–Dioxolane
Heat Capacity 298 K, C_p = 28.2 cal·mol^{-1}·K^{-1}
118.0 J·mol^{-1}·K^{-1}
One temperature
Molecular Weight 74.0792
Wiswesser Line Notation T5O COTJ
Evaluation B

$C_3H_6O_2$ (liq) 33KOL/UDO
Ethyl methanoate; Ethyl formate
Heat Capacity 294.7 K, C_p = 35.4 cal·mol^{-1}·K^{-1}
148.1 J·mol^{-1}·K^{-1}
One temperature
Molecular Weight 74.0792
Wiswesser Line Notation VHO2
Evaluation C

$C_3H_6O_2$ (liq) 34KOL/UDO 2
Ethyl methanoate; Ethyl formate
Heat Capacity 294.7 K, C_p = 35.4 cal·mol^{-1}·K^{-1}
148.1 J·mol^{-1}·K^{-1}
One temperature
Molecular Weight 74.0792
Wiswesser Line Notation VHO2
Evaluation C

$C_3H_6O_2$ (liq) 36KUR/VOS
Ethyl methanoate; Ethyl formate
Heat Capacity 290 K, C_p = 37.8 cal·mol^{-1}·K^{-1}
158.2 J·mol^{-1}·K^{-1}
One temperature
Molecular Weight 74.0792
Wiswesser Line Notation VHO2
Evaluation D

$C_3H_6O_2$ (liq) 79FUC
Ethyl methanoate; Ethyl formate
Heat Capacity 298.15 K, C_p = 34.5 cal·mol^{-1}·K^{-1}
144.3 J·mol^{-1}·K^{-1}
One temperature
Molecular Weight 74.0792
Wiswesser Line Notation VHO2
Evaluation B

$C_3H_6O_2S$ (liq) 35HUF/ELL
3–Thiolpropanoic acid; β–Thiolactic acid
Heat Capacity 299.8 K, C_p = 48.4 cal·mol^{-1}·K^{-1}
202.5 J·mol^{-1}·K^{-1}
Temperature range 85–310 K. Value is unsmoothed experimental datum.
Entropy 298.1 K, S = 54.7 cal·mol^{-1}·K^{-1}
228.9 J·mol^{-1}·K^{-1}
Extrapolation below 90 K, 11.88 cal·mol^{-1}K^{-1}

Phase Changes
c/liq 291.9 K, ΔH = 4057 cal·mol^{-1}
16974 J·mol^{-1}
ΔS = 13.9 cal·mol^{-1}·K^{-1}
58.2 J·mol^{-1}·K^{-1}
Molecular Weight 106.1392
Wiswesser Line Notation SH2VQ
Evaluation B(C_p), C(S)

$C_3H_6O_3$ (liq) 36PAR/THO
2–Hydroxypropanoic acid (DL); Lactic acid (DL)
Heat Capacity 298.1 K, C_p = 50.3 cal·mol^{-1}·K^{-1}
210.4 J·mol^{-1}·K^{-1}
Temperature range 90–300 K. Unsmoothed experimental datum.
Entropy 298.1 K, S = 45.9 cal·mol^{-1}·K^{-1}
192.0 J·mol^{-1}·K^{-1}
Extrapolation below 90 K, 11.51 cal·mol^{-1}K^{-1}
Phase Changes
c/liq 289.9 K, ΔH = 2710 cal·mol^{-1}
11340 J·mol^{-1}
ΔS = 9.35 cal·mol^{-1}·K^{-1}
39.12 J·mol^{-1}·K^{-1}
Molecular Weight 90.0786
Wiswesser Line Notation QY1&VQ –DL
Evaluation B(C_p),C(S)

$C_3H_6O_3$ (c) 40HUF/ELL
2–Hydroxypropanoic acid(D); Lactic acid(D)
Heat Capacity 298.2 K, C_p = 30.51 cal·mol^{-1}·K^{-1}
127.65 J·mol^{-1}·K^{-1}
Temperature range 84–312 K. Value is unsmoothed experimental datum.
Entropy 298.1 K, S = 34.30 cal·mol^{-1}·K^{-1}
143.51 J·mol^{-1}·K^{-1}
Extrapolation below 90 K, 10.43 cal·mol^{-1}K^{-1}
Molecular Weight 90.0786
Wiswesser Line Notation QY1&VQ –D
Evaluation A(C_p),C(S)

$C_3H_6O_3$ (c) 40HUF/ELL
2–Hydroxypropanoic acid(L); Lactic acid(L)
Heat Capacity 297.9 K, C_p = 30.73 cal·mol^{-1}·K^{-1}
128.57 J·mol^{-1}·K^{-1}
Temperature range 84–298 K. Value is unsmoothed experimental datum.
Entropy 298.1 K, S = 34.00 cal·mol^{-1}·K^{-1}
142.26 J·mol^{-1}·K^{-1}
Extrapolation below 90 K, 10.26 cal·mol^{-1}K^{-1}
Molecular Weight 90.0786
Wiswesser Line Notation QY1&VQ –L
Evaluation A(C_p),C(S)

C_3H_6S (liq) 53SCO/FIN
Thiacyclobutane
Heat Capacity 294.37 K, C_p = 26.91 cal·mol^{-1}·K^{-1}
112.59 J·mol^{-1}·K^{-1}
Temperature range 12–321 K. Value is unsmoothed experimental datum.
Entropy 298.15 K, S = 44.72 cal·mol^{-1}·K^{-1}
184.93 J·mol^{-1}·K^{-1}
Phase Changes
c,II/c,I 176.7 K, ΔH = 159.8 cal·mol^{-1}
668.6 J·mol^{-1}
ΔS = 0.90 cal·mol^{-1}·K^{-1}
3.78 J·mol^{-1}·K^{-1}

Table of Heat Capacities, Entropies, and Phase Transition Properties — Continued

c,I/liq 199.91 K, ΔH = 1971.4 cal·mol^{-1}
8248.3 J·mol^{-1}
ΔS = 9.86 cal·mol^{-1}·K^{-1}
41.26 J·mol^{-1}·K^{-1}

Molecular Weight 74.1404
Wiswesser Line Notation T4STJ
Evaluation A

C_3H_7Br (liq) **81REI**
1–Bromopropane; n–Propyl bromide
Heat Capacity 298 K, C_p = 31.2 cal·mol^{-1}·K^{-1}
130.5 J·mol^{-1}·K^{-1}
Temperature range 289–364 K
Molecular Weight 122.9923
Wiswesser Line Notation E3
Evaluation D

C_3H_7Br (liq) 48KUR
1–Bromopropane; n–Propyl bromide
Heat Capacity 298 K, C_p = 33.5 cal·mol^{-1}·K^{-1}
140.2 J·mol^{-1}·K^{-1}
Temperature range –30 to 67 °C, mean C_p, three temperatures.
Molecular Weight 122.9923
Wiswesser Line Notation E3
Evaluation D

C_3H_7Br (liq) **81REI**
2–Bromopropane; Isopropyl bromide
Heat Capacity 298 K, C_p = 31.6 cal·mol^{-1}·K^{-1}
132.2 J·mol^{-1}·K^{-1}
Temperature range 287–344 K
Molecular Weight 122.9923
Wiswesser Line Notation EY1&1
Evaluation D

C_3H_7Br (liq) 50KUS/CRO
2–Bromopropane; Isopropyl bromide
Heat Capacity 209.6 K, C_p = 30.3 cal·mol^{-1}·K^{-1}
126.8 J·mol^{-1}·K^{-1}
Temperature range 117–209 K. Value is unsmoothed experimental datum.
Phase Changes
c/liq 184.1 K, ΔH = 1560 cal·mol^{-1}
6530 J·mol^{-1}
ΔS = 8.5 cal·mol^{-1}·K^{-1}
35.5 J·mol^{-1}·K^{-1}
Molecular Weight 122.9923
Wiswesser Line Notation EY1&1
Evaluation B

C_3H_7Cl (liq) **81REI**
1–Chloropropane; n–Propyl chloride
Heat Capacity 298 K, C_p = 31.4 cal·mol^{-1}·K^{-1}
131.4 J·mol^{-1}·K^{-1}
Temperature range 288–333 K
Molecular Weight 78.5413
Wiswesser Line Notation G3
Evaluation D

C_3H_7Cl (liq) 48EUC
1–Chloropropane; n–Propyl chloride
Heat Capacity 297 K, C_p = 31.2 cal·mol^{-1}·K^{-1}
130.5 J·mol^{-1}·K^{-1}
Temperature range 200–297 K (4 pts). From unpublished measurements of A. Landsberg.
Molecular Weight 78.5413
Wiswesser Line Notation G3
Evaluation C

C_3H_7Cl (liq) 48KUR
1–Chloropropane; n–Propyl chloride
Heat Capacity 298 K, C_p = 31.6 cal·mol^{-1}·K^{-1}
132.2 J·mol^{-1}·K^{-1}
Temperature range –39 to 43 °C, mean C_p, two temperatures.
Molecular Weight 78.5413
Wiswesser Line Notation G3
Evaluation D

C_3H_7I (liq) **81REI**
1–Iodopropane; n–Propyl iodide
Heat Capacity 298 K, C_p = 30.3 cal·mol^{-1}·K^{-1}
126.8 J·mol^{-1}·K^{-1}
Temperature range 293–383 K
Molecular Weight 169.9928
Wiswesser Line Notation I3
Evaluation D

C_3H_7N (liq) 81FIN/MES
Cyclopropylamine
Heat Capacity 298.15 K, C_p = 35.16 cal·mol^{-1}·K^{-1}
147.11 J·mol^{-1}·K^{-1}
Temperature range 12–315 K. Equation also given in temperature range 242–315 K.
Entropy 298.15 K, S = 44.86 cal·mol^{-1}·K^{-1}
187.69 J·mol^{-1}·K^{-1}
Phase Changes
c,I/liq 237.76 K, ΔH = 3150.9 cal·mol^{-1}
13183.4 J·mol^{-1}
ΔS = 13.25 cal·mol^{-1}·K^{-1}
55.45 J·mol^{-1}·K^{-1}
Molecular Weight 57.0950
Wiswesser Line Notation L3TJ AZ
Evaluation A

C_3H_7NO (liq) 66GEL
N,N–Dimethylmethanamide; N,N–Dimethylformamide
Heat Capacity 298 K, C_p = 37.45 cal·mol^{-1}·K^{-1}
156.69 J·mol^{-1}·K^{-1}
Temperature range 273–323 K
Molecular Weight 73.0944
Wiswesser Line Notation VHN1&1
Evaluation C

C_3H_7NO (liq) 74VIS/SOM
N,N–Dimethylmethanamide; N,N–Dimethylformamide
Heat Capacity 298.15 K, C_p = 36.33 cal·mol^{-1}·K^{-1}
152.0 J·mol^{-1}·K^{-1}
One temperature
Molecular Weight 73.0944
Wiswesser Line Notation VHN1&1
Evaluation A

C_3H_7NO (liq) 77VIS/PER
N,N–Dimethylmethanamide; N,N–Dimethylformamide
Heat Capacity 298 K, C_p = 35.4 cal·mol^{-1}·K^{-1}
148 J·mol^{-1}·K^{-1}
One temperature only
Molecular Weight 73.0944
Wiswesser Line Notation VHN1&1
Evaluation B

Table of Heat Capacities, Entropies, and Phase Transition Properties — Continued

C_3H_7NO (liq) 78MAR/CIO
N,N-Dimethylmethanamide; N,N-Dimethylformamide
Heat Capacity 298 K, C_p = 28.8 cal·mol^{-1}·K^{-1}
120.5 J·mol^{-1}·K^{-1}
Temperature range 298–427 K. Mean value over range.
Molecular Weight 73.0944
Wiswesser Line Notation VHN1&1
Evaluation D

C_3H_7NO (liq) 79VIS/SOM
N,N-Dimethylmethanamide; N,N-Dimethylformamide
Heat Capacity 298.15 K, C_p = 36.0 cal·mol^{-1}·K^{-1}
150.8 J·mol^{-1}·K^{-1}
One temperature
Molecular Weight 73.0944
Wiswesser Line Notation VHN1&1
Evaluation B

$C_3H_7NO_2$ (c) 77SAB/LAF
Sarcosine; N-Methylglycine
Heat Capacity 298.15 K, C_p = 30.8 cal·mol^{-1}·K^{-1}
128.9 J·mol^{-1}·K^{-1}
One temperature
Molecular Weight 89.0938
Wiswesser Line Notation QV1M1
Evaluation B

$C_3H_7NO_2$ (c) 32HUF/BOR
2-Aminopropanoic acid(D); Alanine(D)
Heat Capacity 296.8 K, C_p = 28.9 cal·mol^{-1}·K^{-1}
120.8 J·mol^{-1}·K^{-1}
Temperature range 84–297 K. Value is unsmoothed experimental datum.
Entropy 298.1 K, S = 31.6 cal·mol^{-1}·K^{-1}
132.2 J·mol^{-1}·K^{-1}
Extrapolation below 90 K, 8.88 cal·mol^{-1}K^{-1}
Molecular Weight 89.0938
Wiswesser Line Notation ZY1&VQ -D
Evaluation B(C_p),C(S)

$C_3H_7NO_2$ (c) 37HUF/ELL
2-Aminopropanoic acid(DL); Alanine(DL)
Heat Capacity 297.5 K, C_p = 29.09 cal·mol^{-1}·K^{-1}
121.71 J·mol^{-1}·K^{-1}
Temperature range 85–298 K. Value is unsmoothed experimental datum.
Entropy 298.15 K, S = 31.6 cal·mol^{-1}·K^{-1}
132.2 J·mol^{-1}·K^{-1}
Extrapolation below 90 K, 8.84 cal·mol^{-1}K^{-1}
Molecular Weight 89.0938
Wiswesser Line Notation ZY1&VQ -DL
Evaluation B(C_p),C(S)

$C_3H_7NO_2$ (c) 60HUT/COL
2-Aminopropanoic acid(L); Alanine(L)
Heat Capacity 298.15 K, C_p = 29.22 cal·mol^{-1}·K^{-1}
122.26 J·mol^{-1}·K^{-1}
Temperature range 11–305 K
Entropy 298.15 K, S = 30.88 cal·mol^{-1}·K^{-1}
129.21 J·mol^{-1}·K^{-1}
Molecular Weight 89.0938
Wiswesser Line Notation ZY1&VQ -L
Evaluation A

$C_3H_7NO_2$ (c) 75DAU/DEL
2-Aminopropanic acid (L); Alanine (L)
Heat Capacity
Temperature range 1–300 K, C_p data given graphically.
Entropy 273 K, S = 28.4 cal·mol^{-1}·K^{-1}
118.8 J·mol^{-1}·K^{-1}
Molecular Weight 87.1011
Wiswesser Line Notation ZYVQ -L
Evaluation B

$C_3H_7NO_2$ (c) 75SPI/WAD
2-Aminopropanoic acid(DL); Alanine(DL)
Heat Capacity 298.15 K, C_p = 29.1 cal·mol^{-1}·K^{-1}
121.6 J·mol^{-1}·K^{-1}
One temperature
Molecular Weight 89.0938
Wiswesser Line Notation ZY1&VQ -DL
Evaluation B

$C_3H_7NO_2S$ (c) 35HUF/ELL
Cysteine(L)
Heat Capacity 297.6 K, C_p = 38.8 cal·mol^{-1}·K^{-1}
162.3 J·mol^{-1}·K^{-1}
Temperature range 85–298 K. Value is unsmoothed experimental datum.
Entropy 298.1 K, S = 40.6 cal·mol^{-1}·K^{-1}
169.9 J·mol^{-1}·K^{-1}
Extrapolation below 90 K, 11.71 cal·mol^{-1}K^{-1}
Molecular Weight 121.1538
Wiswesser Line Notation SH1YZVQ -L
Evaluation B(C_p),C(S)

$C_3H_7NO_3$ (c) 64HUT/COL 2
2-Amino-3-hydroxypropanoic acid(L); Serine(L)
Heat Capacity 298.15 K, C_p = 32.40 cal·mol^{-1}·K^{-1}
135.56 J·mol^{-1}·K^{-1}
Temperature range 10–310 K
Entropy 298.15 K, S = 35.65 cal·mol^{-1}·K^{-1}
149.16 J·mol^{-1}·K^{-1}
Molecular Weight 105.0932
Wiswesser Line Notation QVYZ1Q -L
Evaluation A

$C_3H_7NO_3$ (c) 75SPI/WAD
2-Amino-3-hydroxypropanoic acid(DL); Serine(DL)
Heat Capacity 298.15 K, C_p = 31.6 cal·mol^{-1}·K^{-1}
132.4 J·mol^{-1}·K^{-1}
One temperature
Molecular Weight 105.0932
Wiswesser Line Notation QVYZ1Q -DL
Evaluation B

$C_3H_7NO_3$ (c) 78SAB/LAF
2-Amino-3-hydroxypropanoic acid(L); Serine(L)
Heat Capacity 298.15 K, C_p = 33.2 cal·mol^{-1}·K^{-1}
138.9 J·mol^{-1}·K^{-1}
One temperature
Molecular Weight 105.0932
Wiswesser Line Notation QVYZ1Q -L
Evaluation B

Table of Heat Capacities, Entropies, and Phase Transition Properties — Continued

C_3H_8 (liq) 38KEM/EGA
Propane
Heat Capacity 230 K, C_p = 23.49 cal·mol^{-1}·K^{-1}
98.28 J·mol^{-1}·K^{-1}
Temperature range 15–230 K
Entropy 231.04 K, S = 40.87 cal·mol^{-1}·K^{-1}
171.0 J·mol^{-1}·K^{-1}
Debye extrapolation, 0–15 K
Phase Changes
c,I/liq 85.45 K, ΔH = 842.2 cal·mol^{-1}
3524 J·mol^{-1}
ΔS = 9.856 cal·mol^{-1}·K^{-1}
41.24 J·mol^{-1}·K^{-1}
liq/g 231.04 K, ΔH = 4487 cal·mol^{-1}
18774 J·mol^{-1}
ΔS = 19.421 cal·mol^{-1}·K^{-1}
81.26 J·mol^{-1}·K^{-1}
Molecular Weight 44.0962
Wiswesser Line Notation 3H
Evaluation A

C_3H_8 (liq) 78GOO
Propane
Heat Capacity 300 K, C_p = 28.59 cal·mol^{-1}·K^{-1}
119.6 J·mol^{-1}·K^{-1}
Temperature range 81–289 K, C_p data reported for an extended data set; unsmoothed experimental datum.
Molecular Weight 44.0962
Wiswesser Line Notation 3H
Evaluation A

$C_3H_8N_2O_3$ (c) 81SHE/KAM
Dimethanolurea
Heat Capacity 300 K, C_p = 37.7 cal·mol^{-1}·K^{-1}
157.7 J·mol^{-1}·K^{-1}
Temperature range 4–300 K
Entropy 300 K, S = 41.5 cal·mol^{-1}·K^{-1}
173.6 J·mol^{-1}·K^{-1}
Molecular Weight 120.1078
Wiswesser Line Notation Q1MVM1Q
Evaluation A

C_3H_8O (liq) 81REI
1–Propanol; n–Propyl alcohol
Heat Capacity 298 K, C_p = 34.6 cal·mol^{-1}·K^{-1}
144.8 J·mol^{-1}·K^{-1}
Temperature range 289–363 K
Molecular Weight 60.0956
Wiswesser Line Notation Q3
Evaluation D

C_3H_8O (liq) 26PAR/HUF
1–Propanol; n–Propyl alcohol
Heat Capacity 275.0 K, C_p = 31.9 cal·mol^{-1}·K^{-1}
133.5 J·mol^{-1}·K^{-1}
Temperature range 86–275 K. Value is unsmoothed experimental datum.
Entropy 298.1 K, S = 51.2 cal·mol^{-1}·K^{-1}
214.2 J·mol^{-1}·K^{-1}
Extrapolation below 90 K, 15.50 cal·mol^{-1}K^{-1}
Phase Changes
c/liq 147.0 K, ΔH = 1241 cal·mol^{-1}
5192 J·mol^{-1}
ΔS = 8.41 cal·mol^{-1}·K^{-1}
35.3 J·mol^{-1}·K^{-1}
Molecular Weight 60.0956
Wiswesser Line Notation Q3
Evaluation B(C_p),C(S)

C_3H_8O (liq) 27PAR/HUF
1–Propanol; n–Propyl alcohol
Heat Capacity 275.4 K, C_p = 31.9 cal·mol^{-1}·K^{-1}
133.5 J·mol^{-1}·K^{-1}
Temperature range 86–275 K. Value is unsmoothed experimental datum.
Molecular Weight 60.0956
Wiswesser Line Notation Q3
Evaluation B

C_3H_8O (liq) 29MIT/HAR
1–Propanol; n–Propyl alcohol
Heat Capacity 270 K, C_p = 32.5 cal·mol^{-1}·K^{-1}
136.0 J·mol^{-1}·K^{-1}
Temperature range 170–270 K
Molecular Weight 60.0956
Wiswesser Line Notation Q3
Evaluation B

C_3H_8O (liq) 29PAR/KEL
1–Propanol; n–Propyl alcohol
Entropy 298.1 K, S = 46.1 cal·mol^{-1}·K^{-1}
192.9 J·mol^{-1}·K^{-1}
Extrapolation below 90 K, 10.4 cal·mol^{-1}K^{-1}. Revision of previous data.
Molecular Weight 60.0956
Wiswesser Line Notation Q3
Evaluation C

C_3H_8O (liq) 39PHI
1–Propanol; n–Propyl alcohol
Heat Capacity 301.2 K, C_p = 39.4 cal·mol^{-1}·K^{-1}
164.8 J·mol^{-1}·K^{-1}
One temperature
Molecular Weight 60.0956
Wiswesser Line Notation Q3
Evaluation C

C_3H_8O (liq) 41ZHD
1–Propanol; n–Propyl alcohol
Heat Capacity 298.1 K, C_p = 34.8 cal·mol^{-1}·K^{-1}
145.6 J·mol^{-1}·K^{-1}
Temperature range 5 to 46°C
Molecular Weight 60.0956
Wiswesser Line Notation Q3
Evaluation C

C_3H_8O (liq) 51EUC/EIG
1–Propanol; n–Propyl alcohol
Heat Capacity 303 K, C_p = 33.51 cal·mol^{-1}·K^{-1}
140.21 J·mol^{-1}·K^{-1}
Temperature range 303–393 K
Molecular Weight 60.0956
Wiswesser Line Notation Q3
Evaluation A

C_3H_8O (liq) 60SWI/ZIE
1–Propanol; n–Propyl alcohol
Heat Capacity 320 K, C_p = 37.2 cal·mol^{-1}·K^{-1}
155.6 J·mol^{-1}·K^{-1}
Mean value 21 to 74°C
Molecular Weight 60.0956
Wiswesser Line Notation Q3
Evaluation C

Table of Heat Capacities, Entropies, and Phase Transition Properties — Continued

C_3H_8O (liq) 68COU/LEE
1–Propanol; n–Propyl alcohol
Heat Capacity 298.15 K, C_p = 34.37 cal·mol^{-1}·K^{-1}
143.8 J·mol^{-1}·K^{-1}
Temperature range 11–350 K
Entropy 298.15 K, S = 46.08 cal·mol^{-1}·K^{-1}
192.8 J·mol^{-1}·K^{-1}
Phase Changes
c/liq 148.75 K, ΔH = 1284 cal·mol^{-1}
5372 J·mol^{-1}
ΔS = 8.63 cal·mol^{-1}·K^{-1}
36.11 J·mol^{-1}·K^{-1}
Molecular Weight 60.0956
Wiswesser Line Notation Q3
Evaluation A

C_3H_8O (liq) 68REC
1–Propanol; n–Propyl alcohol
Heat Capacity 298 K, C_p = 34.9 cal·mol^{-1}·K^{-1}
146.1 J·mol^{-1}·K^{-1}
Temperature range 24 to 40 °C, equation only.
Molecular Weight 60.0956
Wiswesser Line Notation Q3
Evaluation C

C_3H_8O (liq) 70PAZ/PAZ
1–Propanol; n–Propyl alcohol
Heat Capacity 313.2 K, C_p = 37.9 cal·mol^{-1}·K^{-1}
158.6 J·mol^{-1}·K^{-1}
One temperature
Molecular Weight 60.0956
Wiswesser Line Notation Q3
Evaluation B

C_3H_8O (liq) 76FOR/BEN
1–Propanol; n–Propyl alcohol
Heat Capacity 298.15 K, C_p = 34.39 cal·mol^{-1}·K^{-1}
143.87 J·mol^{-1}·K^{-1}
One temperature
Molecular Weight 60.0956
Wiswesser Line Notation Q3
Evaluation B

C_3H_8O (gls) 68COU/LEE
1–Propanol; n–Propyl alcohol
Heat Capacity 150 K, C_p = 25.41 cal·mol^{-1}·K^{-1}
106.3 J·mol^{-1}·K^{-1}
Temperature range 10–150 K
Entropy 150 K, S = 26.94 cal·mol^{-1}·K^{-1}
112.7 J·mol^{-1}·K^{-1}
Molecular Weight 60.0956
Wiswesser Line Notation Q3
Evaluation A

C_3H_8O (liq) 24WIL/DAN
2–Propanol; Isopropyl alcohol
Heat Capacity 303 K, C_p = 40.6 cal·mol^{-1}·K^{-1}
169.9 J·mol^{-1}·K^{-1}
Temperature range 303–323 K, equation only.
Molecular Weight 60.0956
Wiswesser Line Notation QY1&1
Evaluation C

C_3H_8O (liq) 25PAR/KEL
2–Propanol; Isopropyl alcohol
Heat Capacity 293.1 K, C_p = 36.4 cal·mol^{-1}·K^{-1}
152.3 J·mol^{-1}·K^{-1}
Temperature range 71–293 K. Value is unsmoothed experimental datum.
Entropy 298.1 K, S = 45.6 cal·mol^{-1}·K^{-1}
190.8 J·mol^{-1}·K^{-1}
Extrapolation below 90 K, 12.72 cal·mol^{-1}K^{-1}
Phase Changes
c/liq 184.6 K, ΔH = 1266 cal·mol^{-1}
5297 J·mol^{-1}
ΔS = 6.86 cal·mol^{-1}·K^{-1}
28.7 J·mol^{-1}·K^{-1}
Molecular Weight 60.0956
Wiswesser Line Notation QY1&1
Evaluation B(C_p),C(S)

C_3H_8O (liq) 28PAR/KEL
2–Propanol; Isopropyl alcohol
Heat Capacity 293.1 K, C_p = 36.1 cal·mol^{-1}·K^{-1}
151.0 J·mol^{-1}·K^{-1}
Temperature range 71–293 K. Value is unsmoothed experimental datum.
Entropy 298.1 K, S = 46.1 cal·mol^{-1}·K^{-1}
192.9 J·mol^{-1}·K^{-1}
Extrapolation below 70 K, 10.41 cal·mol^{-1}·K^{-1}
Phase Changes
c/liq 184.6 K, ΔH = 1267 cal·mol^{-1}
5301 J·mol^{-1}
ΔS = 6.86 cal·mol^{-1}·K^{-1}
28.72 J·mol^{-1}·K^{-1}
Molecular Weight 60.0956
Wiswesser Line Notation QY1&1
Evaluation B(C_p),C(S)

C_3H_8O (liq) 29KEL 3
2–Propanol; Isopropyl alcohol
Heat Capacity 292.84 K, C_p = 35.79 cal·mol^{-1}·K^{-1}
149.75 J·mol^{-1}·K^{-1}
Temperature range 16–298 K. Value is unsmoothed experimental datum.
Entropy 298.15 K, S = 43.0 cal·mol^{-1}·K^{-1}
179.9 J·mol^{-1}·K^{-1}
Phase Changes
c/liq 184.67 K, ΔH = 1284 cal·mol^{-1}
5372 J·mol^{-1}
ΔS = 6.95 cal·mol^{-1}·K^{-1}
29.09 J·mol^{-1}·K^{-1}
Molecular Weight 60.0956
Wiswesser Line Notation QY1&1
Evaluation A

C_3H_8O (liq) 29PAR/KEL
2–Propanol; Isopropyl alcohol
Entropy 298.1 K, S = 43.1 cal·mol^{-1}·K^{-1}
180.3 J·mol^{-1}·K^{-1}
Extrapolation below 90 K, 10.2 cal·mol^{-1}K^{-1}. Revision of previous data.
Molecular Weight 60.0956
Wiswesser Line Notation QY1&1
Evaluation C

Table of Heat Capacities, Entropies, and Phase Transition Properties — Continued

C_3H_8O (liq) 33TRE/WAT
2–Propanol; Isopropyl alcohol
Heat Capacity 298 K, C_p = 39.1 cal·mol^{-1}·K^{-1}
163.6 J·mol^{-1}·K^{-1}
One temperature
Molecular Weight 60.0956
Wiswesser Line Notation QY1&1
Evaluation B

C_3H_8O (liq) 39PHI
2–Propanol; Isopropyl alcohol
Heat Capacity 303.2 K, C_p = 41.2 cal·mol^{-1}·K^{-1}
172.4 J·mol^{-1}·K^{-1}
One temperature
Molecular Weight 60.0956
Wiswesser Line Notation QY1&1
Evaluation C

C_3H_8O (liq) 45ZHD
2–Propanol; Isopropyl alcohol
Heat Capacity 298.04 K, C_p = 38.24 cal·mol^{-1}·K^{-1}
159.99 J·mol^{-1}·K^{-1}
Temperature range 7 to 41°C. Value is unsmoothed experimental datum.
Molecular Weight 60.0956
Wiswesser Line Notation QY1&1
Evaluation C

C_3H_8O (liq) 48GIN/COR
2–Propanol; Isopropyl alcohol
Heat Capacity 298 K, C_p = 36.8 cal·mol^{-1}·K^{-1}
154.0 J·mol^{-1}·K^{-1}
Temperature range 0 to 200°C
Molecular Weight 60.0956
Wiswesser Line Notation QY1&1
Evaluation B

C_3H_8O (liq) 58SWI/ZIE 2
2–Propanol; Isopropyl alcohol
Heat Capacity 324 K, C_p = 43.1 cal·mol^{-1}·K^{-1}
180.3 J·mol^{-1}·K^{-1}
Mean value 21 to 81°C
Molecular Weight 60.0956
Wiswesser Line Notation QY1&1
Evaluation C

C_3H_8O (liq) 62KAT
2–Propanol; Isopropyl alcohol
Heat Capacity 298.2 K, C_p = 38.9 cal·mol^{-1}·K^{-1}
162.8 J·mol^{-1}·K^{-1}
Temperature range 10 to 60°C
Molecular Weight 60.0956
Wiswesser Line Notation QY1&1
Evaluation B

C_3H_8O (liq) 63AND/COU 2
2–Propanol; Isopropyl alcohol
Heat Capacity 298.15 K, C_p = 36.91 cal·mol^{-1}·K^{-1}
154.43 J·mol^{-1}·K^{-1}
Temperature range 10–330 K
Entropy 298.15 K, S = 43.16 cal·mol^{-1}·K^{-1}
180.58 J·mol^{-1}·K^{-1}
Phase Changes
c/liq 185.20 K, ΔH = 1293 cal·mol^{-1}
5410 J·mol^{-1}
ΔS = 6.98 cal·mol^{-1}·K^{-1}
29.21 J·mol^{-1}·K^{-1}
Molecular Weight 60.0956
Wiswesser Line Notation QY1&1
Evaluation A

C_3H_8O (liq) 79BRO/ZIE
2–Propanol; Isopropyl alcohol
Heat Capacity 298.15 K, C_p = 36.99 cal·mol^{-1}·K^{-1}
154.75 J·mol^{-1}·K^{-1}
Temperature range 185–304 K. Results as equation only.
Molecular Weight 60.0956
Wiswesser Line Notation QY1&1
Evaluation B

$C_3H_8O_2$ (liq) 81REI
Methylal; 2,4–Dioxapentane; Formaldehyde,dimethylacetal; Dimethoxymethane
Heat Capacity 298 K, C_p = 39.1 cal·mol^{-1}·K^{-1}
163.6 J·mol^{-1}·K^{-1}
Temperature range 289–323 K
Molecular Weight 76.0950
Wiswesser Line Notation 1O1O1
Evaluation D

$C_3H_8O_2$ (liq) 64MCE/KIL
Methylal; 2,4–Dioxapentane; Formaldehyde,dimethylacetal; Dimethoxymethane
Heat Capacity 298.15 K, C_p = 38.58 cal·mol^{-1}·K^{-1}
161.42 J·mol^{-1}·K^{-1}
Temperature range 15–300 K
Entropy 298.15 K, S = 58.32 cal·mol^{-1}·K^{-1}
244.01 J·mol^{-1}·K^{-1}
Phase Changes
c/liq 168.01 K, ΔH = 1991.3 cal·mol^{-1}
8331.6 J·mol^{-1}
ΔS = 11.85 cal·mol^{-1}·K^{-1}
49.59 J·mol^{-1}·K^{-1}
liq/g 298.15 K, ΔH = 6904 cal·mol^{-1}
28886 J·mol^{-1}
ΔS = 23.2 cal·mol^{-1}·K^{-1}
96.9 J·mol^{-1}·K^{-1}
P = 53.14 kPa
Molecular Weight 76.0950
Wiswesser Line Notation 1O1O1
Evaluation A

$C_3H_8O_2$ (liq) 73KUS/SUU
3–Oxa–1–butanol; 2–Methoxyethanol
Heat Capacity 298.15 K, C_p = 41.80 cal·mol^{-1}·K^{-1}
174.9 J·mol^{-1}·K^{-1}
One temperature
Molecular Weight 76.0950
Wiswesser Line Notation Q2O1
Evaluation B

$C_3H_8O_2$ (liq) 78ROU/PER
3–Oxa–1–butanol; 2–Methoxyethanol
Heat Capacity 298.15 K, C_p = 42.2 cal·mol^{-1}·K^{-1}
176.6 J·mol^{-1}·K^{-1}
One temperature
Molecular Weight 76.0950
Wiswesser Line Notation Q2O1
Evaluation C

$C_3H_8O_2$ (liq) 27PAR/HUF
1,2–Propanediol; 1,2–Dihydroxypropane; Propylene glycol
Heat Capacity 276.7 K, C_p = 43.1 cal·mol^{-1}·K^{-1}
180.3 J·mol^{-1}·K^{-1}
Temperature range 91–277 K. Value is unsmoothed experimental datum.
Molecular Weight 76.0950
Wiswesser Line Notation QY1&1Q
Evaluation B

Table of Heat Capacities, Entropies, and Phase Transition Properties — Continued

$C_3H_8O_2$ (liq) 72KAW/OTA
1,2–Propanediol; 1,2–Dihydroxypropane; Propylene glycol
Heat Capacity 303 K, C_p = 42.3. cal·mol^{-1}·K^{-1}
177.0 J·mol^{-1}·K^{-1}
One temperature
Molecular Weight 76.0950
Wiswesser Line Notation QY1&1Q
Evaluation B

$C_3H_8O_3$ (c) 37AHL/BLA
1,2,3–Trihydroxypropane; 1,2,3–Propanetriol; Glycerol
Heat Capacity 86.92 K, C_p = 11.90 cal·mol^{-1}·K^{-1}
49.79 J·mol^{-1}·K^{-1}
Temperature range 3–87 K. Value is unsmoothed experimental datum.
Entropy 90 K, S = 9.05 cal·mol^{-1}·K^{-1}
37.87 J·mol^{-1}·K^{-1}
Molecular Weight 92.0944
Wiswesser Line Notation Q1YQ1Q
Evaluation A

$C_3H_8O_3$ (gls) 37AHL/BLA
1,2,3–Trihydroxypropane; 1,2,3–Propanetriol; Glycerol
Heat Capacity 85.12 K, C_p = 12.00 cal·mol^{-1}·K^{-1}
50.21 J·mol^{-1}·K^{-1}
Temperature range 2.3–95 K. Value is unsmoothed experimental datum.
Entropy 90 K, S = 10.12 cal·mol^{-1}·K^{-1}
42.34 J·mol^{-1}·K^{-1}
Value S–S_0; zero point entropy calculated as 4.64 cal·mol^{-1}K^{-1}.
Molecular Weight 92.0944
Wiswesser Line Notation Q1YQ1Q
Evaluation A

$C_3H_8O_3$ (liq) 22SIM
1,2,3–Trihydroxypropane; 1,2,3–Propanetriol; Glycerol
Heat Capacity 289.7 K, C_p = 54.0 cal·mol^{-1}·K^{-1}
225.9 J·mol^{-1}·K^{-1}
Temperature range 19–294 K. Value is unsmoothed experimental datum.
C_p also measured for glass.
Molecular Weight 92.0944
Wiswesser Line Notation Q1YQ1Q
Evaluation C
Contained 1.3% water. Mp 17°C.

$C_3H_8O_3$ (liq) 23GIB/GIA
1,2,3–Trihydroxypropane; 1,2,3–Propanetriol; Glycerol
Heat Capacity 299.4 K, C_p = 53.4 cal·mol^{-1}·K^{-1}
223.4. J·mol^{-1}·K^{-1}
Temperature range 70.2–299.4 K. Value is unsmoothed experimental datum. C_p also measured for glass.
Phase Changes
c/liq 291.0 K, ΔH = 4370 cal·mol^{-1}
18285 J·mol^{-1}
ΔS = 15.0 cal·mol^{-1}·K^{-1}
62.8 J·mol^{-1}·K^{-1}
Molecular Weight 92.0944
Wiswesser Line Notation Q1YQ1Q
Evaluation B

$C_3H_8O_3$ (liq) 29PAR/KEL
1,2,3–Trihydroxypropane; 1,2,3–Propanetriol; Glycerol
Entropy 298.1 K, S = 49.7 cal·mol^{-1}·K^{-1}
207.9 J·mol^{-1}·K^{-1}
Extrapolation below 90 K, 9.9 cal·mol^{-1}K^{-1}.
Revision of previous data.
Molecular Weight 92.0944
Wiswesser Line Notation Q1YQ1Q
Evaluation C

$C_3H_8O_3$ (liq) 36ERN/WAT
1,2,3–Trihydroxypropane; 1,2,3–Propanetriol; Glycerol
Heat Capacity 298 K, C_p = 51.1 cal·mol^{-1}·K^{-1}
213.8 J·mol^{-1}·K^{-1}
One temperature
Molecular Weight 92.0944
Wiswesser Line Notation Q1YQ1Q
Evaluation C

$C_3H_8O_3$ (liq) 62RAB/NIK
1,2,3–Trihydroxypropane; 1,2,3–Propanetriol; Glycerol
Heat Capacity 298 K, C_p = 52.14 cal·mol^{-1}·K^{-1}
218.5 J·mol^{-1}·K^{-1}
Temperature range 10 to 55°C
Molecular Weight 92.0944
Wiswesser Line Notation Q1YQ1Q
Evaluation B

$C_3H_8O_3$ (liq) 70PAZ/PAZ
1,2,3–Trihydroxypropane; 1,2,3–Propanetriol; Glycerol
Heat Capacity 301.2 K, C_p = 53.0 cal·mol^{-1}·K^{-1}
221.7 J·mol^{-1}·K^{-1}
Temperature range 28, 40°C
Molecular Weight 92.0944
Wiswesser Line Notation Q1YQ1Q
Evaluation B

C_3H_8S (liq) 51SCO/FIN
Thiabutane; Ethyl methyl sulfide
Heat Capacity 298.15 K, C_p = 34.57 cal·mol^{-1}·K^{-1}
144.64 J·mol^{-1}·K^{-1}
Temperature range 14–298 K
Entropy 298.15 K, S = 57.14 cal·mol^{-1}·K^{-1}
239.07 J·mol^{-1}·K^{-1}
Phase Changes
c/liq 167.23 K, ΔH = 2333 cal·mol^{-1}
9761 J·mol^{-1}
ΔS = 13.95 cal·mol^{-1}·K^{-1}
58.37 J·mol^{-1}·K^{-1}
liq/g 301.66 K, ΔH = 7563 cal·mol^{-1}
31644 J·mol^{-1}
ΔS = 25.07 cal·mol^{-1}·K^{-1}
104.90 J·mol^{-1}·K^{-1}
P = 24.75 kPa
Molecular Weight 76.1562
Wiswesser Line Notation 2S1
Evaluation A

C_3H_8S (liq) 56PEN/SCO
1–Propanethiol; n–Propyl mercaptan
Heat Capacity 298.15 K, C_p = 34.55 cal·mol^{-1}·K^{-1}
144.56 J·mol^{-1}·K^{-1}
Temperature range 10–320 K
Entropy 298.15 K, S = 57.96 cal·mol^{-1}·K^{-1}
242.50 J·mol^{-1}·K^{-1}

Table of Heat Capacities, Entropies, and Phase Transition Properties — Continued

Phase Changes
c,II/c,I 142.10 K, ΔH = 949.1 cal·mol^{-1}
3971.0 J·mol^{-1}
ΔS = 6.68 cal·mol^{-1}·K^{-1}
27.95 J·mol^{-1}·K^{-1}
c,I/liq 160.00 K, ΔH = 1309 cal·mol^{-1}
5476.9 J·mol^{-1}
ΔS = 8.18 cal·mol^{-1}·K^{-1}
34.23 J·mol^{-1}·K^{-1}
Molecular Weight 76.1562
Wiswesser Line Notation SH3
Evaluation A

C_3H_8S (liq) 54MCC/FIN 2
2–Propanethiol; Isopropyl mercaptan
Heat Capacity 298.15 K, C_p = 34.74 cal·mol^{-1}·K^{-1}
145.35 J·mol^{-1}·K^{-1}
Temperature range 12–322 K
Entropy 298.15 K, S = 55.82 cal·mol^{-1}·K^{-1}
233.55 J·mol^{-1}·K^{-1}
Phase Changes
c,II/c,I 112.5 K, ΔH = 12.63 cal·mol^{-1}
52.84 J·mol^{-1}
ΔS = 0.11 cal·mol^{-1}·K^{-1}
0.47 J·mol^{-1}·K^{-1}
c,I/liq 142.64 K, ΔH = 1371 cal·mol^{-1}
5736 J·mol^{-1}
ΔS = 9.61 cal·mol^{-1}·K^{-1}
40.21 J·mol^{-1}·K^{-1}
liq/g 325.72 K, ΔH = 6670 cal·mol^{-1}
27910 J·mol^{-1}
ΔS = 20.48 cal·mol^{-1}·K^{-1}
85.69 J·mol^{-1}·K^{-1}
P = 101.325 kPa
Molecular Weight 76.1562
Wiswesser Line Notation SHY1&1
Evaluation A

C_3H_9Al (liq) 63MCC/MES
Trimethylaluminum
Heat Capacity 298.15 K, C_p = 37.19 cal·mol^{-1}·K^{-1}
155.60 J·mol^{-1}·K^{-1}
Temperature range 10–380 K
Entropy 298.15 K, S = 50.05 cal·mol^{-1}·K^{-1}
209.41 J·mol^{-1}·K^{-1}
Phase Changes
c/liq 288.43 K, ΔH = 2101.0 cal·mol^{-1}
8790.6 J·mol^{-1}
ΔS = 7.28 cal·mol^{-1}·K^{-1}
30.48 J·mol^{-1}·K^{-1}
Molecular Weight 72.0856
Wiswesser Line Notation 1–AL–1&1
Evaluation A

C_3H_9B (liq) 54FUR/PAR
Trimethylborane
Heat Capacity 210 K, C_p = 27.91 cal·mol^{-1}·K^{-1}
116.79 J·mol^{-1}·K^{-1}
Temperature range 15–220 K
Entropy 210 K, S = 47.55 cal·mol^{-1}·K^{-1}
198.95 J·mol^{-1}·K^{-1}
Phase Changes
c/liq 113.21 K, ΔH = 776.8 cal·mol^{-1}
3250 J·mol^{-1}
ΔS = 6.86 cal·mol^{-1}·K^{-1}
28.71 J·mol^{-1}·K^{-1}
liq/g 199.91 K, ΔH = 5518 cal·mol^{-1}
23089 J·mol^{-1}
ΔS = 27.61 cal·mol^{-1}·K^{-1}
115.5 J·mol^{-1}·K^{-1}
P = 6.20 kPa
Molecular Weight 55.9141
Wiswesser Line Notation 1B1&1
Evaluation A

C_3H_9Ga (liq) 73MAS/NOV
Trimethylgallium
Heat Capacity 298.15 K, C_p = 42.75 cal·mol^{-1}·K^{-1}
178.87 J·mol^{-1}·K^{-1}
Temperature range 60–300 K
Phase Changes
Anomaly in specific heat at 82.4 K
c/liq 257.9 K, ΔH = 2640 cal·mol^{-1}
11046 J·mol^{-1}
ΔS = 10.24 cal·mol^{-1}·K^{-1}
42.83 J·mol^{-1}·K^{-1}
Molecular Weight 114.8241
Wiswesser Line Notation 1–GA–1&1
Evaluation B

C_3H_9N (liq) 44AST/SAG
Trimethylamine
Heat Capacity 280 K, C_p = 31.68 cal·mol^{-1}·K^{-1}
132.55 J·mol^{-1}·K^{-1}
Temperature range 12–280 K
Entropy 276.03 K, S = 47.28 cal·mol^{-1}·K^{-1}
197.82 J·mol^{-1}·K^{-1}
Phase Changes
c/liq 156.08 K, ΔH = 1564 cal·mol^{-1}
6544 J·mol^{-1}
ΔS = 10.02 cal·mol^{-1}·K^{-1}
41.93 J·mol^{-1}·K^{-1}
liq/g 276.03 K, ΔH = 5482 cal·mol^{-1}
22937 J·mol^{-1}
ΔS = 19.86 cal·mol^{-1}·K^{-1}
83.10 J·mol^{-1}·K^{-1}
P = 101.325 kPa
Molecular Weight 59.1108
Wiswesser Line Notation 1N1&1
Evaluation A

C_3H_9N (liq) 67SMI/GOO 2
1–Aminopropane; n–Propylamine
Heat Capacity 298.15 K, C_p = 38.8 cal·mol^{-1}·K^{-1}
162.3 J·mol^{-1}·K^{-1}
One temperature
Molecular Weight 59.1108
Wiswesser Line Notation Z3
Evaluation B

C_3H_9N (liq) 71KON/WAD
1–Aminopropane; n–Propylamine
Heat Capacity 298.15 K, C_p = 38.2 cal·mol^{-1}·K^{-1}
160 J·mol^{-1}·K^{-1}
One temperature
Molecular Weight 59.1108
Wiswesser Line Notation Z3
Evaluation B

Table of Heat Capacities, Entropies, and Phase Transition Properties — Continued

C_3H_9N (liq) 71VAS/PET
1–Aminopropane; n–Propylamine
Heat Capacity 298.15 K, C_p = 39.77 cal·mol^{-1}·K^{-1}
166.4 J·mol^{-1}·K^{-1}
Temperature range 60–300 K. Details deposited VINITI, No. 2530–71, 30 Jan 1971.
Entropy 298.15 K, S = 54.54 cal·mol^{-1}·K^{-1}
228.2 J·mol^{-1}·K^{-1}
Extrapolation below 60 K, 26.7 J·mol^{-1}K^{-1}
Phase Changes
c/liq 188.36 K, ΔH = 2539 cal·mol^{-1}
10625 J·mol^{-1}
ΔS = 13.48 cal·mol^{-1}·K^{-1}
56.4 J·mol^{-1}·K^{-1}
Molecular Weight 59.1108
Wiswesser Line Notation Z3
Evaluation A(C_p), B(S)

C_3H_9N (liq) 72FIN/MES
1–Aminopropane; n–Propylamine
Heat Capacity 298.15 K, C_p = 38.84 cal·mol^{-1}·K^{-1}
162.51 J·mol^{-1}·K^{-1}
Temperature range 12–350 K
Entropy 298.15 K, S = 54.36 cal·mol^{-1}·K^{-1}
227.44 J·mol^{-1}·K^{-1}
Phase Changes
c/liq 188.36 K, ΔH = 2622.9 cal·mol^{-1}
10974.2 J·mol^{-1}
ΔS = 13.92 cal·mol^{-1}·K^{-1}
58.26 J·mol^{-1}·K^{-1}
Molecular Weight 59.1108
Wiswesser Line Notation Z3
Evaluation A

C_3H_9N (liq) 50HOU/MAS
2–Aminopropane; Isopropylamine
Heat Capacity 313 K, C_p = 39.2 cal·mol^{-1}·K^{-1}
164.0 J·mol^{-1}·K^{-1}
Temperature range 313–343 K
Molecular Weight 59.1108
Wiswesser Line Notation ZY1&1
Evaluation B

C_3H_9N (liq) 67SMI/GOO 2
2–Aminopropane; Isopropylamine
Heat Capacity 298.15 K, C_p = 39.5 cal·mol^{-1}·K^{-1}
165.3 J·mol^{-1}·K^{-1}
One temperature
Molecular Weight 59.1108
Wiswesser Line Notation ZY1&1
Evaluation B

C_3H_9N (liq) 71KON/WAD
2–Aminopropane; Isopropylamine
Heat Capacity 298.15 K, C_p = 39.2 cal·mol^{-1}·K^{-1}
164 J·mol^{-1}·K^{-1}
One temperature
Molecular Weight 59.1108
Wiswesser Line Notation ZY1&1
Evaluation B

C_3H_9N (liq) 72FIN/MES
2–Aminopropane; Isopropylamine
Heat Capacity 298.15 K, C_p = 39.16 cal·mol^{-1}·K^{-1}
163.85 J·mol^{-1}·K^{-1}
Temperature range 12–350 K
Entropy 298.15 K, S = 52.18 cal·mol^{-1}·K^{-1}
218.32 J·mol^{-1}·K^{-1}
Phase Changes
c/liq 177.99 K, ΔH = 1750.6 cal·mol^{-1}
7324.5 J·mol^{-1}
ΔS = 9.84 cal·mol^{-1}·K^{-1}
41.15 J·mol^{-1}·K^{-1}
Molecular Weight 59.1108
Wiswesser Line Notation ZY1&1
Evaluation A

$C_3H_{10}N_2$ (liq) 55AST/ZOL
Trimethylhydrazine
Heat Capacity 292.15 K, C_p = 44.44 cal·mol^{-1}·K^{-1}
185.94 J·mol^{-1}·K^{-1}
Temperature range 12–294 K
Entropy 292.15 K, S = 55.44 cal·mol^{-1}·K^{-1}
231.96 J·mol^{-1}·K^{-1}
Phase Changes
c/liq 201.24 K, ΔH = 2267 cal·mol^{-1}
9485 J·mol^{-1}
ΔS = 11.27 cal·mol^{-1}·K^{-1}
47.13 J·mol^{-1}·K^{-1}
liq/g 292.15 K, ΔH = 7949 cal·mol^{-1}
33259 J·mol^{-1}
ΔS = 27.21 cal·mol^{-1}·K^{-1}
113.84 J·mol^{-1}·K^{-1}
P = 19.83 kPa
Molecular Weight 74.1254
Wiswesser Line Notation 1N1&M1
Evaluation A
Corrected for 2 mole % unsym–dimethylhydrazine.

$C_3H_{10}N_2$ (liq) 75MES/FIN
1,2–Diaminopropane; 1,2–Propanediamine
Heat Capacity 298.15 K, C_p = 49.15 cal·mol^{-1}·K^{-1}
205.64 J·mol^{-1}·K^{-1}
Temperature range 11–368 K
Entropy 298.15 K, S = 59.10 cal·mol^{-1}·K^{-1}
247.27 J·mol^{-1}·K^{-1}
Includes 1.38 cal·mol^{-1}K^{-1} for mixing of d & l isomers.
Phase Changes
c,II/c,I 222.0 K, ΔH = 16.1 cal·mol^{-1}
67.4 J·mol^{-1}
ΔS = 0.073 cal·mol^{-1}·K^{-1}
0.304 J·mol^{-1}·K^{-1}
c,I/liq 236.53 K, ΔH = 4403.1 cal·mol^{-1}
18422.6 J·mol^{-1}
ΔS = 18.62 cal·mol^{-1}·K^{-1}
77.89 J·mol^{-1}·K^{-1}
Molecular Weight 74.1254
Wiswesser Line Notation ZY1&1Z
Evaluation A

$C_3H_{12}BN$ (c,III) 70FIN/TOD
Trimethylamineborane
Heat Capacity 298.15 K, C_p = 41.53 cal·mol^{-1}·K^{-1}
173.76 J·mol^{-1}·K^{-1}
Temperature range 12–390 K
Entropy 298.15 K, S = 40.53 cal·mol^{-1}·K^{-1}
169.58 J·mol^{-1}·K^{-1}

Table of Heat Capacities, Entropies, and Phase Transition Properties — Continued

Phase Changes

c,III/c,II 350.1 K, ΔH = 606 cal·mol^{-1}
2535 J·mol^{-1}
ΔS = 1.73 cal·mol^{-1}·K^{-1}
7.24 J·mol^{-1}·K^{-1}

c,II/c,I 360.4 K, ΔH = 1420 cal·mol^{-1}
594 J·mol^{-1}
ΔS = 3.94 cal·mol^{-1}·K^{-1}
16.48 J·mol^{-1}·K^{-1}

c,I/liq 368.70 K, ΔH = 1182.4 cal·mol^{-1}
4947.2 J·mol^{-1}
ΔS = 3.21 cal·mol^{-1}·K^{-1}
13.42 J·mol^{-1}·K^{-1}

Molecular Weight 72.9445
Wiswesser Line Notation 1N1&1 &BHHH
Evaluation A

C_4F_8 (liq) 54FUR/MCC
Octafluorocyclobutane
Heat Capacity 268.52 K, C_p = 50.14 cal·mol^{-1}·K^{-1}
209.77 J·mol^{-1}·K^{-1}
Temperature range 17–270 K. Value is unsmoothed experimental datum.
Entropy 261.25 K, S = 69.6 cal·mol^{-1}·K^{-1}
291.1 J·mol^{-1}·K^{-1}
Phase Changes
Solid state transitions at 141.3 K, 174.6 K, 214.84 K, and 216.99 K. No enthalpies of transition reported, entropy changes calculated from integration of total heat input and temperature measurements. Anomalous heat capacity region about 97 K.

c,I/liq 232.96 K, ΔH = 661.6 cal·mol^{-1}
2768.2 J·mol^{-1}
ΔS = 2.84 cal·mol^{-1}·K^{-1}
11.89 J·mol^{-1}·K^{-1}

liq/g 261.25 K, ΔH = 5669 cal·mol^{-1}
23721 J·mol^{-1}
ΔS = 21.70 cal·mol^{-1}·K^{-1}
90.80 J·mol^{-1}·K^{-1}
P = 78.78 kPa

Molecular Weight 200.0312
Wiswesser Line Notation L4TJ AF AF BF BF CF CF DF DF
Evaluation A

$C_4H_2N_2O_4$ (c) 35STI/HUF
Alloxan
Heat Capacity 297.2 K, C_p = 36.57 cal·mol^{-1}·K^{-1}
153.01 J·mol^{-1}·K^{-1}
Temperature range 85–297 K. Value is unsmoothed experimental datum.
Entropy 298.15 K, S = 44.6 cal·mol^{-1}·K^{-1}
153.0 J·mol^{-1}·K^{-1}
Extrapolation below 90 K, 13.15 cal·mol^{-1}K^{-1}
Molecular Weight 142.0708
Wiswesser Line Notation T6MVMVVVJ
Evaluation B(C_p),C(S)

$C_4H_2O_3$ (c) 52SPE/TAM
Maleic anhydride
Phase Changes
c/liq 325 K, ΔH = 3090 cal·mol^{-1}
12930 J·mol^{-1}
ΔS = 9.5 cal·mol^{-1}·K^{-1}
39.8 J·mol^{-1}·K^{-1}

Molecular Weight 98.0580
Wiswesser Line Notation T5VOVJ
Evaluation C

$C_4H_2O_3$ (c) 78MAR/CIO 2
Maleic anhydride
Heat Capacity 298.15 K, C_p = 16.1 cal·mol^{-1}·K^{-1}
67.4 J·mol^{-1}·K^{-1}
Temperature range 298–480 K
Phase Changes
c/liq 325.65 K, ΔH = 2930 cal·mol^{-1}
12260 J·mol^{-1}
ΔS = 9.0 cal·mol^{-1}·K^{-1}
37.6 J·mol^{-1}·K^{-1}
Molecular Weight 98.0580
Wiswesser Line Notation T5VOVJ
Evaluation D

$C_4H_2O_4$ (c,II) 79BAR/HEL
Squaric acid
Heat Capacity 300 K, C_p = 29 cal·mol^{-1}·K^{-1}
121 J·mol^{-1}·K^{-1}
Temperature range 300–420 K, data graphically only.
Phase Changes
c,II/c,I 373.57 K, ΔH = 83 cal·mol^{-1}
347 J·mol^{-1}
ΔS = 0.22 cal·mol^{-1}·K^{-1}
0.93 J·mol^{-1}·K^{-1}
Molecular Weight 114.0574
Wiswesser Line Notation L4VVJ CQ DQ
Evaluation B

$C_4H_4Cl_2O_3$ (liq) 76MAS/PET
4,5–Dichloro–1,3–dioxolan–2–one
Heat Capacity 298 K, C_p = 46.6 cal·mol^{-1}·K^{-1}
195 J·mol^{-1}·K^{-1}
Temperature range 200–340 K. Data graphically only. Value estimated from graph.
Molecular Weight 170.9798
Wiswesser Line Notation T6OVOTJ DG EG
Evaluation D

$C_4H_4N_2$ (c,I) 63WUL/WES
1,4–Butanedinitrile; Succinonitrile
Heat Capacity 298.15 K, C_p = 34.80 cal·mol^{-1}·K^{-1}
145.60 J·mol^{-1}·K^{-1}
Temperature range 5–350 K
Entropy 298.15 K, S = 45.79 cal·mol^{-1}·K^{-1}
191.59 J·mol^{-1}·K^{-1}
Phase Changes
c,II/c,I 233.3 K, ΔH = 1481.7 cal·mol^{-1}
6199.4 J·mol^{-1}
ΔS = 6.35 cal·mol^{-1}·K^{-1}
26.57 J·mol^{-1}·K^{-1}

c,I/liq 331.16 K, ΔH = 885.1 cal·mol^{-1}
3703.3 J·mol^{-1}
ΔS = 2.68 cal·mol^{-1}·K^{-1}
11.18 J·mol^{-1}·K^{-1}
Molecular Weight 80.0890
Wiswesser Line Notation NC2CN
Evaluation A

$C_4H_4N_2$ (c,III) 79BOY/COM
Pyrazine
Heat Capacity 298 K, C_p = 43 cal·mol^{-1}·K^{-1}
180 J·mol^{-1}·K^{-1}
Temperature range 295–312K; data graphically only.

Table of Heat Capacities, Entropies, and Phase Transition Properties — Continued

Phase Changes
c,III/c,II 300.6 K, ΔH = 232 cal·mol^{-1}
969 J·mol^{-1}
ΔS = 0.77 cal·mol^{-1}·K^{-1}
3.22 J·mol^{-1}·K^{-1}
c,II/c,I 310 K, ΔH = 14 cal·mol^{-1}
60 J·mol^{-1}
ΔS = 0.05 cal·mol^{-1}·K^{-1}
0.2 J·mol^{-1}·K^{-1}
Molecular Weight 80.0890
Wiswesser Line Notation T6N DNJ
Evaluation C(C_p),B(transitions)

$C_4H_4N_2O_2$ (c) 78KIL 4
Uracil
Heat Capacity 298 K, C_p = 28.8 cal·mol^{-1}·K^{-1}
120.5 J·mol^{-1}·K^{-1}
298 K, One temperature
Molecular Weight 112.0878
Wiswesser Line Notation T6MVMVJ
Evaluation B

C_4H_4O (liq) 52GUT/SCO
Furan
Heat Capacity 298.15 K, C_p = 27.38 cal·mol^{-1}·K^{-1}
114.56 J·mol^{-1}·K^{-1}
Temperature range 11–300 K
Entropy 298.15 K, S = 42.22 cal·mol^{-1}·K^{-1}
176.65 J·mol^{-1}·K^{-1}
Phase Changes
c,II/c,I 150.0 K, ΔH = 489.2 cal·mol^{-1}
2046.8 J·mol^{-1}
ΔS = 3.26 cal·mol^{-1}·K^{-1}
13.65 J·mol^{-1}·K^{-1}
c,I/liq 187.55 K, ΔH = 908.8 cal·mol^{-1}
3802.4 J·mol^{-1}
ΔS = 4.85 cal·mol^{-1}·K^{-1}
20.27 J·mol^{-1}·K^{-1}
liq/g 298.15 K, ΔH = 6561 cal·mol^{-1}
27451 J·mol^{-1}
ΔS = 22.01 cal·mol^{-1}·K^{-1}
92.07 J·mol^{-1}·K^{-1}
P = 79.934 kPa
Molecular Weight 68.0750
Wiswesser Line Notation T5OJ
Evaluation A

$C_4H_4O_4$ (c) 30PAR/HUF 2
cis–2–Butenedioic acid; Maleic acid
Heat Capacity 294.4 K, C_p = 32.4 cal·mol^{-1}·K^{-1}
135.6 J·mol^{-1}·K^{-1}
Temperature range 91–294 K. Value is unsmoothed experimental datum.
Entropy 298.15 K, S = 38.1 cal·mol^{-1}·K^{-1}
159.4 J·mol^{-1}·K^{-1}
Extrapolation below 90 K, 12.10 cal·mol^{-1}K^{-1}
Molecular Weight 116.0732
Wiswesser Line Notation QV1U1VQ -C
Evaluation B(C_p),C(S)

$C_4H_4O_4$ (c) 30PAR/HUF 2
trans–2–Butenedioic acid; Fumaric acid
Heat Capacity 297.1 K, C_p = 33.9 cal·mol^{-1}·K^{-1}
141.8 J·mol^{-1}·K^{-1}
Temperature range 91–297.1 K. Value is unsmoothed experimental datum.
Entropy 298.15 K, S = 39.7 cal·mol^{-1}·K^{-1}
166.1 J·mol^{-1}·K^{-1}
Extrapolation below 90 K, 12.31 cal·mol^{-1}K^{-1}
Molecular Weight 116.0732
Wiswesser Line Notation QV1U1VQ -T
Evaluation B(C_p),C(S)

$C_4H_4O_4$ (c,II) 78EVS/BEL
Glycolide; 1,4–Dioxane–2,5–dione
Heat Capacity 298.15 K, C_p = 31.8 cal·mol^{-1}·K^{-1}
133.2 J·mol^{-1}·K^{-1}
Temperature range 14–350 K. Complete data deposited VINITI, No. 2144–77, 2 June 1977.
Entropy 298.15 K, S = 37.6 cal·mol^{-1}·K^{-1}
157.2 J·mol^{-1}·K^{-1}
Phase Changes
c,II/c,I 312.1 K, ΔH = 433 cal·mol^{-1}
1810 J·mol^{-1}
ΔS = 1.39 cal·mol^{-1}·K^{-1}
5.80 J·mol^{-1}·K^{-1}
c,I/liq 356.2 K, ΔH = 3537 cal·mol^{-1}
14800 J·mol^{-1}
ΔS = 9.93 cal·mol^{-1}·K^{-1}
41.5 J·mol^{-1}·K^{-1}
Molecular Weight 116.0732
Wiswesser Line Notation T6OV DOVTJ
Evaluation B

C_4H_4S (liq) 34JAC/PAR
Thiophene
Heat Capacity 289.3 K, C_p = 29.45 cal·mol^{-1}·K^{-1}
123.22 J·mol^{-1}·K^{-1}
Temperature range 93–294 K. Data for solid, 90–237 K, not given (table omitted, apparently). Value is unsmoothed experimental datum.
Entropy 298.1 K, S = 42.2 cal·mol^{-1}·K^{-1}
176.6 J·mol^{-1}·K^{-1}
Details of extrapolation below 90 K not given. Scatter in data for solid introduce uncertainty. Value good to about 1 cal·mol^{-1}·K^{-1}.
Phase Changes
c,II/c,I 171.1 K, ΔH = 289 cal·mol^{-1}
1209 J·mol^{-1}
ΔS = 1.7 cal·mol^{-1}·K^{-1}
7.1 J·mol^{-1}·K^{-1}
c,I/liq 233.7 K, ΔH = 1187 cal·mol^{-1}
4966 J·mol^{-1}
ΔS = 5.1 cal·mol^{-1}·K^{-1}
21.3 J·mol^{-1}·K^{-1}
Molecular Weight 84.1356
Wiswesser Line Notation T5SJ
Evaluation B(C_p),C(S)

Table of Heat Capacities, Entropies, and Phase Transition Properties — Continued

C_4H_4S (liq) 49WAD/KNO
Thiophene
Heat Capacity 297.45 K, C_p = 29.60 cal·mol⁻¹·K⁻¹ / 123.85 J·mol⁻¹·K⁻¹
Temperature range 11–336 K. Value is unsmoothed experimental datum.
Entropy 298.15 K, S = 43.30 cal·mol⁻¹·K⁻¹ / 181.17 J·mol⁻¹·K⁻¹
Phase Changes
c,II/c,I 171.6 K, ΔH = 152.4 cal·mol⁻¹ / 637.6 J·mol⁻¹
ΔS = 0.89 cal·mol⁻¹·K⁻¹ / 3.72 J·mol⁻¹·K⁻¹
Anomalous heat capacity 100–150 K. Apparently two second order transitions at about 112, 138 K, with small energies involved.
c,I/liq 234.95 K, ΔH = 1215.6 cal·mol⁻¹ / 5086.1 J·mol⁻¹
ΔS = 5.17 cal·mol⁻¹·K⁻¹ / 21.65 J·mol⁻¹·K⁻¹
Molecular Weight 84.1356
Wiswesser Line Notation T5SJ
Evaluation A

$C_4H_5ClO_2$ (c) 28SKA/SAX
cis-3-Chloro-2-butenoic acid;
β-Chloroisocrotonic acid
Heat Capacity 298 K, C_p = 33.5 cal·mol⁻¹·K⁻¹ / 140.2 J·mol⁻¹·K⁻¹
Temperature range 295–394 K. Equations only.
Phase Changes
c/liq 333.7 K, ΔH = 3300 cal·mol⁻¹ / 13810 J·mol⁻¹
ΔS = 9.9 cal·mol⁻¹·K⁻¹ / 41.4 J·mol⁻¹·K⁻¹
Molecular Weight 120.5353
Wiswesser Line Notation QV1UYG1 -C
Evaluation C

$C_4H_5ClO_2$ (c) 28SKA/SAX
trans-3-Chloro-2-butenoic acid;
β-Chlorocrotonic acid
Heat Capacity 298 K, C_p = 38.2 cal·mol⁻¹·K⁻¹ / 159.8 J·mol⁻¹·K⁻¹
Temperature range 295–394 K. Equations only.
Phase Changes
c/liq 366.8 K, ΔH = 4950 cal·mol⁻¹ / 20710 J·mol⁻¹
ΔS = 13.5 cal·mol⁻¹·K⁻¹ / 56.5 J·mol⁻¹·K⁻¹
Molecular Weight 120.5353
Wiswesser Line Notation QV1UYG1 -T
Evaluation C

$C_4H_5ClO_3$ (liq) 76MAS/PET
4-Chloro-1,3-dioxolan-2-one
Heat Capacity 298 K, C_p = 56.9 cal·mol⁻¹·K⁻¹ / 238 J·mol⁻¹·K⁻¹
Temperature range 200–325 K. Data graphically only. Value estimated from graph.
Molecular Weight 136.5347
Wiswesser Line Notation T6OVOTJ DG
Evaluation D

$C_4H_5Cl_3O$ (liq) 81REI
2,2,3-Trichlorobutanal; Butylchloral
Heat Capacity 298 K, C_p = 57.8 cal·mol⁻¹·K⁻¹ / 241.8 J·mol⁻¹·K⁻¹
Temperature range 291–457 K
Molecular Weight 175.4419
Wiswesser Line Notation VHXGGYG1
Evaluation D

$C_4H_5Cl_3O_2$ (liq) 81REI
Ethyl trichloroacetate
Heat Capacity 298 K, C_p = 55.0 cal·mol⁻¹·K⁻¹ / 230.1 J·mol⁻¹·K⁻¹
Temperature range 289–457 K
Molecular Weight 191.4413
Wiswesser Line Notation GXGGVO2
Evaluation D

C_4H_5N (liq) 71HAL/BAL
Cyclopropyl cyanide; Cyanocyclopropane
Heat Capacity 297 K, C_p = 27.6 cal·mol⁻¹·K⁻¹ / 115.4 J·mol⁻¹·K⁻¹
One temperature
Molecular Weight 67.0902
Wiswesser Line Notation L3TJ ACN
Evaluation C

C_4H_5N (liq) 67SCO/BER
Pyrrole
Heat Capacity 298.15 K, C_p = 30.53 cal·mol⁻¹·K⁻¹ / 127.74 J·mol⁻¹·K⁻¹
Temperature range 11–365 K
Entropy 298.15 K, S = 37.39 cal·mol⁻¹·K⁻¹ / 156.44 J·mol⁻¹·K⁻¹
Phase Changes
c/liq 249.74 K, ΔH = 1890.0 cal·mol⁻¹ / 7907.8 J·mol⁻¹
ΔS = 7.568 cal·mol⁻¹·K⁻¹ / 31.664 J·mol⁻¹·K⁻¹
Molecular Weight 67.0902
Wiswesser Line Notation T5MJ
Evaluation A

$C_4H_5NO_2$ (c) 41SAT/SOG 4
Succinimide
Heat Capacity 323 K, C_p = 31.4 cal·mol⁻¹·K⁻¹ / 131.4 J·mol⁻¹·K⁻¹
Temperature range 0 to 100°C. Mean value.
Molecular Weight 99.0890
Wiswesser Line Notation T6VMVTJ
Evaluation C
Same data in 40SAT/SOG 5.

C_4H_5NS (liq) 36KUR/VOS
Allyl isothiocyanate
Heat Capacity 290 K, C_p = 37.5 cal·mol⁻¹·K⁻¹ / 156.9 J·mol⁻¹·K⁻¹
One temperature
Molecular Weight 99.1502
Wiswesser Line Notation SCN2U1
Evaluation D

Table of Heat Capacities, Entropies, and Phase Transition Properties — Continued

C_4H_5NS (liq) 68GOU/WES
2–Methylthiazole
Heat Capacity 298.15 K, C_p = 36.01 cal·mol⁻¹·K⁻¹ 150.67 J·mol⁻¹·K⁻¹
Temperature range 5–340 K
Entropy 298.15 K, S = 50.64 cal·mol⁻¹·K⁻¹ 211.88 J·mol⁻¹·K⁻¹
Phase Changes
c,I/liq 248.42 K, ΔH = 2907 cal·mol⁻¹ 12163 J·mol⁻¹
ΔS = 11.70 cal·mol⁻¹·K⁻¹ 48.96 J·mol⁻¹·K⁻¹
c,II/liq 246.5 K, ΔH = 2712 cal·mol⁻¹ 11347 J·mol⁻¹
ΔS = 11.00 cal·mol⁻¹·K⁻¹ 46.03 J·mol⁻¹·K⁻¹
Metastable crystal form
Molecular Weight 99.1502
Wiswesser Line Notation T5N CSJ B1
Evaluation A

C_4H_5NS (liq) 68GOU/WES 3
2–Methylthiazole
Heat Capacity 298.15 K, C_p = 36.01 cal·mol⁻¹·K⁻¹ 150.67 J·mol⁻¹·K⁻¹
Temperature range 15–298 K
Entropy 298.15 K, S = 50.64 cal·mol⁻¹·K⁻¹ 211.88 J·mol⁻¹·K⁻¹
Phase Changes
c,I/liq 248.43 K, ΔH = 2906.9 cal·mol⁻¹ 12162.5 J·mol⁻¹
ΔS = 11.70 cal·mol⁻¹·K⁻¹ 48.96 J·mol⁻¹·K⁻¹
c,II/liq 246.53 K, ΔH = 2712.7 cal·mol⁻¹ 11349.9 J·mol⁻¹
ΔS = 11.00 cal·mol⁻¹·K⁻¹ 46.04 J·mol⁻¹·K⁻¹
Molecular Weight 99.1502
Wiswesser Line Notation T5N CSJ B1
Evaluation A

C_4H_5NS (liq) 69SOU/GOU
2–Methylthiazole
Heat Capacity 298.15 K, C_p = 36.01 cal·mol⁻¹·K⁻¹ 150.7 J·mol⁻¹·K⁻¹
Temperature range 5–300 K
Entropy 298.15 K, S = 50.64 cal·mol⁻¹·K⁻¹ 211.9 J·mol⁻¹·K⁻¹
Phase Changes
c,I/liq 248.43 K, ΔH = 2906 cal·mol⁻¹ 12159 J·mol⁻¹
ΔS = 11.70 cal·mol⁻¹·K⁻¹ 48.94 J·mol⁻¹·K⁻¹
c,II/liq 246.5 K, ΔH = 2713 cal·mol⁻¹ 11351 J·mol⁻¹
ΔS = 11.01 cal·mol⁻¹·K⁻¹ 46.05 J·mol⁻¹·K⁻¹
Metastable crystals
Molecular Weight 99.1502
Wiswesser Line Notation T5N CSJ B1
Evaluation A

$C_4H_5N_3O$ (c) 78KIL 3
Cytosine
Heat Capacity 298 K, C_p = 31.7 cal·mol⁻¹·K⁻¹ 132.6 J·mol⁻¹·K⁻¹
298 K, One temperature
Molecular Weight 111.1030
Wiswesser Line Notation T6MVNJ DZ
Evaluation B

C_4H_6 (liq) 45SCO/MEY
1,3–Butadiene
Heat Capacity 298.15 K, C_p = 29.55 cal·mol⁻¹·K⁻¹ 123.65 J·mol⁻¹·K⁻¹
Temperature range 15–303 K
Entropy 298.15 K, S = 47.56 cal·mol⁻¹·K⁻¹ 199.00 J·mol⁻¹·K⁻¹
At vapor pressure of 2105 Torr
Phase Changes
c/liq 164.24 K, ΔH = 1908.2 cal·mol⁻¹ 7983.9 J·mol⁻¹
ΔS = 11.62 cal·mol⁻¹·K⁻¹ 48.61 J·mol⁻¹·K⁻¹
liq/g 273.15 K, ΔH = 5315 cal·mol⁻¹ 81.41 J·mol⁻¹
ΔS = 19.46 cal·mol⁻¹·K⁻¹ 280.64 J·mol⁻¹·K⁻¹
P = 119.95 kPa
Molecular Weight 54.0914
Wiswesser Line Notation 1U2U1
Evaluation A

C_4H_6 (liq) 47AST/SZA
1,2–Butadiene
Heat Capacity 290 K, C_p = 29.35 cal·mol⁻¹·K⁻¹ 122.80 J·mol⁻¹·K⁻¹
Temperature range 14–282 K
Entropy 290 K, S = 49.28 cal·mol⁻¹·K⁻¹ 206.19 J·mol⁻¹·K⁻¹
Phase Changes
c/liq 136.92 K, ΔH = 1663.8 cal·mol⁻¹ 6961.3 J·mol⁻¹
ΔS = 12.15 cal·mol⁻¹·K⁻¹ 50.84 J·mol⁻¹·K⁻¹
liq/g 273.24 K, ΔH = 5885 cal·mol⁻¹ 24623 J·mol⁻¹
ΔS = 21.54 cal·mol⁻¹·K⁻¹ 90.11 J·mol⁻¹·K⁻¹
P = 66.41 kPa
Molecular Weight 54.0914
Wiswesser Line Notation 2UCU1
Evaluation A
Values of S and C_p for liquid for saturation vapor pressure.

C_4H_6 (liq) 41YOS/OSB
2–Butyne; Dimethylacetylene
Heat Capacity 290 K, C_p = 29.67 cal·mol⁻¹·K⁻¹ 124.14 J·mol⁻¹·K⁻¹
Temperature range 15–290 K
Entropy 298.15 K, S = 46.63 cal·mol⁻¹·K⁻¹ 195.10 J·mol⁻¹·K⁻¹
Extrapolated from 291.0 K. Anomalous heat capacity between 145–160 K. ΔS obtained from total energy divided by average temperature.

Table of Heat Capacities, Entropies, and Phase Transition Properties — Continued

Phase Changes
c/liq 240.92 K, ΔH = 2207.2 cal·mol^{-1}
9234.9 J·mol^{-1}
ΔS = 9.165 cal·mol^{-1}·K^{-1}
38.348 J·mol^{-1}·K^{-1}
liq/g 291.0 K, ΔH = 6440 cal·mol^{-1}
26945 J·mol^{-1}
ΔS = 22.13 cal·mol^{-1}·K^{-1}
92.59 J·mol^{-1}·K^{-1}
P = 71.46 kPa
Molecular Weight 54.0914
Wiswesser Line Notation 2UU2
Evaluation A

C_4H_6 (liq) 50AST/MAS
1–Butyne
Heat Capacity 280 K, C_p = 31.65 cal·mol^{-1}·K^{-1}
132.42 J·mol^{-1}·K^{-1}
Temperature range 13–280 K
Entropy 281.23 K, S = 47.20 cal·mol^{-1}·K^{-1}
197.48 J·mol^{-1}·K^{-1}
Phase Changes
c/liq 147.43 K, ΔH = 1441.0 cal·mol^{-1}
6029.1 J·mol^{-1}
ΔS = 9.77 cal·mol^{-1}·K^{-1}
40.89 J·mol^{-1}·K^{-1}
liq/g 281.23 K, ΔH = 5861 cal·mol^{-1}
24522 J·mol^{-1}
ΔS = 20.84 cal·mol^{-1}·K^{-1}
87.20 J·mol^{-1}·K^{-1}
P = 101.325 kPa
Molecular Weight 54.0914
Wiswesser Line Notation 3UU1
Evaluation A

$(C_4H_6)_n$ (liq) 62DAI/EVA 5
cis–1,4–Polybutadiene
Heat Capacity 298.15 K, C_p = 24.13 cal·mol^{-1}·K^{-1}
100.9 J·mol^{-1}·K^{-1}
Temperature range 20–310 K
Entropy 298.15 K, S = 27.64 cal·mol^{-1}·K^{-1}
115.6 J·mol^{-1}·K^{-1}
Phase Changes
c,I/gls 165 K
c,I/liq 262 K, ΔH = 945 cal·mol^{-1}
3953 J·mol^{-1}
ΔS = 3.61 cal·mol^{-1}·K^{-1}
15.1 J·mol^{-1}·K^{-1}
Molecular Weight 54.0914
Wiswesser Line Notation /*1U2U1 –C*/
Evaluation B

$(C_4H_6)_n$ (c) 62DAI/EVA 5
trans–1,4–Polybutadiene
Heat Capacity 298.15 K, C_p = 31.26 cal·mol^{-1}·K^{-1}
130.8 J·mol^{-1}·K^{-1}
Temperature range 20–345 K
Phase Changes
c,II/c,I 317 K, ΔH = 821 cal·mol^{-1}
3456 J·mol^{-1}
ΔS = 2.59 cal·mol^{-1}·K^{-1}
10.8 J·mol^{-1}·K^{-1}
Solid–solid transition is a linear to helical crystal structure.
Molecular Weight 54.0914
Wiswesser Line Notation /*1U2U1 –T*/
Evaluation B

$C_4H_6Cl_2O_2$ (liq) 81REI
Ethyl dichloroacetate
Heat Capacity 298 K, C_p = 50.1 cal·mol^{-1}·K^{-1}
209.6 J·mol^{-1}·K^{-1}
Temperature range 291–434 K
Molecular Weight 156.9962
Wiswesser Line Notation GYGVO2
Evaluation D

$C_4H_6KNaO_6$•$4H_2O$ (c) 78TAT/MAT
Sodium potassium tartrate tetrahydrate
Heat Capacity 298.15 K, C_p = 92.287 cal·mol^{-1}·K^{-1}
386.13 J·mol^{-1}·K^{-1}
Temperature range 190–308 K
Molecular Weight 284.2367
Wiswesser Line Notation OVYQYQVO . K .NA &QH 4
Evaluation A

$C_4H_6N_2O$ (c) 73HAM/AYE
3–Amino–5–methylisoxazole
Heat Capacity 287.15 K, C_p = 35.0 cal·mol^{-1}·K^{-1}
146.4 J·mol^{-1}·K^{-1}
One value
Molecular Weight 98.1042
Wiswesser Line Notation T5NOJ C1 EZ
Evaluation C

$C_4H_6N_2O_2$ (c) 81LEB/KUL
2,5–Diketopiperazine
Heat Capacity 298.15 K, C_p = 32.0 cal·mol^{-1}·K^{-1}
133.9 J·mol^{-1}·K^{-1}
Temperature range 5–330 K
Entropy 298.15 K, S = 34.8 cal·mol^{-1}·K^{-1}
145.6 J·mol^{-1}·K^{-1}
Molecular Weight 114.1036
Wiswesser Line Notation T6MV DMVTJ
Evaluation A

$C_4H_6N_4O_3$ (c) 35STI/HUF
Allantoin
Heat Capacity 296.6 K, C_p = 43.01 cal·mol^{-1}·K^{-1}
179.95 J·mol^{-1}·K^{-1}
Temperature range 84–297 K. Value is unsmoothed experimental datum.
Entropy 298.15 K, S = 46.6 cal·mol^{-1}·K^{-1}
195.0 J·mol^{-1}·K^{-1}
Extrapolation below 90 K, 14.18 cal·mol^{-1}K^{-1}
Molecular Weight 158.1164
Wiswesser Line Notation T5MVMV EHJ EMVZ
Evaluation B(C_p),C(S)

$C_4H_6O_2$ (liq) 71HAL/BAL
Methyl propenoate; Methyl acrylate
Heat Capacity 297 K, C_p = 31.9 cal·mol^{-1}·K^{-1}
133.4 J·mol^{-1}·K^{-1}
One temperature
Molecular Weight 86.0902
Wiswesser Line Notation 1U1VO1
Evaluation C

$C_4H_6O_2$ (liq) 79FUC
Methyl propenoate; Methyl acrylate
Heat Capacity 298.15 K, C_p = 38.6 cal·mol^{-1}·K^{-1}
161.5 J·mol^{-1}·K^{-1}
One temperature
Molecular Weight 86.0902
Wiswesser Line Notation 1U1VO1
Evaluation B

Table of Heat Capacities, Entropies, and Phase Transition Properties — Continued

$C_4H_6O_2$ (liq) 59BEN/THO
Ethenyl ethanoate; Vinyl acetate
Heat Capacity 298 K, C_p = 40.5 cal·mol⁻¹·K⁻¹
169.5 J·mol⁻¹·K⁻¹
Mean value 23 to 50°C
Molecular Weight 86.0902
Wiswesser Line Notation 1VO1U1
Evaluation C

$C_4H_6O_2$ (liq) 79FUC
γ–Butyrolactone
Heat Capacity 298.15 K, C_p = 33.8 cal·mol⁻¹·K⁻¹
141.4 J·mol⁻¹·K⁻¹
One temperature
Molecular Weight 86.0902
Wiswesser Line Notation T5OVTJ
Evaluation B

$C_4H_6O_3$ (liq) 39PHI
Ethanoic anhydride; Acetic anhydride
Heat Capacity 303.2 K, C_p = 40.2 cal·mol⁻¹·K⁻¹
168.2 J·mol⁻¹·K⁻¹
One temperature
Molecular Weight 102.0896
Wiswesser Line Notation 1VOV1
Evaluation C

$C_4H_6O_3$ (liq) 58PEP
Propylene carbonate
Heat Capacity 323.15 K, C_p = 44.0 cal·mol⁻¹·K⁻¹
184.1 J·mol⁻¹·K⁻¹
One temperature
Molecular Weight 102.0896
Wiswesser Line Notation T5OVOTJ D1
Evaluation C

$C_4H_6O_4$ (c,I) 30PAR/HUF 2
1,4–Butanedioic acid; Succinic acid
Heat Capacity 289.8 K, C_p = 35.8 cal·mol⁻¹·K⁻¹
149.8 J·mol⁻¹·K⁻¹
Temperature range 93–290 K. Value is unsmoothed experimental datum.
Entropy 298.15 K, S = 42.0 cal·mol⁻¹·K⁻¹
175.7 J·mol⁻¹·K⁻¹
Extrapolation below 90 K, 13.17 cal·mol⁻¹K⁻¹
Phase Changes
c,II/c,I 272 K, ΔH = 41 cal·mol⁻¹
172 J·mol⁻¹
ΔS = 0.15 cal·mol⁻¹·K⁻¹
0.63 J·mol⁻¹·K⁻¹
Molecular Weight 118.0890
Wiswesser Line Notation QV2VQ
Evaluation B(C_p),C(S)

$C_4H_6O_4$ (c) 39SAT/SOG
1,4–Butanedioic acid; Succinic acid
Heat Capacity 323 K, C_p = 39.2 cal·mol⁻¹·K⁻¹
164.0 J·mol⁻¹·K⁻¹
Temperature range 0 to 100°C. Mean value.
Molecular Weight 118.0890
Wiswesser Line Notation QV2VQ
Evaluation C

$C_4H_6O_4$ (c) 70VAN/WES 2
1,4–Butanedioic acid; Succinic acid
Heat Capacity 298.15 K, C_p = 36.55 cal·mol⁻¹·K⁻¹
152.93 J·mol⁻¹·K⁻¹
Temperature range 5–328 K
Entropy 298.15 K, S = 39.99 cal·mol⁻¹·K⁻¹
167.32 J·mol⁻¹·K⁻¹
Molecular Weight 118.0890
Wiswesser Line Notation QV2VQ
Evaluation A

$C_4H_6O_6$ (c) 39SAT/SOG
Tartaric acid
Heat Capacity 323 K, C_p = 44.1 cal·mol⁻¹·K⁻¹
184.5 J·mol⁻¹·K⁻¹
Temperature range 0 to 100°C. Mean value.
Molecular Weight 150.0878
Wiswesser Line Notation QVYQYQVQ
Evaluation C

C_4H_7ClO (liq) 81REI
Butanoyl chloride; Butyryl chloride
Heat Capacity 298 K, C_p = 40.8 cal·mol⁻¹·K⁻¹
170.7 J·mol⁻¹·K⁻¹
Temperature range 290–390 K
Molecular Weight 106.5517
Wiswesser Line Notation GV3
Evaluation D

C_4H_7ClO (liq) 81REI
2–Methylpropanoyl chloride; Isobutyryl chloride
Heat Capacity 298 K, C_p = 31.5 cal·mol⁻¹·K⁻¹
131.8 J·mol⁻¹·K⁻¹
Temperature range 293–368 K
Molecular Weight 106.5517
Wiswesser Line Notation GVY1&1
Evaluation D

$C_4H_7CsO_2$ (c) 75FER/SAN
Cesium butyrate
Phase Changes
c,II/c,I 344 K, ΔH = 350 cal·mol⁻¹
1460 J·mol⁻¹
ΔS = 1.0 cal·mol⁻¹·K⁻¹
4.2 J·mol⁻¹·K⁻¹
c,III/c,II 263 K, ΔH = 300 cal·mol⁻¹
1260 J·mol⁻¹
ΔS = 1.1 cal·mol⁻¹·K⁻¹
4.8 J·mol⁻¹·K⁻¹
c,I/liq 628 K, ΔH = 3300 cal·mol⁻¹
13810 J·mol⁻¹
ΔS = 5.3 cal·mol⁻¹·K⁻¹
22.0 J·mol⁻¹·K⁻¹
Molecular Weight 220.0035
Wiswesser Line Notation OV3 .CS
Evaluation C

$C_4H_7KO_2$ (c) 75FER/SAN
Potassium butyrate
Phase Changes
c,VII/c,V 133 K, ΔH = 160 cal·mol⁻¹
670 J·mol⁻¹
ΔS = 1.2 cal·mol⁻¹·K⁻¹
5.0 J·mol⁻¹·K⁻¹
Taken as sum of data for transitions at 123 K and 143 K at average temperature.
Molecular Weight 126.1964
Wiswesser Line Notation OV3 .KA
Evaluation C

Table of Heat Capacities, Entropies, and Phase Transition Properties — Continued

C_4H_7N (liq) 02LOU
Butanenitrile; n–Propyl cyanide
Heat Capacity 340 K, C_p = 38 cal·mol^{-1}·K^{-1}
159 J·mol^{-1}·K^{-1}
Mean value 21 to 113°C
Molecular Weight 69.1060
Wiswesser Line Notation NC3
Evaluation D

C_4H_7N (liq) 71HAL/BAL
2–Methylpropionitrile; 2–Cyanopropane;
Isopropyl cyanide
Heat Capacity 297 K, C_p = 37.3 cal·mol^{-1}·K^{-1}
156.2 J·mol^{-1}·K^{-1}
One temperature
Molecular Weight 69.1060
Wiswesser Line Notation NCY1&1
Evaluation C

$C_4H_7NO_4$ (c) 32HUF/BOR
Aminosuccinic acid(L); Aspartic acid(L)
Heat Capacity 293.9 K, C_p = 36.5 cal·mol^{-1}·K^{-1}
152.7 J·mol^{-1}·K^{-1}
Temperature range 88–293 K. Value is unsmoothed experimental datum.
Entropy 298.1 K, S = 41.5 cal·mol^{-1}·K^{-1}
173.6 J·mol^{-1}·K^{-1}
Extrapolation below 90 K, 12.21 cal·mol^{-1}K^{-1}
Molecular Weight 133.1036
Wiswesser Line Notation QVYZ1VQ -L
Evaluation B(C_p),C(S)

$C_4H_7NO_4$ (c) 63HUT/COL 2
Aminosuccinic acid(L); Aspartic acid(L)
Heat Capacity 298.15 K, C_p = 37.09 cal·mol^{-1}·K^{-1}
155.18 J·mol^{-1}·K^{-1}
Temperature range 10–310 K
Entropy 298.15 K, S = 40.66 cal·mol^{-1}·K^{-1}
170.12 J·mol^{-1}·K^{-1}
Molecular Weight 133.1036
Wiswesser Line Notation ZVYZ1VQ -L
Evaluation A

$C_4H_7N_3O$ (c) 32HUF/BOR
Creatinine
Heat Capacity 296.5 K, C_p = 33.0 cal·mol^{-1}·K^{-1}
138.1 J·mol^{-1}·K^{-1}
Temperature range 87–297 K. Value is unsmoothed experimental datum.
Entropy 298.1 K, S = 40.0 cal·mol^{-1}·K^{-1}
167.4 J·mol^{-1}·K^{-1}
Extrapolation below 90 K, 13.38 cal·mol^{-1}K^{-1}
Molecular Weight 113.1188
Wiswesser Line Notation T5NYMVTJ A1 BUM
Evaluation B(C_p),C(S)

$C_4H_7O_2Rb$ (c) 75FER/SAN
Rubidium butyrate
Phase Changes
c,IV/c,III 191 K, ΔH = 570 cal·mol^{-1}
2385 J·mol^{-1}
ΔS = 3.0 cal·mol^{-1}·K^{-1}
12.5 J·mol^{-1}·K^{-1}
c,III/c,II 346 K, ΔH = 240 cal·mol^{-1}
1005 J·mol^{-1}
ΔS = 0.7 cal·mol^{-1}·K^{-1}
2.9 J·mol^{-1}·K^{-1}
c,II/c,I 466 K, ΔH = 560 cal·mol^{-1}
2340 J·mol^{-1}
ΔS = 1.2 cal·mol^{-1}·K^{-1}
5.0 J·mol^{-1}·K^{-1}
c,I/liq 652 K, ΔH = 3760 cal·mol^{-1}
15730 J·mol^{-1}
ΔS = 5.8 cal·mol^{-1}·K^{-1}
24.1 J·mol^{-1}·K^{-1}
Molecular Weight 172.5659
Wiswesser Line Notation OV3 .RB
Evaluation C

C_4H_8 (liq) 36TOD/PAR
2–Methylpropene; Isobutene
Heat Capacity 253.1 K, C_p = 29.02 cal·mol^{-1}·K^{-1}
121.42 J·mol^{-1}·K^{-1}
Temperature range 93.3–253 K. Value is unsmoothed experimental datum.
Entropy 266.0 K, S = 46.4 cal·mol^{-1}·K^{-1}
194 J·mol^{-1}·K^{-1}
Extrapolation below 90 K, 10.81 cal·mol^{-1}K^{-1}.
Phase Changes
c/liq 132.4 K, ΔH = 1415 cal·mol^{-1}
5920 J·mol^{-1}
ΔS = 10.69 cal·mol^{-1}·K^{-1}
44.71 J·mol^{-1}·K^{-1}
Molecular Weight 56.1072
Wiswesser Line Notation 1Y1&U1
Evaluation B(C_p),C(S)

C_4H_8 (liq) 36TOD/PAR
cis–2–Butene
Heat Capacity 266.6 K, C_p = 28.41 cal·mol^{-1}·K^{-1}
118.87 J·mol^{-1}·K^{-1}
Temperature range 93–267 K. Value is unsmoothed experimental datum.
Entropy 276.8 K, S = 50 cal·mol^{-1}·K^{-1}
212.88 J·mol^{-1}·K^{-1}
Extrapolation below 90 K, 11.70 cal·mol^{-1}K^{-1}.
Phase Changes
c/liq 133.8 K, ΔH = 1749 cal·mol^{-1}
7318 J·mol^{-1}
ΔS = 13.07 cal·mol^{-1}·K^{-1}
54.69 J·mol^{-1}·K^{-1}
Molecular Weight 56.1072
Wiswesser Line Notation 2U2 -C
Evaluation B(C_p),C(S)

C_4H_8 (liq) 44SCO/FER
cis–2–Butene
Heat Capacity 298.15 K, C_p = 30.15 cal·mol^{-1}·K^{-1}
126.15 J·mol^{-1}·K^{-1}
Temperature range 15–300 K
Entropy 298.15 K, S = 52.56 cal·mol^{-1}·K^{-1}
219.91 J·mol^{-1}·K^{-1}
Phase Changes
c/liq 134.26 K, ΔH = 1746.9 cal·mol^{-1}
7309.2 J·mol^{-1}
ΔS = 13.01 cal·mol^{-1}·K^{-1}
54.44 J·mol^{-1}·K^{-1}
Molecular Weight 56.1072
Wiswesser Line Notation 2U2 -C
Evaluation A

Table of Heat Capacities, Entropies, and Phase Transition Properties — Continued

C_4H_8 (liq) 52SCH/SAG
cis–2–Butene
Heat Capacity 299.8 K, C_p = 31.07 cal·mol^{-1}·K^{-1}
130.00 J·mol^{-1}·K^{-1}
Temperature range 80 to 200°F
Molecular Weight 56.1072
Wiswesser Line Notation 2U2 –C
Evaluation B

C_4H_8 (liq) 36TOD/PAR
trans–2–Butene
Heat Capacity 259.6 K, C_p = 29.17 cal·mol^{-1}·K^{-1}
122.05 J·mol^{-1}·K^{-1}
Temperature range 93–260 K. Value is unsmoothed experimental datum.
Entropy 274.1 K, S = 48.99 cal·mol^{-1}·K^{-1}
204.97 J·mol^{-1}·K^{-1}
Extrapolation below 90 K, 10.23 cal·mol^{-1}K^{-1}
Phase Changes
c/liq 167.3 K, ΔH = 2357 cal·mol^{-1}
9861 J·mol^{-1}
ΔS = 14.09 cal·mol^{-1}·K^{-1}
58.94 J·mol^{-1}·K^{-1}
Molecular Weight 56.1072
Wiswesser Line Notation 2U2 –T
Evaluation B(C_p),C(S)

C_4H_8 (liq) 45GUT/PIT
trans–2–Butene
Heat Capacity 270 K, C_p = 29.24 cal·mol^{-1}·K^{-1}
122.34 J·mol^{-1}·K^{-1}
Temperature range 15–274 K
Entropy 274.04 K, S = 49.07 cal·mol^{-1}·K^{-1}
205.31 J·mol^{-1}·K^{-1}
Phase Changes
c/liq 167.61 K, ΔH = 2332 cal·mol^{-1}
9757 J·mol^{-1}
ΔS = 13.91 cal·mol^{-1}·K^{-1}
58.21 J·mol^{-1}·K^{-1}
liq/g 274.04 K, ΔH = 5439 cal·mol^{-1}
22757 J·mol^{-1}
ΔS = 19.85 cal·mol^{-1}·K^{-1}
83.04 J·mol^{-1}·K^{-1}
P = 101.325 kPa
Molecular Weight 56.1072
Wiswesser Line Notation 2U2 –T
Evaluation A

C_4H_8 (liq) 36TOD/PAR
1–Butene
Heat Capacity 253.4 K, C_p = 28.48 cal·mol^{-1}·K^{-1}
119.16 J·mol^{-1}·K^{-1}
Temperature range 81–253 K. Value is unsmoothed experimental datum.
Molecular Weight 56.1072
Wiswesser Line Notation 3U1
Evaluation B

C_4H_8 (liq) 46AST/FIN
1–Butene
Heat Capacity 260 K, C_p = 28.55 cal·mol^{-1}·K^{-1}
119.45 J·mol^{-1}·K^{-1}
Temperature range 11.5–260 K
Entropy 266.91 K, S = 51.11 cal·mol^{-1}·K^{-1}
213.84 J·mol^{-1}·K^{-1}
Phase Changes
c/liq 87.82 K, ΔH = 920 cal·mol^{-1}
3849 J·mol^{-1}
ΔS = 10.48 cal·mol^{-1}·K^{-1}
43.83 J·mol^{-1}·K^{-1}
liq/g 266.91 K, ΔH = 5226 cal·mol^{-1}
21866 J·mol^{-1}
ΔS = 19.58 cal·mol^{-1}·K^{-1}
81.92 J·mol^{-1}·K^{-1}
P = 101.325 kPa
Molecular Weight 56.1072
Wiswesser Line Notation 3U1
Evaluation A

C_4H_8 (liq) 53RAT/GWI
Cyclobutane
Heat Capacity 285 K, C_p = 25.41 cal·mol^{-1}·K^{-1}
106.32 J·mol^{-1}·K^{-1}
Temperature range 15–285 K
Entropy 285.66 K, S = 42.34 cal·mol^{-1}·K^{-1}
175.15 J·mol^{-1}·K^{-1}
Phase Changes
c,II/c,I 145.7 K, ΔH = 1364 cal·mol^{-1}
5707 J·mol^{-1}
ΔS = 9.36 cal·mol^{-1}·K^{-1}
39.17 J·mol^{-1}·K^{-1}
Transition over about 120 to 145.7. Values represent excess over extrapolated C_p curves.
c,I/liq 182.42 K, ΔH = 260 cal·mol^{-1}
1088 J·mol^{-1}
ΔS = 1.43 cal·mol^{-1}·K^{-1}
5.96 J·mol^{-1}·K^{-1}
liq/g 285.66 K, ΔH = 5781 cal·mol^{-1}
24188 J·mol^{-1}
ΔS = 20.24 cal·mol^{-1}·K^{-1}
84.67 J·mol^{-1}·K^{-1}
P = 101.325 kPa
Molecular Weight 56.1072
Wiswesser Line Notation L4TJ
Evaluation A

$(C_4H_8)_n$ (gls) 62DAI/EVA 4
Poly(1–butene), isotactic
Heat Capacity 298.15 K, C_p = 26.82 cal·mol^{-1}·K^{-1}
112.2 J·mol^{-1}·K^{-1}
Temperature range 20–310 K
Entropy 298.15 K, S = 24.61 cal·mol^{-1}·K^{-1}
103.0 J·mol^{-1}·K^{-1}
When extrapolated to 100% crystallinity, the entropy is 20.7 cal·mol^{-1}·K^{-1}.
Phase Changes
c,I/gls 249 K
Molecular Weight 56.1072
Wiswesser Line Notation /*Y2&1*/
Evaluation A

$C_4H_8N_2O_2$ (c) 41SAT/SOG 4
1,4–Butanediamide; Succinamide
Heat Capacity 323 K, C_p = 41.6 cal·mol^{-1}·K^{-1}
174.0 J·mol^{-1}·K^{-1}
Temperature range 0 to 100°C. Mean value.
Molecular Weight 116.1194
Wiswesser Line Notation ZV2VZ
Evaluation C
Same data in 40SAT/SOG 5.

Table of Heat Capacities, Entropies, and Phase Transition Properties — Continued

$C_4H_8N_2O_3$ (c) 41HUF
Glycylglycine
Heat Capacity 293.9 K, C_p = 38.67 cal·mol^{-1}·K^{-1}
161.80 J·mol^{-1}·K^{-1}
Temperature range 87–294 K. Value is unsmoothed experimental datum.
Entropy 298.1 K, S = 45.4 cal·mol^{-1}·K^{-1}
190.0 J·mol^{-1}·K^{-1}
Extrapolation below 90 K, 13.60 cal·mol^{-1}K^{-1}
Molecular Weight 132.1188
Wiswesser Line Notation Z1VM1VQ
Evaluation A(C_p),C(S)

$C_4H_8N_2O_3$ (c) 69HUT/COL 2
Glycylglycine
Heat Capacity 298.15 K, C_p = 39.19 cal·mol^{-1}·K^{-1}
163.97 J·mol^{-1}·K^{-1}
Temperature range 11–305 K
Entropy 298.15 K, S = 43.09 cal·mol^{-1}·K^{-1}
180.3 J·mol^{-1}·K^{-1}
Molecular Weight 132.1188
Wiswesser Line Notation Z1VM1VQ
Evaluation A

$C_4H_8N_2O_3$ (c) 32HUF/BOR
Asparagine(L)
Heat Capacity 296.5 K, C_p = 38.2 cal·mol^{-1}·K^{-1}
159.8 J·mol^{-1}·K^{-1}
Temperature range 85–297 K. Value is unsmoothed experimental datum.
Entropy 298.1 K, S = 41.7 cal·mol^{-1}·K^{-1}
174.5 J·mol^{-1}·K^{-1}
Extrapolation below 90 K, 11.90 cal·mol^{-1}K^{-1}
Molecular Weight 132.1188
Wiswesser Line Notation ZV1YZVQ -L
Evaluation B(C_p),C(S)

$C_4H_8N_2O_3$•H_2O (c) 32HUF/BOR
Asparagine hydrate (L)
Heat Capacity 296.7 K, C_p = 49.2 cal·mol^{-1}·K^{-1}
205.9 J·mol^{-1}·K^{-1}
Temperature range 90–297 K. Value is unsmoothed experimental datum.
Entropy 298.1 K, S = 51.0 cal·mol^{-1}·K^{-1}
213.4 J·mol^{-1}·K^{-1}
Extrapolation below 90 K, 13.57 cal·mol^{-1}K^{-1}
Molecular Weight 150.1340
Wiswesser Line Notation ZV1YZVQ &QH -L
Evaluation B(C_p),C(S)

$C_4H_8N_2O_3$•H_2O (c) 63HUT/COL 2
Asparagine hydrate(L)
Heat Capacity 298.15 K, C_p = 49.69 cal·mol^{-1}·K^{-1}
207.90 J·mol^{-1}·K^{-1}
Temperature range 10–310 K
Entropy 298.15 K, S = 50.10 cal·mol^{-1}·K^{-1}
209.62 J·mol^{-1}·K^{-1}
Molecular Weight 150.1340
Wiswesser Line Notation ZV1YZVQ &QH -L
Evaluation A

$C_4H_8N_8O_8$ (c,α) 73KRI/LIC
1,3,5,7–Tetranitro–1,3,5,7–tetraazacyclooctane(α); Octogen(α)
Heat Capacity 298 K, C_p = 75.1 cal·mol^{-1}·K^{-1}
314.2 J·mol^{-1}·K^{-1}
Temperature range 200–465 K. Equation only.
Phase Changes
c,α/c,δ 466–474 K, ΔH = 1770 cal·mol^{-1}
7406 J·mol^{-1}
Molecular Weight 296.1560
Wiswesser Line Notation T8N CN EN GNTJ ANW CNW ENW GNW
Evaluation C

$C_4H_8N_8O_8$ (c,β) 73KRI/LIC
1,3,5,7–Tetranitro–1,3,5,7–tetraazacyclooctane(β); Octogen(β)
Heat Capacity 298 K, C_p = 73.6 cal·mol^{-1}·K^{-1}
307.9 J·mol^{-1}·K^{-1}
Temperature range 200–452 K. Equation only.
Phase Changes
c,β/c,δ 440–456 K, ΔH = 2340 cal·mol^{-1}
9791 J·mol^{-1}
Molecular Weight 296.1560
Wiswesser Line Notation T8N CN EN GNTJ ANW CNW ENW GNW
Evaluation C

$C_4H_8N_8O_8$ (c,γ) 73KRI/LIC
1,3,5,7–Tetranitro–1,3,5,7–tetraazacyclooctane(γ); Octogen(γ)
Heat Capacity 298 K, C_p = 75.5 cal·mol^{-1}·K^{-1}
315.9 J·mol^{-1}·K^{-1}
Temperature range 200–440 K. Equation only.
Phase Changes
c,γ/c,δ 440–455 K, ΔH = 670 cal·mol^{-1}
2803 J·mol^{-1}
Molecular Weight 296.1560
Wiswesser Line Notation T8N CN EN GNTJ ANW CNW ENW GNW
Evaluation C

$C_4H_8N_8O_8$ (c,δ) 73KRI/LIC
1,3,5,7–Tetranitro–1,3,5,7–tetraazacyclooctane(δ); Octogen(δ)
Heat Capacity 298 K, C_p = 89.5 cal·mol^{-1}·K^{-1}
374.5 J·mol^{-1}·K^{-1}
Temperature range 200–530 K. Equation only.
Molecular Weight 296.1560
Wiswesser Line Notation T8N CN EN GNTJ ANW CNW ENW GNW
Evaluation C

C_4H_8O (liq) 33KOL/UDO
Butanone; Methyl ethyl ketone
Heat Capacity 297.0 K, C_p = 38.4 cal·mol^{-1}·K^{-1}
160.7 J·mol^{-1}·K^{-1}
One temperature
Molecular Weight 72.1066
Wiswesser Line Notation 2V1
Evaluation C

C_4H_8O (liq) 34KOL/UDO 2
Butanone; Methyl ethyl ketone
Heat Capacity 297.0 K, C_p = 38.4 cal·mol^{-1}·K^{-1}
160.7 J·mol^{-1}·K^{-1}
One temperature
Molecular Weight 72.1066
Wiswesser Line Notation 2V1
Evaluation C

Table of Heat Capacities, Entropies, and Phase Transition Properties — Continued

C_4H_8O (liq) 56PAR/KEN
Butanone; Methyl ethyl ketone
Heat Capacity 298.15 K, C_p = 37.86 cal·mol^{-1}·K^{-1}
158.41 J·mol^{-1}·K^{-1}
Temperature range 80–300 K
Entropy 298.1 K, S = 57.7 cal·mol^{-1}·K^{-1}
241.4 J·mol^{-1}·K^{-1}
Extrapolation below 80 K, 12.78 cal·mol^{-1}K^{-1}
Phase Changes
c/liq 186.1 K, ΔH = 2028 cal·mol^{-1}
8485 J·mol^{-1}
ΔS = 10.90 cal·mol^{-1}·K^{-1}
45.59 J·mol^{-1}·K^{-1}
Molecular Weight 72.1066
Wiswesser Line Notation 2V1
Evaluation B(C_p),C(S)

C_4H_8O (liq) 64SIN/OET
Butanone; Methyl ethyl ketone
Heat Capacity 298.15 K, C_p = 37.98 cal·mol^{-1}·K^{-1}
158.91 J·mol^{-1}·K^{-1}
Temperature range 13–308 K
Entropy 298.15 K, S = 57.08 cal·mol^{-1}·K^{-1}
238.82 J·mol^{-1}·K^{-1}
Phase Changes
c/liq 186.48 K, ΔH = 2016.9 cal·mol^{-1}
8438.7 J·mol^{-1}
ΔS = 10.82 cal·mol^{-1}·K^{-1}
45.25 J·mol^{-1}·K^{-1}
Molecular Weight 72.1066
Wiswesser Line Notation 2V1
Evaluation A

C_4H_8O (liq) 67RAS/GAN
Butanone; Methyl ethyl ketone
Heat Capacity 293 K, C_p = 37.8 cal·mol^{-1}·K^{-1}
158.0 J·mol^{-1}·K^{-1}
Temperature range 293–353 K
Molecular Weight 72.1066
Wiswesser Line Notation 2V1
Evaluation C

C_4H_8O (liq) 68AND/COU
Butanone; Methyl ethyl ketone
Heat Capacity 298.15 K, C_p = 37.93 cal·mol^{-1}·K^{-1}
158.7 J·mol^{-1}·K^{-1}
Temperature range 10–320 K
Entropy 298.15 K, S = 57.12 cal·mol^{-1}·K^{-1}
239.0 J·mol^{-1}·K^{-1}
Phase Changes
c/liq 186.47 K, ΔH = 2004 cal·mol^{-1}
8385 J·mol^{-1}
ΔS = 10.75 cal·mol^{-1}·K^{-1}
44.98 J·mol^{-1}·K^{-1}
Molecular Weight 72.1066
Wiswesser Line Notation 2V1
Evaluation A

C_4H_8O (liq) 75GRO/BEN
Butanone; Methyl ethyl ketone
Heat Capacity 298.15 K, C_p = 38.05 cal·mol^{-1}·K^{-1}
159.2 J·mol^{-1}·K^{-1}
One temperature
Molecular Weight 72.1066
Wiswesser Line Notation 2V1
Evaluation B

C_4H_8O (liq) 78ROU/PER 2
Butanone; Methyl ethyl ketone
Heat Capacity 298.1 K, C_p = 37.86 cal·mol^{-1}·K^{-1}
158.4 J·mol^{-1}·K^{-1}
Temperature range 277–313 K
Molecular Weight 72.1066
Wiswesser Line Notation 2V1
Evaluation B

C_4H_8O (liq) 76CON/GIN
Tetrahydrofuran; Oxolane
Heat Capacity 298 K, C_p = 28.8 cal·mol^{-1}·K^{-1}
120.5 J·mol^{-1}·K^{-1}
One temperature
Molecular Weight 72.1066
Wiswesser Line Notation T5OTJ
Evaluation C

C_4H_8O (liq) 78LEB/RAB
Tetrahydrofuran; Oxolane
Heat Capacity 298.15 K, C_p = 29.61 cal·mol^{-1}·K^{-1}
123.9 J·mol^{-1}·K^{-1}
Temperature range 8–322 K
Entropy 298.15 K, S = 48.73 cal·mol^{-1}·K^{-1}
203.9 J·mol^{-1}·K^{-1}
Phase Changes
c/liq 164.76 K, ΔH = 2041 cal·mol^{-1}
8540 J·mol^{-1}
ΔS = 12.4 cal·mol^{-1}·K^{-1}
51.8 J·mol^{-1}·K^{-1}
Molecular Weight 72.1066
Wiswesser Line Notation T5OTJ
Evaluation A

C_4H_8O (liq) 79KIY/D'A
Tetrahydrofuran; Oxolane
Heat Capacity 298.15 K, C_p = 29.53 cal·mol^{-1}·K^{-1}
123.56 J·mol^{-1}·K^{-1}
One temperature
Molecular Weight 72.1066
Wiswesser Line Notation T5OTJ
Evaluation B

C_4H_8O (liq) 79LEB/LIT
Tetrahydrofuran; Oxolane
Heat Capacity 298.15 K, C_p not given
Temperature range 8–322 K, Data deposited in VINITI, No. 1077–78, 30 March 1978.
Entropy 298.15 K, S = 48.7 cal·mol^{-1}·K^{-1}
203.8 J·mol^{-1}·K^{-1}
Phase Changes
c,I/liq 164.76 K, ΔH = 2041 cal·mol^{-1}
8540 J·mol^{-1}
ΔS = 12.39 cal·mol^{-1}·K^{-1}
51.83 J·mol^{-1}·K^{-1}
Molecular Weight 72.1066
Wiswesser Line Notation T5OTJ
Evaluation A

C_4H_8O (liq) 56PAR/KEN
Butanal; n–Butryaldehyde
Heat Capacity 298.15 K, C_p = 39.08 cal·mol^{-1}·K^{-1}
163.51 J·mol^{-1}·K^{-1}
Temperature range 80–300 K
Entropy 298.1 K, S = 59.0 cal·mol^{-1}·K^{-1}
246.9 J·mol^{-1}·K^{-1}
Extrapolation below 80 K, 10.50 cal·mol^{-1}K^{-1}

Table of Heat Capacities, Entropies, and Phase Transition Properties — Continued

Phase Changes
c/liq 176.8 K, ΔH = 2654 cal·mol^{-1}
11104 J·mol^{-1}
ΔS = 15.01 cal·mol^{-1}·K^{-1}
62.81 J·mol^{-1}·K^{-1}
Molecular Weight 72.1066
Wiswesser Line Notation VH3
Evaluation B(C_p),C(S)

$C_4H_8O_2$ (liq) 33PAR/HUF
Ethyl ethanoate; Ethyl acetate
Heat Capacity 293.6 K, C_p = 40.44 cal·mol^{-1}·K^{-1}
169.20 J·mol^{-1}·K^{-1}
Temperature range 92–294 K. Value is unsmoothed experimental datum.
Entropy 298.1 K, S = 62.0 cal·mol^{-1}·K^{-1}
259.4 J·mol^{-1}·K^{-1}
Extrapolation below 90 K, 15.01 cal·mol^{-1}K^{-1}
Phase Changes
c/liq 189.3 K, ΔH = 2505 cal·mol^{-1}
10481 J·mol^{-1}
ΔS = 13.23 cal·mol^{-1}·K^{-1}
55.37 J·mol^{-1}·K^{-1}
Molecular Weight 88.1060
Wiswesser Line Notation 2OV1
Evaluation B(C_p),C(S)

$C_4H_8O_2$ (liq) 36KUR/VOS
Ethyl ethanoate; Ethyl acetate
Heat Capacity 290 K, C_p = 37.7 cal·mol^{-1}·K^{-1}
157.7 J·mol^{-1}·K^{-1}
One temperature
Molecular Weight 88.1060
Wiswesser Line Notation 2OV1
Evaluation D

$C_4H_8O_2$ (liq) 45ZHD
Ethyl ethanoate; Ethyl acetate
Heat Capacity 303.61 K, C_p = 40.35 cal·mol^{-1}·K^{-1}
168.82 J·mol^{-1}·K^{-1}
Temperature range 5 to 46°C. Value is unsmoothed experimental datum.
Molecular Weight 88.1060
Wiswesser Line Notation 2OV1
Evaluation C

$C_4H_8O_2$ (liq) 78ROU/PER 2
Ethyl ethanoate; Ethyl acetate
Heat Capacity 298.1 K, C_p = 40.51 cal·mol^{-1}·K^{-1}
169.5 J·mol^{-1}·K^{-1}
Temperature range 283–313 K
Molecular Weight 88.1060
Wiswesser Line Notation 2OV1
Evaluation B

$C_4H_8O_2$ (liq) 79FUC
Ethyl ethanoate; Ethyl acetate
Heat Capacity 298.15 K, C_p = 40.0 cal·mol^{-1}·K^{-1}
167.4 J·mol^{-1}·K^{-1}
One temperature
Molecular Weight 88.1060
Wiswesser Line Notation 2OV1
Evaluation B

$C_4H_8O_2$ (liq) 79FUC
Methyl propanoate; Methyl propionate
Heat Capacity 298.15 K, C_p = 41.6 cal·mol^{-1}·K^{-1}
174.2 J·mol^{-1}·K^{-1}
One temperature
Molecular Weight 88.1060
Wiswesser Line Notation 2VO1
Evaluation B

$C_4H_8O_2$ (liq) 81REI
Butanoic acid; n–Butyric acid
Heat Capacity 298 K, C_p = 42.1 cal·mol^{-1}·K^{-1}
176.1 J·mol^{-1}·K^{-1}
Temperature range 292–448 K
Molecular Weight 88.1060
Wiswesser Line Notation QV3
Evaluation D

$C_4H_8O_2$ (liq) 26PAR/AND
Butanoic acid; n–Butyric acid
Heat Capacity 290.7 K, C_p = 42.1 cal·mol^{-1}·K^{-1}
176.1 J·mol^{-1}·K^{-1}
Temperature range 89–291 K. Value is unsmoothed experimental datum.
Entropy 298.1 K, S = 61.0 cal·mol^{-1}·K^{-1}
255.2 J·mol^{-1}·K^{-1}
Extrapolation below 90 K, 19.70 cal·mol^{-1}K^{-1}
Phase Changes
c/liq 267.4 K, ΔH = 2646 cal·mol^{-1}
11071 J·mol^{-1}
ΔS = 9.90 cal·mol^{-1}·K^{-1}
41.40 J·mol^{-1}·K^{-1}
Molecular Weight 88.1060
Wiswesser Line Notation QV3
Evaluation B(C_p),C(S)

$C_4H_8O_2$ (liq) 29PAR/KEL
Butanoic acid; n–Butyric acid
Entropy 298.1 K, S = 54.1 cal·mol^{-1}·K^{-1}
226.4 J·mol^{-1}·K^{-1}
Extrapolation below 90 K, 12.8 cal·mol^{-1}K^{-1}. Revision of previous data.
Molecular Weight 88.1060
Wiswesser Line Notation QV3
Evaluation C

$C_4H_8O_2$ (liq) 71KON/WAD
Butanoic acid; n–Butyric acid
Heat Capacity 298.15 K, C_p = 42.5 cal·mol^{-1}·K^{-1}
178 J·mol^{-1}·K^{-1}
One temperature
Molecular Weight 88.1060
Wiswesser Line Notation QV3
Evaluation B

$C_4H_8O_2$ (liq) 82MAR/AND
Butanoic acid; n–Butyric acid
Heat Capacity 298.15 K, C_p = 42.47 cal·mol^{-1}·K^{-1}
177.7 J·mol^{-1}·K^{-1}
Temperature range 13–450 K. Data also given by equation.
Entropy 298.15 K, S = 53.83 cal·mol^{-1}·K^{-1}
225.3 J·mol^{-1}·K^{-1}
Phase Changes
c,II/c,I 155–230 K, ΔH = 248.6 cal·mol^{-1}
1040 J·mol^{-1}
ΔS = 1.21 cal·mol^{-1}·K^{-1}
5.06 J·mol^{-1}·K^{-1}

Table of Heat Capacities, Entropies, and Phase Transition Properties — Continued

c,I/liq 268.03 K, ΔH = 2770 cal·mol^{-1}
11590 J·mol^{-1}
ΔS = 10.33 cal·mol^{-1}·K^{-1}
43.24 J·mol^{-1}·K^{-1}

Molecular Weight 88.1060
Wiswesser Line Notation QV3
Evaluation A

$C_4H_8O_2$ (liq) 81REI
2–Methylpropanoic acid; Isobutyric acid
Heat Capacity 298 K, C_p = 40.9 cal·mol^{-1}·K^{-1}
171.1 J·mol^{-1}·K^{-1}
Temperature range 291–448 K
Molecular Weight 88.1060
Wiswesser Line Notation QVY1&1
Evaluation D

$C_4H_8O_2$ (liq) 71KON/WAD
2–Methylpropanoic acid; Isobutyric acid
Heat Capacity 298.15 K, C_p = 41.3 cal·mol^{-1}·K^{-1}
173 J·mol^{-1}·K^{-1}
One temperature
Molecular Weight 88.1060
Wiswesser Line Notation QVY1&1
Evaluation B

$C_4H_8O_2$ (liq) 76CON/GIN
1,3–Dioxane
Heat Capacity 298 K, C_p = 34.4 cal·mol^{-1}·K^{-1}
143.9 J·mol^{-1}·K^{-1}
One temperature
Molecular Weight 88.1060
Wiswesser Line Notation T6O COTJ
Evaluation B

$C_4H_8O_2$ (liq) 29HER/LOR
1,4–Dioxane
Heat Capacity 296 K, C_p = 37.0 cal·mol^{-1}·K^{-1}
154.8 J·mol^{-1}·K^{-1}
One temperature
Molecular Weight 88.1060
Wiswesser Line Notation T6O DOTJ
Evaluation C

$C_4H_8O_2$ (liq) 34JAC/PAR
1,4–Dioxane
Heat Capacity 298.2 K, C_p = 36.56 cal·mol^{-1}·K^{-1}
152.97 J·mol^{-1}·K^{-1}
Temperature range 92–299 K. Value is unsmoothed experimental datum.
Entropy 298.1 K, S = 47.0 cal·mol^{-1}·K^{-1}
196.6 J·mol^{-1}·K^{-1}
Extrapolation below 90 K, 11.12 cal·mol^{-1}K^{-1}
Phase Changes
c,II/c,I 272.9 K, ΔH = 562 cal·mol^{-1}
2351 J·mol^{-1}
ΔS = 2.1 cal·mol^{-1}·K^{-1}
8.6 J·mol^{-1}·K^{-1}
c,I/liq 284.1 K, ΔH = 3070 cal·mol^{-1}
12845 J·mol^{-1}
ΔS = 10.8 cal·mol^{-1}·K^{-1}
45.2 J·mol^{-1}·K^{-1}
Molecular Weight 88.1060
Wiswesser Line Notation T6O DOTJ
Evaluation B(C_p),C(S)

$C_4H_8O_2$ (liq) 71DES/BHA
1,4–Dioxane
Heat Capacity 298 K, C_p = 37.2 cal·mol^{-1}·K^{-1}
155.6 J·mol^{-1}·K^{-1}
Temperature range 298–318 K
Molecular Weight 88.1060
Wiswesser Line Notation T6O DOTJ
Evaluation B

$C_4H_8O_2$ (liq) 33ROT/MEY
1,4–Dioxane
Heat Capacity 291 K, C_p = 34.9 cal·mol^{-1}·K^{-1}
146.0 J·mol^{-1}·K^{-1}
Temperature range 8 to 28°C
Phase Changes
c/liq 283.2 K, ΔH = 2840 cal·mol^{-1}
11880 J·mol^{-1}
ΔS = 10.0 cal·mol^{-1}·K^{-1}
41.9 J·mol^{-1}·K^{-1}
Molecular Weight 88.1060
Wiswesser Line Notation T6O DOTJ
Evaluation C

$C_4H_8O_2$ (liq) 71HYD/SUB
1,4–Dioxane
Heat Capacity 298.15 K, C_p = 35.35 cal·mol^{-1}·K^{-1}
147.9 J·mol^{-1}·K^{-1}
Temperature range 298; 313 K
Molecular Weight 88.1060
Wiswesser Line Notation T6O DOTJ
Evaluation B

$C_4H_8O_2$ (liq) 76CON/GIN
1,4–Dioxane
Heat Capacity 298 K, C_p = 33.5 cal·mol^{-1}·K^{-1}
140.2 J·mol^{-1}·K^{-1}
One temperature
Molecular Weight 88.1060
Wiswesser Line Notation T6O DOTJ
Evaluation B

$C_4H_8O_4$ (c) 69CLE/MEL
Tetroxan
Heat Capacity 298.15 K, C_p = 33.99 cal·mol^{-1}·K^{-1}
142.22 J·mol^{-1}·K^{-1}
Temperature range 80–420 K
Entropy 298.15 K, S = 40.03 cal·mol^{-1}·K^{-1}
167.5 J·mol^{-1}·K^{-1}
Extrapolation below 80 K, 50.3 J·mol^{-1}K^{-1}
Phase Changes
c/liq 385 K, ΔH = 5400 cal·mol^{-1}
22600 J·mol^{-1}
ΔS = 14.0 cal·mol^{-1}·K^{-1}
58.7 J·mol^{-1}·K^{-1}
Molecular Weight 120.1048
Wiswesser Line Notation T8O CO EO GOTJ
Evaluation B(C_p),C(S)

C_4H_8S (liq) 52HUB/FIN
Thiacyclopentane
Heat Capacity 298.15 K, C_p = 33.50 cal·mol^{-1}·K^{-1}
140.16 J·mol^{-1}·K^{-1}
Temperature range 13–333 K
Entropy 298.15 K, S = 49.67 cal·mol^{-1}·K^{-1}
207.82 J·mol^{-1}·K^{-1}

Table of Heat Capacities, Entropies, and Phase Transition Properties — Continued

Phase Changes
c/liq 176.98 K, ΔH = 1757.2 cal·mol^{-1}
7352.1 J·mol^{-1}
ΔS = 9.93 cal·mol^{-1}·K^{-1}
41.54 J·mol^{-1}·K^{-1}
Molecular Weight 88.1672
Wiswesser Line Notation T5STJ
Evaluation A

C_4H_9Br (liq) 48KUR
1-Bromo-2-methylpropane; Isobutyl bromide
Heat Capacity 298 K, C_p = 36.9 cal·mol^{-1}·K^{-1}
154.4 J·mol^{-1}·K^{-1}
Temperature range 11 to 80 °C, mean C_p, two temperatures.
Molecular Weight 137.0191
Wiswesser Line Notation E1Y1&1
Evaluation D

C_4H_9Br (liq) 31DEE
1-Bromobutane; n-Butyl bromide
Heat Capacity 292.3 K, C_p = 36.38 cal·mol^{-1}·K^{-1}
152.21 J·mol^{-1}·K^{-1}
Temperature range 94–293 K. Value is unsmoothed experimental datum.
Entropy 298.15 K, S = 78.16 cal·mol^{-1}·K^{-1}
327.02 J·mol^{-1}·K^{-1}
Extrapolation below 100 K, 10.24 cal·mol^{-1}K^{-1}
Phase Changes
c/liq 160.4 K, ΔH = 2207 cal·mol^{-1}
9234 J·mol^{-1}
ΔS = 13.76 cal·mol^{-1}·K^{-1}
57.57 J·mol^{-1}·K^{-1}
Molecular Weight 137.0191
Wiswesser Line Notation E4
Evaluation B(C_p),C(S)

C_4H_9Br (liq) 48KUR
1-Bromobutane; n-Butyl bromide
Heat Capacity 298 K, C_p = 36.1 cal·mol^{-1}·K^{-1}
151.0 J·mol^{-1}·K^{-1}
Temperature range 13 to 100 °C, mean C_p, two temperatures.
Molecular Weight 137.0191
Wiswesser Line Notation E4
Evaluation D

C_4H_9Br (liq) 50KUS/CRO
2-Bromo-2-methylpropane; tert-Butyl bromide
Heat Capacity 265.1 K, C_p = 36.1 cal·mol^{-1}·K^{-1}
151.0 J·mol^{-1}·K^{-1}
Temperature range 117–265 K. Value is unsmoothed experimental datum.
Phase Changes
c,III/c,II 208.6 K, ΔH = 1350 cal·mol^{-1}
5650 J·mol^{-1}
ΔS = 6.5 cal·mol^{-1}·K^{-1}
27.1 J·mol^{-1}·K^{-1}
c,II/c,I 231.5 K, ΔH = 250 cal·mol^{-1}
1045 J·mol^{-1}
ΔS = 1.1 cal·mol^{-1}·K^{-1}
4.5 J·mol^{-1}·K^{-1}
c,I/liq 256.1 K, ΔH = 470 cal·mol^{-1}
1965 J·mol^{-1}
ΔS = 1.8 cal·mol^{-1}·K^{-1}
7.7 J·mol^{-1}·K^{-1}
Molecular Weight 137.0191
Wiswesser Line Notation EX1&1&1
Evaluation B

C_4H_9Cl (liq) 48KUR
1-Chloro-2-methylpropane; Isobutyl chloride
Heat Capacity 298 K, C_p = 37.9 cal·mol^{-1}·K^{-1}
158.6 J·mol^{-1}·K^{-1}
Temperature range 14 to 59 °C, mean C_p, two temperatures.
Molecular Weight 92.5681
Wiswesser Line Notation G1Y1&1
Evaluation D

C_4H_9Cl (liq) 50KUS/CRO
2-Chloro-2-methylpropane; tert-Butyl chloride
Heat Capacity 259.6 K, C_p = 36.5 cal·mol^{-1}·K^{-1}
152.7 J·mol^{-1}·K^{-1}
Temperature range 122–260 K. Value is unsmoothed experimental datum.
Phase Changes
c,III/c,II 183.1 K, ΔH = 410 cal·mol^{-1}
1715 J·mol^{-1}
ΔS = 2.2 cal·mol^{-1}·K^{-1}
9.4 J·mol^{-1}·K^{-1}
c,II/c,I 219.6 K, ΔH = 1390 cal·mol^{-1}
5815 J·mol^{-1}
ΔS = 6.3 cal·mol^{-1}·K^{-1}
26.5 J·mol^{-1}·K^{-1}
c,I/liq 248.1 K, ΔH = 480 cal·mol^{-1}
2010 J·mol^{-1}
ΔS = 1.9 cal·mol^{-1}·K^{-1}
8.1 J·mol^{-1}·K^{-1}
Molecular Weight 92.5681
Wiswesser Line Notation GX1&1&1
Evaluation B

C_4H_9Cl (liq) 66DWO/GUI
2-Chloro-2-methylpropane; tert-Butyl chloride
Heat Capacity 272.73 K, C_p = 41.30 cal·mol^{-1}·K^{-1}
172.80 J·mol^{-1}·K^{-1}
Temperature range 82–273 K. Value is unsmoothed experimental datum.
Phase Changes
c,III/c,II 182.91 K, ΔH = 448 cal·mol^{-1}
1874 J·mol^{-1}
ΔS = 2.45 cal·mol^{-1}·K^{-1}
10.25 J·mol^{-1}·K^{-1}
c,II/c,I 219.25 K, ΔH = 1406 cal·mol^{-1}
5883 J·mol^{-1}
ΔS = 6.41 cal·mol^{-1}·K^{-1}
26.83 J·mol^{-1}·K^{-1}
c,I/liq 247.53 K, ΔH = 495 cal·mol^{-1}
2071 J·mol^{-1}
ΔS = 2.00 cal·mol^{-1}·K^{-1}
8.37 J·mol^{-1}·K^{-1}
Molecular Weight 92.5681
Wiswesser Line Notation GX1&1&1
Evaluation A

C_4H_9I (liq) **81REI**
1-Iodo-2-methylpropane; Isobutyl iodide
Heat Capacity 298 K, C_p = 38.8 cal·mol^{-1}·K^{-1}
162.3 J·mol^{-1}·K^{-1}
Temperature range 290–417 K
Molecular Weight 184.0196
Wiswesser Line Notation I1Y1&1
Evaluation D

Table of Heat Capacities, Entropies, and Phase Transition Properties — Continued

C_4H_9N (liq) 59MCC/DOU
Pyrrolidine
Heat Capacity 298.15 K, C_p = 37.42 cal·mol^{-1}·K^{-1}
156.57 J·mol^{-1}·K^{-1}
Temperature range 13–350 K
Entropy 298.15 K, S = 48.76 cal·mol^{-1}·K^{-1}
204.01 J·mol^{-1}·K^{-1}
Phase Changes
c,II/c,I 207.14 K, ΔH = 129.1 cal·mol^{-1}
540.1 J·mol^{-1}
ΔS = 0.62 cal·mol^{-1}·K^{-1}
2.61 J·mol^{-1}·K^{-1}
c,I/liq 215.31 K, ΔH = 2050 cal·mol^{-1}
8577 J·mol^{-1}
ΔS = 9.52 cal·mol^{-1}·K^{-1}
39.84 J·mol^{-1}·K^{-1}
Molecular Weight 71.1218
Wiswesser Line Notation T5MTJ
Evaluation A

C_4H_9N (liq) 59HIL/SIN
Pyrrolidine
Heat Capacity 298.15 K, C_p = 37.42 cal·mol^{-1}·K^{-1}
156.57 J·mol^{-1}·K^{-1}
Temperature range 14–312 K
Entropy 298.15 K, S = 48.78 cal·mol^{-1}·K^{-1}
204.10 J·mol^{-1}·K^{-1}
Phase Changes
c,II/c,I 207.14 K, ΔH = 127 cal·mol^{-1}
531 J·mol^{-1}
ΔS = 0.61 cal·mol^{-1}·K^{-1}
2.56 J·mol^{-1}·K^{-1}
c,I/liq 215.31 K, ΔH = 2053 cal·mol^{-1}
8590 J·mol^{-1}
ΔS = 9.54 cal·mol^{-1}·K^{-1}
39.90 J·mol^{-1}·K^{-1}
Molecular Weight 71.1218
Wiswesser Line Notation T5MTJ
Evaluation A

C_4H_9N (liq) 76CON/GIN
Pyrollidine
Heat Capacity 298 K, C_p = 38.3 cal·mol^{-1}·K^{-1}
160.2 J·mol^{-1}·K^{-1}
One temperature
Molecular Weight 71.1218
Wiswesser Line Notation T5MTJ
Evaluation B

C_4H_9NO (liq) 71KON/WAD
N–Ethylethanamide; N–Ethylacetamide
Heat Capacity 298.15 K, C_p = 43.0 cal·mol^{-1}·K^{-1}
180 J·mol^{-1}·K^{-1}
One temperature
Molecular Weight 87.1212
Wiswesser Line Notation 2MV1
Evaluation B

C_4H_9NO (liq) 71KON/WAD
N–Methylpropanamide
Heat Capacity 298.15 K, C_p = 42.8 cal·mol^{-1}·K^{-1}
179 J·mol^{-1}·K^{-1}
One temperature
Molecular Weight 87.1212
Wiswesser Line Notation 2VM1
Evaluation B

C_4H_9NO (liq) 02LOU
Methyl ethyl ketoxime
Heat Capacity 350 K, C_p = 57 cal·mol^{-1}·K^{-1}
238 J·mol^{-1}·K^{-1}
Mean value 21 to 151°C
Molecular Weight 87.1212
Wiswesser Line Notation QNUY2&1
Evaluation D

C_4H_9NO (liq) 42TRI/ENG
Morpholine; Diethyleneimide oxide
Heat Capacity 298 K, C_p = 41.56 cal·mol^{-1}·K^{-1}
173.89 J·mol^{-1}·K^{-1}
Temperature range 273–403 K
Molecular Weight 87.1212
Wiswesser Line Notation T6M DOTJ
Evaluation C

$C_4H_9NO_2$ (c) 41SAT/SOG 3
2–Aminobutanoic acid; α–Aminobutyric acid
Heat Capacity 323 K, C_p = 38.4 cal·mol^{-1}·K^{-1}
160.7 J·mol^{-1}·K^{-1}
Temperature range 0 to 100°C. Mean value.
Molecular Weight 103.1206
Wiswesser Line Notation ZY2&QV
Evaluation C
Same data as 40SAT/SOG 4.

$C_4H_9NO_2$ (c) 75SPI/WAD
2–Aminobutanoic acid(DL); α–Aminobutyric acid(DL)
Heat Capacity 298.15 K, C_p = 35.0 cal·mol^{-1}·K^{-1}
146.4 J·mol^{-1}·K^{-1}
One temperature
Molecular Weight 103.1206
Wiswesser Line Notation ZY2&VQ -DL
Evaluation B

$C_4H_9NO_4$ (c) 39SAT/SOG
Ammonium acid succinate
Heat Capacity 323 K, C_p = 48.7 cal·mol^{-1}·K^{-1}
203.8 J·mol^{-1}·K^{-1}
Temperature range 0 to 100°C. Mean value.
Molecular Weight 135.1194
Wiswesser Line Notation QV2VQ &ZH
Evaluation C

$C_4H_9NO_6$ (c) 39SAT/SOG
Ammonium acid tartrate
Heat Capacity 323 K, C_p = 54.2 cal·mol^{-1}·K^{-1}
226.8 J·mol^{-1}·K^{-1}
Temperature range 0 to 100°C. Mean value.
Molecular Weight 167.1182
Wiswesser Line Notation QVYQYQVQ &ZH
Evaluation C

$C_4H_9N_3O_2$ (c) 41SAT/SOG 3
Creatine
Heat Capacity 323 K, C_p = 44.1 cal·mol^{-1}·K^{-1}
184.5 J·mol^{-1}·K^{-1}
Temperature range 0 to 100°C. Mean value.
Molecular Weight 131.1340
Wiswesser Line Notation QV1N1&YZUM
Evaluation C
Same data as 40SAT/SUG 4.

Table of Heat Capacities, Entropies, and Phase Transition Properties — Continued

$C_4H_9N_3O_2$ (c) 32HUF/BOR
Creatine
Heat Capacity 296.3 K, C_p = 40.9 cal·mol^{-1}·K^{-1}
171.1 J·mol^{-1}·K^{-1}
Temperature range 87–295 K. Value is unsmoothed experimental datum.
Entropy 298.1 K, S = 45.3 cal·mol^{-1}·K^{-1}
189.5 J·mol^{-1}·K^{-1}
Extrapolation below 90 K, 13.31 cal·mol^{-1}K^{-1}
Molecular Weight 131.1340
Wiswesser Line Notation QV1N1&YZUM
Evaluation B(C_p),C(S)

$C_4H_9N_3O_2{\cdot}H_2O$ (c) 40HUF/FOX
Creatine hydrate
Heat Capacity 298.4 K, C_p = 51.05 cal·mol^{-1}·K^{-1}
213.59 J·mol^{-1}·K^{-1}
Temperature range 90–298 K. Value is unsmoothed experimental datum.
Entropy 298.15 K, S = 56.0 cal·mol^{-1}·K^{-1}
234.3 J·mol^{-1}·K^{-1}
Extrapolation below 90 K, 16.39 cal·mol^{-1}K^{-1}
Molecular Weight 149.1492
Wiswesser Line Notation QV1N1&YZUM &QH
Evaluation B(C_p),C(S)

C_4H_{10} (liq) 37PAR/SHO
2–Methylpropane; Isobutane
Heat Capacity 258.3 K, C_p = 30.68 cal·mol^{-1}·K^{-1}
128.4 J·mol^{-1}·K^{-1}
Temperature range 79–261 K. Value is unsmoothed experimental datum.
Entropy 260.9 K, S = 47.5 cal·mol^{-1}·K^{-1}
198.7 J·mol^{-1}·K^{-1}
Extrapolation below 67 K, 10.52 cal·mol^{-1}K^{-1}
Phase Changes
c/liq 113.2 K, ΔH = 1075 cal·mol^{-1}
4498 J·mol^{-1}
ΔS = 9.50 cal·mol^{-1}·K^{-1}
39.73 J·mol^{-1}·K^{-1}
Molecular Weight 58.1230
Wiswesser Line Notation 1Y1&1
Evaluation B(C_p),C(S)

C_4H_{10} (liq) 40AST/KEN
2–Methylpropane; Isobutane
Heat Capacity 260 K, C_p = 31.00 cal·mol^{-1}·K^{-1}
129.70 J·mol^{-1}·K^{-1}
Temperature range 20–260 K
Entropy 261.44 K, S = 47.99 cal·mol^{-1}·K^{-1}
200.79 J·mol^{-1}·K^{-1}
Phase Changes
c/liq 113.74 K, ΔH = 1085 cal·mol^{-1}
4540 J·mol^{-1}
ΔS = 9.54 cal·mol^{-1}·K^{-1}
39.92 J·mol^{-1}·K^{-1}
liq/g 261.44 K, ΔH = 5090 cal·mol^{-1}
21297 J·mol^{-1}
ΔS = 19.47 cal·mol^{-1}·K^{-1}
81.46 J·mol^{-1}·K^{-1}
P = 101.325 kPa
Molecular Weight 58.1230
Wiswesser Line Notation 1Y1&1
Evaluation A

C_4H_{10} (liq) 31HUF/PAR
n–Butane
Heat Capacity 261.8 K, C_p = 31.0 cal·mol^{-1}·K^{-1}
129.7 J·mol^{-1}·K^{-1}
Temperature range 69–262 K. Value is unsmoothed experimental datum.
Entropy 298.1 K, S = 54.9 cal·mol^{-1}·K^{-1}
229.7 J·mol^{-1}·K^{-1}
Extrapolation below 90 K, 11.7 cal·mol^{-1}K^{-1}. Extrapolated above 262 K.
Phase Changes
c,II/c,I 107.0 K, ΔH = 506 cal·mol^{-1}
2117 J·mol^{-1}
ΔS = 4.7 cal·mol^{-1}·K^{-1}
19.8 J·mol^{-1}·K^{-1}
c,I/liq 134.1 K, ΔH = 1045 cal·mol^{-1}
4372 J·mol^{-1}
ΔS = 7.8 cal·mol^{-1}·K^{-1}
32.6 J·mol^{-1}·K^{-1}
Molecular Weight 58.1230
Wiswesser Line Notation 4H
Evaluation B(C_p),C(S)

C_4H_{10} (liq) 37PAR/SHO
n–Butane
Entropy 272.5 K, S = 54.2 cal·mol^{-1}·K^{-1}
226.8 J·mol^{-1}·K^{-1}
Calculated from heat capacity data reported by 31HUF/PAR. Extrapolation below 67 K, 9.88 cal·mol^{-1}K^{-1}.
Molecular Weight 58.1230
Wiswesser Line Notation 4H
Evaluation C

C_4H_{10} (liq) 40AST/MES
n–Butane
Heat Capacity 270 K, C_p = 31.65 cal·mol^{-1}·K^{-1}
13242 J·mol^{-1}·K^{-1}
Temperature range 11–270 K
Entropy 298.15 K, S = 55.2 cal·mol^{-1}·K^{-1}
231.0 J·mol^{-1}·K^{-1}
Using extrapolated values of C_p 273–298 K for the superheated liquid.
Phase Changes
c,II/c,I 107.55 K, ΔH = 494 cal·mol^{-1}
2067 J·mol^{-1}
ΔS = 4.59 cal·mol^{-1}·K^{-1}
19.22 J·mol^{-1}·K^{-1}
c,I/liq 134.86 K, ΔH = 1114 cal·mol^{-1}
4661 J·mol^{-1}
ΔS = 8.26 cal·mol^{-1}·K^{-1}
34.56 J·mol^{-1}·K^{-1}
liq/g 272.05 K, ΔH = 5351 cal·mol^{-1}
22389 J·mol^{-1}
ΔS = 19.67 cal·mol^{-1}·K^{-1}
82.30 J·mol^{-1}·K^{-1}
P = 101.325 kPa
Molecular Weight 58.1230
Wiswesser Line Notation 4H
Evaluation A

$C_4H_{10}O$ (liq) 75AND/MAR
2–Oxa–3–methylbutane; Methyl isopropyl ether
Heat Capacity 298.15 K, C_p = 38.70 cal·mol^{-1}·K^{-1}
161.9 J·mol^{-1}·K^{-1}
Temperature range 12–350 K
Entropy 298.15 K, S = 60.64 cal·mol^{-1}·K^{-1}
253.7 J·mol^{-1}·K^{-1}

Table of Heat Capacities, Entropies, and Phase Transition Properties — Continued

Phase Changes
c,I/liq 127.93 K, ΔH = 1398 cal·mol^{-1}
5850 J·mol^{-1}
ΔS = 10.93 cal·mol^{-1}·K^{-1}
45.73 J·mol^{-1}·K^{-1}
c,II/liq 123.06 K, ΔH = 1219 cal·mol^{-1}
5100 J·mol^{-1}
ΔS = 9.90 cal·mol^{-1}·K^{-1}
41.44 J·mol^{-1}·K^{-1}
Metastable crystals
Molecular Weight 74.1224
Wiswesser Line Notation 1Y1&O1
Evaluation A

$C_4H_{10}O$ (liq) 24KEY/BEA
3–Oxapentane; Diethyl ether
Heat Capacity 286.6 K, C_p = 42.8 cal·mol^{-1}·K^{-1}
179.1 J·mol^{-1}·K^{-1}
Temperature range 274, 286 K
Phase Changes
liq/g 285.0 K, ΔH = 6580 cal·mol^{-1}
27530 J·mol^{-1}
ΔS = 23.09 cal·mol^{-1}·K^{-1}
96.60 J·mol^{-1}·K^{-1}
p = 101.325 kPa
Molecular Weight 74.1224
Wiswesser Line Notation 2O2
Evaluation B

$C_4H_{10}O$ (liq) 26PAR/HUF
3–Oxapentane; Diethyl ether
Heat Capacity 290.0 K, C_p = 40.8 cal·mol^{-1}·K^{-1}
170.7 J·mol^{-1}·K^{-1}
Temperature range 76–290 K. Value is unsmoothed experimental datum.
Entropy 298.1 K, S = 67.7 cal·mol^{-1}·K^{-1}
283.3 J·mol^{-1}·K^{-1}
Extrapolation below 90 K, 21.20 cal·mol^{-1}K^{-1}
Phase Changes
c/liq 156.8 K, ΔH = 1745 cal·mol^{-1}
7301 J·mol^{-1}
ΔS = 11.1 cal·mol^{-1}·K^{-1}
46.6 J·mol^{-1}·K^{-1}
Molecular Weight 74.1224
Wiswesser Line Notation 2O2
Evaluation B(C_p),C(S)

$C_4H_{10}O$ (liq) 27BEN/WEN
3–Oxapentane; Diethyl ether
Heat Capacity 308 K, C_p = 43.0 cal·mol^{-1}·K^{-1}
179.9 J·mol^{-1}·K^{-1}
Temperature range 308–488 K. Value is unsmoothed experimental datum. Pressure 40 atmospheres.
Molecular Weight 74.1224
Wiswesser Line Notation 2O2
Evaluation B

$C_4H_{10}O$ (liq) 29PAR/KEL
3–Oxapentane; Diethyl ether
Entropy 298.1 K, S = 60.4 cal·mol^{-1}·K^{-1}
252.7 J·mol^{-1}·K^{-1}
Extrapolation below 90 K, 14.0 cal·mol^{-1}K^{-1}. Revision of previous data.
Molecular Weight 74.1224
Wiswesser Line Notation 2O2
Evaluation C

$C_4H_{10}O$ (liq) 35AOY/KAN
3–Oxapentane; Diethyl ether
Heat Capacity 255.2 K, C_p = 39.4 cal·mol^{-1}·K^{-1}
164.8 J·mol^{-1}·K^{-1}
Temperature range 80–255 K. Value is unsmoothed experimental datum.
Molecular Weight 74.1224
Wiswesser Line Notation 2O2
Evaluation B

$C_4H_{10}O$ (liq) 36KUR/VOS
3–Oxapentane; Diethyl ether
Heat Capacity 290 K, C_p = 40.0 cal·mol^{-1}·K^{-1}
167.4 J·mol^{-1}·K^{-1}
One temperature
Molecular Weight 74.1224
Wiswesser Line Notation 2O2
Evaluation D

$C_4H_{10}O$ (liq) 39MAZ 2
3–Oxapentane; Diethyl ether
Heat Capacity 293.15 K, C_p = 41.08 cal·mol^{-1}·K^{-1}
171.88 J·mol^{-1}·K^{-1}
Temperature range –112 to 20°C
Molecular Weight 74.1224
Wiswesser Line Notation 2O2
Evaluation B

$C_4H_{10}O$ (liq) 39MAZ 3
3–Oxapentane; Diethyl ether
Heat Capacity 293 K, C_p = 41.1 cal·mol^{-1}·K^{-1}
172.0 J·mol^{-1}·K^{-1}
Temperature range –110 to 20°C
Molecular Weight 74.1224
Wiswesser Line Notation 2O2
Evaluation B

$C_4H_{10}O$ (liq) 71COU/LEE
3–Oxapentane; Diethyl ether
Heat Capacity 298.15 K, C_p = 41.23 cal·mol^{-1}·K^{-1}
172.5 J·mol^{-1}·K^{-1}
Temperature range 15–300 K
Entropy 298.15 K, S = 60.59 cal·mol^{-1}·K^{-1}
253.5 J·mol^{-1}·K^{-1}
Phase Changes
c,II/liq 149.86 K, ΔH = 1630 cal·mol^{-1}
6820 J·mol^{-1}
ΔS = 10.88 cal·mol^{-1}·K^{-1}
45.5 J·mol^{-1}·K^{-1}
c,I/liq 156.92 K, ΔH = 1718 cal·mol^{-1}
7190 J·mol^{-1}
ΔS = 10.95 cal·mol^{-1}·K^{-1}
45.82 J·mol^{-1}·K^{-1}
Metastable crystal
Molecular Weight 74.1224
Wiswesser Line Notation 2O2
Evaluation A

$C_4H_{10}O$ (liq) 75AND/MAR
2–Oxapentane; Methyl n-propyl ether
Heat Capacity 298.15 K, C_p = 39.53 cal·mol^{-1}·K^{-1}
165.4 J·mol^{-1}·K^{-1}
Temperature range 12–350 K
Entropy 298.15 K, S = 62.83 cal·mol^{-1}·K^{-1}
262.9 J·mol^{-1}·K^{-1}

Table of Heat Capacities, Entropies, and Phase Transition Properties — Continued

Phase Changes
c/liq 133.97 K, ΔH = 1833 cal·mol^{-1}
7670 J·mol^{-1}
ΔS = 13.68 cal·mol^{-1}·K^{-1}
57.25 J·mol^{-1}·K^{-1}
Molecular Weight 74.1224
Wiswesser Line Notation 3O1
Evaluation A

$C_4H_{10}O$ (gls) 68COU/LEE
2–Methyl–1–propanol; Isobutyl alcohol
Heat Capacity 180 K, C_p = 30.88 cal·mol^{-1}·K^{-1}
129.2 J·mol^{-1}·K^{-1}
Temperature range 10–180 K
Entropy 180 K, S = 33.63 cal·mol^{-1}·K^{-1}
140.7 J·mol^{-1}·K^{-1}
Molecular Weight 74.1224
Wiswesser Line Notation Q1Y1&1
Evaluation A

$C_4H_{10}O$ (liq) 24WIL/DAN
2–Methyl–1–propanol; Isobutyl alcohol
Heat Capacity 303 K, C_p = 44.7 cal·mol^{-1}·K^{-1}
187.0 J·mol^{-1}·K^{-1}
Temperature range 303–343 K, equation only.
Molecular Weight 74.1224
Wiswesser Line Notation Q1Y1&1
Evaluation C

$C_4H_{10}O$ (liq) 41ZHD
2–Methyl–1–propanol; Isobutyl alcohol
Heat Capacity 298.1 K, C_p = 44.0 cal·mol^{-1}·K^{-1}
184.1 J·mol^{-1}·K^{-1}
Temperature range 5 to 46°C
Molecular Weight 74.1224
Wiswesser Line Notation Q1Y1&1
Evaluation C

$C_4H_{10}O$ (liq) 58SWI/ZIE 2
2–Methyl–1–propanol; Isobutyl alcohol
Heat Capacity 333 K, C_p = 51.4 cal·mol^{-1}·K^{-1}
215.1 J·mol^{-1}·K^{-1}
Mean value 21 to 99°C
Molecular Weight 74.1224
Wiswesser Line Notation Q1Y1&1
Evaluation C

$C_4H_{10}O$ (liq) 60SWI/ZIE
2–Methyl–1–propanol; Isobutyl alcohol
Heat Capacity 323 K, C_p = 48.1 cal·mol^{-1}·K^{-1}
201.3 J·mol^{-1}·K^{-1}
Mean value 21 to 78°C
Molecular Weight 74.1224
Wiswesser Line Notation Q1Y1&1
Evaluation C

$C_4H_{10}O$ (liq) 68COU/LEE
2–Methyl–1–propanol; Isobutyl alcohol
Heat Capacity 298.15 K, C_p = 43.26 cal·mol^{-1}·K^{-1}
181.0 J·mol^{-1}·K^{-1}
Temperature range 10–350 K
Entropy 298.15 K, S = 51.27 cal·mol^{-1}·K^{-1}
214.5 J·mol^{-1}·K^{-1}
Phase Changes
c/liq 171.18 K, ΔH = 1511 cal·mol^{-1}
6322 J·mol^{-1}
ΔS = 8.83 cal·mol^{-1}·K^{-1}
36.93 J·mol^{-1}·K^{-1}
Molecular Weight 74.1224
Wiswesser Line Notation Q1Y1&1
Evaluation A

$C_4H_{10}O$ (liq) 70PAZ/PAZ
2–Methyl–1–propanol; Isobutyl alcohol
Heat Capacity 301.2 K, C_p = 44.3 cal·mol^{-1}·K^{-1}
185.4 J·mol^{-1}·K^{-1}
Temperature range 28, 40°C
Molecular Weight 74.1224
Wiswesser Line Notation Q1Y1&1
Evaluation B

$C_4H_{10}O$ (liq) 81REI
1–Butanol; n–Butyl alcohol
Heat Capacity 298 K, C_p = 41.7 cal·mol^{-1}·K^{-1}
174.5 J·mol^{-1}·K^{-1}
Temperature range 290–390 K
Molecular Weight 74.1224
Wiswesser Line Notation Q4
Evaluation D

$C_4H_{10}O$ (liq) 24WIL/DAN
1–Butanol; n–Butyl alcohol
Heat Capacity 303 K, C_p = 43.1 cal·mol^{-1}·K^{-1}
180.3 J·mol^{-1}·K^{-1}
Temperature range 303–343 K, equation only.
Molecular Weight 74.1224
Wiswesser Line Notation Q4
Evaluation C

$C_4H_{10}O$ (liq) 25PAR
1–Butanol; n–Butyl alcohol
Heat Capacity 294.0 K, C_p = 41.9 cal·mol^{-1}·K^{-1}
175.3 J·mol^{-1}·K^{-1}
Temperature range 90–294 K. Value is unsmoothed experimental datum.
Entropy 298.1 K, S = 60.2 cal·mol^{-1}·K^{-1}
251.9 J·mol^{-1}·K^{-1}
Extrapolation below 90 K, 17.64 cal·mol^{-1}K^{-1}
Phase Changes
c/liq 183.9 K, ΔH = 2218 cal·mol^{-1}
9280 J·mol^{-1}
ΔS = 12.06 cal·mol^{-1}·K^{-1}
50.46 J·mol^{-1}·K^{-1}
Molecular Weight 74.1224
Wiswesser Line Notation Q4
Evaluation B(C_p),C(S)

$C_4H_{10}O$ (liq) 29PAR/KEL
1–Butanol; n–Butyl alcohol
Entropy 298.1 K, S = 54.5 cal·mol^{-1}·K^{-1}
228.0 J·mol^{-1}·K^{-1}
Extrapolation below 90 K, 11.9 cal·mol^{-1}K^{-1}.
Revision of previous data.
Molecular Weight 74.1224
Wiswesser Line Notation Q4
Evaluation C

$C_4H_{10}O$ (liq) 33TRE/WAT
1–Butanol; n–Butyl alcohol
Heat Capacity 298 K, C_p = 43.8 cal·mol^{-1}·K^{-1}
183.3 J·mol^{-1}·K^{-1}
One temperature
Molecular Weight 74.1224
Wiswesser Line Notation Q4
Evaluation B

Table of Heat Capacities, Entropies, and Phase Transition Properties — Continued

$C_4H_{10}O$ (liq) 39PHI
1–Butanol; n–Butyl alcohol
Heat Capacity 302.6 K, C_p = 51.5 cal·mol^{-1}·K^{-1}
215.5 J·mol^{-1}·K^{-1}
One temperature
Molecular Weight 74.1224
Wiswesser Line Notation Q4
Evaluation C
Isomer not specified; normal assumed.

$C_4H_{10}O$ (liq) 60SWI/ZIE
1–Butanol; n–Butyl alcohol
Heat Capacity 323 K, C_p = 45.2 cal·mol^{-1}·K^{-1}
189.1 J·mol^{-1}·K^{-1}
Mean value 21 to 78°C
Molecular Weight 74.1224
Wiswesser Line Notation Q4
Evaluation C

$C_4H_{10}O$ (liq) 65COU/HAL
1–Butanol; n–Butyl alcohol
Heat Capacity 298.15 K, C_p = 42.31 cal·mol^{-1}·K^{-1}
177.03 J·mol^{-1}·K^{-1}
Temperature range 11–323 K
Entropy 298.15 K, S = 53.95 cal·mol^{-1}·K^{-1}
225.73 J·mol^{-1}·K^{-1}
Phase Changes
c/liq 184.5 K, ΔH = 2240 cal·mol^{-1}
9372 J·mol^{-1}
ΔS = 12.14 cal·mol^{-1}·K^{-1}
50.79 J·mol^{-1}·K^{-1}
Molecular Weight 74.1224
Wiswesser Line Notation Q4
Evaluation A

$C_4H_{10}O$ (liq) 70PAZ/PAZ
1–Butanol; n–Butyl alcohol
Heat Capacity 301.2 K, C_p = 42.9 cal·mol^{-1}·K^{-1}
179.5 J·mol^{-1}·K^{-1}
Temperature range 28, 40°C
Molecular Weight 74.1224
Wiswesser Line Notation Q4
Evaluation B

$C_4H_{10}O$ (c,I) 63OET
2–Methyl–2–propanol; tert–Butyl alcohol
Heat Capacity 298.15 K, C_p = 34.92 cal·mol^{-1}·K^{-1}
146.11 J·mol^{-1}·K^{-1}
Temperature range 15–330 K
Entropy 298.15 K, S = 40.84 cal·mol^{-1}·K^{-1}
170.87 J·mol^{-1}·K^{-1}
Phase Changes
c,II/c,I 286.14 K, ΔH = 198 cal·mol^{-1}
828 J·mol^{-1}
ΔS = 0.69 cal·mol^{-1}·K^{-1}
2.89 J·mol^{-1}·K^{-1}
c,III/c,I 294.47 K, ΔH = 117 cal·mol^{-1}
490 J·mol^{-1}
ΔS = 0.40 cal·mol^{-1}·K^{-1}
1.66 J·mol^{-1}·K^{-1}
Metastable transition, not always reproducible, c,III, metastable form.
c,I/liq 298.97 K, ΔH = 1602.0 cal·mol^{-1}
6702.8 J·mol^{-1}
ΔS = 5.36 cal·mol^{-1}·K^{-1}
22.42 J·mol^{-1}·K^{-1}
Molecular Weight 74.1224
Wiswesser Line Notation QX1&1&1
Evaluation A

$C_4H_{10}O$ (liq) 26PAR/AND
2–Methyl–2–propanol; tert–Butyl alcohol
Heat Capacity 300 K, C_p = 53.7 cal·mol^{-1}·K^{-1}
224.7 J·mol^{-1}·K^{-1}
Temperature range 87–300 K. Value is unsmoothed experimental datum.
Entropy 298.15 K, S = 47.2 cal·mol^{-1}·K^{-1}
197.5 J·mol^{-1}·K^{-1}
Extrapolation below 90 K, 12.75 cal·mol^{-1}K^{-1}. Value for supercooled liquid.
Phase Changes
c/liq 298.5 K, ΔH = 1621 cal·mol^{-1}
6782 J·mol^{-1}
ΔS = 5.43 cal·mol^{-1}·K^{-1}
22.72 J·mol^{-1}·K^{-1}
Molecular Weight 74.1224
Wiswesser Line Notation QX1&1&1
Evaluation B(C_p),C(S)

$C_4H_{10}O$ (liq) 29PAR/KEL
2–Methyl–2–propanol; tert–Butyl alcohol
Entropy 298.1 K, S = 45.3 cal·mol^{-1}·K^{-1}
189.5 J·mol^{-1}·K^{-1}
Extrapolation below 90 K, 10.8 cal·mol^{-1}K^{-1}.
Revision of previous data.
Molecular Weight 74.1224
Wiswesser Line Notation QX1&1&1
Evaluation C

$C_4H_{10}O$ (liq) 77VIS/PER
2–Methyl–2–propanol; tert–Butyl alcohol
Heat Capacity 298 K, C_p = 50.2 cal·mol^{-1}·K^{-1}
210 J·mol^{-1}·K^{-1}
One temperature
Molecular Weight 74.1224
Wiswesser Line Notation QX1&1&1
Evaluation B

$C_4H_{10}O$ (liq) 36PAR/THO
2–Butanol; sec–Butyl alcohol
Heat Capacity 281.7 K, C_p = 44.2 cal·mol^{-1}·K^{-1}
184.9 J·mol^{-1}·K^{-1}
Temperature range 103–282 K. Glass at lower temp. Unsmoothed experimental datum.
Molecular Weight 74.1224
Wiswesser Line Notation QY2&1
Evaluation B

$C_4H_{10}O$ (liq) 76CON/GIN
2–Butanol; sec–Butyl alcohol
Heat Capacity 298 K, C_p = 47.6 cal·mol^{-1}·K^{-1}
199.2 J·mol^{-1}·K^{-1}
One temperature
Molecular Weight 74.1224
Wiswesser Line Notation QY2&1
Evaluation B

Table of Heat Capacities, Entropies, and Phase Transition Properties — Continued

$C_4H_{10}O$ (liq) 71AND/CON
2–Butanol; sec–Butyl alcohol
Heat Capacity 298.15 K, C_p = 47.18 cal·mol^{-1}·K^{-1}
197.4 J·mol^{-1}·K^{-1}
Temperature range 11–350 K
Entropy 298.15 K, S = 50.93 cal·mol^{-1}·K^{-1}
213.1 J·mol^{-1}·K^{-1}
Phase Changes
c/liq 177.38 K, ΔH = 1434 cal·mol^{-1}
6000 J·mol^{-1}
ΔS = 8.08 cal·mol^{-1}·K^{-1}
33.83 J·mol^{-1}·K^{-1}
Molecular Weight 74.1224
Wiswesser Line Notation QY2&1
Evaluation A
Optically active form.

$C_4H_{10}O$ (liq) 71AND/CON
2–Butanol; sec–Butyl alcohol
Heat Capacity 298.15 K, C_p = 47.04 cal·mol^{-1}·K^{-1}
196.8 J·mol^{-1}·K^{-1}
Temperature range 11–350 K
Entropy 298.15 K, S = 51.31 cal·mol^{-1}·K^{-1}
214.7 J·mol^{-1}·K^{-1}
Phase Changes
c/liq 184.70 K, ΔH = 1427 cal·mol^{-1}
5970 J·mol^{-1}
ΔS = 7.73 cal·mol^{-1}·K^{-1}
32.32 J·mol^{-1}·K^{-1}
Molecular Weight 74.1224
Wiswesser Line Notation QY2&1
Evaluation A
Optically inactive form.

$C_4H_{10}O_2$ (liq) 73KUS/SUU
2,5–Dioxahexane; 1,2–Dimethoxyethane
Heat Capacity 298.15 K, C_p = 46.20 cal·mol^{-1}·K^{-1}
193.3 J·mol^{-1}·K^{-1}
One temperature
Molecular Weight 90.1218
Wiswesser Line Notation 1O2O1
Evaluation B

$C_4H_{10}O_2$ (liq) 73KUS/SUU
3–Oxa–1–pentanol; 2–Ethoxyethanol
Heat Capacity 298.15 K, C_p = 50.38 cal·mol^{-1}·K^{-1}
210.8 J·mol^{-1}·K^{-1}
One temperature
Molecular Weight 90.1218
Wiswesser Line Notation Q2O2
Evaluation B

$C_4H_{10}O_2$ (liq) 78ROU/PER
3–Oxa–1–pentanol; 2–Ethoxyethanol
Heat Capacity 298.15 K, C_p = 50.3 cal·mol^{-1}·K^{-1}
210.5 J·mol^{-1}·K^{-1}
Temperature range 283.15, 298.15, 313.15 K. Data at three temperatures.
Molecular Weight 90.1218
Wiswesser Line Notation Q2O2
Evaluation C

$C_4H_{10}O_2$ (liq) 72KAW/OTA
1,3–Butanediol; 1,3–Dihydroxybutane
Heat Capacity 303 K, C_p = 54.3 cal·mol^{-1}·K^{-1}
227.2 J·mol^{-1}·K^{-1}
One temperature
Molecular Weight 90.1218
Wiswesser Line Notation QY1&2Q
Evaluation B

$C_4H_{10}O_2$ (liq) 74PET/TER
1,4–Butanediol; 1,4–Dihydroxybutane
Heat Capacity 297.79 K, C_p = 43 cal·mol^{-1}·K^{-1}
178 J·mol^{-1}·K^{-1}
Temperature range 297–470 K. Value is unsmoothed experimental datum.
Molecular Weight 90.1218
Wiswesser Line Notation Q4Q
Evaluation B

$C_4H_{10}O_2$ (liq) 36KHO/KAL
2,3–Butanediol; 2,3–Dihydroxybutane
Heat Capacity 298.2 K, C_p = 50.9 cal·mol^{-1}·K^{-1}
213.0 J·mol^{-1}·K^{-1}
Temperature range 26 to 140°C
Molecular Weight 90.1218
Wiswesser Line Notation QY1&YQ1
Evaluation C
Extracted from Chem. Abst. *30*, 4080 (1936).

$C_4H_{10}O_3$ (liq) 79STE/TAM
1,5–Dihydroxy–3–oxapentane; Diethylene glycol
Heat Capacity 298 K, C_p = 58.5 cal·mol^{-1}·K^{-1}
287.0 J·mol^{-1}·K^{-1}
Temperature range 273–513 K
Molecular Weight 106.1212
Wiswesser Line Notation Q2O2Q
Evaluation B

$C_4H_{10}O_4$ (c) 26PAR/AND
1,2,3,4–Tetrahydroxybutane;
1,2,3,4–Butanetetrol; Erythritol
Heat Capacity 291.7 K, C_p = 38.7 cal·mol^{-1}·K^{-1}
161.9 J·mol^{-1}·K^{-1}
Temperature range 87–292 K. Value is unsmoothed experimental datum.
Entropy 298.1 K, S = 42.5 cal·mol^{-1}·K^{-1}
177.8 J·mol^{-1}·K^{-1}
Extrapolation below 90 K, 13.26 cal·mol^{-1}K^{-1}
Molecular Weight 122.1206
Wiswesser Line Notation Q1YQYQ1Q
Evaluation B(C_p),C(S)

$C_4H_{10}O_4$ (c) 29PAR/KEL
1,2,3,4–Tetrahydroxybutane;
1,2,3,4–Butanetetrol; Erythritol
Entropy 298.1 K, S = 39.8 cal·mol^{-1}·K^{-1}
166.5 J·mol^{-1}·K^{-1}
Extrapolation below 90 K, 10.6 cal·mol^{-1}K^{-1}.
Revision of previous data.
Molecular Weight 122.1206
Wiswesser Line Notation Q1YQYQ1Q
Evaluation C

Table of Heat Capacities, Entropies, and Phase Transition Properties — Continued

$C_4H_{10}O_4$ (c) 32SPA/THO
1,2,3,4–Tetrahydroxybutane;
1,2,3,4–Butanetetrol; Erythritol
Heat Capacity 303 K, C_p = 40.8 cal·mol⁻¹·K⁻¹ / 170.7 J·mol⁻¹·K⁻¹
Temperature range 30 to 150°C
Phase Changes
c/liq 381.6 K, ΔH = 10124 cal·mol⁻¹ / 42359 J·mol⁻¹
ΔS = 26.5 cal·mol⁻¹·K⁻¹ / 111.0 J·mol⁻¹·K⁻¹
Molecular Weight 122.1206
Wiswesser Line Notation Q1YQYQ1Q
Evaluation B

$C_4H_{10}O_6S_3$ (c) 56DAV/STA
Trimethylsulfonylmethane
Heat Capacity 298.15 K, C_p = 64.44 cal·mol⁻¹·K⁻¹ / 269.62 J·mol⁻¹·K⁻¹
Temperature range 22–293 K
Entropy 298.15 K, S = 72.8 cal·mol⁻¹·K⁻¹ / 304.60 J·mol⁻¹·K⁻¹
Molecular Weight 250.2994
Wiswesser Line Notation 1SWYSW1&SW1
Evaluation A

$C_4H_{10}S$ (liq) 55MCC/FIN
3–Methyl–2–thiabutane; Isopropyl methyl sulfide
Heat Capacity 298.15 K, C_p = 41.20 cal·mol⁻¹·K⁻¹ / 172.38 J·mol⁻¹·K⁻¹
Temperature range 12–344 K
Entropy 298.15 K, S = 62.88 cal·mol⁻¹·K⁻¹ / 263.09 J·mol⁻¹·K⁻¹
Phase Changes
c/liq 171.65 K, ΔH = 2236 cal·mol⁻¹ / 9355 J·mol⁻¹
ΔS = 13.03 cal·mol⁻¹·K⁻¹ / 54.50 J·mol⁻¹·K⁻¹
Molecular Weight 90.1830
Wiswesser Line Notation 1Y1&S1
Evaluation A

$C_4H_{10}S$ (liq) 52SCO/FIN 2
3–Thiapentane; Diethyl sulfide
Heat Capacity 298.15 K, C_p = 40.97 cal·mol⁻¹·K⁻¹ / 171.42 J·mol⁻¹·K⁻¹
Temperature range 16–316 K
Entropy 298.15 K, S = 64.36 cal·mol⁻¹·K⁻¹ / 269.28 J·mol⁻¹·K⁻¹
Phase Changes
c/liq 169.21 K, ΔH = 2845 cal·mol⁻¹ / 11903 J·mol⁻¹
ΔS = 16.81 cal·mol⁻¹·K⁻¹ / 70.35 J·mol⁻¹·K⁻¹
Molecular Weight 90.1830
Wiswesser Line Notation 2S2
Evaluation A

$C_4H_{10}S$ (liq) 57SCO/FIN
2–Thiapentane; Methyl n–propyl sulfide
Heat Capacity 298.15 K, C_p = 41.02 cal·mol⁻¹·K⁻¹ / 171.63 J·mol⁻¹·K⁻¹
Temperature range 12–326 K
Entropy 298.15 K, S = 65.14 cal·mol⁻¹·K⁻¹ / 272.55 J·mol⁻¹·K⁻¹
Phase Changes
c/liq 160.17 K, ΔH = 2369 cal·mol⁻¹ / 9912 J·mol⁻¹
ΔS = 14.79 cal·mol⁻¹·K⁻¹ / 61.88 J·mol⁻¹·K⁻¹
Molecular Weight 90.1830
Wiswesser Line Notation 3S1
Evaluation A

$C_4H_{10}S$ (liq) 58SCO/MCC
2–Methyl–1–propanethiol; Isobutyl mercaptan
Heat Capacity 298.15 K, C_p = 41.08 cal·mol⁻¹·K⁻¹ / 171.88 J·mol⁻¹·K⁻¹
Temperature range 10–350 K
Entropy 298.15 K, S = 63.66 cal·mol⁻¹·K⁻¹ / 266.35 J·mol⁻¹·K⁻¹
Phase Changes
c/liq 128.31 K, ΔH = 1190.8 cal·mol⁻¹ / 4982.3 J·mol⁻¹
ΔS = 9.28 cal·mol⁻¹·K⁻¹ / 388.3 J·mol⁻¹·K⁻¹
Molecular Weight 90.1830
Wiswesser Line Notation SH1Y1&1
Evaluation A

$C_4H_{10}S$ (liq) 57SCO/FIN
1–Butanethiol; n–Butyl mercaptan
Heat Capacity 298.15 K, C_p = 41.18 cal·mol⁻¹·K⁻¹ / 172.30 J·mol⁻¹·K⁻¹
Temperature range 12–314 K
Entropy 298.15 K, S = 65.96 cal·mol⁻¹·K⁻¹ / 275.98 J·mol⁻¹·K⁻¹
Phase Changes
c/liq 157.47 K, ΔH = 2500 cal·mol⁻¹ / 10460 J·mol⁻¹
ΔS = 15.88 cal·mol⁻¹·K⁻¹ / 66.43 J·mol⁻¹·K⁻¹
Molecular Weight 90.1830
Wiswesser Line Notation SH4
Evaluation A

$C_4H_{10}S$ (liq) 53MCC/SCO
2–Methyl–2–propanethiol; tert–Butyl mercaptan
Heat Capacity 298.15 K, C_p = 41.84 cal·mol⁻¹·K⁻¹ / 175.06 J·mol⁻¹·K⁻¹
Temperature range 12–329 K
Entropy 298.15 K, S = 58.90 cal·mol⁻¹·K⁻¹ / 246.44 J·mol⁻¹·K⁻¹
Phase Changes
c,IV/c,III 151.6 K, ΔH = 9720 cal·mol⁻¹ / 4066.8 J·mol⁻¹
ΔS = 6.41 cal·mol⁻¹·K⁻¹ / 26.83 J·mol⁻¹·K⁻¹
c,III/c,II. 157.0 K, ΔH = 154.9 cal·mol⁻¹ / 648.1 J·mol⁻¹
ΔS = 0.99 cal·mol⁻¹·K⁻¹ / 4.13 J·mol⁻¹·K⁻¹
c,II/c,I 199.4 K, ΔH = 232.0 cal·mol⁻¹ / 970.7 J·mol⁻¹
ΔS = 1.16 cal·mol⁻¹·K⁻¹ / 4.87 J·mol⁻¹·K⁻¹
c,I/liq 274.42 K, ΔH = 593.2 cal·mol⁻¹ / 2481.9 J·mol⁻¹
ΔS = 2.16 cal·mol⁻¹·K⁻¹ / 9.04 J·mol⁻¹·K⁻¹
Molecular Weight 90.1830
Wiswesser Line Notation SHX1&1&1
Evaluation A

Table of Heat Capacities, Entropies, and Phase Transition Properties — Continued

$C_4H_{10}S$ (liq) 58MCC/FIN
2–Butanethiol; sec–Butyl mercaptan
Heat Capacity 298.15 K, C_p = 40.92 cal·mol⁻¹·K⁻¹ 171.2 J·mol⁻¹·K⁻¹
Temperature range 12–309 K
Entropy 298.15 K, S = 64.87 cal·mol⁻¹·K⁻¹ 271.42 J·mol⁻¹·K⁻¹
Phase Changes
c/liq 133.02 K, ΔH = 1548 cal·mol⁻¹ 6477 J·mol⁻¹
ΔS = 11.64 cal·mol⁻¹·K⁻¹ 48.69 J·mol⁻¹·K⁻¹
Molecular Weight 90.1830
Wiswesser Line Notation SHY2&1
Evaluation A

$C_4H_{10}S_2$ (liq) 52SCO/FIN
Diethyl disulfide; 3,4–Dithiahexane
Heat Capacity 298.15 K, C_p = 48.76 cal·mol⁻¹·K⁻¹ 204.01 J·mol⁻¹·K⁻¹
Temperature range 13–300 K
Entropy 298.15 K, S = 72.90 cal·mol⁻¹·K⁻¹ 305.01 J·mol⁻¹·K⁻¹
Phase Changes
c/liq 171.64 K, ΔH = 2247.7 cal·mol⁻¹ 9404.4 J·mol⁻¹
ΔS = 13.09 cal·mol⁻¹·K⁻¹ 54.79 J·mol⁻¹·K⁻¹
Molecular Weight 122.2430
Wiswesser Line Notation 2SS2
Evaluation A

$C_4H_{11}N$ (liq) 36KUR/VOS
Diethylamine
Heat Capacity 290 K, C_p = 25.4 cal·mol⁻¹·K⁻¹ 106.3 J·mol⁻¹·K⁻¹
One temperature
Molecular Weight 73.1376
Wiswesser Line Notation 2M2
Evaluation D

$C_4H_{11}N$ (liq) 71KON/WAD
2–Methyl–1–aminopropane; Isobutylamine
Heat Capacity 298.15 K, C_p = 46.4 cal·mol⁻¹·K⁻¹ 194 J·mol⁻¹·K⁻¹
One temperature
Molecular Weight 73.1376
Wiswesser Line Notation Z1Y1&1
Evaluation B

$C_4H_{11}N$ (liq) 71KON/WAD
1–Aminobutane; n–Butylamine
Heat Capacity 298.15 K, C_p = 44.9 cal·mol⁻¹·K⁻¹ 188 J·mol⁻¹·K⁻¹
One temperature
Molecular Weight 73.1376
Wiswesser Line Notation Z4
Evaluation B

$C_4H_{11}N$ (liq) 67SMI/GOO 2
2–Methyl–2–aminopropane; tert–Butylamine
Heat Capacity 298.15 K, C_p = 45.9 cal·mol⁻¹·K⁻¹ 192.0 J·mol⁻¹·K⁻¹
One temperature
Molecular Weight 73.1376
Wiswesser Line Notation ZX1&1&1
Evaluation B

$C_4H_{11}N$ (liq) 71KON/WAD
2–Methyl–2–aminopropane; tert–Butylamine
Heat Capacity 298.15 K, C_p = 45.4 cal·mol⁻¹·K⁻¹ 190 J·mol⁻¹·K⁻¹
One temperature
Molecular Weight 73.1376
Wiswesser Line Notation ZX1&1&1
Evaluation B

$C_4H_{11}N$ (liq) 72FIN/MES
2–Methyl–2–aminopropane; tert–Butylamine
Heat Capacity 298.15 K, C_p = 45.82 cal·mol⁻¹·K⁻¹ 191.71 J·mol⁻¹·K⁻¹
Temperature range 12–350 K
Entropy 298.15 K, S = 55.84 cal·mol⁻¹·K⁻¹ 233.63 J·mol⁻¹·K⁻¹
Phase Changes
c,III/c,II 91.30 K, ΔH = 27.13 cal·mol⁻¹ 113.51 J·mol⁻¹
ΔS = 0.30 cal·mol⁻¹·K⁻¹ 1.24 J·mol⁻¹·K⁻¹
c,II/c,I 202.27 K, ΔH = 1446.6 cal·mol⁻¹ 6052.6 J·mol⁻¹
ΔS = 7.15 cal·mol⁻¹·K⁻¹ 29.91 J·mol⁻¹·K⁻¹
c,I/liq 206.19 K, ΔH = 210.8 cal·mol⁻¹ 882.0 J·mol⁻¹
ΔS = 1.02 cal·mol⁻¹·K⁻¹ 4.28 J·mol⁻¹·K⁻¹
Molecular Weight 73.1376
Wiswesser Line Notation ZX1&1&1
Evaluation A

$C_4H_{11}NO_3$ (c) 72ARV/WES
Tris(hydroxymethyl)aminomethane; TRIS; THAM
Heat Capacity 298.15 K, C_p = 39.96 cal·mol⁻¹·K⁻¹ 167.19 J·mol⁻¹·K⁻¹
Temperature range 5–300 K
Entropy 298.15 K, S = 41.93 cal·mol⁻¹·K⁻¹ 175.44 J·mol⁻¹·K⁻¹
Molecular Weight 121.1358
Wiswesser Line Notation Q1XZ1Q1Q
Evaluation A

$C_4H_{12}BrN$ (c) 62CHA/WES 2
Tetramethylammonium bromide
Heat Capacity 298.15 K, C_p = 38.64 cal·mol⁻¹·K⁻¹ 161.67 J·mol⁻¹·K⁻¹
Temperature range 5–350 K
Entropy 298.15 K, S = 47.99 cal·mol⁻¹·K⁻¹ 200.79 J·mol⁻¹·K⁻¹
Molecular Weight 154.0495
Wiswesser Line Notation 1K1&1&1 &E
Evaluation A

$C_4H_{12}BrN$ (c) 74BUR/VER
Tetramethylammonium bromide
Heat Capacity 298 K, C_p = 38 cal·mol⁻¹·K⁻¹ 159.4 J·mol⁻¹·K⁻¹
Temperature range 273–373 K
Molecular Weight 154.0495
Wiswesser Line Notation 1K1&1&1 &E
Evaluation B

Table of Heat Capacities, Entropies, and Phase Transition Properties — Continued

$C_4H_{12}ClN$ (c) 62CHA/WES 2

Tetramethylammonium chloride

Heat Capacity 298.15 K, C_p = 37.51 cal·mol^{-1}·K^{-1}
156.94 J·mol^{-1}·K^{-1}

Temperature range 5–350 K

Entropy 298.15 K, S = 45.58 cal·mol^{-1}·K^{-1}
190.71 J·mol^{-1}·K^{-1}

Phase Changes

c,III/c,II 75.76 K, ΔH = 27.8 cal·mol^{-1}
116.3 J·mol^{-1}
ΔS = 0.37 cal·mol^{-1}·K^{-1}
1.54 J·mol^{-1}·K^{-1}

c,II/c,I 184.85 K, ΔH = 25.9 cal·mol^{-1}
108.4 J·mol^{-1}
ΔS = 0.14 cal·mol^{-1}·K^{-1}
0.59 J·mol^{-1}·K^{-1}

Transition over range 180–190 K.

Molecular Weight 109.5985

Wiswesser Line Notation 1K1&1&1 &G

Evaluation A

$C_4H_{12}IN$ (c) 40COU/PIT

Tetramethylammonium iodide

Heat Capacity 298.15 K, C_p = 38.43 cal·mol^{-1}·K^{-1}
160.79 J·mol^{-1}·K^{-1}

Temperature range 15–300 K.

Entropy 298.15 K, S = 49.7 cal·mol^{-1}·K^{-1}
207.9 J·mol^{-1}·K^{-1}

Molecular Weight 201.0500

Wiswesser Line Notation 1K1&1&1 &I

Evaluation A

$C_4H_{12}N_2$ (liq) 75MES/FIN

1,2–Diamino–2–methylpropane

Heat Capacity 298.15 K, C_p = 56.06 cal·mol^{-1}·K^{-1}
234.56 J·mol^{-1}·K^{-1}

Temperature range 11–375 K

Entropy 298.15 K, S = 62.03 cal·mol^{-1}·K^{-1}
259.53 J·mol^{-1}·K^{-1}

Phase Changes

c,II/c,I 237.5 K, ΔH = 3696 cal·mol^{-1}
15464 J·mol^{-1}
ΔS = 15.56 cal·mol^{-1}·K^{-1}
65.11 J·mol^{-1}·K^{-1}

c,I/liq 256.1 K, ΔH = 533 cal·mol^{-1}
2230 J·mol^{-1}
ΔS = 2.08 cal·mol^{-1}·K^{-1}
8.71 J·mol^{-1}·K^{-1}

Molecular Weight 88.1423

Wiswesser Line Notation ZX1&1&1Z

Evaluation A

$C_4H_{12}N_2O_4$ (c) 39SAT/SOG

Ammonium succinate

Heat Capacity 323 K, C_p = 61.7 cal·mol^{-1}·K^{-1}
258.2 J·mol^{-1}·K^{-1}

Temperature range 0 to 100°C. Mean value.

Molecular Weight 152.1498

Wiswesser Line Notation QV2VQ &ZH 2

Evaluation C

$C_4H_{12}N_2O_6$ (c) 39SAT/SOG

Ammonium tartrate

Heat Capacity 323 K, C_p = 68.0 cal·mol^{-1}·K^{-1}
284.5 J·mol^{-1}·K^{-1}

Temperature range 0 to 100°C. Mean value.

Molecular Weight 184.1486

Wiswesser Line Notation QVYQYQVQ &ZH 2

Evaluation C

$C_4H_{12}Pb$ (c) 59GOO/SCO

Tetramethyl lead

Heat Capacity 298.15 K, C_p = 48.4 cal·mol^{-1}·K^{-1}
202.5 J·mol^{-1}·K^{-1}

One temperature

Entropy 298.16 K, S = 76.48 cal·mol^{-1}·K^{-1}
319.99 J·mol^{-1}·K^{-1}

Molecular Weight 267.3388

Wiswesser Line Notation 1–PB–1&1&1

Evaluation B

$C_4H_{12}Si$ (liq) 41AST/KEN

Tetramethylsilane; Silicon tetramethyl

Heat Capacity 290 K, C_p = 47.30 cal·mol^{-1}·K^{-1}
197.90 J·mol^{-1}·K^{-1}

Temperature range 11–290 K

Entropy 298.15 K, S = 66.27 cal·mol^{-1}·K^{-1}
277.27 J·mol^{-1}·K^{-1}

Phase Changes

c,I/liq 171.04 K, ΔH = 1426.8 cal·mol^{-1}
5969.7 J·mol^{-1}
ΔS = 8.34 cal·mol^{-1}·K^{-1}
34.90 J·mol^{-1}·K^{-1}

c,II/liq 174.12 K, ΔH = 1648 cal·mol^{-1}
6895 J·mol^{-1}
ΔS = 9.46 cal·mol^{-1}·K^{-1}
39.60 J·mol^{-1}·K^{-1}

liq/g 299.80 K, ΔH = 5785 cal·mol^{-1}
24204 J·mol^{-1}
ΔS = 19.29 cal·mol^{-1}·K^{-1}
80.73 J·mol^{-1}·K^{-1}
P = 101.325 kPa

Molecular Weight 88.2243

Wiswesser Line Notation 1–SI–1&1&1

Evaluation A

$C_4H_{12}Si$ (liq) 73SHI/ENO

Tetramethylsilane; Silicon tetramethyl

Heat Capacity 177.45 K, C_p = 39.33 cal·mol^{-1}·K^{-1}
164.56 J·mol^{-1}·K^{-1}

Temperature range 2–26 K and 106–177.5 K

Phase Changes

c,II/liq 70.983 K, ΔH = 1396.0 cal·mol^{-1}
5840.9 J·mol^{-1}
ΔS = 8.16 cal·mol^{-1}·K^{-1}
34.16 J·mol^{-1}·K^{-1}

Metastable form

c,I/liq 174.049 K, ΔH = 1611.2 cal·mol^{-1}
6741.3 J·mol^{-1}
ΔS = 9.26 cal·mol^{-1}·K^{-1}
38.73 J·mol^{-1}·K^{-1}

Stable form

Molecular Weight 88.2243

Wiswesser Line Notation 1–SI–1&1&1

Evaluation A

Table of Heat Capacities, Entropies, and Phase Transition Properties — Continued

$C_4H_{12}Si$ (liq) 77HAR/ATA

Tetramethylsilane; Silicon tetramethyl

Heat Capacity 290 K, C_p = 45.37 cal·mol^{-1}·K^{-1}
189.84 J·mol^{-1}·K^{-1}

Temperature range 3–300 K

Entropy 290 K, S = 65.02 cal·mol^{-1}·K^{-1}
272.04 J·mol^{-1}·K^{-1}

Phase Changes

c,α/liq	165.920 K,	ΔH =	168 cal·mol^{-1}
			703 J·mol^{-1}
		ΔS =	1.01 cal·mol^{-1}·K^{-1}
			4.24 J·mol^{-1}·K^{-1}
c,β/liq	171.016 K,	ΔH =	1405 cal·mol^{-1}
			5878 J·mol^{-1}
		ΔS =	8.21 cal·mol^{-1}·K^{-1}
			34.37 J·mol^{-1}·K^{-1}
c,γ/liq	174.074 K,	ΔH =	1643 cal·mol^{-1}
			6874 J·mol^{-1}
		ΔS =	9.44 cal·mol^{-1}·K^{-1}
			39.51 J·mol^{-1}·K^{-1}

Molecular Weight 88.2243
Wiswesser Line Notation 1-SI-1&1&1
Evaluation A

$C_4H_{13}Cl_2N$ (c) 62CHA/WES 3

Tetramethylammonium hydrogen dichloride

Heat Capacity 298.15 K, C_p = 49.07 cal·mol^{-1}·K^{-1}
205.31 J·mol^{-1}·K^{-1}

Temperature range 5–350 K. Corrected for decomposition above 250 K.

Entropy 298.15 K, S = 60.63 cal·mol^{-1}·K^{-1}
253.68 J·mol^{-1}·K^{-1}

Molecular Weight 146.0594
Wiswesser Line Notation 1K1&1&1 &G &GH
Evaluation A

$C_5F_{11}N$ (liq) 63GOO/TOD

Perfluoropiperidine

Heat Capacity 298.15 K, C_p = 70.93 cal·mol^{-1}·K^{-1}
296.77 J·mol^{-1}·K^{-1}

Temperature range 12–320 K

Entropy 298.15 K, S = 94.02 cal·mol^{-1}·K^{-1}
393.38 J·mol^{-1}·K^{-1}

Phase Changes

c,III/c,II	161.0 K,	ΔH =	1584.0 cal·mol^{-1}
			6627.5 J·mol^{-1}
		ΔS =	9.84 cal·mol^{-1}·K^{-1}
			41.16 J·mol^{-1}·K^{-1}
c,II/c,I	171.9 K,	ΔH =	439.5 cal·mol^{-1}
			1838.9 J·mol^{-1}
		ΔS =	2.56 cal·mol^{-1}·K^{-1}
			10.70 J·mol^{-1}·K^{-1}
c,I/liq	274.12 K,	ΔH =	673 cal·mol^{-1}
			2816 J·mol^{-1}
		ΔS =	2.45 cal·mol^{-1}·K^{-1}
			10.27 J·mol^{-1}·K^{-1}

Molecular Weight 283.0441
Wiswesser Line Notation T6NTJ AF BF BF CF CF DF DF EF EF FF FF
Evaluation A

$C_5F_{13}N$ (liq) 80ZHO/KOS

Perfluoromethyldiethylamine

Heat Capacity 298.15 K, C_p = 80.63 cal·mol^{-1}·K^{-1}
337.46 J·mol^{-1}·K^{-1}

Temperature range 6–300 K

Entropy 298.15 K, S = 113.55 cal·mol^{-1}·K^{-1}
475.09 J·mol^{-1}·K^{-1}

Phase Changes

c,III/c,II 102.4 K

Metastable phase is a supercooled liquid in the range 102.4–126.0 K and forms a glass at 102.4 K.

c,II/c,I	126.0 K,	ΔH =	1322.2 cal·mol^{-1}
			5532.1 J·mol^{-1}
		ΔS =	10.49 cal·mol^{-1}·K^{-1}
			43.89 J·mol^{-1}·K^{-1}
c,I/liq	149.64 K,	ΔH =	1710.7 cal·mol^{-1}
			7157.6 J·mol^{-1}
		ΔS =	11.44 cal·mol^{-1}·K^{-1}
			47.86 J·mol^{-1}·K^{-1}

Monotropic transition from metastable to stable phase.

Molecular Weight 321.0409
Wiswesser Line Notation FXFFXFF 2NXFFF
Evaluation A

$C_5H_4N_4O$ (c) 35STI/HUF

Hypoxanthine

Heat Capacity 298.5 K, C_p = 32.14 cal·mol^{-1}·K^{-1}
134.52 J·mol^{-1}·K^{-1}

Temperature range 85–298 K. Value is unsmoothed experimental datum.

Entropy 298.15 K, S = 34.8 cal·mol^{-1}·K^{-1}
145.6 J·mol^{-1}·K^{-1}

Extrapolation below 90 K, 10.89 cal·mol^{-1}K^{-1}

Molecular Weight 136.1128
Wiswesser Line Notation T56 BM DN FVM INJ
Evaluation B(C_p),C(S)

$C_5H_4N_4O_2$ (c) 35STI/HUF

Xanthine

Heat Capacity 298.5 K, C_p = 36.17 cal·mol^{-1}·K^{-1}
151.34 J·mol^{-1}·K^{-1}

Temperature range 85–299 K. Value is unsmoothed experimental datum.

Entropy 298.15 K, S = 38.5 cal·mol^{-1}·K^{-1}
161.1 J·mol^{-1}·K^{-1}

Extrapolation below 90 K, 11.68 cal·mol^{-1}K^{-1}

Molecular Weight 152.1122
Wiswesser Line Notation T56 BM DN FMVMVJ
Evaluation B(C_p), C(S)

$C_5H_4N_4O_3$ (c) 35STI/HUF

Uric acid

Heat Capacity 297.1 K, C_p = 39.71 cal·mol^{-1}·K^{-1}
166.15 J·mol^{-1}·K^{-1}

Temperature range 85–297 K. Value is unsmoothed experimental datum.

Entropy 298.15 K, S = 41.4 cal·mol^{-1}·K^{-1}
173.2 J·mol^{-1}·K^{-1}

Extrapolation below 90 K, 12.01 cal·mol^{-1}K^{-1}

Molecular Weight 168.1116
Wiswesser Line Notation T56 BMVM FMVMVJ
Evaluation B(C_p), C(S)

Table of Heat Capacities, Entropies, and Phase Transition Properties — Continued

$C_5H_4O_2$ (liq) 81REI
Furfural; Furfuraldehyde
Heat Capacity 298 K, C_p = 37.8 cal·mol^{-1}·K^{-1}
158.2 J·mol^{-1}·K^{-1}
Temperature range 290–347 K
Molecular Weight 96.0854
Wiswesser Line Notation T5OJ BVH
Evaluation D

$C_5H_4O_2$ (liq) 35MIL
Furfural; Furfuraldehyde
Heat Capacity
Temperature range 100–298 K. C_p data in thesis only.
Entropy 298.15 K, S = 52.1 cal·mol^{-1}·K^{-1}
218.0 J·mol^{-1}·K^{-1}
Extrapolation below 90 K, 12.52 cal·mol^{-1}K^{-1}
Phase Changes
c/liq 235.1 K, ΔH = 3434 cal·mol^{-1}
14368 J·mol^{-1}
ΔS = 14.6 cal·mol^{-1}·K^{-1}
61.1 J·mol^{-1}·K^{-1}
Molecular Weight 96.0854
Wiswesser Line Notation T5OJ BVH
Evaluation C

$C_5H_4O_2$ (liq) 62OME
Furfural; Furfuraldehyde
Heat Capacity 298.15 K, C_p = 38.9 cal·mol^{-1}·K^{-1}
162.8 J·mol^{-1}·K^{-1}
Temperature range 288–412 K
Molecular Weight 96.0854
Wiswesser Line Notation T5OJ BVH
Evaluation B

$C_5H_4O_2$ (liq) 67RAS/GAN
Furfural; Furfuraldehyde
Heat Capacity 293 K, C_p = 38.1 cal·mol^{-1}·K^{-1}
159.5 J·mol^{-1}·K^{-1}
Temperature range 293–373 K
Molecular Weight 96.0854
Wiswesser Line Notation T5OJ BVH
Evaluation C

C_5H_5N (liq) 71HAL/BAL
1–Bicyclobutyl cyanide; 1–Cyanobicyclobutane
Heat Capacity 297 K, C_p = 31.5 cal·mol^{-1}·K^{-1}
132.1 J·mol^{-1}·K^{-1}
One temperature
Molecular Weight 79.1012
Wiswesser Line Notation L33TJ ACN
Evaluation C

C_5H_5N (liq) 16BRA
Pyridine
Heat Capacity 283 K, C_p = 31.2 cal·mol^{-1}·K^{-1}
130.5 J·mol^{-1}·K^{-1}
Mean value, 0 to 20°C
Molecular Weight 79.1012
Wiswesser Line Notation T6NJ
Evaluation C

C_5H_5N (liq) 17MAT/KRA
Pyridine
Heat Capacity 294 K, C_p = 30.91 cal·mol^{-1}·K^{-1}
129.33 J·mol^{-1}·K^{-1}
One temperature
Molecular Weight 79.1012
Wiswesser Line Notation T6NJ
Evaluation B

C_5H_5N (liq) 31SWI/RYB
Pyridine
Heat Capacity 290 K, C_p = 32.4 cal·mol^{-1}·K^{-1}
135.6 J·mol^{-1}·K^{-1}
One temperature
Molecular Weight 79.1012
Wiswesser Line Notation T6NJ
Evaluation B

C_5H_5N (liq) 31SWI/RYB 2
Pyridine
Heat Capacity 273.4 K, C_p = 32.35 cal·mol^{-1}·K^{-1}
135.35 J·mol^{-1}·K^{-1}
One temperature
Molecular Weight 79.1012
Wiswesser Line Notation T6NJ
Evaluation B

C_5H_5N (liq) 34RAD/JUL
Pyridine
Heat Capacity 289 K, C_p = 30.9 cal·mol^{-1}·K^{-1}
129.3 J·mol^{-1}·K^{-1}
One temperature
Molecular Weight 79.1012
Wiswesser Line Notation T6NJ
Evaluation C

C_5H_5N (liq) 36PAR/TOD
Pyridine
Heat Capacity 298.1 K, C_p = 32.25 cal·mol^{-1}·K^{-1}
134.93 J·mol^{-1}·K^{-1}
Temperature range 90–300 K
Entropy 298.1 K, S = 42.8 cal·mol^{-1}·K^{-1}
179.1 J·mol^{-1}·K^{-1}
Extrapolation below 90 K, 11.96 cal·mol^{-1}K^{-1}
Phase Changes
c/liq 231.1 K, ΔH = 1977 cal·mol^{-1}
8272 J·mol^{-1}
ΔS = 8.55 cal·mol^{-1}·K^{-1}
35.79 J·mol^{-1}·K^{-1}
Molecular Weight 79.1012
Wiswesser Line Notation T6NJ
Evaluation B(C_p),C(S)

C_5H_5N (liq) 36PEA/BAK
Pyridine
Heat Capacity 298.1 K, C_p = 31.86 cal·mol^{-1}·K^{-1}
133.30 J·mol^{-1}·K^{-1}
Temperature range 90–298 K. Value is unsmoothed expermental datum.
Entropy 298.15 K, S = 50.29 cal·mol^{-1}·K^{-1}
210.41 J·mol^{-1}·K^{-1}
Extrapolation below 90 K, 21.35 cal·mol^{-1}K^{-1}
Phase Changes
c/liq 230.38 K, ΔH = 741 cal·mol^{-1}
3100 J·mol^{-1}
ΔS = 3.22 cal·mol^{-1}·K^{-1}
13.46 J·mol^{-1}·K^{-1}

Table of Heat Capacities, Entropies, and Phase Transition Properties — Continued

Molecular Weight 79.1012
Wiswesser Line Notation T6NJ
Evaluation B(C_p),C(S)

C_5H_5N (liq) 57MCC/DOU
Pyridine
Heat Capacity 298.15 K, C_p = 31.72 cal·mol^{-1}·K^{-1}
132.72 J·mol^{-1}·K^{-1}
Temperature range 10–350 K
Entropy 298.15 K, S = 42.52 cal·mol^{-1}·K^{-1}
177.90 J·mol^{-1}·K^{-1}
Phase Changes
c/liq 231.49 K, ΔH = 1978.6 cal·mol^{-1}
8278.5 J·mol^{-1}
ΔS = 8.55 cal·mol^{-1}·K^{-1}
35.76 J·mol^{-1}·K^{-1}
Includes energy of anomaly at about 210 K.
Molecular Weight 79.1012
Wiswesser Line Notation T6NJ
Evaluation A

C_5H_5N (liq) 58SWI/ZIE
Pyridine
Heat Capacity 332 K, C_p = 35.1 cal·mol^{-1}·K^{-1}
146.9 J·mol^{-1}·K^{-1}
Mean value 22 to 96°C
Molecular Weight 79.1012
Wiswesser Line Notation T6NJ
Evaluation C

C_5H_5N (liq) 67RAS/GAN
Pyridine
Heat Capacity 293 K, C_p = 46.2 cal·mol^{-1}·K^{-1}
193.4 J·mol^{-1}·K^{-1}
Temperature range 293–353 K
Molecular Weight 79.1012
Wiswesser Line Notation T6NJ
Evaluation C

$C_5H_5N_5$ (c) 35STI/HUF
Adenine
Heat Capacity 298.1 K, C_p = 34.21 cal·mol^{-1}·K^{-1}
143.13 J·mol^{-1}·K^{-1}
Temperature range 88–298 K. Value is unsmoothed experimental datum.
Entropy 298.15 K, S = 36.1 cal·mol^{-1}·K^{-1}
151.01 J·mol^{-1}·K^{-1}
Extrapolation below 90 K, 11.16 cal·mol^{-1}K^{-1}
Molecular Weight 135.1280
Wiswesser Line Notation T56 BM DN FN HNJ IZ
Evaluation B(C_p), C(S)

$C_5H_5N_5$ (c) 78KIL
Adenine
Heat Capacity 298 K, C_p = 35.13 cal·mol^{-1}·K^{-1}
147.0 J·mol^{-1}·K^{-1}
One temperature
Molecular Weight 135.1280
Wiswesser Line Notation T56 BM DN FN HNJ IZ
Evaluation C

$C_5H_5N_5O$ (c) 35STI/HUF
Guanine
Heat Capacity 296.7 K, C_p = 37.51 cal·mol^{-1}·K^{-1}
156.94 J·mol^{-1}·K^{-1}
Temperature range 84–297 K. Value is unsmoothed experimental datum.
Entropy 298.15 K, S = 38.3 cal·mol^{-1}·K^{-1}
160.2 J·mol^{-1}·K^{-1}
Extrapolation below 90 K, 10.77 cal·mol^{-1}K^{-1}
Molecular Weight 151.1274
Wiswesser Line Notation T56 BN DM FN HNJ GZ IQ
Evaluation B(C_p),C(S)

$C_5H_5N_5O$ (c) 81KIL
Guanine
Heat Capacity 298 K, C_p = 38.4 cal·mol^{-1}·K^{-1}
160.7 J·mol^{-1}·K^{-1}
One temperature
Molecular Weight 151.1274
Wiswesser Line Notation T56 BN DM FN HNJ GZ IQ
Evaluation C

$C_5H_6N_2$ (liq) 65CLE/WUL
Glutaronitrile; 1,3–Dicyanopropane
Heat Capacity 298.15 K, C_p = 43.80 cal·mol^{-1}·K^{-1}
186.26 J·mol^{-1}·K^{-1}
Temperature range 5–350 K
Entropy 298.15 K, S = 57.23 cal·mol^{-1}·K^{-1}
239.45 J·mol^{-1}·K^{-1}
Phase Changes
c/liq 244.21 K, ΔH = 3008 cal·mol^{-1}
12585 J·mol^{-1}
ΔS = 12.32 cal·mol^{-1}·K^{-1}
51.53 J·mol^{-1}·K^{-1}
Molecular Weight 94.1158
Wiswesser Line Notation NC3CN
Evaluation A

$C_5H_6N_2$ (c,II) 67RIB/WES
Dimethylmalonitrile; 2,2–Dicyanopropane
Heat Capacity 298.15 K, C_p = 42.90 cal·mol^{-1}·K^{-1}
179.49 J·mol^{-1}·K^{-1}
Temperature range 5–350 K
Entropy 298.15 K, S = 44.92 cal·mol^{-1}·K^{-1}
187.95 J·mol^{-1}·K^{-1}
Phase Changes
c,II/c,I 302.60 K, ΔH = 2358 cal·mol^{-1}
9866 J·mol^{-1}
ΔS = 7.79 cal·mol^{-1}·K^{-1}
32.60 J·mol^{-1}·K^{-1}
c,I/liq 307.47 K, ΔH = 969 cal·mol^{-1}
4054 J·mol^{-1}
ΔS = 3.15 cal·mol^{-1}·K^{-1}
13.19 J·mol^{-1}·K^{-1}
Molecular Weight 94.1158
Wiswesser Line Notation NCX1&1&CN
Evaluation A

C_5H_6O (liq) 65CAR/WES 2
2–Methylfuran
Heat Capacity 298.15 K, C_p = 34.35 cal·mol^{-1}·K^{-1}
143.72 J·mol^{-1}·K^{-1}
Temperature range 5–310 K
Entropy 298.15 K, S = 51.12 cal·mol^{-1}·K^{-1}
213.89 J·mol^{-1}·K^{-1}

Table of Heat Capacities, Entropies, and Phase Transition Properties — Continued

Phase Changes
c/liq 181.90 K, ΔH = 2044 cal·mol^{-1}
8552 J·mol^{-1}
ΔS = 11.24 cal·mol^{-1}·K^{-1}
47.01 J·mol^{-1}·K^{-1}
Molecular Weight 82.1018
Wiswesser Line Notation T5OJ B1
Evaluation A

$C_5H_6O_2$ (liq) 35MIL
Furfuryl alcohol
Temperature range 100–298 K. Data in thesis only.
Entropy 298.15 K, S = 51.5 cal·mol^{-1}·K^{-1}
215.5 J·mol^{-1}·K^{-1}
Extrapolation below 90 K, 10.27 cal·mol^{-1}K^{-1}
Phase Changes
c/liq 253.5 K, ΔH = 3540 cal·mol^{-1}
14811 J·mol^{-1}
ΔS = 14.0 cal·mol^{-1}·K^{-1}
58.4 J·mol^{-1}·K^{-1}
Molecular Weight 98.1012
Wiswesser Line Notation T5OJ B1Q
Evaluation C

$C_5H_6O_2$ (liq) 50HOU/MAS
Furfuryl alcohol
Heat Capacity 303 K, C_p = 46.2 cal·mol^{-1}·K^{-1}
193.3 J·mol^{-1}·K^{-1}
Temperature range 303–333 K
Molecular Weight 98.1012
Wiswesser Line Notation T5OJ B1Q
Evaluation B

$C_5H_6O_2$ (liq) 56PAR/KEN
Furfuryl alcohol
Heat Capacity 298.15 K, C_p = 48.76 cal·mol^{-1}·K^{-1}
204.01 J·mol^{-1}·K^{-1}
Temperature range 90–300 K
Entropy 298.1 K, S = 51.6 cal·mol^{-1}·K^{-1}
215.9 J·mol^{-1}·K^{-1}
Extrapolation below 80 K, 11.71 cal·mol^{-1}K^{-1}
Phase Changes
c/liq 258.6 K, ΔH = 3138 cal·mol^{-1}
13129 J·mol^{-1}
ΔS = 12.13 cal·mol^{-1}·K^{-1}
50.77 J·mol^{-1}·K^{-1}
Molecular Weight 98.1012
Wiswesser Line Notation T5OJ B1Q
Evaluation B(C_p),C(S)

C_5H_6S (gls) 68CAR/WES
2–Methylthiophene
Heat Capacity 199.70 K, C_p = 23.40 cal·mol^{-1}·K^{-1}
97.91 J·mol^{-1}·K^{-1}
Temperature range 110–200 K. Value is unsmoothed experimental datum.
Molecular Weight 98.1624
Wiswesser Line Notation T5SJ B1
Evaluation A
Data for glass transition region.

C_5H_6S (liq) 56PEN/FIN
2–Methylthiophene
Heat Capacity 298.15 K, C_p = 35.81 cal·mol^{-1}·K^{-1}
149.83 J·mol^{-1}·K^{-1}
Temperature range 12–340 K
Entropy 298.15 K, S = 52.22 cal·mol^{-1}·K^{-1}
218.49 J·mol^{-1}·K^{-1}

Phase Changes
c/liq 207.79 K, ΔH = 2263 cal·mol^{-1}
11142 J·mol^{-1}
ΔS = 12.82 cal·mol^{-1}·K^{-1}
53.62 J·mol^{-1}·K^{-1}
Molecular Weight 98.1624
Wiswesser Line Notation T5SJ B1
Evaluation A

$C_5H_7ClO_3$ (liq) 76MAS/PET
4–Chloromethyl–1,3–dioxolan–2–one
Heat Capacity 298 K, C_p = 59.3 cal·mol^{-1}·K^{-1}
248 J·mol^{-1}·K^{-1}
Temperature range 200–320 K. Data graphically only. Value estimated from graph.
Molecular Weight 150.5615
Wiswesser Line Notation T6OVOTJ D1G
Evaluation D

C_5H_7N (liq) 71HAL/BAL
Cyclobutyl cyanide; Cyanocyclobutane
Heat Capacity 297 K, C_p = 34.9 cal·mol^{-1}·K^{-1}
146.0 J·mol^{-1}·K^{-1}
One temperature
Molecular Weight 81.1170
Wiswesser Line Notation L4TJ ACN
Evaluation C

$C_5H_7N_2O_2$ (c) 73ALV/BIL
Thymine
Heat Capacity 298.15 K, C_p = 36.19 cal·mol^{-1}·K^{-1}
151.4 J·mol^{-1}·K^{-1}
One temperature
Molecular Weight 127.1225
Wiswesser Line Notation T6MVMVJ E1
Evaluation C

$C_5H_7N_2O_2$ (c) 78KIL 2
Thymine
Heat Capacity 298 K, C_p = 35.9 cal·mol^{-1}·K^{-1}
150.2 J·mol^{-1}·K^{-1}
One temperature
Molecular Weight 127.1225
Wiswesser Line Notation T6MVMVJ E1
Evaluation C

C_5H_8 (liq) 36PAR/TOD 2
1,4–Pentadiene
Heat Capacity 292.5 K, C_p = 35.04 cal·mol^{-1}·K^{-1}
146.61 J·mol^{-1}·K^{-1}
Temperature range 82–293 K. Value is unsmoothed experimental datum.
Entropy 298.15 K, S = 58.2 cal·mol^{-1}·K^{-1}
338.7 J·mol^{-1}·K^{-1}
Extrapolation below 80 K, 11.26 cal·mol^{-1}K^{-1}
Phase Changes
c/liq 124.3 K, ΔH = 1468 cal·mol^{-1}
6142 J·mol^{-1}
ΔS = 11.81 cal·mol^{-1}·K^{-1}
49.41 J·mol^{-1}·K^{-1}
Molecular Weight 68.1182
Wiswesser Line Notation 1U3U1
Evaluation B(C_p),C(S)

Table of Heat Capacities, Entropies, and Phase Transition Properties — Continued

C_5H_8 (liq) 70MES/TOD
1,4–Pentadiene
Heat Capacity 298.15 K, C_p = 35.09 cal·mol^{-1}·K^{-1}
146.82 J·mol^{-1}·K^{-1}
Temperature range 12–320 K
Entropy 298.15 K, S = 59.48 cal·mol^{-1}·K^{-1}
248.86 J·mol^{-1}·K^{-1}
Phase Changes
c/liq 124.9 K, ΔH = 1461.5 cal·mol^{-1}
6073.1 J·mol^{-1}
ΔS = 11.620 cal·mol^{-1}·K^{-1}
48.620 J·mol^{-1}·K^{-1}
Molecular Weight 68.1182
Wiswesser Line Notation 1U3U1
Evaluation A

C_5H_8 (liq) 37BEK/WOO
2–Methyl–1,3–butadiene; Isoprene
Heat Capacity 298.2 K, C_p = 36.48 cal·mol^{-1}·K^{-1}
152.6 J·mol^{-1}·K^{-1}
Temperature range 20–300 K
Entropy 298.2 K, S = 54.81 cal·mol^{-1}·K^{-1}
229.3 J·mol^{-1}·K^{-1}
Phase Changes
c/liq 126.4 K, ΔH = 1155 cal·mol^{-1}
4830 J·mol^{-1}
ΔS = 9.14 cal·mol^{-1}·K^{-1}
38.21 J·mol^{-1}·K^{-1}
Molecular Weight 68.1182
Wiswesser Line Notation 1UY1&1U1
Evaluation A

C_5H_8 (liq) 70MES/TOD
2–Methyl–1,3–butadiene; Isoprene
Heat Capacity 298.15 K, C_p = 36.11 cal·mol^{-1}·K^{-1}
151.08 J·mol^{-1}·K^{-1}
Temperature range 12–320 K
Entropy 298.15 K, S = 54.56 cal·mol^{-1}·K^{-1}
228.28 J·mol^{-1}·K^{-1}
Phase Changes
c/liq 127.27 K, ΔH = 1177.0 cal·mol^{-1}
4924.6 J·mol^{-1}
ΔS = 9.248 cal·mol^{-1}·K^{-1}
38.694 J·mol^{-1}·K^{-1}
Molecular Weight 68.1182
Wiswesser Line Notation 1UY1&1U1
Evaluation A

C_5H_8 (liq) 70MES/TOD
3–Methyl–1,2–butadiene
Heat Capacity 298.15 K, C_p = 36.43 cal·mol^{-1}·K^{-1}
152.42 J·mol^{-1}·K^{-1}
Temperature range 12–320 K
Entropy 298.15 K, S = 55.40 cal·mol^{-1}·K^{-1}
231.79 J·mol^{-1}·K^{-1}
Phase Changes
c/liq 159.53 K, ΔH = 1901.6 cal·mol^{-1}
7956.3 J·mol^{-1}
ΔS = 11.920 cal·mol^{-1}·K^{-1}
49.873 J·mol^{-1}·K^{-1}
Molecular Weight 68.1182
Wiswesser Line Notation 1Y1&UCU1
Evaluation A

C_5H_8 (liq) 70MES/TOD
1–cis–3–Pentadiene
Heat Capacity 298.15 K, C_p = 35.03 cal·mol^{-1}·K^{-1}
146.57 J·mol^{-1}·K^{-1}
Temperature range 12–320 K
Entropy 298.15 K, S = 55.75 cal·mol^{-1}·K^{-1}
233.25 J·mol^{-1}·K^{-1}
Phase Changes
c/liq 132.35 K, ΔH = 1347.7 cal·mol^{-1}
5638.8 J·mol^{-1}
ΔS = 10.183 cal·mol^{-1}·K^{-1}
42.605 J·mol^{-1}·K^{-1}
Molecular Weight 68.1182
Wiswesser Line Notation 2U2U1 –C
Evaluation A

C_5H_8 (liq) 70MES/TOD
1–trans–3–Pentadiene
Heat Capacity 298.15 K, C_p = 35.69 cal·mol^{-1}·K^{-1}
149.33 J·mol^{-1}·K^{-1}
Temperature range 12–320 K
Entropy 298.15 K, S = 54.28 cal·mol^{-1}·K^{-1}
227.11 J·mol^{-1}·K^{-1}
Phase Changes
c/liq 185.71 K, ΔH = 1707.4 cal·mol^{-1}
7143.8 J·mol^{-1}
ΔS = 9.194 cal·mol^{-1}·K^{-1}
38.468 J·mol^{-1}·K^{-1}
Molecular Weight 68.1182
Wiswesser Line Notation 2U2U1 –T
Evaluation A

C_5H_8 (liq) 70MES/TOD
2,3–Pentadiene
Heat Capacity 298.15 K, C_p = 36.41 cal·mol^{-1}·K^{-1}
152.34 J·mol^{-1}·K^{-1}
Temperature range 12–320 K
Entropy 298.15 K, S = 56.72 cal·mol^{-1}·K^{-1}
237.32 J·mol^{-1}·K^{-1}
Phase Changes
c/liq 147.52 K, ΔH = 1584.1 cal·mol^{-1}
6127.9 J·mol^{-1}
ΔS = 10.738 cal·mol^{-1}·K^{-1}
44.929 J·mol^{-1}·K^{-1}
Molecular Weight 68.1182
Wiswesser Line Notation 2UCU2
Evaluation A

C_5H_8 (liq) 70MES/TOD
1,2–Pentadiene
Heat Capacity 298.15 K, C_p = 36.05 cal·mol^{-1}·K^{-1}
150.83 J·mol^{-1}·K^{-1}
Temperature range 12–320 K
Entropy 298.15 K, S = 58.55 cal·mol^{-1}·K^{-1}
244.97 J·mol^{-1}·K^{-1}
Phase Changes
c/liq 135.89 K, ΔH = 1806.7 cal·mol^{-1}
7559.2 J·mol^{-1}
ΔS = 13.245 cal·mol^{-1}·K^{-1}
55.628 J·mol^{-1}·K^{-1}
Molecular Weight 68.1182
Wiswesser Line Notation 3UCU1
Evaluation A

Table of Heat Capacities, Entropies, and Phase Transition Properties — Continued

C_5H_8 (liq) 50SCO/FIN 2
Spiropentane
Heat Capacity 298.15 K, C_p = 32.15 cal·mol^{-1}·K^{-1}
134.52 J·mol^{-1}·K^{-1}
Temperature range 12–312 K
Entropy 298.15 K, S = 46.29 cal·mol^{-1}·K^{-1}
193.68 J·mol^{-1}·K^{-1}
Phase Changes
c/liq 166.14 K, ΔH = 1537.6 cal·mol^{-1}
6433.3 J·mol^{-1}
ΔS = 9.25 cal·mol^{-1}·K^{-1}
38.72 J·mol^{-1}·K^{-1}
liq/g 312.13 K, ΔH = 6393 cal·mol^{-1}
26748 J·mol^{-1}
ΔS = 20.48 cal·mol^{-1}·K^{-1}
85.70 J·mol^{-1}·K^{-1}
P = 101.325 kPa
Molecular Weight 68.1182
Wiswesser Line Notation L3XT J A-& AL3XTJ
Evaluation A

C_5H_8 (liq) 48HUF/EAT
Cyclopentene
Heat Capacity 298.15 K, C_p = 29.25 cal·mol^{-1}·K^{-1}
122.38 J·mol^{-1}·K^{-1}
Temperature range 12–300 K
Entropy 298.15 K, S = 48.10 cal·mol^{-1}·K^{-1}
201.25 J·mol^{-1}·K^{-1}
Phase Changes
c,II/c,I 87.07 K, ΔH = 114.6 cal·mol^{-1}
479.5 J·mol^{-1}
ΔS = 1.32 cal·mol^{-1}·K^{-1}
9.51 J·mol^{-1}·K^{-1}
c,I/liq 138.13 K, ΔH = 803.9 cal·mol^{-1}
3363.5 J·mol^{-1}
ΔS = 5.82 cal·mol^{-1}·K^{-1}
24.35 J·mol^{-1}·K^{-1}
Molecular Weight 68.1182
Wiswesser Line Notation L5UTJ
Evaluation A

C_5H_8 (liq) 81FIN/MES
Methylenecyclobutane
Heat Capacity 298.15 K, C_p = 31.33 cal·mol^{-1}·K^{-1}
131.1 J·mol^{-1}·K^{-1}
Temperature range 12–301 K; equation also given for temperature range 146–301 K.
Entropy 298.15 K, S = 50.24 cal·mol^{-1}·K^{-1}
210.2 J·mol^{-1}·K^{-1}
Phase Changes
c,I/liq 138.535 K, ΔH = 1400.9 cal·mol^{-1}
5861.4 J·mol^{-1}
ΔS = 10.11 cal·mol^{-1}·K^{-1}
42.31 J·mol^{-1}·K^{-1}
Molecular Weight 68.1182
Wiswesser Line Notation L4YTJ AU1
Evaluation A

$(C_5H_8)_n$ (c) 35BEK/MAT
Rubber; Latex
Heat Capacity 280 K, C_p = 29.37 cal·mol^{-1}·K^{-1}
122.9 J·mol^{-1}·K^{-1}
Temperature range 15–280 K. Values per C_5H_8 unit.
Phase Changes
c/amorp 284 K, ΔH = 1138 cal·mol^{-1}
4761 J·mol^{-1}
ΔS = 4.0 cal·mol^{-1}·K^{-1}
16.8 J·mol^{-1}·K^{-1}
Corrected for premelting, 255–284 K. Value per C_5H_8 unit. "Fusion" actually transition to amorphous form.
Molecular Weight 68.1182
Wiswesser Line Notation /*Y1&U3* -C/
Evaluation C
Latex digested with steam at 190 °C, and extracted with alcohol and water.

$(C_5H_8)_n$ (amorp) 35BEK/MAT
Rubber; Latex
Heat Capacity 298.15 K, C_p = 30.62 cal·mol^{-1}·K^{-1}
128.1 J·mol^{-1}·K^{-1}
Temperature range 14–320 K, Value per C_5H_8 unit.
Entropy 298.15 K, S = 30.6 cal·mol^{-1}·K^{-1}
128.1 J·mol^{-1}·K^{-1}
From graphical integration of C_p data of crystal, entropy of fusion, and C_p data of amorphous above 284 K.
Molecular Weight 68.1182
Wiswesser Line Notation /*Y1&U3* -C/
Evaluation C
Latex digested with steam at 190 °C, and extracted with alcohol and water.

$C_5H_8Br_4$ (c) 59WES
2,2-Bis(bromomethyl)-1,3-dibromopropane;
Pentaerythrityl tetrabromide
Heat Capacity 298.15 K, C_p = 51.10 cal·mol^{-1}·K^{-1}
213.80 J·mol^{-1}·K^{-1}
Based on data 10–350 K, to be reported elsewhere.
Entropy 298.15 K, S = 69.58 cal·mol^{-1}·K^{-1}
291.12 J·mol^{-1}·K^{-1}
Molecular Weight 387.7342
Wiswesser Line Notation E1X1E1E1E
Evaluation B

$C_5H_8Br_4$ (c) 62PAY/WES
2,2-Bis(bromomethyl)-1,3-dibromopropane;
Pentaerythrityl tetrabromide
Heat Capacity 298.15 K, C_p = 51.10 cal·mol^{-1}·K^{-1}
213.80 J·mol^{-1}·K^{-1}
Temperature range 6–300 K. Anomalous region, 260–290 K
Entropy 298.15 K, S = 69.58 cal·mol^{-1}·K^{-1}
291.12 J·mol^{-1}·K^{-1}
Molecular Weight 387.7342
Wiswesser Line Notation E1X1E1E1E
Evaluation A

$C_5H_8Br_4$ (c) 65CLE/WON
2,2-Bis(bromomethyl)-1,3-dibromopropane;
Pentaerythrityl tetrabromide
Heat Capacity 298.15 K, C_p = 52.24 cal·mol^{-1}·K^{-1}
218.57 J·mol^{-1}·K^{-1}
Temperature range 250–460 K
Entropy 298.15 K, S = 69.58 cal·mol^{-1}·K^{-1}
297.12 J·mol^{-1}·K^{-1}
Based on low tempertaure data of 62PAY/WES.
Phase Changes
c/liq 433.45 K, ΔH = 6684 cal·mol^{-1}
27966 J·mol^{-1}
ΔS = 15.42 cal·mol^{-1}·K^{-1}
64.52 J·mol^{-1}·K^{-1}
Molecular Weight 387.7342
Wiswesser Line Notation E1X1E1E1E
Evaluation A

Table of Heat Capacities, Entropies, and Phase Transition Properties — Continued

$C_5H_8Cl_2O$ (liq) 62DAI/EVA 2
3,3–Bis–(chloromethyl)oxacyclobutane
Heat Capacity 300 K, C_p = 52.27 cal·mol⁻¹·K⁻¹
218.7 J·mol⁻¹·K⁻¹
Temperature range 20–310 K
Entropy 300 K, S = 65.58 cal·mol⁻¹·K⁻¹
274.4 J·mol⁻¹·K⁻¹
Phase Changes
c,I/liq 292.16 K, ΔH = 4049.7 cal·mol⁻¹
16944 J·mol⁻¹
ΔS = 13.86 cal·mol⁻¹·K⁻¹
58.00 J·mol⁻¹·K⁻¹
Premelting occurs above 250 K.
Molecular Weight 155.0236
Wiswesser Line Notation T4OTJ C1G C1G
Evaluation A

$(C_5H_8Cl_2O)_n$ (c) 62DAI/EVA 2
3,3–Bis–(chloromethyl)polyoxacyclobutane; Penton
Heat Capacity 300 K, C_p = 42.97 cal·mol⁻¹·K⁻¹
179.8 J·mol⁻¹·K⁻¹
Temperature range 20–310 K
Entropy 300 K, S = 45.65 cal·mol⁻¹·K⁻¹
191.0 J·mol⁻¹·K⁻¹
Molecular Weight 155.0236
Wiswesser Line Notation /*T40TJ C1G C1G*/
Evaluation A

$C_5H_8Cl_2O_2$ (liq) 81REI
Ethyl 2,3–dichloropropanoate; Ethyl α,β–dichloropropionate
Heat Capacity 298 K, C_p = 59.5 cal·mol⁻¹·K⁻¹
248.9 J·mol⁻¹·K⁻¹
Temperature range 290–465 K
Molecular Weight 171.0230
Wiswesser Line Notation G1YGVO2
Evaluation D

$C_5H_8Cl_4$ (c) 59WES
2,2–Bis(chloromethyl)–1,3–dichloropropane;
Pentaerythrityl tetrachloride
Heat Capacity 298.15 K, C_p = 47.44 cal·mol⁻¹·K⁻¹
198.49 J·mol⁻¹·K⁻¹
Based on data 10–350 K, to be reported elsewhere.
Entropy 298.15 K, S = 61.54 cal·mol⁻¹·K⁻¹
257.48 J·mol⁻¹·K⁻¹
Molecular Weight 209.9302
Wiswesser Line Notation G1X1G1G1G
Evaluation B

$C_5H_8Cl_4$ (c) 62PAY/WES
2,2–Bis(chloromethyl)–1,3–dichloropropane;
Pentaerythrityl tetrachloride
Heat Capacity 298.15 K, C_p = 47.44 cal·mol⁻¹·K⁻¹
198.49 J·mol⁻¹·K⁻¹
Temperature range 6–300 K. Anomalous region, 220–240 K.
Entropy 298.15 K, S = 61.54 cal·mol⁻¹·K⁻¹
257.48 J·mol⁻¹·K⁻¹
Molecular Weight 209.9302
Wiswesser Line Notation G1X1G1G1G
Evaluation A

$C_5H_8Cl_4$ (c,II) 65CLE/WON
2,2–Bis(chloromethyl)–1,3–dichloropropane;
Pentaerythrityl tetrachloride
Heat Capacity 298.15 K, C_p = 44.68 cal·mol⁻¹·K⁻¹
186.94 J·mol⁻¹·K⁻¹
Temperature range 290–420 K
Entropy 298.15 K, S = 57.51 cal·mol⁻¹·K⁻¹
240.62 J·mol⁻¹·K⁻¹
Based on low temp data of 62PAY/WES.
Phase Changes
c,II/c,I 368.23 K, ΔH = 5320 cal·mol⁻¹
22259 J·mol⁻¹
ΔS = 14.45 cal·mol⁻¹·K⁻¹
60.45 J·mol⁻¹·K⁻¹
Molecular Weight 209.9302
Wiswesser Line Notation G1X1G1G1G
Evaluation A

$C_5H_8F_4$ (c,I) 59WES
2,2–Bis(fluoromethyl)–1,3–difluoropropane;
Pentaerythrityl tetrafluoride
Heat Capacity 298.15 K, C_p = 50.80 cal·mol⁻¹·K⁻¹
212.55 J·mol⁻¹·K⁻¹
Based on data 10–350 K to be reported elsewhere.
Entropy 298.15 K, S = 69.35 cal·mol⁻¹·K⁻¹
290.16 J·mol⁻¹·K⁻¹
Phase Changes
c,II/c,I 249.40 K, ΔH = 3157.5 cal·mol⁻¹
13210 J·mol⁻¹
ΔS = 12.7 cal·mol⁻¹·K⁻¹
53.0 J·mol⁻¹·K⁻¹
Lambda transition, no details.
Molecular Weight 144.1118
Wiswesser Line Notation F1X1F1F1F
Evaluation B

$C_5H_8F_4$ (c) 64TRO/WES
2,2–Bis(fluoromethyl)–1,3–difluoropropane;
Pentaerythrityl tetrafluoride
Heat Capacity Data on solid and liquid 295–385 K, supplementing previous work of Westrum and Payne.
Phase Changes
c,I/liq 367.43 K, ΔH = 1229 cal·mol⁻¹
5142 J·mol⁻¹
ΔS = 3.34 cal·mol⁻¹·K⁻¹
13.99 J·mol⁻¹·K⁻¹
Molecular Weight 144.1118
Wiswesser Line Notation F1X1F1F1F
Evaluation A

$C_5H_8I_4$ (c) 59WES
2,2–Bis(iodomethyl)–1,3–diiodopropane;
Pentaerythrityl tetraiodide
Heat Capacity 298.15 K, C_p = 49.64 cal·mol⁻¹·K⁻¹
207.69 J·mol⁻¹·K⁻¹
Based on data 10–350 K, to be reported elsewhere.
Entropy 298.15 K, S = 75.70 cal·mol⁻¹·K⁻¹
316.73 J·mol⁻¹·K⁻¹
Molecular Weight 575.7362
Wiswesser Line Notation I1X1I1I1I
Evaluation B

Table of Heat Capacities, Entropies, and Phase Transition Properties — Continued

$C_5H_8I_4$ (c) 62PAY/WES
2,2–Bis(iodomethyl)–1,3–diiodopropane;
Pentaerythrityl tetraiodide
Heat Capacity 298.15 K, C_p = 49.64 cal·mol⁻¹·K⁻¹ 207.70 J·mol⁻¹·K⁻¹
Temperature range 6–300 K
Entropy 298.15 K, S = 75.70 cal·mol⁻¹·K⁻¹ 316.73 J·mol⁻¹·K⁻¹
Molecular Weight 575.7362
Wiswesser Line Notation I1X1I1I1I
Evaluation A

$C_5H_8I_4$ (c) 65CLE/WON
2,2–Bis(iodomethyl)–1,3–diiodopropane;
Pentaerythrityl tetraiodide
Heat Capacity 298.15 K, C_p = 50.12 cal·mol⁻¹·K⁻¹ 209.62 J·mol⁻¹·K⁻¹
Temperature range 260–420 K
Entropy 298.15 K, S = 75.75 cal·mol⁻¹·K⁻¹ 316.94 J·mol⁻¹·K⁻¹
Based on low temp data of 62PAY/WES.
Molecular Weight 575.7362
Wiswesser Line Notation I1X1I1I1I
Evaluation A

$C_5H_8O_2$ (liq) **81REI**
Ethenyl ethanoate; Allyl acetate
Heat Capacity 298 K, C_p = 44.0 cal·mol⁻¹·K⁻¹ 184.1 J·mol⁻¹·K⁻¹
Temperature range 292–382 K
Molecular Weight 100.1170
Wiswesser Line Notation 1VO2U1
Evaluation D

$C_5H_8O_2$ (liq) 52ERD/JAG
Methyl 2–methylpropenoate; Methyl methacrylate
Heat Capacity 293 K, C_p = 45.05 cal·mol⁻¹·K⁻¹ 188.49 J·mol⁻¹·K⁻¹
Temperature range 20 to 50°C
Molecular Weight 100.1170
Wiswesser Line Notation 1UY1&VO1
Evaluation C

$C_5H_8O_2$ (c) 58SOC/TRA
Methyl 2–methylpropenoate; Methyl methacrylate
Heat Capacity 210 K, C_p = 35.9 cal·mol⁻¹·K⁻¹ 150.2 J·mol⁻¹·K⁻¹
Temperature range 60–210 K
Molecular Weight 100.1170
Wiswesser Line Notation 1UY1&VO1
Evaluation B

$(C_5H_8O_2)_n$ (c) 58SOC/TRA
Poly(methyl methacrylate)
Heat Capacity 260 K, C_p = 26.7 cal·mol⁻¹·K⁻¹ 111.7 J·mol⁻¹·K⁻¹
Temperature range 60–260 K. Values per monomer unit.
Molecular Weight 100.1170
Wiswesser Line Notation /*X1&1*VO1/
Evaluation B

$C_5H_8O_4$ (c) 39SAT/SOG
Pyrotartaric acid; 2–Methylsuccinic acid
Heat Capacity 323 K, C_p = 47.7 cal·mol⁻¹·K⁻¹ 199.6 J·mol⁻¹·K⁻¹
Temperature range 0 to 100°C. Mean value.
Molecular Weight 132.1158
Wiswesser Line Notation QVY1&1VQ
Evaluation C

C_5H_9ClO (liq) **81REI**
Pentanoyl chloride; Valeryl chloride
Heat Capacity 298 K, C_p = 44.9 cal·mol⁻¹·K⁻¹ 187.9 J·mol⁻¹·K⁻¹
Temperature range 291–400 K
Molecular Weight 120.5785
Wiswesser Line Notation GV4
Evaluation D

$C_5H_9ClO_2$ (liq) **81REI**
Ethyl 2-chloropropanoate; Ethyl α–chloropropionate
Heat Capacity 298 K, C_p = 52.7 cal·mol⁻¹·K⁻¹ 220.5 J·mol⁻¹·K⁻¹
Temperature range 291–431 K
Molecular Weight 136.5779
Wiswesser Line Notation GY1&VO2
Evaluation D

C_5H_9N (liq) 71HAL/BAL
2,2–Dimethylpropionitrile; 2–Cyano–2–methylpropane;
tert–Butyl cyanide
Heat Capacity 297 K, C_p = 36.6 cal·mol⁻¹·K⁻¹ 153.3 J·mol⁻¹·K⁻¹
One temperature
Molecular Weight 83.1328
Wiswesser Line Notation NCX1&1&1
Evaluation C

C_5H_9N (liq) 67WES/RIB
2,2–Dimethylpropionitrile; 2–Cyano–2–methylpropane;
tert–Butyl cyanide
Heat Capacity 298.15 K, C_p = 42.87 cal·mol⁻¹·K⁻¹ 179.37 J·mol⁻¹·K⁻¹
Temperature range 5–350 K
Entropy 298.15 K, S = 55.45 cal·mol⁻¹·K⁻¹ 232.00 J·mol⁻¹·K⁻¹

Phase Changes

c,III/c,II	213 K,	ΔH =	55 cal·mol⁻¹ 230 J·mol⁻¹
		ΔS =	0.26 cal·mol⁻¹·K⁻¹ 1.08 J·mol⁻¹·K⁻¹
c,II/c,I	232.74 K,	ΔH =	457 cal·mol⁻¹ 1912 J·mol⁻¹
		ΔS =	1.86 cal·mol⁻¹·K⁻¹ 7.78 J·mol⁻¹·K⁻¹

ΔS not equal to $\Delta H/T$, see article.

c,I/liq	292.13 K,	ΔH =	2220 cal·mol⁻¹ 9288 J·mol⁻¹
		ΔS =	7.60 cal·mol⁻¹·K⁻¹ 31.79 J·mol⁻¹·K⁻¹

Molecular Weight 83.1328
Wiswesser Line Notation NCX1&1&1
Evaluation A

Table of Heat Capacities, Entropies, and Phase Transition Properties — Continued

C_5H_9NO (liq) 78MAR/CIO
1–Methyl–2–pyrrolidone
Heat Capacity 298 K, C_p = 73.6 cal·mol^{-1}·K^{-1}
307.8 J·mol^{-1}·K^{-1}
Temperature range 298–461 K
Molecular Weight 99.1322
Wiswesser Line Notation T5NVJ A1
Evaluation D

$C_5H_9NO_2$ (c) 40HUF/FOX
Proline(L); Pyrrolidine–2–carboxylic acid(L)
Heat Capacity 300.4 K, C_p = 35.64 cal·mol^{-1}·K^{-1}
149.12 J·mol^{-1}·K^{-1}
Temperature range 90–298 K. Value is unsmoothed experimental datum.
Entropy 298.15 K, S = 40.8 cal·mol^{-1}·K^{-1}
170.7 J·mol^{-1}·K^{-1}
Extrapolation below 90 K, 13.40 cal·mol^{-1}K^{-1}
Molecular Weight 115.1316
Wiswesser Line Notation T5MTJ BVQ -L
Evaluation C

$C_5H_9NO_2$ (c) 63COL/HUT
Proline(L); Pyrrolidine–2–carboxylic acid(L)
Heat Capacity 298.15 K, C_p = 36.13 cal·mol^{-1}·K^{-1}
151.17 J·mol^{-1}·K^{-1}
Temperature range 11–305 K
Entropy 298.15 K, S = 39.21 cal·mol^{-1}·K^{-1}
164.06 J·mol^{-1}·K^{-1}
Molecular Weight 115.1316
Wiswesser Line Notation T5MTJ BVQ -L
Evaluation A

$C_5H_9NO_2$ (c) 75SPI/WAD
Proline(L); Pyrrolidine–2–carboxylic acid(L)
Heat Capacity 298.15 K, C_p = 35.9 cal·mol^{-1}·K^{-1}
150.4 J·mol^{-1}·K^{-1}
One temperature
Molecular Weight 115.1316
Wiswesser Line Notation T5MTJ BVQ -L
Evaluation B

$C_5H_9NO_4$ (c) 32HUF/BOR
Glutamic acid(D)
Heat Capacity 294.6 K, C_p = 41.3 cal·mol^{-1}·K^{-1}
172.8 J·mol^{-1}·K^{-1}
Temperature range 91–295 K. Value is unsmoothed experimental datum.
Entropy 298.1 K, S = 45.7 cal·mol^{-1}·K^{-1}
191.2 J·mol^{-1}·K^{-1}
Extrapolation below 90 K, 13.00 cal·mol^{-1}K^{-1}
Molecular Weight 147.1304
Wiswesser Line Notation QVYZ2VQ -D
Evaluation B(C_p),C(S)

$C_5H_9NO_4$ (c) 63HUT/COL 2
Glutamic acid(L)
Heat Capacity 298.15 K, C_p = 41.84 cal·mol^{-1}·K^{-1}
175.06 J·mol^{-1}·K^{-1}
Temperature range 10–310 K
Entropy 298.15 K, S = 44.98 cal·mol^{-1}·K^{-1}
188.20 J·mol^{-1}·K^{-1}
Molecular Weight 147.1304
Wiswesser Line Notation QVYZ2VQ -L
Evaluation A

C_5H_{10} (liq) 47TOD/OLI
3–Methyl–1–butene
Heat Capacity 298.15 K, C_p = 37.30 cal·mol^{-1}·K^{-1}
156.06 J·mol^{-1}·K^{-1}
Temperature range 12–300 K
Entropy 298.15 K, S = 60.54 cal·mol^{-1}·K^{-1}
253.30 J·mol^{-1}·K^{-1}
Phase Changes
c/liq 104.72 K, ΔH = 1280.9 cal·mol^{-1}
5359.3 J·mol^{-1}
ΔS = 12.23 cal·mol^{-1}·K^{-1}
51.18 J·mol^{-1}·K^{-1}
Molecular Weight 70.1340
Wiswesser Line Notation 1Y1&1U1
Evaluation A

C_5H_{10} (liq) 30PAR/HUF 2
2–Methyl–2–butene
Heat Capacity 293.9 K, C_p = 35.9 cal·mol^{-1}·K^{-1}
146.4 J·mol^{-1}·K^{-1}
Temperature range 93–294 K. Value is unsmoothed experimental datum.
Entropy 298.15 K, S = 59.5 cal·mol^{-1}·K^{-1}
248.9 J·mol^{-1}·K^{-1}
Extrapolation below 90 K, 13.12 cal·mol^{-1}K^{-1}
Phase Changes
c/liq 138.9 K, ΔH = 1777 cal·mol^{-1}
7435 J·mol^{-1}
ΔS = 1.27 cal·mol^{-1}·K^{-1}
5.35 J·mol^{-1}·K^{-1}
Molecular Weight 70.1340
Wiswesser Line Notation 2UY1&1
Evaluation B(C_p),C(S)

C_5H_{10} (liq) 47TOD/OLI
2–Methyl–2–butene
Heat Capacity 298.15 K, C_p = 36.52 cal·mol^{-1}·K^{-1}
152.80 J·mol^{-1}·K^{-1}
Temperature range 12–300 K
Entropy 298.15 K, S = 60.00 cal·mol^{-1}·K^{-1}
251.04 J·mol^{-1}·K^{-1}
Phase Changes
c/liq 139.42 K, ΔH = 1815.8 cal·mol^{-1}
7597.3 J·mol^{-1}
ΔS = 13.02 cal·mol^{-1}·K^{-1}
54.49 J·mol^{-1}·K^{-1}
Molecular Weight 70.1340
Wiswesser Line Notation 2UY1&1
Evaluation A

C_5H_{10} (liq) 47TOD/OLI
2–Methyl–1–butene
Heat Capacity 298.15 K, C_p = 37.57 cal·mol^{-1}·K^{-1}
157.19 J·mol^{-1}·K^{-1}
Temperature range 12–300 K
Entropy 298.15 K, S = 60.70 cal·mol^{-1}·K^{-1}
253.97 J·mol^{-1}·K^{-1}
Phase Changes
c/liq 135.62 K, ΔH = 1890.6 cal·mol^{-1}
7910.3 J·mol^{-1}
ΔS = 13.94 cal·mol^{-1}·K^{-1}
58.33 J·mol^{-1}·K^{-1}
Molecular Weight 70.1340
Wiswesser Line Notation 2Y1&U1
Evaluation A

Table of Heat Capacities, Entropies, and Phase Transition Properties — Continued

C_5H_{10} (liq) 81REI
2-Pentene
Heat Capacity 298 K, C_p = 46.8 cal·mol^{-1}·K^{-1}
195.8 J·mol^{-1}·K^{-1}
Temperature range 290–322 K
Molecular Weight 70.1340
Wiswesser Line Notation 3U2
Evaluation D

C_5H_{10} (liq) 47TOD/OLI
cis-2-Pentene
Heat Capacity 298.15 K, C_p = 36.26 cal·mol^{-1}·K^{-1}
151.71 J·mol^{-1}·K^{-1}
Temperature range 12–300 K
Entropy 298.15 K, S = 61.81 cal·mol^{-1}·K^{-1}
258.61 J·mol^{-1}·K^{-1}
Phase Changes
c/liq 121.80 K, ΔH = 1699.7 cal·mol^{-1}
7111.5 J·mol^{-1}
ΔS = 13.95 cal·mol^{-1}·K^{-1}
58.39 J·mol^{-1}·K^{-1}
Molecular Weight 70.1340
Wiswesser Line Notation 3U2 -C
Evaluation A

C_5H_{10} (liq) 47TOD/OLI
trans-2-Pentene
Heat Capacity 298.15 K, C_p = 37.52 cal·mol^{-1}·K^{-1}
156.98 J·mol^{-1}·K^{-1}
Temperature range 12–300 K
Entropy 298.15 K, S = 61.31 cal·mol^{-1}·K^{-1}
256.52 J·mol^{-1}·K^{-1}
Phase Changes
c/liq 132.95 K, ΔH = 1996.0 cal·mol^{-1}
8351 J·mol^{-1}
ΔS = 15.01 cal·mol^{-1}·K^{-1}
62.81 J·mol^{-1}·K^{-1}
Molecular Weight 70.1340
Wiswesser Line Notation 3U2 -T
Evaluation A

C_5H_{10} (liq) 30PAR/HUF 2
1-Pentene
Heat Capacity 289.1 K, C_p = 36.12 cal·mol^{-1}·K^{-1}
151.13 J·mol^{-1}·K^{-1}
Temperature range 136–289 K. Value is unsmoothed experimental datum.
Molecular Weight 70.1340
Wiswesser Line Notation 4U1
Evaluation B

C_5H_{10} (liq) 47TOD/OLI
1-Pentene
Heat Capacity 298.15 K, C_p = 37.12 cal·mol^{-1}·K^{-1}
155.31 J·mol^{-1}·K^{-1}
Temperature range 12–300 K
Entropy 298.15 K, S = 62.75 cal·mol^{-1}·K^{-1}
262.55 J·mol^{-1}·K^{-1}
Phase Changes
c/liq 107.9 K, ΔH = 1388 cal·mol^{-1}
5807 J·mol^{-1}
ΔS = 12.86 cal·mol^{-1}·K^{-1}
53.82 J·mol^{-1}·K^{-1}
Molecular Weight 70.1340
Wiswesser Line Notation 4U1
Evaluation A

C_5H_{10} (liq) 34JAC/PAR
Cyclopentane
Heat Capacity 293.7 K, C_p = 30.09 cal·mol^{-1}·K^{-1}
125.90 J·mol^{-1}·K^{-1}
Temperature range 93–294 K. Value is unsmoothed experimental datum.
Entropy 298 K, S = 49.4 cal·mol^{-1}·K^{-1}
206.7 J·mol^{-1}·K^{-1}
Extrapolation below 90 K, 12.69 calmol^{-1}K^{-1}
Phase Changes
c,III/c,II 121.6 K, ΔH = 1134 cal·mol^{-1}
4745 J·mol^{-1}
ΔS = 9.33 cal·mol^{-1}·K^{-1}
39.02 J·mol^{-1}·K^{-1}
c,II/c,I 137.1 K, ΔH = 85.6 cal·mol^{-1}
358.2 J·mol^{-1}
ΔS = 0.62 cal·mol^{-1}·K^{-1}
2.61 J·mol^{-1}·K^{-1}
c,I/liq 179.0 K, ΔH = 144.5 cal·mol^{-1}
604.6 J·mol^{-1}
ΔS = 0.81 cal·mol^{-1}·K^{-1}
3.38 J·mol^{-1}·K^{-1}
Molecular Weight 70.1340
Wiswesser Line Notation L5TJ
Evaluation B(C_p),C(S)

C_5H_{10} (liq) 43AST/FIN
Cyclopentane
Heat Capacity 298.15 K, C_p = 30.79 cal·mol^{-1}·K^{-1}
128.83 J·mol^{-1}·K^{-1}
Temperature range 15–300 K
Entropy 298.15 K, S = 48.87 cal·mol^{-1}·K^{-1}
204.47 J·mol^{-1}·K^{-1}
Phase Changes
c,III/c,II 122.39 K, ΔH = 1165 cal·mol^{-1}
4874 J·mol^{-1}
ΔS = 9.52 cal·mol^{-1}·K^{-1}
39.82 J·mol^{-1}·K^{-1}
c,II/c,I 138.07 K, ΔH = 82.8 cal·mol^{-1}
346.4 J·mol^{-1}
ΔS = 0.60 cal·mol^{-1}·K^{-1}
2.51 J·mol^{-1}·K^{-1}
c,I/liq 179.69 K, ΔH = 144.0 cal·mol^{-1}
602 J·mol^{-1}
ΔS = 0.80 cal·mol^{-1}·K^{-1}
3.35 J·mol^{-1}·K^{-1}
liq/g 298.15 K, ΔH = 6982 cal·mol^{-1}
29213 J·mol^{-1}
ΔS = 23.42 cal·mol^{-1}·K^{-1}
97.98 J·mol^{-1}·K^{-1}
P = 41.10 kPa
Molecular Weight 70.1340
Wiswesser Line Notation L5TJ
Evaluation A

C_5H_{10} (liq) 46DOU/HUF 2
Cyclopentane
Heat Capacity 298.15 K, C_p = 30.30 cal·mol^{-1}·K^{-1}
126.78 J·mol^{-1}·K^{-1}
Temperature range 12–300 K
Entropy 298.15 K, S = 48.79 cal·mol^{-1}·K^{-1}
204.14 J·mol^{-1}·K^{-1}

Table of Heat Capacities, Entropies, and Phase Transition Properties — Continued

Phase Changes
c,III/c,II 122.39 K, ΔH = 1167.3 cal·mol^{-1}
4884.0 J·mol^{-1}
ΔS = 9.54 cal·mol^{-1}·K^{-1}
39.91 J·mol^{-1}·K^{-1}
c,II/c,I 138.09 K, ΔH = 82.32 cal·mol^{-1}
344.43 J·mol^{-1}
ΔS = 0.60 cal·mol^{-1}·K^{-1}
2.49 J·mol^{-1}·K^{-1}
c,I/liq 179.71 K, ΔH = 145.54 cal·mol^{-1}
608.94 J·mol^{-1}
ΔS = 0.81 cal·mol^{-1}·K^{-1}
3.38 J·mol^{-1}·K^{-1}
Molecular Weight 70.1340
Wiswesser Line Notation L5TJ
Evaluation A

C_5H_{10} (liq) 47SZA/MOR
Cyclopentane
Heat Capacity 300 K, C_p = 30.46 cal·mol^{-1}·K^{-1}
127.44 J·mol^{-1}·K^{-1}
Temperature range 14–300 K
Phase Changes
c,III/c,II 122.36 K, ΔH = 1167.4 cal·mol^{-1}
4884.4 J·mol^{-1}
ΔS = 9.54 cal·mol^{-1}·K^{-1}
39.92 J·mol^{-1}·K^{-1}
c,II/c,I 138.07 K, ΔH = 81.9 cal·mol^{-1}
342.7 J·mol^{-1}
ΔS = 0.59 cal·mol^{-1}·K^{-1}
2.48 J·mol^{-1}·K^{-1}
Temperature from 43AST/FIN.
c,I/liq 179.69 K, ΔH = 144.3 cal·mol^{-1}
603.8 J·mol^{-1}
ΔS = 0.80 cal·mol^{-1}·K^{-1}
3.36 J·mol^{-1}·K^{-1}
Temperature from 43AST/FIN.
Molecular Weight 70.1340
Wiswesser Line Notation L5TJ
Evaluation A

$C_5H_{10}ClNO_4$ (c) 40HUF/ELL
Glutamic acid hydrochloride
Heat Capacity 296.8 K, C_p = 49.83 cal·mol^{-1}·K^{-1}
208.49 J·mol^{-1}·K^{-1}
Temperature range 85–297 K. Value is unsmoothed experimental datum.
Entropy 298.1 K, S = 59.33 cal·mol^{-1}·K^{-1}
248.24 J·mol^{-1}·K^{-1}
Extrapolation below 90 K, 18.39 cal·mol^{-1}K^{-1}
Molecular Weight 183.5913
Wiswesser Line Notation QVYZ2VQ &GH
Evaluation A(C_p),C(S)

$C_5H_{10}N_2$ (liq) 82DZH/KAR
β–Dimethylaminopropionitrile
Heat Capacity 298.15 K, C_p = 50.79 cal·mol^{-1}·K^{-1}
212.5 J·mol^{-1}·K^{-1}
Temperature range 12–300 K
Entropy 298.15 K, S = 63.67 cal·mol^{-1}·K^{-1}
266.4 J·mol^{-1}·K^{-1}
Molecular Weight 98.1474
Wiswesser Line Notation NC2N1&1
Evaluation A

$C_5H_{10}N_2O_3$ (c) 63HUT/COL 2
Glutamine(L)
Heat Capacity 298.15 K, C_p = 44.02 cal·mol^{-1}·K^{-1}
184.18 J·mol^{-1}·K^{-1}
Temperature range 10–310 K
Entropy 298.15 K, S = 46.62 cal·mol^{-1}·K^{-1}
195.06 J·mol^{-1}·K^{-1}
Molecular Weight 146.1456
Wiswesser Line Notation ZV2YZVQ -L
Evaluation A

$C_5H_{10}N_2O_3$ (c) 41HUF
Alanylglycine(DL)
Heat Capacity 296.4 K, C_p = 43.45 cal·mol^{-1}·K^{-1}
181.79 J·mol^{-1}·K^{-1}
Temperature range 85–296 K. Value is unsmoothed experimental datum.
Entropy 298.1 K, S = 51.0 cal·mol^{-1}·K^{-1}
213.4 J·mol^{-1}·K^{-1}
Extrapolation below 90 K, 15.84 cal·mol^{-1}K^{-1}
Molecular Weight 146.1456
Wiswesser Line Notation ZY1&VM1VQ -DL
Evaluation A(C_p),C(S)

$C_5H_{10}O$ (liq) 68AND/COU
3–Methylbutanone; Isopropyl methyl ketone
Heat Capacity 298.15 K, C_p = 43.00 cal·mol^{-1}·K^{-1}
179.9 J·mol^{-1}·K^{-1}
Temperature range 10–320 K
Entropy 298.15 K, S = 64.17 cal·mol^{-1}·K^{-1}
268.5 J·mol^{-1}·K^{-1}
Phase Changes
c/liq 180.01 K, ΔH = 2233 cal·mol^{-1}
9343 J·mol^{-1}
ΔS = 12.40 cal·mol^{-1}·K^{-1}
51.90 J·mol^{-1}·K^{-1}
Molecular Weight 86.1334
Wiswesser Line Notation 1Y1&V1
Evaluation A

$C_5H_{10}O$ (liq) 68AND/COU
3–Pentanone; Diethyl ketone
Heat Capacity 298.15 K, C_p = 45.63 cal·mol^{-1}·K^{-1}
190.9 J·mol^{-1}·K^{-1}
Temperature range 10–320 K
Entropy 298.15 K, S = 63.58 cal·mol^{-1}·K^{-1}
266.0 J·mol^{-1}·K^{-1}
Phase Changes
c,III/c,II 118.5 K, ΔH = 26.5 cal·mol^{-1}
110.9 J·mol^{-1}
ΔS = 0.23 cal·mol^{-1}·K^{-1}
0.96 J·mol^{-1}·K^{-1}
c,II/c,I 180 K, ΔH = 2.3 cal·mol^{-1}
9.6 J·mol^{-1}
ΔS = 0.01 cal·mol^{-1}·K^{-1}
0.04 J·mol^{-1}·K^{-1}
c,I/liq 234.16 K, ΔH = 2771 cal·mol^{-1}
11594 J·mol^{-1}
ΔS = 11.83 cal·mol^{-1}·K^{-1}
49.51 J·mol^{-1}·K^{-1}
Molecular Weight 86.1334
Wiswesser Line Notation 2V2
Evaluation A

Table of Heat Capacities, Entropies, and Phase Transition Properties — Continued

$C_5H_{10}O$ (liq) 79SAL/PEA
3–Pentanone; Diethyl ketone
Heat Capacity 298.15 K, C_p = 45.4 cal·mol^{-1}·K^{-1}
190.0 J·mol^{-1}·K^{-1}
One temperature
Molecular Weight 86.1334
Wiswesser Line Notation 2V2
Evaluation B

$C_5H_{10}O$ (liq) 65OET
2–Pentanone; n–Propyl methyl ketone
Heat Capacity 298.15 K, C_p = 44.06 cal·mol^{-1}·K^{-1}
184.35 J·mol^{-1}·K^{-1}
Temperature range 12–330 K
Entropy 298.15 K, S = 65.11 cal·mol^{-1}·K^{-1}
272.42 J·mol^{-1}·K^{-1}
Phase Changes
c/liq 196.35 K, ΔH = 2539 cal·mol^{-1}
10623 J·mol^{-1}
ΔS = 12.93 cal·mol^{-1}·K^{-1}
54.10 J·mol^{-1}·K^{-1}
Molecular Weight 86.1334
Wiswesser Line Notation 3V1
Evaluation A

$C_5H_{10}O$ (liq) 68AND/COU
2–Pentanone; n–Propyl methyl ketone
Heat Capacity 298.15 K, C_p = 44.02 cal·mol^{-1}·K^{-1}
184.2 J·mol^{-1}·K^{-1}
Temperature range 10–360 K
Entropy 298.15 K, S = 65.51 cal·mol^{-1}·K^{-1}
274.1 J·mol^{-1}·K^{-1}
Phase Changes
c,II/c,I 110 K, ΔH = 50.7 cal·mol^{-1}
237.7 J·mol^{-1}
ΔS = 0.52 cal·mol^{-1}·K^{-1}
2.18 J·mol^{-1}·K^{-1}
Apparently a typographic error in ΔH in paper; given as 137.7 J·mol^{-1}.
c,I/liq 196.31 K, ΔH = 2541 cal·mol^{-1}
10632 J·mol^{-1}
ΔS = 12.94 cal·mol^{-1}·K^{-1}
54.16 J·mol^{-1}·K^{-1}
Molecular Weight 86.1334
Wiswesser Line Notation 3V1
Evaluation A

$C_5H_{10}O$ (liq) 75GRO/BEN
2–Pentanone; n–Propyl methyl ketone
Heat Capacity 298.15 K, C_p = 44.24 cal·mol^{-1}·K^{-1}
185.1 J·mol^{-1}·K^{-1}
One temperature
Molecular Weight 86.1334
Wiswesser Line Notation 3V1
Evaluation B

$C_5H_{10}O$ (liq) 79SAL/PEA
2–Pentanone; n–Propyl methyl ketone
Heat Capacity 298.15 K, C_p = 44.3 cal·mol^{-1}·K^{-1}
185.4 J·mol^{-1}·K^{-1}
One temperature
Molecular Weight 86.1334
Wiswesser Line Notation 3V1
Evaluation B

$C_5H_{10}O$ (liq) 56PAR/KEN
Cyclopentanol
Heat Capacity 298.15 K, C_p = 44.01 cal·mol^{-1}·K^{-1}
184.14 J·mol^{-1}·K^{-1}
Temperature range 80–300 K
Entropy 298.1 K, S = 49.3 cal·mol^{-1}·K^{-1}
206.3 J·mol^{-1}·K^{-1}
Extrapolation below 80 K, 11.66 cal·mol^{-1}K^{-1}
Phase Changes
c,II/c,I 202.8 K, ΔH = 886 cal·mol^{-1}
3707 J·mol^{-1}
ΔS = 4.37 cal·mol^{-1}·K^{-1}
18.28 J·mol^{-1}·K^{-1}
c,I/liq 257.4 K, ΔH = 367 cal·mol^{-1}
1536 J·mol^{-1}
ΔS = 1.43 cal·mol^{-1}·K^{-1}
5.97 J·mol^{-1}·K^{-1}
Molecular Weight 86.1334
Wiswesser Line Notation L5TJ AQ
Evaluation B(C_p),C(S)

$C_5H_{10}O$ (liq) 76CON/GIN
Cyclopentanol
Heat Capacity 298 K, C_p = 44.3 cal·mol^{-1}·K^{-1}
185.4 J·mol^{-1}·K^{-1}
One temperature
Molecular Weight 86.1334
Wiswesser Line Notation L5TJ AQ
Evaluation B

$C_5H_{10}O$ (liq) 64MOE/THO
Tetrahydropyran; Oxane
Heat Capacity 297.62 K, C_p = 36.12 cal·mol^{-1}·K^{-1}
151.13 J·mol^{-1}·K^{-1}
Temperature range 297–327 K
Molecular Weight 86.1334
Wiswesser Line Notation T6OTJ
Evaluation B

$C_5H_{10}O$ (liq) 76CON/GIN
Tetrahydropyran; Oxane
Heat Capacity 298 K, C_p = 33.6 cal·mol^{-1}·K^{-1}
140.6 J·mol^{-1}·K^{-1}
One temperature
Molecular Weight 86.1334
Wiswesser Line Notation T6OTJ
Evaluation B

$C_5H_{10}O$ (liq) 81REI
Valeral; n–Pentanal; Valeraldehyde
Heat Capacity 298 K, C_p = 41.0 cal·mol^{-1}·K^{-1}
171.5 J·mol^{-1}·K^{-1}
Temperature range 290–385 K
Molecular Weight 86.1334
Wiswesser Line Notation VH4
Evaluation D

$C_5H_{10}O_2$ (liq) 79FUC
1–Methylethyl ethanoate; Isopropyl acetate
Heat Capacity 298.15 K, C_p = 47.0 cal·mol^{-1}·K^{-1}
196.6 J·mol^{-1}·K^{-1}
One temperature
Molecular Weight 102.1328
Wiswesser Line Notation 1Y1&OV1
Evaluation B

Table of Heat Capacities, Entropies, and Phase Transition Properties — Continued

$C_5H_{10}O_2$ (liq) 81REI
Propyl ethanoate; n-Propyl acetate
Heat Capacity 298 K, C_p = 46.4 cal·mol^{-1}·K^{-1}
194.1 J·mol^{-1}·K^{-1}
Temperature range 292–382 K
Molecular Weight 102.1328
Wiswesser Line Notation 3OV1
Evaluation D

$C_5H_{10}O_2$ (liq) 79FUC
Methyl butanoate; Methyl butyrate
Heat Capacity 298.15 K, C_p = 48.0 cal·mol^{-1}·K^{-1}
200.8 J·mol^{-1}·K^{-1}
One temperature
Molecular Weight 102.1328
Wiswesser Line Notation 3VO1
Evaluation B

$C_5H_{10}O_2$ (liq) 81REI
3-Methylbutanoic acid; Isovaleric acid
Heat Capacity 298 K, C_p = 47.1 cal·mol^{-1}·K^{-1}
197.1 J·mol^{-1}·K^{-1}
Temperature range 290–470 K
Molecular Weight 102.1328
Wiswesser Line Notation QV1Y1&1
Evaluation D

$C_5H_{10}O_2$ (liq) 65MCD/KIL
Pentanoic acid; n-Valeric acid
Heat Capacity 298.15 K, C_p = 50.27 cal·mol^{-1}·K^{-1}
210.33 J·mol^{-1}·K^{-1}
Temperature range 15–300 K
Entropy 298.15 K, S = 62.10 cal·mol^{-1}·K^{-1}
259.83 J·mol^{-1}·K^{-1}
Phase Changes
c/liq 239.49 K, ΔH = 3384.7 cal·mol^{-1}
14161.6 J·mol^{-1}
ΔS = 14.13 cal·mol^{-1}·K^{-1}
59.13 J·mol^{-1}·K^{-1}
Molecular Weight 102.1328
Wiswesser Line Notation QV4
Evaluation A

$C_5H_{10}O_2$ (liq) 71KON/WAD
Pentanoic acid; n-Valeric acid
Heat Capacity 298.15 K, C_p = 47.1 cal·mol^{-1}·K^{-1}
197 J·mol^{-1}·K^{-1}
One temperature
Molecular Weight 102.1328
Wiswesser Line Notation QV4
Evaluation B

$C_5H_{10}O_2$ (c) 71KON/WAD
2,2-Dimethylpropanoic acid; Pivalic acid
Heat Capacity 298.15 K, C_p = 42.5 cal·mol^{-1}·K^{-1}
178 J·mol^{-1}·K^{-1}
One temperature
Molecular Weight 102.1328
Wiswesser Line Notation QVX1&1&1
Evaluation B

$C_5H_{10}O_2$ (liq) 35MIL
Tetrahydrofurfuryl alcohol
Heat Capacity (C_p)
Temperature range 100–298 K. Data in thesis only.
Entropy 298.15 K, S = 52.4 cal·mol^{-1}·K^{-1}
219.2 J·mol^{-1}·K^{-1}
Extrapolation below 90 K, 10.81 cal·mol^{-1}K^{-1}
Molecular Weight 102.1328
Wiswesser Line Notation T5OTJ B1Q
Evaluation C

$C_5H_{10}O_2$ (liq) 76CON/GIN
1,3-Dioxepane
Heat Capacity 298 K, C_p = 40 cal·mol^{-1}·K^{-1}
167 J·mol^{-1}·K^{-1}
One temperature
Molecular Weight 102.1328
Wiswesser Line Notation T7O COTJ
Evaluation C

$C_5H_{10}O_2$ (liq) 36KUR/VOS
2-Methylpropyl methanoate; Isobutyl formate
Heat Capacity 290 K, C_p = 51.2 cal·mol^{-1}·K^{-1}
214.2 J·mol^{-1}·K^{-1}
One temperature
Molecular Weight 102.1328
Wiswesser Line Notation VHO1Y1&1
Evaluation D

$C_5H_{10}O_3$ (liq) 81REI
Diethyl carbonate
Heat Capacity 298 K, C_p = 43.5 cal·mol^{-1}·K^{-1}
182.0 J·mol^{-1}·K^{-1}
Temperature range 297–416 K
Molecular Weight 118.1322
Wiswesser Line Notation 2OVO2
Evaluation D

$C_5H_{10}O_3$ (liq) 33KOL/UDO
Diethyl carbonate
Heat Capacity 294.7 K, C_p = 50.4 cal·mol^{-1}·K^{-1}
210.9 J·mol^{-1}·K^{-1}
One temperature
Molecular Weight 118.1322
Wiswesser Line Notation 2OVO2
Evaluation C

$C_5H_{10}O_3$ (liq) 34KOL/UDO 2
Diethyl carbonate
Heat Capacity 294.2 K, C_p = 50.4 cal·mol^{-1}·K^{-1}
210.9 J·mol^{-1}·K^{-1}
One temperature
Molecular Weight 118.1322
Wiswesser Line Notation 2OVO2
Evaluation C

$C_5H_{10}O_3$ (liq) 76MAS/PET
4-Methyl-1,3-dioxolan-2-one
Heat Capacity 298 K, C_p = 41.8 cal·mol^{-1}·K^{-1}
175 J·mol^{-1}·K^{-1}
Temperature range 200–325 K. Data graphically only. Value estimated from graph.
Molecular Weight 118.1322
Wiswesser Line Notation T6OVOTJ D1
Evaluation D

Table of Heat Capacities, Entropies, and Phase Transition Properties — Continued

$C_5H_{10}O_5$ (c) 69CLE/MEL 3
Pentoxan; Pentacycloformaldehyde
Heat Capacity 298.15 K, C_p = 41.0 cal·mol^{-1}·K^{-1}
171.4 J·mol^{-1}·K^{-1}
Temperature range 180–400 K
Entropy 298.15 K, S = 44.9 cal·mol^{-1}·K^{-1}
187.7 J·mol^{-1}·K^{-1}
Extrapolation below 180 K, 119.0 J·mol^{-1}K^{-1}
Phase Changes
c/liq 334 K, ΔH = 5230 cal·mol^{-1}
21900 J·mol^{-1}
ΔS = 15.7 cal·mol^{-1}·K^{-1}
65.6 J·mol^{-1}·K^{-1}
Molecular Weight 150.1310
Wiswesser Line Notation T-10-O CO EO GO IOTJ
Evaluation B(C_p), C(S)

$C_5H_{10}O_5$ (c) 35MIL
α-Xylose(D)
Heat Capacity C_p data in thesis only.
Temperature range 100–298 K.
Entropy 298.15 K, S = 34.3 cal·mol^{-1}·K^{-1}
143.5 J·mol^{-1}·K^{-1}
Extrapolation below 90 K, 7.50 cal·mol^{-1}K^{-1}
Molecular Weight 150.1310
Wiswesser Line Notation T6OTJ BQ CQ DQ EQ -A&BCE -B&D
Evaluation C

$C_5H_{10}S$ (liq) 61BER/SCO
Cyclopentanethiol; Cyclopentyl mercaptan
Heat Capacity 298.15 K, C_p = 39.49 cal·mol^{-1}·K^{-1}
165.23 J·mol^{-1}·K^{-1}
Temperature range 12–367 K. For metastable crystals 12–137 K.
Entropy 298.15 K, S = 61.39 cal·mol^{-1}·K^{-1}
256.86 J·mol^{-1}·K^{-1}
Phase Changes
c,I/liq 155.39 K, ΔH = 1871.6 cal·mol^{-1}
7830.8 J·mol^{-1}
ΔS = 12.04 cal·mol^{-1}·K^{-1}
50.39 J·mol^{-1}·K^{-1}
Molecular Weight 102.1940
Wiswesser Line Notation L5TJ ASH
Evaluation A
From enthalpy data at 102–162 K calculate enthalpy of fusion of metastable crystals at 155.39 K as 1764 cal·mol^{-1}K^{-1}. Adiabatic transition from metastable to stable crystals, ΔH(155.39K) = 108 cal·mol^{-1}. Sum gives ΔH_{fusion} = 1872 cal·mol^{-1}.

$C_5H_{10}S$ (liq) 74MES/FIN
2-Methylthiolane; 2-Methylcyclothiapentane
Heat Capacity 298.15 K, C_p = 41.06 cal·mol^{-1}·K^{-1}
171.80 J·mol^{-1}·K^{-1}
Temperature range 10–380 K
Entropy 298.15 K, S = 58.63 cal·mol^{-1}·K^{-1}
245.31 J·mol^{-1}·K^{-1}
Phase Changes
c/liq 172.39 K, ΔH = 2121.4 cal·mol^{-1}
8875.9 J·mol^{-1}
ΔS = 12.31 cal·mol^{-1}·K^{-1}
51.49 J·mol^{-1}·K^{-1}
Molecular Weight 102.1940
Wiswesser Line Notation T5STJ B1
Evaluation A

$C_5H_{10}S$ (liq) 74MES/FIN
3-Methylthiolane; 3-Methylcyclothiapentane
Heat Capacity 298.15 K, C_p = 41.06 cal·mol^{-1}·K^{-1}
171.80 J·mol^{-1}·K^{-1}
Temperature range 10–340 K
Entropy 298.15 K, S = 57.60 cal·mol^{-1}·K^{-1}
241.00 J·mol^{-1}·K^{-1}
Phase Changes
c/liq 192.00 K, ΔH = 2478.6 cal·mol^{-1}
10370.5 J·mol^{-1}
ΔS = 12.91 cal·mol^{-1}·K^{-1}
54.01 J·mol^{-1}·K^{-1}
Molecular Weight 102.1940
Wiswesser Line Notation T5STJ C1
Evaluation A

$C_5H_{10}S$ (liq) 54MCC/FIN
Thiacyclohexane
Heat Capacity 298.15 K, C_p = 39.03 cal·mol^{-1}·K^{-1}
163.30 J·mol^{-1}·K^{-1}
Temperature range 13–340 K
Entropy 298.15 K, S = 52.16 cal·mol^{-1}·K^{-1}
218.24 J·mol^{-1}·K^{-1}
Phase Changes
c,III/c,II 201.4 K, ΔH = 262.4 cal·mol^{-1}
1097.9 J·mol^{-1}
ΔS = 1.30 cal·mol^{-1}·K^{-1}
5.45 J·mol^{-1}·K^{-1}
c,II/c,I 240.02 K, ΔH = 1858.3 cal·mol^{-1}
7775.1 J·mol^{-1}
ΔS = 7.74 cal·mol^{-1}·K^{-1}
32.39 J·mol^{-1}·K^{-1}
c,I/liq 292.25 K, ΔH = 585.2 cal·mol^{-1}
2448.5 J·mol^{-1}
ΔS = 2.00 cal·mol^{-1}·K^{-1}
8.38 J·mol^{-1}·K^{-1}
Molecular Weight 102.1940
Wiswesser Line Notation T6STJ
Evaluation A

$C_5H_{11}Br$ (liq) 48KUR
1-Bromo-3-methylbutane; Isoamyl bromide
Heat Capacity 298 K, C_p = 44.7 cal·mol^{-1}·K^{-1}
187.0 J·mol^{-1}·K^{-1}
Temperature range 12 to 100 °C, mean C_p, two temperatures.
Molecular Weight 151.0459
Wiswesser Line Notation E2Y1&1
Evaluation D

$C_5H_{11}Br$ (liq) 31DEE
1-Bromopentane; n-Amyl bromide; n-Pentyl bromide
Heat Capacity 290.7 K, C_p = 41.01 cal·mol^{-1}·K^{-1}
171.59 J·mol^{-1}·K^{-1}
Temperature range 96–291 K. Value is unsmoothed experimental datum.
Entropy 298.15 K, S = 97.22 cal·mol^{-1}·K^{-1}
406.77 J·mol^{-1}·K^{-1}
Extrapolation below 100 K, 13.58 cal·mol^{-1}K^{-1}
Phase Changes
c/liq 185.1 K, ΔH = 3433 cal·mol^{-1}
14364 J·mol^{-1}
ΔS = 18.55 cal·mol^{-1}·K^{-1}
77.60 J·mol^{-1}·K^{-1}
Molecular Weight 151.0459
Wiswesser Line Notation E5
Evaluation B(C_p),C(S)

Table of Heat Capacities, Entropies, and Phase Transition Properties — Continued

$C_5H_{11}Br$ (liq) 50KUS/CRO
1–Bromopentane; n–Amyl bromide; n–Pentyl bromide
Heat Capacity 206.6 K, C_p = 41.8 cal·mol^{-1}·K^{-1}
174.9 J·mol^{-1}·K^{-1}
Temperature range 122–207 K. Value is unsmoothed experimental datum.
Phase Changes
c/liq 185.1 K, ΔH = 2740 cal·mol^{-1}
11465 J·mol^{-1}
ΔS = 14.8 cal·mol^{-1}·K^{-1}
62.0 J·mol^{-1}·K^{-1}
Molecular Weight 151.0459
Wiswesser Line Notation E5
Evaluation B

$C_5H_{11}Cl$ (liq) 48KUR
1–Chloro–3–methylbutane; Isoamyl chloride
Heat Capacity 298 K, C_p = 42.9 cal·mol^{-1}·K^{-1}
179.5 J·mol^{-1}·K^{-1}
Temperature range 14 to 98 °C, mean C_p, two temperatures.
Molecular Weight 106.5949
Wiswesser Line Notation G2Y1&1
Evaluation D

$C_5H_{11}I$ (liq) 48KUR
1–Iodo–3–methylbutane; Isoamyl iodide
Heat Capacity 298 K, C_p = 42.7 cal·mol^{-1}·K^{-1}
178.7 J·mol^{-1}·K^{-1}
Temperature range 13 to 137 °C, mean C_p, three temperatures.
Molecular Weight 198.0464
Wiswesser Line Notation I2Y1&1
Evaluation D

$C_5H_{11}N$ (liq) 76CON/GIN
1–Methylpyrrolidine
Heat Capacity 298 K, C_p = 38.5 cal·mol^{-1}·K^{-1}
161.1 J·mol^{-1}·K^{-1}
One temperature
Molecular Weight 85.1486
Wiswesser Line Notation T5NTJ A1
Evaluation B

$C_5H_{11}N$ (liq) 34RAD/JUL
Piperidine
Heat Capacity 290 K, C_p = 40.8 cal·mol^{-1}·K^{-1}
170.7 J·mol^{-1}·K^{-1}
One temperature
Molecular Weight 85.1486
Wiswesser Line Notation T6MTJ
Evaluation C

$C_5H_{11}N$ (liq) 36KUR/VOS
Piperidine
Heat Capacity 290 K, C_p = 44.5 cal·mol^{-1}·K^{-1}
186.2 J·mol^{-1}·K^{-1}
One temperature
Molecular Weight 85.1486
Wiswesser Line Notation T6MTJ
Evaluation D

$C_5H_{11}N$ (liq) 64MOE/THO
Piperidine
Heat Capacity 297.39 K, C_p = 43.68 cal·mol^{-1}·K^{-1}
182.76 J·mol^{-1}·K^{-1}
Temperature range 297–327 K
Molecular Weight 85.1486
Wiswesser Line Notation T6MTJ
Evaluation B

$C_5H_{11}N$ (liq) 76CON/GIN
Piperidine
Heat Capacity 298 K, C_p = 43.4 cal·mol^{-1}·K^{-1}
181.6 J·mol^{-1}·K^{-1}
One temperature
Molecular Weight 85.1486
Wiswesser Line Notation T6MTJ
Evaluation B

$C_5H_{11}N$ (liq) 81FIN/MES
Cyclopentylamine
Heat Capacity 298 K, C_p = 43.31 cal·mol^{-1}·K^{-1}
181.21 J·mol^{-1}·K^{-1}
Temperature range 12–349 K. Equation also given for temperature range 197–349 K.
Entropy 298.15 K, S = 57.61 cal·mol^{-1}·K^{-1}
241.04 J·mol^{-1}·K^{-1}
Phase Changes
c,II/c,I 184.5 K, ΔH = 113.6 cal·mol^{-1}
475.3 J·mol^{-1}
ΔS = .616 cal·mol^{-1}·K^{-1}
2.58 J·mol^{-1}·K^{-1}
Lambda type transition.
c,I/liq 190.45 K, ΔH = 1986.8 cal·mol^{-1}
8312.8 J·mol^{-1}
ΔS = 10.43 cal·mol^{-1}·K^{-1}
43.65 J·mol^{-1}·K^{-1}
Molecular Weight 85.1486
Wiswesser Line Notation L5TJ AZ
Evaluation A

$C_5H_{11}NO$ (liq) 71KON/WAD
N–(1–Methylethyl)ethanamide; N–Isopropylacetamide
Heat Capacity 298.15 K, C_p = 50.4 cal·mol^{-1}·K^{-1}
211 J·mol^{-1}·K^{-1}
One temperature
Molecular Weight 101.1480
Wiswesser Line Notation 1Y1&MV1
Evaluation B

$C_5H_{11}NO$ (liq) 71KON/WAD
2,N–Dimethylpropanamide
Heat Capacity 298.15 K, C_p = 50.0 cal·mol^{-1}·K^{-1}
209 J·mol^{-1}·K^{-1}
One temperature
Molecular Weight 101.1480
Wiswesser Line Notation 1Y1&VM1
Evaluation B

$C_5H_{11}NO$ (liq) 71KON/WAD
N–(n–Propyl)ethanamide; N–(n–Propyl)acetamide
Heat Capacity 298.15 K, C_p = 49.5 cal·mol^{-1}·K^{-1}
207 J·mol^{-1}·K^{-1}
One temperature
Molecular Weight 101.1480
Wiswesser Line Notation 3MV1
Evaluation B

$C_5H_{11}NO$ (liq) 71KON/WAD
N–Methylbutanamide
Heat Capacity 298.15 K, C_p = 49.5 cal·mol^{-1}·K^{-1}
207 J·mol^{-1}·K^{-1}
One temperature
Molecular Weight 101.1480
Wiswesser Line Notation 3VM1
Evaluation B

Table of Heat Capacities, Entropies, and Phase Transition Properties — Continued

$C_5H_{11}NO_2$ (c) 63HUT/COL
2-Amino-3-methylbutanoic acid(L); Valine(L);
α-Aminoisovaleric acid
Heat Capacity 298.15 K, C_p = 40.35 cal·mol⁻¹·K⁻¹
168.82 J·mol⁻¹·K⁻¹
Temperature range 11–310 K
Entropy 298.15 K, S = 42.75 cal·mol⁻¹·K⁻¹
178.87 J·mol⁻¹·K⁻¹
Molecular Weight 117.1474
Wiswesser Line Notation QVYZY1&1 -L
Evaluation A

$C_5H_{11}NO_2$ (c) 75SPI/WAD
2-Amino-3-methylbutanoic acid(L); Valine(L);
α-Aminoisovaleric acid(L)
Heat Capacity 298.15 K, C_p = 40.3 cal·mol⁻¹·K⁻¹
168.5 J·mol⁻¹·K⁻¹
One temperature
Molecular Weight 117.1474
Wiswesser Line Notation QVYZY1&1 -L
Evaluation B

$C_5H_{11}NO_2S$ (c) 64HUT/COL
Methionine
Heat Capacity 298.15 K, C_p = 69.32 cal·mol⁻¹·K⁻¹
290.03 J·mol⁻¹·K⁻¹
Temperature range 11–348 K
Entropy 298.15 K, S = 55.32 cal·mol⁻¹·K⁻¹
231.46 J·mol⁻¹·K⁻¹
Phase Changes
c,II/c,I 305.5 K, ΔH = 1300 cal·mol⁻¹
5440 J·mol⁻¹
ΔS = 4.2 cal·mol⁻¹·K⁻¹
17.6 J·mol⁻¹·K⁻¹
Lambda transition over the temperature range 250–350 K with a maximum at 305.5 K.
Molecular Weight 149.2074
Wiswesser Line Notation QVYZ2S1
Evaluation A

$C_5H_{11}NO_4$ (c) 39SAT/SOG
Ammonium acid pyrotartrate; Ammonium acid
2-methylsuccinate
Heat Capacity 323 K, C_p = 56.1 cal·mol⁻¹·K⁻¹
234.7 J·mol⁻¹·K⁻¹
Temperature range 0 to 100°C. Mean value.
Molecular Weight 149.1462
Wiswesser Line Notation QVY1&1VQ &ZH
Evaluation C

C_5H_{12} (liq) 36AST/MES
2,2-Dimethylpropane; Neopentane
Heat Capacity 278.92 K, C_p = 39.17 cal·mol⁻¹·K⁻¹
163.89 J·mol⁻¹·K⁻¹
Temperature range 13–283 K. Value is unsmoothed experimental datum.
Entropy 282.61 K, S = 52.3 cal·mol⁻¹·K⁻¹
218.8 J·mol⁻¹·K⁻¹
Phase Changes
c,II/c,I 140.0 K, ΔH = 616 cal·mol⁻¹
2577 J·mol⁻¹
ΔS = 4.40 cal·mol⁻¹·K⁻¹
18.41 J·mol⁻¹·K⁻¹
c,I/liq 256.53 K, ΔH = 778 cal·mol⁻¹
3255 J·mol⁻¹
ΔS = 3.03 cal·mol⁻¹·K⁻¹
12.68 J·mol⁻¹·K⁻¹
liq/g 282.61 K, ΔH = 5438 cal·mol⁻¹
22753 J·mol⁻¹
ΔS = 19.24 cal·mol⁻¹·K⁻¹
80.50 J·mol⁻¹·K⁻¹
P = 101.325 kPa
Molecular Weight 72.1498
Wiswesser Line Notation 1X1&1&1
Evaluation A

C_5H_{12} (liq) 69ENO/SHI
2,2-Dimethylpropane; Neopentane
Heat Capacity 259.93 K, C_p = 36.59 cal·mol⁻¹·K⁻¹
153.09 J·mol⁻¹·K⁻¹
Temperature range 4–260 K. Value is unsmoothed experimental datum.
Entropy 282.61 K, S = 51.82 cal·mol⁻¹·K⁻¹
216.81 J·mol⁻¹·K⁻¹
At normal boiling point.
Phase Changes
c,II/c,I 140–142 K, ΔH = 628.7 cal·mol⁻¹
2630.5 J·mol⁻¹
ΔS = 4.47 cal·mol⁻¹·K⁻¹
18.70 J·mol⁻¹·K⁻¹
Second-order transition, 140–142 K. ΔS for total change between 140 and 142 K.
c,I/liq 256.76 K, ΔH = 740.0 cal·mol⁻¹
3096.2 J·mol⁻¹
ΔS = 2.88 cal·mol⁻¹·K⁻¹
12.06 J·mol⁻¹·K⁻¹
Molecular Weight 72.1498
Wiswesser Line Notation 1X1&1&1
Evaluation A

C_5H_{12} (liq) 30PAR/HUF
2-Methylbutane; Isopentane
Heat Capacity 275.8 K, C_p = 37.6 cal·mol⁻¹·K⁻¹
157.3 J·mol⁻¹·K⁻¹
Temperature range 80–276 K. Value is unsmoothed experimental datum.
Entropy 298.15 K, S = 60.8 cal·mol⁻¹·K⁻¹
254.4 J·mol⁻¹·K⁻¹
Extrapolation below 90 K, 13.74 cal·mol⁻¹K⁻¹
Phase Changes
c/liq 112.6 K, ΔH = 1222 cal·mol⁻¹
5113 J·mol⁻¹
ΔS = 10.85 cal·mol⁻¹·K⁻¹
45.41 J·mol⁻¹·K⁻¹
Molecular Weight 72.1498
Wiswesser Line Notation 2Y1&1
Evaluation B(C_p)(,C(S)

Table of Heat Capacities, Entropies, and Phase Transition Properties — Continued

C_5H_{12} (liq) 42SCH/AST
2–Methylbutane; Isopentane
Heat Capacity 290 K, C_p = 40.49 cal·mol^{-1}·K^{-1}
169.41 J·mol^{-1}·K^{-1}
Temperature range 20–290 K
Entropy 298.15 K, S = 62.39 cal·mol^{-1}·K^{-1}
261.04 J·mol^{-1}·K^{-1}
Phase Changes
c/liq 113.39 K, ΔH = 1226 cal·mol^{-1}
5130 J·mol^{-1}
ΔS = 10.81 cal·mol^{-1}·K^{-1}
45.24 J·mol^{-1}·K^{-1}
liq/g 293.95 K, ΔH = 5935 cal·mol^{-1}
24832 J·mol^{-1}
ΔS = 20.19 cal·mol^{-1}·K^{-1}
84.48 J·mol^{-1}·K^{-1}
P = 79.15 kPa
Molecular Weight 72.1498
Wiswesser Line Notation 2Y1&1
Evaluation A

C_5H_{12} (liq) 43GUT/HUF
2–Methylbutane; Isopentane
Heat Capacity 298.15 K, C_p = 39.40 cal·mol^{-1}·K^{-1}
164.85 J·mol^{-1}·K^{-1}
Temperature range 13–300 K
Entropy 298.15 K, S = 62.24 cal·mol^{-1}·K^{-1}
260.41 J·mol^{-1}·K^{-1}
Phase Changes
Small second order transition 69–77 K.
c/liq 113.37 K, ΔH = 1232.2 cal·mol^{-1}
5155.5 J·mol^{-1}
ΔS = 10.87 cal·mol^{-1}·K^{-1}
45.48 J·mol^{-1}·K^{-1}
Molecular Weight 72.1498
Wiswesser Line Notation 2Y1&1
Evaluation A

C_5H_{12} (liq) 30PAR/HUF 2
n–Pentane
Heat Capacity 290.0 K, C_p = 39.0 cal·mol^{-1}·K^{-1}
163.2 J·mol^{-1}·K^{-1}
Temperature range 93–290 K. Value is unsmoothed experimental datum.
Entropy 298.15 K, S = 62.0 cal·mol^{-1}·K^{-1}
259.4 J·mol^{-1}·K^{-1}
Extrapolation below 90 K, 13.53 cal·mol^{-1}K^{-1}
Phase Changes
c/liq 143.4 K, ΔH = 2002 cal·mol^{-1}
8376 J·mol^{-1}
ΔS = 13.96 cal·mol^{-1}·K^{-1}
58.41 J·mol^{-1}·K^{-1}
Molecular Weight 72.1498
Wiswesser Line Notation 5H
Evaluation B(C_p),C(S)

C_5H_{12} (liq) 40MES/KEN
n–Pentane
Heat Capacity 290 K, C_p = 40.15 cal·mol^{-1}·K^{-1}
167.99 J·mol^{-1}·K^{-1}
Temperature range 12–290 K
Entropy 298.15 K, S = 62.78 cal·mol^{-1}·K^{-1}
262.67 J·mol^{-1}·K^{-1}
Phase Changes
c/liq 143.46 K, ΔH = 2011.2 cal·mol^{-1}
8414.9 J·mol^{-1}
ΔS = 14.02 cal·mol^{-1}·K^{-1}
58.66 J·mol^{-1}·K^{-1}
liq/g 298.15 K, ΔH = 6262 cal·mol^{-1}
26200 J·mol^{-1}
ΔS = 21.00 cal·mol^{-1}·K^{-1}
87.88 J·mol^{-1}·K^{-1}
P = 68.68 kPa
Molecular Weight 72.1498
Wiswesser Line Notation 5H
Evaluation A

C_5H_{12} (liq) 67MES/GUT
n–Pentane
Heat Capacity 298.15 K, C_p = 39.96 cal·mol^{-1}·K^{-1}
167.19 J·mol^{-1}·K^{-1}
Temperature range 12–300 K
Entropy 298.15 K, S = 62.97 cal·mol^{-1}·K^{-1}
263.47 J·mol^{-1}·K^{-1}
Phase Changes
c/liq 143.47 K, ΔH = 2008 cal·mol^{-1}
8401 J·mol^{-1}
ΔS = 14.00 cal·mol^{-1}·K^{-1}
58.56 J·mol^{-1}·K^{-1}
Molecular Weight 72.1498
Wiswesser Line Notation 5H
Evaluation A

C_5H_{12} (liq) 75GRI/RAS
n–Pentane
Heat Capacity 298 K, C_p = 40.3 cal·mol^{-1}·K^{-1}
168.6 J·mol^{-1}·K^{-1}
Temperature range 300–463 K
Molecular Weight 72.1498
Wiswesser Line Notation 5H
Evaluation B

$C_5H_{12}N_2O_2$ (c) 40HUF/FOX
Ornithine(DL)
Heat Capacity 298.1 K, C_p = 45.73 cal·mol^{-1}·K^{-1}
191.33 J·mol^{-1}·K^{-1}
Temperature range 90–298 K. Value is unsmoothed experimental datum.
Entropy 298.15 K, S = 46.2 cal·mol^{-1}·K^{-1}
193.3 J·mol^{-1}·K^{-1}
Extrapolation below 90 K, 13.16 cal·mol^{-1}K^{-1}
Molecular Weight 132.1620
Wiswesser Line Notation Z3YZVQ -DL
Evaluation C

$C_5H_{12}O$ (liq) 36EVA/EDL
3,3–Dimethyl–2–oxabutane; Methyl tert–butyl ether
Heat Capacity 298 K, C_p = 45 cal·mol^{-1}·K^{-1}
188 J·mol^{-1}·K^{-1}
One temperature
Molecular Weight 88.1492
Wiswesser Line Notation 1X1&1&O1
Evaluation C

Table of Heat Capacities, Entropies, and Phase Transition Properties — Continued

$C_5H_{12}O$ (liq) 75AND/MAR
3,3-Dimethyl-2-oxabutane; Methyl tert-butyl ether
Heat Capacity 298.15 K, C_p = 44.81 cal·mol^{-1}·K^{-1}
187.5 J·mol^{-1}·K^{-1}
Temperature range 12–350 K
Entropy 298.15 K, S = 63.41 cal·mol^{-1}·K^{-1}
265.3 J·mol^{-1}·K^{-1}
Phase Changes
c/liq 164.56 K, ΔH = 1816 cal·mol^{-1}
7600 J·mol^{-1}
ΔS = 11.04 cal·mol^{-1}·K^{-1}
46.18 J·mol^{-1}·K^{-1}
Molecular Weight 88.1492
Wiswesser Line Notation 1X1&1&O1
Evaluation A

$C_5H_{12}O$ (liq) 75AND/MAR
3-Oxahexane; Ethyl n-propyl ether
Heat Capacity 298.15 K, C_p = 47.13 cal·mol^{-1}·K^{-1}
197.2 J·mol^{-1}·K^{-1}
Temperature range 10–350 K
Entropy 298.15 K, S = 70.51 cal·mol^{-1}·K^{-1}
295.0 J·mol^{-1}·K^{-1}
Phase Changes
c/liq 145.65 K, ΔH = 2006 cal·mol^{-1}
8395 J·mol^{-1}
ΔS = 13.77 cal·mol^{-1}·K^{-1}
57.64 J·mol^{-1}·K^{-1}
Molecular Weight 88.1492
Wiswesser Line Notation 3O2
Evaluation A

$C_5H_{12}O$ (liq) 75AND/MAR
2-Oxahexane; Methyl n-butyl ether
Heat Capacity 298.15 K, C_p = 46.06 cal·mol^{-1}·K^{-1}
192.7 J·mol^{-1}·K^{-1}
Temperature range 12–350 K
Entropy 298.15 K, S = 70.58 cal·mol^{-1}·K^{-1}
295.3 J·mol^{-1}·K^{-1}
Phase Changes
c/liq 157.48 K, ΔH = 2593 cal·mol^{-1}
10850 J·mol^{-1}
ΔS = 16.47 cal·mol^{-1}·K^{-1}
68.90 J·mol^{-1}·K^{-1}
Molecular Weight 88.1492
Wiswesser Line Notation 4O1
Evaluation A

$C_5H_{12}O$ (liq) 33PAR/HUF
2,2-Dimethyl-1-propanol; tert-Amyl alcohol
Heat Capacity 294.4 K, C_p = 58.35 cal·mol^{-1}·K^{-1}
244.14 J·mol^{-1}·K^{-1}
Temperature range 92–294 K. Value is unsmoothed experimental datum.
Entropy 298.1 K, S = 54.8 cal·mol^{-1}·K^{-1}
229.3 J·mol^{-1}·K^{-1}
Extrapolation below 90 K, 11.18 cal·mol^{-1}K^{-1}
Phase Changes
c,III/c,II 146.0 K, ΔH = 469 cal·mol^{-1}
1962 J·mol^{-1}
ΔS = 3.21 cal·mol^{-1}·K^{-1}
13.44 J·mol^{-1}·K^{-1}
c,II/c,I 213 K, ΔH = 40 cal·mol^{-1}
167 J·mol^{-1}
ΔS = 0.19 cal·mol^{-1}·K^{-1}
0.79 J·mol^{-1}·K^{-1}
c,I/liq 264.0 K, ΔH = 1065 cal·mol^{-1}
4456 J·mol^{-1}
ΔS = 4.03 cal·mol^{-1}·K^{-1}
16.88 J·mol^{-1}·K^{-1}
Molecular Weight 88.1492
Wiswesser Line Notation Q1X1&1&1
Evaluation B(C_p),C(S)

$C_5H_{12}O$ (liq) 24WIL/DAN
3-Methyl-1-butanol; Isoamyl alcohol
Heat Capacity 303 K, C_p = 50.2 cal·mol^{-1}·K^{-1}
210.0 J·mol^{-1}·K^{-1}
Temperature range 303–343 K, equation only.
Molecular Weight 88.1492
Wiswesser Line Notation Q2Y1&1
Evaluation C

$C_5H_{12}O$ (liq) 45ZHD
3-Methyl-1-butanol; Isoamyl alcohol
Heat Capacity 295.52 K, C_p = 50.08 cal·mol^{-1}·K^{-1}
209.52 J·mol^{-1}·K^{-1}
Temperature range 7 to 47°C. Value is unsmoothed experimental datum.
Molecular Weight 88.1492
Wiswesser Line Notation Q2Y1&1
Evaluation C

$C_5H_{12}O$ (liq) 58SWI/ZIE 2
3-Methyl-1-butanol; Isoamyl alcohol
Heat Capacity 347 K, C_p = 61.6 cal·mol^{-1}·K^{-1}
257.7 J·mol^{-1}·K^{-1}
Mean value 22 to 126°C
Molecular Weight 88.1492
Wiswesser Line Notation Q2Y1&1
Evaluation C

$C_5H_{12}O$ (liq) 81REI
1-Pentanol; n-Amyl alcohol; n-Pentyl alcohol
Heat Capacity 298 K, C_p = 43.8 cal·mol^{-1}·K^{-1}
183.3 J·mol^{-1}·K^{-1}
Temperature range 298–400 K
Molecular Weight 88.1492
Wiswesser Line Notation Q5
Evaluation D

$C_5H_{12}O$ (liq) 33PAR/HUF
1-Pentanol; n-Amyl alcohol; n-Pentyl alcohol
Heat Capacity 298.0 K, C_p = 49.98 cal·mol^{-1}·K^{-1}
209.12 J·mol^{-1}·K^{-1}
Temperature range 94–298 K. Value is unsmoothed experimental datum.
Entropy 298.1 K, S = 60.9 cal·mol^{-1}·K^{-1}
254.8 J·mol^{-1}·K^{-1}
Extrapolation below 90 K, 13.78 cal·mol^{-1}K^{-1}
Phase Changes
c/liq 194.2 K, ΔH = 2349 cal·mol^{-1}
9828 J·mol^{-1}
ΔS = 12.10 cal·mol^{-1}·K^{-1}
50.61 J·mol^{-1}·K^{-1}
Molecular Weight 88.1492
Wiswesser Line Notation Q5
Evaluation B(C_p),C(S)

Table of Heat Capacities, Entropies, and Phase Transition Properties — Continued

$C_5H_{12}O$ (liq) 39PHI
1-Pentanol; n-Amyl alcohol; n-Pentyl alcohol
Heat Capacity 302.4 K, C_p = 48.2 cal·mol^{-1}·K^{-1}
201.7 J·mol^{-1}·K^{-1}
One temperature
Molecular Weight 88.1492
Wiswesser Line Notation Q5
Evaluation C
Isomer not specified; normal assumed.

$C_5H_{12}O$ (liq) 68COU/LEE
1-Pentanol; n-Amyl alcohol; n-Pentyl alcohol
Heat Capacity 298.15 K, C_p = 49.78 cal·mol^{-1}·K^{-1}
208.3 J·mol^{-1}·K^{-1}
Temperature range 10–390 K
Entropy 298.15 K, S = 61.88 cal·mol^{-1}·K^{-1}
258.9 J·mol^{-1}·K^{-1}
Phase Changes
c/liq 195.56 K, ΔH = 2510 cal·mol^{-1}
10502 J·mol^{-1}
ΔS = 12.84 cal·mol^{-1}·K^{-1}
53.70 J·mol^{-1}·K^{-1}
Molecular Weight 88.1492
Wiswesser Line Notation Q5
Evaluation A

$C_5H_{12}O$ (liq) 70PAZ/PAZ
1-Pentanol; n-Amyl alcohol; n-Pentyl alcohol
Heat Capacity 313.2 K, C_p = 57.5 cal·mol^{-1}·K^{-1}
240.6 J·mol^{-1}·K^{-1}
One temperature
Molecular Weight 88.1492
Wiswesser Line Notation Q5
Evaluation B

$C_5H_{12}O$ (liq) 76CON/GIN
3-Pentanol
Heat Capacity 298 K, C_p = 57.3 cal·mol^{-1}·K^{-1}
239.7 J·mol^{-1}·K^{-1}
One temperature
Molecular Weight 88.1492
Wiswesser Line Notation QY2&2
Evaluation B

$C_5H_{12}O_2$ (liq) 73KUS/SUU
2,5-Dioxaheptane; 1-Ethoxy-2-methoxyethane
Heat Capacity 298.15 K, C_p = 53.70 cal·mol^{-1}·K^{-1}
224.7 J·mol^{-1}·K^{-1}
One temperature
Molecular Weight 104.1486
Wiswesser Line Notation 2O2O1
Evaluation B

$C_5H_{12}O_2$ (liq) 73KUS/SUU
3-Oxa-1-hexanol; 2-n-Propoxyethanol
Heat Capacity 298.15 K, C_p = 57.74 cal·mol^{-1}·K^{-1}
241.6 J·mol^{-1}·K^{-1}
One temperature
Molecular Weight 104.1486
Wiswesser Line Notation Q2O3
Evaluation B

$C_5H_{12}O_2$ (liq) 78ROU/PER
3-Oxa-1-hexanol; 2-n-Propoxyethanol
Heat Capacity 298.15 K, C_p = 58.4 cal·mol^{-1}·K^{-1}
244.3 J·mol^{-1}·K^{-1}
One temperature
Molecular Weight 104.1486
Wiswesser Line Notation Q2O3
Evaluation C

$C_5H_{12}O_2$ (liq) 73KUS/SUU
4-Methyl-3-oxa-1-pentanol; 2-Isopropoxyethanol
Heat Capacity 298.15 K, C_p = 57.07 cal·mol^{-1}·K^{-1}
238.8 J·mol^{-1}·K^{-1}
One temperature
Molecular Weight 104.1486
Wiswesser Line Notation Q2OY1&1
Evaluation B

$C_5H_{12}O_2$ (liq) 35MIL
1,5-Pentanediol
Heat Capacity C_p data in thesis only.
Temperature range 100–298 K. Data in thesis only.
Entropy 298.15 K, S = 76.8 cal·mol^{-1}·K^{-1}
321.3 J·mol^{-1}·K^{-1}
Extrapolation below 90 K, 18.76 cal·mol^{-1}K^{-1}
Phase Changes
c/liq 248.0 K, ΔH = 3759 cal·mol^{-1}
15728 J·mol^{-1}
ΔS = 15.2 cal·mol^{-1}·K^{-1}
63.4 J·mol^{-1}·K^{-1}
Molecular Weight 104.1486
Wiswesser Line Notation Q5Q
Evaluation C

$C_5H_{12}O_4$ (c,II) 50NIT/SEK
2,2-Bis(hydroxymethyl)-1,3-dihydroxypropane;
Pentaerythritol
Heat Capacity 373.2 K, C_p = 60.8 cal·mol^{-1}·K^{-1}
254.4 J·mol^{-1}·K^{-1}
Temperature range 373–567 K
Phase Changes
c,II/c,I 460.4 K, ΔH = 10500 cal·mol^{-1}
43930 J·mol^{-1}
ΔS = 22.8 cal·mol^{-1}·K^{-1}
95.4 J·mol^{-1}·K^{-1}
c,I/liq 538.7 K, ΔH = 1700 cal·mol^{-1}
7110 J·mol^{-1}
ΔS = 3.2 cal·mol^{-1}·K^{-1}
13.2 J·mol^{-1}·K^{-1}
Molecular Weight 136.1474
Wiswesser Line Notation Q1X1Q1Q1Q
Evaluation B

$C_5H_{12}O_4$ (c) 59WES
2,2-Bis(hydroxymethyl)-1,3-dihydroxypropane;
Pentaerythritol
Heat Capacity 298.15 K, C_p = 45.51 cal·mol^{-1}·K^{-1}
190.41 J·mol^{-1}·K^{-1}
Based on data 10–350 K to be reported elsewhere.
Entropy 298.15 K, S = 47.34 cal·mol^{-1}·K^{-1}
198.07 J·mol^{-1}·K^{-1}
Molecular Weight 136.1474
Wiswesser Line Notation Q1X1Q1Q1Q
Evaluation B

Table of Heat Capacities, Entropies, and Phase Transition Properties — Continued

$C_5H_{12}S$ (liq) 62SCO/GOO
3,3–Dimethyl–2–thiabutane; Methyl tert–butyl sulfide
Heat Capacity 298.15 K, C_p = 47.79 cal·mol^{-1}·K^{-1}
199.95 J·mol^{-1}·K^{-1}
Temperature range 15–364 K
Entropy 298.15 K, S = 66.00 cal·mol^{-1}·K^{-1}
276.14 J·mol^{-1}·K^{-1}
Phase Changes
c/liq 190.84 K, ΔH = 2011 cal·mol^{-1}
8414 J·mol^{-1}
ΔS = 10.54 cal·mol^{-1}·K^{-1}
44.09 J·mol^{-1}·K^{-1}
Molecular Weight 104.2098
Wiswesser Line Notation 1X1&1&S1
Evaluation A

$C_5H_{12}S$ (liq) 61MCC/FIN
3–Thiahexane; Ethyl n–propyl sulfide
Heat Capacity 298.15 K, C_p = 47.42 cal·mol^{-1}·K^{-1}
198.41 J·mol^{-1}·K^{-1}
Temperature range 11–370 K
Entropy 298.15 K, S = 73.98 cal·mol^{-1}·K^{-1}
309.53 J·mol^{-1}·K^{-1}
Phase Changes
c/liq 156.10 K, ΔH = 2529 cal·mol^{-1}
10581 J·mol^{-1}
ΔS = 16.20 cal·mol^{-1}·K^{-1}
67.78 J·mol^{-1}·K^{-1}
Molecular Weight 104.2098
Wiswesser Line Notation 3S2
Evaluation A

$C_5H_{12}S$ (liq) 61MCC/FIN
2–Thiahexane; n–Butyl methyl sulfide
Heat Capacity 298.15 K, C_p = 48.02 cal·mol^{-1}·K^{-1}
200.92 J·mol^{-1}·K^{-1}
Temperature range 11–370 K
Entropy 298.15 K, S = 73.49 cal·mol^{-1}·K^{-1}
307.48 J·mol^{-1}·K^{-1}
Phase Changes
c/liq 175.30 K, ΔH = 2976 cal·mol^{-1}
12452 J·mol^{-1}
ΔS = 16.98 cal·mol^{-1}·K^{-1}
71.03 J·mol^{-1}·K^{-1}
Molecular Weight 104.2098
Wiswesser Line Notation 4S1
Evaluation A

$C_5H_{12}S$ (gls) 74MES/FIN
3–Methyl–1–butanethiol; Isoamyl mercaptan
Heat Capacity 103 K, C_p = 22.40 cal·mol^{-1}·K^{-1}
93.72 J·mol^{-1}·K^{-1}
Temperature range 10–103 K
Molecular Weight 104.2098
Wiswesser Line Notation SH2Y1&1
Evaluation A

$C_5H_{12}S$ (liq) 74MES/FIN
3–Methyl–1–butanethiol; Isoamyl mercaptan
Heat Capacity 298.15 K, C_p = 47.88 cal·mol^{-1}·K^{-1}
200.33 J·mol^{-1}·K^{-1}
Temperature range 12–370 K
Entropy 298.15 K, S = 71.34 cal·mol^{-1}·K^{-1}
298.49 J·mol^{-1}·K^{-1}
Phase Changes
c/liq 139.635 K, ΔH = 1770 cal·mol^{-1}
7406 J·mol^{-1}
ΔS = 12.68 cal·mol^{-1}·K^{-1}
53.04 J·mol^{-1}·K^{-1}
Molecular Weight 104.2098
Wiswesser Line Notation SH2Y1&1
Evaluation A

$C_5H_{12}S$ (liq) 52FIN/SCO
1–Pentanethiol; n–Amyl mercaptan
Heat Capacity 296.21 K, C_p = 48.08 cal·mol^{-1}·K^{-1}
201.17 J·mol^{-1}·K^{-1}
Temperature range 12–320 K. Value is unsmoothed experimental datum.
Entropy 298.15 K, S = 74.18 cal·mol^{-1}·K^{-1}
310.37 J·mol^{-1}·K^{-1}
Phase Changes
c/liq 197.46 K, ΔH = 4190 cal·mol^{-1}
17531 J·mol^{-1}
ΔS = 21.22 cal·mol^{-1}·K^{-1}
88.78 J·mol^{-1}·K^{-1}
Molecular Weight 104.2098
Wiswesser Line Notation SH5
Evaluation A

$C_5H_{12}S$ (liq) 62SCO/DOU
2–Methyl–2–butanethiol; tert–Amyl mercaptan
Heat Capacity 298.15 K, C_p = 47.36 cal·mol^{-1}·K^{-1}
198.15 J·mol^{-1}·K^{-1}
Temperature range 10–350 K
Entropy 298.15 K, S = 69.34 cal·mol^{-1}·K^{-1}
290.12 J·mol^{-1}·K^{-1}
Phase Changes
c,II/c,I 159.1 K, ΔH = 1907.1 cal·mol^{-1}
7979.3 J·mol^{-1}
ΔS = 11.99 cal·mol^{-1}·K^{-1}
50.15 J·mol^{-1}·K^{-1}
Lambda transition at about 145 K.
c,I/liq 169.3 K, ΔH = 145.4 cal·mol^{-1}
608.4 J·mol^{-1}
ΔS = 0.86 cal·mol^{-1}·K^{-1}
3.59 J·mol^{-1}·K^{-1}
Molecular Weight 104.2098
Wiswesser Line Notation SHX1&1&2
Evaluation A

$C_5H_{12}S$ (liq) 74MES/FIN
2–Methyl–2–butanethiol; tert–Amyl mercaptan
Heat Capacity 298.15 K, C_p = 47.55 cal·mol^{-1}·K^{-1}
198.95 J·mol^{-1}·K^{-1}
Temperature range 10–390 K
Entropy 298.15 K, S = 70.65 cal·mol^{-1}·K^{-1}
295.60 J·mol^{-1}·K^{-1}
Phase Changes
c,II/c,I 144.47 K, ΔH = 1688 cal·mol^{-1}
7063 J·mol^{-1}
ΔS = 11.68 cal·mol^{-1}·K^{-1}
48.89 J·mol^{-1}·K^{-1}
c,I/liq 146.05 K, ΔH = 145.1 cal·mol^{-1}
607.1 J·mol^{-1}
ΔS = 0.99 cal·mol^{-1}·K^{-1}
4.16 J·mol^{-1}·K^{-1}
Molecular Weight 104.2098
Wiswesser Line Notation SHY1&Y1&1
Evaluation A

Table of Heat Capacities, Entropies, and Phase Transition Properties — Continued

$C_5H_{12}S_4$ (c,II) 43BAC/PER

Tetrakis(methylthia)methane

Heat Capacity 307 K, C_p = 35.0 cal·mol^{-1}·K^{-1} 146.4 J·mol^{-1}·K^{-1}

Mean value 23.2–45.5°C. Value for c,III, at 23.2°C = 32.0 cal·mol^{-1}K^{-1}; c,I, between 45.5 and 65.5°C = 47.2 cal·mol^{-1}K^{-1}; liquid at 65.5°C = 51.8 cal·mol^{-1}K^{-1}.

Phase Changes

c,III/c,II 296.4 K, ΔH = 1460 cal·mol^{-1} 6110 J·mol^{-1}

ΔS = 4.9 cal·mol^{-1}·K^{-1} 20.6 J·mol^{-1}·K^{-1}

c,II/c,I 318.7 K, ΔH = 1820 cal·mol^{-1} 7610 J·mol^{-1}

ΔS = 5.7 cal·mol^{-1}·K^{-1} 23.9 J·mol^{-1}·K^{-1}

c,I/liq 338.7 K, ΔH = 990 cal·mol^{-1} 4140 J·mol^{-1}

ΔS = 2.9 cal·mol^{-1}·K^{-1} 12.2 J·mol^{-1}·K^{-1}

Molecular Weight 200.3898

Wiswesser Line Notation 1SXS1&S1&S1

Evaluation C

$C_5H_{13}N$ (liq) 71KON/WAD

1–Aminopentane; n–Amylamine; n–Pentylamine

Heat Capacity 298.15 K, C_p = 52.1 cal·mol^{-1}·K^{-1} 218 J·mol^{-1}·K^{-1}

One temperature

Molecular Weight 87.1644

Wiswesser Line Notation Z5

Evaluation B

$C_5H_{14}ClN$ (c,I) 33SOU/MIL

n–Amylammonium chloride; n–Pentylammonium chloride

Heat Capacity 278.19 K, C_p = 51.00 cal·mol^{-1}·K^{-1} 213.38 J·mol^{-1}·K^{-1}

Temperature range 20–280 K. Value is unsmoothed experimental datum.

Entropy 298.15 K, S = 63.74 cal·mol^{-1}·K^{-1} 266.69 J·mol^{-1}·K^{-1}

Phase Changes

c,III/c,II 221.5 K, ΔH = 283 cal·mol^{-1} 1184 J·mol^{-1}

ΔS = 1.28 cal·mol^{-1}·K^{-1} 5.35 J·mol^{-1}·K^{-1}

c,II/c,I 246.5 K, ΔH = 32 cal·mol^{-1} 134 J·mol^{-1}

ΔS = 0.13 cal·mol^{-1}·K^{-1} 0.54 J·mol^{-1}·K^{-1}

Molecular Weight 123.6253

Wiswesser Line Notation M5 &GH

Evaluation A

$C_5H_{14}Cl_2N_2O_2$ (c) 40HUF/ELL

Ornithine dihydrochloride

Heat Capacity 292.8 K, C_p = 56.97 cal·mol^{-1}·K^{-1} 238.36 J·mol^{-1}·K^{-1}

Temperature range 85–293 K. Value is unsmoothed experimental datum.

Entropy 298.1 K, S = 70.25 cal·mol^{-1}·K^{-1} 293.93 J·mol^{-1}·K^{-1}

Extrapolation below 90 K, 21.31 cal·mol^{-1}K^{-1}

Molecular Weight 205.0838

Wiswesser Line Notation Z3YZVQ &GH 2

Evaluation A(C_p),C(S)

$C_5H_{14}N_2$ (liq) 82DZH/KAR 2

N,N–dimethyl–1,3–propanediamine

Heat Capacity 298.15 K, C_p = 61.1 cal·mol^{-1}·K^{-1} 255.7 J·mol^{-1}·K^{-1}

Temperature range 12–300 K

Entropy 298.15 K, S = 77.2 cal·mol^{-1}·K^{-1} 323.0 J·mol^{-1}·K^{-1}

Phase Changes

c/liq 194.43 K, ΔH = 2960 cal·mol^{-1} 12385 J·mol^{-1}

ΔS = 15.22 cal·mol^{-1}·K^{-1}, 63.70 J·mol^{-1}·K^{-1}

Glass to crystal transition at 135 K shown graphically.

Molecular Weight 130.1924

Wiswesser Line Notation Z3NI&1

Evaluation A

C_6ClF_5 (liq) 68AND/COU 2

Pentafluorochlorobenzene

Heat Capacity 298.15 K, C_p = 52.92 cal·mol^{-1}·K^{-1} 221.4 J·mol^{-1}·K^{-1}

Temperature range 12–395 K

Entropy 298.15 K, S = 71.87 cal·mol^{-1}·K^{-1} 300.7 J·mol^{-1}·K^{-1}

Phase Changes

c,III/c,II 191 K, ΔH = 869 cal·mol^{-1} 3636 J·mol^{-1}

ΔS = 4.55 cal·mol^{-1}·K^{-1} 19.04 J·mol^{-1}·K^{-1}

Entropy change reported as 17.91 J·mol^{-1}K^{-1} from integration of excess heat capacity. Value given assumes isothermal transition.

c,II/c,I 245 K, ΔH = 235 cal·mol^{-1} 983 J·mol^{-1}

ΔS = 0.96 cal·mol^{-1}·K^{-1} 4.01 J·mol^{-1}·K^{-1}

c,I/liq 257.49 K, ΔH = 1997 cal·mol^{-1} 8355 J·mol^{-1}

ΔS = 7.76 cal·mol^{-1}·K^{-1} 32.45 J·mol^{-1}·K^{-1}

Molecular Weight 202.5110

Wiswesser Line Notation GR BF CF DF EF FF

Evaluation A

C_6ClF_5 (liq) 69PAU/GLU 2

Pentafluorochlorobenzene

Heat Capacity 298.15 K, C_p = 53.34 cal·mol^{-1}·K^{-1} 223.17 J·mol^{-1}·K^{-1}

Temperature range 13–303 K

Entropy 298.15 K, S = 72.56 cal·mol^{-1}·K^{-1} 303.59 J·mol^{-1}·K^{-1}

Phase Changes

c,II/c,I 191.2 K, ΔH = 297 cal·mol^{-1} 1243 J·mol^{-1}

ΔS = 1.55 cal·mol^{-1}·K^{-1} 6.50 J·mol^{-1}·K^{-1}

c,I/liq 257.29 K, ΔH = 2007 cal·mol^{-1} 8397 J·mol^{-1}

ΔS = 7.80 cal·mol^{-1}·K^{-1} 32.64 J·mol^{-1}·K^{-1}

Molecular Weight 202.5110

Wiswesser Line Notation GR BF CF DF EF FF

Evaluation A

Table of Heat Capacities, Entropies, and Phase Transition Properties — Continued

$C_6Cl_3F_3$ (c) 69PAU/GLU
1,3,5–Trichloro–2,4,6–trifluorobenzene
Heat Capacity 298.15 K, C_p = 47.31 cal·mol^{-1}·K^{-1}
197.95 J·mol^{-1}·K^{-1}
Temperature range 13–355 K
Entropy 298.15 K, S = 58.64 cal·mol^{-1}·K^{-1}
245.35 J·mol^{-1}·K^{-1}
Phase Changes
c/liq 334.16 K, ΔH = 4744 cal·mol^{-1}
19849 J·mol^{-1}
ΔS = 14.20 cal·mol^{-1}·K^{-1}
59.40 J·mol^{-1}·K^{-1}
Smoothed table gives ΔH = 4712 cal·mol^{-1}, ΔS = 14.12 cal·mol^{-1}K^{-1}.
Molecular Weight 235.4202
Wiswesser Line Notation GR CG EG BF DF FF
Evaluation B

$C_6Cl_3F_3$ (c) 73AND/MAR 2
1,3,5–Trichloro–2,4,6–trifluorobenzene
Heat Capacity 298.15 K, C_p = 47.08 cal·mol^{-1}·K^{-1}
197.0 J·mol^{-1}·K^{-1}
Temperature range 14–347 K
Entropy 298.15 K, S = 58.15 cal·mol^{-1}·K^{-1}
243.3 J·mol^{-1}·K^{-1}
Phase Changes
Anomalous heat capacity between 285–305 K. Enthalpy of transition 18.1 J·mol^{-1} (excess over extrapolated heat capacities).
c/liq 335.01 K, ΔH = 4739 cal·mol^{-1}
19830 J·mol^{-1}
ΔS = 14.15 cal·mol^{-1}·K^{-1}
59.19 J·mol^{-1}·K^{-1}
Molecular Weight 235.4202
Wiswesser Line Notation GR CG EG BF DF FF
Evaluation A

$C_6Cl_4KO_2$ (c,I) 77KOS/SOR
p–Chloranil potassium
Heat Capacity 300 K, C_p = 57.28 cal·mol^{-1}·K^{-1}
239.67 J·mol^{-1}·K^{-1}
Temperature range 14–331 K
Entropy 300 K, S = 75.04 cal·mol^{-1}·K^{-1}
313.97 J·mol^{-1}·K^{-1}
Phase Changes
c,II/c,I 260.01 K, ΔH = 668 cal·mol^{-1}
2796 J·mol^{-1}
ΔS = 2.64 cal·mol^{-1}·K^{-1}
11.06 J·mol^{-1}·K^{-1}
ΔS obtained by summation of experimental C_p/T data.
Molecular Weight 284.9751
Wiswesser Line Notation L6V DVJ BG CG EG FG .KA
Evaluation A

$C_6Cl_4O_2$ (c,I) 73CHI/MAS
Chloranil; Tetrachloro–p–benzoquinone
Heat Capacity 298.15 K, C_p = 46.3 cal·mol^{-1}·K^{-1}
193.8 J·mol^{-1}·K^{-1}
Temperature range 11–300 K
Entropy 298.15 K, S = 61.8 cal·mol^{-1}·K^{-1}
258.4 J·mol^{-1}·K^{-1}
Phase Changes
c,II/c,I 92 K, ΔH = 9 cal·mol^{-1}
38 J·mol^{-1}
ΔS = 0.1 cal·mol^{-1}·K^{-1}
0.4 J·mol^{-1}·K^{-1}
Lambda transition, 70–100 K.
Molecular Weight 245.8768
Wiswesser Line Notation L6V DVJ BG CG EG FG
Evaluation A

C_6Cl_6 (c) 28AND/HAW
Hexachlorobenzene; Perchlorobenzene
Heat Capacity 299.8 K, C_p = 61.6 cal·mol^{-1}·K^{-1}
257.7 J·mol^{-1}·K^{-1}
Temperature range 101–336 K. Value is unsmoothed experimental datum.
Molecular Weight 284.7840
Wiswesser Line Notation GR BG CG DG EG FG
Evaluation C

C_6Cl_6 (c) 58HIL/KRA
Hexachlorobenzene; Perchlorobenzene
Heat Capacity 298.15 K, C_p = 48.11 cal·mol^{-1}·K^{-1}
201.29 J·mol^{-1}·K^{-1}
Temperature range 15–300 K
Entropy 298.15 K, S = 62.20 cal·mol^{-1}·K^{-1}
260.24 J·mol^{-1}·K^{-1}
Molecular Weight 284.7840
Wiswesser Line Notation GR BG CG DG EG FG
Evaluation A

$C_6F_5NO_2$ (liq) 71PAU 2
Pentafluoronitrobenzene
Heat Capacity 298.15 K, C_p = 65.24 cal·mol^{-1}·K^{-1}
272.96 J·mol^{-1}·K^{-1}
Temperature range 12–300 K
Entropy 298.15 K, S = 77.23 cal·mol^{-1}·K^{-1}
323.13 J·mol^{-1}·K^{-1}
Phase Changes
c/liq 250.5 K, ΔH = 2822 cal·mol^{-1}
11807 J·mol^{-1}
ΔS = 11.27 cal·mol^{-1}·K^{-1}
47.13 J·mol^{-1}·K^{-1}
Molecular Weight 213.0635
Wiswesser Line Notation WNR BF CF DF EF FF
Evaluation A

C_6F_6 (liq) 65COU/GRE
Hexafluorobenzene; Perfluorobenzene
Heat Capacity 298.15 K, C_p = 52.96 cal·mol^{-1}·K^{-1}
221.58 J·mol^{-1}·K^{-1}
Temperature range 10–310 K
Entropy 298.15 K, S = 66.90 cal·mol^{-1}·K^{-1}
279.91 J·mol^{-1}·K^{-1}
Phase Changes
c/liq 278.25 K, ΔH = 2770 cal·mol^{-1}
11590 J·mol^{-1}
ΔS = 9.96 cal·mol^{-1}·K^{-1}
41.65 J·mol^{-1}·K^{-1}
Molecular Weight 186.0564
Wiswesser Line Notation FR BF CF DF EF FF
Evaluation A

C_6F_6 (liq) 70MES/FIN
Hexafluorobenzene; Perfluorobenzene
Heat Capacity 298.15 K, C_p = 52.96 cal·mol^{-1}·K^{-1}
221.58 J·mol^{-1}·K^{-1}
Temperature range 13–342 K

Table of Heat Capacities, Entropies, and Phase Transition Properties — Continued

Entropy 298.15 K, S = 67.11 cal·mol^{-1}·K^{-1}
280.79 J·mol^{-1}·K^{-1}
Phase Changes
c/liq 278.30 K, ΔH = 2769 cal·mol^{-1}
11585 J·mol^{-1}
ΔS = 9.95 cal·mol^{-1}·K^{-1}
41.63 J·mol^{-1}·K^{-1}
Molecular Weight 186.0564
Wiswesser Line Notation FR BF CF DF EF FF
Evaluation A

C_6F_6 (liq) 82GOR/GRI
Hexafluorobenzene; Perfluorobenzene
Heat Capacity 300 K, C_p = 53.06 cal·mol^{-1}·K^{-1}
222.0 J·mol^{-1}·K^{-1}
Temperature range 280–353 K. Data also given by equation.
Molecular Weight 186.0564
Wiswesser Line Notation FR BF CF DF EF FF
Evaluation B

C_6F_{14} (liq) 82CAM/REY
n-Perfluorohexane
Heat Capacity C_p data is given graphically only.
Temperature range 4.2–300 K
Phase Changes
c,II/c,I 103 K, ΔH = 231 cal·mol^{-1}
967 J·mol^{-1}
ΔS = 2.39 cal·mol^{-1}·K^{-1}
10.00 J·mol^{-1}·K^{-1}
Transition between 90 and 130 K.
c,,I/liq 185 K, ΔH = 1634 cal·mol^{-1}
6837 J·mol^{-1}
ΔS = 8.8 cal·mol^{-1}·K^{-1}
36.8 J·mol^{-1}·K^{-1}
Molecular Weight 338.0436
Wiswesser Line Notation FXFFXFFXFFXFFXFFXFFF
Evaluation A

$C_6F_{15}N$ (liq) 79ZHO/KOS
Perfluorotriethylamine
Heat Capacity 298.15 K, C_p = 90.70 cal·mol^{-1}·K^{-1}
379.49 J·mol^{-1}·K^{-1}
Temperature range 9–300 K
Entropy 298.15 K, S = 125.98 cal·mol^{-1}·K^{-1}
527.10 J·mol^{-1}·K^{-1}
Phase Changes
c,III/c,II 126.0 K, ΔH = −896.2 cal·mol^{-1}
−3749.7 J·mol^{-1}
Glassy (G type) transition at 108.7 K.
Monotropic transition at 126.0 K with the liberation of heat.
c,II/c,I 146.4 K, ΔH = 373.8 cal·mol^{-1}
1564 J·mol^{-1}
ΔS = 2.55 cal·mol^{-1}·K^{-1}
10.67 J·mol^{-1}·K^{-1}
Enantiotropic transition at 146.4 K.
c,I/liq 156.2 K, ΔH = 1328.9 cal·mol^{-1}
5560.1 J·mol^{-1}
ΔS = 8.51 cal·mol^{-1}
35.61 J·mol^{-1}·K^{-1}
Molecular Weight 371.0487
Wiswesser Line Notation FXFFXFF 3N
Evaluation A

C_6HCl_5O (c) 58HIL/KRA
Pentachlorophenol; Perchlorophenol
Heat Capacity 298.15 K, C_p = 48.27 cal·mol^{-1}·K^{-1}
201.96 J·mol^{-1}·K^{-1}
Temperature range 15–300 K
Entropy 298.15 K, S = 60.51 cal·mol^{-1}·K^{-1}
253.17 J·mol^{-1}·K^{-1}
Molecular Weight 266.3383
Wiswesser Line Notation QR BG CG DG EG FG
Evaluation A

C_6HF_5 (liq) 68COU/HAL
Pentafluorobenzene
Heat Capacity 298.15 K, C_p = 48.92 cal·mol^{-1}·K^{-1}
204.7 J·mol^{-1}·K^{-1}
Temperature range 12–324 K
Entropy 298.15 K, S = 65.94 cal·mol^{-1}·K^{-1}
275.9 J·mol^{-1}·K^{-1}
Phase Changes
c/liq 225.83 K, ΔH = 2594 cal·mol^{-1}
10853 J·mol^{-1}
ΔS = 11.49 cal·mol^{-1}·K^{-1}
48.06 J·mol^{-1}·K^{-1}
Molecular Weight 168.0659
Wiswesser Line Notation FR BF CF DF EF
Evaluation A

C_6HF_5 (liq) 69PAU/LAV
Pentafluorobenzene
Heat Capacity 298.15 K, C_p = 50.38 cal·mol^{-1}·K^{-1}
210.79 J·mol^{-1}·K^{-1}
Temperature range 12–300 K
Entropy 298.15 K, S = 66.80 cal·mol^{-1}·K^{-1}
279.49 J·mol^{-1}·K^{-1}
Phase Changes
c/liq 225.67 K, ΔH = 2601 cal·mol^{-1}
10883 J·mol^{-1}
ΔS = 11.53 cal·mol^{-1}·K^{-1}
48.22 J·mol^{-1}·K^{-1}
Molecular Weight 168.0659
Wiswesser Line Notation FR BF CF DF EF
Evaluation A

C_6HF_5O (c,I) 68AND/COU 2
Pentafluorophenol
Heat Capacity 298.15 K, C_p = 48.11 cal·mol^{-1}·K^{-1}
201.3 J·mol^{-1}·K^{-1}
Temperature range 12–377 K
Entropy 298.15 K, S = 54.28 cal·mol^{-1}·K^{-1}
227.1 J·mol^{-1}·K^{-1}
Phase Changes
c,II/c,I 287 K, ΔH = 277 cal·mol^{-1}
1134 J·mol^{-1}
ΔS = 0.94 cal·mol^{-1}·K^{-1}
3.95 J·mol^{-1}·K^{-1}
c,I/liq 310.62 K, ΔH = 3922 cal·mol^{-1}
16410 J·mol^{-1}
ΔS = 12.63 cal·mol^{-1}·K^{-1}
52.82 J·mol^{-1}·K^{-1}
Molecular Weight 184.0653
Wiswesser Line Notation QR BF CF DF EF FF
Evaluation A

Table of Heat Capacities, Entropies, and Phase Transition Properties — Continued

C_6HF_5O (c,I) 69PAU/LAV 2

Pentafluorophenol

Heat Capacity 298.15 K, C_p = 62.30 cal·mol^{-1}·K^{-1}
260.66 J·mol^{-1}·K^{-1}

Temperature range 12–329 K

Entropy 298.15 K, S = 58.04 cal·mol^{-1}·K^{-1}
242.84 J·mol^{-1}·K^{-1}

Phase Changes

c,II/c,I	248.15 K,	ΔH =	355.0 cal·mol^{-1}
			1485.3 J·mol^{-1}
		ΔS =	1.43 cal·mol^{-1}·K^{-1}
			5.99 J·mol^{-1}·K^{-1}
c/liq	305.18 K,	ΔH =	3070 cal·mol^{-1}
			12845 J·mol^{-1}
		ΔS =	10.06 cal·mol^{-1}·K^{-1}
			42.09 J·mol^{-1}·K^{-1}

Molecular Weight 184.0653
Wiswesser Line Notation QR BF CF DF EF FF
Evaluation A

$C_6H_2Cl_4$ (c) 28AND/HAW

1,2,4,5–Tetrachlorobenzene

Heat Capacity 299.8 K, C_p = 48.4 cal·mol^{-1}·K^{-1}
202.5 J·mol^{-1}·K^{-1}

Temperature range 101–336 K. Value is unsmoothed experimental datum.

Molecular Weight 215.8938
Wiswesser Line Notation GR BG DG EG
Evaluation C

$C_6H_2F_4$ (liq) 73AND/MAR

1,2,3,4–Tetrafluorobenzene

Heat Capacity 298.15 K, C_p = 45.39 cal·mol^{-1}·K^{-1}
189.9 J·mol^{-1}·K^{-1}

Temperature range 10–320 K

Entropy 298.15 K, S = 61.21 cal·mol^{-1}·K^{-1}
256.1 J·mol^{-1}·K^{-1}

Phase Changes

c,II/c,I	221 K,	ΔH =	1170 cal·mol^{-1}
			4900 J·mol^{-1}
		ΔS =	5.3 cal·mol^{-1}·K^{-1}
			22.2 J·mol^{-1}·K^{-1}
c,II/liq	231.25 K,	ΔH =	1477 cal·mol^{-1}
			6180 J·mol^{-1}
		ΔS =	6.4 cal·mol^{-1}·K^{-1}
			26.7 J·mol^{-1}·K^{-1}
c,I/liq	233.26 K,	ΔH =	2612 cal·mol^{-1}
			10930 J·mol^{-1}
		ΔS =	11.2 cal·mol^{-1}·K^{-1}
			46.9 J·mol^{-1}·K^{-1}

Molecular Weight 150.0754
Wiswesser Line Notation FR BF CF DF
Evaluation A

$C_6H_2F_4$ (liq) 73AND/MAR

1,2,3,5–Tetrafluorobenzene

Heat Capacity 298.15 K, C_p = 45.48 cal·mol^{-1}·K^{-1}
190.3 J·mol^{-1}·K^{-1}

Temperature range 10–310 K

Entropy 298.15 K, S = 61.50 cal·mol^{-1}·K^{-1}
257.3 J·mol^{-1}·K^{-1}

Phase Changes

c,II/c,I	224.2 K,	ΔH =	1030 cal·mol^{-1}
			4300 J·mol^{-1}
		ΔS =	4.6 cal·mol^{-1}·K^{-1}
			19.2 J·mol^{-1}·K^{-1}
c,I/liq	226.90 K,	ΔH =	1520 cal·mol^{-1}
			6360 J·mol^{-1}
		ΔS =	6.7 cal·mol^{-1}·K^{-1}
			28.0 J·mol^{-1}·K^{-1}

Molecular Weight 150.0754
Wiswesser Line Notation FR BF CF EF
Evaluation A

$C_6H_2F_4$ (liq) 73AND/MAR

1,2,4,5–Tetrafluorobenzene

Heat Capacity 298.15 K, C_p = 45.94 cal·mol^{-1}·K^{-1}
192.2 J·mol^{-1}·K^{-1}

Temperature range 10–350 K

Entropy 298.15 K, S = 59.85 cal·mol^{-1}·K^{-1}
250.4 J·mol^{-1}·K^{-1}

Phase Changes

c/liq	277.03 K,	ΔH =	3597 cal·mol^{-1}
			15050 J·mol^{-1}
		ΔS =	12.98 cal·mol^{-1}·K^{-1}
			54.33 J·mol^{-1}·K^{-1}

Molecular Weight 150.0754
Wiswesser Line Notation FR BF DF EF
Evaluation A

$C_6H_2F_5N$ (c,I) 69PAU/LAV 3

Pentafluoroaniline

Heat Capacity 298.15 K, C_p = 55.16 cal·mol^{-1}·K^{-1}
230.79 J·mol^{-1}·K^{-1}

Temperature range 12–320 K

Entropy 298.15 K, S = 58.85 cal·mol^{-1}·K^{-1}
246.23 J·mol^{-1}·K^{-1}

Phase Changes

c,II/c,I	287.4 K,	ΔH =	942 cal·mol^{-1}
			3941 J·mol^{-1}
		ΔS =	3.28 cal·mol^{-1}·K^{-1}
			13.71 J·mol^{-1}·K^{-1}
c,I/liq	306.75 K,	ΔH =	3410 cal·mol^{-1}
			14267 J·mol^{-1}
		ΔS =	11.12 cal·mol^{-1}·K^{-1}
			46.51 J·mol^{-1}·K^{-1}

Molecular Weight 183.0805
Wiswesser Line Notation ZR BF CF DF EF FF
Evaluation A

$C_6H_3Cl_3$ (liq) 74PET/TER

1,2,4–Trichlorobenzene

Heat Capacity 297.95 K, C_p = 45 cal·mol^{-1}·K^{-1}
188 J·mol^{-1}·K^{-1}

Temperature range 297–454 K. Value is unsmoothed experimental datum.

Molecular Weight 181.4487
Wiswesser Line Notation GR BG DG
Evaluation B

$C_6H_3Cl_3$ (liq) 82WIL/ING

1,2,4–Trichlorobenzene

Heat Capacity 298.15 K, C_p = 46.50 cal·mol^{-1}·K^{-1}
194.6 J·mol^{-1}·K^{-1}

One temperature

Molecular Weight 181.4487
Wiswesser Line Notation GR BG DG
Evaluation A

Table of Heat Capacities, Entropies, and Phase Transition Properties — Continued

$C_6H_3N_3O_6$ (c) 80RAD/RAD
1,3,5–Trinitrobenzene
Heat Capacity 298.15 K, C_p = 51.3 cal·mol⁻¹·K⁻¹ 214.6 J·mol⁻¹·K⁻¹
Temperature range 180–400 K, Data given graphically; C_p calculated from equation. Thermodynamically stable modification I.
Phase Changes
c,I/liq 398.4 K, ΔH = 3585 cal·mol^{-1} 15000 J·mol^{-1}
ΔS = 9.00 cal·mol^{-1}·K^{-1} 37.7 J·mol^{-1}·K^{-1}
Molecular Weight 181.4487
Wiswesser Line Notation WNR CNW ENW
Evaluation B

$C_6H_3N_3O_6$ (c) 80RAD/RAD
1,3,5–Trinitrobenzene
Heat Capacity 298.15 K, C_p = 53.2 cal·mol^{-1}·K^{-1} 222.7 J·mol^{-1}·K^{-1}
Temperature range 180–380 K. Data given graphically. C_p calculated from equation. Metastable modification II.
Phase Changes
c,II/c,I 370 K, ΔH = 454 cal·mol^{-1} 1900 J·mol^{-1}
ΔS = 1.23 cal·mol^{-1}·K^{-1} 5.13 J·mol^{-1}·K^{-1}
c,I/liq 380.3 K, ΔH = 3537 cal·mol^{-1} 14800 J·mol^{-1}
ΔS = 9.30 cal·mol^{-1}·K^{-1} 38.9 J·mol^{-1}·K^{-1}
Molecular Weight 213.1062
Wiswesser Line Notation WNR CNW ENW
Evaluation B

$C_6H_3N_3O_6$ (c) 80RAD/RAD
1,3,5–Trinitrobenzene
Phase Changes
c,I/liq 383.0 K, ΔH = 3150 cal·mol^{-1} 13200 J·mol^{-1}
ΔS = 8.25 cal·mol^{-1}·K^{-1} 34.5 J·mol^{-1}·K^{-1}
Metastable modification III
Molecular Weight 213.1062
Wiswesser Line Notation WNR CNW ENW
Evaluation B

$C_6H_3N_3O_7$ (c) 24TAY/RIN
2,4,6–Trinitrophenol; Picric acid
Heat Capacity 293 K, C_p = 57.3 cal·mol^{-1}·K^{-1} 223.9 J·mol^{-1}·K^{-1}
Temperature range 90–395 K
Molecular Weight 229.1056
Wiswesser Line Notation WNR BQ CNW ENW
Evaluation C

$C_6H_4Br_2$ (c) 28AND/HAW
1,4–Dibromobenzene; p–Dibromobenzene
Heat Capacity 299.8 K, C_p = 41.7 cal·mol^{-1}·K^{-1} 174.5 J·mol^{-1}·K^{-1}
Temperature range 101–336 K. Value is unsmoothed experimental datum.
Molecular Weight 235.9056
Wiswesser Line Notation ER DE
Evaluation C

$C_6H_4Br_2$ (c) 50UEB/ORT
1,4–Dibromobenzene; p–Dibromobenzene
Heat Capacity 298.15 K, C_p = 34.8 cal·mol^{-1}·K^{-1} 145.6 J·mol^{-1}·K^{-1}
Temperature range 293–368 K. Equation only.
Phase Changes
c/liq 360 K, ΔH = 4790 cal·mol^{-1} 20040 J·mol^{-1}
ΔS = 13.3 cal·mol^{-1}·K^{-1} 55.7 J·mol^{-1}·K^{-1}
Molecular Weight 235.9056
Wiswesser Line Notation ER DE
Evaluation C

$C_6H_4Br_2O$ (liq) **84WER**
2,4–Dibromophenol
Heat Capacity ~313 K, C_p = 62 cal·mol^{-1}·K^{-1} 260 J·mol^{-1}·K^{-1}
Temperature range 18.5–73°C, mean value for supercooled liquid.
Phase Changes
c/liq 313 K, ΔH = 3500 cal·mol^{-1} 15000 J·mol^{-1}
ΔS = 11 cal·mol^{-1}·K^{-1} 46 J·mol^{-1}·K^{-1}
Molecular Weight 252.4990
Wiswesser Line Notation QR BE DE
Evaluation D

$C_6H_4ClNO_2$ (c) 78MAR/CIO
4–Nitrochlorobenzene
Heat Capacity 298 K, C_p = 59.8 cal·mol^{-1}·K^{-1} 250.2 J·mol^{-1}·K^{-1}
Temperature range 298–467 K
Phase Changes
c/liq 354.6 K, ΔH = 2832 cal·mol^{-1} 11850 J·mol^{-1}
ΔS = 8.0 cal·mol^{-1}·K^{-1} 33.4 J·mol^{-1}·K^{-1}
Molecular Weight 157.5561
Wiswesser Line Notation WNR DG
Evaluation C

$C_6H_4Cl_2$ (c) 28AND/HAW
1,4–Dichlorobenzene; p–Dichlorobenzene
Heat Capacity 299.8 K, C_p = 41.1 cal·mol^{-1}·K^{-1} 172.0 J·mol^{-1}·K^{-1}
Temperature range 101–336 K. Value is unsmoothed experimental datum.
Molecular Weight 147.0036
Wiswesser Line Notation GR DG
Evaluation C

$C_6H_4Cl_2$ (c) 50UEB/ORT
1,4–Dichlorobenzene; p–Dichlorobenzene
Heat Capacity 298.15 K, C_p = 34.1 cal·mol^{-1}·K^{-1} 142.7 J·mol^{-1}·K^{-1}
Temperature range 293–368 K. Equation only.
Phase Changes
c/liq 326 K, ΔH = 4340 cal·mol^{-1} 18160 J·mol^{-1}
ΔS = 13.3 cal·mol^{-1}·K^{-1} 55.7 J·mol^{-1}·K^{-1}
Molecular Weight 147.0036
Wiswesser Line Notation GR DG
Evaluation C

Table of Heat Capacities, Entropies, and Phase Transition Properties — Continued

$C_6H_4Cl_2$ (c,II) 76DWO/FIG
1,4–Dichlorobenzene; p–Dichlorobenzene
Heat Capacity 298.15 K, C_p = 35.32 cal·mol^{-1}·K^{-1}
147.76 J·mol^{-1}·K^{-1}
Temperature range 20–330 K
Entropy 298.15 K, S = 41.92 cal·mol^{-1}·K^{-1}
175.41 J·mol^{-1}·K^{-1}
Phase Changes
c,III/c,II 271.77 K, ΔH = 300 cal·mol^{-1}
1256 J·mol^{-1}
ΔS = 1.10 cal·mol^{-1}·K^{-1}
4.62 J·mol^{-1}·K^{-1}
c,II/c,I 304.35 K, ΔH = 51.3 cal·mol^{-1}
214.5 J·mol^{-1}
ΔS = 0.169 cal·mol^{-1}·K^{-1}
0.705 J·mol^{-1}·K^{-1}
c,I/liq 326.15 K, ΔH = 4347 cal·mol^{-1}
18187 J·mol^{-1}
ΔS = 13.33 cal·mol^{-1}·K^{-1}
55.76 J·mol^{-1}·K^{-1}
Molecular Weight 147.0036
Wiswesser Line Notation GR DG
Evaluation A

$C_6H_4F_2$ (liq) 63SCO/MES
1,2–Difluorobenzene
Heat Capacity 298.15 K, C_p = 38.01 cal·mol^{-1}·K^{-1}
159.03 J·mol^{-1}·K^{-1}
Temperature range 14–357 K
Entropy 298.15 K, S = 53.20 cal·mol^{-1}·K^{-1}
222.59 J·mol^{-1}·K^{-1}
Phase Changes
c/liq 226.01 K, ΔH = 2640 cal·mol^{-1}
11046 J·mol^{-1}
ΔS = 11.68 cal·mol^{-1}·K^{-1}
48.87 J·mol^{-1}·K^{-1}
Molecular Weight 114.0944
Wiswesser Line Notation FR BF
Evaluation A

$C_6H_4F_2$ (liq) 70MES/FIN
1,3–Difluorobenzene
Heat Capacity 298.15 K, C_p = 38.03 cal·mol^{-1}·K^{-1}
159.11 J·mol^{-1}·K^{-1}
Temperature range 11–355 K
Entropy 298.15 K, S = 53.50 cal·mol^{-1}·K^{-1}
223.84 J·mol^{-1}·K^{-1}
Phase Changes
c,II/c,I 186.77 K, ΔH = 197.67 cal·mol^{-1}
827.05 J·mol^{-1}
ΔS = 1.058 cal·mol^{-1}·K^{-1}
4.428 J·mol^{-1}·K^{-1}
c,I/liq 204.03 K, ΔH = 2050.9 cal·mol^{-1}
8581.0 J·mol^{-1}
ΔS = 10.05 cal·mol^{-1}·K^{-1}
42.06 J·mol^{-1}·K^{-1}
Molecular Weight 114.0944
Wiswesser Line Notation FR CF
Evaluation A

$C_6H_4I_2$ (c) 50UEB/ORT
1,4–Diiodobenzene
Heat Capacity 298.15 K, C_p = 38.4 cal·mol^{-1}·K^{-1}
160.7 J·mol^{-1}·K^{-1}
Temperature range 293–368 K. Equation only.
Phase Changes
c/liq 402 K, ΔH = 5340 cal·mol^{-1}
22340 J·mol^{-1}
ΔS = 13.3 cal·mol^{-1}·K^{-1}
55.6 J·mol^{-1}·K^{-1}
Molecular Weight 329.9066
Wiswesser Line Notation IR DI
Evaluation C

$C_6H_4NNaO_3{\bullet}2H_2O$ (c) 78MAR/CIO
Sodium p–nitrophenoxide dihydrate
Heat Capacity 310 K, C_p = 69.4 cal·mol^{-1}·K^{-1}
290.4 J·mol^{-1}·K^{-1}
Temperature range 309–393 K
Molecular Weight 197.1227
Wiswesser Line Notation WNR DO .NA &QH 2
Evaluation D

$C_6H_4N_2O_4$ (c) 26AND
1,2–Dinitrobenzene
Heat Capacity 297.9 K, C_p = 44.5 cal·mol^{-1}·K^{-1}
186.2 J·mol^{-1}·K^{-1}
Temperature range 110–344 K. Value is unsmoothed experimental datum.
Molecular Weight 168.1086
Wiswesser Line Notation WNR BNW
Evaluation C

$C_6H_4N_2O_4$ (c) 26AND/LYN
1,2–Dinitrobenzene
Heat Capacity 298 K, C_p = 46.6 cal·mol^{-1}·K^{-1}
195.0 J·mol^{-1}·K^{-1}
Temperature range 22 to 240°C
Phase Changes
c/liq 396.1 K, ΔH = 5460 cal·mol^{-1}
22840 J·mol^{-1}
ΔS = 14.0 cal·mol^{-1}·K^{-1}
58.6 J·mol^{-1}·K^{-1}
Molecular Weight 168.1086
Wiswesser Line Notation WNR BNW
Evaluation C

$C_6H_4N_2O_4$ (c) 50UEB/ORT
1,2–Dinitrobenzene
Heat Capacity 298.15 K, C_p = 47.9 cal·mol^{-1}·K^{-1}
200.4 J·mol^{-1}·K^{-1}
Temperature range 293–368 K. Equation only.
Molecular Weight 168.1086
Wiswesser Line Notation WNR BNW
Evaluation C

$C_6H_4N_2O_4$ (c) 26AND
1,3–Dinitrobenzene
Heat Capacity 297.9 K, C_p = 45.0 cal·mol^{-1}·K^{-1}
188.3 J·mol^{-1}·K^{-1}
Temperature range 110–332 K. Value is unsmoothed experimental datum.
Molecular Weight 168.1086
Wiswesser Line Notation WNR CNW
Evaluation C

Table of Heat Capacities, Entropies, and Phase Transition Properties — Continued

$C_6H_4N_2O_4$ (c) 26AND/LYN
1,3–Dinitrobenzene
Heat Capacity 298 K, C_p = 43.1 cal·mol^{-1}·K^{-1}
180.3 J·mol^{-1}·K^{-1}
Temperature range 22 to 245°C
Phase Changes
c/liq 363.2 K, ΔH = 4150 cal·mol^{-1}
17360 J·mol^{-1}
ΔS = 11.4 cal·mol^{-1}·K^{-1}
47.8 J·mol^{-1}·K^{-1}
Molecular Weight 168.1086
Wiswesser Line Notation WNR CNW
Evaluation C

$C_6H_4N_2O_4$ (c) 50UEB/ORT
1,3–Dinitrobenzene
Heat Capacity 298.15 K, C_p = 47.2 cal·mol^{-1}·K^{-1}
197.5 J·mol^{-1}·K^{-1}
Temperature range 293–368 K. Equation only.
Molecular Weight 168.1086
Wiswesser Line Notation WNR CNW
Evaluation C

$C_6H_4N_2O_4$ (c) 26AND/LYN
1,4–Dinitrobenzene
Heat Capacity 298 K, C_p = 45.9 cal·mol^{-1}·K^{-1}
192.0 J·mol^{-1}·K^{-1}
Temperature range 22 to 210°C
Phase Changes
c/liq 446.7 K, ΔH = 6720 cal·mol^{-1}
28120 J·mol^{-1}
ΔS = 15.0 cal·mol^{-1}·K^{-1}
62.9 J·mol^{-1}·K^{-1}
Molecular Weight 168.1086
Wiswesser Line Notation WNR DNW
Evaluation C

$C_6H_4N_2O_4$ (c) 50UEB/ORT
1,4–Dinitrobenzene
Heat Capacity 298.15 K, C_p = 47.8 cal·mol^{-1}·K^{-1}
200.0 J·mol^{-1}·K^{-1}
Temperature range 293–368 K. Equation only.
Molecular Weight 168.1086
Wiswesser Line Notation WNR DNW
Evaluation C

$C_6H_4N_2O_4$ (c) 64DAV
1,4–Dinitrobenzene
Heat Capacity 325 K, C_p = 40 cal·mol^{-1}·K^{-1}
167 J·mol^{-1}·K^{-1}
Temperature range 298–373 K. Mean value.
Temperature range uncertain.
Molecular Weight 168.1086
Wiswesser Line Notation WNR DNW
Evaluation D

$C_6H_4O_2$ (c) 24LAN
Quinone; p–Benzoquinone
Heat Capacity 291.2 K, C_p = 31.0 cal·mol^{-1}·K^{-1}
129.7 J·mol^{-1}·K^{-1}
Temperature range 22–291 K. Value is unsmoothed experimental datum.
Molecular Weight 108.0964
Wiswesser Line Notation L6V DVJ
Evaluation B

$C_6H_4O_2$ (c) 26AND/LYN
Quinone; p–Benzoquinone
Heat Capacity 298 K, C_p = 31.6 cal·mol^{-1}·K^{-1}
132.2 J·mol^{-1}·K^{-1}
Temperature range 22 to 160°C
Phase Changes
c/liq 386.0 K, ΔH = 4410 cal·mol^{-1}
18450 J·mol^{-1}
ΔS = 11.4 cal·mol^{-1}·K^{-1}
47.8 J·mol^{-1}·K^{-1}
Molecular Weight 108.0964
Wiswesser Line Notation L6V DVJ
Evaluation C

$C_6H_4O_2$ (c) 50UEB/ORT
Quinone; p–Benzoquinone
Heat Capacity 298.15 K, C_p = 30.2 cal·mol^{-1}·K^{-1}
126.4 J·mol^{-1}·K^{-1}
Temperature range 293–368 K. Equation only.
Molecular Weight 108.0964
Wiswesser Line Notation L6V DVJ
Evaluation C

$C_6H_4O_3$ (c) 36PAR/TOD
Phthalic anhydride
Heat Capacity 298.1 K, C_p = 38.67 cal·mol^{-1}·K^{-1}
161.80 J·mol^{-1}·K^{-1}
Temperature range 90–300 K
Entropy 298.1 K, S = 42.9 cal·mol^{-1}·K^{-1}
179.5 J·mol^{-1}·K^{-1}
Extrapolation below 90 K, 13.89 cal·mol^{-1}K^{-1}
Molecular Weight 124.0958
Wiswesser Line Notation T56 BVOVJ
Evaluation B(C_p),C(S)

C_6H_5Br (c) 28AND/HAW
Bromobenzene
Heat Capacity 231.7 K, C_p = 30.5 cal·mol^{-1}·K^{-1}
127.6 J·mol^{-1}·K^{-1}
Temperature range 101–232 K. Value is unsmoothed experimental datum.
Molecular Weight 157.0095
Wiswesser Line Notation ER
Evaluation C

C_6H_5Br (liq) **81REI**
Bromobenzene
Heat Capacity 298 K, C_p = 36.1 cal·mol^{-1}·K^{-1}
151.0 J·mol^{-1}·K^{-1}
Temperature range 291–444 K
Molecular Weight 157.0095
Wiswesser Line Notation BR
Evaluation D

C_6H_5Br (liq) 25WIL/DAN
Bromobenzene
Heat Capacity 293.2 K, C_p = 36.2 cal·mol^{-1}·K^{-1}
151.5 J·mol^{-1}·K^{-1}
Temperature range 20 to 80°C
Molecular Weight 157.0095
Wiswesser Line Notation ER
Evaluation B

Table of Heat Capacities, Entropies, and Phase Transition Properties — Continued

C_6H_5Br (liq) 34KOL/UDO
Bromobenzene
Heat Capacity 302.6 K, C_p = 34.8 cal·mol^{-1}·K^{-1}
145.6 J·mol^{-1}·K^{-1}
One temperature
Molecular Weight 157.0095
Wiswesser Line Notation ER
Evaluation C

C_6H_5Br (liq) 34KOL/UDO 2
Bromobenzene
Heat Capacity 302.6 K, C_p = 34.8 cal·mol^{-1}·K^{-1}
145.6 J·mol^{-1}·K^{-1}
One temperature
Molecular Weight 157.0095
Wiswesser Line Notation ER
Evaluation C

C_6H_5Br (liq) 37STU
Bromobenzene
Heat Capacity 298.1 K, C_p = 37.14 cal·mol^{-1}·K^{-1}
155.39 J·mol^{-1}·K^{-1}
Temperature range 90–320 K
Entropy 298.1 K, S = 49.7 cal·mol^{-1}·K^{-1}
207.9 J·mol^{-1}·K^{-1}
Extrapolation below 91 K, 13.35 cal·mol^{-1}K^{-}1
Phase Changes
c/liq 242.43 K, ΔH = 3540 cal·mol^{-1}
10627 J·mol^{-1}
ΔS = 10.48 cal·mol^{-1}·K^{-1}
43.84 J·mol^{-1}·K^{-1}
Molecular Weight 157.0095
Wiswesser Line Notation ER
Evaluation B(C_p),C(S)

C_6H_5Br (liq) 75MAS/SCO
Bromobenzene
Heat Capacity 298.15 K, C_p = 36.88 cal·mol^{-1}·K^{-1}
154.29 J·mol^{-1}·K^{-1}
Temperature range 11–300 K
Entropy 298.15 K, S = 52.39 cal·mol^{-1}·K^{-1}
21922 J·mol^{-1}·K^{-1}
Phase Changes
c/liq 242.40 K, ΔH = 2557.8 cal·mol^{-1}
10702 J·mol^{-1}
ΔS = 10.55 cal·mol^{-1}·K^{-1}
44.15 J·mol^{-1}·K^{-1}
Molecular Weight 157.0095
Wiswesser Line Notation ER
Evaluation A

C_6H_5BrO (liq) **84WER**
4–Bromophenol
Heat Capacity ~337 K, C_p = 55 cal·mol^{-1}·K^{-1}
230 J·mol^{-1}·K^{-1}
Temperature range 18–77°C, mean value for supercooled liquid.
Phase Changes
c/liq 337 K, ΔH = 3000 cal·mol^{-1}
13000 J·mol^{-1}
ΔS = 9 cal·mol^{-1}·K^{-1}
39 J·mol^{-1}·K^{-1}
Molecular Weight 173.6029
Wiswesser Line Notation QR DE
Evaluation D

C_6H_5BrO (c) **84WER**
4–Bromophenol
Heat Capacity ~300 K, C_p = 46 cal·mol^{-1}·K^{-1}
192 J·mol^{-1}·K^{-1}
Temperature range 13.5–51.5°C, mean value.
Molecular Weight 173.6029
Wiswesser Line Notation QR DE
Evaluation D

C_6H_5Cl (c) 28AND/HAW
Chlorobenzene
Heat Capacity 216.8 K, C_p = 25.4 cal·mol^{-1}·K^{-1}
106.3 J·mol^{-1}·K^{-1}
Temperature range 101–217 K. Value is unsmoothed experimental datum.
Molecular Weight 112.5585
Wiswesser Line Notation GR
Evaluation C

C_6H_5Cl (liq) **81REI**
Chlorobenzene
Heat Capacity 298 K, C_p = 33.7 cal·mol^{-1}·K^{-1}
141.0 J·mol^{-1}·K^{-1}
Temperature range 294–425 K
Molecular Weight 112.5585
Wiswesser Line Notation GR
Evaluation D

C_6H_5Cl (liq) 25WIL/DAN
Chlorobenzene
Heat Capacity 293.2 K, C_p = 34.8 cal·mol^{-1}·K^{-1}
145.6 J·mol^{-1}·K^{-1}
Temperature range 20 to 80°C
Molecular Weight 112.5585
Wiswesser Line Notation GR
Evaluation B

C_6H_5Cl (liq) 37STU
Chlorobenzene
Heat Capacity 298.1 K, C_p = 35.87 cal·mol^{-1}·K^{-1}
150.08 J·mol^{-1}·K^{-1}
Temperature range 90–320 K
Entropy 298.1 K, S = 47.2 cal·mol^{-1}·K^{-1}
197.5 J·mol^{-1}·K^{-1}
Extrapolation below 91 K, 10.52 cal·mol^{-1}K^{-1}
Phase Changes
c/liq 227.89 K, ΔH = 2284 cal·mol^{-1}
9556 J·mol^{-1}
ΔS = 10.02 cal·mol^{-1}·K^{-1}
41.93 J·mol^{-1}·K^{-1}
Molecular Weight 112.5585
Wiswesser Line Notation GR
Evaluation B(C_p),C(S)

C_6H_5Cl (liq) 39PHI
Chlorobenzene
Heat Capacity 305.6 K, C_p = 37.6 cal·mol^{-1}·K^{-1}
157.3 J·mol^{-1}·K^{-1}
One temperature
Molecular Weight 112.5585
Wiswesser Line Notation GR
Evaluation C

Table of Heat Capacities, Entropies, and Phase Transition Properties — Continued

C_6H_5Cl (liq) 71DES/BHA
Chlorobenzene
Heat Capacity 298 K, C_p = 35.3 cal·mol⁻¹·K⁻¹ 147.7 J·mol⁻¹·K⁻¹
Temperature range 298–318 K
Molecular Weight 112.5585
Wiswesser Line Notation GR
Evaluation B

C_6H_5ClO (liq) 16BRA
o–Chlorophenol
Heat Capacity 283 K, C_p = 45.1 cal·mol⁻¹·K⁻¹ 188.7 J·mol⁻¹·K⁻¹
Mean value, 0 to 20°C
Molecular Weight 128.5579
Wiswesser Line Notation QR BG
Evaluation C

$C_6H_5Cl_3Si$ (liq) 65GUM/KOS
Phenyltrichlorosilane
Heat Capacity 298.15 K, C_p = 52.7 cal·mol⁻¹·K⁻¹ 220.7 J·mol⁻¹·K⁻¹
Temperature range 14–289 K
Entropy 298.15 K, S = 75.2 cal·mol⁻¹·K⁻¹ 314.7 J·mol⁻¹·K⁻¹
Phase Changes
c/liq 233.4 K, ΔH = 2787 cal·mol⁻¹ 11660 J·mol⁻¹
ΔS = 11.94 cal·mol⁻¹·K⁻¹ 49.96 J·mol⁻¹·K⁻¹
Smoothed table gives ΔH = 11650 J·mol⁻¹.
Molecular Weight 211.5500
Wiswesser Line Notation G–Si–GGR
Evaluation A

C_6H_5F (liq) 37STU
Fluorobenzene
Heat Capacity 298.1 K, C_p = 35.03 cal·mol⁻¹·K⁻¹ 146.57 J·mol⁻¹·K⁻¹
Temperature range 90–320 K
Entropy 298.1 K, S = 46.6 cal·mol⁻¹·K⁻¹ 195.0 J·mol⁻¹·K⁻¹
Extrapolation below 91 K, 10.17 cal·mol⁻¹K⁻¹
Phase Changes
c/liq 231.10 K, ΔH = 2485 cal·mol⁻¹ 10397 J·mol⁻¹
ΔS = 10.75 cal·mol⁻¹·K⁻¹ 44.99 J·mol⁻¹·K⁻¹
Molecular Weight 96.1039
Wiswesser Line Notation FR
Evaluation B(C_p),C(S)

C_6H_5F (liq) 56SCO/MCC
Fluorobenzene
Heat Capacity 298.15 K, C_p = 34.98 cal·mol⁻¹·K⁻¹ 146.36 J·mol⁻¹·K⁻¹
Temperature range 14–350 K.
Entropy 298.15 K, S = 49.22 cal·mol⁻¹·K⁻¹ 205.94 J·mol⁻¹·K⁻¹
Phase Changes
c/liq 230.94 K, ΔH = 2702 cal·mol⁻¹ 11305 J·mol⁻¹
ΔS = 11.70 cal·mol⁻¹·K⁻¹ 48.95 J·mol⁻¹·K⁻¹
Molecular Weight 96.1039
Wiswesser Line Notation FR
Evaluation A

C_6H_5I (c) 35AOY/KAN
Iodobenzene
Heat Capacity 226.1 K, C_p = 26.8 cal·mol⁻¹·K⁻¹ 112.1 J·mol⁻¹·K⁻¹
Temperature range 81–226 K. Value is unsmoothed experimental datum.
Molecular Weight 204.0100
Wiswesser Line Notation IR
Evaluation B

C_6H_5I (liq) 37STU
Iodobenzene
Heat Capacity 298.1 K, C_p = 37.93 cal·mol⁻¹·K⁻¹ 158.70 J·mol⁻¹·K⁻¹
Temperature range 90–320 K
Entropy 298.1 K, S = 49.1 cal·mol⁻¹·K⁻¹ 205.4 J·mol⁻¹·K⁻¹
Extrapolation below 91 K, 12.70 cal·mol⁻¹K⁻¹
Phase Changes
c/liq 241.83 K, ΔH = 2330 cal·mol⁻¹ 9749 J·mol⁻¹
ΔS = 9.63 cal·mol⁻¹·K⁻¹ 40.31 J·mol⁻¹·K⁻¹
Molecular Weight 204.0100
Wiswesser Line Notation IR
Evaluation B(C_p),C(S)

$C_6H_5NO_2$ (liq) 81REI
Nitrobenzene
Heat Capacity 298 K, C_p = 42.5 cal·mol⁻¹·K⁻¹ 177.8 J·mol⁻¹·K⁻¹
Temperature range 291–486 K
Molecular Weight 123.1110
Wiswesser Line Notation WNR
Evaluation D

$C_6H_5NO_2$ (liq) 02LOU
Nitrobenzene
Heat Capacity 380 K, C_p = 49 cal·mol⁻¹·K⁻¹ 205 J·mol⁻¹·K⁻¹
Mean value 21 to 199°C
Molecular Weight 123.1110
Wiswesser Line Notation WNR
Evaluation D

$C_6H_5NO_2$ (liq) 07WAL
Nitrobenzene
Heat Capacity 293 K, C_p = 48 cal·mol⁻¹·K⁻¹ 201 J·mol⁻¹·K⁻¹
One temperature
Molecular Weight 123.1110
Wiswesser Line Notation WNR
Evaluation D

$C_6H_5NO_2$ (liq) 24WIL/DAN
Nitrobenzene
Heat Capacity 303 K, C_p = 42.4 cal·mol⁻¹·K⁻¹ 177.4 J·mol⁻¹·K⁻¹
Temperature range 303–358 K, equation only.
Molecular Weight 123.1110
Wiswesser Line Notation WNR
Evaluation C

Table of Heat Capacities, Entropies, and Phase Transition Properties — Continued

$C_6H_5NO_2$ (liq) 34PAR/TOD
Nitrobenzene
Heat Capacity 298 K, C_p = 44.63 cal·mol^{-1}·K^{-1}
186.73 J·mol^{-1}·K^{-1}
Temperature range 273–299 K
Molecular Weight 123.1110
Wiswesser Line Notation WNR
Evaluation B

$C_6H_5NO_2$ (liq) 36PAR/TOD
Nitrobenzene
Heat Capacity 298.1 K, C_p = 44.62 cal·mol^{-1}·K^{-1}
186.69 J·mol^{-1}·K^{-1}
Temperature range 90–300 K
Entropy 298.1 K, S = 53.6 cal·mol^{-1}·K^{-1}
224.3 J·mol^{-1}·K^{-1}
Extrapolation below 90 K, 14.85 cal·mol^{-1}K^{-1}
Phase Changes
c/liq 278.8 K, ΔH = 2897 cal·mol^{-1}
12121 J·mol^{-1}
ΔS = 10.39 cal·mol^{-1}·K^{-1}
43.48 J·mol^{-1}·K^{-1}
Molecular Weight 123.1110
Wiswesser Line Notation WNR
Evaluation B(C_p),C(S)

$C_6H_5NO_2$ (liq) 39MAZ
Nitrobenzene
Heat Capacity 293.15 K, C_p = 43.02 cal·mol^{-1}·K^{-1}
179.95 J·mol^{-1}·K^{-1}
Temperature range 5 to 20°C
Molecular Weight 123.1110
Wiswesser Line Notation WNR
Evaluation B

$C_6H_5NO_2$ (liq) 39MAZ 3
Nitrobenzene
Heat Capacity 293 K, C_p = 43.0 cal·mol^{-1}·K^{-1}
179.9 J·mol^{-1}·K^{-1}
Temperature range 5 to 20°C
Molecular Weight 123.1110
Wiswesser Line Notation WNR
Evaluation B

$C_6H_5NO_2$ (liq) 58LUT/PAN
Nitrobenzene
Heat Capacity 335.5 K, C_p = 45.1 cal·mol^{-1}·K^{-1}
188.7 J·mol^{-1}·K^{-1}
Temperature range 62 to 141°C. Value is unsmoothed experimental datum.
Molecular Weight 123.1110
Wiswesser Line Notation WNR
Evaluation B

$C_6H_5NO_2$ (liq) 67PAC
Nitrobenzene
Heat Capacity 303 K, C_p = 42 cal·mol^{-1}·K^{-1}
176 J·mol^{-1}·K^{-1}
One temperature
Phase Changes
c/liq 278.9 K, ΔH = 2585 cal·mol^{-1}
10815 J·mol^{-1}
ΔS = 9.3 cal·mol^{-1}·K^{-1}
38.8 J·mol^{-1}·K^{-1}
Molecular Weight 123.1110
Wiswesser Line Notation WNR
Evaluation C

$C_6H_5NO_2$ (liq) 67RAS/GAN
Nitrobenzene
Heat Capacity 293 K, C_p = 43.1 cal·mol^{-1}·K^{-1}
180.2 J·mol^{-1}·K^{-1}
Temperature range 293–373 K
Molecular Weight 123.1110
Wiswesser Line Notation WNR
Evaluation C

C_6H_6 (c) 25MAA/WAL
Benzene
Heat Capacity 273 K, C_p = 28.3 cal·mol^{-1}·K^{-1}
118.4 J·mol^{-1}·K^{-1}
Temperature range 93–273 K
Phase Changes
c/liq 278.64 K, ΔH = 2390 cal·mol^{-1}
10000 J·mol^{-1}
ΔS = 8.6 cal·mol^{-1}·K^{-1}
35.9 J·mol^{-1}·K^{-1}
Molecular Weight 78.1134
Wiswesser Line Notation R
Evaluation C

C_6H_6 (c) 35AOY/KAN
Benzene
Heat Capacity 223.9 K, C_p = 23.4 cal·mol^{-1}·K^{-1}
97.9 J·mol^{-1}·K^{-1}
Temperature range 82–224 K. Value is unsmoothed experimental datum.
Molecular Weight 78.1134
Wiswesser Line Notation R
Evaluation B

C_6H_6 (c) 37AHL/BLA
Benzene
Heat Capacity 90 K, C_p = 11.44 cal·mol^{-1}·K^{-1}
47.86 J·mol^{-1}·K^{-1}
Temperature range 4–93 K
Entropy 90 K, S = 10.89 cal·mol^{-1}·K^{-1}
45.56 J·mol^{-1}·K^{-1}
Molecular Weight 78.1134
Wiswesser Line Notation R
Evaluation A

C_6H_6 (liq) **81REI**
Benzene
Heat Capacity 298 K, C_p = 31.9 cal·mol^{-1}·K^{-1}
133.5 J·mol^{-1}·K^{-1}
Temperature range 292–364 K
Molecular Weight 78.1134
Wiswesser Line Notation R
Evaluation D

C_6H_6 (liq) 19DEJ
Benzene
Heat Capacity 298 K, C_p = 32.8 cal·mol^{-1}·K^{-1}
137.2 J·mol^{-1}·K^{-1}
Temperature range 24 to 50°C
Molecular Weight 78.1134
Wiswesser Line Notation R
Evaluation B

Table of Heat Capacities, Entropies, and Phase Transition Properties — Continued

C_6H_6 (liq) 24WIL/DAN
Benzene
Heat Capacity 303 K, C_p = 32.0 cal·mol^{-1}·K^{-1}
133.9 J·mol^{-1}·K^{-1}
Temperature range 303–333 K, equation only.
Molecular Weight 78.1134
Wiswesser Line Notation R
Evaluation C

C_6H_6 (liq) 25WIL/DAN
Benzene
Heat Capacity 293.2 K, C_p = 31.8 cal·mol^{-1}·K^{-1}
133.1 J·mol^{-1}·K^{-1}
Temperature range 20 to 60°C
Molecular Weight 78.1134
Wiswesser Line Notation R
Evaluation B

C_6H_6 (liq) 26AND/LYN
Benzene
Heat Capacity 298 K, C_p = 31.6 cal·mol^{-1}·K^{-1}
132.2 J·mol^{-1}·K^{-1}
Temperature range –18 to 110°C
Phase Changes
c/liq 278.55 K, ΔH = 2360 cal·mol^{-1}
9875 J·mol^{-1}
ΔS = 8.5 cal·mol^{-1}·K^{-1}
35.5 J·mol^{-1}·K^{-1}
Molecular Weight 78.1134
Wiswesser Line Notation R
Evaluation C

C_6H_6 (liq) 30HUF/PAR
Benzene
Heat Capacity 300.0 K, C_p = 32.3 cal·mol^{-1}·K^{-1}
135.1 J·mol^{-1}·K^{-1}
Temperature range 93–300 K. Value is unsmoothed experimental datum.
Entropy 298.1 K, S = 41.9 cal·mol^{-1}·K^{-1}
175.3 J·mol^{-1}·K^{-1}
Extrapolation below 90 K, 11.35 cal·mol^{-1}K^{-1}
Phase Changes
c/liq 278.6 K, ΔH = 2343 cal·mol^{-1}
9803 J·mol^{-1}
ΔS = 8.41 cal·mol^{-1}·K^{-1}
35.19 J·mol^{-1}·K^{-1}
Molecular Weight 78.1134
Wiswesser Line Notation R
Evaluation B(C_p),C(S)

C_6H_6 (liq) 31FIO/GIN
Benzene
Heat Capacity 323.15 K, C_p = 34.31 cal·mol^{-1}·K^{-1}
143.57 J·mol^{-1}·K^{-1}
Temperature range 50 to 110°C
Molecular Weight 78.1134
Wiswesser Line Notation R
Evaluation A

C_6H_6 (liq) 32RIC/WAL
Benzene
Heat Capacity 298.1 K, C_p = 32.3 cal·mol^{-1}·K^{-1}
135.1 J·mol^{-1}·K^{-1}
Temperature range 293–333 K
Molecular Weight 78.1134
Wiswesser Line Notation R
Evaluation C

C_6H_6 (liq) 33FER/MIL
Benzene
Heat Capacity 298.15 K, C_p = 31.41 cal·mol^{-1}·K^{-1}
131.4 J·mol^{-1}·K^{-1}
Temperature range 293–323 K, Data calculated from equation.
Molecular Weight 78.1134
Wiswesser Line Notation R
Evaluation B

C_6H_6 (liq) 33KOL/UDO
Benzene
Heat Capacity 287.8 K, C_p = 31.4 cal·mol^{-1}·K^{-1}
131.4 J·mol^{-1}·K^{-1}
One temperature
Molecular Weight 78.1134
Wiswesser Line Notation R
Evaluation C

C_6H_6 (liq) 34KOL/UDO 2
Benzene
Heat Capacity 287.8 K, C_p = 31.4 cal·mol^{-1}·K^{-1}
131.4 J·mol^{-1}·K^{-1}
One temperature.
Molecular Weight 78.1134
Wiswesser Line Notation R
Evaluation C

C_6H_6 (liq) 40BUR
Benzene
Heat Capacity 298.2 K, C_p = 32.37 cal·mol^{-1}·K^{-1}
135.44 J·mol^{-1}·K^{-1}
Temperature range 281–353 K
Molecular Weight 78.1134
Wiswesser Line Notation R
Evaluation B

C_6H_6 (liq) 41ZHD
Benzene
Heat Capacity 298.1 K, C_p = 32.5 cal·mol^{-1}·K^{-1}
136.0 J·mol^{-1}·K^{-1}
Temperature range 8 to 46°C
Molecular Weight 78.1134
Wiswesser Line Notation R
Evaluation C

C_6H_6 (liq) 42ZIE/AND
Benzene
Phase Changes
c/liq 278.65 K, ΔH = 2370 cal·mol^{-1}
9916 J·mol^{-1}
ΔS = 8.51 cal·mol^{-1}·K^{-1}
35.59 J·mol^{-1}·K^{-1}
Molecular Weight 78.1134
Wiswesser Line Notation R
Evaluation B

C_6H_6 (liq) 47KUR
Benzene
Heat Capacity 298 K, C_p = 31.9 cal·mol^{-1}·K^{-1}
133.5 J·mol^{-1}·K^{-1}
Temperature range 9 to 80 °C, mean C_p, five temperatures.
Molecular Weight 78.1134
Wiswesser Line Notation R
Evaluation D

Table of Heat Capacities, Entropies, and Phase Transition Properties — Continued

C_6H_6 (liq) 48OLI/EAT
Benzene
Heat Capacity 298.15 K, C_p = 32.52 cal·mol^{-1}·K^{-1}
136.06 J·mol^{-1}·K^{-1}
Temperature range 13–337 K
Entropy 298.15 K, S = 41.41 cal·mol^{-1}·K^{-1}
173.26 J·mol^{-1}·K^{-1}
Phase Changes
c/liq 278.69 K, ΔH = 2358.1 cal·mol^{-1}
9866.3 J·mol^{-1}
ΔS = 8.46 cal·mol^{-1}·K^{-1}
35.40 J·mol^{-1}·K^{-1}
Molecular Weight 78.1134
Wiswesser Line Notation R
Evaluation A

C_6H_6 (liq) 51SIE/CRU
Benzene
Heat Capacity 293 K, C_p = 31.8 cal·mol^{-1}·K^{-1}
133.1 J·mol^{-1}·K^{-1}
One temperature
Molecular Weight 78.1134
Wiswesser Line Notation R
Evaluation C

C_6H_6 (liq) 55STA/TUP
Benzene
Heat Capacity 298 K, C_p = 32.32 cal·mol^{-1}·K^{-1}
135.23 J·mol^{-1}·K^{-1}
Temperature range 288–347 K
Molecular Weight 78.1134
Wiswesser Line Notation R
Evaluation B

C_6H_6 (liq) 56DUF/EVE
Benzene
Heat Capacity 303 K, C_p = 32.6 cal·mol^{-1}·K^{-1}
136.4 J·mol^{-1}·K^{-1}
Temperature range 303–353 K
Molecular Weight 78.1134
Wiswesser Line Notation R
Evaluation B

C_6H_6 (liq) 60SWI/ZIE
Benzene
Heat Capacity 316 K, C_p = 32.3 cal·mol^{-1}·K^{-1}
135.1 J·mol^{-1}·K^{-1}
Mean value 21 to 66°C
Molecular Weight 78.1134
Wiswesser Line Notation R
Evaluation C

C_6H_6 (liq) 62RAB/NIK
Benzene
Heat Capacity 298 K, C_p = 32.26 cal·mol^{-1}·K^{-1}
134.98 J·mol^{-1}·K^{-1}
Temperature range 10 to 35°C
Molecular Weight 78.1134
Wiswesser Line Notation R
Evaluation B

C_6H_6 (liq) 67PAC
Benzene
Heat Capacity 298 K, C_p = 31 cal·mol^{-1}·K^{-1}
130 J·mol^{-1}·K^{-1}
One temperature
Phase Changes
c/liq 278.8 K, ΔH = 2140 cal·mol^{-1}
8950 J·mol^{-1}
ΔS = 7.7 cal·mol^{-1}·K^{-1}
32.1 J·mol^{-1}·K^{-1}
Molecular Weight 78.1134
Wiswesser Line Notation R
Evaluation C

C_6H_6 (liq) 67RAS/GAN
Benzene
Heat Capacity 293 K, C_p = 32.2 cal·mol^{-1}·K^{-1}
134.6 J·mol^{-1}·K^{-1}
Temperature range 293–353 K
Molecular Weight 78.1134
Wiswesser Line Notation R
Evaluation C

C_6H_6 (liq) 68REC
Benzene
Heat Capacity 298 K, C_p = 32.4 cal·mol^{-1}·K^{-1}
135.4 J·mol^{-1}·K^{-1}
Temperature range 24 to 40 °C, equation only.
Molecular Weight 78.1134
Wiswesser Line Notation R
Evaluation C

C_6H_6 (liq) 71DES/BHA
Benzene
Heat Capacity 298 K, C_p = 32.1 cal·mol^{-1}·K^{-1}
134.3 J·mol^{-1}·K^{-1}
Temperature range 298–318 K
Molecular Weight 78.1134
Wiswesser Line Notation R
Evaluation B

C_6H_6 (liq) 71HYD/SUB
Benzene
Heat Capacity 298.15 K, C_p = 32.48 cal·mol^{-1}·K^{-1}
135.9 J·mol^{-1}·K^{-1}
Temperature range 298; 313 K
Molecular Weight 78.1134
Wiswesser Line Notation R
Evaluation C

C_6H_6 (liq) 76FOR/BEN
Benzene
Heat Capacity 298.15 K, C_p = 32.45 cal·mol^{-1}·K^{-1}
135.76 J·mol^{-1}·K^{-1}
One temperature
Molecular Weight 78.1134
Wiswesser Line Notation R
Evaluation B

C_6H_6 (liq) 82GOR/GRI
Benzene
Heat Capacity 300 K, C_p = 32.62 cal·mol^{-1}·K^{-1}
136.5 J·mol^{-1}·K^{-1}
Temperature range 280–353 K, Data also given by equation.
Molecular Weight 78.1134
Wiswesser Line Notation R
Evaluation A

Table of Heat Capacities, Entropies, and Phase Transition Properties — Continued

C_6H_6 (liq) 82GRO/ING
Benzene
Heat Capacity 298.15 K, C_p = 32.44 cal·mol^{-1}·K^{-1}
135.7 J·mol^{-1}·K^{-1}
Temperature range 298.15 K, One temperature
Molecular Weight 78.1134
Wiswesser Line Notation R
Evaluation A

C_6H_6 (liq) 82TAN
Benzene
Heat Capacity 298.15 K, C_p = 32.44 cal·mol^{-1}·K^{-1}
135.74 J·mol^{-1}·K^{-1}
Temperatures 293.15, 298.15, 303.15 K.
Molecular Weight 78.1134
Wiswesser Line Notation R
Evaluation A

C_6H_6ClN (liq) 65ZAL/KOC
m–Chloroaniline
Heat Capacity 294.7 K, C_p = 47.5 cal·mol^{-1}·K^{-1}
198.7 J·mol^{-1}·K^{-1}
Temperature range 295, 323 K
Molecular Weight 127.5731
Wiswesser Line Notation ZR CG
Evaluation C

C_6H_6ClN (c) 65ZAL/KOC
p–Chloroaniline
Heat Capacity 305 K, C_p = 35.2 cal·mol^{-1}·K^{-1}
147.3 J·mol^{-1}·K^{-1}
Temperature range 292–323 K. Mean value.
Molecular Weight 127.5731
Wiswesser Line Notation ZR DG
Evaluation C

$C_6H_6N_2O_2$ (c) 26AND
2–Nitroaniline
Heat Capacity 297.9 K, C_p = 40.2 cal·mol^{-1}·K^{-1}
168.2 J·mol^{-1}·K^{-1}
Temperature range 110–332 K. Value is unsmoothed experimental datum.
Molecular Weight 138.1256
Wiswesser Line Notation ZR BNW
Evaluation C

$C_6H_6N_2O_2$ (c) 26AND/LYN
2–Nitroaniline
Heat Capacity 298 K, C_p = 39.3 cal·mol^{-1}·K^{-1}
164.4 J·mol^{-1}·K^{-1}
Temperature range 22 to 150°C
Phase Changes
c/liq 342.5 K, ΔH = 3850 cal·mol^{-1}
16110 J·mol^{-1}
ΔS = 11.2 cal·mol^{-1}·K^{-1}
47.0 J·mol^{-1}·K^{-1}
Molecular Weight 138.1256
Wiswesser Line Notation ZR BNW
Evaluation C

$C_6H_6N_2O_2$ (c) 26AND
3–Nitroaniline
Heat Capacity 297.9 K, C_p = 40.0 cal·mol^{-1}·K^{-1}
167.4 J·mol^{-1}·K^{-1}
Temperature range 110–344 K. Value is unsmoothed experimental datum.
Molecular Weight 138.1256
Wiswesser Line Notation ZR CNW
Evaluation C

$C_6H_6N_2O_2$ (c) 26AND/LYN
3–Nitroaniline
Heat Capacity 298 K, C_p = 40.2 cal·mol^{-1}·K^{-1}
168.2 J·mol^{-1}·K^{-1}
Temperature range 22 to 210°C
Phase Changes
c/liq 285.0 K, ΔH = 5660 cal·mol^{-1}
23680 J·mol^{-1}
ΔS = 14.7 cal·mol^{-1}·K^{-1}
61.5 J·mol^{-1}·K^{-1}
Molecular Weight 138.1256
Wiswesser Line Notation ZR CNW
Evaluation C

$C_6H_6N_2O_2$ (c) 41SAT/SOG 3
3–Nitroaniline
Heat Capacity 323 K, C_p = 44.6 cal·mol^{-1}·K^{-1}
186.6 J·mol^{-1}·K^{-1}
Temperature range 0 to 100°C. Mean value.
Molecular Weight 138.1256
Wiswesser Line Notation ZR CNW
Evaluation C
Same data as 40SAT/SOG 4.

$C_6H_6N_2O_2$ (c) 26AND
4–Nitroaniline
Heat Capacity 297.9 K, C_p = 39.6 cal·mol^{-1}·K^{-1}
165.7 J·mol^{-1}·K^{-1}
Temperature range 110–344 K. Value is unsmoothed experimental datum.
Molecular Weight 138.1256
Wiswesser Line Notation ZR DNW
Evaluation C

$C_6H_6N_2O_2$ (c) 26AND/LYN
4–Nitroaniline
Heat Capacity 298 K, C_p = 40.4 cal·mol^{-1}·K^{-1}
169.0 J·mol^{-1}·K^{-1}
Temperature range 22 to 195°C
Phase Changes
c/liq 420.7 K, ΔH = 5040 cal·mol^{-1}
21090 J·mol^{-1}
ΔS = 12.0 cal·mol^{-1}·K^{-1}
50.1 J·mol^{-1}·K^{-1}
Molecular Weight 138.1256
Wiswesser Line Notation ZR DNW
Evaluation C

$C_6H_6N_2O_2$ (c) 41SAT/SOG 3
4–Nitroaniline
Heat Capacity 323 K, C_p = 44.2 cal·mol^{-1}·K^{-1}
184.9 J·mol^{-1}·K^{-1}
Temperature range 0 to 100°C. Mean value.
Molecular Weight 138.1256
Wiswesser Line Notation ZR DNW
Evaluation C
Same data as 40SAT/SOG 4.

C_6H_6O (c) 33PAR/HUF
Phenol; Hydroxybenzene
Heat Capacity 295.8 K, C_p = 31.81 cal·mol^{-1}·K^{-1}
133.09 J·mol^{-1}·K^{-1}
Temperature range 93–296 K. Value is unsmoothed experimental datum.

Table of Heat Capacities, Entropies, and Phase Transition Properties — Continued

Entropy 298.1 K, S = 34.1 cal·mol^{-1}·K^{-1}
142.7 J·mol^{-1}·K^{-1}
Extrapolation below 90 K, 11.72 cal·mol^{-1}K^{-1}
Molecular Weight 94.1128
Wiswesser Line Notation QR
Evaluation B(C_p),C(S)

C_6H_6O (c) 35AOY/KAN
Phenol; Hydroxybenzene
Heat Capacity 229.3 K, C_p = 24.8 cal·mol^{-1}·K^{-1}
103.8 J·mol^{-1}·K^{-1}
Temperature range 78–229 K. Value is unsmoothed experimental datum.
Molecular Weight 94.1128
Wiswesser Line Notation QR
Evaluation B

C_6H_6O (c) 40CAM/CAM
Phenol; Hydroxybenzene
Heat Capacity 293 K, C_p = 22.4 cal·mol^{-1}·K^{-1}
93.7 J·mol^{-1}·K^{-1}
One temperature
Molecular Weight 94.1128
Wiswesser Line Notation QR
Evaluation C

C_6H_6O (c) 63AND/COU
Phenol; Hydroxybenzene
Heat Capacity 298.15 K, C_p = 30.46 cal·mol^{-1}·K^{-1}
127.44 J·mol^{-1}·K^{-1}
Temperature range 13–336 K
Entropy 298.15 K, S = 34.42 cal·mol^{-1}·K^{-1}
144.01 J·mol^{-1}·K^{-1}
Phase Changes
c/liq 314.06 K, ΔH = 2752 cal·mol^{-1}
11514 J·mol^{-1}
ΔS = 8.76 cal·mol^{-1}·K^{-1}
36.66 J·mol^{-1}·K^{-1}
Molecular Weight 94.1128
Wiswesser Line Notation QR
Evaluation A

C_6H_6O (liq) 67RAS/GAN
Phenol; Hydroxybenzene
Heat Capacity 313 K, C_p = 47.8 cal·mol^{-1}·K^{-1}
199.8 J·mol^{-1}·K^{-1}
Temperature range 313–373 K
Molecular Weight 94.1128
Wiswesser Line Notation QR
Evaluation C

C_6H_6O (c) 75NIC/WAD
Phenol; Hydroxybenzene
Heat Capacity 298.15 K, C_p = 30.40 cal·mol^{-1}·K^{-1}
127.21 J·mol^{-1}·K^{-1}
One temperature
Molecular Weight 94.1128
Wiswesser Line Notation QR
Evaluation B

$C_6H_6O_2$ (c) 26AND
1,2–Dihydroxybenzene; Catechol
Heat Capacity 297.9 K, C_p = 33.3 cal·mol^{-1}·K^{-1}
139.3 J·mol^{-1}·K^{-1}
Temperature range 110–344 K. Value is unsmoothed experimental datum.
Molecular Weight 110.1122
Wiswesser Line Notation QR BQ
Evaluation C

$C_6H_6O_2$ (c) 26AND/LYN
1,2–Dihydroxybenzene; Catechol
Heat Capacity 298 K, C_p = 31.6 cal·mol^{-1}·K^{-1}
132.2 J·mol^{-1}·K^{-1}
Temperature range 22 to 200°C
Phase Changes
c/liq 337.5 K, ΔH = 5440 cal·mol^{-1}
22760 J·mol^{-1}
ΔS = 14.4 cal·mol^{-1}·K^{-1}
60.3 J·mol^{-1}·K^{-1}
Molecular Weight 110.1122
Wiswesser Line Notation QR BQ
Evaluation C

$C_6H_6O_2$ (c) 41SAT/SOG 3
1,2–Dihydroxybenzene; Catechol
Heat Capacity 323 K, C_p = 37.5 cal·mol^{-1}·K^{-1}
156.9 J·mol^{-1}·K^{-1}
Temperature range 0 to 100°C. Mean value.
Molecular Weight 110.1122
Wiswesser Line Notation QR BQ
Evaluation C
Same data as 40SAT/SOG 4.

$C_6H_6O_2$ (c) 50UEB/ORT
1,2–Dihydroxybenzene; Catechol
Heat Capacity 298.15 K, C_p = 33.6 cal·mol^{-1}·K^{-1}
140.6 J·mol^{-1}·K^{-1}
Temperature range 293–368 K. Equation only.
Molecular Weight 110.1122
Wiswesser Line Notation QR BQ
Evaluation C

$C_6H_6O_2$ (c) 26AND
1,3–Dihydroxybenzene; Resorcinol
Heat Capacity 297.9 K, C_p = 31.4 cal·mol^{-1}·K^{-1}
131.4 J·mol^{-1}·K^{-1}
Temperature range 110–344 K. Value is unsmoothed experimental datum.
Molecular Weight 110.1122
Wiswesser Line Notation QR CQ
Evaluation C

$C_6H_6O_2$ (c) 26AND/LYN
1,3–Dihydroxybenzene; Resorcinol
Heat Capacity 298 K, C_p = 31.3 cal·mol^{-1}·K^{-1}
131.0 J·mol^{-1}·K^{-1}
Temperature range 22 to 200°C
Phase Changes
c/liq 382.8 K, ΔH = 5090 cal·mol^{-1}
21300 J·mol^{-1}
ΔS = 13.3 cal·mol^{-1}·K^{-1}
55.6 J·mol^{-1}·K^{-1}
Molecular Weight 110.1122
Wiswesser Line Notation QR CQ
Evaluation C

$C_6H_6O_2$ (c) 41SAT/SOG 3
1,3–Dihydroxybenzene; Resorcinol
Heat Capacity 323 K, C_p = 36.1 cal·mol^{-1}·K^{-1}
151.0 J·mol^{-1}·K^{-1}
Temperature range 0 to 100°C. Mean value.
Molecular Weight 110.1122
Wiswesser Line Notation QR CQ
Evaluation C
Same data as 40SAT/SOG 4.

Table of Heat Capacities, Entropies, and Phase Transition Properties — Continued

$C_6H_6O_2$ (c) 50UEB/ORT
1,3–Dihydroxybenzene; Resorcinol
Heat Capacity 298.15 K, C_p = 33.3 $cal{\cdot}mol^{-1}{\cdot}K^{-1}$
139.3 $J{\cdot}mol^{-1}{\cdot}K^{-1}$
Temperature range 293–368 K. Equation only.
Molecular Weight 110.1122
Wiswesser Line Notation QR CQ
Evaluation C

$C_6H_6O_2$ (c) 24LAN
1,4–Dihydroxybenzene; Hydroquinone
Heat Capacity 274.3 K, C_p = 31.2 $cal{\cdot}mol^{-1}{\cdot}K^{-1}$
130.5 $J{\cdot}mol^{-1}{\cdot}K^{-1}$
Temperature range 28–275 K. Value is unsmoothed experimental datum.
Molecular Weight 110.1122
Wiswesser Line Notation QR DQ
Evaluation B

$C_6H_6O_2$ (c) 26AND
1,4–Dihydroxybenzene; Hydroquinone
Heat Capacity 297.9 K, C_p = 31.9 $cal{\cdot}mol^{-1}{\cdot}K^{-1}$
133.5 $J{\cdot}mol^{-1}{\cdot}K^{-1}$
Temperature range 110–344 K. Value is unsmoothed experimental datum.
Molecular Weight 110.1122
Wiswesser Line Notation QR DQ
Evaluation C

$C_6H_6O_2$ (c) 26AND/LYN
1,4–Dihydroxybenzene; Hydroquinone
Heat Capacity 298 K, C_p = 33.4 $cal{\cdot}mol^{-1}{\cdot}K^{-1}$
139.7 $J{\cdot}mol^{-1}{\cdot}K^{-1}$
Temperature range 22 to 200°C
Phase Changes
c/liq 445.5 K, ΔH = 6480 $cal{\cdot}mol^{-1}$
27110 $J{\cdot}mol^{-1}$
ΔS = 14.5 $cal{\cdot}mol^{-1}{\cdot}K^{-1}$
60.9 $J{\cdot}mol^{-1}{\cdot}K^{-1}$
Molecular Weight 110.1122
Wiswesser Line Notation QR DQ
Evaluation C

$C_6H_6O_2$ (c) 41SAT/SOG 3
1,4–Dihydroxybenzene; Hydroquinone
Heat Capacity 323 K, C_p = 35.9 $cal{\cdot}mol^{-1}{\cdot}K^{-1}$
150.2 $J{\cdot}mol^{-1}{\cdot}K^{-1}$
Temperature range 0 to 100°C. Mean value.
Molecular Weight 110.1122
Wiswesser Line Notation QR DQ
Evaluation C
Same data as 40SAT/SOG 4.

$C_6H_6O_2$ (c) 50UEB/ORT
1,4–Dihydroxybenzene; Hydroquinone
Heat Capacity 298.15 K, C_p = 32.6 $cal{\cdot}mol^{-1}{\cdot}K^{-1}$
136.4 $J{\cdot}mol^{-1}{\cdot}K^{-1}$
Temperature range 293–368 K. Equation only.
Molecular Weight 110.1122
Wiswesser Line Notation QR DQ
Evaluation C

C_6H_6S (liq) 36PAR/TOD
Thiophenol; Phenyl mercaptan
Heat Capacity 298.1 K, C_p = 42.20 $cal{\cdot}mol^{-1}{\cdot}K^{-1}$
176.56 $J{\cdot}mol^{-1}{\cdot}K^{-1}$
Temperature range 90–300 K
Entropy 298.1 K, S = 52.6 $cal{\cdot}mol^{-1}{\cdot}K^{-1}$
220.1 $J{\cdot}mol^{-1}{\cdot}K^{-1}$
Extrapolation below 90 K, 13.07 $cal{\cdot}mol^{-1}K^{-1}$
Phase Changes
c/liq 258.2 K, ΔH = 2743 $cal{\cdot}mol^{-1}$
11478 $J{\cdot}mol^{-1}$
ΔS = 10.62 $cal{\cdot}mol^{-1}{\cdot}K^{-1}$
44.45 $J{\cdot}mol^{-1}{\cdot}K^{-1}$
Molecular Weight 110.1734
Wiswesser Line Notation SHR
Evaluation B(C_p),C(S)

C_6H_6S (liq) 56SCO/MCC 2
Thiophenol; Phenyl mercaptan
Heat Capacity 298.15 K, C_p = 41.40 $cal{\cdot}mol^{-1}{\cdot}K^{-1}$
173.22 $J{\cdot}mol^{-1}{\cdot}K^{-1}$
Temperature range 10–380 K
Entropy 298.15 K, S = 53.25 $cal{\cdot}mol^{-1}{\cdot}K^{-1}$
222.80 $J{\cdot}mol^{-1}{\cdot}K^{-1}$
Phase Changes
Lambda transition of about 40 $cal{\cdot}mol^{-1}$ excess enthalpy at 110–135 K.
c/liq 258.27 K, ΔH = 2736 $cal{\cdot}mol^{-1}$
11447 $J{\cdot}mol^{-1}$
ΔS = 10.59 $cal{\cdot}mol^{-1}{\cdot}K^{-1}$
44.32 $J{\cdot}mol^{-1}{\cdot}K^{-1}$
Molecular Weight 110.1734
Wiswesser Line Notation SHR
Evaluation A

C_6H_7N (liq) 71HAL/BAL
1–Bicyclo[2.1.0]pentyl cyanide;
1–Cyanobicyclo[2.1.0]pentane
Heat Capacity 297 K, C_p = 39.2 $cal{\cdot}mol^{-1}{\cdot}K^{-1}$
163.9 $J{\cdot}mol^{-1}{\cdot}K^{-1}$
One temperature
Molecular Weight 93.1280
Wiswesser Line Notation L34TJ ACN
Evaluation C

C_6H_7N (liq) 71HAL/BAL
3–Methylenecyclobutyl cyanide;
1–Cyano–3–methylenecyclobutane
Heat Capacity 297 K, C_p = 45.6 $cal{\cdot}mol^{-1}{\cdot}K^{-1}$
190.9 $J{\cdot}mol^{-1}{\cdot}K^{-1}$
One temperature
Molecular Weight 93.1280
Wiswesser Line Notation L4YTJ AU1 CCN
Evaluation C

C_6H_7N (liq) 63SCO/HUB
2–Methylpyridine; α–Picoline
Heat Capacity 298.15 K, C_p = 37.86 $cal{\cdot}mol^{-1}{\cdot}K^{-1}$
158.41 $J{\cdot}mol^{-1}{\cdot}K^{-1}$
Temperature range 12–370 K
Entropy 298.15 K, S = 52.07 $cal{\cdot}mol^{-1}{\cdot}K^{-1}$
217.86 $J{\cdot}mol^{-1}{\cdot}K^{-1}$
Phase Changes
c/liq 206.45 K, ΔH = 2324.1 $cal{\cdot}mol^{-1}$
9724.0 $J{\cdot}mol^{-1}$
ΔS = 11.26 $cal{\cdot}mol^{-1}{\cdot}K^{-1}$
47.10 $J{\cdot}mol^{-1}{\cdot}K^{-1}$
Molecular Weight 93.1280
Wiswesser Line Notation T6NJ B1
Evaluation A

Table of Heat Capacities, Entropies, and Phase Transition Properties — Continued

C_6H_7N (liq) 63SCO/GOO
3-Methylpyridine; β-Picoline
Heat Capacity 298.15 K, C_p = 37.93 cal·mol^{-1}·K^{-1}
158.69 J·mol^{-1}·K^{-1}
Temperature range 12–400 K
Entropy 298.15 K, S = 51.70 cal·mol^{-1}·K^{-1}
216.31 J·mol^{-1}·K^{-1}
Phase Changes
c/liq 255.01 K, ΔH = 3389 cal·mol^{-1}
14180 J·mol^{-1}
ΔS = 13.29 cal·mol^{-1}·K^{-1}
55.61 J·mol^{-1}·K^{-1}
Molecular Weight 93.1280
Wiswesser Line Notation T6NJ C1
Evaluation A

C_6H_7N (liq) 81REI
Aniline
Heat Capacity 298 K, C_p = 46.0 cal·mol^{-1}·K^{-1}
192.5 J·mol^{-1}·K^{-1}
Temperature range 290–465 K
Molecular Weight 93.1280
Wiswesser Line Notation ZR
Evaluation D

C_6H_7N (liq) 02LOU
Aniline
Heat Capacity 370 K, C_p = 51 cal·mol^{-1}·K^{-1}
213 J·mol^{-1}·K^{-1}
Mean value 20 to 176°C
Molecular Weight 93.1280
Wiswesser Line Notation ZR
Evaluation D

C_6H_7N (liq) 28LAN
Aniline
Heat Capacity 298.2 K, C_p = 46.22 cal·mol^{-1}·K^{-1}
193.38 J·mol^{-1}·K^{-1}
Temperature range 5 to 60°C
Molecular Weight 93.1280
Wiswesser Line Notation ZR
Evaluation B

C_6H_7N (liq) 33FER/MIL
Aniline
Heat Capacity 298.15 K, C_p = 42.74 cal·mol^{-1}·K^{-1}
178.8 J·mol^{-1}·K^{-1}
Temperature range 293–323 K, Data calculated from equation.
Molecular Weight 93.1280
Wiswesser Line Notation ZR
Evaluation B

C_6H_7N (liq) 33PAR/HUF
Aniline
Heat Capacity 298.2 K, C_p = 45.63 cal·mol^{-1}·K^{-1}
190.92 J·mol^{-1}·K^{-1}
Temperature range 94–298 K. Value is unsmoothed experimental datum.
Entropy 298.1 K, S = 45.8 cal·mol^{-1}·K^{-1}
191.6 J·mol^{-1}·K^{-1}
Extrapolation below 90 K, 10.82 cal·mol^{-1}K^{-1}
Phase Changes
c/liq 266.8 K, ΔH = 2523 cal·mol^{-1}
10556 J·mol^{-1}
ΔS = 9.46 cal·mol^{-1}·K^{-1}
39.57 J·mol^{-1}·K^{-1}
Molecular Weight 93.1280
Wiswesser Line Notation ZR
Evaluation B(C_p),C(S)

C_6H_7N (liq) 34RAD/JUL
Aniline
Heat Capacity 288 K, C_p = 43.9 cal·mol^{-1}·K^{-1}
183.7 J·mol^{-1}·K^{-1}
One temperature
Molecular Weight 93.1280
Wiswesser Line Notation ZR
Evaluation C

C_6H_7N (liq) 42ZIE/AND
Aniline
Phase Changes
c/liq 267.3 K, ΔH = 2610 cal·mol^{-1}
10920 J·mol^{-1}
ΔS = 9.76 cal·mol^{-1}·K^{-1}
40.84 J·mol^{-1}·K^{-1}
Molecular Weight 93.1280
Wiswesser Line Notation ZR
Evaluation B

C_6H_7N (liq) 50HOU/MAS
Aniline
Heat Capacity 323 K, C_p = 47.2 cal·mol^{-1}·K^{-1}
197.5 J·mol^{-1}·K^{-1}
Temperature range 323–453 K
Molecular Weight 93.1280
Wiswesser Line Notation ZR
Evaluation B

C_6H_7N (liq) 57CRU/JOS
Aniline
Heat Capacity 293 K, C_p = 45.9 cal·mol^{-1}·K^{-1}
192.0 J·mol^{-1}·K^{-1}
One temperature
Molecular Weight 93.1280
Wiswesser Line Notation ZR
Evaluation B

C_6H_7N (liq) 71DES/BHA
Aniline
Heat Capacity 298 K, C_p = 46.3 cal·mol^{-1}·K^{-1}
193.7 J·mol^{-1}·K^{-1}
Temperature range 298–318 K
Molecular Weight 93.1280
Wiswesser Line Notation ZR
Evaluation B

C_6H_7N (liq) 75NIC/WAD
Aniline
Heat Capacity 298.15 K, C_p = 45.65 cal·mol^{-1}·K^{-1}
191.01 J·mol^{-1}·K^{-1}
One temperature
Molecular Weight 93.1280
Wiswesser Line Notation ZR
Evaluation B

Table of Heat Capacities, Entropies, and Phase Transition Properties — Continued

$C_6H_7NO_2S$ (c) 41SAT/SOG 2
Benzenesulfonamide
Heat Capacity 323 K, C_p = 46.2 cal·mol^{-1}·K^{-1}
193.3 J·mol^{-1}·K^{-1}
Temperature range 0 to 100°C. Mean value.
Molecular Weight 157.1868
Wiswesser Line Notation ZSWR
Evaluation C
Same data as 40SAT/SOG3.

C_6H_8 (liq) 76GEI/WOL
1,3–Cyclohexadiene
Heat Capacity 298.15 K, C_p = 34.55 cal·mol^{-1}·K^{-1}
144.56 J·mol^{-1}·K^{-1}
Temperature range 10–300 K
Entropy 298.15 K, S = 47.15 cal·mol^{-1}·K^{-1}
197.28 J·mol^{-1}·K^{-1}
Phase Changes
c/liq 161.0 K, ΔH = 1004.9 cal·mol^{-1}
4204.5 J·mol^{-1}
ΔS = 6.24 cal·mol^{-1}·K^{-1}
26.11 J·mol^{-1}·K^{-1}
Molecular Weight 80.1292
Wiswesser Line Notation L6U CUTJ
Evaluation B

C_6H_8 (liq) 76GEI/WOL
1,4–Cyclohexadiene
Heat Capacity 298.15 K, C_p = 34.88 cal·mol^{-1}·K^{-1}
145.94 J·mol^{-1}·K^{-1}
Temperature range 10–300 K
Entropy 298.15 K, S = 45.26 cal·mol^{-1}·K^{-1}
189.37 J·mol^{-1}·K^{-1}
Phase Changes
c,II/c,I 192.0 K, ΔH = 195 cal·mol^{-1}
816 J·mol^{-1}
ΔS = 1.02 cal·mol^{-1}·K^{-1}
4.25 J·mol^{-1}·K^{-1}
c,I/liq 224.0 K, ΔH = 1366.0 cal·mol^{-1}
5715.3 J·mol^{-1}
ΔS = 6.10 cal·mol^{-1}·K^{-1}
25.51 J·mol^{-1}·K^{-1}
Molecular Weight 80.1292
Wiswesser Line Notation L6U DUTJ
Evaluation B

C_6H_8BrN (c,I) 51SUG
Aniline hydrobromide
Heat Capacity 298.75 K, C_p = 38.15 cal·mol^{-1}·K^{-1}
159.62 J·mol^{-1}·K^{-1}
Temperature range –74 to 67°C. Value is unsmoothed experimental datum.
Phase Changes
c,II/c,I 293 K, ΔH = 297 cal·mol^{-1}
1243 J·mol^{-1}
ΔS = 1.10 cal·mol^{-1}·K^{-1}
4.60 J·mol^{-1}·K^{-1}
Values obtained by summing excess specific heat between –20°C and transition temperature.
Molecular Weight 174.0399
Wiswesser Line Notation ZR &EH
Evaluation B

$C_6H_8N_2O_2S$ (c) 41SAT/SOG 2
p–Aminobenzenesulfonamide
Heat Capacity 323 K, C_p = 52.8 cal·mol^{-1}·K^{-1}
220.9 J·mol^{-1}·K^{-1}
Temperature range 0 to 100°C. Mean value.
Molecular Weight 172.2014
Wiswesser Line Notation ZSWR DZ
Evaluation C
Same data as 40SAT/SOG 3.

$C_6H_8O_2$ (liq) 71HAL/BAL
Methyl bicyclobutane–1–carboxylate
Heat Capacity 297 K, C_p = 46.1 cal·mol^{-1}·K^{-1}
192.9 J·mol^{-1}·K^{-1}
One temperature
Molecular Weight 112.1280
Wiswesser Line Notation L33TJ AVO1
Evaluation C

$C_6H_8O_4$ (liq) 30WAS
Dimethyl maleate
Heat Capacity 298 K, C_p = 62.9 cal·mol^{-1}·K^{-1}
263.2 J·mol^{-1}·K^{-1}
Temperature range 0 to 99°C.
Phase Changes
c/liq 254 K, ΔH = 3500 cal·mol^{-1}
14640 J·mol^{-1}
ΔS = 14 cal·mol^{-1}·K^{-1}
58 J·mol^{-1}·K^{-1}
Molecular Weight 144.1268
Wiswesser Line Notation 1OV1U1VO1 –C
Evaluation C

$C_6H_8O_4$ (c) 30WAS
Dimethyl fumarate
Heat Capacity 298 K, C_p = 47.6 cal·mol^{-1}·K^{-1}
199.2 J·mol^{-1}·K^{-1}
Temperature range –14 to 99 °C
Phase Changes
c/liq 375 K, ΔH = 8400 cal·mol^{-1}
35150 J·mol^{-1}
ΔS = 22 cal·mol^{-1}·K^{-1}
94 J·mol^{-1}·K^{-1}
Molecular Weight 144.1268
Wiswesser Line Notation 1OV1U1VO1 –T
Evaluation C

$C_6H_8O_7 \cdot H_2O$ (c) 62EVA/HOA
Citric acid monohydrate
Heat Capacity 298.15 K, C_p = 64.1 cal·mol^{-1}·K^{-1}
268.1 J·mol^{-1}·K^{-1}
Temperature range 21–303 K
Entropy 298.15 K, S = 67.74 cal·mol^{-1}·K^{-1}
283.5 J·mol^{-1}·K^{-1}
Molecular Weight 210.1402
Wiswesser Line Notation QV1XQVQ1VQ &QH
Evaluation A

C_6H_8S (liq) 65CAR/WES
2,5–Dimethylthiophene
Heat Capacity 298.15 K, C_p = 42.62 cal·mol^{-1}·K^{-1}
178.32 J·mol^{-1}·K^{-1}
Temperature range 5–305 K
Entropy 298.15 K, S = 58.49 cal·mol^{-1}·K^{-1}
244.72 J·mol^{-1}·K^{-1}

Table of Heat Capacities, Entropies, and Phase Transition Properties — Continued

Phase Changes
c,I/liq 210.58 K, ΔH = 1957.7 cal·mol^{-1}
8191.0 J·mol^{-1}
ΔS = 9.30 cal·mol^{-1}·K^{-1}
38.90 J·mol^{-1}·K^{-1}
c,II/liq 204.87 K, ΔH = 1768.6 cal·mol^{-1}
7401.1 J·mol^{-1}
ΔS = 8.63 cal·mol^{-1}·K^{-1}
36.12 J·mol^{-1}·K^{-1}
Metastable crystals. Note error in table on ΔH_m and T_m.
Molecular Weight 112.1892
Wiswesser Line Notation T5SJ B1 E1
Evaluation A

C_6H_9N (liq) 71HAL/BAL
Cyclopentyl cyanide; Cyanocyclopentane
Heat Capacity 297 K, C_p = 40.0 cal·mol^{-1}·K^{-1}
167.5 J·mol^{-1}·K^{-1}
One temperature
Molecular Weight 95.1438
Wiswesser Line Notation L5TJ ACN
Evaluation C

C_6H_{10} (liq) 81REI
1,5–Hexadiene; Diallyl
Heat Capacity 298 K, C_p = 31.8 cal·mol^{-1}·K^{-1}
133.1 J·mol^{-1}·K^{-1}
Temperature range 291–328 K
Molecular Weight 82.1450
Wiswesser Line Notation 1U4U1
Evaluation D

C_6H_{10} (liq) 79PUC/PEA
1–Methylcyclopentene
Heat Capacity 298.15 K, C_p = 36.6 cal·mol^{-1}·K^{-1}
153.1 J·mol^{-1}·K^{-1}
One temperature
Molecular Weight 82.1450
Wiswesser Line Notation L5UTJ A1
Evaluation B

C_6H_{10} (liq) 79PUC/PEA
3–Methylcyclopentene
Heat Capacity 298.15 K, C_p = 36.4 cal·mol^{-1}·K^{-1}
152.3 J·mol^{-1}·K^{-1}
One temperature
Molecular Weight 82.1450
Wiswesser Line Notation L5UTJ C1
Evaluation B

C_6H_{10} (liq) 30PAR/HUF 2
Cyclohexene
Heat Capacity 293.2 K, C_p = 34.8 cal·mol^{-1}·K^{-1}
145.6 J·mol^{-1}·K^{-1}
Temperature range 92–293 K. Value is unsmoothed experimental datum.
Entropy 298.15 K, S = 51.8 cal·mol^{-1}·K^{-1}
216.7 J·mol^{-1}·K^{-1}
Extrapolation below 90 K, 11.76 cal·mol^{-1}K^{-1}
Phase Changes
c,II/c,I 138.7 K, ΔH = 974 cal·mol^{-1}
4075 J·mol^{-1}
ΔS = 7.02 cal·mol^{-1}·K^{-1}
29.38 J·mol^{-1}·K^{-1}
c,I/liq 169.0 K, ΔH = 786 cal·mol^{-1}
3289 J·mol^{-1}
ΔS = 4.65 cal·mol^{-1}·K^{-1}
19.46 J·mol^{-1}·K^{-1}
Molecular Weight 82.1450
Wiswesser Line Notation L6UTJ
Evaluation B(C_p),C(S)

C_6H_{10} (liq) 48HUF/EAT
Cyclohexene
Heat Capacity 298.15 K, C_p = 35.65 cal·mol^{-1}·K^{-1}
149.16 J·mol^{-1}·K^{-1}
Temperature range 12–300 K
Entropy 298.15 K, S = 51.67 cal·mol^{-1}·K^{-1}
216.19 J·mol^{-1}·K^{-1}
Phase Changes
c,II/c,I 138.7 K, ΔH = 1015.9 cal·mol^{-1}
4250.5 J·mol^{-1}
ΔS = 7.32 cal·mol^{-1}·K^{-1}
30.65 J·mol^{-1}·K^{-1}
c,I/liq 169.67 K, ΔH = 787.1 cal·mol^{-1}
3293.2 J·mol^{-1}
ΔS = 4.64 cal·mol^{-1}·K^{-1}
19.41 J·mol^{-1}·K^{-1}
Molecular Weight 82.1450
Wiswesser Line Notation L6UTJ
Evaluation A

C_6H_{10} (liq) 77HAI/SUG 2
Cyclohexene
Heat Capacity 298.15 K, C_p = 35.46 cal·mol^{-1}·K^{-1}
148.35 J·mol^{-1}·K^{-1}
Temperature range 15–293 K
Entropy 298.15 K, S = 51.29 cal·mol^{-1}·K^{-1}
214.60 J·mol^{-1}·K^{-1}
Phase Changes
c,III/c,I 112.3 K, ΔH = 354 cal·mol^{-1}
1483 J·mol^{-1}
ΔS = 3.15 cal·mol^{-1}·K^{-1}
13.2 J·mol^{-1}·K^{-1}
c,II/c,I 138.63 K, ΔH = 1011 cal·mol^{-1}
4231 J·mol^{-1}
ΔS = 7.29 cal·mol^{-1}·K^{-1}
30.52 J·mol^{-1}·K^{-1}
c,I/liq 169.66 K, ΔH = 785 cal·mol^{-1}
3284 J·mol^{-1}
ΔS = 4.63 cal·mol^{-1}·K^{-1}
19.36 J·mol^{-1}·K^{-1}
Molecular Weight 82.1450
Wiswesser Line Notation L6UTJ
Evaluation A

$C_6H_{10}ClN_3O_2$ (c) 63COL/HUT 2
Histidine hydrochloride(L)
Heat Capacity 298.15 K, C_p = 59.64 cal·mol^{-1}·K^{-1}
249.53 J·mol^{-1}·K^{-1}
Temperature range 11–305 K
Entropy 298.15 K, S = 65.99 cal·mol^{-1}·K^{-1}
276.10 J·mol^{-1}·K^{-1}
Molecular Weight 191.6169
Wiswesser Line Notation T5M DNJ B1YZVQ &GH –L
Evaluation A

Table of Heat Capacities, Entropies, and Phase Transition Properties — Continued

$C_6H_{10}O$ (liq) 81REI
4–Methylpenten–3–one–2; Mesityl oxide
Heat Capacity 298 K, C_p = 44.9 cal·mol^{-1}·K^{-1}
187.9 J·mol^{-1}·K^{-1}
Temperature range 290–415 K
Molecular Weight 98.1444
Wiswesser Line Notation 1Y1&U1V1
Evaluation D

$C_6H_{10}O$ (liq) 24HER/BLO
Cyclohexanone
Heat Capacity 290 K, C_p = 42.5 cal·mol^{-1}·K^{-1}
177.8 J·mol^{-1}·K^{-1}
One temperature
Molecular Weight 98.1444
Wiswesser Line Notation L6VTJ
Evaluation C

$C_6H_{10}O$ (liq) 39PHI
Cyclohexanone
Heat Capacity 304.2 K, C_p = 47.9 cal·mol^{-1}·K^{-1}
200.4 J·mol^{-1}·K^{-1}
One temperature
Molecular Weight 98.1444
Wiswesser Line Notation L6VTJ
Evaluation C

$C_6H_{10}O_2$ (liq) 79FUC
Ethyl cyclopropanecarboxylate
Heat Capacity 298.15 K, C_p = 50.9 cal·mol^{-1}·K^{-1}
213.0 J·mol^{-1}·K^{-1}
One temperature
Molecular Weight 114.1438
Wiswesser Line Notation L3TJ AVO2
Evaluation B

$C_6H_{10}O_2$ (liq) 71HAL/BAL
Methyl cyclobutanecarboxylate
Heat Capacity 297 K, C_p = 45.6 cal·mol^{-1}·K^{-1}
190.6 J·mol^{-1}·K^{-1}
One temperature
Molecular Weight 114.1438
Wiswesser Line Notation L4TJ AVO1
Evaluation C

$C_6H_{10}O_3$ (liq) 81REI
Ethyl acetoacetate; Acetoacetic ester
Heat Capacity 298 K, C_p = 57.8 cal·mol^{-1}·K^{-1}
241.8 J·mol^{-1}·K^{-1}
Temperature range 288–455 K
Molecular Weight 130.1432
Wiswesser Line Notation 2OV1V1
Evaluation D

$C_6H_{10}O_3$ (liq) 33KOL/UDO
Ethyl acetoacetate; Acetoacetic ester
Heat Capacity 297.5 K, C_p = 59.0 cal·mol^{-1}·K^{-1}
246.9 J·mol^{-1}·K^{-1}
One temperature
Molecular Weight 130.1432
Wiswesser Line Notation 2OV1V1
Evaluation C

$C_6H_{10}O_3$ (liq) 34KOL/UDO 2
Ethyl acetoacetate; Acetoacetic ester
Heat Capacity 297.5 K, C_p = 59.8 cal·mol^{-1}·K^{-1}
250.2 J·mol^{-1}·K^{-1}
One temperature
Molecular Weight 130.1432
Wiswesser Line Notation 2OV1V1
Evaluation C

$C_6H_{10}O_4$ (liq) 81REI
Diethyl ethanedioate; Diethyl oxalate
Heat Capacity 298 K, C_p = 62.3 cal·mol^{-1}·K^{-1}
260.7 J·mol^{-1}·K^{-1}
Temperature range 294–472 K
Molecular Weight 146.1426
Wiswesser Line Notation 2OVVO2
Evaluation D

$C_6H_{10}O_4S_2$ (c) 35HUF/ELL
4,5–Dithia–1,8–octanedioic acid;
β,β'–Dithiodilactic acid
Heat Capacity 296.8 K, C_p = 57.2 cal·mol^{-1}·K^{-1}
239.3 J·mol^{-1}·K^{-1}
Temperature range 85–305 K. Value is unsmoothed experimental datum.
Entropy 298.1 K, S = 65.5 cal·mol^{-1}·K^{-1}
274.1 J·mol^{-1}·K^{-1}
Extrapolation below 90 K, 20.18 cal·mol^{-1}K^{-1}
Molecular Weight 210.2626
Wiswesser Line Notation QV2SS2VQ
Evaluation B(C_p),C(S)

$C_6H_{11}NO$ (c) 59PAU/KOL
ϵ–Caprolactam
Heat Capacity 298.15 K, C_p = 37.47 cal·mol^{-1}·K^{-1}
156.77 J·mol^{-1}·K^{-1}
Temperature range 60–373 K
Entropy 298.15 K, S = 40.3 cal·mol^{-1}·K^{-1}
168.6 J·mol^{-1}·K^{-1}
Debye–Einstein extrapolation below 60 K.
Molecular Weight 113.1590
Wiswesser Line Notation T7MVTJ
Evaluation B

C_6H_{12} (liq) 38KEN/SHO
3,3–Dimethyl–1–butene; tert–Butylethylene
Heat Capacity 295.9 K, C_p = 45.00 cal·mol^{-1}·K^{-1}
188.28 J·mol^{-1}·K^{-1}
Temperature range 80–298 K. Value is unsmoothed experimental datum.
Entropy 298.1 K, S = 61.3 cal·mol^{-1}·K^{-1}
256.5 J·mol^{-1}·K^{-1}
Extrapolation below 80 K, 11.39 cal·mol^{-1}K^{-1}
Phase Changes
c,II/c,I 124.9 K, ΔH = 1039 cal·mol^{-1}
4347 J·mol^{-1}
ΔS = 8.32 cal·mol^{-1}·K^{-1}
34.80 J·mol^{-1}·K^{-1}
c,II/liq 158.4 K, ΔH = 262 cal·mol^{-1}
1096 J·mol^{-1}
ΔS = 1.65 cal·mol^{-1}·K^{-1}
6.92 J·mol^{-1}·K^{-1}
Molecular Weight 84.1608
Wiswesser Line Notation 1X1&1&1U1
Evaluation B(C_p),C(S)

Table of Heat Capacities, Entropies, and Phase Transition Properties — Continued

C_6H_{12} (liq) 36PAR/TOD 2
2,3–Dimethyl–2–butene; Tetramethylethylene
Heat Capacity 295.5 K, C_p = 42.00 cal·mol^{-1}·K^{-1}
175.73 J·mol^{-1}·K^{-1}
Temperature range 83–296 K. Value is unsmoothed datum.
Entropy 298.15 K, S = 65.1 cal·mol^{-1}·K^{-1}
272.4 J·mol^{-1}·K^{-1}
Extrapolation below 90 K, 13.97 cal·mol^{-1}K^{-1}
Phase Changes
c,II/c,I 196.6 K, ΔH = 1094 cal·mol^{-1}
4577 J·mol^{-1}
ΔS = 5.56 cal·mol^{-1}·K^{-1}
23.28 J·mol^{-1}·K^{-1}
c,I/liq 198.5 K, ΔH = 1305 cal·mol^{-1}
5460 J·mol^{-1}
ΔS = 6.57 cal·mol^{-1}·K^{-1}
27.51 J·mol^{-1}·K^{-1}
Molecular Weight 84.1608
Wiswesser Line Notation 1Y1&UY1&1
Evaluation B(C_p)C(S)

C_6H_{12} (liq) 55SCO/FIN
2,3–Dimethyl–2–butene; Tetramethylethylene
Heat Capacity 298.15 K, C_p = 41.75 cal·mol^{-1}·K^{-1}
174.68 J·mol^{-1}·K^{-1}
Temperature range 10–320 K
Entropy 298.15 K, S = 64.58 cal·mol^{-1}·K^{-1}
270.20 J·mol^{-1}·K^{-1}
Phase Changes
c,II/c,I 196.82 K, ΔH = 844 cal·mol^{-1}
3531 J·mol^{-1}
ΔS = 4.29 cal·mol^{-1}·K^{-1}
17.94 J·mol^{-1}·K^{-1}
c,I/liq 198.92 K, ΔH = 1542 cal·mol^{-1}
6452 J·mol^{-1}
ΔS = 7.75 cal·mol^{-1}·K^{-1}
32.44 J·mol^{-1}·K^{-1}
Enthalpies of transition and fusion calculated from total enthalpy change 196.82–198.92 K and assumed C_p of 36 cal·mol^{-1}K^{-1} for c,I.
Molecular Weight 84.1608
Wiswesser Line Notation 1Y1&UY1&1
Evaluation A

C_6H_{12} (liq) 57MCC/FIN 2
1–Hexene
Heat Capacity 298.15 K, C_p = 43.81 cal·mol^{-1}·K^{-1}
183.30 J·mol^{-1}·K^{-1}
Temperature range 11–360 K
Entropy 298.15 K, S = 70.55 cal·mol^{-1}·K^{-1}
295.18 J·mol^{-1}·K^{-1}
Phase Changes
c/liq 133.39 K, ΔH = 2334 cal·mol^{-1}
9347 J·mol^{-1}
ΔS = 16.75 cal·mol^{-1}·K^{-1}
70.07 J·mol^{-1}·K^{-1}
Molecular Weight 84.1608
Wiswesser Line Notation 5U1
Evaluation A

C_6H_{12} (liq) 31HUF/PAR
Methylcyclopentane
Heat Capacity 295.7 K, C_p = 37.6 cal·mol^{-1}·K^{-1}
157.3 J·mol^{-1}·K^{-1}
Temperature range 92–294 K. Value is unsmoothed experimental datum.
Entropy 298.1 K, S = 59.2 cal·mol^{-1}·K^{-1}
247.7 J·mol^{-1}·K^{-1}
Extrapolation below 90 K, 13.83 cal·mol^{-1}K^{-1}
Phase Changes
c/liq 130.1 K, ΔH = 1645 cal·mol^{-1}
6883 J·mol^{-1}
ΔS = 12.6 cal·mol^{-1}·K^{-1}
52.9 J·mol^{-1}·K^{-1}
Molecular Weight 84.1608
Wiswesser Line Notation L5TJ A1
Evaluation B(C_p),C(S)

C_6H_{12} (liq) 46DOU/HUF 2
Methylcyclopentane
Heat Capacity 298.15 K, C_p = 37.93 cal·mol^{-1}·K^{-1}
158.70 J·mol^{-1}·K^{-1}
Temperature range 12–300 K
Entropy 298.15 K, S = 59.22 cal·mol^{-1}·K^{-1}
247.78 J·mol^{-1}·K^{-1}
Phase Changes
c/liq 130.73 K, ΔH = 1656.0 cal·mol^{-1}
6928.7 J·mol^{-1}
ΔS = 12.67 cal·mol^{-1}·K^{-1}
53.00 J·mol^{-1}·K^{-1}
Molecular Weight 84.1608
Wiswesser Line Notation L5TJ A1
Evaluation A

C_6H_{12} (liq) 51CON/SAG
Methylcyclopentane
Heat Capacity 299.8 K, C_p = 38.03 cal·mol^{-1}·K^{-1}
159.12 J·mol^{-1}·K^{-1}
Temperature range 80 to 200°F
Molecular Weight 84.1608
Wiswesser Line Notation L5TJ A1
Evaluation B

C_6H_{12} (liq) 19DEJ
Cyclohexane
Heat Capacity 298 K, C_p = 42.1 cal·mol^{-1}·K^{-1}
176.1 J·mol^{-1}·K^{-1}
Temperature range 22 to 50°C
Molecular Weight 84.1608
Wiswesser Line Notation L6TJ
Evaluation B

C_6H_{12} (liq) 30PAR/HUF
Cyclohexane
Heat Capacity 298.9 K, C_p = 34.4 cal·mol^{-1}·K^{-1}
143.9 J·mol^{-1}·K^{-1}
Temperature range 92–299 K. Value is unsmoothed experimental datum.
Entropy 298.15 K, S = 49.2 cal·mol^{-1}·K^{-1}
205.9 J·mol^{-1}·K^{-1}
Extrapolation below 90 K, 12.08 cal·mol^{-1}K^{-1}
Phase Changes
c,II/c,I 185.9 K, ΔH = 1490 cal·mol^{-1}
6234 J·mol^{-1}
ΔS = 8.02 cal·mol^{-1}·K^{-1}
33.53 J·mol^{-1}·K^{-1}
c,I/liq 279.3 K, ΔH = 579 cal·mol^{-1}
2423 J·mol^{-1}
ΔS = 2.07 cal·mol^{-1}·K^{-1}
8.68 J·mol^{-1}·K^{-1}
Molecular Weight 84.1608
Wiswesser Line Notation L6TJ
Evaluation B(C_p),C(S)

Table of Heat Capacities, Entropies, and Phase Transition Properties — Continued

C_6H_{12} (liq) 39PHI
Cyclohexane
Heat Capacity 304.2 K, C_p = 24.0 cal·mol⁻¹·K⁻¹
100.4 J·mol⁻¹·K⁻¹
One temperature
Molecular Weight 84.1608
Wiswesser Line Notation L6TJ
Evaluation C

C_6H_{12} (liq) 42ZIE/AND
Cyclohexane
Phase Changes
c,II/c,I 186.4 K, ΔH = 1630 cal·mol^{-1}
6820 J·mol^{-1}
ΔS = 8.74 cal·mol^{-1}·K^{-1}
36.59 J·mol^{-1}·K^{-1}
c,I/liq 279.4 K, ΔH = 652 cal·mol^{-1}
2728 J·mol^{-1}
ΔS = 2.33 cal·mol^{-1}·K^{-1}
9.76 J·mol^{-1}·K^{-1}
Molecular Weight 84.1608
Wiswesser Line Notation L6TJ
Evaluation B

C_6H_{12} (liq) 43AST/SZA
Cyclohexane
Heat Capacity 295 K, C_p = 37.25 cal·mol^{-1}·K^{-1}
155.85 J·mol^{-1}·K^{-1}
Temperature range 12–293 K
Entropy 298.15 K, S = 48.73 cal·mol^{-1}·K^{-1}
203.89 J·mol^{-1}·K^{-1}
Phase Changes
c,II/c,I 186.09 K, ΔH = 1598 cal·mol^{-1}
6686 J·mol^{-1}
ΔS = 8.59 cal·mol^{-1}·K^{-1}
35.93 J·mol^{-1}·K^{-1}
c,I/liq 279.84 K, ΔH = 628 cal·mol^{-1}
2628 J·mol^{-1}
ΔS = 2.24 cal·mol^{-1}·K^{-1}
9.39 J·mol^{-1}·K^{-1}
liq/g 298.15 K, ΔH = 7967 cal·mol^{-1}
33334 J·mol^{-1}
ΔS = 26.72 cal·mol^{-1}·K^{-1}
111.80 J·mol^{-1}·K^{-1}
P = 13.18 kPa
Molecular Weight 84.1608
Wiswesser Line Notation L6TJ
Evaluation A

C_6H_{12} (liq) 43RUE/HUF
Cyclohexane
Heat Capacity 298.15 K, C_p = 37.36 cal·mol^{-1}·K^{-1}
156.31 J·mol^{-1}·K^{-1}
Temperature range 13–302 K
Entropy 298.15 K, S = 48.84 cal·mol^{-1}·K^{-1}
204.35 J·mol^{-1}·K^{-1}
Phase Changes
c,II/c,I 186.1 K, ΔH = 1610.8 cal·mol^{-1}
6739.6 J·mol^{-1}
ΔS = 8.66 cal·mol^{-1}·K^{-1}
36.21 J·mol^{-1}·K^{-1}
c,I/liq 279.82 K, ΔH = 639.8 cal·mol^{-1}
2676.9 J·mol^{-1}
ΔS = 2.29 cal·mol^{-1}·K^{-1}
9.57 J·mol^{-1}·K^{-1}
Molecular Weight 84.1608
Wiswesser Line Notation L6TJ
Evaluation A

C_6H_{12} (liq) 60SWI/ZIE
Cyclohexane
Heat Capacity 311 K, C_p = 37.1 cal·mol^{-1}·K^{-1}
155.2 J·mol^{-1}·K^{-1}
Mean value 20 to 56°C.
Molecular Weight 84.1608
Wiswesser Line Notation L6TJ
Evaluation C

C_6H_{12} (liq) 64MOE/THO
Cyclohexane
Heat Capacity 298.00 K, C_p = 37.12 cal·mol^{-1}·K^{-1}
155.31 J·mol^{-1}·K^{-1}
Temperature range 297–327 K
Molecular Weight 84.1608
Wiswesser Line Notation L6TJ
Evaluation B

C_6H_{12} (liq) 66NIK/RAB
Cyclohexane
Heat Capacity 298 K, C_p = 36.55 cal·mol^{-1}·K^{-1}
152.93 J·mol^{-1}·K^{-1}
Temperature range 10 to 50°C
Molecular Weight 84.1608
Wiswesser Line Notation L6TJ
Evaluation B

C_6H_{12} (liq) 68REC
Cyclohexane
Heat Capacity 298 K, C_p = 37.2 cal·mol^{-1}·K^{-1}
155.5 J·mol^{-1}·K^{-1}
Temperature range 24 to 40 °C, equation only.
Molecular Weight 84.1608
Wiswesser Line Notation L6TJ
Evaluation C

C_6H_{12} (liq) 69WIL/SCH
Cyclohexane
Heat Capacity 298.15 K, C_p = 37.1 cal·mol^{-1}·K^{-1}
155.2 J·mol^{-1}·K^{-1}
Temperature 20, 30, and 40°C.
Molecular Weight 84.1608
Wiswesser Line Notation L6TJ
Evaluation B

C_6H_{12} (liq) 73SUB/RAS
Cyclohexane
Heat Capacity 298.15 K, C_p = 38.1 cal·mol^{-1}·K^{-1}
159.6 J·mol^{-1}·K^{-1}
Temperature range 298–323 K
Molecular Weight 84.1608
Wiswesser Line Notation L6TJ
Evaluation B

C_6H_{12} (liq) 76FOR/BEN
Cyclohexane
Heat Capacity 298.15 K, C_p = 37.30 cal·mol^{-1}·K^{-1}
156.07 J·mol^{-1}·K^{-1}
One temperature
Molecular Weight 84.1608
Wiswesser Line Notation L6TJ
Evaluation B

Table of Heat Capacities, Entropies, and Phase Transition Properties — Continued

C_6H_{12} (liq) 79WIL/GRO
Cyclohexane
Heat Capacity 298.15 K, C_p = 37.4 cal·mol^{-1}·K^{-1}
156.4 J·mol^{-1}·K^{-1}
One temperature
Molecular Weight 84.1608
Wiswesser Line Notation L6TJ
Evaluation B

C_6H_{12} (liq) 82GRO/ING
Cyclohexane
Heat Capacity 298.15 K, C_p = 37.39 cal·mol^{-1}·K^{-1}
156.4 J·mol^{-1}·K^{-1}
One temperature.
Molecular Weight 84.1608
Wiswesser Line Notation L6TJ
Evaluation A

C_6H_{12} (liq) 82TAN
Cyclohexane
Heat Capacity 298.15 K, C_p = 37.29 cal·mol^{-1}·K^{-1}
156.0 J·mol^{-1}·K^{-1}
Temperature range 293.15, 298.15, 303.15 K. Data at three temperatures.
Molecular Weight 84.1608
Wiswesser Line Notation L6TJ
Evaluation A

$(C_6H_{12})_n$ (c) 67MEL/TYS
Poly(4–methyl–1–pentene)
Heat Capacity 298.15 K, C_p = 34.8 cal·mol^{-1}·K^{-1}
145.6 J·mol^{-1}·K^{-1}
Temperature range 80–310 K. Values per unit formula weight.
Entropy 298.15 K, S = 36.2 cal·mol^{-1}·K^{-1}
151.5 J·mol^{-1}·K^{-1}
Extrapolation below 80 K, 38.2 J·mol^{-1}K^{-1}. Values do not include zero–point entropy.
Molecular Weight 84.1608
Wiswesser Line Notation /*Y1*1Y1&1/
Evaluation B(C_p), C(S)
65% crystalline, isotactic material.

$C_6H_{12}N_2$ (c) 60CHA/WES 2
Triethylenediamine; 1,4–Diazabicyclo[2.2.2]octane
Heat Capacity 298.15 K, C_p = 36.56 cal·mol^{-1}·K^{-1}
152.97 J·mol^{-1}·K^{-1}
Temperature range 5–353 K
Entropy 298.15 K, S = 37.67 cal·mol^{-1}·K^{-1}
157.61 J·mol^{-1}·K^{-1}
Molecular Weight 112.1742
Wiswesser Line Notation T66 A B CN FNTJ
Evaluation A

$C_6H_{12}N_2$ (c) 61WES
Triethylenediamine; 1,4–Diazabicyclo[2.2.2]octane
Heat Capacity 298.15 K, C_p = 36.61 cal·mol^{-1}·K^{-1}
153.18 J·mol^{-1}·K^{-1}
Temperature range 5–350 K. Only value at 298.15 K given.
Entropy 298.15 K, S = 37.67 cal·mol^{-1}·K^{-1}
157.61 J·mol^{-1}·K^{-1}
Molecular Weight 112.1742
Wiswesser Line Notation T66 A B CN FNTJ
Evaluation A
Details reported in other papers.

$C_6H_{12}N_2$ (c,II) 63TRO/WES
Triethylenediamine; 1,4–Diazabicyclo[2,2,2]octane
Heat Capacity 300 K, C_p = 36.80 cal·mol^{-1}·K^{-1}
153.97 J·mol^{-1}·K^{-1}
Temperature range 300–450 K
Phase Changes
c,II/c,I 351.08 K, ΔH = 2524 cal·mol^{-1}
10560 J·mol^{-1}
ΔS = 7.19 cal·mol^{-1}·K^{-1}
30.08 J·mol^{-1}·K^{-1}
Transformation to plastic crystalline phase.
c,I/liq 432.99 K, ΔH = 1776 cal·mol^{-1}
7431 J·mol^{-1}
ΔS = 4.10 cal·mol^{-1}·K^{-1}
17.16 J·mol^{-1}·K^{-1}
Molecular Weight 112.1742
Wiswesser Line Notation T66 A B CN FNTJ
Evaluation A

$C_6H_{12}N_2O_4S_2$ (c) 35HUF/ELL
Cystine(L)
Heat Capacity 297.3 K, C_p = 64.2 cal·mol^{-1}·K^{-1}
268.6 J·mol^{-1}·K^{-1}
Temperature range 85–298 K. Value is unsmoothed experimental datum.
Entropy 298.1 K, S = 68.5 cal·mol^{-1}·K^{-1}
286.6 J·mol^{-1}·K^{-1}
Extrapolation below 90 K, 18.99 cal·mol^{-1}K^{-1}
Molecular Weight 240.2918
Wiswesser Line Notation QVYZ1SS1YZVQ -L
Evaluation B(C_p),C(S)

$C_6H_{12}N_2O_4S_2$ (c) 64HUT/COL
Cystine(L)
Heat Capacity 298.15 K, C_p = 62.60 cal·mol^{-1}·K^{-1}
261.92 J·mol^{-1}·K^{-1}
Temperature range 10–310 K
Entropy 298.15 K, S = 67.06 cal·mol^{-1}·K^{-1}
280.58 J·mol^{-1}·K^{-1}
Molecular Weight 240.2918
Wiswesser Line Notation QVYZ1SS1YZVQ -L
Evaluation A

$C_6H_{12}N_4$ (c) 60CHA/WES
Hexamethylenetetramine;
1,3,5,7–Tetraazatricyclo[3.3.1.1^{3,7}]decane
Heat Capacity 298.15 K, C_p = 36.40 cal·mol^{-1}·K^{-1}
152.29 J·mol^{-1}·K^{-1}
Temperature range 5–350 K
Entropy 298.15 K, S = 39.05 cal·mol^{-1}·K^{-1}
163.38 J·mol^{-1}·K^{-1}
Molecular Weight 140.1876
Wiswesser Line Notation T66 B6/B–H/DI A B– C 1B I BN DN FN HNTJ
Evaluation A

$C_6H_{12}N_4$ (c) 61WES
Hexamethylenetetramine;
1,3,5,7–Tetraazatricyclo[3.3.1.1^{3,7}]decane
Heat Capacity 298.15 K, C_p = 36.40 cal·mol^{-1}·K^{-1}
152.29 J·mol^{-1}·K^{-1}
Temperature range 5–350 K. Only value at 298.15 K given.
Entropy 298.15 K, S = 39.05 cal·mol^{-1}·K^{-1}
163.38 J·mol^{-1}·K^{-1}
Molecular Weight 140.1876
Wiswesser Line Notation T66 B6/B–H/DI A B– C 1B I BN DN FN HNTJ
Evaluation A
Details reported in other papers.

Table of Heat Capacities, Entropies, and Phase Transition Properties — Continued

$C_6H_{12}O$ (liq) 70AND/COU
3,3–Dimethyl–2–butanone; Methyl tert–butyl ketone
Heat Capacity 298.15 K, C_p = 49.45 cal·mol^{-1}·K^{-1}
206.9 J·mol^{-1}·K^{-1}
Temperature range 10–360 K
Entropy 298.15 K, S = 67.50 cal·mol^{-1}·K^{-1}
282.4 J·mol^{-1}·K^{-1}
Phase Changes
c/liq 221.74 K, ΔH = 2710 cal·mol^{-1}
11330 J·mol^{-1}
ΔS = 12.22 cal·mol^{-1}·K^{-1}
51.10 J·mol^{-1}·K^{-1}
Molecular Weight 100.1602
Wiswesser Line Notation 1X1&1&V1
Evaluation A

$C_6H_{12}O$ (liq) 47SCH/ZOS
5–Methyl–3–oxahex–1–ene; Vinyl isobutyl ether
Heat Capacity 298.15 K, C_p = 55.4 cal·mol^{-1}·K^{-1}
231.8 J·mol^{-1}·K^{-1}
One value, no details.
Molecular Weight 100.1602
Wiswesser Line Notation 1Y1&1O1U1
Evaluation D

$C_6H_{12}O$ (liq) 70AND/COU
3–Hexanone; Ethyl n–propyl ketone
Heat Capacity 298.15 K, C_p = 51.84 cal·mol^{-1}·K^{-1}
216.9 J·mol^{-1}·K^{-1}
Temperature range 10–320 K
Entropy 298.15 K, S = 72.97 cal·mol^{-1}·K^{-1}
305.3 J·mol^{-1}·K^{-1}
Phase Changes
c,II/c,I 145 K, ΔH = 163 cal·mol^{-1}
682 J·mol^{-1}
ΔS = 1.12 cal·mol^{-1}·K^{-1}
4.73 J·mol^{-1}·K^{-1}
c,I/liq 217.72 K, ΔH = 3225 cal·mol^{-1}
13490 J·mol^{-1}
ΔS = 14.81 cal·mol^{-1}·K^{-1}
61.98 J·mol^{-1}·K^{-1}
Molecular Weight 100.1602
Wiswesser Line Notation 3V2
Evaluation A

$C_6H_{12}O$ (liq) 47SCH/ZOS
3–Oxahept–1–ene; Vinyl n–butyl ether
Heat Capacity 298.15 K, C_p = 55.4 cal·mol^{-1}·K^{-1}
231.8 J·mol^{-1}·K^{-1}
One value, no details.
Molecular Weight 100.1602
Wiswesser Line Notation 4O1U1
Evaluation D

$C_6H_{12}O$ (liq) 70AND/COU
2–Hexanone; Methyl n–butyl ketone
Heat Capacity 298.15 K, C_p = 51.00 cal·mol^{-1}·K^{-1}
213.4 J·mol^{-1}·K^{-1}
Temperature range 10–380 K
Entropy 298.15 K, S = 73.64 cal·mol^{-1}·K^{-1}
308.1 J·mol^{-1}·K^{-1}
Phase Changes
c/liq 217.69 K, ΔH = 3560 cal·mol^{-1}
14900 J·mol^{-1}
ΔS = 16.35 cal·mol^{-1}·K^{-1}
68.44 J·mol^{-1}·K^{-1}
Molecular Weight 100.1602
Wiswesser Line Notation 4V1
Evaluation A

$C_6H_{12}O$ (liq) 24HER/BLO
Cyclohexanol; Cyclohexyl alcohol
Heat Capacity 290 K, C_p = 41.8 cal·mol^{-1}·K^{-1}
174.9 J·mol^{-1}·K^{-1}
One temperature
Molecular Weight 100.1602
Wiswesser Line Notation L6TJ AQ
Evaluation C

$C_6H_{12}O$ (liq) 29KEL
Cyclohexanol; Cyclohexyl alcohol
Heat Capacity 298.15 K, C_p = 49.96 cal·mol^{-1}·K^{-1}
209.03 J·mol^{-1}·K^{-1}
Temperature range 13–300 K. Value is unsmoothed experimental datum.
Entropy 298.15 K, S = 47.7 cal·mol^{-1}·K^{-1}
199.6 J·mol^{-1}·K^{-1}
Average of values derived from measurements on both low and high–temperature crystal forms down to 13 K, plus entropy of transition and fusion. Debye extrapolation below 13.5 K.
Phase Changes
c,II/c,I 263.5 K, ΔH = 1961 cal·mol^{-1}
8205 J·mol^{-1}
ΔS = 7.44 cal·mol^{-1}·K^{-1}
31.14 J·mol^{-1}·K^{-1}
Excess enthalpy over extrapolated heat capacity curves.
c,I/liq 297.0 K, ΔH = 406 cal·mol^{-1}
1699 J·mol^{-1}
ΔS = 1.37 cal·mol^{-1}·K^{-1}
5.72 J·mol^{-1}·K^{-1}
Tm is 23.87 °C from 16RIC/SHI.
Molecular Weight 100.1602
Wiswesser Line Notation L6TJ AQ
Evaluation B
Sample may have contained a trace of water.

$C_6H_{12}O$ (liq) 39PHI
Cyclohexanol; Cyclohexyl alcohol
Heat Capacity 305.1 K, C_p = 48.4 cal·mol^{-1}·K^{-1}
202.5 J·mol^{-1}·K^{-1}
One temperature
Molecular Weight 100.1602
Wiswesser Line Notation L6TJ AQ
Evaluation C

$C_6H_{12}O$ (liq) 68ADA/SUG
Cyclohexanol; Cyclohexyl alcohol
Heat Capacity 300 K, C_p = 51.05 cal·mol^{-1}·K^{-1}
273.60 J·mol^{-1}·K^{-1}
Temperature range 14–320 K
Entropy 300 K, S = 48.73 cal·mol^{-1}·K^{-1}
203.87 J·mol^{-1}·K^{-1}

Table of Heat Capacities, Entropies, and Phase Transition Properties — Continued

Phase Changes

c,III/c,I	244.8 K,	ΔH =	2065 cal·mol^{-1}
			8640 J·mol^{-1}
		ΔS =	8.44 cal·mol^{-1}·K^{-1}
			35.29 J·mol^{-1}·K^{-1}
c,II/c,I	265.50 K,	ΔH =	2110 cal·mol^{-1}
			8827 J·mol^{-1}
		ΔS =	7.95 cal·mol^{-1}·K^{-1}
			33.25 J·mol^{-1}·K^{-1}
c,I/liq	299.09 K,	ΔH =	426 cal·mol^{-1}
			1783 J·mol^{-1}
		ΔS =	1.42 cal·mol^{-1}·K^{-1}
			5.96 J·mol^{-1}·K^{-1}

Molecular Weight 100.1602
Wiswesser Line Notation L6TJ AQ
Evaluation B

$C_6H_{12}O$ (liq) 74PET/TER
Cyclohexanol; Cyclohexyl alcohol
Heat Capacity 297.95 K, C_p = 51 cal·mol^{-1}·K^{-1}
212 J·mol^{-1}·K^{-1}
Temperature range 297–428 K. Value is unsmoothed experimental datum.
Molecular Weight 100.1602
Wiswesser Line Notation L6TJ AQ
Evaluation B

$C_6H_{12}O$ (liq) 76CON/GIN
Cyclohexanol; Cyclohexyl alcohol
Heat Capacity 298 K, C_p = 52.6 cal·mol^{-1}·K^{-1}
220.1 J·mol^{-1}·K^{-1}
One temperature
Molecular Weight 100.1602
Wiswesser Line Notation L6TJ AQ
Evaluation B

$C_6H_{12}O_2$ (liq) 79FUC
2–Methyl–2–propyl ethanoate; tert–Butyl acetate
Heat Capacity 298.15 K, C_p = 55.2 cal·mol^{-1}·K^{-1}
231.0 J·mol^{-1}·K^{-1}
One temperature
Molecular Weight 116.1596
Wiswesser Line Notation 1X1&1&OV1
Evaluation B

$C_6H_{12}O_2$ (liq) 71HAL/BAL
Methyl 2,2–dimethylpropanoate; Methyl pivalate
Heat Capacity 297 K, C_p = 53.3 cal·mol^{-1}·K^{-1}
223.0 J·mol^{-1}·K^{-1}
One temperature
Molecular Weight 116.1596
Wiswesser Line Notation 1X1&1&VO1
Evaluation C

$C_6H_{12}O_2$ (liq) 79FUC
Methyl 2,2–dimethylpropanoate; Methyl pivalate
Heat Capacity 298.15 K, C_p = 61.6 cal·mol^{-1}·K^{-1}
257.9 J·mol^{-1}·K^{-1}
One temperature
Molecular Weight 116.1596
Wiswesser Line Notation 1X1&1&VO1
Evaluation B

$C_6H_{12}O_2$ (liq) 36KUR/VOS
2–Methyl–1–propyl ethanoate; Isobutyl acetate
Heat Capacity 290 K, C_p = 57.4 cal·mol^{-1}·K^{-1}
240.2 J·mol^{-1}·K^{-1}
One temperature
Molecular Weight 116.1596
Wiswesser Line Notation 1Y1&1OV1
Evaluation D

$C_6H_{12}O_2$ (liq) 33KOL/UDO
Ethyl butanoate; Ethyl butyrate
Heat Capacity 297.2 K, C_p = 52.6 cal·mol^{-1}·K^{-1}
220.1 J·mol^{-1}·K^{-1}
One temperature
Molecular Weight 116.1596
Wiswesser Line Notation 3VO2
Evaluation C

$C_6H_{12}O_2$ (liq) 34KOL/UDO 2
Ethyl butanoate; Ethyl butyrate
Heat Capacity 297.2 K, C_p = 52.6 cal·mol^{-1}·K^{-1}
220.1 J·mol^{-1}·K^{-1}
One temperature
Molecular Weight 116.1596
Wiswesser Line Notation 3VO2
Evaluation C

$C_6H_{12}O_2$ (liq) 36KUR/VOS
Ethyl butanoate; Ethyl butyrate
Heat Capacity 290 K, C_p = 54.9 cal·mol^{-1}·K^{-1}
229.7 J·mol^{-1}·K^{-1}
One temperature
Molecular Weight 116.1596
Wiswesser Line Notation 3VO2
Evaluation D

$C_6H_{12}O_2$ (liq) 79FUC
Ethyl butanoate; Ethyl butyrate
Heat Capacity 298.15 K, C_p = 54.5 cal·mol^{-1}·K^{-1}
228.0 J·mol^{-1}·K^{-1}
One temperature
Molecular Weight 116.1596
Wiswesser Line Notation 3VO2
Evaluation B

$C_6H_{12}O_2$ (liq) 33KOL/UDO
n–Butyl ethanoate; n–Butyl acetate
Heat Capacity 292.5 K, C_p = 57.9 cal·mol^{-1}·K^{-1}
242.3 J·mol^{-1}·K^{-1}
One temperature
Molecular Weight 116.1596
Wiswesser Line Notation 4OV1
Evaluation C

$C_6H_{12}O_2$ (liq) 34KOL/UDO 2
n–Butyl ethanoate; n–Butyl acetate
Heat Capacity 292.5 K, C_p = 57.9 cal·mol^{-1}·K^{-1}
242.3 J·mol^{-1}·K^{-1}
One temperature
Molecular Weight 116.1596
Wiswesser Line Notation 4OV1
Evaluation C

Table of Heat Capacities, Entropies, and Phase Transition Properties — Continued

$C_6H_{12}O_2$ (liq) 79FUC
n–Butyl ethanoate; n–Butyl acetate
Heat Capacity 298.15 K, C_p = 54.6 cal·mol^{-1}·K^{-1}
228.4 J·mol^{-1}·K^{-1}
One temperature
Molecular Weight 116.1596
Wiswesser Line Notation 4OV1
Evaluation B

$C_6H_{12}O_2$ (liq) 79FUC
Methyl pentanoate; Methyl valerate
Heat Capacity 298.15 K, C_p = 54.8 cal·mol^{-1}·K^{-1}
229.3 J·mol^{-1}·K^{-1}
One temperature
Molecular Weight 116.1596
Wiswesser Line Notation 4VO1
Evaluation B

$C_6H_{12}O_2$ (liq) 81REI
Hexanoic acid; n–Caproic acid
Heat Capacity 298 K, C_p = 53.8 cal·mol^{-1}·K^{-1}
225.1 J·mol^{-1}·K^{-1}
Temperature range 292–483 K
Molecular Weight 116.1596
Wiswesser Line Notation QV5
Evaluation D

$C_6H_{12}O_3$ (liq) 81REI
2,4,6–Trimethyl–1,3,5–trioxane; Paraldehyde
Heat Capacity 298 K, C_p = 59.8 cal·mol^{-1}·K^{-1}
250.2 J·mol^{-1}·K^{-1}
Temperature range 293–400 K
Molecular Weight 132.1590
Wiswesser Line Notation T6O CO EOTJ B1 D1 F1
Evaluation D

$C_6H_{12}O_3$ (liq) 39PHI
2,4,6–Trimethyl–1,3,5–trioxane; Paraldehyde
Heat Capacity 306.6 K, C_p = 60.7 cal·mol^{-1}·K^{-1}
254.0 J·mol^{-1}·K^{-1}
One temperature
Molecular Weight 132.1590
Wiswesser Line Notation T6O CO EOTJ B1 D1 F1
Evaluation C

$C_6H_{12}O_3$ (liq) 69CLE/MEL 2
2,4,6–Trimethyl–1,3,5–trioxane; Paraldehyde
Heat Capacity 298.15 K, C_p = 61.50 cal·mol^{-1}·K^{-1}
257.3 J·mol^{-1}·K^{-1}
Temperature range 80–310 K
Entropy 298.15 K, S = 69.16 cal·mol^{-1}·K^{-1}
289.4 J·mol^{-1}·K^{-1}
Extrapolation below 80 K, 61.1 J·mol^{-1}K^{-1}
Phase Changes
c,IV/c,III 142.7 K, ΔH = 61.6 cal·mol^{-1}
257.7 J·mol^{-1}
ΔS = 0.43 cal·mol^{-1}·K^{-1}
1.81 J·mol^{-1}·K^{-1}
c,III/c,II 147.5 K, ΔH = 184.8 cal·mol^{-1}
773.1 J·mol^{-1}
ΔS = 1.25 cal·mol^{-1}·K^{-1}
5.24 J·mol^{-1}·K^{-1}
ΔH and ΔS of transition for c,II/c,I at 230.3 K has been added into ΔH and ΔS of fusion at 285.7 K.
c,I/liq 285.7 K, ΔH = 3231 cal·mol^{-1}
13520 J·mol^{-1}
ΔS = 11.31 cal·mol^{-1}·K^{-1}
47.32 J·mol^{-1}·K^{-1}
Molecular Weight 132.1590
Wiswesser Line Notation T6O CO EOTJ B1 D1 F1
Evaluation B(C_p),C(S)

$C_6H_{12}O_6$ (c,I) 41JAC/STE
Sorbose(L)
Heat Capacity 295.9 K, C_p = 54.64 cal·mol^{-1}·K^{-1}
228.61 J·mol^{-1}·K^{-1}
Temperature range 64–296 K. Value is unsmoothed experimental datum.
Entropy 298.15 K, S = 52.8 cal·mol^{-1}·K^{-1}
220.9 J·mol^{-1}·K^{-1}
Extrapolation below 90 K, 12.0 cal·mol^{-1}K^{-1}
Phase Changes
c,II/c,I 199.22 K, ΔH = 143.7 cal·mol^{-1}
601.2 J·mol^{-1}
ΔS = 0.72 cal·mol^{-1}·K^{-1}
3.02 J·mol^{-1}·K^{-1}
Molecular Weight 180.1572
Wiswesser Line Notation T6OTJ BQ CQ DQ EQ F1Q -A&E -B&BCDF
Evaluation B(C_p),C(S)

$C_6H_{12}O_6$ (c) 81KAW/NIS
Fructose
Heat Capacity 300 K, C_p = 55.1 cal·mol^{-1}·K^{-1}
230.5 J·mol^{-1}·K^{-1}
Temperature range 270–325 K; C_p given as 1.28 J·g^{-1}·K^{-1}
Molecular Weight 180.1572
Wiswesser Line Notation T6OTJ BQ B1Q CQ DQ EQ -A&DE -B&BC
Evaluation A

$C_6H_{12}O_6$ (c) 22SIM
α–Glucose(D), Dextrose
Heat Capacity 300K K, C_p = 54.8 cal·mol^{-1}·K^{-1}
229.3 J·mol^{-1}·K^{-1}
Temperature range 20–287 K
Molecular Weight 180.1572
Wiswesser Line Notation T6OTJ BQ CQ DQ EQ F1Q -A&BCE -B&DF
Evaluation B

$C_6H_{12}O_6$ (c) 29PAR/KEL
α–Glucose(D), Dextrose
Entropy 298.1 K, S = 50.5 cal·mol^{-1}·K^{-1}
211.3 J·mol^{-1}·K^{-1}
Extrapolation below 90 K, 13.2 cal·mol^{-1}K^{-1}.
Revision of previous data.
Molecular Weight 180.1572
Wiswesser Line Notation T6OTJ BQ CQ DQ EQ F1Q -A&BCE -B&DF
Evaluation C

$C_6H_{12}O_6$ (c) 34PAR/THO
α–Glucose(D), Dextrose
Heat Capacity 298 K, C_p = 52.8 cal·mol^{-1}·K^{-1}
220.9 J·mol^{-1}·K^{-1}
Temperature range 273–318K; curve given also for undercooled liquid.
Phase Changes
c/liq 414 K, ΔH = 7510 cal·mol^{-1}
31420 J·mol^{-1}
ΔS = 18.1 cal·mol^{-1}·K^{-1}
75.9 J·mol^{-1}·K^{-1}
Molecular Weight 180.1572
Wiswesser Line Notation T6OTJ BQ CQ DQ EQ F1Q -A&BCE -B&DF
Evaluation B

Table of Heat Capacities, Entropies, and Phase Transition Properties — Continued

$C_6H_{12}O_6$ (c) 41NEL/NEW
α–Glucose(D), Dextrose
Heat Capacity 298 K, C_p = 52.1 cal·mol⁻¹·K⁻¹
218.0 J·mol⁻¹·K⁻¹
Temperature range 0 to 60 °C. Equation only.
Molecular Weight 180.1572
Wiswesser Line Notation T6OTJ BQ CQ DQ EQ F1Q -A&BCE -B&DF
Evaluation B

$C_6H_{12}O_6$ (c) 51DOU/BAL
α–Glucose(D), Dextrose
Heat Capacity 298.15 K, C_p = 52.3 cal·mol⁻¹·K⁻¹
218.8 J·mol⁻¹·K⁻¹
Temperature range 273–368 K
Molecular Weight 180.1572
Wiswesser Line Notation T6OTJ BQ CQ DQ EQ F1Q -A&BCE -B&DF
Evaluation B

$C_6H_{12}O_6$ (c) 81KAW/NIS
Glucose
Heat Capacity 300 K, C_p = 52.96 cal·mol⁻¹·K⁻¹
221.6 J·mol⁻¹ K⁻¹
Temperature range 270–325 K; C_p given as 1.23 J·g⁻¹·K⁻¹
Molecular Weight 180.1572
Wiswesser Line Notation T6OTJ BQ CQ DQ EQ F1Q -A&BCE -B&DF
Evaluation A

$C_6H_{12}O_6$ (c,α) 41JAC/STE
α–Galactose(D)
Heat Capacity 296.9 K, C_p = 52.71 cal·mol⁻¹·K⁻¹
220.54 J·mol⁻¹·K⁻¹
Temperature range 64–297 K. Value is unsmoothed experimental datum.
Entropy 298.15 K, S = 49.1 cal·mol⁻¹·K⁻¹
205.4 J·mol⁻¹·K⁻¹
Extrapolation below 90 K, 11.7 cal·mol⁻¹K⁻¹
Molecular Weight 180.1572
Wiswesser Line Notation T6OTJ BQ CQ DQ EQ F1Q -A&BC -B&DEF
Evaluation B(C_p),C(S)

$C_6H_{12}O_6$ (c) 81KAW/NIS
Galactose
Heat Capacity 300 K, C_p = 51.7 cal·mol⁻¹·K⁻¹
216.3 J·mol⁻¹·K⁻¹
Temperature range 270–325 K, C_p given as 1.20 J·g⁻¹·K⁻¹.
Molecular Weight 180.1572
Wiswesser Line Notation T6OTJ BQ CQ DQ EQ F1Q -A&BC -B&DEF
Evaluation A

$C_6H_{12}O_6$ (c) 81KAW/NIS
Mannose
Heat Capacity 300 K, C_p = 51.2 cal·mol⁻¹·K⁻¹
214.2 J·mol⁻¹·K⁻¹
Temperature range 270–325 K, C_p given as 1.19 J·g⁻¹·K⁻¹.
Molecular Weight 180.1572
Wiswesser Line Notation T6OTJ BQ CQ DQ EQ F1Q -A&E -B&CDF
Evaluation A

$C_6H_{12}S$ (liq) 74MES/FIN
Cyclopentyl–1–thiaethane; Cyclopentyl methyl sulfide
Heat Capacity 298.15 K, C_p = 46.11 cal·mol⁻¹·K⁻¹
192.92 J·mol⁻¹·K⁻¹
Temperature range 10–370 K
Entropy 298.15 K, S = 68.23 cal·mol⁻¹·K⁻¹
285.47 J·mol⁻¹·K⁻¹
Phase Changes
c,II/c,I 165.0 K, ΔH = 214 cal·mol⁻¹
895 J·mol⁻¹
ΔS = 1.30 cal·mol⁻¹·K⁻¹
5.42 J·mol⁻¹·K⁻¹
c,I/liq 169.85 K, ΔH = 2205 cal·mol⁻¹
9226 J·mol⁻¹
ΔS = 12.98 cal·mol⁻¹·K⁻¹
54.32 J·mol⁻¹·K⁻¹
c,II/liq 169.34 K, ΔH = 2415 cal·mol⁻¹
10104 J·mol⁻¹
ΔS = 14.26 cal·mol⁻¹·K⁻¹
59.67 J·mol⁻¹·K⁻¹
Molecular Weight 116.2208
Wiswesser Line Notation L5TJ AS1
Evaluation A

$C_6H_{12}S$ (liq) 67MES/TOD
Cyclohexanethiol; Cyclohexyl mercaptan
Heat Capacity 298.15 K, C_p = 6.04 cal·mol⁻¹·K⁻¹
192.63 J·mol⁻¹·K⁻¹
Temperature range 10–370 K
Entropy 298.15 K, S = 61.80 cal·mol⁻¹·K⁻¹
255.57 J·mol⁻¹·K⁻¹
Phase Changes
c/liq 189.64 K, ΔH = 2390 cal·mol⁻¹
10000 J·mol⁻¹
ΔS = 12.60 cal·mol⁻¹·K⁻¹
52.73 J·mol⁻¹·K⁻¹
Molecular Weight 116.2208
Wiswesser Line Notation L6TJ ASH
Evaluation A

$C_6H_{13}Br$ (liq) 31DEE
1–Bromohexane; n–Hexyl bromide
Heat Capacity 298.8 K, C_p = 48.65 cal·mol⁻¹·K⁻¹
203.55 J·mol⁻¹·K⁻¹
Temperature range 95–290 K. Value is unsmoothed experimental datum.
Entropy 298.15 K, S = 108.25 cal·mol⁻¹·K⁻¹
452.92 J·mol⁻¹·K⁻¹
Extrapolation below 100 K, 13.89 cal·mol⁻¹K⁻¹
Phase Changes
c/liq 188.1 K, ΔH = 4315 cal·mol⁻¹
18054 J·mol⁻¹
ΔS = 22.94 cal·mol⁻¹·K⁻¹
95.98 J·mol⁻¹·K⁻¹
Molecular Weight 165.0727
Wiswesser Line Notation E6
Evaluation B(C_p),C(S)

$C_6H_{13}N$ (liq) 76CON/GIN
2–Methylpiperidine
Heat Capacity 298 K, C_p = 49 cal·mol⁻¹·K⁻¹
205 J·mol⁻¹·K⁻¹
One temperature
Molecular Weight 99.1754
Wiswesser Line Notation T6MTJ B1
Evaluation C

Table of Heat Capacities, Entropies, and Phase Transition Properties — Continued

$C_6H_{13}N$ (liq) 76CON/GIN
4–Methylpiperidine
Heat Capacity 298 K, C_p = 50 cal·mol^{-1}·K^{-1}
209 J·mol^{-1}·K^{-1}
One temperature
Molecular Weight 99.1754
Wiswesser Line Notation T6MTJ D1
Evaluation C

$C_6H_{13}N$ (liq) 76CON/GIN
N–Methylpiperidine
Heat Capacity 298 K, C_p = 44.2 cal·mol^{-1}·K^{-1}
184.9 J·mol^{-1}·K^{-1}
One temperature
Molecular Weight 99.1754
Wiswesser Line Notation T6NTJ A1
Evaluation B

$C_6H_{13}N$ (liq) 76CON/GIN
Hexamethyleneimine; Perhydroazepine
Heat Capacity 298 K, C_p = 49 cal·mol^{-1}·K^{-1}
205 J·mol^{-1}·K^{-1}
One temperature
Molecular Weight 99.1754
Wiswesser Line Notation T7MTJ
Evaluation C

$C_6H_{13}NO$ (c) 71KON/WAD
2,2,N–Trimethylpropanamide
Heat Capacity 298.15 K, C_p = 43.7 cal·mol^{-1}·K^{-1}
183 J·mol^{-1}·K^{-1}
One temperature
Molecular Weight 115.1748
Wiswesser Line Notation 1X1&1&VM1
Evaluation B

$C_6H_{13}NO$ (c) 71KON/WAD
N–(2–Methyl–2–propyl)ethanamide; N–tert–Butylacetamide
Heat Capacity 298.15 K, C_p = 45.4 cal·mol^{-1}·K^{-1}
190 J·mol^{-1}·K^{-1}
One temperature
Molecular Weight 115.1748
Wiswesser Line Notation 1X1&1&MV1
Evaluation B

$C_6H_{13}NO$ (liq) 71KON/WAD
N–n–Butylethanamide; N–n–Butylacetamide
Heat Capacity 298.15 K, C_p = 56.4 cal·mol^{-1}·K^{-1}
236 J·mol^{-1}·K^{-1}
One temperature
Molecular Weight 115.1748
Wiswesser Line Notation 4MV1
Evaluation B

$C_6H_{13}NO$ (liq) 71KON/WAD
N–Methylpentanamide; N–Methylvaleramide
Heat Capacity 298.15 K, C_p = 54.7 cal·mol^{-1}·K^{-1}
229 J·mol^{-1}·K^{-1}
One temperature
Molecular Weight 115.1748
Wiswesser Line Notation 4VM1
Evaluation B

$C_6H_{13}NO_2$ (c) 37HUF/ELL
2–Amino–4–methylpentanoic acid(DL); Leucine(DL)
Heat Capacity 296.6 K, C_p = 46.44 cal·mol^{-1}·K^{-1}
194.30 J·mol^{-1}·K^{-1}
Temperature range 86–297 K. Value is unsmoothed experimental datum.
Entropy 298.15 K, S = 49.5 cal·mol^{-1}·K^{-1}
207.1 J·mol^{-1}·K^{-1}
Extrapolation below 90 K, 13.91 cal·mol^{-1}K^{-1}
Molecular Weight 131.1742
Wiswesser Line Notation QVYZ1Y1&1 -DL
Evaluation B(C_p),C(S)

$C_6H_{13}NO_2$ (c) 63HUT/COL
2–Amino–4–methylpentanoic acid(L); Leucine(L)
Heat Capacity 298.15 K, C_p = 48.03 cal·mol^{-1}·K^{-1}
200.96 J·mol^{-1}·K^{-1}
Temperature range 11–310 K
Entropy 298.15 K, S = 50.62 cal·mol^{-1}·K^{-1}
211.79 J·mol^{-1}·K^{-1}
Molecular Weight 131.1742
Wiswesser Line Notation QVYZ1Y1&1 -L
Evaluation A

$C_6H_{13}NO_2$ (c) 75SPI/WAD
2–Amino–4–methylpentanoic acid(L); Leucine(L)
Heat Capacity 298.15 K, C_p = 48.1 cal·mol^{-1}·K^{-1}
201.4 J·mol^{-1}·K^{-1}
One temperature
Molecular Weight 131.1742
Wiswesser Line Notation QVYZ1Y1&1 -L
Evaluation B

$C_6H_{13}NO_2$ (c) 63HUT/COL
2–Amino–3–methylpentanoic acid(L); Isoleucine(L)
Heat Capacity 298.15 K, C_p = 45.00 cal·mol^{-1}·K^{-1}
188.28 J·mol^{-1}·K^{-1}
Temperature range 11–310 K
Entropy 298.15 K, S = 49.71 cal·mol^{-1}·K^{-1}
207.99 J·mol^{-1}·K^{-1}
Molecular Weight 131.1742
Wiswesser Line Notation QVYZY2&1 -L
Evaluation A

$C_6H_{13}N_3O_3$ (c) 40HUF/FOX
Citrulline(DL)
Heat Capacity 300.8 K, C_p = 55.64 cal·mol^{-1}·K^{-1}
232.80 J·mol^{-1}·K^{-1}
Temperature range 90–298 K. Value is unsmoothed experimental datum.
Entropy 298.15 K, S = 60.8 cal·mol^{-1}·K^{-1}
254.4 J·mol^{-1}·K^{-1}
Extrapolation below 90 K, 18.15 cal·mol^{-1}K^{-1}
Molecular Weight 175.1870
Wiswesser Line Notation ZVM3YZVQ
Evaluation B(C_p),C(S)

$C_6H_{13}N_3O_8S$ (c,II) 68AGU/TEL
Triglycine sulfate
Heat Capacity C_p data only graphically.
Temperature range 12 to 60°C.
Phase Changes
c,II/c,I 322 K, ΔH = 258 cal·mol^{-1}
1079 J·mol^{-1}
ΔS = 0.78 cal·mol^{-1}·K^{-1}
3.26 J·mol^{-1}·K^{-1}
Molecular Weight 287.2440
Wiswesser Line Notation Z1VM1VM1VQ &WSQQ
Evaluation C

Table of Heat Capacities, Entropies, and Phase Transition Properties — Continued

$C_6H_{13}N_3O_8S$ (c) 80RAM/CER
Triglycine sulfate
Heat Capacity 322 K, C_p = 90.6 cal·mol⁻¹·K⁻¹
379.1 J·mol⁻¹·K⁻¹
Temperature 322 K. One temperature near the critical temperature.
Phase Changes
c,II/c,I 322 K
Specific heat anomaly at 322 K equal to 0.28 J·g⁻¹·K⁻¹.
Molecular Weight 287.2440
Wiswesser Line Notation Z1VM1VM1VQ &WSQQ
Evaluation A

C_6H_{14} (liq) 37STU
2,3–Dimethylbutane
Heat Capacity 298.1 K, C_p = 44.06 cal·mol⁻¹·K⁻¹
184.35 J·mol⁻¹·K⁻¹
Temperature range 140–320 K
Molecular Weight 86.1766
Wiswesser Line Notation 1Y1&Y1&1
Evaluation B(C_p)

C_6H_{14} (liq) 46DOU/HUF
2,3–Dimethylbutane
Heat Capacity 298.15 K, C_p = 45.11 cal·mol⁻¹·K⁻¹
188.74 J·mol⁻¹·K⁻¹
Temperature range 13–300 K
Entropy 298.15 K, S = 66.33 cal·mol⁻¹·K⁻¹
277.52 J·mol⁻¹·K⁻¹
Phase Changes
c,II/c,I 136.07 K, ΔH = 1552 cal·mol⁻¹
6494 J·mol⁻¹
ΔS = 11.41 cal·mol⁻¹·K⁻¹
47.72 J·mol⁻¹·K⁻¹
c,I/liq 145.19 K, ΔH = 191.4 cal·mol⁻¹
800.8 J·mol⁻¹
ΔS = 1.32 cal·mol⁻¹·K⁻¹
5.52 J·mol⁻¹·K⁻¹
Molecular Weight 86.1766
Wiswesser Line Notation 1Y1&Y1&1
Evaluation A

C_6H_{14} (liq) 71ADA/SUG
2,3–Dimethylbutane
Heat Capacity 298.15 K, C_p = 45.34 cal·mol⁻¹·K⁻¹
189.70 J·mol⁻¹·K⁻¹
Temperature range 13–300 K
Entropy 298.15 K, S = 66.65 cal·mol⁻¹·K⁻¹
278.85 J·mol⁻¹·K⁻¹
Phase Changes
c,III/c,I 107 K, ΔH = 566 cal·mol⁻¹
2370 J·mol⁻¹
ΔS = 5.29 cal·mol⁻¹·K⁻¹
22.15 J·mol⁻¹·K⁻¹
c,III has residual entropy of 2.7 J·mol⁻¹K⁻¹
c,II/c,I 136.02 K, ΔH = 1536 cal·mol⁻¹
6425 J·mol⁻¹
ΔS = 11.29 cal·mol⁻¹·K⁻¹
47.24 J·mol⁻¹·K⁻¹
c,II stable form to 10K; apparently has no zero point entropy.
c,I/liq 145.04 K, ΔH = 189.7 cal·mol⁻¹
793.7 J·mol⁻¹
ΔS = 1.31 cal·mol⁻¹·K⁻¹
5.47 J·mol⁻¹·K⁻¹
Molecular Weight 86.1766
Wiswesser Line Notation 1Y1&Y1&1
Evaluation A

C_6H_{14} (liq) 37STU
2,2–Dimethylbutane
Heat Capacity 298.1 K, C_p = 43.78 cal·mol⁻¹·K⁻¹
183.18 J·mol⁻¹·K⁻¹
Temperature range 90–320 K
Entropy 298.1 K, S = 64.4 cal·mol⁻¹·K⁻¹
269.4 J·mol⁻¹·K⁻¹
Extrapolation below 90 K, 17.76 cal·mol⁻¹K⁻¹
Phase Changes
c,II/c,I 127.11 K, ΔH = 1095 cal·mol⁻¹
4581 J·mol⁻¹
ΔS = 8.61 cal·mol⁻¹·K⁻¹
36.04 J·mol⁻¹·K⁻¹
c,I/liq 172.13 K, ΔH = 111 cal·mol⁻¹
464 J·mol⁻¹
ΔS = 0.64 cal·mol⁻¹·K⁻¹
2.70 J·mol⁻¹·K⁻¹
Molecular Weight 86.1766
Wiswesser Line Notation 2X1&1&1
Evaluation B(C_p),C(S)

C_6H_{14} (liq) 46DOU/HUF
2,2–Dimethylbutane
Heat Capacity 298.15 K, C_p = 45.11 cal·mol⁻¹·K⁻¹
188.74 J·mol⁻¹·K⁻¹
Temperature range 13–300 K
Entropy 298.15 K, S = 65.01 cal·mol⁻¹·K⁻¹
272.00 J·mol⁻¹·K⁻¹
Phase Changes
c,III/c,II 126.81 K, ΔH = 1293 cal·mol⁻¹
5410 J·mol⁻¹
ΔS = 10.20 cal·mol⁻¹·K⁻¹
42.66 J·mol⁻¹·K⁻¹
c,II/c,I 140.79 K, ΔH = 68.2 cal·mol⁻¹
285.3 J·mol⁻¹
ΔS = 0.48 cal·mol⁻¹·K⁻¹
2.03 J·mol⁻¹·K⁻¹
c,I/liq 174.28 K, ΔH = 138.4 cal·mol⁻¹
579.1 J·mol⁻¹
ΔS = 0.79 cal·mol⁻¹·K⁻¹
3.32 J·mol⁻¹·K⁻¹
Molecular Weight 86.1766
Wiswesser Line Notation 2X1&1&1
Evaluation A

C_6H_{14} (c) 46DOU/HUF
3–Methylpentane
Heat Capacity 298.15 K, C_p = 45.61 cal·mol⁻¹·K⁻¹
190.83 J·mol⁻¹·K⁻¹
Temperature range 13–300 K
Molecular Weight 86.1766
Wiswesser Line Notation 2Y2&1
Evaluation A

C_6H_{14} (liq) 37STU
3–Methylpentane
Heat Capacity 298.1 K, C_p = 44.78 cal·mol⁻¹·K⁻¹
187.36 J·mol⁻¹·K⁻¹
Temperature range 90–320 K
Molecular Weight 86.1766
Wiswesser Line Notation 2Y2&1
Evaluation B(C_p),C(S)

Table of Heat Capacities, Entropies, and Phase Transition Properties — Continued

C_6H_{14} (liq) 73FIN/MES
3–Methylpentane
Heat Capacity 298.15 K, C_p = 45.572 cal·mol^{-1}·K^{-1}
190.67 J·mol^{-1}·K^{-1}
Temperature range 10–330 K
Entropy 298.15 K, S = 69.92 cal·mol^{-1}·K^{-1}
292.5 J·mol^{-1}·K^{-1}
Thermodynamic properties calculated from a Debye function at 10 K.
Phase Changes
c,I/liq 110.26 K, ΔH = 1267.5 cal·mol^{-1}
5303.2 J·mol^{-1}
ΔS = 11.496 cal·mol^{-1}·K^{-1}
48.101 J·mol^{-1}·K^{-1}
Molecular Weight 86.1766
Wiswesser Line Notation 2Y2
Evaluation A

C_6H_{14} (liq) 37STU
2–Methylpentane
Heat Capacity 298.1 K, C_p = 47.43 cal·mol^{-1}·K^{-1}
198.45 J·mol^{-1}·K^{-1}
Temperature range 90–320 K
Entropy 298.1 K, S = 69.9 cal·mol^{-1}·K^{-1}
292.5 J·mol^{-1}·K^{-1}
Extrapolation below 90 K, 24.79 cal·mol^{-1}K^{-1}
Phase Changes
c/liq 119.69 K, ΔH = 89 cal·mol^{-1}
372 J·mol^{-1}
ΔS = 0.74 cal·mol^{-1}·K^{-1}
3.12 J·mol^{-1}·K^{-1}
Molecular Weight 86.1766
Wiswesser Line Notation 3Y1&1
Evaluation B(C_p),C(S)

C_6H_{14} (liq) 46DOU/HUF
2–Methylpentane
Heat Capacity 298.15 K, C_p = 46.30 cal·mol^{-1}·K^{-1}
193.7 J·mol^{-1}·K^{-1}
Temperature range 13–300 K
Entropy 298.15 K, S = 69.45 cal·mol^{-1}·K^{-1}
290.58 J·mol^{-1}·K^{-1}
Phase Changes
c/liq 119.55 K, ΔH = 1498 cal·mol^{-1}
6268 J·mol^{-1}
ΔS = 12.53 cal·mol^{-1}·K^{-1}
52.43 J·mol^{-1}·K^{-1}
Molecular Weight 86.1766
Wiswesser Line Notation 3Y1&1
Evaluation A

C_6H_{14} (liq) 81REI
n–Hexane
Heat Capacity 298 K, C_p = 46.4 cal·mol^{-1}·K^{-1}
194.1 J·mol^{-1}·K^{-1}
Temperature range 290–363 K
Molecular Weight 86.1766
Wiswesser Line Notation 6H
Evaluation D

C_6H_{14} (liq) 30PAR/HUF
n–Hexane
Heat Capacity 295.1 K, C_p = 45.8 cal·mol^{-1}·K^{-1}
191.6 J·mol^{-1}·K^{-1}
Temperature range 90–295 K. Value is unsmoothed experimental datum.
Entropy 298.15 K, S = 71.1 cal·mol^{-1}·K^{-1}
297.5 J·mol^{-1}·K^{-1}
Extrapolation below 90 K, 15.64 cal·mol^{-1}K^{-1}
Phase Changes
c/liq 178.6 K, ΔH = 3007 cal·mol^{-1}
12581 J·mol^{-1}
ΔS = 16.84 cal·mol^{-1}·K^{-1}
70.44 J·mol^{-1}·K^{-1}
Molecular Weight 86.1766
Wiswesser Line Notation 6H
Evaluation B(C_p),C(S)

C_6H_{14} (liq) 31HUF/PAR
n–Hexane
Heat Capacity 293.5 K, C_p = 46.2 cal·mol^{-1}·K^{-1}
193.3 J·mol^{-1}·K^{-1}
Temperature range 140–294 K. Value is unsmoothed experimental datum.
Entropy 298.1 K, S = 70.6 cal·mol^{-1}·K^{-1}
295.4 J·mol^{-1}·K^{-1}
Extrapolation below 90 K, 15.3 cal·mol^{-1}K^{-1}
Phase Changes
c/liq 177.9 K, ΔH = 3115 cal·mol^{-1}
13033 J·mol^{-1}
ΔS = 17.5 cal·mol^{-1}·K^{-1}
73.3 J·mol^{-1}·K^{-1}
Molecular Weight 86.1766
Wiswesser Line Notation 6H
Evaluation B(C_p),C(S)

C_6H_{14} (liq) 37STU
n–Hexane
Heat Capacity 298.1 K, C_p = 45.2 cal·mol^{-1}·K^{-1}
189.1 J·mol^{-1}·K^{-1}
Temperature range 90–320 K. Hump about 262 K with abnormal curve to 320 K.
Entropy 298.1 K, S = 69.2 cal·mol^{-1}·K^{-1}
289.5 J·mol^{-1}·K^{-1}
Extrapolation below 91 K, 13.07 cal·mol^{-1}K^{-1}
Phase Changes
c/liq 177.90 K, ΔH = 2950 cal·mol^{-1}
12343 J·mol^{-1}
ΔS = 16.58 cal·mol^{-1}·K^{-1}
69.38 J·mol^{-1}·K^{-1}
Molecular Weight 86.1766
Wiswesser Line Notation 6H
Evaluation B(C_p),C(S)

C_6H_{14} (liq) 39PHI
n–Hexane
Heat Capacity 300.7 K, C_p = 44.5 cal·mol^{-1}·K^{-1}
186.2 J·mol^{-1}·K^{-1}
One temperature
Molecular Weight 86.1766
Wiswesser Line Notation 6H
Evaluation C

C_6H_{14} (liq) 46DOU/HUF
n–Hexane
Heat Capacity 298.15 K, C_p = 46.60 cal·mol^{-1}·K^{-1}
194.97 J·mol^{-1}·K^{-1}
Temperature range 13–300 K
Entropy 298.15 K, S = 70.76 cal·mol^{-1}·K^{-1}
296.06 J·mol^{-1}·K^{-1}

Table of Heat Capacities, Entropies, and Phase Transition Properties — Continued

Phase Changes
c/liq 177.84 K, ΔH = 3126 cal·mol^{-1}
13079 J·mol^{-1}
ΔS = 17.5 cal·mol^{-1}·K^{-1}
73.54 J·mol^{-1}·K^{-1}
Molecular Weight 86.1766
Wiswesser Line Notation 6H
Evaluation A

C_6H_{14} (liq) 51CON/SAG
n–Hexane
Heat Capacity 299.8 K, C_p = 46.87 cal·mol^{-1}·K^{-1}
196.10 J·mol^{-1}·K^{-1}
Temperature range 80 to 200°F
Molecular Weight 86.1766
Wiswesser Line Notation 6H
Evaluation B

C_6H_{14} (liq) 74DIA/REN
n–Hexane
Heat Capacity 298.15 K, C_p = 47.44 cal·mol^{-1}·K^{-1}
198.5 J·mol^{-1}·K^{-1}
Temperature range 298–325 K
Molecular Weight 86.1766
Wiswesser Line Notation 6H
Evaluation A

C_6H_{14} (liq) 75GRI/RAS
n–Hexane
Heat Capacity 298 K, C_p = 46.9 cal·mol^{-1}·K^{-1}
196.2 J·mol^{-1}·K^{-1}
Temperature range 300–463 K
Molecular Weight 86.1766
Wiswesser Line Notation 6H
Evaluation B

C_6H_{14} (liq) 82WIL/ING
n–Hexane
Heat Capacity 298.15 K, C_p = 46.68 cal·mol^{-1}·K^{-1}
195.33 J·mol^{-1}·K^{-1}
One temperature
Molecular Weight 86.1766
Wiswesser Line Notation 6H
Evaluation A

$C_6H_{14}N_4O_2$ (c) 37HUF/ELL
Arginine(D)
Heat Capacity 296.8 K, C_p = 55.54 cal·mol^{-1}·K^{-1}
232.8 J·mol^{-1}·K^{-1}
Temperature range 86–297 K. Value is unsmoothed experimental datum.
Entropy 298.15 K, S = 59.9 cal·mol^{-1}·K^{-1}
250.6 J·mol^{-1}·K^{-1}
Extrapolation below 90 K, 17.52 cal·mol^{-1}K^{-1}
Molecular Weight 174.2022
Wiswesser Line Notation QVYZ3MYZUM -D
Evaluation B(C_p),C(S)

$C_6H_{14}O$ (liq) 33PAR/HUF
2,4–Dimethyl–3–oxapentane; Isopropyl ether
Heat Capacity 293.1 K, C_p = 51.70 cal·mol^{-1}·K^{-1}
216.31 J·mol^{-1}·K^{-1}
Temperature range 92–213 K. Value is unsmoothed experimental datum.
Entropy 298.1 K, S = 70.4 cal·mol^{-1}·K^{-1}
294.6 J·mol^{-1}·K^{-1}
Extrapolation below 90 K, 14.65 cal·mol^{-1}K^{-1}
Phase Changes
c/liq 186.3 K, ΔH = 2635 cal·mol^{-1}
11025 J·mol^{-1}
ΔS = 14.14 cal·mol^{-1}·K^{-1}
59.18 J·mol^{-1}·K^{-1}
Molecular Weight 102.1760
Wiswesser Line Notation 1Y1&OY1&1
Evaluation B(C_p),C(S)

$C_6H_{14}O$ (liq) 74AND/COU
2,4–Dimethyl–3–oxapentane; Isopropyl ether
Heat Capacity 298.15 K, C_p = 51.65 cal·mol^{-1}·K^{-1}
216.1 J·mol^{-1}·K^{-1}
Temperature range 10–340 K
Entropy 298.15 K, S = 72.80 cal·mol^{-1}·K^{-1}
304.6 J·mol^{-1}·K^{-1}
Phase Changes
c/liq 187.77 K, ΔH = 2876 cal·mol^{-1}
12035 J·mol^{-1}
ΔS = 15.32 cal·mol^{-1}·K^{-1}
64.09 J·mol^{-1}·K^{-1}
Molecular Weight 102.1760
Wiswesser Line Notation 1Y1&OY1&1
Evaluation A

$C_6H_{14}O$ (liq) 36EVA/EDL
4,4–Dimethyl–3–oxapentane; tert–Butyl ethyl ether
Heat Capacity 298 K, C_p = 52 cal·mol^{-1}·K^{-1}
218 J·mol^{-1}·K^{-1}
One temperature
Molecular Weight 102.1760
Wiswesser Line Notation 2OX1&1&1
Evaluation C

$C_6H_{14}O$ (liq) 36EVA/EDL
3,3–Dimethyl–2–oxapentane; tert–Amyl methyl ether
Heat Capacity 298 K, C_p = 53 cal·mol^{-1}·K^{-1}
222 J·mol^{-1}·K^{-1}
One temperature
Molecular Weight 102.1760
Wiswesser Line Notation 2X1&1&O1
Evaluation C

$C_6H_{14}O$ (liq) 75AND/COU
4–Oxaheptane; Di–n–propyl ether
Heat Capacity 298.15 K, C_p = 52.96 cal·mol^{-1}·K^{-1}
221.6 J·mol^{-1}·K^{-1}
Temperature range 10–330 K
Entropy 298.15 K, S = 77.41 cal·mol^{-1}·K^{-1}
323.9 J·mol^{-1}·K^{-1}
Phase Changes
c,I/liq 149.40 K, ΔH = 2280 cal·mol^{-1}
9540 J·mol^{-1}
ΔS = 15.26 cal·mol^{-1}·K^{-1}
63.86 J·mol^{-1}·K^{-1}
c,II/liq 158.36 K, ΔH = 2574 cal·mol^{-1}
10770 J·mol^{-1}
ΔS = 16.25 cal·mol^{-1}·K^{-1}
68.01 J·mol^{-1}·K^{-1}
liq/g 363.22 K, ΔH = 7475 cal·mol^{-1}
31274 J·mol^{-1}
ΔS = 20.58 cal·mol^{-1}·K^{-1}
86.10 J·mol^{-1}·K^{-1}
P = 101.30 kPa

Table of Heat Capacities, Entropies, and Phase Transition Properties — Continued

Molecular Weight 102.1760
Wiswesser Line Notation 3O3
Evaluation A

$C_6H_{14}O$ (liq) 29KEL 2
1-Hexanol; n-Hexyl alcohol
Heat Capacity 290.01 K, C_p = 55.56 cal·mol^{-1}·K^{-1}
232.46. J·mol^{-1}·K^{-1}
Temperature range 16–298 K. Value is unsmoothed experimental datum.
Entropy 298.15 K, S = 68.7 cal·mol^{-1}·K^{-1}
287.4 J·mol^{-1}·K^{-1}
Phase Changes
c/liq 225.8 K, ΔH = 3676 cal·mol^{-1}
15380 J·mol^{-1}
ΔS = 16.28 cal·mol^{-1}·K^{-1}
68.11 J·mol^{-1}·K^{-1}
Molecular Weight 102.1760
Wiswesser Line Notation Q6
Evaluation B

$C_6H_{14}O$ (liq) 59HUT/BAI
1-Hexanol; n-Hexyl alcohol
Heat Capacity 298 K, C_p = 58.5 cal·mol^{-1}·K^{-1}
244.8 J·mol^{-1}·K^{-1}
One temperature
Molecular Weight 102.1760
Wiswesser Line Notation Q6
Evaluation C

$C_6H_{14}O$ (liq) 76CON/GIN
3-Hexanol
Heat Capacity 298 K, C_p = 68.4 cal·mol^{-1}·K^{-1}
286.2 J·mol^{-1}·K^{-1}
One temperature
Molecular Weight 102.1760
Wiswesser Line Notation QY3&2
Evaluation B

$C_6H_{14}O_2$ (liq) 73KUS/SUU
3,6-Dioxaoctane; 1,2-Diethoxyethane
Heat Capacity 298.15 K, C_p = 62.00 cal·mol^{-1}·K^{-1}
259.4 J·mol^{-1}·K^{-1}
One temperature
Molecular Weight 118.1754
Wiswesser Line Notation 2O2O2
Evaluation B

$C_6H_{14}O_2$ (liq) **81REI**
4-Methyl-3,5-dioxaheptane; Acetal; 1,1-Diethoxyethane
Heat Capacity 298 K, C_p = 56.8 cal·mol^{-1}·K^{-1}
237.7 J·mol^{-1}·K^{-1}
Temperature range 289–382 K
Molecular Weight 118.1754
Wiswesser Line Notation 2OY1&O2
Evaluation D

$C_6H_{14}O_2$ (liq) 73KUS/SUU
2,5-Dioxaoctane; 1-n-Propoxy-2-methoxyethane
Heat Capacity 298.15 K, C_p = 59.49 cal·mol^{-1}·K^{-1}
248.9 J·mol^{-1}·K^{-1}
One temperature
Molecular Weight 118.1754
Wiswesser Line Notation 3O2O1
Evaluation B

$C_6H_{14}O_2$ (liq) 73KUS/SUU
3-Oxa-1-heptanol; 2-n-Butoxyethanol
Heat Capacity 298.15 K, C_p = 65.27 cal·mol^{-1}·K^{-1}
273.1 J·mol^{-1}·K^{-1}
One temperature
Molecular Weight 118.1754
Wiswesser Line Notation Q2O4
Evaluation B

$C_6H_{14}O_2$ (liq) 78ROU/PER
3-Oxa-1-heptanol; 2-Butoxyethanol
Heat Capacity 298.15 K, C_p = 64.7 cal·mol^{-1}·K^{-1}
270.6 J·mol^{-1}·K^{-1}
Temperatures 278.15, 283.15, 298.15, 313.15, 328.15 K.
Molecular Weight 118.1754
Wiswesser Line Notation Q2O4
Evaluation C

$C_6H_{14}O_3$ (liq) 66BEA/CLE
2,5,8-Trioxanonane; Diglyme
Heat Capacity 298.15 K, C_p = 65.5 cal·mol^{-1}·K^{-1}
274.1 J·mol^{-1}·K^{-1}
Temperature range 90–350 K
Entropy 298.15 K, S = 84.3 cal·mol^{-1}·K^{-1}
352.7 J·mol^{-1}·K^{-1}
Extrapolation below 90 K, 19.1 cal.mol^{-1}.K^{-1}
Phase Changes
c/liq 209.1 K, ΔH = 4253 cal·mol^{-1}
17795 J·mol^{-1}
ΔS = 20.3 cal·mol^{-1}·K^{-1}
85.1 J·mol^{-1}·K^{-1}
Molecular Weight 134.1748
Wiswesser Line Notation 1O2O2O1
Evaluation A(C_p),C(S)

$C_6H_{14}O_4$ (liq) 79STE/TAM
1,8-Dihydroxy-3,6-dioxaoctane; Triethylene glycol
Heat Capacity 298 K, C_p = 78.3 cal·mol^{-1}·K^{-1}
327.6 J·mol^{-1}·K^{-1}
Temperature range 273–533 K
Molecular Weight 150.1742
Wiswesser Line Notation Q2O2O2Q
Evaluation B

$C_6H_{14}O_6$ (c) 26PAR/HUF
Dulcitol; Galactitol
Heat Capacity 292.8 K, C_p = 57.0 cal·mol^{-1}·K^{-1}
238.5 J·mol^{-1}·K^{-1}
Temperature range 88–293 K. Value is unsmoothed experimental datum.
Entropy 298.1 K, S = 59.2 cal·mol^{-1}·K^{-1}
247.7 J·mol^{-1}·K^{-1}
Extrapolation below 90 K, 17.53 cal·mol^{-1}K^{-1}
Molecular Weight 182.1730
Wiswesser Line Notation Q1YQYQYQYQ1Q -DLLD
Evaluation B(C_p),C(S)
Meso form

$C_6H_{14}O_6$ (c) 29PAR/KEL
Dulcitol; Galactitol
Entropy 298.1 K, S = 56.0 cal·mol^{-1}·K^{-1}
234.3 J·mol^{-1}·K^{-1}
Extrapolation below 90 K, 14.4 cal·mol^{-1}K^{-1}.
Revision of previous data.
Molecular Weight 182.1730
Wiswesser Line Notation Q1YQYQYQYQ1Q -DLLD
Evaluation C
Meso form

Table of Heat Capacities, Entropies, and Phase Transition Properties — Continued

$C_6H_{14}O_6$ (c) 32SPA/THO
Mannitol
Heat Capacity 303 K, C_p = 58.5 cal·mol^{-1}·K^{-1}
244.8 J·mol^{-1}·K^{-1}
Temperature range 30 to 200°C
Phase Changes
c/liq 433.2 K, ΔH = 12806 cal·mol^{-1}
53580 J·mol^{-1}
ΔS = 29.6 cal·mol^{-1}·K^{-1}
123.7 J·mol^{-1}·K^{-1}
Molecular Weight 182.1730
Wiswesser Line Notation Q1YQYQYQYQ1Q -DDLL
Evaluation B

$C_6H_{14}O_6$ (c) 26PAR/AND
Mannitol(D)
Heat Capacity 294.1 K, C_p = 56.4 cal·mol^{-1}·K^{-1}
236.0 J·mol^{-1}·K^{-1}
Temperature range 88–294 K. Value is unsmoothed experimental datum.
Entropy 298 K, S = 60.5 cal·mol^{-1}·K^{-1}
253.1 J·mol^{-1}·K^{-1}
Extrapolation below 90 K, 18.80 cal·mol^{-1}K^{-1}
Molecular Weight 182.1730
Wiswesser Line Notation Q1YQYQYQYQ1Q -DDLL -D
Evaluation B(C_p),C(S)

$C_6H_{14}O_6$ (c) 29PAR/KEL
Mannitol(D)
Entropy 298.1 K, S = 57.0 cal·mol^{-1}·K^{-1}
238.5 J·mol^{-1}·K^{-1}
Extrapolation below 90 K, 15.3 cal·mol^{-1}K^{-1}.
Revision of previous data.
Molecular Weight 182.1730
Wiswesser Line Notation Q1YQYQYQYQ1Q -DDLL -D
Evaluation C

$C_6H_{14}S$ (liq) 67MES/TOD
2,4–Dimethyl–3–thiapentane; Diisopropyl sulfide
Heat Capacity 298.15 K, C_p = 55.45 cal·mol^{-1}·K^{-1}
232.00 J·mol^{-1}·K^{-1}
Temperature range 10–390 K
Entropy 298.15 K, S = 74.82 cal·mol^{-1}·K^{-1}
313.05 J·mol^{-1}·K^{-1}
Phase Changes
c/liq 195.07 K, ΔH = 2489 cal·mol^{-1}
10414 J·mol^{-1}
ΔS = 12.76 cal·mol^{-1}·K^{-1}
53.39 J·mol^{-1}·K^{-1}
Molecular Weight 118.2366
Wiswesser Line Notation 1Y1&SY1&1
Evaluation A

$C_6H_{14}S$ (liq) 61MCC/FIN
4–Thiaheptane; Dipropyl sulfide
Heat Capacity 298.15 K, C_p = 53.89 cal·mol^{-1}·K^{-1}
225.48 J·mol^{-1}·K^{-1}
Temperature range 11–370 K
Entropy 298.15 K, S = 80.85 cal·mol^{-1}·K^{-1}
338.28 J·mol^{-1}·K^{-1}
Phase Changes
c/liq 170.44 K, ΔH = 2902 cal·mol^{-1}
12142 J·mol^{-1}
ΔS = 17.03 cal·mol^{-1}·K^{-1}
71.24 J·mol^{-1}·K^{-1}
Molecular Weight 118.2366
Wiswesser Line Notation 3S3
Evaluation A

$C_6H_{14}S$ (liq) 70FIN/MCC
1–Hexanethiol; n–Hexyl mercaptan
Heat Capacity 298.15 K, C_p = 55.14 cal·mol^{-1}·K^{-1}
230.71 J·mol^{-1}·K^{-1}
Temperature range 10–370 K
Entropy 298.15 K, S = 82.03 cal·mol^{-1}·K^{-1}
343.21 J·mol^{-1}·K^{-1}
Phase Changes
c/liq 192.62 K, ΔH = 4305 cal·mol^{-1}
18012 J·mol^{-1}
ΔS = 22.35 cal·mol^{-1}·K^{-1}
93.51 J·mol^{-1}·K^{-1}
Molecular Weight 118.2366
Wiswesser Line Notation SH6
Evaluation A

$C_6H_{14}S_2$ (liq) 58HUB/DOU
4,5–Dithiaoctane; Dipropyl disulfide
Heat Capacity 298.15 K, C_p = 62.73 cal·mol^{-1}·K^{-1}
262.46 J·mol^{-1}·K^{-1}
Temperature range 10–360 K
Entropy 298.15 K, S = 89.28 cal·mol^{-1}·K^{-1}
373.55 J·mol^{-1}·K^{-1}
Phase Changes
c/liq 187.66 K, ΔH = 3300 cal·mol^{-1}
13807 J·mol^{-1}
ΔS = 17.58 cal·mol^{-1}·K^{-1}
73.57 J·mol^{-1}·K^{-1}
Molecular Weight 150.2966
Wiswesser Line Notation 3SS3
Evaluation A

$C_6H_{15}As$ (liq) 72MAS/FAM
Triethylarsine
Heat Capacity 298.15 K, C_p = 56.90 cal·mol^{-1}·K^{-1}
234.30 J·mol^{-1}·K^{-1}
Temperature range 60–300 K
Phase Changes
c/liq 181.8 K, ΔH = 2643 cal·mol^{-1}
11058 J·mol^{-1}
ΔS = 14.54 cal·mol^{-1}·K^{-1}
60.83 J·mol^{-1}·K^{-1}
Molecular Weight 162.1061
Wiswesser Line Notation 2–AS–2&2
Evaluation B

$C_6H_{15}B$ (c) 55FUR
Triethylborane
Heat Capacity 300 K, C_p = 57.7 cal·mol^{-1}·K^{-1}
241.4 J·mol^{-1}·K^{-1}
Temperature range 15–300 K
Entropy 300 K, S = 80.8 cal·mol^{-1}·K^{-1}
338.1 J·mol^{-1}·K^{-1}
Below 15°. Debye extrapolation.
Phase Changes
c,I/liq 180.3 K, ΔH = 2833 cal·mol^{-1}
11853 J·mol^{-1}
ΔS = 15.7 cal·mol^{-1}·K^{-1}
65.7 J·mol^{-1}·K^{-1}
liq/g 300 K, ΔH = 877 cal·mol^{-1}
3669 J·mol^{-1}
ΔS = 29.2 cal·mol^{-1}·K^{-1}
122 J·mol^{-1}·K^{-1}
P = 56.27 mmHg
Molecular Weight 97.9945
Wiswesser Line Notation 2B2&2
Evaluation B

Table of Heat Capacities, Entropies, and Phase Transition Properties — Continued

$C_6H_{15}ClN_4O_2$ (c) 63COL/HUT 2
Arginine hydrochloride(L)
Heat Capacity 298.15 K, C_p = 62.37 cal·mol^{-1}·K^{-1}
260.96 J·mol^{-1}·K^{-1}
Temperature range 11–305 K
Entropy 298.15 K, S = 68.43 cal·mol^{-1}·K^{-1}
286.31 J·mol^{-1}·K^{-1}
Molecular Weight 210.6631
Wiswesser Line Notation QVYZ3MYZUM &GH -L
Evaluation A

$C_6H_{15}Ga$ (liq) 72MAS/FAM
Triethylgallium; Gallium triethyl
Heat Capacity 298.15 K, C_p = 66.45 cal·mol^{-1}·K^{-1}
278.03 J·mol^{-1}·K^{-1}
Temperature range 60–300 K
Phase Changes
c/liq 193.5 K, ΔH = 2783 cal·mol^{-1}
11644 J·mol^{-1}
ΔS = 14.38 cal·mol^{-1}·K^{-1}
60.18 J·mol^{-1}·K^{-1}
Molecular Weight 156.9045
Wiswesser Line Notation 2–GA–2&2
Evaluation B

$C_6H_{15}In$ (liq) 73MAS/NOV
Triethylindium
Heat Capacity 298.15 K, C_p = 71.05 cal·mol^{-1}·K^{-1}
297.27 J·mol^{-1}·K^{-1}
Temperature range 60–300 K
Phase Changes
Anomaly in specific heat at 194.8 K
c/liq 237.6 K, ΔH = 3110 cal·mol^{-1}
13012 J·mol^{-1}
ΔS = 13.09 cal·mol^{-1}·K^{-1}
54.77 J·mol^{-1}·K^{-1}
Molecular Weight 202.0045
Wiswesser Line Notation 2–IN–2&2
Evaluation B

$C_6H_{15}In$ (liq) 73MAS/NOV
Triethylindium
Heat Capacity 298.15 K, C_p = 71.05 cal·mol^{-1}·K^{-1}
297.27 J·mol^{-1}·K^{-1}
Temperature range 60–300 K
Phase Changes
Anomaly in specific heat at 194.8 K
c/liq 237.6 K, ΔH = 3110 cal·mol^{-1}
13012 J·mol^{-1}
ΔS = 13.09 cal·mol^{-1}·K^{-1}
54.77 J·mol^{-1}·K^{-1}
Molecular Weight 202.0045
Wiswesser Line Notation 2–IN–2&2
Evaluation B

$C_6H_{15}N$ (liq) **81REI**
Triethylamine
Heat Capacity 298 K, C_p = 42.3 cal·mol^{-1}·K^{-1}
177.0 J·mol^{-1}·K^{-1}
Temperature range 293–378 K
Molecular Weight 101.1912
Wiswesser Line Notation 2N2&2
Evaluation D

$C_6H_{15}N$ (liq) 71KON/WAD
1–Aminohexane; n–Hexylamine
Heat Capacity 298.15 K, C_p = 60.2 cal·mol^{-1}·K^{-1}
252 J·mol^{-1}·K^{-1}
One temperature
Molecular Weight 101.1912
Wiswesser Line Notation Z6
Evaluation B

$C_6H_{15}Sb$ (liq) 73MAS/NOV
Triethylstibine; Triethylantimony
Heat Capacity 298.15 K, C_p = 57.90 cal·mol^{-1}·K^{-1}
242.25 J·mol^{-1}·K^{-1}
Temperature range 60–300 K
Phase Changes
c/liq 153.9 K, ΔH = 2259 cal·mol^{-1}
9452 J·mol^{-1}
ΔS = 14.68 cal·mol^{-1}·K^{-1}
61.42 J·mol^{-1}·K^{-1}
Molecular Weight 208.9345
Wiswesser Line Notation 2–SB–2&2
Evaluation B

$C_6H_{18}Si_2$ (liq) 59SUG/SEK
Hexamethyldisilane
Heat Capacity 295.67 K, C_p = 61.16 cal·mol^{-1}·K^{-1}
255.89 J·mol^{-1}·K^{-1}
Temperature range 200–300 K. Value is unsmoothed experimental datum.
Phase Changes
c,II/c,I 221.8 K, ΔH = 2330 cal·mol^{-1}
9750 J·mol^{-1}
ΔS = 10.50 cal·mol^{-1}·K^{-1}
43.96 J·mol^{-1}·K^{-1}
c,I/liq 287.7 K, ΔH = 721 cal·mol^{-1}
3017 J·mol^{-1}
ΔS = 2.51 cal·mol^{-1}·K^{-1}
10.49 J·mol^{-1}·K^{-1}
Molecular Weight 146.3792
Wiswesser Line Notation 1–SI–1&1&–SI–1&1&1
Evaluation B

$C_6H_{16}CdCl_4N_2$ (c) 82WHI/STA
Tetrachlorobis–(2–propeneammonium) cadmium II
Heat Capacity 298.15 K, C_p = 86.71 cal·mol^{-1}·K^{-1}
362.8 J·mol^{-1}·K^{-1}
Temperature range 10–300 K
Entropy 298.15 K, S = 118.8 cal·mol^{-1}·K^{-1}
497.0 J·mol^{-1}·K^{-1}
Phase Changes
c,III/c,II 206.9 K, ΔH = 511.5 cal·mol^{-1}
2140 J·mol^{-1}
ΔS = 2.51 cal·mol^{-1}·K^{-1}
7.87 J·mol^{-1}·K^{-1}
c,II/c,I 266.7 K, ΔH = 478.0 cal·mol^{-1}
2000 J·mol^{-1}
ΔS = 1.88 cal·mol^{-1}·K^{-1}
7.87 J·mol^{-1}·K^{-1}
Molecular Weight 370.4278
Wiswesser Line Notation CD Z2U1&2 G4
Evaluation A

Table of Heat Capacities, Entropies, and Phase Transition Properties — Continued

$C_6H_{16}Si_2$ (liq) 73DZH/GUS
Tetramethyldisiletan; Tetramethyldisilacyclobutane
Temperature range 12–300K; C_p data only in complete paper deposited at VINITI, No. 5024–72, 3 Nov 1972.
Entropy 298.15 K, S = 70.81 cal·mol^{-1}·K^{-1}
296.27 J·mol^{-1}·K^{-1}
Phase Changes
c/liq 266.02 K, ΔH = 2452 cal·mol^{-1}
10259 J·mol^{-1}
ΔS = 9.22 cal·mol^{-1}·K^{-1}
38.56 J·mol^{-1}·K^{-1}
Molecular Weight 144.3634
Wiswesser Line Notation T4–SI–C–SI–TJ A1 A1 C1 C1
Evaluation B

$C_6H_{18}BN$ (liq) 67SMI/GOO
Triethylamineborane
Heat Capacity 298.15 K, C_p = 61.4 cal·mol^{-1}·K^{-1}
256.9 J·mol^{-1}·K^{-1}
One temperature
Molecular Weight 115.0249
Wiswesser Line Notation 2N2&2 &BHHH
Evaluation B

$C_6H_{18}BN$ (liq) 70FIN/TOD
Triethylamineborane
Heat Capacity 298.15 K, C_p = 61.45 cal·mol^{-1}·K^{-1}
257.11 J·mol^{-1}·K^{-1}
Temperature range 10–310 K
Entropy 298.15 K, S = 72.11 cal·mol^{-1}·K^{-1}
301.71 J·mol^{-1}·K^{-1}
Phase Changes
c/liq 269.48 K, ΔH = 3562.7 cal·mol^{-1}
14906.3 J·mol^{-1}
ΔS = 13.22 cal·mol^{-1}·K^{-1}
55.32 J·mol^{-1}·K^{-1}
Molecular Weight 115.0249
Wiswesser Line Notation 2N2&2 &BHHH
Evaluation A

$C_6H_{18}OSi_2$ (liq) 61SCO/MES
Hexamethyldisiloxane
Heat Capacity 298.15 K, C_p = 74.42 cal·mol^{-1}·K^{-1}
311.37 J·mol^{-1}·K^{-1}
Temperature range 12–371 K
Entropy 298.15 K, S = 103.69 cal·mol^{-1}·K^{-1}
433.84 J·mol^{-1}·K^{-1}
Phase Changes
c/liq 204.93 K, ΔH = 2849.4 cal·mol^{-1}
11921.9 J·mol^{-1}
ΔS = 13.90 cal·mol^{-1}·K^{-1}
58.18 J·mol^{-1}·K^{-1}
Molecular Weight 162.3786
Wiswesser Line Notation 1–SI–1&1&O–SI–1&1&1
Evaluation A

$C_6H_{18}O_3Si_3$ (liq) 77KUL/DZH 2
Hexamethylcyclotrisiloxane
Temperature range 12–350 K. Data deposited in VINITI, No. 987–77, 14 March 1977. Includes C_p, S, ΔH phase transitions.
Molecular Weight 222.4629
Wiswesser Line Notation T6–SI–O–SI–O–SI–OTJ A1 A1 C1 C1 E1 E1
Evaluation C

$C_6H_{18}O_3Si_3$ (c) 82KUL/DZH
Hexamethylcyclotrisiloxane
Heat Capacity 298.15 K, C_p = 86.0 cal·mol^{-1}·K^{-1}
360.0 J·mol^{-1}·K^{-1}
Temperature range 4.2–370 K. Data given graphically except for data at 298.15 K.
Entropy 298.15 K, S = 98.5 cal·mol^{-1}·K^{-1}
412.1 J·mol^{-1}·K^{-1}
Phase Changes
c,I/liq 335.22 K, ΔH = 3970 cal·mol^{-1}
16611 J·mol^{-1}
ΔS = 11.8 cal·mol^{-1}·K^{-1}
49.5 J·mol^{-1}·K^{-1}
Molecular Weight 222.4629
Wiswesser Line Notation T6–SI–O–SI–O–SI–OTJ A1 A1 C1 C1 E1 E1
Evaluation B

$C_6H_{20}CdCl_4N_2$ (c) 81WHI/GRA
Tetrachlorobis–(n–propylammonium) cadmium II
Heat Capacity 298.15 K, C_p = 90.9 cal·mol^{-1}·K^{-1}
380.3 J·mol^{-1}·K^{-1}
Temperature range 10–300 K
Entropy 298.15 K, S = 124.2 cal·mol^{-1}·K^{-1}
519.7 J·mol^{-1}·K^{-1}
Phase Changes
c,IV/c,III 105.5 K, ΔH = 351.9 cal·mol^{-1}
1472 J·mol^{-1}
ΔS = 3.336 cal·mol^{-1}·K^{-1}
13.96 J·mol^{-1}·K^{-1}
c,III/c,II 156.8 K, ΔH = 143 cal·mol^{-1}
598 J·mol^{-1}
ΔS = 0.92 cal·mol^{-1}·K^{-1}
3.85 J·mol^{-1}·K^{-1}
c,II/c,I 178.7 K, ΔH = 244 cal·mol^{-1}
1021 J·mol^{-1}
ΔS = 1.47 cal·mol^{-1}·K^{-1}
6.15 J·mol^{-1}·K^{-1}
Molecular Weight 374.4594
Wiswesser Line Notation CD Z3&2 G4
Evaluation A

$C_6H_{20}Cl_4MnN_2$ (c) 81WHI/GRA
Tetrachlorobis–(n–propylammonium) manganese II
Heat Capacity 298.15 K, C_p = 93.5 cal·mol^{-1}·K^{-1}
391.2 J·mol^{-1}·K^{-1}
Temperature range 10–300 K
Entropy 298.15 K, S = 123.4 cal·mol^{-1}·K^{-1}
516.3 J·mol^{-1}·K^{-1}
Phase Changes
c,III/c,II 112.8 K, ΔH = 140 cal·mol^{-1}
586 J·mol^{-1}
ΔS = 1.31 cal·mol^{-1}·K^{-1}
5.48 J·mol^{-1}·K^{-1}
c,II/c,I 164.3 K, ΔH = 119 cal·mol^{-1}
498 J·mol^{-1}
ΔS = 0.79 cal·mol^{-1}·K^{-1}
3.31 J·mol^{-1}·K^{-1}
Molecular Weight 316.9874
Wiswesser Line Notation MN Z3&2 G4
Evaluation A

Table of Heat Capacities, Entropies, and Phase Transition Properties — Continued

C_6D_6 (liq) 42ZIE/AND
Benzene–d_6
Heat Capacity 298.5 K, C_p = 35.7 cal·mol^{-1}·K^{-1}
149.4 J·mol^{-1}·K^{-1}
Temperature range 100–320 K. Value is unsmoothed experimental datum.
Phase Changes
c/liq 279.85 K, ΔH = 2340 cal·mol^{-1}
9791 J·mol^{-1}
ΔS = 8.36 cal·mol^{-1}·K^{-1}
34.99 J·mol^{-1}·K^{-1}
Molecular Weight 84.1506
Wiswesser Line Notation R &1A–F/H–2 6
Evaluation B

C_6D_6 (liq) 62RAB/NIK
Benzene–d_6
Heat Capacity 298 K, C_p = 36.44 cal·mol^{-1}·K^{-1}
152.46 J·mol^{-1}·K^{-1}
Temperature range 10 to 35°C
Molecular Weight 84.1506
Wiswesser Line Notation R &1A–F/H–2 6
Evaluation B

C_6D_{12} (liq) 66NIK/RAB
Cyclohexane–d_{12}
Heat Capacity 298 K, C_p = 43.94 cal·mol^{-1}·K^{-1}
183.84 J·mol^{-1}·K^{-1}
Temperature range 10 to 50°C
Molecular Weight 96.2352
Wiswesser Line Notation L6TJ &1A–F/H–2 12
Evaluation B

C_7F_8 (liq) 74AND/MAR
Octafluorotoluene; Perfluorotoluene
Heat Capacity 298.15 K, C_p = 62.69 cal·mol^{-1}·K^{-1}
262.3 J·mol^{-1}·K^{-1}
Temperature range 10–315 K
Entropy 298.15 K, S = 84.97 cal·mol^{-1}·K^{-1}
355.5 J·mol^{-1}·K^{-1}
Phase Changes
c/liq 207.69 K, ΔH = 2746 cal·mol^{-1}
11490 J·mol^{-1}
ΔS = 13.22 cal·mol^{-1}·K^{-1}
55.32 J·mol^{-1}·K^{-1}
Molecular Weight 236.0642
Wiswesser Line Notation FXFFR BF CF DF EF FF
Evaluation A

C_7F_{14} (liq) 57YAR/KAY
Perfluoromethylcyclohexane
Heat Capacity 298 K, C_p = 84.4 cal·mol^{-1}·K^{-1}
353.1 J·mol^{-1}·K^{-1}
Temperature range 298–373 K. Equation only.
Molecular Weight 350.0546
Wiswesser Line Notation L6TJ AXFFF AF BF BF CF CF DF DF EF EF FF FF
Evaluation B

C_7F_{16} (liq) 57YAR/KAY
Perfluoroheptane; Hexadecafluoroheptane
Heat Capacity 298 K, C_p = 98.3 cal·mol^{-1}·K^{-1}
411.3 J·mol^{-1}·K^{-1}
Temperature range 298–373 K. Equation only.
Molecular Weight 388.0514
Wiswesser Line Notation FXFFXFFXFFXFFXFFXFFXFFF
Evaluation B

$C_7H_3F_5$ (liq) 68COU/HAL 2
2,3,4,5,6–Pentafluorotoluene
Heat Capacity 298.15 K, C_p = 53.97 cal·mol^{-1}·K^{-1}
225.8 J·mol^{-1}·K^{-1}
Temperature range 10–376 K
Entropy 298.15 K, S = 73.23 cal·mol^{-1}·K^{-1}
306.4 J·mol^{-1}·K^{-1}
Phase Changes
c,II/c,I 70.3 K, ΔH = 50.3 cal·mol^{-1}
210.3 J·mol^{-1}
ΔS = 0.70 cal·mol^{-1}·K^{-1}
2.94 J·mol^{-1}·K^{-1}
c,II/c,I 243.35 K, ΔH = 3105 cal·mol^{-1}
12990 J·mol^{-1}
ΔS = 12.76 cal·mol^{-1}·K^{-1}
53.38 J·mol^{-1}·K^{-1}
Molecular Weight 182.0927
Wiswesser Line Notation FR BF CF DF EF F1
Evaluation A

$C_7H_3F_5$ (liq) 71PAU
2,3,4,5,6–Pentafluorotoluene
Heat Capacity 298.15 K, C_p = 55.24 cal·mol^{-1}·K^{-1}
231.12 J·mol^{-1}·K^{-1}
Temperature range 12–305 K. Data deposited VINITI, No. 2538–71, 20 Jan 1971.
Entropy 298.15 K, S = 74.30 cal·mol^{-1}·K^{-1}
310.87 J·mol^{-1}·K^{-1}
Phase Changes
c/liq 243.7 K, ΔH = 3174 cal·mol^{-1}
13280 J·mol^{-1}
ΔS = 13.02 cal·mol^{-1}·K^{-1}
54.49 J·mol^{-1}·K^{-1}
Transition 0.3 K below mp. ΔH is sum of transition and fusion.
Molecular Weight 182.0927
Wiswesser Line Notation FR BF CF DF EF F1
Evaluation A

C_7H_4ClNO (liq) 65ZAL/KOC
m–Chlorophenylisocyanate
Heat Capacity 294.2 K, C_p = 44.7 cal·mol^{-1}·K^{-1}
187.0 J·mol^{-1}·K^{-1}
Temperature range 294, 323 K
Molecular Weight 153.5677
Wiswesser Line Notation OCNR CG
Evaluation C

C_7H_4ClNO (c) 65ZAL/KOC
p–Chlorophenylisocyanate
Heat Capacity 282 K, C_p = 50.4 cal·mol^{-1}·K^{-1}
210.9 J·mol^{-1}·K^{-1}
Temperature range 273–290 K, mean value
Molecular Weight 153.5677
Wiswesser Line Notation OCNR DG
Evaluation C

C_7H_5ClO (liq) **81REI**
Benzoyl chloride
Heat Capacity 298 K, C_p = 44.7 cal·mol^{-1}·K^{-1}
187.0 J·mol^{-1}·K^{-1}
Temperature range 290–477 K
Molecular Weight 140.5689
Wiswesser Line Notation GVR
Evaluation D

Table of Heat Capacities, Entropies, and Phase Transition Properties — Continued

$C_7H_5ClO_2$ (c) 26AND/LYN
2–Chlorobenzoic acid
Heat Capacity 298 K, C_p = 39.0 cal·mol^{-1}·K^{-1}
163.2 J·mol^{-1}·K^{-1}
Temperature range 22 to 200°C
Phase Changes
c/liq 413.4 K, ΔH = 6150 cal·mol^{-1}
25730 J·mol^{-1}
ΔS = 14.9 cal·mol^{-1}·K^{-1}
62.2 J·mol^{-1}·K^{-1}
Molecular Weight 156.5683
Wiswesser Line Notation QVR BG
Evaluation C

$C_7H_5ClO_2$ (c) 26AND/LYN
3–Chlorobenzoic acid
Heat Capacity 298 K, C_p = 39.1 cal·mol^{-1}·K^{-1}
163.6 J·mol^{-1}·K^{-1}
Temperature range 22 to 205°C
Phase Changes
c/liq 427.4 K, ΔH = 5700 cal·mol^{-1}
23850 J·mol^{-1}
ΔS = 13.3 cal·mol^{-1}·K^{-1}
55.8 J·mol^{-1}·K^{-1}
Molecular Weight 156.5683
Wiswesser Line Notation QVR CG
Evaluation C

$C_7H_5ClO_2$ (c) 26AND/LYN
4–Chlorobenzoic acid
Heat Capacity 298 K, C_p = 40.1 cal·mol^{-1}·K^{-1}
167.8 J·mol^{-1}·K^{-1}
Temperature range 22 to 265°C
Phase Changes
c/liq 512.9 K, ΔH = 7710 cal·mol^{-1}
32260 J·mol^{-1}
ΔS = 15.0 cal·mol^{-1}·K^{-1}
62.9 J·mol^{-1}·K^{-1}
Molecular Weight 156.5683
Wiswesser Line Notation QVR DG
Evaluation C

$C_7H_5F_3$ (liq) 59SCO/DOU
Benzotrifluoride; α,α,α–Trifluorotoluene
Heat Capacity 298.15 K, C_p = 45.04 cal·mol^{-1}·K^{-1}
188.45 J·mol^{-1}·K^{-1}
Temperature range 12–365 K
Entropy 298.15 K, S = 64.89 cal·mol^{-1}·K^{-1}
271.50 J·mol^{-1}·K^{-1}
Phase Changes
c/liq 244.14 K, ΔH = 3294 cal·mol^{-1}
13782 J·mol^{-1}
ΔS = 13.49 cal·mol^{-1}·K^{-1}
56.45 J·mol^{-1}·K^{-1}
Molecular Weight 146.1117
Wiswesser Line Notation FXFFR
Evaluation A

$C_7H_5NO_4$ (c) 26AND
2–Nitrobenzoic acid
Heat Capacity 297.9 K, C_p = 45.4 cal·mol^{-1}·K^{-1}
190.0 J·mol^{-1}·K^{-1}
Temperature range 110–344 K. Value is unsmoothed experimental datum.
Molecular Weight 167.1208
Wiswesser Line Notation WNR BVQ
Evaluation C

$C_7H_5NO_4$ (c) 26AND/LYN
2–Nitrobenzoic acid
Heat Capacity 298 K, C_p = 45.8 cal·mol^{-1}·K^{-1}
191.6 J·mol^{-1}·K^{-1}
Temperature range 22 to 240°C
Phase Changes
c/liq 419.0 K, ΔH = 6690 cal·mol^{-1}
27990 J·mol^{-1}
ΔS = 16.0 cal·mol^{-1}·K^{-1}
66.8 J·mol^{-1}·K^{-1}
Molecular Weight 167.1208
Wiswesser Line Notation WNR BVQ
Evaluation C

$C_7H_5NO_4$ (c) 41SAT/SOG
2–Nitrobenzoic acid
Heat Capacity 323 K, C_p = 48.3 cal·mol^{-1}·K^{-1}
202.1 J·mol^{-1}·K^{-1}
Temperature range 0 to 100°C. Mean value.
Molecular Weight 167.1208
Wiswesser Line Notation WNR BVQ
Evaluation C
Same data in 40SAT/SOG 2.

$C_7H_5NO_4$ (c) 26AND
3–Nitrobenzoic acid
Heat Capacity 297.9 K, C_p = 43.0 cal·mol^{-1}·K^{-1}
179.9 J·mol^{-1}·K^{-1}
Temperature range 110–344 K. Value is unsmoothed experimental datum.
Molecular Weight 167.1208
Wiswesser Line Notation WNR CVQ
Evaluation C

$C_7H_5NO_4$ (c) 26AND/LYN
3–Nitrobenzoic acid
Heat Capacity 298 K, C_p = 41.4 cal·mol^{-1}·K^{-1}
173.2 J·mol^{-1}·K^{-1}
Temperature range 22 to 225°C
Phase Changes
c/liq 414.3 K, ΔH = 4620 cal·mol^{-1}
19330 J·mol^{-1}
ΔS = 11.2 cal·mol^{-1}·K^{-1}
46.7 J·mol^{-1}·K^{-1}
Molecular Weight 167.1208
Wiswesser Line Notation WNR CVQ
Evaluation C

$C_7H_5NO_4$ (c) 41SAT/SOG
3–Nitrobenzoic acid
Heat Capacity 323 K, C_p = 48.2 cal·mol^{-1}·K^{-1}
201.7 J·mol^{-1}·K^{-1}
Temperature range 0 to 100°C. Mean value.
Molecular Weight 167.1208
Wiswesser Line Notation WNR CVQ
Evaluation C
Same data in 40SAT/SOG 2.

$C_7H_5NO_4$ (c) 26AND
4–Nitrobenzoic acid
Heat Capacity 297.9 K, C_p = 43.7 cal·mol^{-1}·K^{-1}
182.8 J·mol^{-1}·K^{-1}
Temperature range 110–344 K. Value is unsmoothed expermential datum.
Molecular Weight 167.1208
Wiswesser Line Notation WNR DVQ
Evaluation C

Table of Heat Capacities, Entropies, and Phase Transition Properties — Continued

$C_7H_5NO_4$ (c) 26AND/LYN
4–Nitrobenzoic acid
Heat Capacity 298 K, C_p = 43.1 cal·mol^{-1}·K^{-1}
180.3 J·mol^{-1}·K^{-1}
Temperature range 22 to 245°C
Phase Changes
c/liq 512.4 K, ΔH = 8820 cal·mol^{-1}
36900 J·mol^{-1}
ΔS = 17.2 cal·mol^{-1}·K^{-1}
72.0 J·mol^{-1}·K^{-1}
Molecular Weight 167.1208
Wiswesser Line Notation WNR DVQ
Evaluation C

$C_7H_5NO_4$ (c) 41SAT/SOG
4–Nitrobenzoic acid
Heat Capacity 323 K, C_p = 47.3 cal·mol^{-1}·K^{-1}
197.9 J·mol^{-1}·K^{-1}
Temperature range 0 to 100°C. Mean value.
Molecular Weight 167.1208
Wiswesser Line Notation WNR DVQ
Evaluation C
Same data in 40SAT/SOG 2.

C_7H_5NS (liq) 36KUR/VOS
Phenyl isothiocyanate
Heat Capacity 290 K, C_p = 44.5 cal·mol^{-1}·K^{-1}
186.2 J·mol^{-1}·K^{-1}
One temperature
Molecular Weight 135.1832
Wiswesser Line Notation SCNR
Evaluation D

C_7H_5NS (liq) 69GOU/WES
Benzothiazole
Heat Capacity 298.15 K, C_p = 45.30 cal·mol^{-1}·K^{-1}
189.54 J·mol^{-1}·K^{-1}
Temperature range 5–325 K
Entropy 298.15 K, S = 50.15 cal·mol^{-1}·K^{-1}
209.83 J·mol^{-1}·K^{-1}
Phase Changes
c/liq 275.60 K, ΔH = 3055 cal·mol^{-1}
12782 J·mol^{-1}
ΔS = 11.08 cal·mol^{-1}·K^{-1}
46.38 J·mol^{-1}·K^{-1}
Molecular Weight 135.1832
Wiswesser Line Notation T56 BN DSJ
Evaluation A

$C_7H_5N_3O_6$ (c) 24TAY/RIN
2,4,6–Trinitrotoluene
Heat Capacity 293 K, C_p = 74.5 cal·mol^{-1}·K^{-1}
311.7 J·mol^{-1}·K^{-1}
Temperature range 90–352 K
Molecular Weight 227.1330
Wiswesser Line Notation WNR B1 CNW ENW
Evaluation C

$C_7H_5N_3O_6$ (c,II) 71CHI/THO
2,4,5–Trinitrotoluene; γ–TNT
Heat Capacity 313.2 K, C_p = 74.6 cal·mol^{-1}·K^{-1}
312 J·mol^{-1}·K^{-1}
Temperature range 313–333 K. At 313.2 K, C_p of c,I is 248 J·mol^{-1}K^{-1} (59.3 cal·mol^{-1}K^{-1}).
Phase Changes
c,II/c,I 319.2 K, ΔH = 1360 cal·mol^{-1}
5700 J·mol^{-1}
ΔS = 4 cal·mol^{-1}·K^{-1}
18 J·mol^{-1}·K^{-1}
c,I/liq 376.2 K, ΔH = 5900 cal·mol^{-1}
24700 J·mol^{-1}
ΔS = 16 cal·mol^{-1}·K^{-1}
66 J·mol^{-1}·K^{-1}
c,II/liq 347.2 K, ΔH = 3180 cal·mol^{-1}
13300 J·mol^{-1}
ΔS = 9 cal·mol^{-1}·K^{-1}
38 J·mol^{-1}·K^{-1}
Molecular Weight 227.1330
Wiswesser Line Notation WNR B1 DNW ENW
Evaluation C

$C_7H_5N_5O_8$ (c) 24TAY/RIN
2,4,6–Trinitrophenylmethyl nitramine; Tetryl;
N–Methyl–2,4,6,N–tetranitroaniline
Heat Capacity 293 K, C_p = 62.3 cal·mol^{-1}·K^{-1}
260.7 J·mol^{-1}·K^{-1}
Temperature range 90–370 K
Molecular Weight 287.1452
Wiswesser Line Notation WNN1&R BNW DNW FNW
Evaluation C

$C_7H_5N_5O_8$ (c) 73KRI/LIC
2,4,6–Trinitrophenylmethyl nitramine; Tetryl;
N–Methyl–2,4,6,N–tetranitroaniline
Heat Capacity 298 K, C_p = 72.2 cal·mol^{-1}·K^{-1}
302.1 J·mol^{-1}·K^{-1}
Temperature range 200–403 K. Equation only.
Phase Changes
c/liq 402.6 K, ΔH = 5480 cal·mol^{-1}
22930 J·mol^{-1}
ΔS = 13.6 cal·mol^{-1}·K^{-1}
57.0 J·mol^{-1}·K^{-1}
Molecular Weight 287.1452
Wiswesser Line Notation WNN1&R BNW DNW FNW
Evaluation C

$C_7H_6F_8O_3$ (liq) 80LEB/DOB
Bis–(tetrafluoropropyl)carbonate
Heat Capacity 298.15 K, C_p = 93.18 cal·mol^{-1}·K^{-1}
389.87 J·mol^{-1}·K^{-1}
Temperature range 12–360 K, Data given in tables and by equations.
Entropy 298.15 K, S = 128.83 cal·mol^{-1}·K^{-1}
539.02 J·mol^{-1}·K^{-1}
Phase Changes
c,I/liq 253.35 K, ΔH = 9811 cal·mol^{-1}
41049 J·mol^{-1}
ΔS = 38.73 cal·mol^{-1}·K^{-1}
162.0 J·mol^{-1}·K^{-1}
Molecular Weight 290.1098
Wiswesser Line Notation 1XFFXFF&O 2V
Evaluation A

$C_7H_6N_2O_4$ (c) 64DAV
2,4–Dinitrotoluene
Heat Capacity 325 K, C_p = 61 cal·mol^{-1}·K^{-1}
255 J·mol^{-1}·K^{-1}
Temperature range 298–340 K. Mean value.
Temperature range uncertain.

Table of Heat Capacities, Entropies, and Phase Transition Properties — Continued

Phase Changes
c/liq 345 K, ΔH = 5000 cal·mol^{-1}
20900 J·mol^{-1}
ΔS = 15 cal·mol^{-1}·K^{-1}
61 J·mol^{-1}·K^{-1}

Temperature not measured
Molecular Weight 182.1354
Wiswesser Line Notation WNR B1 ENW
Evaluation D

C_7H_6O (liq) 34KOL/UDO
Benzaldehyde
Heat Capacity 302.4 K, C_p = 40.9 cal·mol^{-1}·K^{-1}
171.1 J·mol^{-1}·K^{-1}

One temperature
Molecular Weight 106.1238
Wiswesser Line Notation VHR
Evaluation C

C_7H_6O (liq) 34KOL/UDO 2
Benzaldehyde
Heat Capacity 302.3 K, C_p = 40.9 cal·mol^{-1}·K^{-1}
171.1 J·mol^{-1}·K^{-1}

One temperature
Molecular Weight 106.1238
Wiswesser Line Notation VHR
Evaluation C

C_7H_6O (liq) 75AMB/CON
Benzaldehyde
Heat Capacity 298.15 K, C_p = 41.11 cal·mol^{-1}·K^{-1}
172.0 J·mol^{-1}·K^{-1}

Temperature range 13–425 K
Entropy 298.15 K, S = 52.87 cal·mol^{-1}·K^{-1}
221.2 J·mol^{-1}·K^{-1}

Phase Changes
c/liq 216.02 K, ΔH = 2228 cal·mol^{-1}
9320 J·mol^{-1}
ΔS = 10.31 cal·mol^{-1}·K^{-1}
43.14 J·mol^{-1}·K^{-1}

Molecular Weight 106.1238
Wiswesser Line Notation VHR
Evaluation A

$C_7H_6O_2$ (c) 26AND/LYN
Benzoic acid
Heat Capacity 298 K, C_p = 37.1 cal·mol^{-1}·K^{-1}
155.2 J·mol^{-1}·K^{-1}

Temperature range 22 to 200°C
Phase Changes
c/liq 395.0 K, ΔH = 4140 cal·mol^{-1}
17320 J·mol^{-1}
ΔS = 10.5 cal·mol^{-1}·K^{-1}
43.8 J·mol^{-1}·K^{-1}

Molecular Weight 122.1232
Wiswesser Line Notation QVR
Evaluation C

$C_7H_6O_2$ (c) 33PAR/HUF
Benzoic acid
Heat Capacity 295.1 K, C_p = 34.68 cal·mol^{-1}·K^{-1}
145.10 J·mol^{-1}·K^{-1}

Temperature range 93–295 K. Value is unsmoothed experimental datum.
Entropy 298.1 K, S = 40.8 cal·mol^{-1}·K^{-1}
170.7 J·mol^{-1}·K^{-1}

Extrapolation below 90 K, 14.16 cal·mol^{-1}K^{-1}
Molecular Weight 122.1232
Wiswesser Line Notation QVR
Evaluation B(C_p),C(S)

$C_7H_6O_2$ (c) 39SAT/SOG 2
Benzoic acid
Heat Capacity 323 K, C_p = 38.3 cal·mol^{-1}·K^{-1}
160.2 J·mol^{-1}·K^{-1}

Temperature range 0 to 100°C. Mean value.
Molecular Weight 122.1232
Wiswesser Line Notation QVR
Evaluation C
Same data in 40SAT/SOG.

$C_7H_6O_2$ (c) 51FUR/MCC
Benzoic acid
Heat Capacity 298.15 K, C_p = 35.09 cal·mol^{-1}·K^{-1}
146.81 J·mol^{-1}·K^{-1}

Temperature range 13–410 K
Entropy 298.15 K, S = 40.05 cal·mol^{-1}·K^{-1}
167.59 J·mol^{-1}·K^{-1}

Phase Changes
c/liq 395.52 K, ΔH = 4300 cal·mol^{-1}
18000 J·mol^{-1}
ΔS = 10.87 cal·mol^{-1}·K^{-1}
45.51 J·mol^{-1}·K^{-1}

Molecular Weight 122.1232
Wiswesser Line Notation QVR
Evaluation A

$C_7H_6O_2$ (c) 53GIN/FUR
Benzoic acid
Heat Capacity 298.15 K, C_p = 35.09 cal·mol^{-1}·K^{-1}
146.81 J·mol^{-1}·K^{-1}

Temperature range 14–570 K
Phase Changes
c/liq 395.52 K, ΔH = 4304 cal·mol^{-1}
18006 J·mol^{-1}
ΔS = 10.88 cal·mol^{-1}·K^{-1}
45.52 J·mol^{-1}·K^{-1}

Molecular Weight 122.1232
Wiswesser Line Notation QVR
Evaluation A

$C_7H_6O_2$ (c) 56POP/KOL
Benzoic acid
Heat Capacity 298.15 K, C_p = 35.18 cal·mol^{-1}·K^{-1}
149.79 J·mol^{-1}·K^{-1}

Temperature range 80–300 K
Molecular Weight 122.1232
Wiswesser Line Notation QVR
Evaluation A

$C_7H_6O_2$ (c) 62KOL/SER
Benzoic acid
Heat Capacity 298.15 K, C_p = 34.97 cal·mol^{-1}·K^{-1}
146.31 J·mol^{-1}·K^{-1}

Temperature range 22–310 K
Molecular Weight 122.1232
Wiswesser Line Notation QVR
Evaluation A

Table of Heat Capacities, Entropies, and Phase Transition Properties — Continued

$C_7H_6O_2$ (c) 64DAV
Benzoic acid
Heat Capacity 340 K, C_p = 31 cal·mol^{-1}·K^{-1}
130 J·mol^{-1}·K^{-1}
Temperature range 298–373 K. Mean value.
Temperature range uncertain.
Phase Changes
c/liq 395 K, ΔH = 4160 cal·mol^{-1}
17400 J·mol^{-1}
ΔS = 11 cal·mol^{-1}·K^{-1}
44 J·mol^{-1}·K^{-1}
Temperature not measured.
Molecular Weight 122.1232
Wiswesser Line Notation QVR
Evaluation D

$C_7H_6O_2$ (c) 65SUG/SEK
Benzoic acid
Heat Capacity 299.99 K, C_p = 35.17 cal·mol^{-1}·K^{-1}
147.14 J·mol^{-1}·K^{-1}
Temperature range 13–300 K. Value is unsmoothed experimental datum.
Molecular Weight 122.1232
Wiswesser Line Notation QVR
Evaluation A

$C_7H_6O_2$ (c) 69JUS
Benzoic acid
Entropy 298.15 K, S = 40.01 cal·mol^{-1}·K^{-1}
167.40 J·mol^{-1}·K^{-1}
As check on system. Only value at 298 K given.
Molecular Weight 122.1232
Wiswesser Line Notation QVR
Evaluation B

$C_7H_6O_2$ (c) 74MOS/MOU
Benzoic acid
Heat Capacity 301 K, C_p = 35.6 cal·mol^{-1}·K^{-1}
149 J·mol^{-1}·K^{-1}
One temperature, ΔT = 5 K. Value ±5 J·mol^{-1}K^{-1}.
Molecular Weight 122.1232
Wiswesser Line Notation QVR
Evaluation B

$C_7H_6O_2$ (c) 76ARV/FAL
Benzoic acid
Heat Capacity 298.15 K, C_p = 35.08 cal·mol^{-1}·K^{-1}
146.79 J·mol^{-1}·K^{-1}
Temperature range 6–341 K
Entropy 298.15 K, S = 40.09 cal·mol^{-1}·K^{-1}
167.73 J·mol^{-1}·K^{-1}
Molecular Weight 122.1232
Wiswesser Line Notation QVR
Evaluation A

$C_7H_6O_2$ (c) 80AND/CON
Benzoic acid
Phase Changes
c,I/liq 395.527 K, ΔH = 4317 cal·mol^{-1}
18062 J·mol^{-1}
ΔS = 10.91 cal·mol^{-1}·K^{-1}
45.67 J·mol^{-1}·K^{-1}
Molecular Weight 122.1232
Wiswesser Line Notation QVR
Evaluation A

$C_7H_6O_2$ (c) 82MOR/MAT
Benzoic acid
Heat Capacity 296.29 K, C_p = 34.95 cal·mol^{-1}·K^{-1}
146.23 J·mol^{-1}·K^{-1}
Temperature range 13–355 K, NBS SRM 39.
Molecular Weight 122.1232
Wiswesser Line Notation QVR
Evaluateion A

$C_7H_6O_2$ (liq) 67PAC
Benzoic acid
Heat Capacity 413 K, C_p = 62 cal·mol^{-1}·K^{-1}
259 J·mol^{-1}·K^{-1}
One temperature
Phase Changes
c/liq 395 K, ΔH = 3880 cal·mol^{-1}
16230 J·mol^{-1}
ΔS = 9.8 cal·mol^{-1}·K^{-1}
41.1 J·mol^{-1}·K^{-1}
Molecular Weight 122.1232
Wiswesser Line Notation QVR
Evaluation C

$C_7H_6O_2$ (liq) 07WAL
2–Hydroxybenzaldehyde; Salicylaldehyde
Heat Capacity 291 K, C_p = 53 cal·mol^{-1}·K^{-1}
222 J·mol^{-1}·K^{-1}
One temperature
Molecular Weight 122.1232
Wiswesser Line Notation VHR BQ
Evaluation D

$C_7H_6O_3$ (c) 34PAR/LIG
Salicylic acid; o–Hydroxybenzoic acid
Heat Capacity 288.6 K, C_p = 38.1 cal·mol^{-1}·K^{-1}
159.4 J·mol^{-1}·K^{-1}
Temperature range 96–289 K. Value is unsmoothed experimental datum.
Entropy 298.15 K, S = 42.5 cal·mol^{-1}·K^{-1}
178.2 J·mol^{-1}·K^{-1}
Extrapolation below 90 K, 13.70 cal·mol^{-1}K^{-1}
Molecular Weight 138.1226
Wiswesser Line Notation QVR BQ
Evaluation B(C_p),C(S)

$C_7H_6O_3$ (c) 40CAM/CAM
Salicylic acid; o–Hydroxybenzoic acid
Heat Capacity 293 K, C_p = 24.9 cal·mol^{-1}·K^{-1}
104.2 J·mol^{-1}·K^{-1}
One temperature
Molecular Weight 138.1226
Wiswesser Line Notation QVR BQ
Evaluation C

$C_7H_6O_3$ (c) 34PAR/LIG
m–Hydroxybenzoic acid
Heat Capacity 288.4 K, C_p = 37.6 cal·mol^{-1}·K^{-1}
157.3 J·mol^{-1}·K^{-1}
Temperature range 94–288 K. Value is unsmoothed experimental datum.
Entropy 298.15 K, S = 42.3 cal·mol^{-1}·K^{-1}
177.0 J·mol^{-1}·K^{-1}
Extrapolation below 90 K, 13.50 cal·mol^{-1}K^{-1}
Molecular Weight 138.1226
Wiswesser Line Notation QVR CQ
Evaluation B(C_p),C(S)

Table of Heat Capacities, Entropies, and Phase Transition Properties — Continued

$C_7H_6O_3$ (c) 34PAR/LIG
p–Hydroxybenzoic acid
Heat Capacity 283.8 K, C_p = 37.1 cal·mol^{-1}·K^{-1}
155.2 J·mol^{-1}·K^{-1}
Temperature range 95–284 K. Value is unsmoothed experimental datum.
Entropy 298.15 K, S = 42.0 cal·mol^{-1}·K^{-1}
175.7 J·mol^{-1}·K^{-1}
Extrapolation below 90 K, 13.23 cal·mol^{-1}K^{-1}
Molecular Weight 138.1226
Wiswesser Line Notation QVR DQ
Evaluation B(C_p),C(S)

C_7H_7Cl (liq) 31SMI/AND
Benzyl chloride; α–Chlorotoluene; Phenylchloromethane
Heat Capacity 298.5 K, C_p = 43.6 cal·mol^{-1}·K^{-1}
182.4 J·mol^{-1}·K^{-1}
Temperature range 102–299 K. Value is unsmoothed experimental datum
Molecular Weight 126.5853
Wiswesser Line Notation G1R
Evaluation C

C_7H_7F (liq) 62SCO/MES
4–Fluorotoluene
Heat Capacity 298.15 K, C_p = 40.91 cal·mol^{-1}·K^{-1}
171.17 J·mol^{-1}·K^{-1}
Temperature range 13–361 K
Entropy 298.15 K, S = 56.67 cal·mol^{-1}·K^{-1}
237.11 J·mol^{-1}·K^{-1}
Phase Changes
c/liq 216.49 K, ΔH = 2235 cal·mol^{-1}
9351 J·mol^{-1}
ΔS = 10.32 cal·mol^{-1}·K^{-1}
43.19 J·mol^{-1}·K^{-1}
Molecular Weight 110.1307
Wiswesser Line Notation FR D1
Evaluation A

$C_7H_7NO_2$ (liq) 34KOL/UDO 2
2–Nitrotoluene
Heat Capacity 302.3 K, C_p = 48.4 cal·mol^{-1}·K^{-1}
202.5 J·mol^{-1}·K^{-1}
One temperature
Molecular Weight 137.1378
Wiswesser Line Notation WNR C1
Evaluation C

$C_7H_7NO_2$ (liq) 34KOL/UDO
3–Nitrotoluene
Heat Capacity 302.4 K, C_p = 48.3 cal·mol^{-1}·K^{-1}
202.1 J·mol^{-1}·K^{-1}
One temperature
Molecular Weight 137.1378
Wiswesser Line Notation WNR C1
Evaluation C

$C_7H_7NO_2$ (c,I) 79RIC/SAV
4–Nitrotoluene
Heat Capacity 298.15 K, C_p = 41.2 cal·mol^{-1}·K^{-1}
172.3 J·mol^{-1}·K^{-1}
Temperature range 270–340 K. Equations only. Data for stable and metastable crystal form.
Phase Changes
c,II/liq 318.5 K, ΔH = 16917 cal·mol^{-1}
70781 J·mol^{-1}
ΔS = 53.1 cal·mol^{-1}·K^{-1}
222.2 J·mol^{-1}·K^{-1}
metastable form
c,I/liq 324.7 K, ΔH = 16683 cal·mol^{-1}
69802 J·mol^{-1}
ΔS = 51.4 cal·mol^{-1}·K^{-1}
215.0 J·mol^{-1}·K^{-1}
Molecular Weight 137.1378
Wiswesser Line Notation WNR D1
Evaluation B

$C_7H_7NO_2$ (c) 80AND/CON
4–Nitrotoluene
Phase Changes
c,I/liq 324.788 K, ΔH = 4018 cal·mol^{-1}
16811 J·mol^{-1}
ΔS = 12.37 cal·mol^{-1}·K^{-1}
51.76 J·mol^{-1}·K^{-1}
Molecular Weight 137.1378
Wiswesser Line Notation WNR D1
Evaluation A

$C_7H_7NO_2$ (c) 26AND/LYN
2–Aminobenzoic acid
Heat Capacity 298 K, C_p = 39.5 cal·mol^{-1}·K^{-1}
165.3 J·mol^{-1}·K^{-1}
Temperature range 22 to 160°C
Phase Changes
c/liq 417.8 K, ΔH = 4870 cal·mol^{-1}
20380 J·mol^{-1}
ΔS = 11.7 cal·mol^{-1}·K^{-1}
48.8 J·mol^{-1}·K^{-1}
Molecular Weight 137.1378
Wiswesser Line Notation ZR BVQ
Evaluation C

$C_7H_7NO_2$ (c) 41SAT/SOG 2
2–Aminobenzoic acid
Heat Capacity 323 K, C_p = 44.7 cal·mol^{-1}·K^{-1}
187.0 J·mol^{-1}·K^{-1}
Temperature range 0 to 100°C. Mean value.
Molecular Weight 137.1378
Wiswesser Line Notation ZR BVQ
Evaluation C
Same data as 40SAT/SOG 3.

$C_7H_7NO_2$ (c) 26AND/LYN
3–Aminobenzoic acid
Heat Capacity 298 K, C_p = 38.9 cal·mol^{-1}·K^{-1}
162.8 J·mol^{-1}·K^{-1}
Temperature range 22 to 180°C
Phase Changes
c/liq 452.9 K, ΔH = 5220 cal·mol^{-1}
21840 J·mol^{-1}
ΔS = 11.5 cal·mol^{-1}·K^{-1}
48.2 J·mol^{-1}·K^{-1}
Molecular Weight 137.1378
Wiswesser Line Notation ZR CVQ
Evaluation C

Table of Heat Capacities, Entropies, and Phase Transition Properties — Continued

$C_7H_7NO_2$ (c) 41SAT/SOG 2
3–Aminobenzoic acid
Heat Capacity 323 K, C_p = 43.0 cal·mol^{-1}·K^{-1}
179.9 J·mol^{-1}·K^{-1}
Temperature range 0 to 100°C. Mean value.
Molecular Weight 137.1378
Wiswesser Line Notation ZR CVQ
Evaluation C
Same data as 40SAT/SOG 3.

$C_7H_7NO_2$ (c) 26AND/LYN
4–Aminobenzoic acid
Heat Capacity 298 K, C_p = 42.5 cal·mol^{-1}·K^{-1}
177.8 J·mol^{-1}·K^{-1}
Temperature range 22 to 190°C
Phase Changes
c/liq 461.4 K, ΔH = 5000 cal·mol^{-1}
20920 J·mol^{-1}
ΔS = 10.8 cal·mol^{-1}·K^{-1}
45.3 J·mol^{-1}·K^{-1}
Molecular Weight 137.1378
Wiswesser Line Notation ZR DVQ
Evaluation C

$C_7H_7NO_2$ (c) 41SAT/SOG 2
4–Aminobenzoic acid
Heat Capacity 323 K, C_p = 44.8 cal·mol^{-1}·K^{-1}
187.4 J·mol^{-1}·K^{-1}
Temperature range 0 to 100°C. Mean value.
Molecular Weight 137.1378
Wiswesser Line Notation ZR DVQ
Evaluation C
Same data as 40SAT/SOG 3.

C_7H_8 (liq) 81REI
Toluene
Heat Capacity 298 K, C_p = 37.8 cal·mol^{-1}·K^{-1}
158.2 J·mol^{-1}·K^{-1}
Temperature range 292–390 K
Molecular Weight 92.1402
Wiswesser Line Notation 1R
Evaluation D

C_7H_8 (liq) 24WIL/DAN
Toluene
Heat Capacity 303 K, C_p = 36.7 cal·mol^{-1}·K^{-1}
153.6 J·mol^{-1}·K^{-1}
Temperature range 303–343 K. Equation only.
Molecular Weight 92.1402
Wiswesser Line Notation 1R
Evaluation C

C_7H_8 (liq) 25WIL/DAN
Toluene
Heat Capacity 293.2 K, C_p = 36.1 cal·mol^{-1}·K^{-1}
151.0 J·mol^{-1}·K^{-1}
Temperature range 20 to 60°C
Molecular Weight 92.1402
Wiswesser Line Notation 1R
Evaluation B

C_7H_8 (liq) 29KEL 4
Toluene
Heat Capacity 284.44 K, C_p = 36.59 cal·mol^{-1}·K^{-1}
153.09 J·mol^{-1}·K^{-1}
Temperature range 14–284 K. Value is unsmoothed experimental datum.
Entropy 298.15 K, S = 52.4 cal·mol^{-1}·K^{-1}
219.2 J·mol^{-1}·K^{-1}
Phase Changes
c/liq 177.95 K, ΔH = 1582 cal·mol^{-1}
6619 J·mol^{-1}
ΔS = 8.89 cal·mol^{-1}·K^{-1}
37.20 J·mol^{-1}·K^{-1}
Molecular Weight 92.1402
Wiswesser Line Notation 1R
Evaluation A

C_7H_8 (liq) 31SMI/AND
Toluene
Heat Capacity 298.15 K, C_p = 38.7 cal·mol^{-1}·K^{-1}
161.9 J·mol^{-1}·K^{-1}
Temperature range 102–299 K. Value is unsmoothed experimental datum.
Molecular Weight 92.1402
Wiswesser Line Notation 1R
Evaluation C

C_7H_8 (liq) 32RIC/WAL
Toluene
Heat Capacity 298.1 K, C_p = 37.4 cal·mol^{-1}·K^{-1}
156.5 J·mol^{-1}·K^{-1}
Temperature range 293–333 K
Molecular Weight 92.1402
Wiswesser Line Notation 1R
Evaluation C

C_7H_8 (liq) 35AOY/KAN
Toluene
Heat Capacity 227.8 K, C_p = 34.1 cal·mol^{-1}·K^{-1}
142.7 J·mol^{-1}·K^{-1}
Temperature range 78–228 K. Value is unsmoothed experimental datum.
Molecular Weight 92.1402
Wiswesser Line Notation 1R
Evaluation B

C_7H_8 (liq) 37VOL
Toluene
Heat Capacity 298 K, C_p = 37.4 cal·mol^{-1}·K^{-1}
156.5 J·mol^{-1}·K^{-1}
One temperature
Molecular Weight 92.1402
Wiswesser Line Notation 1R
Evaluation B

C_7H_8 (liq) 40BUR
Toluene
Heat Capacity 298.2 K, C_p = 37.54 cal·mol^{-1}·K^{-1}
157.07 J·mol^{-1}·K^{-1}
Temperature range 281–383 K
Molecular Weight 92.1402
Wiswesser Line Notation 1R
Evaluation B

C_7H_8 (liq) 41ZHD
Toluene
Heat Capacity 298.1 K, C_p = 37.5 cal·mol^{-1}·K^{-1}
156.9 J·mol^{-1}·K^{-1}
Temperature range 5 to 47°C
Molecular Weight 92.1402
Wiswesser Line Notation 1R
Evaluation C

Table of Heat Capacities, Entropies, and Phase Transition Properties — Continued

C_7H_8 (liq) 42ZIE/AND
Toluene
Phase Changes
c/liq 178.0 K, ΔH = 1565 cal·mol^{-1}
6548 J·mol^{-1}
ΔS = 8.79 cal·mol^{-1}·K^{-1}
36.79 J·mol^{-1}·K^{-1}
Molecular Weight 92.1402
Wiswesser Line Notation 1R
Evaluation B

C_7H_8 (liq) 47KUR
Toluene
Heat Capacity 298 K, C_p = 37.9 cal·mol^{-1}·K^{-1}
158.6 J·mol^{-1}·K^{-1}
Temperature range –76 to 60 °C, mean C_p, four temperatures.
Molecular Weight 92.1402
Wiswesser Line Notation 1R
Evaluation D

C_7H_8 (liq) 58SWI/ZIE 2
Toluene
Heat Capacity 324 K, C_p = 39.9 cal·mol^{-1}·K^{-1}
166.9 J·mol^{-1}·K^{-1}
Mean value 21 to 81°C
Molecular Weight 92.1402
Wiswesser Line Notation 1R
Evaluation C

C_7H_8 (liq) 62SCO/GUT
Toluene
Heat Capacity 298.15 K, C_p = 37.58 cal·mol^{-1}·K^{-1}
157.23 J·mol^{-1}·K^{-1}
Temperature range 10–360 K
Entropy 298.15 K, S = 52.81 cal·mol^{-1}·K^{-1}
220.96 J·mol^{-1}·K^{-1}
Phase Changes
c/liq 178.15 K, ΔH = 1586 cal·mol^{-1}
6636 J·mol^{-1}
ΔS = 8.90 cal·mol^{-1}·K^{-1}
37.25 J·mol^{-1}·K^{-1}
Molecular Weight 92.1402
Wiswesser Line Notation 1R
Evaluation A

C_7H_8 (liq) 67RAS/GAN
Toluene
Heat Capacity 293 K, C_p = 37.9 cal·mol^{-1}·K^{-1}
158.6 J·mol^{-1}·K^{-1}
Temperature range 293–373 K
Molecular Weight 92.1402
Wiswesser Line Notation 1R
Evaluation C

C_7H_8 (liq) 71DES/BHA
Toluene
Heat Capacity 298 K, C_p = 37.4 cal·mol^{-1}·K^{-1}
156.5 J·mol^{-1}·K^{-1}
Temperature range 298–318 K
Molecular Weight 92.1402
Wiswesser Line Notation 1R
Evaluation B

C_7H_8 (liq) 75HOL/ZIE
Toluene
Heat Capacity 298.15 K, C_p = 37.5 cal·mol^{-1}·K^{-1}
157.0 J·mol^{-1}·K^{-1}
Temperature range 162–312 K. Equation only; experimental data deposited with journal as supplementary material.
Molecular Weight 92.1402
Wiswesser Line Notation 1R
Evaluation B

C_7H_8 (liq) 82GRO/ING
Toluene
Heat Capacity 298.15 K, C_p = 37.53 cal·mol^{-1}·K^{-1}
157.0 J·mol^{-1}·K^{-1}
One temperature
Molecular Weight 92.1402
Wiswesser Line Notation 1R
Evaluation A

C_7H_8 (liq) 73HAL/SMI
Quadricyclane
Heat Capacity 297 K, C_p = 33.2 cal·mol^{-1}·K^{-1}
139.1 J·mol^{-1}·K^{-1}
One temperature
Molecular Weight 92.1402
Wiswesser Line Notation L435 B3 2AB GTJ
Evaluation C

C_7H_8 (liq) 73HAL/SMI
Norbornadiene
Heat Capacity 297 K, C_p = 27.7 cal·mol^{-1}·K^{-1}
116.1 J·mol^{-1}·K^{-1}
One temperature
Molecular Weight 92.1402
Wiswesser Line Notation L55 A CU FUTJ
Evaluation C

C_7H_8 (liq) 56FIN/SCO
Cycloheptatriene
Heat Capacity 298.15 K, C_p = 38.90 cal·mol^{-1}·K^{-1}
162.76 J·mol^{-1}·K^{-1}
Temperature range 10–320 K
Entropy 298.15 K, S = 51.30 cal·mol^{-1}·K^{-1}
214.64 J·mol^{-1}·K^{-1}
Phase Changes
c,II/c,I 153.98 K, ΔH = 560.9 cal·mol^{-1}
2346.8 J·mol^{-1}
ΔS = 3.64 cal·mol^{-1}·K^{-1}
15.24 J·mol^{-1}·K^{-1}
c,I/liq 197.92 K, ΔH = 277.4 cal·mol^{-1}
1160.6 J·mol^{-1}
ΔS = 1.40 cal·mol^{-1}·K^{-1}
5.86 J·mol^{-1}·K^{-1}
Molecular Weight 92.1402
Wiswesser Line Notation L7HJ
Evaluation A

$C_7H_8N_2O_2$ (c) 41SAT/SOG 3
5–Nitro–2–aminotoluene
Heat Capacity 323 K, C_p = 51.9 cal·mol^{-1}·K^{-1}
217.1 J·mol^{-1}·K^{-1}
Temperature range 0 to 100°C. Mean value.
Molecular Weight 152.1524
Wiswesser Line Notation ZR B1 DNW
Evaluation C
Same data at 40SAT/SOG 4.

Table of Heat Capacities, Entropies, and Phase Transition Properties — Continued

$C_7H_8N_2O_2$ (c) 41SAT/SOG 3
3–Nitro–4–aminotoluene
Heat Capacity 323 K, C_p = 49.1 cal·mol^{-1}·K^{-1}
205.4 J·mol^{-1}·K^{-1}
Temperature range 0 to 100°C. Mean value.
Molecular Weight 152.1524
Wiswesser Line Notation ZR D1 BNW
Evaluation C
Same data as 40SAT/SOG 4.

C_7H_8O (liq) 02LOU
Anisole; Methyl phenyl ether; Methoxybenzene
Heat Capacity 360 K, C_p = 52 cal·mol^{-1}·K^{-1}
218 J·mol^{-1}·K^{-1}
Mean value 20 to 152°C
Molecular Weight 108.1396
Wiswesser Line Notation 1OR
Evaluation D

C_7H_8O (liq) 33KOL/UDO
Anisole; Methyl phenyl ether; Methoxybenzene; Phenyl methyl ether
Heat Capacity 297.2 K, C_p = 45.7 cal·mol^{-1}·K^{-1}
191.2 J·mol^{-1}·K^{-1}
One temperature
Molecular Weight 108.1396
Wiswesser Line Notation 1OR
Evaluation C

C_7H_8O (liq) 39PHI
Anisole; Methyl phenyl ether; Methoxybenzene
Heat Capacity 304.8 K, C_p = 49.9 cal·mol^{-1}·K^{-1}
208.8 J·mol^{-1}·K^{-1}
One temperature
Molecular Weight 108.1396
Wiswesser Line Notation 1OR
Evaluation C

C_7H_8O (liq) 81REI
Benzyl alcohol
Heat Capacity 298 K, C_p = 52.7 cal·mol^{-1}·K^{-1}
220.5 J·mol^{-1}·K^{-1}
Temperature range 290–485 K
Molecular Weight 108.1396
Wiswesser Line Notation Q1R
Evaluation D

C_7H_8O (liq) 31SMI/AND
Benzyl alcohol
Heat Capacity 298.5 K, C_p = 51.6 cal·mol^{-1}·K^{-1}
215.9 J·mol^{-1}·K^{-1}
Temperature range 102–299 K. Value is unsmoothed experimental datum.
Molecular Weight 108.1396
Wiswesser Line Notation Q1R
Evaluation C

C_7H_8O (liq) 36PAR/TOD
Benzyl alcohol
Heat Capacity 298.1 K, C_p = 52.07 cal·mol^{-1}·K^{-1}
217.86 J·mol^{-1}·K^{-1}
Temperature range 90–300 K
Entropy 298.1 K, S = 51.8 cal·mol^{-1}·K^{-1}
216.7 J·mol^{-1}·K^{-1}
Extrapolation below 90 K, 12.92 cal·mol^{-1}K^{-1}
Phase Changes
c/liq 257.6 K, ΔH = 2144 cal·mol^{-1}
8970 J·mol^{-1}
ΔS = 8.32 cal·mol^{-1}·K^{-1}
34.82 J·mol^{-1}·K^{-1}
Molecular Weight 108.1396
Wiswesser Line Notation Q1R
Evaluation B(C_p),C(S)

C_7H_8O (liq) 75NIC/WAD
Benzyl alcohol
Heat Capacity 298.15 K, C_p = 51.61 cal·mol^{-1}·K^{-1}
215.94 J·mol^{-1}·K^{-1}
One temperature
Molecular Weight 108.1396
Wiswesser Line Notation Q1R
Evaluation B

C_7H_8O (c) 67AND/COU
o–Hydroxytoluene
Heat Capacity 298.15 K, C_p = 36.94 cal·mol^{-1}·K^{-1}
154.56 J·mol^{-1}·K^{-1}
Temperature range 10–400 K
Entropy 298.15 K, S = 39.54 cal·mol^{-1}·K^{-1}
165.44 J·mol^{-1}·K^{-1}
Phase Changes
c/liq 304.20 K, ΔH = 3781 cal·mol^{-1}
15820 J·mol^{-1}
ΔS = 12.43 cal·mol^{-1}·K^{-1}
52.00 J·mol^{-1}·K^{-1}
Molecular Weight 108.1396
Wiswesser Line Notation QR B1
Evaluation A

C_7H_8O (liq) 16BRA
o–Hydroxytoluene
Heat Capacity 283 K, C_p = 54.0 cal·mol^{-1}·K^{-1}
225.9 J·mol^{-1}·K^{-1}
Mean value, 0 to 20°C
Molecular Weight 108.1396
Wiswesser Line Notation QR B1
Evaluation C

C_7H_8O (liq) 67RAS/GAN
o–Hydroxytoluene
Heat Capacity 313 K, C_p = 55.8 cal·mol^{-1}·K^{-1}
233.6 J·mol^{-1}·K^{-1}
Temperature range 313–373 K
Molecular Weight 108.1396
Wiswesser Line Notation QR B1
Evaluation C

C_7H_8O (liq) 16BRA
m–Hydroxytoluene
Heat Capacity 283 K, C_p = 51.8 cal·mol^{-1}·K^{-1}
216.7 J·mol^{-1}·K^{-1}
Mean value, 0 to 20°C
Molecular Weight 108.1396
Wiswesser Line Notation QR C1
Evaluation C

Table of Heat Capacities, Entropies, and Phase Transition Properties — Continued

C_7H_8O (liq) 51TSC/KRI
m–Hydroxytoluene
Heat Capacity 298 K, C_p = 52.3 cal·mol^{-1}·K^{-1}
218.8 J·mol^{-1}·K^{-1}
One temperature
Molecular Weight 108.1396
Wiswesser Line Notation QR C1
Evaluation C

C_7H_8O (liq) 67RAS/GAN
m–Hydroxytoluene
Heat Capacity 93 K, C_p = 52.8 cal·mol^{-1}·K^{-1}
220.9 J·mol^{-1}·K^{-1}
Temperature range 293–373 K
Molecular Weight 108.1396
Wiswesser Line Notation QR C1
Evaluation C

C_7H_8O (liq) 67AND/COU
m–Hydroxytoluene
Heat Capacity 298.15 K, C_p = 53.76 cal·mol^{-1}·K^{-1}
224.93 J·mol^{-1}·K^{-1}
Temperature range 10–400 K
Entropy 298.15 K, S = 50.81 cal·mol^{-1}·K^{-1}
212.59 J·mol^{-1}·K^{-1}
Phase Changes
c/liq 285.40 K, ΔH = 2559 cal·mol^{-1}
10707 J·mol^{-1}
ΔS = 8.97 cal·mol^{-1}·K^{-1}
27.53 J·mol^{-1}·K^{-1}
Molecular Weight 108.1396
Wiswesser Line Notation QR C1
Evaluation A

C_7H_8O (c) 67AND/COU
p–Hydroxytoluene
Heat Capacity 298.15 K, C_p = 35.91 cal·mol^{-1}·K^{-1}
150.25 J·mol^{-1}·K^{-1}
Temperature range 10–400 K
Entropy 298.15 K, S = 39.99 cal·mol^{-1}·K^{-1}
167.32 J·mol^{-1}·K^{-1}
Phase Changes
c/liq 307.94 K, ΔH = 3037 cal·mol^{-1}
12707 J·mol^{-1}
ΔS = 9.86 cal·mol^{-1}·K^{-1}
41.26 J·mol^{-1}·K^{-1}
Molecular Weight 108.1396
Wiswesser Line Notation QR D1
Evaluation A

C_7H_8O (liq) 67RAS/GAN
p–Hydroxytoluene
Heat Capacity 313 K, C_p = 54.1 cal·mol^{-1}·K^{-1}
226.2 J·mol^{-1}·K^{-1}
Temperature range 313–373 K
Molecular Weight 108.1396
Wiswesser Line Notation QR D1
Evaluation C

C_7H_8O (liq) 75NIC/WAD
p–Hydroxytoluene
Heat Capacity 298.15 K, C_p = 52.83 cal·mol^{-1}·K^{-1}
221.03 J·mol^{-1}·K^{-1}
One temperature
Molecular Weight 108.1396
Wiswesser Line Notation QR D1
Evaluation B

$C_7H_8O_2$ (c) 41SAT/SOG 3
2,5–Dihydroxytoluene
Heat Capacity 323 K, C_p = 41.8 cal·mol^{-1}·K^{-1}
174.9 J·mol^{-1}·K^{-1}
Temperature range 0 to 100°C. Mean value.
Molecular Weight 124.1390
Wiswesser Line Notation QR DQ B1
Evaluation C
Same data as 40SAT/SOG 4.

C_7H_8S (liq) 74MES/FIN
Methyl phenyl sulfide
Heat Capacity 298.15 K, C_p = 49.24 cal·mol^{-1}·K^{-1}
206.02 J·mol^{-1}·K^{-1}
Temperature range 10–330 K
Entropy 298.15 K, S = 60.35 cal·mol^{-1}·K^{-1}
252.50 J·mol^{-1}·K^{-1}
Phase Changes
c/liq 256.44 K, ΔH = 3545.8 cal·mol^{-1}
14835.8 J·mol^{-1}
ΔS = 13.83 cal·mol^{-1}·K^{-1}
57.85 J·mol^{-1}·K^{-1}
Molecular Weight 124.2002
Wiswesser Line Notation 1SR
Evaluation A

C_7H_9N (liq) 02LOU
N–Methylaniline
Heat Capacity 380 K, C_p = 55 cal·mol^{-1}·K^{-1}
230 J·mol^{-1}·K^{-1}
Mean value 20 to 196°C
Molecular Weight 107.1548
Wiswesser Line Notation 1MR
Evaluation D

C_7H_9N (liq) 36KUR/VOS
N–Methylaniline
Heat Capacity 290 K, C_p = 55.0 cal·mol^{-1}·K^{-1}
230.1 J·mol^{-1}·K^{-1}
One temperature
Molecular Weight 107.1548
Wiswesser Line Notation 1MR
Evaluation D

C_7H_9N (liq) 71HAL/BAL
1–Bicyclo[3.1.0]hexyl cyanide;
1–Cyanobicyclo[3.1.0]hexane
Heat Capacity 297 K, C_p = 40.7 cal·mol^{-1}·K^{-1}
170.4 J·mol^{-1}·K^{-1}
One temperature
Molecular Weight 107.1548
Wiswesser Line Notation L35TJ ACN
Evaluation C

C_7H_9N (liq) 75NIC/WAD
Benzylamine; Phenylmethylamine
Heat Capacity 298.15 K, C_p = 49.52 cal·mol^{-1}·K^{-1}
207.18 J·mol^{-1}·K^{-1}
One temperature
Molecular Weight 107.1548
Wiswesser Line Notation Z1R
Evaluation B

Table of Heat Capacities, Entropies, and Phase Transition Properties — Continued

C_7H_9N (liq) **81REI**
2–Methylaniline; o–Toluidine
Heat Capacity 298 K, C_p = 50.5 cal·mol⁻¹·K⁻¹ 211.3 J·mol⁻¹·K⁻¹
Temperature range 294–485 K
Molecular Weight 107.1548
Wiswesser Line Notation ZR B1
Evaluation D

C_7H_9N (liq) 02LOU
2–Methylaniline; o–Toluidine
Heat Capacity 380 K, C_p = 56 cal·mol⁻¹·K⁻¹ 234 J·mol⁻¹·K⁻¹
Mean value 20 to 196°C
Molecular Weight 107.1548
Wiswesser Line Notation ZR B1
Evaluation D

C_7H_9N (liq) 34KOL/UDO
2–Methylaniline; o–Toluidine
Heat Capacity 302.5 K, C_p = 50.1 cal·mol⁻¹·K⁻¹ 209.6 J·mol⁻¹·K⁻¹
One temperature
Molecular Weight 107.1548
Wiswesser Line Notation ZR B1
Evaluation C

C_7H_9N (liq) 34KOL/UDO 2
2–Methylaniline; o–Toluidine
Heat Capacity 302.5 K, C_p = 50.1 cal·mol⁻¹·K⁻¹ 209.6 J·mol⁻¹·K⁻¹
One temperature
Molecular Weight 107.1548
Wiswesser Line Notation ZR B1
Evaluation C

C_7H_9N (liq) 34RAD/JUL
2–Methylaniline; o–Toluidine
Heat Capacity 288 K, C_p = 48.2 cal·mol⁻¹·K⁻¹ 201.7 J·mol⁻¹·K⁻¹
One temperature
Molecular Weight 107.1548
Wiswesser Line Notation ZR B1
Evaluation C

C_7H_9N (liq) 34KOL/UDO
3–Methylaniline; m–Toluidine
Heat Capacity 302.7 K, C_p = 51.8 cal·mol⁻¹·K⁻¹ 216.7 J·mol⁻¹·K⁻¹
One temperature
Molecular Weight 107.1548
Wiswesser Line Notation ZR C1
Evaluation C

C_7H_9N (liq) 34KOL/UDO 2
3–Methylaniline; m–Toluidine
Heat Capacity 302.7 K, C_p = 51.8 cal·mol⁻¹·K⁻¹ 216.7 J·mol⁻¹·K⁻¹
One temperature
Molecular Weight 107.1548
Wiswesser Line Notation ZR C1
Evaluation C

C_7H_9N (c) 40CAM/CAM
4–Methylaniline; p–Toluidine
Heat Capacity 293 K, C_p = 29.7 cal·mol⁻¹·K⁻¹ 124.3 J·mol⁻¹·K⁻¹
One temperature
Molecular Weight 107.1548
Wiswesser Line Notation ZR D1
Evaluation C

$C_7H_9NO_2$ (c) 39SAT/SOG 2
Ammonium benzoate
Heat Capacity 323 K, C_p = 50.8 cal·mol⁻¹·K⁻¹ 212.5 J·mol⁻¹·K⁻¹
Temperature range 0 to 100°C. Mean value.
Molecular Weight 139.1536
Wiswesser Line Notation QVR &ZH
Evaluation C
Same data in 40SAT/SOG.

C_7H_{10} (c) 73HAL/SMI
Nortricyclene; Tricyclo[2.2.1.0^{2,6}]heptane
Heat Capacity 297 K, C_p = 26.3 cal·mol⁻¹·K⁻¹ 110.2 J·mol⁻¹·K⁻¹
One temperature
Molecular Weight 94.1560
Wiswesser Line Notation L535 B 1A GTJ
Evaluation C

C_7H_{10} (c) 73HAL/SMI
Norbornene
Heat Capacity 297 K, C_p = 31.0 cal·mol⁻¹·K⁻¹ 129.9 J·mol⁻¹·K⁻¹
One temperature
Molecular Weight 94.1560
Wiswesser Line Notation L55 A CUTJ
Evaluation C

$C_7H_{11}N$ (liq) 71HAL/BAL
Cyclohexyl cyanide; Cyanocyclohexane
Heat Capacity 297 K, C_p = 42.5 cal·mol⁻¹·K⁻¹ 177.9 J·mol⁻¹·K⁻¹
One temperature
Molecular Weight 109.1706
Wiswesser Line Notation L6TJ ACN
Evaluation C

C_7H_{12} (liq) 70CHA/MCC
cis–Bicyclo[4.1.0]heptane
Heat Capacity 315 K, C_p = 44.9 cal·mol⁻¹·K⁻¹ 187.9 J·mol⁻¹·K⁻¹
One temperature
Molecular Weight 96.1718
Wiswesser Line Notation L36TJ –C
Evaluation B

C_7H_{12} (liq) 79PUC/PEA
1–Ethylcyclopentene
Heat Capacity 298.15 K, C_p = 45.0 cal·mol⁻¹·K⁻¹ 188.3 J·mol⁻¹·K⁻¹
One temperature
Molecular Weight 96.1718
Wiswesser Line Notation L5UTJ A2
Evaluation B

Table of Heat Capacities, Entropies, and Phase Transition Properties — Continued

C_7H_{12} (liq) 79PUC/PEA
Ethylidenecyclopentane
Heat Capacity 298.15 K, C_p = 43.3 cal·mol^{-1}·K^{-1}
181.2 J·mol^{-1}·K^{-1}
One temperature
Molecular Weight 96.1718
Wiswesser Line Notation L5YTJ AU2
Evaluation B

C_7H_{12} (liq) 79PUC/PEA
Methylenecyclohexane
Heat Capacity 298.15 K, C_p = 42.4 cal·mol^{-1}·K^{-1}
177.4 J·mol^{-1}·K^{-1}
One temperature
Molecular Weight 96.1718
Wiswesser Line Notation L6YTJ AU1
Evaluation B

$C_7H_{12}O$ (liq) 24HER/BLO
2–Methylcyclohexanone
Heat Capacity 290 K, C_p = 48.9 cal·mol^{-1}·K^{-1}
204.6 J·mol^{-1}·K^{-1}
One temperature
Molecular Weight 112.1712
Wiswesser Line Notation L6VTJ B1
Evaluation C

$C_7H_{12}O$ (liq) 24HER/BLO
3–Methylcyclohexanone
Heat Capacity 290 K, C_p = 49.5 cal·mol^{-1}·K^{-1}
207.1 J·mol^{-1}·K^{-1}
One temperature
Molecular Weight 112.1712
Wiswesser Line Notation L6VTJ C1
Evaluation C

$C_7H_{12}O$ (liq) 24HER/BLO
4–Methylcyclohexanone
Heat Capacity 290 K, C_p = 49.5 cal·mol^{-1}·K^{-1}
207.1 J·mol^{-1}·K^{-1}
One temperature
Molecular Weight 112.1712
Wiswesser Line Notation L6VTJ D1
Evaluation C

$C_7H_{12}O_4$ (liq) 33KOL/UDO
Diethyl malonate
Heat Capacity 294.6 K, C_p = 68.1 cal·mol^{-1}·K^{-1}
284.9 J·mol^{-1}·K^{-1}
One temperature
Molecular Weight 160.1694
Wiswesser Line Notation 2OV1VO2
Evaluation C

$C_7H_{12}O_4$ (liq) 34KOL/UDO 2
Diethyl malonate
Heat Capacity 294.6 K, C_p = 68.1 cal·mol^{-1}·K^{-1}
284.9 J·mol^{-1}·K^{-1}
One temperature
Molecular Weight 160.1694
Wiswesser Line Notation 2OV1VO2
Evaluation C

$C_7H_{13}N$ (c,I) 70WES/WON
1–Azabicyclo[2.2.2]octane; Quinuclidine
Heat Capacity 298.15 K, C_p = 40.48 cal·mol^{-1}·K^{-1}
169.37 J·mol^{-1}·K^{-1}
Temperature range 5–433 K
Entropy 298.15 K, S = 49.47 cal·mol^{-1}·K^{-1}
206.98 J·mol^{-1}·K^{-1}
Phase Changes
c,II/c,I 196 K, ΔH = 1249 cal·mol^{-1}
5226 J·mol^{-1}
ΔS = 6.34 cal·mol^{-1}·K^{-1}
26.52 J·mol^{-1}·K^{-1}
Transition region 190–200K; maximum at 196 K. Entropy change obtained by difference of integrated heat input and lattice (extrapolated C_p) contribution.
c,I/liq 430 K, ΔH = 1400 cal·mol^{-1}
5860 J·mol^{-1}
ΔS = 3.3 cal·mol^{-1}·K^{-1}
13.6 J·mol^{-1}·K^{-1}
Molecular Weight 111.1864
Wiswesser Line Notation T66 A B CNTJ
Evaluation A

C_7H_{14} (liq) 36PAR/TOD 2
1–Heptene
Heat Capacity 295.1 K, C_p = 50.87 cal·mol^{-1}·K^{-1}
212.84 J·mol^{-1}·K^{-1}
Temperature range 80–295 K. Value is unsmoothed experimental datum.
Entropy 298.15 K, S = 78.6 cal·mol^{-1}·K^{-1}
328.9 J·mol^{-1}·K^{-1}
Extrapolation below 80 K, 14.0 cal·mol^{-1}K^{-1}
Phase Changes
c/liq 153.4 K, ΔH = 3026 cal·mol^{-1}
12661 J·mol^{-1}
ΔS = 19.73 cal·mol^{-1}·K^{-1}
82.54 J·mol^{-1}·K^{-1}
Molecular Weight 98.1876
Wiswesser Line Notation 6U1
Evaluation B(C_p),C(S)

C_7H_{14} (liq) 57MCC/FIN 2
1–Heptene
Heat Capacity 298.15 K, C_p = 50.62 cal·mol^{-1}·K^{-1}
211.79 J·mol^{-1}·K^{-1}
Temperature range 11–360 K
Entropy 298.15 K, S = 78.31 cal·mol^{-1}·K^{-1}
327.65 J·mol^{-1}·K^{-1}
Phase Changes
c,II/liq 153.89 K, ΔH = 3021 cal·mol^{-1}
12640 J·mol^{-1}
ΔS = 19.63 cal·mol^{-1}·K^{-1}
82.14 J·mol^{-1}·K^{-1}
c,I/liq 154.30 K, ΔH = 2964 cal·mol^{-1}
12401 J·mol^{-1}
ΔS = 19.21 cal·mol^{-1}·K^{-1}
80.37 J·mol^{-1}·K^{-1}
Molecular Weight 98.1876
Wiswesser Line Notation 6U1
Evaluation A

C_7H_{14} (liq) 53GRO/OLI
1,1–Dimethylcyclopentane
Heat Capacity 299.81 K, C_p = 44.78 cal·mol^{-1}·K^{-1}
187.36 J·mol^{-1}·K^{-1}
Temperature range 13–300 K. Unsmoothed experimental datum.

Table of Heat Capacities, Entropies, and Phase Transition Properties — Continued

Entropy 298.15 K, S = 63.34 cal·mol^{-1}·K^{-1}
265.01 J·mol^{-1}·K^{-1}

Phase Changes
c,II/c,I 146.80 K, ΔH = 1551.0 cal·mol^{-1}
6489.4 J·mol^{-1}
ΔS = 10.57 cal·mol^{-1}·K^{-1}
44.21 J·mol^{-1}·K^{-1}
c,I/liq 203.68 K, ΔH = 257.8 cal·mol^{-1}
1078.6 J·mol^{-1}
ΔS = 1.27 cal·mol^{-1}·K^{-1}
5.30 J·mol^{-1}·K^{-1}

Molecular Weight 98.1876
Wiswesser Line Notation L5TJ A1 A1
Evaluation A

C_7H_{14} (liq) 31HUF/PAR
1,2–Dimethylcyclopentane
Heat Capacity 294.2 K, C_p = 44.8 cal·mol^{-1}·K^{-1}
187.4 J·mol^{-1}·K^{-1}
Temperature range 93–294 K. Value is unsmoothed experimental datum.
Entropy 298.1 K, S = 64.5 cal·mol^{-1}·K^{-1}
269.9 J·mol^{-1}·K^{-1}
Extrapolation below 90 K, 16.67 cal·mol^{-1}K^{-1}
Phase Changes
c/liq 154.1 K, ΔH = 1532 cal·mol^{-1}
6410 J·mol^{-1}
ΔS = 9.9 cal·mol^{-1}·K^{-1}
41.6 J·mol^{-1}·K^{-1}

Molecular Weight 98.1876
Wiswesser Line Notation L5TJ A1 B1
Evaluation B(C_p),C(S)

C_7H_{14} (liq) 53GRO/OLI
1–cis–2–Dimethylcyclopentane
Heat Capacity 302.84 K, C_p = 45.57 cal·mol^{-1}·K^{-1}
190.66 J·mol^{-1}·K^{-1}
Temperature range 13–300 K. Unsmoothed experimental datum.
Entropy 298.15 K, S = 64.33 cal·mol^{-1}·K^{-1}
269.16 J·mol^{-1}·K^{-1}
Phase Changes
c,II/c,I 141.50 K, ΔH = 1593.9 cal·mol^{-1}
6668.9 J·mol^{-1}
ΔS = 11.26 cal·mol^{-1}·K^{-1}
47.13 J·mol^{-1}·K^{-1}
c,I/liq 219.45 K, ΔH = 396.1 cal·mol^{-1}
1657.3 J·mol^{-1}
ΔS = 1.80 cal·mol^{-1}·K^{-1}
7.55 J·mol^{-1}·K^{-1}

Molecular Weight 98.1876
Wiswesser Line Notation L5TJ A1 B1 –C
Evaluation A

C_7H_{14} (liq) 53GRO/OLI
1–trans–3–Dimethylcyclopentane
Heat Capacity 304.03 K, C_p = 45.62 cal·mol^{-1}·K^{-1}
190.87 J·mol^{-1}·K^{-1}
Temperature range 13–300 K. Unsmoothed experimental datum.
Entropy 298.15 K, S = 64.90 cal·mol^{-1}·K^{-1}
271.54 J·mol^{-1}·K^{-1}

Phase Changes
c/liq 139.48 K, ΔH = 1768.2 cal·mol^{-1}
7398.1 J·mol^{-1}
ΔS = 12.68 cal·mol^{-1}·K^{-1}
53.04 J·mol^{-1}·K^{-1}

Molecular Weight 98.1876
Wiswesser Line Notation L5TJ A1 C1 –T
Evaluation A

C_7H_{14} (liq) 53GRO/OLI
Ethylcyclopentane
Heat Capacity 301.83 K, C_p = 44.84 cal·mol^{-1}·K^{-1}
187.61 J·mol^{-1}·K^{-1}
Temperature range 13–300 K. Unsmoothed experimental datum.
Entropy 298.15 K, S = 66.90 cal·mol^{-1}·K^{-1}
279.91 J·mol^{-1}·K^{-1}
Phase Changes
c,I/liq 134.73 K, ΔH = 1641.8 cal·mol^{-1}
6869.3 J·mol^{-1}
ΔS = 12.19 cal·mol^{-1}·K^{-1}
50.99 J·mol^{-1}·K^{-1}
Form stable above 129.5 K.
c,II/liq 134.03 K, ΔH = 1889.2 cal·mol^{-1}
2904.4 J·mol^{-1}
ΔS = 14.10 cal·mol^{-1}·K^{-1}
58.97 J·mol^{-1}·K^{-1}
Form stable below 129.5 K.
Molecular Weight 98.1876
Wiswesser Line Notation L5TJ A2
Evaluation A

C_7H_{14} (liq) 30PAR/HUF 2
Methylcyclohexane
Heat Capacity 294.2 K, C_p = 43.5 cal·mol^{-1}·K^{-1}
182.0 J·mol^{-1}·K^{-1}
Temperature range 93–294 K. Value is unsmoothed experimental datum.
Entropy 298.15 K, S = 59.3 cal·mol^{-1}·K^{-1}
248.1 J·mol^{-1}·K^{-1}
Extrapolation below 90 K, 13.08 cal·mol^{-1}K^{-1}
Phase Changes
c/liq 146.2 K, ΔH = 1595 cal·mol^{-1}
6673 J·mol^{-1}
ΔS = 10.91 cal·mol^{-1}·K^{-1}
45.14 J·mol^{-1}·K^{-1}

Molecular Weight 98.1876
Wiswesser Line Notation L6TJ A1
Evaluation B(C_p),C(S)

C_7H_{14} (liq) 46DOU/HUF 2
Methylcyclohexane
Heat Capacity 298.15 K, C_p = 44.10 cal·mol^{-1}·K^{-1}
184.51 J·mol^{-1}·K^{-1}
Temperature range 12–300 K
Entropy 298.15 K, S = 59.25 cal·mol^{-1}·K^{-1}
247.90 J·mol^{-1}·K^{-1}
Phase Changes
c/liq 146.58 K, ΔH = 1613.4 cal·mol^{-1}
6750.5 J·mol^{-1}
ΔS = 11.01 cal·mol^{-1}·K^{-1}
46.05 J·mol^{-1}·K^{-1}

Molecular Weight 98.1876
Wiswesser Line Notation L6TJ A1
Evaluation A

Table of Heat Capacities, Entropies, and Phase Transition Properties — Continued

C_7H_{14} (liq) 75HOL/ZIE
Methylcyclohexane
Heat Capacity 298.15 K, C_p = 44.2 cal·mol⁻¹·K⁻¹
184.8 J·mol⁻¹·K⁻¹
Temperature range 144–312 K. Equation only. Experimental data deposited with journal as supplementary material.
Molecular Weight 98.1876
Wiswesser Line Notation L6TJ A1
Evaluation B

C_7H_{14} (liq) 79WIL/GRO
Methylcyclohexane
Heat Capacity 298.15 K, C_p = 44.2 cal·mol⁻¹·K⁻¹
184.8 J·mol⁻¹·K⁻¹
One temperature
Molecular Weight 98.1876
Wiswesser Line Notation L6TJ A1
Evaluation B

C_7H_{14} (liq) 82GRO/ING
Methylcyclohexane
Heat Capacity 298.15 K, C_p = 44.29 cal·mol⁻¹·K⁻¹
185.29 J·mol⁻¹·K⁻¹
Temperature range 298.15 K, One temperature
Molecular Weight 98.1876
Wiswesser Line Notation L6TJ A1
Evaluation A

C_7H_{14} (liq) 56FIN/SCO
Cycloheptane
Heat Capacity 298.15 K, C_p = 43.20 cal·mol⁻¹·K⁻¹
180.75 J·mol⁻¹·K⁻¹
Temperature range 12–300 K
Entropy 298.15 K, S = 57.97 cal·mol⁻¹·K⁻¹
242.55 J·mol⁻¹·K⁻¹
Phase Changes
c,IV/c,III 134.8 K, ΔH = 1187.0 cal·mol⁻¹
4966.4 J·mol⁻¹
ΔS = 8.81 cal·mol⁻¹·K⁻¹
36.84 J·mol⁻¹·K⁻¹
c,III/c,II 198.2 K, ΔH = 69.2 cal·mol⁻¹
289.5 J·mol⁻¹
ΔS = 0.35 cal·mol⁻¹·K⁻¹
1.46 J·mol⁻¹·K⁻¹
c,II/c,I 212.4 K, ΔH = 107.5 cal·mol⁻¹
449.8 J·mol⁻¹
ΔS = 0.51 cal·mol⁻¹·K⁻¹
2.12 J·mol⁻¹·K⁻¹
c,I/liq 265.12 K, ΔH = 449.8 cal·mol⁻¹
1882.0 J·mol⁻¹
ΔS = 1.70 cal·mol⁻¹·K⁻¹
7.10 J·mol⁻¹·K⁻¹
Molecular Weight 98.1876
Wiswesser Line Notation L7TJ
Evaluation A

$C_7H_{14}O$ (liq) 70AND/COU
2,4–Dimethyl–3–pentanone; Diisopropyl ketone
Heat Capacity 298.15 K, C_p = 55.86 cal·mol⁻¹·K⁻¹
233.7 J·mol⁻¹·K⁻¹
Temperature range 10–320 K
Entropy 298.15 K, S = 76.00 cal·mol⁻¹·K⁻¹
318.0 J·mol⁻¹·K⁻¹
Phase Changes
c/liq 204.81 K, ΔH = 2670 cal·mol⁻¹
11180 J·mol⁻¹
ΔS = 13.05 cal·mol⁻¹·K⁻¹
54.58 J·mol⁻¹·K⁻¹
Molecular Weight 114.1870
Wiswesser Line Notation 1Y1&VY1&1
Evaluation A

$C_7H_{14}O$ (liq) 24HER/BLO
2–Methylcyclohexanol
Heat Capacity 290 K, C_p = 47.7 cal·mol⁻¹·K⁻¹
199.6 J·mol⁻¹·K⁻¹
One temperature
Molecular Weight 114.1870
Wiswesser Line Notation L6TJ AQ B1
Evaluation C

$C_7H_{14}O$ (liq) 24HER/BLO
3–Methylcyclohexanol
Heat Capacity 290 K, C_p = 48.2 cal·mol⁻¹·K⁻¹
201.7 J·mol⁻¹·K⁻¹
One temperature
Molecular Weight 114.1870
Wiswesser Line Notation L6TJ AQ C1
Evaluation C

$C_7H_{14}O$ (liq) 24HER/BLO
4–Methylcyclohexanol
Heat Capacity 290 K, C_p = 48.3 cal·mol⁻¹·K⁻¹
202.1 J·mol⁻¹·K⁻¹
One temperature
Molecular Weight 114.1870
Wiswesser Line Notation L6TJ AQ D1
Evaluation C

$C_7H_{14}O$ (liq) 72ADA/SUG
Cycloheptanol
Heat Capacity 298.15 K, C_p = 59.80 cal·mol⁻¹·K⁻¹
250.22 J·mol⁻¹·K⁻¹
Temperature range 13–300 K
Entropy 298.15 K, S = 57.75 cal·mol⁻¹·K⁻¹
241.638 J·mol⁻¹·K⁻¹
Phase Changes
c,II/c,I 258.45 K, ΔH = 209 cal·mol⁻¹
875 J·mol⁻¹
ΔS = 0.81 cal·mol⁻¹·K⁻¹
3.39 J·mol⁻¹·K⁻¹
c,II′/c,II 172.23 K,
ΔH = 2925 J·mol⁻¹ between 162.50 and 176.50 K.
c,III/c,II 227.26 K, ΔH = 132 cal·mol⁻¹
554 J·mol⁻¹
ΔS = 0.58 cal·mol⁻¹·K⁻¹
2.44 J·mol⁻¹·K⁻¹
c,III′/c,III 152.30 K, ΔH = 13.88 cal·mol⁻¹
58.07 J·mol⁻¹
ΔS = 0.09 cal·mol⁻¹·K⁻¹
0.38 J·mol⁻¹·K⁻¹
c,III″/c,III 128.2 K,
c,I/liq 280.30 K, ΔH = 383 cal·mol⁻¹
1604 J·mol⁻¹
ΔS = 1.37 cal·mol⁻¹·K⁻¹
5.72 J·mol⁻¹·K⁻¹
Molecular Weight 114.1870
Wiswesser Line Notation L7TJ AQ
Evaluation A

Table of Heat Capacities, Entropies, and Phase Transition Properties — Continued

$C_7H_{14}O$ (liq) 76CON/GIN
Cycloheptanol
Heat Capacity 298 K, C_p = 58.4 cal·mol⁻¹·K⁻¹
244.3 J·mol⁻¹·K⁻¹
One temperature
Molecular Weight 114.1870
Wiswesser Line Notation L7TJ AQ
Evaluation B

$C_7H_{14}O$ (liq) 81REI
Heptanal; Oenanthal; Enanthal; n–Heptaldehyde
Heat Capacity 298 K, C_p = 48.9 cal·mol⁻¹·K⁻¹
204.6 J·mol⁻¹·K⁻¹
Temperature range 298–450 K
Molecular Weight 114.1870
Wiswesser Line Notation VH6
Evaluation D

$C_7H_{14}O$ (liq) 56PAR/KEN
Heptanal; Oenanthal; Enanthal; n–Heptaldehyde
Heat Capacity 298.15 K, C_p = 59.83 cal·mol⁻¹·K⁻¹
250.33 J·mol⁻¹·K⁻¹
Temperature range 80–300 K
Entropy 298.1 K, S = 83.3 cal·mol⁻¹·K⁻¹
348.5 J·mol⁻¹·K⁻¹
Extrapolation below 80 K, 13.55 cal·mol⁻¹K⁻¹
Phase Changes
c/liq 229.8 K, ΔH = 5637 cal·mol⁻¹
23585 J·mol⁻¹
ΔS = 24.53 cal·mol⁻¹·K⁻¹
102.63 J·mol⁻¹·K⁻¹
Molecular Weight 114.1870
Wiswesser Line Notation VH6
Evaluation B(C_p),C(S)

$C_7H_{14}O_2$ (liq) 82SCH/MIL
Heptanoic acid
Heat Capacity 298.15 K, C_p = 63.439 cal·mol⁻¹·K⁻¹
265.43 J·mol⁻¹·K⁻¹
Temperature range 80–305 K
Phase Changes
c,II/c,I 224.8 K, ΔH = 487.1 cal·mol⁻¹
2038 J·mol⁻¹
ΔS = 2.17 cal·mol⁻¹·K⁻¹
9.07 J·mol⁻¹·K⁻¹
c,I/liq 265.83 K, ΔH = 3689.6 cal·mol⁻¹
15437 J·mol⁻¹
ΔS = 13.88 cal·mol⁻¹·K⁻¹
58.07 J·mol⁻¹·K⁻¹
Molecular Weight 130.1864
Wiswesser Line Notation QV6
Evaluation B

$C_7H_{14}O_2$ (liq) 39PHI
Pentyl ethanoate; n–Amyl acetate
Heat Capacity 304.0 K, C_p = 66.0 cal·mol⁻¹·K⁻¹
276.1 J·mol⁻¹·K⁻¹
One temperature
Molecular Weight 130.1864
Wiswesser Line Notation 5OV1
Evaluation C

$C_7H_{15}N$ (liq) 76CON/GIN
Octahydroazocine
Heat Capacity 298 K, C_p = 55 cal·mol⁻¹·K⁻¹
230 J·mol⁻¹·K⁻¹
One temperature
Molecular Weight 113.2022
Wiswesser Line Notation T8MTJ
Evaluation C

C_7H_{16} (liq) 30HUF/PAR 2
2,4–Dimethylpentane
Heat Capacity 294.4 K, C_p = 52.8 cal·mol⁻¹·K⁻¹
220.9 J·mol⁻¹·K⁻¹
Temperature range 92–294 K. Value is unsmoothed experimental datum.
Entropy 298.1 K, S = 69.7 cal·mol⁻¹·K⁻¹
291.6 J·mol⁻¹·K⁻¹
Extrapolation below 90 K, 15.75 cal·mol⁻¹K⁻¹
Phase Changes
c/liq 152.5 K, ΔH = 1600 cal·mol⁻¹
6694 J·mol⁻¹
ΔS = 10.49 cal·mol⁻¹·K⁻¹
43.90 J·mol⁻¹·K⁻¹
Molecular Weight 100.2034
Wiswesser Line Notation 1Y1&1Y1&1
Evaluation B(C_p),C(S)

C_7H_{16} (liq) 61HUF/GRO
2,4–Dimethylpentane
Heat Capacity 298.15 K, C_p = 53.59 cal·mol⁻¹·K⁻¹
224.22 J·mol⁻¹·K⁻¹
Temperature range 10–300 K
Entropy 298.15 K, S = 72.46 cal·mol⁻¹·K⁻¹
303.17 J·mol⁻¹·K⁻¹
Phase Changes
c/liq 153.97 K, ΔH = 1636 cal·mol⁻¹
6845 J·mol⁻¹
ΔS = 10.36 cal·mol⁻¹·K⁻¹
44.46 J·mol⁻¹·K⁻¹
Molecular Weight 100.2034
Wiswesser Line Notation 1Y1&1Y1&1
Evaluation A

C_7H_{16} (liq) 61HUF/GRO
2,2,3–Trimethylbutane
Heat Capacity 298.15 K, C_p = 51.03 cal·mol⁻¹·K⁻¹
213.51 J·mol⁻¹·K⁻¹
Temperature range 10–300 K
Entropy 298.15 K, S = 69.85 cal·mol⁻¹·K⁻¹
292.25 J·mol⁻¹·K⁻¹
Phase Changes
Lambda–type transitions at 87, 108 K
c,II/c,I 121.4 K, ΔH = 535.8 cal·mol⁻¹
2242 J·mol⁻¹
ΔS = 4.41 cal·mol⁻¹·K⁻¹
18.47 J·mol⁻¹·K⁻¹
Molecular Weight 100.2034
Wiswesser Line Notation 1Y1&X1&1&1
Evaluation A

C_7H_{16} (liq) 30HUF/PAR 2
2,2,3–Trimethylbutane
Heat Capacity 293.9 K, C_p = 49.8 cal·mol⁻¹·K⁻¹
208.4 J·mol⁻¹·K⁻¹
Temperature range 89–294 K. Value is unsmoothed experimental datum.

Table of Heat Capacities, Entropies, and Phase Transition Properties — Continued

Entropy 298.1 K, S = 64.8 cal·mol^{-1}·K^{-1}
271.1 J·mol^{-1}·K^{-1}
Extrapolation below 90 K, 14.0 cal·mol^{-1}K^{-1}
Phase Changes
c,II/c,I 121.0 K, ΔH = 568 cal·mol^{-1}
2377 J·mol^{-1}
ΔS = 4.69 cal·mol^{-1}·K^{-1}
19.64 J·mol^{-1}·K^{-1}
Hump in heat capacity curve at about 105 K, with excess enthalpy of 58 cal·mol^{-1}.
c,I/liq 247.7 K, ΔH = 526 cal·mol^{-1}
2201 J·mol^{-1}
ΔS = 2.12 cal·mol^{-1}·K^{-1}
8.89 J·mol^{-1}·K^{-1}
Molecular Weight 100.2034
Wiswesser Line Notation 1Y1&X1&1&1
Evaluation B(C_p),C(S)

C_7H_{16} (liq) 30HUF/PAR 2
3,3–Dimethylpentane
Heat Capacity 292.9 K, C_p = 50.6 cal·mol^{-1}·K^{-1}
211.7 J·mol^{-1}·K^{-1}
Temperature range 92–293 K. Value is unsmoothed experimental datum.
Entropy 298.1 K, S = 70.1 cal·mol^{-1}·K^{-1}
293.3 J·mol^{-1}·K^{-1}
Extrapolation below 90 K, 14.50 cal·mol^{-1}K^{-1}
Phase Changes
c/liq 138.2 K, ΔH = 1689 cal·mol^{-1}
7067 J·mol^{-1}
ΔS = 12.22 cal·mol^{-1}·K^{-1}
51.14 J·mol^{-1}·K^{-1}
Molecular Weight 100.2034
Wiswesser Line Notation 2X2&1&1
Evaluation B(C_p),C(S)

C_7H_{16} (liq) 30HUF/PAR 2
2,3–Dimethylpentane
Heat Capacity 291.5 K, C_p = 51.6 cal·mol^{-1}·K^{-1}
215.9 J·mol^{-1}·K^{-1}
Temperature range 68–292 K. Value is unsmoothed experimental datum.
Entropy 298.1 K, S = 73.2 cal·mol^{-1}·K^{-1}
306.3 J·mol^{-1}·K^{-1}
Extrapolation below 70 K, 16.6 cal·mol^{-1}K^{-1}. Forms glass at low temperatures. Value includes estimated zero point entropy of 4 cal·mol^{-1}K^{-1}.
Molecular Weight 100.2034
Wiswesser Line Notation 2Y1&Y1&1
Evaluation B(C_p),C(S)

C_7H_{16} (liq) 30HUF/PAR 2
3–Ethylpentane
Heat Capacity 294.8 K, C_p = 52.0 cal·mol^{-1}·K^{-1}
217.6 J·mol^{-1}·K^{-1}
Temperature range 92–295 K. Value is unsmoothed experimental datum.
Entropy 298.1 K, S = 74.6 cal·mol^{-1}·K^{-1}
312.1 J·mol^{-1}·K^{-1}
Extrapolation below 90 K, 16.46 cal·mol^{-1}K^{-1}
Phase Changes
c/liq 154.3 K, ΔH = 2261 cal·mol^{-1}
9460 J·mol^{-1}
ΔS = 14.65 cal·mol^{-1}·K^{-1}
61.31 J·mol^{-1}·K^{-1}
Molecular Weight 100.2034
Wiswesser Line Notation 2Y2&2
Evaluation B(C_p),C(S)

C_7H_{16} (liq) 61HUF/GRO
3–Ethylpentane
Heat Capacity 298.15 K, C_p = 52.48 cal·mol^{-1}·K^{-1}
219.58 J·mol^{-1}·K^{-1}
Temperature range 10–300 K
Entropy 298.15 K, S = 75.18 cal·mol^{-1}·K^{-1}
314.55 J·mol^{-1}·K^{-1}
Phase Changes
c/liq 154.58 K, ΔH = 2282 cal·mol^{-1}
9548 J·mol^{-1}
ΔS = 14.76 cal·mol^{-1}·K^{-1}
61.77 J·mol^{-1}·K^{-1}
Molecular Weight 100.2034
Wiswesser Line Notation 2Y2&2
Evaluation A

C_7H_{16} (liq) 30HUF/PAR 2
2,2–Dimethylpentane
Heat Capacity 294.1 K, C_p = 51.9 cal·mol^{-1}·K^{-1}
217.1 J·mol^{-1}·K^{-1}
Temperature range 92–294 K. Value is unsmoothed experimental datum.
Entropy 298.1 K, S = 68.1 cal·mol^{-1}·K^{-1}
284.9 J·mol^{-1}·K^{-1}
Extrapolation below 90 K, 15.32 cal·mol^{-1}K^{-1}
Phase Changes
c/liq 148.1 K, ΔH = 1401 cal·mol^{-1}
5862 J·mol^{-1}
ΔS = 9.46 cal·mol^{-1}·K^{-1}
39.58 J·mol^{-1}·K^{-1}
Molecular Weight 100.2034
Wiswesser Line Notation 3X1&1&1
Evaluation B(C_p),C(S)

C_7H_{16} (liq) 61HUF/GRO
2,2–Dimethylpentane
Heat Capacity 298.15 K, C_p = 52.85 cal·mol^{-1}·K^{-1}
221.12 J·mol^{-1}·K^{-1}
Temperature range 10–300 K
Entropy 298.15 K, S = 71.77 cal·mol^{-1}·K^{-1}
300.29 J·mol^{-1}·K^{-1}
Phase Changes
Lambda–type transition at 83 K
c/liq 149.43 K, ΔH = 1392 cal·mol^{-1}
5824 J·mol^{-1}
ΔS = 9.32 cal·mol^{-1}·K^{-1}
38.98 J·mol^{-1}·K^{-1}
Molecular Weight 100.2034
Wiswesser Line Notation 3X1&1&1
Evaluation A

C_7H_{16} (liq) 30HUF/PAR 2
3–Methylhexane
Heat Capacity 289.2 K, C_p = 51.2 cal·mol^{-1}·K^{-1}
214.2 J·mol^{-1}·K^{-1}
Temperature range 71–289 K. Value is unsmoothed experimental datum.
Entropy 298.1 K, S = 74.0 cal·mol^{-1}·K^{-1}
309.6 J·mol^{-1}·K^{-1}
Extrapolation below 70 K, 16.9 cal·mol^{-1}K^{-1}. Forms glass at low temperatures. Value includes estimated zero point entropy of 4 cal·mol^{-1}K^{-1}.
Molecular Weight 100.2034
Wiswesser Line Notation 3Y2&1
Evaluation B(C_p),C(S)

Table of Heat Capacities, Entropies, and Phase Transition Properties — Continued

C_7H_{16} (liq) 36PAR/THO
3–Methylhexane
Heat Capacity 289.2 K, C_p = 51.8 cal·mol⁻¹·K⁻¹ 216.7 J·mol⁻¹·K⁻¹
Temperature range 71–290 K. Glass at lower temperatures. Unsmoothed experimental datum.
Molecular Weight 100.2034
Wiswesser Line Notation 3Y2&1
Evaluation B

C_7H_{16} (liq) 30PAR/HUF
2–Methylhexane
Heat Capacity 292.4 K, C_p = 52.4 cal·mol^{-1}·K^{-1} 219.2 J·mol^{-1}·K^{-1}
Temperature range 86–293 K. Value is unsmoothed experimental datum.
Entropy 298.15 K, S = 75.2 cal·mol^{-1}·K^{-1} 314.6 J·mol^{-1}·K^{-1}
Extrapolation below 90 K
Phase Changes
c/liq 154.0 K, ΔH = 2120 cal·mol^{-1} 8870 J·mol^{-1}
ΔS = 13.77 cal·mol^{-1}·K^{-1} 57.60 J·mol^{-1}·K^{-1}
Molecular Weight 100.2034
Wiswesser Line Notation 4Y1&1
Evaluation B(C_p),C(S)

C_7H_{16} (liq) 30HUF/PAR 2
2–Methylhexane
Entropy 298.1 K, S = 75.3 cal·mol^{-1}·K^{-1} 315.1 J·mol^{-1}·K^{-1}
Extrapolation below 90 K, 16.76 cal·mol^{-1}K^{-1}. Based on previously published specific heat data, 30PAR/HUF.
Phase Changes
c/liq 154.0 K, ΔH = 2120 cal·mol^{-1} 8870 J·mol^{-1}
ΔS = 13.77 cal·mol^{-1}·K^{-1} 57.60 J·mol^{-1}·K^{-1}
Molecular Weight 100.2034
Wiswesser Line Notation 4Y1&1
Evaluation C(S)

C_7H_{16} (liq) 61HUF/GRO
2–Methylhexane
Heat Capacity 298.15 K, C_p = 53.28 cal·mol^{-1}·K^{-1} 222.92 J·mol^{-1}·K^{-1}
Temperature range 10–300 K
Entropy 298.15 K, S = 77.28 cal·mol^{-1}·K^{-1} 323.34 J·mol^{-1}·K^{-1}
Phase Changes
Lambda–type transition at 72 K
c/liq 154.90 K, ΔH = 2195 cal·mol^{-1} 9184 J·mol^{-1}
ΔS = 14.17 cal·mol^{-1}·K^{-1} 59.29 J·mol^{-1}·K^{-1}
Molecular Weight 100.2034
Wiswesser Line Notation 4Y1&1
Evaluation A

C_7H_{16} (liq) 24WIL/DAN
n–Heptane
Heat Capacity 303 K, C_p = 51.9 cal·mol^{-1}·K^{-1} 217.0 J·mol^{-1}·K^{-1}
Temperature range 303–350 K. Equation only.
Molecular Weight 100.2034
Wiswesser Line Notation 7H
Evaluation C

C_7H_{16} (liq) 30PAR/HUF
n–Heptane
Heat Capacity 299.2 K, C_p = 53.1 cal·mol^{-1}·K^{-1} 222.2 J·mol^{-1}·K^{-1}
Temperature range 90–300 K. Value is unsmoothed experimental datum.
Entropy 298.15 K, S = 78.9 cal·mol^{-1}·K^{-1} 326.4 J·mol^{-1}·K^{-1}
Extrapolation below 90 K, 16.97 cal·mol^{-1}K^{-1}
Phase Changes
c/liq 182.2 K, ΔH = 3385 cal·mol^{-1} 14163 J·mol^{-1}
ΔS = 18.58 cal·mol^{-1}·K^{-1} 77.73 J·mol^{-1}·K^{-1}
Molecular Weight 100.2034
Wiswesser Line Notation 7H
Evaluation B(C_p),C(S)

C_7H_{16} (liq) 30HUF/PAR 2
n–Heptane
Entropy 298.1 K, S = 78.9 cal·mol^{-1}·K^{-1} 330.1 J·mol^{-1}·K^{-1}
Extrapolation below 90 K, 16.97 cal·mol^{-1}K^{-1}. Based on previously published specific heat data, 30PAR/HUF.
Phase Changes
c/liq 182.2 K, ΔH = 3385 cal·mol^{-1} 14163 J·mol^{-1}
ΔS = 18.58 cal·mol^{-1}·K^{-1} 77.73 J·mol^{-1}·K^{-1}
Molecular Weight 100.2034
Wiswesser Line Notation 7H
Evaluation C(S)

C_7H_{16} (liq) 32RIC/WAL
n–Heptane
Heat Capacity 298.1 K, C_p = 53.6 cal·mol^{-1}·K^{-1} 224.3 J·mol^{-1}·K^{-1}
Temperature range 293–323 K
Molecular Weight 100.2034
Wiswesser Line Notation 7H
Evaluation C

C_7H_{16} (liq) 37VOL
n–Heptane
Heat Capacity 298 K, C_p = 53.2 cal·mol^{-1}·K^{-1} 222.6 J·mol^{-1}·K^{-1}
One temperature
Molecular Weight 100.2034
Wiswesser Line Notation 7H
Evaluation B

C_7H_{16} (liq) 39PHI
n–Heptane
Heat Capacity 300.8 K, C_p = 50.4 cal·mol^{-1}·K^{-1} 210.9 J·mol^{-1}·K^{-1}
One temperature
Molecular Weight 100.2034
Wiswesser Line Notation 7H
Evaluation C

C_7H_{16} (liq) 39BYK
n–Heptane
Heat Capacity 298 K, C_p = 50.5 cal·mol^{-1}·K^{-1} 211.3 J·mol^{-1}·K^{-1}
One temperature
Molecular Weight 100.2034
Wiswesser Line Notation 7H
Evaluation C

Table of Heat Capacities, Entropies, and Phase Transition Properties — Continued

C_7H_{16} (liq) 40PIT
n–Heptane
Heat Capacity 296.5 K, C_p = 53.68 cal·mol^{-1}·K^{-1}
224.60 J·mol^{-1}·K^{-1}
Temperature range 15–318 K. Value is unsmoothed experimental datum.
Entropy 298.15 K, S = 78.60 cal·mol^{-1}·K^{-1}
328.86 J·mol^{-1}·K^{-1}
Phase Changes
c/liq 182.52 K, ΔH = 3355.8 cal·mol^{-1}
14040.7 J·mol^{-1}
ΔS = 18.39 cal·mol^{-1}·K^{-1}
76.93 J·mol^{-1}·K^{-1}
Molecular Weight 100.2034
Wiswesser Line Notation 7H
Evaluation A

C_7H_{16} (liq) 47OSB/GIN
n–Heptane
Heat Capacity 298.15 K, C_p = 53.74 cal·mol^{-1}·K^{-1}
224.85 J·mol^{-1}·K^{-1}
Temperature range 278–318 K
Molecular Weight 100.2034
Wiswesser Line Notation 7H
Evaluation A

C_7H_{16} (liq) 53GIN/FUR
n–Heptane
Heat Capacity 298.15 K, C_p = 53.71 cal·mol^{-1}·K^{-1}
224.74 J·mol^{-1}·K^{-1}
Temperature range 25–520 K
Phase Changes
c/liq 182.56 K, ΔH = 3351 cal·mol^{-1}
14022 J·mol^{-1}
ΔS = 18.36 cal·mol^{-1}·K^{-1}
76.81 J·mol^{-1}·K^{-1}
Molecular Weight 100.2034
Wiswesser Line Notation 7H
Evaluation A
See correction in 53GIN/FUR 2.

C_7H_{16} (liq) 54DOU/FUR
n–Heptane
Heat Capacity 298.15 K, C_p = 53.71 cal·mol^{-1}·K^{-1}
224.74 J·mol^{-1}·K^{-1}
Temperature range 20–520 K
Entropy 298.15 K, S = 78.39 cal·mol^{-1}·K^{-1}
327.98 J·mol^{-1}·K^{-1}
Phase Changes
c/liq 182.56 K, ΔH = 3351 cal·mol^{-1}
14022 J·mol^{-1}
ΔS = 18.36 cal·mol^{-1}·K^{-1}
78.81 J·mol^{-1}·K^{-1}
Molecular Weight 100.2034
Wiswesser Line Notation 7H
Evaluation A

C_7H_{16} (liq) 55HEL/HEI
n–Heptane
Heat Capacity 299.8 K, C_p = 53.33 cal·mol^{-1}·K^{-1}
233.13 J·mol^{-1}·K^{-1}
Temperature range 70–220°F
Molecular Weight 100.2034
Wiswesser Line Notation 7H
Evaluation B

C_7H_{16} (liq) 58SWI/ZIE
n–Heptane
Heat Capacity 332 K, C_p = 59.1 cal·mol^{-1}·K^{-1}
247.3 J·mol^{-1}·K^{-1}
Mean value 22 to 96°C
Molecular Weight 100.2034
Wiswesser Line Notation 7H
Evaluation C

C_7H_{16} (liq) 61HUF/GRO
n–Heptane
Heat Capacity 298.15 K, C_p = 53.76 cal·mol^{-1}·K^{-1}
224.93 J·mol^{-1}·K^{-1}
Temperature range 10–300 K
Entropy 298.15 K, S = 78.53 cal·mol^{-1}·K^{-1}
328.57 J·mol^{-1}·K^{-1}
Phase Changes
c/liq 182.55 K, ΔH = 3355 cal·mol^{-1}
14037 J·mol^{-1}
ΔS = 18.38 cal·mol^{-1}·K^{-1}
76.89 J·mol^{-1}·K^{-1}
Molecular Weight 100.2034
Wiswesser Line Notation 7H
Evaluation A

C_7H_{16} (liq) 63OET
n–Heptane
Entropy 298.15 K, S = 78.54 cal·mol^{-1}·K^{-1}
328.61 J·mol^{-1}·K^{-1}
Phase Changes
c/liq 182.56 K, ΔH = 3353.4 cal·mol^{-1}
14030.6 J·mol^{-1}
ΔS = 18.37 cal·mol^{-1}·K^{-1}
76.81 J·mol^{-1}·K^{-1}
Molecular Weight 100.2034
Wiswesser Line Notation 7H
Evaluation B
Run as check on calorimeter. No details.

C_7H_{16} (liq) 72MIL
n–Heptane
Heat Capacity 250 K, C_p = 50.1 cal·mol^{-1}·K^{-1}
209.6 J·mol^{-1}·K^{-1}
Temperature range 130–263 K
Phase Changes
c/liq 182.56 K, ΔH = 3361 cal·mol^{-1}
14061 J·mol^{-1}
ΔS = 18.41 cal·mol^{-1}·K^{-1}
77.02 J·mol^{-1}·K^{-1}
Molecular Weight 100.2034
Wiswesser Line Notation 7H
Evaluation B

C_7H_{16} (liq) 74DIA/REN
n–Heptane
Heat Capacity 298.15 K, C_p = 53.85 cal·mol^{-1}·K^{-1}
225.3 J·mol^{-1}·K^{-1}
Temperature range 298–323 K
Molecular Weight 100.2034
Wiswesser Line Notation 7H
Evaluation A

Table of Heat Capacities, Entropies, and Phase Transition Properties — Continued

C_7H_{16} (liq) 75HOL/ZIE
n–Heptane
Heat Capacity 298.15 K, C_p = 53.6 cal·mol^{-1}·K^{-1}
224.4 J·mol^{-1}·K^{-1}
Temperature range 182–312 K. Equation only. Experimental data deposited with journal as supplementary material.
Molecular Weight 100.2034
Wiswesser Line Notation 7H
Evaluation B

C_7H_{16} (liq) 75GRI/RAS
n–Heptane
Heat Capacity 298 K, C_p = 53.4 cal·mol^{-1}·K^{-1}
223.4 J·mol^{-1}·K^{-1}
Temperature range 300–463 K
Molecular Weight 100.2034
Wiswesser Line Notation 7H
Evaluation B

C_7H_{16} (liq) 77MEI/BLO
n–Heptane
Heat Capacity 298.15 K, C_p = 60.9 cal·mol^{-1}·K^{-1}
255.0 J·mol^{-1}·K^{-1}
Temperature range 160–350 K
Phase Changes
c/liq 182.7 K, ΔH = 3360 cal·mol^{-1}
14059 J·mol^{-1}
ΔS = 18.4 cal·mol^{-1}·K^{-1}
77.0 J·mol^{-1}·K^{-1}
Molecular Weight 100.2034
Wiswesser Line Notation 7H
Evaluation B

C_7H_{16} (liq) 79SCH/OFF
n–Heptane
Heat Capacity 285 K, C_p = 52.58 cal·mol^{-1}·K^{-1}
220.00 J·mol^{-1}·K^{-1}
Temperature range 90–285 K
Phase Changes
c/liq 182.59 K, ΔH = 3359 cal·mol^{-1}
14053 J·mol^{-1}
ΔS = 18.40 cal·mol^{-1}·K^{-1}
76.96 J·mol^{-1}·K^{-1}
Molecular Weight 100.2034
Wiswesser Line Notation 7H
Evaluation A

C_7H_{16} (liq) 79GRO/HAM
n–Heptane
Heat Capacity 298.15 K, C_p = 53.7 cal·mol^{-1}·K^{-1}
224.6 J·mol^{-1}·K^{-1}
One temperature
Molecular Weight 100.2034
Wiswesser Line Notation 7H
Evaluation B

C_7H_{16} (liq) 79BRO/ZIE
n–Heptane
Heat Capacity 298.15 K, C_p = 53.70 cal·mol^{-1}·K^{-1}
224.66 J·mol^{-1}·K^{-1}
Temperature range 183–302 K. Results as equation only.
Molecular Weight 100.2034
Wiswesser Line Notation 7H
Evaluation B

C_7H_{16} (liq) 82TAN
n–Heptane
Heat Capacity 298.15 K, C_p = 53.72 cal·mol^{-1}·K^{-1}
224.8 J·mol^{-1}·K^{-1}
Temperature range 298.15 K, One temperature
Molecular Weight 100.2034
Wiswesser Line Notation 7H
Evaluation A

$C_7H_{16}O$ (liq) 36EVA/EDL
4,4–Dimethyl–3–oxahexane; tert–Amyl ethyl ether
Heat Capacity 298 K, C_p = 58 cal·mol^{-1}·K^{-1}
243 J·mol^{-1}·K^{-1}
One temperature
Molecular Weight 116.2028
Wiswesser Line Notation 2X1&1&O2
Evaluation C

$C_7H_{16}O$ (liq) 56PAR/KEN
1–Heptanol; n–Heptyl alcohol
Heat Capacity 298.15 K, C_p = 66.58 cal·mol^{-1}·K^{-1}
278.57 J·mol^{-1}·K^{-1}
Temperature range 80–300 K
Entropy 298.1 K, S = 77.9 cal·mol^{-1}·K^{-1}
325.9 J·mol^{-1}·K^{-1}
Extrapolation below 80 K, 15.55 cal·mol^{-1}K^{-1}
Phase Changes
c/liq 240.4 K, ΔH = 4344 cal·mol^{-1}
18175 J·mol^{-1}
ΔS = 18.07 cal·mol^{-1}·K^{-1}
75.60 J·mol^{-1}·K^{-1}
Molecular Weight 116.2028
Wiswesser Line Notation Q7
Evaluation B(C_p),C(S)

$C_7H_{16}O$ (liq) 59HUT/BAI
1–Heptanol; n–Heptyl alcohol
Heat Capacity 298 K, C_p = 65.5 cal·mol^{-1}·K^{-1}
274.1 J·mol^{-1}·K^{-1}
One temperature
Molecular Weight 116.2028
Wiswesser Line Notation Q7
Evaluation C

$C_7H_{16}O$ (liq) 76CON/GIN
4–Heptanol
Heat Capacity 298 K, C_p = 75.9 cal·mol^{-1}·K^{-1}
317.6 J·mol^{-1}·K^{-1}
One temperature
Molecular Weight 116.2028
Wiswesser Line Notation QY3&3
Evaluation B

$C_7H_{16}O_2$ (liq) 73KUS/SUU
2,5–Dioxanonane; 1–n–Butoxy–2–methoxyethane
Heat Capacity 298.15 K, C_p = 67.40 cal·mol^{-1}·K^{-1}
282.0 J·mol^{-1}·K^{-1}
One temperature
Molecular Weight 132.2022
Wiswesser Line Notation 4O2O1
Evaluation B

Table of Heat Capacities, Entropies, and Phase Transition Properties — Continued

$C_7H_{16}S$ (liq) 70FIN/MCC
1-Heptanethiol; n-Heptyl mercaptan
Heat Capacity 298.15 K, C_p = 61.98 cal·mol^{-1}·K^{-1}
259.32 J·mol^{-1}·K^{-1}
Temperature range 10–370 K
Entropy 298.15 K, S = 89.71 cal·mol^{-1}·K^{-1}
375.35 J·mol^{-1}·K^{-1}
Phase Changes
c/liq 229.92 K, ΔH = 6067 cal·mol^{-1}
25384 J·mol^{-1}
ΔS = 26.39 cal·mol^{-1}·K^{-1}
110.40 J·mol^{-1}·K^{-1}
Molecular Weight 132.2634
Wiswesser Line Notation SH7
Evaluation A

C_8F_{16} (liq) 57YAR/KAY
Perfluorodimethylcyclohexane;
Hexadecafluorodimethylcyclohexane
Heat Capacity 298 K, C_p = 96.8 cal·mol^{-1}·K^{-1}
405.0 J·mol^{-1}·K^{-1}
Temperature range 298–373 K. Equation only.
Molecular Weight 400.0624
Wiswesser Line Notation L6TJ AXFFF AF BF CF DF EF FF XXFFF XF
Evaluation C
Unspecified isomer.

$C_8F_{16}O$ (liq) 57YAR/KAY
Perfluoro-3-butyltetrahydrofuran;
Hexadecafluoro-3-butyltetrahydrofuran
Heat Capacity 298 K, C_p = 103.0 cal·mol^{-1}·K^{-1}
431.0 J·mol^{-1}·K^{-1}
Temperature range 298–373 K. Equation only.
Molecular Weight 416.0618
Wiswesser Line Notation T5OTJ BF BF CXFFXFFXFFXFFF CF DF DF EF EF
Evaluation B

C_8F_{18} (liq) 82CAM/REY
n-Perfluorooctane
Heat Capacity C_p is given graphically only.
Temperature range 4.2–300 K
Phase Changes
c,II/c,I 190–200 K, ΔH = 1726 cal·mol^{-1}
7222 J·mol^{-1}
ΔS = 8.59 cal·mol^{-1}·K^{-1}
35.94 J·mol^{-1}·K^{-1}
Data given for two solid-solid transitions at 190 and 200 K.
Molecular Weight 438.0592
Wiswesser Line Notation
FXFFXFFXFFXFFXFFXFFXFFXFFF
Evaluation A

$C_8H_4Cl_2O_2$ (liq) 81REI
Phthalyl dichloride
Heat Capacity 298 K, C_p = 59.4 cal·mol^{-1}·K^{-1}
248.5 J·mol^{-1}·K^{-1}
Temperature range 290–475 K
Molecular Weight 203.0244
Wiswesser Line Notation GVR BVG
Evaluation D

$C_8H_4N_2O_2$ (c) 62STR/BAR
1,4-Phenylenediisocyanate; 1,4-Diisocyanatobenzene
Heat Capacity 298 K, C_p = 50.6 cal·mol^{-1}·K^{-1}
2117 J·mol^{-1}·K^{-1}
One temperature
Molecular Weight 160.1318
Wiswesser Line Notation OCNR DNCO
Evaluation D

C_8H_6 (liq) 31SMI/AND
Phenylacetylene
Heat Capacity 298.5 K, C_p = 42.9 cal·mol^{-1}·K^{-1}
179.5 J·mol^{-1}·K^{-1}
Temperature range 102–298 K. Value is unsmoothed experimental datum.
Molecular Weight 102.1354
Wiswesser Line Notation 1UU1R
Evaluation C

$C_8H_6O_4$ (c) 36PAR/TOD
o-Phthalic acid; Phthalic acid
Heat Capacity 298.1 K, C_p = 44.96 cal·mol^{-1}·K^{-1}
188.11 J·mol^{-1}·K^{-1}
Temperature range 90–300 K
Entropy 298.1 K, S = 49.7 cal·mol^{-1}·K^{-1}
207.94 J·mol^{-1}·K^{-1}
Extrapolation below 90 K, 15.82 cal·mol^{-1}K^{-1}
Molecular Weight 166.1330
Wiswesser Line Notation QVR BVQ
Evaluation B(C_p),C(S)

$C_8H_6O_4$ (c) 39SAT/SOG 2
o-Phthalic acid; Phthalic acid
Heat Capacity 323 K, C_p = 48.2 cal·mol^{-1}·K^{-1}
201.7 J·mol^{-1}·K^{-1}
Temperature range 0 to 100°C. Mean value.
Molecular Weight 166.1330
Wiswesser Line Notation QVR BVQ
Evaluation C
Same data in 40SAT/SOG.

$C_8H_6O_4$ (c) 41SAT/SOG 4
o-Phthalic acid; Phthalic acid
Heat Capacity 323 K, C_p = 48.2 cal·mol^{-1}·K^{-1}
201.7 J·mol^{-1}·K^{-1}
Temperature range 0 to 100°C. Mean value.
Molecular Weight 166.1330
Wiswesser Line Notation QVR BVQ
Evaluation C
Same data in 40SAT/SOG 5.

$C_8H_6O_4$ (c) 39SAT/SOG 2
m-Phthalic acid; Isophthalic acid
Heat Capacity 323 K, C_p = 48.2 cal·mol^{-1}·K^{-1}
201.7 J·mol^{-1}·K^{-1}
Temperature range 0 to 100°C. Mean value.
Molecular Weight 166.1330
Wiswesser Line Notation QVR CVQ
Evaluation C
Same data in 40SAT/SOG.

Table of Heat Capacities, Entropies, and Phase Transition Properties — Continued

$C_8H_6O_4$ (c) 41SAT/SOG 4
m–Phthalic acid; Isophthalic acid
Heat Capacity 323 K, C_p = 48.2 cal·mol^{-1}·K^{-1}
201.7 J·mol^{-1}·K^{-1}
Temperature range 0 to 100°C. Mean value.
Molecular Weight 166.1330
Wiswesser Line Notation QVR CVQ
Evaluation C
Same data in 40SAT/SOG 5.

$C_8H_6O_4$ (c) 41SAT/SOG 4
p–Phthalic acid; Terephthalic acid
Heat Capacity 323 K, C_p = 47.7 cal·mol^{-1}·K^{-1}
199.6 J·mol^{-1}·K^{-1}
Temperature range 0 to 100°C. Mean value.
Molecular Weight 166.1330
Wiswesser Line Notation QVR DVQ
Evaluation C
Same data in 40SAT/SOG 5.

C_8H_6S (c,II) 54FIN/GRO
Benzothiophene
Heat Capacity 298.15 K, C_p = 38.96 cal·mol^{-1}·K^{-1}
163.01 J·mol^{-1}·K^{-1}
Temperature range 12–335 K
Entropy 298.15 K, S = 42.33 cal·mol^{-1}·K^{-1}
177.11 J·mol^{-1}·K^{-1}
Phase Changes
Lambda transition between 250–261.6 K
c/liq 304.50 K, ΔH = 2826.8 cal·mol^{-1}
11827 J·mol^{-1}
ΔS = 9.28 cal·mol^{-1}·K^{-1}
38.84 J·mol^{-1}·K^{-1}
Molecular Weight 134.1954
Wiswesser Line Notation T56 BSJ
Evaluation A

$C_8H_7N_5O_8$ (c) 73KRI/LIC
2,4,6–Trinitro–N–(methylnitro)–m–toluidine; Methyltetryl;
2,4,6,N–Tetranitro–N–methyltoluidine
Heat Capacity 298 K, C_p = 78.0 cal·mol^{-1}·K^{-1}
326.4 J·mol^{-1}·K^{-1}
Temperature range 200–376 K. Equation only.
Phase Changes
c/liq 375.6 K, ΔH = 4620 cal·mol^{-1}
19330 J·mol^{-1}
ΔS = 12.3 cal·mol^{-1}·K^{-1}
51.5 J·mol^{-1}·K^{-1}
Molecular Weight 301.1720
Wiswesser Line Notation WNN1&R C1 BNW DNW FNW
Evaluation C

$C_8H_7N_5O_8$ (c) 73KRI/LIC
2,4,6–Trinitrophenylethyl nitramine; Ethyltetryl;
2,4,6,N–Tetranitro–ethylaniline
Heat Capacity 298 K, C_p = 79.4 cal·mol^{-1}·K^{-1}
332.2 J·mol^{-1}·K^{-1}
Temperature range 200–369 K. Equation only.
Phase Changes
c/liq 369.0 K, ΔH = 5620 cal·mol^{-1}
23510 J·mol^{-1}
ΔS = 15.2 cal·mol^{-1}·K^{-1}
63.7 J·mol^{-1}·K^{-1}
Molecular Weight 301.1720
Wiswesser Line Notation WNN2&R BNW DNW FNW
Evaluation C

C_8H_8 (liq) 31SMI/AND
Styrene
Heat Capacity 298.5 K, C_p = 43.0 cal·mol^{-1}·K^{-1}
179.9 J·mol^{-1}·K^{-1}
Temperature range 102–299 K. Value is unsmoothed experimental datum.
Molecular Weight 104.1512
Wiswesser Line Notation 1U1R
Evaluation C

C_8H_8 (liq) 46PIT/GUT
Styrene
Heat Capacity 298.15 K, C_p = 43.70 cal·mol^{-1}·K^{-1}
182.84 J·mol^{-1}·K^{-1}
Temperature range 15–300 K
Entropy 298.15 K, S = 56.78 cal·mol^{-1}·K^{-1}
237.57 J·mol^{-1}·K^{-1}
Phase Changes
c/liq 242.27 K, ΔH = 2617 cal·mol^{-1}
10949 J·mol^{-1}
ΔS = 10.79 cal·mol^{-1}·K^{-1}
45.16 J·mol^{-1}·K^{-1}
Molecular Weight 104.1512
Wiswesser Line Notation 1U1R
Evaluation A

C_8H_8 (liq) 50KUR
Styrene
Heat Capacity 298 K, C_p = 56.3 cal·mol^{-1}·K^{-1}
23.56 J·mol^{-1}·K^{-1}
Temperature range 21 to 139°C
Molecular Weight 104.1512
Wiswesser Line Notation 1U1R
Evaluation B

C_8H_8 (liq) 49SCO/GRO
Cyclooctatetraene
Heat Capacity 298.15 K, C_p = 44.26 cal·mol^{-1}·K^{-1}
185.18 J·mol^{-1}·K^{-1}
Temperature range 12–340 K
Entropy 298.15 K, S = 52.65 cal·mol^{-1}·K^{-1}
220.29 J·mol^{-1}·K^{-1}
Phase Changes
c/liq 268.48 K, ΔH = 2694.6 cal·mol^{-1}
11274.2 J·mol^{-1}
ΔS = 10.037 cal·mol^{-1}·K^{-1}
41.493 J·mol^{-1}·K^{-1}
Molecular Weight 104.1512
Wiswesser Line Notation L8J
Evaluation A

$(C_8H_8)_n$ (c) 62DAI/EVA 4
Polystyrene, isotactic
Heat Capacity 298.15 K, C_p = 30.49 cal·mol^{-1}·K^{-1}
127.6 J·mol^{-1}·K^{-1}
Temperature range 20–310 K
Entropy 298.15 K, S = 31.40 cal·mol^{-1}·K^{-1}
131.4 J·mol^{-1}·K^{-1}
When extrapolated to 100% crystallinity, the entropy is 30.7 cal·mol^{-1}·K^{-1}.
Molecular Weight 104.1512
Wiswesser Line Notation /*YR&1*/
Evaluation A

Table of Heat Capacities, Entropies, and Phase Transition Properties — Continued

$(C_8H_8)_n$ (c) 65KAR/BAI
Polystyrene
Heat Capacity 298.15 K, C_p = 30.43 cal·mol^{-1}·K^{-1}
127.3 J·mol^{-1}·K^{-1}
Temperature range 305–525 K. Glass transition at 355 K. Values per C_8H_8 unit.
Molecular Weight 104.1512
Wiswesser Line Notation /*Y1&R*/
Evaluation B

$(C_8H_8)_n$ (amorp) 65KAR/BAI
Polystyrene, isotactic
Heat Capacity 298.15 K, C_p = 30.47 cal·mol^{-1}·K^{-1}
127.5 J·mol^{-1}·K^{-1}
Temperature range 300–522 K. Glass transition at 360 K. Values per C_8H_8 unit.
Molecular Weight 104.1512
Wiswesser Line Notation /*Y1&R*/
Evaluation B

$(C_8H_8)_n$ (amorp) 65KAR/BAI
Polystyrene, atactic
Heat Capacity 298.15 K, C_p = 30.52 cal·mol^{-1}·K^{-1}
127.7 J·mol^{-1}·K^{-1}
Temperature range 80–480 K. Also annealed sample, 293–378 K. Glass transition at about 367 K. Values per C_8H_8 unit.
Molecular Weight 104.1512
Wiswesser Line Notation /*Y1&R*/
Evaluation B
NBS broad molecular weight distribution. Sample NS706.

$(C_8H_8)_n$ (c) 65ABU/DOL
Polystyrene, atactic
Heat Capacity 298 K, C_p = 29.7 cal·mol^{-1}·K^{-1}
124.3 J·mol^{-1}·K^{-1}
Temperature range 223–553K; values per monomer unit.
Molecular Weight 104.1512
Wiswesser Line Notation /*YR&1*/
Evaluation B
Also data above the glass transition.

$(C_8H_8)_n$ (c) 65ABU/DOL
Polystyrene, isotactic, annealed
Heat Capacity 298 K, C_p = 29.6 cal·mol^{-1}·K^{-1}
123.8 J·mol^{-1}·K^{-1}
Temperature range 298–348K; values per monomer unit.
Molecular Weight 104.1512
Wiswesser Line Notation /*YR&1*/
Evaluation B
Also data for amorphous and semicrystalline isotactic, and above glass transition.

$(C_8H_8)_n$ (c) 68CHA/BES
Polystyrene, atactic
Heat Capacity 298.15 K, C_p = 30.468 cal·mol^{-1}·K^{-1}
127.48 J·mol^{-1}·K^{-1}
Temperature range 10–360 K, National Bureau of Standards Standard Sample 705
Entropy 298.15 K, S = 32.225 cal·mol^{-1}·K^{-1}
134.83 J·mol^{-1}·K^{-1}
Molecular Weight 104.1512
Wiswesser Line Notation /*YR&1*/
Evaluation A

C_8H_8O (liq) 39PHI
Acetophenone; Methyl phenyl ketone
Heat Capacity 303.2 K, C_p = 54.4 cal·mol^{-1}·K^{-1}
227.6 J·mol^{-1}·K^{-1}
One temperature
Molecular Weight 120.1506
Wiswesser Line Notation 1VR
Evaluation C

$C_8H_8O_2$ (liq) 71HAL/BAL
Methyl benzoate
Heat Capacity 297 K, C_p = 51.7 cal·mol^{-1}·K^{-1}
216.5 J·mol^{-1}·K^{-1}
One temperature
Molecular Weight 136.1500
Wiswesser Line Notation 1OVR
Evaluation C

$C_8H_8O_2$ (liq) 79FUC
Methyl benzoate
Heat Capacity 298.15 K, C_p = 52.9 cal·mol^{-1}·K^{-1}
221.3 J·mol^{-1}·K^{-1}
One temperature
Molecular Weight 136.1500
Wiswesser Line Notation 1OVR
Evaluation B

$C_8H_8O_2$ (c) 26AND/LYN
o–Toluic acid; 2–Methylbenzoic acid
Heat Capacity 298 K, C_p = 41.8 cal·mol^{-1}·K^{-1}
174.9 J·mol^{-1}·K^{-1}
Temperature range 22 to 200°C
Phase Changes
c/liq 376.9 K, ΔH = 4820 cal·mol^{-1}
20170 J·mol^{-1}
ΔS = 12.8 cal·mol^{-1}·K^{-1}
53.5 J·mol^{-1}·K^{-1}
Molecular Weight 136.1500
Wiswesser Line Notation QVR B1
Evaluation C

$C_8H_8O_2$ (c) 26AND/LYN
m–Toluic acid; 3–Methylbenzoic acid
Heat Capacity 298 K, C_p = 39.1 cal·mol^{-1}·K^{-1}
163.6 J·mol^{-1}·K^{-1}
Temperature range 22 to 170°C
Phase Changes
c/liq 381.9 K, ΔH = 3760 cal·mol^{-1}
15730 J·mol^{-1}
ΔS = 9.8 cal·mol^{-1}·K^{-1}
41.2 J·mol^{-1}·K^{-1}
Molecular Weight 136.1500
Wiswesser Line Notation QVR C1
Evaluation C

$C_8H_8O_2$ (c) 26AND/LYN
p–Toluic acid; 4–Methylbenzoic acid
Heat Capacity 298 K, C_p = 40.4 cal·mol^{-1}·K^{-1}
169.0 J·mol^{-1}·K^{-1}
Temperature range 22 to 225°C
Phase Changes
c/liq 452.8 K, ΔH = 5430 cal·mol^{-1}
22720 J·mol^{-1}
ΔS = 12.0 cal·mol^{-1}·K^{-1}
50.2 J·mol^{-1}·K^{-1}
Molecular Weight 136.1500
Wiswesser Line Notation QVR D1
Evaluation C

Table of Heat Capacities, Entropies, and Phase Transition Properties — Continued

$C_8H_8O_3$ (liq) 33KOL/UDO
Methyl salicylate
Heat Capacity 295.2 K, C_p = 64.2 cal·mol^{-1}·K^{-1}
268.6 J·mol^{-1}·K^{-1}
One temperature
Molecular Weight 152.1494
Wiswesser Line Notation QR BVO1
Evaluation C

$C_8H_8O_3$ (liq) 34KOL/UDO 2
Methyl salicylate
Heat Capacity 295.2 K, C_p = 59.5 cal·mol^{-1}·K^{-1}
248.9 J·mol^{-1}·K^{-1}
One temperature
Molecular Weight 152.1494
Wiswesser Line Notation QR BVO1
Evaluation C

$C_8H_8O_3$ (c) 41SAT/SOG 3
4–Methoxybenzoic acid; p–Anisic acid
Heat Capacity 323 K, C_p = 49.6 cal·mol^{-1}·K^{-1}
205.0 J·mol^{-1}·K^{-1}
Temperature range 0 to 100°C. Mean value.
Molecular Weight 152.1494
Wiswesser Line Notation QVR DO1
Evaluation C
Same data as 40SAT/SOG 4.

$C_8H_8O_3$ (c) 41SAT/SOG 3
Mandelic acid
Heat Capacity 323 K, C_p = 47.6 cal·mol^{-1}·K^{-1}
199.2 J·mol^{-1}·K^{-1}
Temperature range 0 to 100°C. Mean value.
Molecular Weight 152.1494
Wiswesser Line Notation QVYQR
Evaluation C
Same data as 40SAT/SOG 4.

$C_8H_8O_3$ (c) 78GEI/KAR 2
Tetrahydrophthalic anhydride
Heat Capacity 12–390 K. Data deposited VINITI, No 3882–77, 5 October 1977. Includes C_p, S, ΔHm, Tm.
Molecular Weight 152.1494
Wiswesser Line Notation T666 1A M CVOVT&&J
Evaluation B(for original data)

C_8H_9NO (c) 41SAT/SOG
Acetanilide
Heat Capacity 323 K, C_p = 45.8 cal·mol^{-1}·K^{-1}
191.6 J·mol^{-1}·K^{-1}
Temperature range 0 to 100°C. Mean value.
Molecular Weight 135.1652
Wiswesser Line Notation 1VMR
Evaluation C
Same data in 40SAT/SOG 2.

C_8H_9NO (c) 80AND/CON
Acetanilide
Phase Changes
c,I/liq 387.525 K, ΔH = 5175 cal·mol^{-1}
21653 J·mol^{-1}
ΔS = 13.35 cal·mol^{-1}·K^{-1}
55.87 J·mol^{-1}·K^{-1}
Molecular Weight 135.1652
Wiswesser Line Notation 1VMR
Evaluation A

$C_8H_9NO_2$ (c) 71PRI
Methyl phenylcarbamate
Heat Capacity 298 K, C_p = 48.6 cal·mol^{-1}·K^{-1}
203.3 J·mol^{-1}·K^{-1}
Temperature range 200–390 K. Complete data deposited VINITI, No. 2713–71, 25 March 1971.
Phase Changes
c/liq 325 K, ΔH = 3477 cal·mol^{-1}
14548 J·mol^{-1}
ΔS = 10.7 cal·mol^{-1}·K^{-1}
44.8 J·mol^{-1}·K^{-1}
Molecular Weight 151.1646
Wiswesser Line Notation 1OVMR
Evaluation B

$C_8H_9NO_2$ (c) 26AND/LYN
o–Hydroxyacetanilide
Heat Capacity 298 K, C_p = 43.6 cal·mol^{-1}·K^{-1}
182.4 J·mol^{-1}·K^{-1}
Temperature range 22 to 140°C
Phase Changes
c/liq 364.5 K, ΔH = 5080 cal·mol^{-1}
21250 J·mol^{-1}
ΔS = 13.9 cal·mol^{-1}·K^{-1}
58.3 J·mol^{-1}·K^{-1}
Molecular Weight 151.1646
Wiswesser Line Notation QR BMV1
Evaluation C

$C_8H_9NO_2$ (c) 80SAB/SKO
N–Phenylglycine
Heat Capacity 298.15 K, C_p = 42.2 cal·mol^{-1}·K^{-1}
176.6 J·mol^{-1}·K^{-1}
Temperature range 298.15 K, One temperature
Phase Changes
c/g 298.15 K, ΔH = 30.6 cal·mol^{-1}
128.0 J·mol^{-1}
ΔS = 0.10 Cal·mol^{-1}·K^{-1}
0.43 J·mol^{-1}·K^{-1}
Sublimation measurements made over 350–380 K; data corrected to 298.15 K.
Molecular Weight 151.1646
Wiswesser Line Notation QV1MR
Evaluation A

$C_8H_9NO_2$ (c) 80SAB/SKO
α–Phenylglycine (D)
Heat Capacity 298.15 K, C_p = 42.4 cal·mol^{-1}·K^{-1}
177.4 J·mol^{-1}·K^{-1}
Temperature range 298.15 K, One temperature
Phase Changes
c/g 298.15 K, ΔH = 39.4 cal·mol^{-1}
165 J·mol^{-1}
ΔS = 0.13 cal·mol^{-1}·K^{-1}
0.55 J·mol^{-1}·K^{-1}
Sublimation measurements made over 435–455 K; data corrected to 298.15 K.
Molecular Weight 151.1646
Wiswesser Line Notation ZYRVQ
Evaluation A

Table of Heat Capacities, Entropies, and Phase Transition Properties — Continued

$C_8H_9NO_4$ (c) 39SAT/SOG 2
Ammonium acid o-phthalate
Heat Capacity 323 K, C_p = 66.7 cal·mol^{-1}·K^{-1}
279.1 J·mol^{-1}·K^{-1}
Temperature range 0 to 100°C. Mean value.
Molecular Weight 183.1634
Wiswesser Line Notation QVR BVQ &ZH
Evaluation C
Same data in 40SAT/SOG.

$C_8H_9NO_4$ (c) 39SAT/SOG 2
Ammonium acid m-phthalate
Heat Capacity 323 K, C_p = 60.1 cal·mol^{-1}·K^{-1}
251.5 J·mol^{-1}·K^{-1}
Temperature range 0 to 100°C. Mean value.
Molecular Weight 183.1634
Wiswesser Line Notation QVR CVQ &ZH
Evaluation C
Same data in 40SAT/SOG.

$C_8H_9N_2O$ (c) 41SAT/SOG
p-Nitroacetanilide
Heat Capacity 323 K, C_p = 55.1 cal·mol^{-1}·K^{-1}
230.5 J·mol^{-1}·K^{-1}
Temperature range 0 to 100°C. Mean value.
Molecular Weight 181.1707
Wiswesser Line Notation WNR DMV1
Evaluation C
Same data in 40SAT/SOG 2.

C_8H_{10} (liq) 24WIL/DAN
1,2-Dimethylbenzene; o-Xylene
Heat Capacity 303 K, C_p = 43.6 cal·mol^{-1}·K^{-1}
182.4 J·mol^{-1}·K^{-1}
Temperature range 303–348 K. Equation only.
Molecular Weight 106.1670
Wiswesser Line Notation 1R B1
Evaluation C

C_8H_{10} (liq) 30HUF/PAR
1,2-Dimethylbenzene; o-Xylene
Heat Capacity 298.1 K, C_p = 43.95 cal·mol^{-1}·K^{-1}
183.89 J·mol^{-1}·K^{-1}
Temperature range 90–295 K. Value is unsmoothed experimental datum.
Entropy 298.1 K, S = 59.3 cal·mol^{-1}·K^{-1}
248.1 J·mol^{-1}·K^{-1}
Extrapolation below 90 K, 14.53 cal·mol^{-1}K^{-1}
Phase Changes
c,II/c,I 208 K, ΔH = 7.4 cal·mol^{-1}
31 J·mol^{-1}
ΔS = 0.036 cal·mol^{-1}·K^{-1}
0.15 J·mol^{-1}·K^{-1}
c,I/liq 247.8 K, ΔH = 3116 cal·mol^{-1}
13037 J·mol^{-1}
ΔS = 12.57 cal·mol^{-1}·K^{-1}
52.61 J·mol^{-1}·K^{-1}
Molecular Weight 106.1670
Wiswesser Line Notation 1R B1
Evaluation B(C_p),C(S)

C_8H_{10} (liq) 43PIT/SCO
1,2-Dimethylbenzene; o-Xylene
Heat Capacity 298.15 K, C_p = 44.89 cal·mol^{-1}·K^{-1}
187.82 J·mol^{-1}·K^{-1}
Temperature range 14–301 K
Entropy 298.15 K, S = 58.80 cal·mol^{-1}·K^{-1}
246.02 J·mol^{-1}·K^{-1}
Phase Changes
c/liq 247.82 K, ΔH = 3250 cal·mol^{-1}
13598 J·mol^{-1}
ΔS = 13.11 cal·mol^{-1}·K^{-1}
54.87 J·mol^{-1}·K^{-1}
Molecular Weight 106.1670
Wiswesser Line Notation 1R B1
Evaluation A

C_8H_{10} (liq) 47KUR
1,2-Dimethylbenzene; o-Xylene
Heat Capacity 298 K, C_p = 44.7 cal·mol^{-1}·K^{-1}
187.0 J·mol^{-1}·K^{-1}
Temperature range 15 to 132 °C, mean C_p, three temperatures.
Molecular Weight 106.1670
Wiswesser Line Notation 1R B1
Evaluation D

C_8H_{10} (liq) 58SWI/ZIE 2
1,2-Dimethylbenzene; o-Xylene
Heat Capacity 347 K, C_p = 49.4 cal·mol^{-1}·K^{-1}
206.7 J·mol^{-1}·K^{-1}
Mean value 22 to 126°C
Molecular Weight 106.1670
Wiswesser Line Notation 1R B1
Evaluation C

C_8H_{10} (liq) 24WIL/DAN
1,3-Dimethylbenzene; m-Xylene
Heat Capacity 303 K, C_p = 42.6 cal·mol^{-1}·K^{-1}
178.2 J·mol^{-1}·K^{-1}
Temperature range 303–348 K. Equation only.
Molecular Weight 106.1670
Wiswesser Line Notation 1R C1
Evaluation C

C_8H_{10} (liq) 30HUF/PAR
1,3-Dimethylbenzene; m-Xylene
Heat Capacity 275.3 K, C_p = 41.9 cal·mol^{-1}·K^{-1}
175.3 J·mol^{-1}·K^{-1}
Temperature range 96–275 K. Value is unsmoothed experimental datum.
Entropy 298.1 K, S = 60.3 cal·mol^{-1}·K^{-1}
252.3 J·mol^{-1}·K^{-1}
Extrapolation below 90 K, 16.00 cal·mol^{-1}K^{-1}
Phase Changes
c,II/c,I 166 K, ΔH = 50 cal·mol^{-1}
208 J·mol^{-1}
ΔS = 0.30 cal·mol^{-1}·K^{-1}
1.25 J·mol^{-1}·K^{-1}
c,I/liq 219.6 K, ΔH = 2735 cal·mol^{-1}
11443 J·mol^{-1}
ΔS = 12.45 cal·mol^{-1}·K^{-1}
52.11 J·mol^{-1}·K^{-1}
Molecular Weight 106.1670
Wiswesser Line Notation 1R C1
Evaluation B(C_p),C(S)

Table of Heat Capacities, Entropies, and Phase Transition Properties — Continued

C_8H_{10} (liq) 43PIT/SCO
1,3–Dimethylbenzene; m–Xylene
Heat Capacity 298.15 K, C_p = 43.78 cal·mol⁻¹·K⁻¹
183.18 J·mol⁻¹·K⁻¹
Temperature range 14–320 K
Entropy 298.15 K, S = 60.66 cal·mol⁻¹·K⁻¹
253.80 J·mol⁻¹·K⁻¹
Phase Changes
c/liq 225.27 K, ΔH = 2765 cal·mol⁻¹
11569 J·mol⁻¹
ΔS = 12.27 cal·mol⁻¹·K⁻¹
51.36 J·mol⁻¹·K⁻¹
Molecular Weight 106.1670
Wiswesser Line Notation 1R C1
Evaluation A

C_8H_{10} (liq) 47KUR
1,3–Dimethylbenzene; m–Xylene
Heat Capacity 298 K, C_p = 44.1 cal·mol⁻¹·K⁻¹
184.5 J·mol⁻¹·K⁻¹
Temperature range 16 to 132 °C, mean C_p, three temperatures.
Molecular Weight 106.1670
Wiswesser Line Notation 1R C1
Evaluation D

C_8H_{10} (liq) 58SWI/ZIE 2
1,3–Dimethylbenzene; m–Xylene
Heat Capacity 336 K, C_p = 47.6 cal·mol⁻¹·K⁻¹
199.2 J·mol⁻¹·K⁻¹
Mean value 21 to 106°C
Molecular Weight 106.1670
Wiswesser Line Notation 1R C1
Evaluation C

C_8H_{10} (liq) 24WIL/DAN
1,4–Dimethylbenzene; p–Xylene
Heat Capacity 303 K, C_p = 42.2 cal·mol⁻¹·K⁻¹
176.6 J·mol⁻¹·K⁻¹
Temperature range 303–348 K. Equation only.
Molecular Weight 106.1670
Wiswesser Line Notation 1R D1
Evaluation C

C_8H_{10} (liq) 30HUF/PAR
1,4–Dimethylbenzene; p–Xylene
Heat Capacity 299.0 K, C_p = 43.1 cal·mol⁻¹·K⁻¹
180.3 J·mol⁻¹·K⁻¹
Temperature range 92–299 K. Value is unsmoothed experimental datum.
Entropy 298.1 K, S = 60.5 cal·mol⁻¹·K⁻¹
253.1 J·mol⁻¹·K⁻¹
Extrapolation below 90 K, 15.58 cal·mol⁻¹K⁻¹
Phase Changes
c/liq 286.3 K, ΔH = 4047 cal·mol⁻¹
16933 J·mol⁻¹
ΔS = 14.14 cal·mol⁻¹·K⁻¹
59.14 J·mol⁻¹·K⁻¹
Molecular Weight 106.1670
Wiswesser Line Notation 1R D1
Evaluation B(C_p),C(S)

C_8H_{10} (liq) 43PIT/SCO
1,4–Dimethylbenzene; p–Xylene
Heat Capacity 298.15 K, C_p = 43.92 cal·mol⁻¹·K⁻¹
183.76 J·mol⁻¹·K⁻¹
Temperature range 14–360 K
Entropy 298.15 K, S = 59.20 cal·mol⁻¹·K⁻¹
243.51 J·mol⁻¹·K⁻¹
Phase Changes
c/liq 286.39 K, ΔH = 4090 cal·mol⁻¹
17113 J·mol⁻¹
ΔS = 14.28 cal·mol⁻¹·K⁻¹
59.75 J·mol⁻¹·K⁻¹
Molecular Weight 106.1670
Wiswesser Line Notation 1R D1
Evaluation A

C_8H_{10} (liq) 47KUR
1,4–Dimethylbenzene; p–Xylene
Heat Capacity 298 K, C_p = 44.2 cal·mol⁻¹·K⁻¹
184.9 J·mol⁻¹·K⁻¹
Temperature range 15 to 132 °C, mean C_p, three temperatures.
Molecular Weight 106.1670
Wiswesser Line Notation 1R D1
Evaluation D

C_8H_{10} (liq) 47COR/GIN
1,4–Dimethylbenzene; p–Xylene
Heat Capacity 298 K, C_p = 43.4 cal·mol⁻¹·K⁻¹
181.6 J·mol⁻¹·K⁻¹
Temperature range 273–573 K
Phase Changes
c/liq 286.3 K, ΔH = 4087 cal·mol⁻¹
17100 J·mol⁻¹
ΔS = 14.27 cal·mol⁻¹·K⁻¹
59.73 J·mol⁻¹·K⁻¹
Molecular Weight 106.1670
Wiswesser Line Notation 1R D1
Evaluation A

C_8H_{10} (liq) 58SWI/ZIE 2
1,4–Dimethylbenzene; p–Xylene
Heat Capacity 336 K, C_p = 47.5 cal·mol⁻¹·K⁻¹
198.7 J·mol⁻¹·K⁻¹
Mean value 21 to 106°C
Molecular Weight 106.1670
Wiswesser Line Notation 1R D1
Evaluation C

C_8H_{10} (liq) 71HYD/SUB
1,4–Dimethylbenzene; p–Xylene
Heat Capacity 298.15 K, C_p = 43.43 cal·mol⁻¹·K⁻¹
181.7 J·mol⁻¹·K⁻¹
Temperature range 298; 313 K
Molecular Weight 106.1670
Wiswesser Line Notation 1R D1
Evaluation B

C_8H_{10} (liq) 79OTT/GOA
1,4–Xylene; p–Xylene
Heat Capacity 298.15 K, C_p = 43.5 cal·mol⁻¹·K⁻¹
181.9 J·mol⁻¹·K⁻¹
Temperature range 288.15–328.15 K
Molecular Weight 106.1670
Wiswesser Line Notation 1R D1
Evaluation B

Table of Heat Capacities, Entropies, and Phase Transition Properties — Continued

C_8H_{10} (liq) 81REI
Ethylbenzene
Heat Capacity 298 K, C_p = 44.1 cal·mol^{-1}·K^{-1}
184.5 J·mol^{-1}·K^{-1}
Temperature range 292–425 K
Molecular Weight 106.1670
Wiswesser Line Notation 2R
Evaluation D

C_8H_{10} (liq) 24WIL/DAN
Ethylbenzene
Heat Capacity 303 K, C_p = 43.4 cal·mol^{-1}·K^{-1}
181.6 J·mol^{-1}·K^{-1}
Temperature range 303–343 K. Equation only.
Molecular Weight 106.1670
Wiswesser Line Notation 2R
Evaluation C

C_8H_{10} (liq) 30HUF/PAR
Ethylbenzene
Heat Capacity 297.4 K, C_p = 43.4 cal·mol^{-1}·K^{-1}
181.6 J·mol^{-1}·K^{-1}
Temperature range 93–305 K. Value is unsmoothed experimental datum.
Entropy 298.1 K, S = 61.2 cal·mol^{-1}·K^{-1}
256.1 J·mol^{-1}·K^{-1}
Extrapolation below 90 K, 14.60 cal·mol^{-1}K^{-1}
Phase Changes
c/liq 178.0 K, ΔH = 2190 cal·mol^{-1}
9163 J·mol^{-1}
ΔS = 12.30 cal·mol^{-1}·K^{-1}
51.48 J·mol^{-1}·K^{-1}
Molecular Weight 106.1670
Wiswesser Line Notation 2R
Evaluation B(C_p),C(S)

C_8H_{10} (liq) 31BLA/LET
Ethylbenzene
Heat Capacity 298.15 K, C_p = 44.6 cal·mol^{-1}·K^{-1}
186.6 J·mol^{-1}·K^{-1}
Temperature range 286–368 K
Heat capacity reported as 0.420 cal·g^{-1}·K^{-1} at 25° C.
Molecular Weight 106.1670
Wiswesser Line Notation 2R
Evaluation B

C_8H_{10} (liq) 31SMI/AND
Ethylbenzene
Heat Capacity 298.5 K, C_p = 43.9 cal·mol^{-1}·K^{-1}
183.7 J·mol^{-1}·K^{-1}
Temperature range 102–299 K. Value is unsmoothed experimental datum.
Molecular Weight 106.1670
Wiswesser Line Notation 2R
Evaluation C

C_8H_{10} (liq) 34KOL/UDO
Ethylbenzene
Heat Capacity 302.8 K, C_p = 42.7 cal·mol^{-1}·K^{-1}
178.7 J·mol^{-1}·K^{-1}
One temperature
Molecular Weight 106.1670
Wiswesser Line Notation 2R
Evaluation C

C_8H_{10} (liq) 34KOL/UDO 2
Ethylbenzene
Heat Capacity 302.7 K, C_p = 42.7 cal·mol^{-1}·K^{-1}
178.7 J·mol^{-1}·K^{-1}
One temperature
Molecular Weight 106.1670
Wiswesser Line Notation 2R
Evaluation C

C_8H_{10} (liq) 44GUT/SPI
Ethylbenzene
Heat Capacity 298.15 K, C_p = 44.41 cal·mol^{-1}·K^{-1}
185.81 J·mol^{-1}·K^{-1}
Temperature range 13–305 K
Entropy 298.15 K, S = 60.95 cal·mol^{-1}·K^{-1}
255.01 J·mol^{-1}·K^{-1}
Phase Changes
c/liq 178.17 K, ΔH = 2190 cal·mol^{-1}
9163 J·mol^{-1}
ΔS = 12.29 cal·mol^{-1}·K^{-1}
51.43 J·mol^{-1}·K^{-1}
Molecular Weight 106.1670
Wiswesser Line Notation 2R
Evaluation A

C_8H_{10} (liq) 45SCO/BRI
Ethylbenzene
Heat Capacity 298.15 K, C_p = 44.46 cal·mol^{-1}·K^{-1}
186.04 J·mol^{-1}·K^{-1}
Temperature range 15–300 K.
Phase Changes
c/liq 178.15 K, ΔH = 2194.5 cal·mol^{-1}
9181.8 J·mol^{-1}
ΔS = 12.32 cal·mol^{-1}·K^{-1}
51.54 J·mol^{-1}·K^{-1}
liq/g 294.01 K, ΔH = 10155 cal·mol^{-1}
42490 J·mol^{-1}
ΔS = 34.54 cal·mol^{-1}·K^{-1}
144.5 J·mol^{-1}·K^{-1}
Molecular Weight 106.1670
Wiswesser Line Notation 2R
Evaluation A

C_8H_{10} (liq) 47KUR
Ethylbenzene
Heat Capacity 298 K, C_p = 44.4 cal·mol^{-1}·K^{-1}
185.8 J·mol^{-1}·K^{-1}
Temperature range 15 to 18 °C, mean C_p, four temperatures.
Molecular Weight 106.1670
Wiswesser Line Notation 2R
Evaluation D

C_8H_{10} (liq) 76FOR/BEN
Ethylbenzene
Heat Capacity 298.15 K, C_p = 44.40 cal·mol^{-1}·K^{-1}
185.78 J·mol^{-1}·K^{-1}
One temperature
Molecular Weight 106.1670
Wiswesser Line Notation 2R
Evaluation B

Table of Heat Capacities, Entropies, and Phase Transition Properties — Continued

$C_8H_{10}N_2O$ (c) 40CAM/CAM
p-Nitrosodimethylaniline
Heat Capacity 293 K, C_p = 49.4 cal·mol^{-1}·K^{-1}
206.7 J·mol^{-1}·K^{-1}
One temperature
Molecular Weight 150.1798
Wiswesser Line Notation ONR DN1&1
Evaluation C

$C_8H_{10}N_4O_2$ (c) 79BOT/CAM
Caffeine, anhydrous
Heat Capacity 298.15 K, C_p = 55.4 cal·mol^{-1}·K^{-1}
232 J·mol^{-1}·K^{-1}
One temperature
Phase Changes
c,II/c,I 414 K, ΔH = 963.2 cal·mol^{-1}
4030 J·mol^{-1}
ΔS = 6.83 cal·mol^{-1}·K^{-1}
28.6 J·mol^{-1}·K^{-1}
Low temperature β to a high temperature α transition.
c,I/liq 509.3 K, ΔH = 5163 cal·mol^{-1}
21600 J·mol^{-1}
ΔS = 10.13 cal·mol^{-1}·K^{-1}
42.4 J·mol^{-1}·K^{-1}
Molecular Weight 194.1926
Wiswesser Line Notation T56 BN DN FNVNVJ B1 F1 H1
Evaluation B

$C_8H_{10}O$ (liq) 75NIC/WAD
2-Phenylethanol
Heat Capacity 298.15 K, C_p = 60.38 cal·mol^{-1}·K^{-1}
252.64 J·mol^{-1}·K^{-1}
One temperature
Molecular Weight 122.1664
Wiswesser Line Notation Q2R
Evaluation B

$C_8H_{10}O$ (c) 75NIC/WAD
4-Ethylphenol
Heat Capacity 298.15 K, C_p = 49.45 cal·mol^{-1}·K^{-1}
206.9 J·mol^{-1}·K^{-1}
One temperature
Molecular Weight 122.1664
Wiswesser Line Notation QR D2
Evaluation B

$C_8H_{10}O_2$ (liq) 75NIC/WAD
2-Phenoxyethanol
Heat Capacity 298.15 K, C_p = 70.42 cal·mol^{-1}·K^{-1}
294.63 J·mol^{-1}·K^{-1}
One temperature
Molecular Weight 138.1658
Wiswesser Line Notation Q2OR
Evaluation B

$C_8H_{11}N$ (liq) 81REI
N,N-Dimethylaniline
Heat Capacity 298 K, C_p = 50.7 cal·mol^{-1}·K^{-1}
212.1 J·mol^{-1}·K^{-1}
Temperature range 292–478 K
Molecular Weight 121.1816
Wiswesser Line Notation 1N1&R
Evaluation D

$C_8H_{11}N$ (liq) 02LOU
N,N-Dimethylaniline
Heat Capacity 375 K, C_p = 58 cal·mol^{-1}·K^{-1}
243 J·mol^{-1}·K^{-1}
Mean value 21 to 186°C
Molecular Weight 121.1816
Wiswesser Line Notation 1N1&R
Evaluation D

$C_8H_{11}N$ (liq) 16BRA
N,N-Dimethylaniline
Heat Capacity 283 K, C_p = 50.7 cal·mol^{-1}·K^{-1}
212.1 J·mol^{-1}·K^{-1}
Mean value, 0 to 20°C
Molecular Weight 121.1816
Wiswesser Line Notation 1N1&R
Evaluation C

$C_8H_{11}N$ (liq) 34KOL/UDO
N,N-Dimethylaniline
Heat Capacity 302.4 K, C_p = 51.3 cal·mol^{-1}·K^{-1}
214.6 J·mol^{-1}·K^{-1}
One temperature
Molecular Weight 121.1816
Wiswesser Line Notation 1N1&R
Evaluation C

$C_8H_{11}N$ (liq) 34KOL/UDO 2
N,N-Dimethylaniline
Heat Capacity 302.3 K, C_p = 51.3 cal·mol^{-1}·K^{-1}
214.6 J·mol^{-1}·K^{-1}
One temperature
Molecular Weight 121.1816
Wiswesser Line Notation 1N1&R
Evaluation C

$C_8H_{11}N$ (liq) 34RAD/JUL
N,N-Dimethylaniline
Heat Capacity 289 K, C_p = 50.1 cal·mol^{-1}·K^{-1}
209.6 J·mol^{-1}·K^{-1}
One temperature
Molecular Weight 121.1816
Wiswesser Line Notation 1N1&R
Evaluation C

$C_8H_{11}N$ (c) 70KOL/SER
2-Cyanobicyclo[2.2.1]heptane(exo)
Heat Capacity 285 K, C_p = 47.81 cal·mol^{-1}·K^{-1}
200.0 J·mol^{-1}·K^{-1}
Temperature range 12–340 K
Entropy 298.15 K, S = 57.77 cal·mol^{-1}·K^{-1}
241.7 J·mol^{-1}·K^{-1}
Phase Changes
c,II/c,I 237.7 K, ΔH = 1895 cal·mol^{-1}
7929 J·mol^{-1}
ΔS = 8.0 cal·mol^{-1}·K^{-1}
33.4 J·mol^{-1}·K^{-1}
c,I/liq 298.8 K, ΔH = 703.5 cal·mol^{-1}
2943.4 J·mol^{-1}
ΔS = 2.35 cal·mol^{-1}·K^{-1}
9.83 J·mol^{-1}·K^{-1}
Molecular Weight 121.1816
Wiswesser Line Notation L55 A TJ CCN -EXO
Evaluation A

Table of Heat Capacities, Entropies, and Phase Transition Properties — Continued

$C_8H_{11}N$ (c) 70KOL/SER
2–Cyanobicyclo[2.2.1]heptane(endo)
Heat Capacity 298.15 K, C_p = 46.35 cal·mol^{-1}·K^{-1}
193.93 J·mol^{-1}·K^{-1}
Temperature range 12–340 K
Entropy 298.15 K, S = 53.70 cal·mol^{-1}·K^{-1}
224.68 J·mol^{-1}·K^{-1}
Phase Changes
c,II/c,I 177.3 K, ΔH = 538.2 cal·mol^{-1}
2251.8 J·mol^{-1}
ΔS = 3.04 cal·mol^{-1}·K^{-1}
12.70 J·mol^{-1}·K^{-1}
c,I/liq 331.2 K, ΔH = 707.7 cal·mol^{-1}
2961.0 J·mol^{-1}
ΔS = 2.14 cal·mol^{-1}·K^{-1}
8.94 J·mol^{-1}·K^{-1}
Molecular Weight 121.1816
Wiswesser Line Notation L55 A TJ CCN -ENDO
Evaluation A

$C_8H_{11}N$ (liq) 75NIC/WAD
2–Phenylethylamine
Heat Capacity 298.15 K, C_p = 57.18 cal·mol^{-1}·K^{-1}
239.24 J·mol^{-1}·K^{-1}
One temperature
Molecular Weight 121.1816
Wiswesser Line Notation Z2R
Evaluation B

C_8H_{12} (c,I) 70WON/WES
Bicyclo[2.2.2]octene–2
Heat Capacity 298.15 K, C_p = 37.46 cal·mol^{-1}·K^{-1}
156.73 J·mol^{-1}·K^{-1}
Temperature range 5–500 K
Entropy 298.15 K, S = 50.30 cal·mol^{-1}·K^{-1}
210.46 J·mol^{-1}·K^{-1}
Phase Changes
c,III/c,II 110.50 K, ΔH = 45 cal·mol^{-1}
188 J·mol^{-1}
ΔS = 0.32 cal·mol^{-1}·K^{-1}
1.33 J·mol^{-1}·K^{-1}
Second order transition
c,II/c,I 176.47 K, ΔH = 1350 cal·mol^{-1}
5648 J·mol^{-1}
ΔS = 7.66 cal·mol^{-1}·K^{-1}
24.52 J·mol^{-1}·K^{-1}
ΔH and ΔS obtained separately by graphical integration of observed C_p (or total enthalpy input) subtracting contributions from extrapolated normal C_p. ΔH, ΔS, and T not self–consistent.
c,I/liq 389.75 K, ΔH = 1286 cal·mol^{-1}
5381 J·mol^{-1}
ΔS = 3.44 cal·mol^{-1}·K^{-1}
14.39 J·mol^{-1}·K^{-1}
ΔH and ΔS obtained separately by graphical integration of observed C_p (or total enthalpy input) subtracting contributions from extrapolated normal C_p. ΔH, ΔS, and T not self–consistent.
Molecular Weight 108.1828
Wiswesser Line Notation L66 A B AUTJ
Evaluation A(C_p),B(Phase changes)

C_8H_{12} (liq) 75LEB/TSV
Cycloocta–1,5–diene
Heat Capacity 298.15 K, C_p = 49.74 cal·mol^{-1}·K^{-1}
208.1 J·mol^{-1}·K^{-1}
Temperature range 10.6–322.7 K, Data deposited in VINITI, No. 1324–75, 15 May 1975.
Entropy 298.15 K, S = 63.18 cal·mol^{-1}·K^{-1}
264.3 J·mol^{-1}·K^{-1}
Phase Changes
c,II/c,I 194.4 K, ΔH = –91 cal·mol^{-1}
–381 J·mol^{-1}
Irreversible, exothermic transition
c,I/liq 203.98 K, ΔH = 2349 cal·mol^{-1}
9828 J·mol^{-1}
ΔS = 11.53 cal·mol^{-1}·K^{-1}
48.24 J·mol^{-1}·K^{-1}
Molecular Weight 108.1828
Wiswesser Line Notation L8 AU EUTJ
Evaluation A

$(C_8H_{12})_n$ (liq) 76LEB/LIT
Polyoctadiene
Heat Capacity 298.15 K, C_p = 49.1 cal·mol^{-1}·K^{-1}
205.4 J·mol^{-1}·K^{-1}
Temperature range 15–328 K, Data deposited in VINITI No. 2097–76, 10 June 1976.
Entropy 298.15 K, S = 52.7 cal·mol^{-1}·K^{-1}
220.5 J·mol^{-1}·K^{-1}
Entropy at zero kelvin of amorphous polyoctadiene was assumed to be equal to the sum of the configurational entropy and the entropy of mixing the cis and the trans forms. $S_0°$ = 15.7 J·mol^{-1}·K^{-1}.
Phase Changes
c,I/gls 168.5 K
Heat capacity increment following conversion to glass is 41.3 J·mol^{-1}·K^{-1}.
c,I/liq 255 K, ΔH = 64.5 cal·mol^{-1}
270 J·mol^{-1}
ΔS = 0.253 cal·mol^{-1}·K^{-1}
1.06 J·mol^{-1}·K^{-1}
Temperature range of anomaly, probably due to fusion of crystal of the cis–form, is 200–255 K.
Molecular Weight 108.1828
Wiswesser Line Notation /*L8 AU EUTJ*/
Evaluation B

$C_8H_{12}N_2O_2$ (liq) 62STR/BAR
Hexamethylenediisocyanate; 1,6–Diisocyanatohexane
Heat Capacity 298 K, C_p = 71.5 cal·mol^{-1}·K^{-1}
299.2 J·mol^{-1}·K^{-1}
One temperature
Molecular Weight 168.1950
Wiswesser Line Notation OCN6NCO
Evaluation D

$C_8H_{12}N_2O_4$ (c) 39SAT/SOG 2
Ammonium o–phthalate
Heat Capacity 323 K, C_p = 70.3 cal·mol^{-1}·K^{-1}
294.1 J·mol^{-1}·K^{-1}
Temperature range 0 to 100°C. Mean value.
Molecular Weight 200.1938
Wiswesser Line Notation QVR BVQ &ZH 2
Evaluation C
Same data in 40SAT/SOG.

Table of Heat Capacities, Entropies, and Phase Transition Properties — Continued

$C_8H_{12}N_2O_4$ (c) 39SAT/SOG 2
Ammonium m-phthalate; Ammonium isophthalate
Heat Capacity 323 K, C_p = 68.2 cal·mol^{-1}·K^{-1}
285.3 J·mol^{-1}·K^{-1}
Temperature range 0 to 100°C. Mean value.
Molecular Weight 200.1938
Wiswesser Line Notation QVR CVQ &ZH 2
Evaluation C
Same data in 40SAT/SOG.

$C_8H_{12}S_6$ (c) 62CHA/WES
1,3,5,7-Tetramethyl-2,4,6,8,9,10-hexathiaadamantane
Heat Capacity 298.15 K, C_p = 72.09 cal·mol^{-1}·K^{-1}
301.62 J·mol^{-1}·K^{-1}
Temperature range 5-300 K
Entropy 298.15 K, S = 76.75 cal·mol^{-1}·K^{-1}
321.12 J·mol^{-1}·K^{-1}
Molecular Weight 300.5428
Wiswesser Line Notation T66 B6/B-H/DI A B- C 1B I AS B-S CS ES GS ISTJ B1 D1 F1 H1
Evaluation A

C_8H_{14} (liq) 70CHA/MCC
cis-Bicyclo[4.2.0]octane
Heat Capacity 345 K, C_p = 61.8 cal·mol^{-1}·K^{-1}
258.6 J·mol^{-1}·K^{-1}
One temperature
Molecular Weight 110.1986
Wiswesser Line Notation L46TJ -C
Evaluation B

C_8H_{14} (liq) 64SER/GOR
2-Methylbicyclo[2.2.1]heptane(exo)
Heat Capacity 298.15 K, C_p = 44.42 cal·mol^{-1}·K^{-1}
185.8 J·mol^{-1}·K^{-1}
Temperature range 12-310 K
Entropy 298.15 K, S = 58.85 cal·mol^{-1}·K^{-1}
246.2 J·mol^{-1}·K^{-1}
Phase Changes
c/liq 164.10 K, ΔH = 2003.3 cal·mol^{-1}
8381.8 J·mol^{-1}
ΔS = 12.2 cal·mol^{-1}·K^{-1}
51.0 J·mol^{-1}·K^{-1}
Includes effects of transition just below melting point.
Molecular Weight 110.1986
Wiswesser Line Notation L55 ATJ C1 -EXO
Evaluation A

C_8H_{14} (liq) 64SER/GOR
2-Methylbicyclo[2.2.1]heptane(endo)
Heat Capacity 298.15 K, C_p = 44.04 cal·mol^{-1}·K^{-1}
184.3 J·mol^{-1}·K^{-1}
Temperature range 12-310 K
Entropy 298.15 K, S = 56.93 cal·mol^{-1}·K^{-1}
238.2 J·mol^{-1}·K^{-1}
Phase Changes
c,II/c,I 152.42 K, ΔH = 1125.0 cal·mol^{-1}
4707.0 J·mol^{-1}
ΔS = 7.38 cal·mol^{-1}·K^{-1}
30.88 J·mol^{-1}·K^{-1}
c,I/liq 278.25 K, ΔH = 387.3 cal·mol^{-1}
1620.5 J·mol^{-1}
ΔS = 1.4 cal·mol^{-1}·K^{-1}
5.8 J·mol^{-1}·K^{-1}
Molecular Weight 110.1986
Wiswesser Line Notation L55 ATJ C1 -ENDO
Evaluation A

C_8H_{14} (liq) 70CHA/MCC
cis-Bicyclo[3.3.0]octane
Heat Capacity 308 K, C_p = 51.0 cal·mol^{-1}·K^{-1}
212.4 J·mol^{-1}·K^{-1}
Temperature range 308, 334 K
Molecular Weight 110.1986
Wiswesser Line Notation L55TJ -C
Evaluation B

C_8H_{14} (liq) 70CHA/MCC
trans-Bicyclo[3.3.0]octane
Heat Capacity 308 K, C_p = 43.1 cal·mol^{-1}·K^{-1}
180.3 J·mol^{-1}·K^{-1}
Temperature range 308,334 K
Molecular Weight 110.1986
Wiswesser Line Notation L55TJ -T
Evaluation B

C_8H_{14} (liq) 79PUC/PEA
Allylcyclopentane
Heat Capacity 298.15 K, C_p = 48.5 cal·mol^{-1}·K^{-1}
202.9 J·mol^{-1}·K^{-1}
One temperature
Molecular Weight 110.1986
Wiswesser Line Notation L5TJ A2U1
Evaluation B

C_8H_{14} (c,I) 70WON/WES
Bicyclo[2.2.2]octane
Heat Capacity 298.15 K, C_p = 37.69 cal·mol^{-1}·K^{-1}
157.69 J·mol^{-1}·K^{-1}
Temperature range 5-470 K
Entropy 298.15 K, S = 50.18 cal·mol^{-1}·K^{-1}
209.95 J·mol^{-1}·K^{-1}
Phase Changes
c,II/c,I 164.25 K, ΔH = 1096 cal·mol^{-1}
4586 J·mol^{-1}
ΔS = 6.66 cal·mol^{-1}·K^{-1}
27.87 J·mol^{-1}·K^{-1}
ΔH and ΔS obtained separately by graphical integration of observed C_p (or total enthalpy input) subtracting contributions from extrapolated normal C_p. ΔH, ΔS, and T not self-consistent.
c,I/liq 447.48 K, ΔH = 1995 cal·mol^{-1}
8347 J·mol^{-1}
ΔS = 4.48 cal·mol^{-1}·K^{-1}
18.74 J·mol^{-1}·K^{-1}
ΔH and ΔS obtained separately by graphical integration of observed C_p (or total enthalpy input) subtracting contributions from extrapolated normal C_p. ΔH, ΔS, and T not self-consistent.
Molecular Weight 110.1986
Wiswesser Line Notation L66 A BTJ
Evaluation A(C_p), B(Phase changes)

C_8H_{14} (liq) 79PUC/PEA
Ethylidenecyclohexane
Heat Capacity 298.15 K, C_p = 48.7 cal·mol^{-1}·K^{-1}
203.8 J·mol^{-1}·K^{-1}
One temperature
Molecular Weight 110.1986
Wiswesser Line Notation L6YTJ AU2
Evaluation B

Table of Heat Capacities, Entropies, and Phase Transition Properties — Continued

$C_8H_{14}O$ (liq) 70WES/WON
3–Oxabicyclo[3.2.2]nonane
Heat Capacity 298.15 K, C_p = 44.23 cal·mol^{-1}·K^{-1}
185.06 J·mol^{-1}·K^{-1}
Temperature range 5–477 K
Entropy 298.15 K, S = 56.47 cal·mol^{-1}·K^{-1}
236.27 J·mol^{-1}·K^{-1}
Phase Changes
c,II/c,I 208.5 K, ΔH = 1677 cal·mol^{-1}
7017 J·mol^{-1}
ΔS = 8.22 cal·mol^{-1}·K^{-1}
34.39 J·mol^{-1}·K^{-1}
Transition region 160–220 K, maximum at 208.5. Entropy change obtained by difference of integrated heat input and lattice (extrapolated C_p) contribution.
c,I/liq 448.43 K, ΔH = 1614 cal·mol^{-1}
6753 J·mol^{-1}
ΔS = 3.60 cal·mol^{-1}·K^{-1}
15.06 J·mol^{-1}·K^{-1}
Molecular Weight 126.1980
Wiswesser Line Notation T67 A B EOTJ
Evaluation A

$C_8H_{14}O_2$ (liq) 52ERD/JAG
Butyl 2–methylpropenoate; Butyl methacrylate
Heat Capacity 293 K, C_p = 64.70 cal·mol^{-1}·K^{-1}
270.70 J·mol^{-1}·K^{-1}
Temperature range 20 to 40°C
Molecular Weight 142.1974
Wiswesser Line Notation 4OVY1&U1
Evaluation C

$C_8H_{14}O_4$ (liq) 33KOL/UDO
Diethyl succinate
Heat Capacity 292.6 K, C_p = 80.8 cal·mol^{-1}·K^{-1}
338.1 J·mol^{-1}·K^{-1}
One temperature
Molecular Weight 174.1962
Wiswesser Line Notation 2OV2VO2
Evaluation C

$C_8H_{14}O_4$ (liq) 34KOL/UDO 2
Diethyl succinate
Heat Capacity 292.6 K, C_p = 80.8 cal·mol^{-1}·K^{-1}
338.1 J·mol^{-1}·K^{-1}
One temperature
Molecular Weight 174.1962
Wiswesser Line Notation 2OV2VO2
Evaluation C

$C_8H_{14}O_4$ (liq) 79FUC
Diethyl succinate
Heat Capacity 298.15 K, C_p = 79.0 cal·mol^{-1}·K^{-1}
330.5 J·mol^{-1}·K^{-1}
One temperature
Molecular Weight 174.1962
Wiswesser Line Notation 2OV2VO2
Evaluation B

$C_8H_{15}N$ (c) 63BAR/WES
3–Azabicyclo[3.2.2]nonane
Heat Capacity 310 K, C_p = 57.13 cal·mol^{-1}·K^{-1}
239.03 J·mol^{-1}·K^{-1}
Temperature range 5–350 K. Transition too close to 298.15 K to allow meaningful value of C_p.
Entropy 310 K, S = 58.73 cal·mol^{-1}·K^{-1}
245.73 J·mol^{-1}·K^{-1}
Phase Changes
c,II/c,I 297.78 K, ΔH = 3461 cal·mol^{-1}
14481 J·mol^{-1}
ΔS = 11.63 cal·mol^{-1}·K^{-1}
48.63 J·mol^{-1}·K^{-1}
Molecular Weight 125.2132
Wiswesser Line Notation T67 A B HMTJ
Evaluation A

$C_8H_{15}N$ (c,I) 64WUL/WES
3–Azabicyclo[3.2.2]nonane
Heat Capacity 350 K, C_p = 56.83 cal·mol^{-1}·K^{-1}
237.8 J·mol^{-1}·K^{-1}
Temperature range 330–490 K
Phase Changes
c,I/liq 467.12 K, ΔH = 1653 cal·mol^{-1}
6916 J·mol^{-1}
ΔS = 3.54 cal·mol^{-1}·K^{-1}
14.81 J·mol^{-1}·K^{-1}
Molecular Weight 125.2132
Wiswesser Line Notation T67 A B HMTJ
Evaluation A

C_8H_{16} (liq) 36PAR/TOD 2
2,4,4–Trimethyl–1–pentene; Diisobutylene
Heat Capacity 296.0 K, C_p = 56.25 cal·mol^{-1}·K^{-1}
235.35 J·mol^{-1}·K^{-1}
Temperature range 81–296 K. Value is unsmoothed experimental datum.
Entropy 298.15 K, S = 73.2 cal·mol^{-1}·K^{-1}
306.3 J·mol^{-1}·K^{-1}
Extrapolation below 90 K, 14.28 cal·mol^{-1}K^{-1}
Phase Changes
c/liq 178.9 K, ΔH = 2095 cal·mol^{-1}
8765 J·mol^{-1}
ΔS = 11.71 cal·mol^{-1}·K^{-1}
48.99 J·mol^{-1}·K^{-1}
Molecular Weight 112.2144
Wiswesser Line Notation 1X1&1&1Y1&U1
Evaluation B(C_p),C(S)
Low boiling isomer.

C_8H_{16} (liq) 30PAR/HUF 2
2,4,4–Trimethyl–2–pentene; Diisobutylene
Heat Capacity 296.0 K, C_p = 55.8 cal·mol^{-1}·K^{-1}
233.5 J·mol^{-1}·K^{-1}
Temperature range 92–296 K. Value is unsmoothed experimental datum.
Entropy 298.15 K, S = 71.4 cal·mol^{-1}·K^{-1}
298.7 J·mol^{-1}·K^{-1}
Extrapolation below 90 K, 15.24 cal·mol^{-1}K^{-1}
Molecular Weight 112.2144
Wiswesser Line Notation 1Y1&U1X1&1&1
Evaluation B(C_p),C(S)

C_8H_{16} (liq) 36PAR/TOD 2
2,4,4–Trimethyl–2–pentene; Diisobutylene
Heat Capacity 298.6 K, C_p = 57.41 cal·mol^{-1}·K^{-1}
240.20 J·mol^{-1}·K^{-1}
Extrapolation below 80 K, 14.94 cal·mol^{-1}K^{-1}
Entropy 298.15 K, S = 74.5 cal·mol^{-1}·K^{-1}
311.7 J·mol^{-1}·K^{-1}

Table of Heat Capacities, Entropies, and Phase Transition Properties — Continued

Phase Changes
c/liq 166 K, ΔH = 1624 cal·mol^{-1}
6795 J·mol^{-1}
ΔS = 9.8 cal·mol^{-1}·K^{-1}
40.9 J·mol^{-1}·K^{-1}
Molecular Weight 112.2144
Wiswesser Line Notation 1Y1&U1X1&1&1
Evaluation B(C_p),C(S)
High boiling isomer

C_8H_{16} (liq) **81REI**
2–Octene; Caprylene
Heat Capacity 298 K, C_p = 57.2 cal·mol^{-1}·K^{-1}
239.3 J·mol^{-1}·K^{-1}
Temperature range 291–365 K
Molecular Weight 112.2144
Wiswesser Line Notation 6U2
Evaluation D
Uncertain isomeric structure.

C_8H_{16} (liq) 57MCC/FIN 2
1–Octene
Heat Capacity 298.15 K, C_p = 57.65 cal·mol^{-1}·K^{-1}
241.21 J·mol^{-1}·K^{-1}
Temperature range 11–360 K
Entropy 298.15 K, S = 86.15 cal·mol^{-1}·K^{-1}
360.45 J·mol^{-1}·K^{-1}
Phase Changes
c/liq 171.46 K, ΔH = 3660 cal·mol^{-1}
15313 J·mol^{-1}
ΔS = 21.35 cal·mol^{-1}·K^{-1}
89.31 J·mol^{-1}·K^{-1}
Molecular Weight 112.2144
Wiswesser Line Notation 7U1
Evaluation A

C_8H_{16} (liq) 65MES/TOD 2
n–Propylcyclopentane
Heat Capacity 298.15 K, C_p = 51.69 cal·mol^{-1}·K^{-1}
216.27 J·mol^{-1}·K^{-1}
Temperature range 12–370 K
Entropy 298.15 K, S = 74.29 cal·mol^{-1}·K^{-1}
310.83 J·mol^{-1}·K^{-1}
Phase Changes
c/liq 155.79 K, ΔH = 2398 cal·mol^{-1}
10033 J·mol^{-1}
ΔS = 15.40 cal·mol^{-1}·K^{-1}
64.44 J·mol^{-1}·K^{-1}
Molecular Weight 112.2144
Wiswesser Line Notation L5TJ A3
Evaluation A

C_8H_{16} (liq) 49HUF/TOD
1,1–Dimethylcyclohexane
Heat Capacity 298.15 K, C_p = 50.01 cal·mol^{-1}·K^{-1}
209.24 J·mol^{-1}·K^{-1}
Temperature range 12–310 K
Entropy 298.15 K, S = 63.87 cal·mol^{-1}·K^{-1}
267.23 J·mol^{-1}·K^{-1}
Phase Changes
c,II/c,I 153.15 K, ΔH = 1430.3 cal·mol^{-1}
5984.4 J·mol^{-1}
ΔS = 9.34 cal·mol^{-1}·K^{-1}
39.00 J·mol^{-1}·K^{-1}
c,I/liq 239.81 K, ΔH = 484 cal·mol^{-1}
2025 J·mol^{-1}
ΔS = 2.02 cal·mol^{-1}·K^{-1}
8.44 J·mol^{-1}·K^{-1}
Molecular Weight 112.2144
Wiswesser Line Notation L6TJ A1 A1
Evaluation A

C_8H_{16} (liq) 49HUF/TOD
1–trans–2–Dimethylcyclohexane
Heat Capacity 298.15 K, C_p = 50.05 cal·mol^{-1}·K^{-1}
209.41 J·mol^{-1}·K^{-1}
Temperature range 12–310 K
Entropy 298.15 K, S = 65.30 cal·mol^{-1}·K^{-1}
273.22 J·mol^{-1}·K^{-1}
Phase Changes
c/liq 184.99 K, ΔH = 2507.6 cal·mol^{-1}
10491.8 J·mol^{-1}
ΔS = 13.56 cal·mol^{-1}·K^{-1}
56.72 J·mol^{-1}·K^{-1}
Molecular Weight 112.2144
Wiswesser Line Notation L6TJ A1 B1 –A&A –B&B
Evaluation A

C_8H_{16} (liq) 49HUF/TOD
1–cis–2–Dimethylcyclohexane
Heat Capacity 298.15 K, C_p = 50.24 cal·mol^{-1}·K^{-1}
210.20 J·mol^{-1}·K^{-1}
Temperature range 12–310 K
Entropy 298.15 K, S = 65.52 cal·mol^{-1}·K^{-1}
274.14 J·mol^{-1}·K^{-1}
Phase Changes
c,I/liq 172.5 K, ΔH = 1973.4 cal·mol^{-1}
8256.7 J·mol^{-1}
ΔS = 11.44 cal·mol^{-1}·K^{-1}
47.86 J·mol^{-1}·K^{-1}
c,I/liq 223.28 K, ΔH = 393.2 cal·mol^{-1}
1645.1 J·mol^{-1}
ΔS = 1.76 cal·mol^{-1}·K^{-1}
7.37 J·mol^{-1}·K^{-1}
Molecular Weight 112.2144
Wiswesser Line Notation L6TJ A1 B1 –A&AB
Evaluation A

C_8H_{16} (liq) 49HUF/TOD
1–trans–3–Dimethylcyclohexane
Heat Capacity 298.15 K, C_p = 50.87 cal·mol^{-1}·K^{-1}
212.84 J·mol^{-1}·K^{-1}
Temperature range 12–310 K
Entropy 298.15 K, S = 66.03 cal·mol^{-1}·K^{-1}
276.27 J·mol^{-1}·K^{-1}
Phase Changes
c/liq 183.06 K, ΔH = 2358 cal·mol^{-1}
9865.9 J·mol^{-1}
ΔS = 12.88 cal·mol^{-1}·K^{-1}
53.89 J·mol^{-1}·K^{-1}
Molecular Weight 112.2144
Wiswesser Line Notation L6TJ A1 C1 –A&A –B&C
Evaluation A

C_8H_{16} (liq) 49HUF/TOD
1–cis–3–Dimethylcyclohexane
Heat Capacity 298.15 K, C_p = 50.04 cal·mol^{-1}·K^{-1}
209.37 J·mol^{-1}·K^{-1}
Temperature range 12–310 K
Entropy 298.15 K, S = 65.16 cal·mol^{-1}·K^{-1}
272.63 J·mol^{-1}·K^{-1}

Table of Heat Capacities, Entropies, and Phase Transition Properties — Continued

Phase Changes
c/liq 197.59 K, ΔH = 2586.1 cal·mol^{-1}
10820.2 J·mol^{-1}
ΔS = 13.09 cal·mol^{-1}·K^{-1}
54.77 J·mol^{-1}·K^{-1}
Molecular Weight 112.2144
Wiswesser Line Notation L6TJ A1 C1 -A&AC
Evaluation A

C_8H_{16} (liq) 49HUF/TOD
1-trans-4-Dimethylcyclohexane
Heat Capacity 298.15 K, C_p = 50.25 cal·mol^{-1}·K^{-1}
210.25 J·mol^{-1}·K^{-1}
Temperature range 12–310 K
Entropy 298.15 K, S = 64.06 cal·mol^{-1}·K^{-1}
268.03 J·mol^{-1}·K^{-1}
Phase Changes
c/liq 236.22 K, ΔH = 2947.2 cal·mol^{-1}
12331.1 J·mol^{-1}
ΔS = 12.48 cal·mol^{-1}·K^{-1}
52.21 J·mol^{-1}·K^{-1}
Molecular Weight 112.2144
Wiswesser Line Notation L6TJ A1 D1 -A&A -B&D
Evaluation A

C_8H_{16} (liq) 49HUF/TOD
1-cis-4-Dimethylcyclohexane
Heat Capacity 298.15 K, C_p = 50.69 cal·mol^{-1}·K^{-1}
212.09 J·mol^{-1}·K^{-1}
Temperature range 12–310 K
Entropy 298.15 K, S = 64.80 cal·mol^{-1}·K^{-1}
271.12 J·mol^{-1}·K^{-1}
Phase Changes
c/liq 185.73 K, ΔH = 2224.4 cal·mol^{-1}
9306.9 J·mol^{-1}
ΔS = 11.98 cal·mol^{-1}·K^{-1}
50.11 J·mol^{-1}·K^{-1}
Molecular Weight 112.2144
Wiswesser Line Notation L6TJ A1 D1 -A&AD
Evaluation A

C_8H_{16} (liq) 49HUF/TOD
Ethylcyclohexane
Heat Capacity 298.15 K, C_p = 50.62 cal·mol^{-1}·K^{-1}
211.79 J·mol^{-1}·K^{-1}
Temperature range 12–310 K
Entropy 298.15 K, S = 67.14 cal·mol^{-1}·K^{-1}
280.91 J·mol^{-1}·K^{-1}
Phase Changes
c/liq 161.84 K, ΔH = 1991.7 cal·mol^{-1}
8333.3 J·mol^{-1}
ΔS = 12.31 cal·mol^{-1}·K^{-1}
51.49 J·mol^{-1}·K^{-1}
Molecular Weight 112.2144
Wiswesser Line Notation L6TJ A2
Evaluation A

C_8H_{16} (liq) 49PAR/MOO
Ethylcyclohexane
Heat Capacity 298.15 K, C_p = 51.2 cal·mol^{-1}·K^{-1}
214.2 J·mol^{-1}·K^{-1}
Temperature range 80–300 K
Entropy 298.15 K, S = 67.3 cal·mol^{-1}·K^{-1}
281.6 J·mol^{-1}·K^{-1}
Extrapolation below 80 K, 13.80 cal·mol^{-1}K^{-1}

Phase Changes
c/liq 161.4 K, ΔH = 1978 cal·mol^{-1}
8276 J·mol^{-1}
ΔS = 12.3 cal·mol^{-1}·K^{-1}
51.3 J·mol^{-1}·K^{-1}
Molecular Weight 112.2144
Wiswesser Line Notation L6TJ A2
Evaluation B(C_p),C(S)

C_8H_{16} (liq) 56FIN/SCO
Cycloöctane
Heat Capacity 298.15 K, C_p = 51.50 cal·mol^{-1}·K^{-1}
215.48 J·mol^{-1}·K^{-1}
Temperature range 12–330 K
Entropy 298.15 K, S = 62.62 cal·mol^{-1}·K^{-1}
262.00 J·mol^{-1}·K^{-1}
Phase Changes
c,III/c,II 166.5 K, ΔH = 1507.1 cal·mol^{-1}
6305.7 J·mol^{-1}
ΔS = 9.05 cal·mol^{-1}·K^{-1}
37.87 J·mol^{-1}·K^{-1}
c,II/c,I 183.8 K, ΔH = 114.3 cal·mol^{-1}
478.2 J·mol^{-1}
ΔS = 0.62 cal·mol^{-1}·K^{-1}
2.60 J·mol^{-1}·K^{-1}
c,I/liq 287.98 K, ΔH = 575.9 cal·mol^{-1}
2409.6 J·mol^{-1}
ΔS = 2.00 cal·mol^{-1}·K^{-1}
8.37 J·mol^{-1}·K^{-1}
Molecular Weight 112.2144
Wiswesser Line Notation L8TJ
Evaluation A

C_8H_{16} (liq) 79WIL/FAR
Cycloöctane
Heat Capacity 298.15 K, C_p = 51.27 cal·mol^{-1}·K^{-1}
214.53 J·mol^{-1}·K^{-1}
One temperature.
Molecular Weight 112.2144
Wiswesser Line Notation L8TJ
Evaluation B

$C_8H_{16}N_2O_3$ (c) 41HUF
Leucylglycine(DL)
Heat Capacity 297.1 K, C_p = 61.10 cal·mol^{-1}·K^{-1}
214.53 J·mol^{-1}·K^{-1}
Temperature range 86–297 K. Value is unsmoothed experimental datum.
Entropy 298.1 K, S = 67.2 cal·mol^{-1}·K^{-1}
281.2 J·mol^{-1}·K^{-1}
Extrapolation below 90 K, 19.71 cal·mol^{-1}K^{-1}
Molecular Weight 188.2260
Wiswesser Line Notation QV1MVYZ1Y1&1-DL
Evaluation A(C_p),C(S)

$C_8H_{16}O$ (liq) **81REI**
2-Octanone; Methyl hexyl ketone
Heat Capacity 298 K, C_p = 65.6 cal·mol^{-1}·K^{-1}
274.5 J·mol^{-1}·K^{-1}
Temperature range 291–463 K
Molecular Weight 128.2138
Wiswesser Line Notation 6V1
Evaluation D

Table of Heat Capacities, Entropies, and Phase Transition Properties — Continued

$C_8H_{16}O$ (liq) 65OET
2–Octanone; Methyl hexyl ketone
Heat Capacity 298.15 K, C_p = 65.31 cal·mol⁻¹·K⁻¹
273.26 J·mol⁻¹·K⁻¹
Temperature range 13–330 K
Entropy 298.15 K, S = 89.35 cal·mol⁻¹·K⁻¹
373.84 J·mol⁻¹·K⁻¹
Phase Changes
c/liq 252.86 K, ΔH = 5836.3 cal·mol⁻¹
24419 J·mol⁻¹
ΔS = 23.08 cal·mol⁻¹·K⁻¹
96.57 J·mol⁻¹·K⁻¹
Molecular Weight 128.2138
Wiswesser Line Notation 6V1
Evaluation A

$C_8H_{16}O_2$ (liq) 79FUC
Methyl heptanoate; Methyl oenanthoate;
Methyl enanthoate
Heat Capacity 298.15 K, C_p = 68.1 cal·mol⁻¹·K⁻¹
285.1 J·mol⁻¹·K⁻¹
One temperature
Molecular Weight 144.2132
Wiswesser Line Notation 6VO1
Evaluation B

$C_8H_{16}O_2$ (liq) 24GAR/RAN
Octanoic acid; Caprylic acid
Heat Capacity 305 K, C_p = 72.8 cal·mol⁻¹·K⁻¹
304.6 J·mol⁻¹·K⁻¹
Temperature range 0 to 46°C. Mean value 18 to 46°C.
Phase Changes
c/liq 289.5 K, ΔH = 5110 cal·mol⁻¹
21380 J·mol⁻¹
ΔS = 17.7 cal·mol⁻¹·K⁻¹
73.9 J·mol⁻¹·K⁻¹
Molecular Weight 144.2132
Wiswesser Line Notation QV7
Evaluation B

$C_8H_{16}O_2$ (liq) 82SCH/MIL 2
Octanoic acid; Caprylic acid
Heat Capacity 298.15 K, C_p = 71.205 cal·mol⁻¹·K⁻¹
297.92 J·mol⁻¹·K⁻¹
Temperature range 80–300 K
Phase Changes
c,I/liq 289.66 K, ΔH = 5102.8 cal·mol⁻¹
21350 J·mol⁻¹
ΔS = 17.62 cal·mol⁻¹·K⁻¹
73.71 J·mol⁻¹·K⁻¹
Molecular Weight 144.2132
Wiswesser Line Notation QV7
Evaluation B

C_8H_{18} (c) 30PAR/HUF
2,2,3,3–Tetramethylbutane
Heat Capacity 295.4 K, C_p = 55.51 cal·mol⁻¹·K⁻¹
232.2 J·mol⁻¹·K⁻¹
Temperature range 89–295 K. Value is unsmoothed experimental datum.
Entropy 298.15 K, S = 61.4 cal·mol⁻¹·K⁻¹
256.9 J·mol⁻¹·K⁻¹
Extrapolation below 90 K, 15.46 cal·mol⁻¹K⁻¹
Phase Changes
c,II/c,I 148.1 K, ΔH = 480 cal·mol⁻¹
2008 J·mol⁻¹
ΔS = 3.24 cal·mol⁻¹·K⁻¹
13.56 J·mol⁻¹·K⁻¹
Molecular Weight 114.2302
Wiswesser Line Notation 1X1&1&X1&1&1
Evaluation B(C_p),C(S)

C_8H_{18} (c) 52SCO/DOU
2,2,3,3–Tetramethylbutane
Heat Capacity 301.60 K, C_p = 57.27 cal·mol⁻¹·K
239.62 J·mol⁻¹·K⁻¹
Temperature range 12–374 K Value is unsmoothed experimental datum.
Entropy 298.15 K, S = 65.43 cal·mol⁻¹·K⁻¹
273.76 J·mol⁻¹·K⁻¹
Phase Changes
c,II/c,I 152.5 K, ΔH = 478 cal·mol⁻¹
2000 J·mol⁻¹
ΔS = 3.13 cal·mol⁻¹·K⁻¹
13.11 J·mol⁻¹·K⁻¹
c,I/liq 373.9 K, ΔH = 1802 cal·mol⁻¹
7540 J·mol⁻¹
ΔS = 4.82 cal·mol⁻¹·K⁻¹
20.17 J·mol⁻¹·K⁻¹
Molecular Weight 114.2302
Wiswesser Line Notation 1X1&1&X1&1&1
Evaluation A

C_8H_{18} (liq) 30PAR/HUF
2,2,4–Trimethylpentane; Isoöctane
Heat Capacity 295.2 K, C_p = 55.9 cal·mol⁻¹·K⁻¹
233.9 J·mol⁻¹·K⁻¹
Temperature range 88–295 K. Value is unsmoothed experimental datum.
Entropy 298.15 K, S = 75.2 cal·mol⁻¹·K⁻¹
314.6 J·mol⁻¹·K⁻¹
Extrapolation below 90 K, 15.90 cal·mol⁻¹K⁻¹
Phase Changes
c/liq 165.3 K, ΔH = 2161 cal·mol⁻¹
9042 J·mol⁻¹
ΔS = 13.07 cal·mol⁻¹·K⁻¹
54.70 J·mol⁻¹·K⁻¹
Molecular Weight 114.2302
Wiswesser Line Notation 1Y1&1X1&1&1
Evaluation B(C_p),C(S)

C_8H_{18} (liq) 40PIT
2,2,4–Trimethylpentane; Isoöctane
Heat Capacity 301.9 K, C_p = 57.60 cal·mol⁻¹·K⁻¹
241.00 J·mol⁻¹·K⁻¹
Temperature range 15–318 K. Value is unsmoothed experimental datum.
Entropy 298.15 K, S = 78.40 cal·mol⁻¹·K⁻¹
328.03 J·mol⁻¹·K⁻¹
Phase Changes
c/liq 165.79 K, ΔH = 2201.6 cal·mol⁻¹
9211.5 J·mol⁻¹
ΔS = 13.28 cal·mol⁻¹·K⁻¹
55.56 J·mol⁻¹·K⁻¹
Molecular Weight 114.2302
Wiswesser Line Notation 1Y1&1X1&1&1
Evaluation A

Table of Heat Capacities, Entropies, and Phase Transition Properties — Continued

C_8H_{18} (liq) 47OSB/GIN
2,2,4–Trimethylpentane; Isoöctane
Heat Capacity 298.15 K, C_p = 57.02 cal·mol^{-1}·K^{-1}
238.57 J·mol^{-1}·K^{-1}
Temperature range 283–318 K
Molecular Weight 114.2302
Wiswesser Line Notation 1Y1&1X&1&1
Evaluation A

C_8H_{18} (liq) 73SUB/RAS
2,2,4–Trimethylpentane; Isoöctane
Heat Capacity 298.15 K, C_p = 56.8 cal·mol^{-1}·K^{-1}
237.8 J·mol^{-1}·K^{-1}
Temperature range 298–323 K
Molecular Weight 114.2302
Wiswesser Line Notation 1Y1&1X1&1&1
Evaluation B

C_8H_{18} (liq) 47OSB/GIN
2,5–Dimethylhexane
Heat Capacity 298.15 K, C_p = 59.56 cal·mol^{-1}·K^{-1}
249.20 J·mol^{-1}·K^{-1}
Temperature range 278–318 K
Molecular Weight 114.2302
Wiswesser Line Notation 1Y1&2Y1&1
Evaluation A

C_8H_{18} (liq) 41PIT/SCO
2,3,4–Trimethylpentane
Heat Capacity 293.79 K, C_p = 58.85 cal·mol^{-1}·K^{-1}
246.23 J·mol^{-1}·K^{-1}
Temperature range 14–325 K. Value is unsmoothed experimental datum.
Entropy 298.15 K, S = 78.71 cal·mol^{-1}·K^{-1}
329.32 J·mol^{-1}·K^{-1}
Phase Changes
c/liq 163.63 K, ΔH = 2215 cal·mol^{-1}
9268 J·mol^{-1}
ΔS = 13.54 cal·mol^{-1}·K^{-1}
56.64 J·mol^{-1}·K^{-1}
Molecular Weight 114.2302
Wiswesser Line Notation 1Y1&Y1&Y1&1
Evaluation B

C_8H_{18} (liq) 47OSB/GIN
2,3,4–Trimethylpentane
Heat Capacity 298.15 K, C_p = 59.11 cal·mol^{-1}·K^{-1}
247.32 J·mol^{-1}·K^{-1}
Temperature range 278–318 K
Molecular Weight 114.2302
Wiswesser Line Notation 1Y1&Y1&Y1&1
Evaluation A

C_8H_{18} (liq) 47OSB/GIN
2,3,3–Trimethylpentane
Heat Capacity 298.15 K, C_p = 58.69 cal·mol^{-1}·K^{-1}
245.56 J·mol^{-1}·K^{-1}
Temperature range 278–318 K
Molecular Weight 114.2302
Wiswesser Line Notation 2X1&1&Y1&1
Evaluation A

C_8H_{18} (liq) 47OSB/GIN
3,3–Dimethylhexane
Heat Capacity 298.15 K, C_p = 58.94 cal·mol^{-1}·K^{-1}
246.60 J·mol^{-1}·K^{-1}
Temperature range 278–318 K
Molecular Weight 114.2302
Wiswesser Line Notation 3X2&1&1
Evaluation A

C_8H_{18} (liq) 47OSB/GIN
4–Methylheptane
Heat Capacity 298.15 K, C_p = 60.01 cal·mol^{-1}·K^{-1}
251.08 J·mol^{-1}·K^{-1}
Temperature range 278–318 K
Molecular Weight 114.2302
Wiswesser Line Notation 3Y3&1
Evaluation A

C_8H_{18} (liq) 47OSB/GIN
3–Methylheptane
Heat Capacity 298.15 K, C_p = 59.67 cal·mol^{-1}·K^{-1}
249.66 J·mol^{-1}·K^{-1}
Temperature range 283–318 K
Molecular Weight 114.2302
Wiswesser Line Notation 4Y2&1
Evaluation A

C_8H_{18} (liq) 73FIN/MES
3–Methylheptane
Heat Capacity 298.15 K, C_p = 59.799 cal·mol^{-1}·K^{-1}
250.20 J·mol^{-1}·K^{-1}
Temperature range 10–380 K
Entropy 298.15 K, S = 86.67 cal·mol^{-1}·K^{-1}
362.6 J·mol^{-1}·K^{-1}
Thermodynamic properties calculated from a Debye function at 10 K.
Phase Changes
c,I/liq 152.6574 K, ΔH = 2794.9 cal·mol^{-1}
11694 J·mol^{-1}
ΔS = 1.832 cal·mol^{-1}·K^{-1}
7.665 J·mol^{-1}·K^{-1}
Molecular Weight 114.2302
Wiswesser Line Notation 4Y2
Evaluation A

C_8H_{18} (liq) 47OSB/GIN
2–Methylheptane
Heat Capacity 298.15 K, C_p = 60.13 cal·mol^{-1}·K^{-1}
251.58 J·mol^{-1}·K^{-1}
Temperature range 283–318 K
Molecular Weight 114.2302
Wiswesser Line Notation 5Y1&1
Evaluation A

C_8H_{18} (liq) 71MES/FIN
2–Methylheptane
Heat Capacity 298.15 K, C_p = 60.23 cal·mol^{-1}·K^{-1}
252.00 J·mol^{-1}·K^{-1}
Temperature range 11–370 K
Entropy 298.15 K, S = 85.18 cal·mol^{-1}·K^{-1}
356.39 J·mol^{-1}·K^{-1}
Phase Changes
c/liq 64.19 K, ΔH = 2849 cal·mol^{-1}
11920 J·mol^{-1}
ΔS = 17.35 cal·mol^{-1}·K^{-1}
72.60 J·mol^{-1}·K^{-1}
Molecular Weight 114.2302
Wiswesser Line Notation 5Y1&1
Evaluation A

Table of Heat Capacities, Entropies, and Phase Transition Properties — Continued

C_8H_{18} (liq) 30PAR/HUF
n–Octane
Heat Capacity 293.7 K, C_p = 59.2 cal·mol^{-1}·K^{-1}
247.7 J·mol^{-1}·K^{-1}
Temperature range 85–294 K. Value is unsmoothed experimental datum.
Entropy 298.15 K, S = 86.0 cal·mol^{-1}·K^{-1}
359.8 J·mol^{-1}·K^{-1}
Extrapolation below 90 K, 18.45 cal·mol^{-1}K^{-1}
Phase Changes
c/liq 215.6 K, ΔH = 4802 cal·mol^{-1}
20092 J·mol^{-1}
ΔS = 22.27 cal·mol^{-1}·K^{-1}
93.19 J·mol^{-1}·K^{-1}
Molecular Weight 114.2302
Wiswesser Line Notation 8H
Evaluation B(C_p),C(S)

C_8H_{18} (liq) 31HUF/PAR
n–Octane
Heat Capacity 298.3 K, C_p = 60.1 cal·mol^{-1}·K^{-1}
251.5 J·mol^{-1}·K^{-1}
Temperature range 92–298 K. Value is unsmoothed experimental datum.
Entropy 298.1 K, S = 86.0 cal·mol^{-1}·K^{-1}
359.8 J·mol^{-1}·K^{-1}
Extrapolation below 90 K, 18.1 cal·mol^{-1}K^{-1}
Phase Changes
c/liq 215.8 K, ΔH = 4936 cal·mol^{-1}
20652 J·mol^{-1}
ΔS = 22.9 cal·mol^{-1}·K^{-1}
95.7 J·mol^{-1}·K^{-1}
Molecular Weight 114.2302
Wiswesser Line Notation 8H
Evaluation B(C_p),C(S)

C_8H_{18} (liq) 47OSB/GIN
n–Octane
Heat Capacity 298.15 K, C_p = 60.68 cal·mol^{-1}·K^{-1}
253.89 J·mol^{-1}·K^{-1}
Temperature range 283–318 K
Molecular Weight 114.2302
Wiswesser Line Notation 8H
Evaluation A

C_8H_{18} (liq) 51CON/SAG
n–Octane
Heat Capacity 299.8 K, C_p = 60.69 cal·mol^{-1}·K^{-1}
253.93 J·mol^{-1}·K^{-1}
Temperature range 80 to 200°F
Molecular Weight 114.2302
Wiswesser Line Notation 8H
Evaluation B

C_8H_{18} (liq) 54FIN/GRO 2
n–Octane
Heat Capacity 298.15 K, C_p = 60.74 cal·mol^{-1}·K^{-1}
254.14 J·mol^{-1}·K^{-1}
Temperature range 12–300 K
Entropy 298.15 K, S = 86.33 cal·mol^{-1}·K^{-1}
361.20 J·mol^{-1}·K^{-1}
Phase Changes
c/liq 216.38 K, ΔH = 4957 cal·mol^{-1}
20740 J·mol^{-1}
ΔS = 22.91 cal·mol^{-1}·K^{-1}
95.85 J·mol^{-1}·K^{-1}
Molecular Weight 114.2302
Wiswesser Line Notation 8H
Evaluation A

C_8H_{18} (liq) 75GRI/RAS
n–Octane
Heat Capacity 298 K, C_p = 60.5 cal·mol^{-1}·K^{-1}
253.2 J·mol^{-1}·K^{-1}
Temperature range 305–463 K
Molecular Weight 114.2302
Wiswesser Line Notation 8H
Evaluation B

$C_8H_{18}N_2O$ (liq) 81BYS
di–t–Butyldiazene N–oxide
Heat Capacity 298.15 K, C_p = 75.6 cal·mol^{-1}·K^{-1}
316.3 J·mol^{-1}·K^{-1}
Temperature range 290–310 K
Phase Changes
c,I/liq 288.4 K
liq/g 298.15 K, ΔH = 12357 cal·mol^{-1}
51702 J·mol^{-1}
ΔS = 41.45 cal·mol^{-1}·K^{-1}
173.4 J·mol^{-1}·K^{-1}
Molecular Weight 158.2430
Wiswesser Line Notation 1X1&1&NO&UNX1&1&1
Evaluation A

$C_8H_{18}O$ (liq) 31CLI/AND
5–Methyl–1–heptanol
Heat Capacity 298.5 K, C_p = 72.7 cal·mol^{-1}·K^{-1}
304.2 J·mol^{-1}·K^{-1}
Temperature range 102–298 K. Value is unsmoothed experimental datum.
Molecular Weight 130.2296
Wiswesser Line Notation Q4Y2&1
Evaluation C

$C_8H_{18}O$ (liq) 81REI
1–Octanol; n–Octyl alcohol; Capryl alcohol
Heat Capacity 298 K, C_p = 77.5 cal·mol^{-1}·K^{-1}
324.3 J·mol^{-1}·K^{-1}
Temperature range 291–470 K
Molecular Weight 130.2296
Wiswesser Line Notation Q8
Evaluation D

$C_8H_{18}O$ (liq) 31CLI/AND
1–Octanol; n–Octyl alcohol; Capryl alcohol
Heat Capacity 286.0 K, C_p = 68.0 cal·mol^{-1}·K^{-1}
284.5 J·mol^{-1}·K^{-1}
Temperature range 102–286 K. Value is unsmoothed experimental datum.
Molecular Weight 130.2296
Wiswesser Line Notation Q8
Evaluation C

$C_8H_{18}O$ (liq) 59HUT/BAI
1–Octanol; n–Octyl alcohol; Capryl alcohol
Heat Capacity 298 K, C_p = 74.6 cal·mol^{-1}·K^{-1}
312.1 J·mol^{-1}·K^{-1}
One temperature
Molecular Weight 130.2296
Wiswesser Line Notation Q8
Evaluation C

Table of Heat Capacities, Entropies, and Phase Transition Properties — Continued

$C_8H_{18}O$ (liq) 31CLI/AND
4–Methyl–4–heptanol
Heat Capacity 298.5 K, C_p = 87.8 cal·mol^{-1}·K^{-1}
367.4 J·mol^{-1}·K^{-1}
Temperature range 102–298 K. Value is unsmoothed experimental datum.
Molecular Weight 130.2296
Wiswesser Line Notation QX3&3&1
Evaluation C

$C_8H_{18}O$ (liq) 31CLI/AND
2–Methyl–2–heptanol
Heat Capacity 298.5 K, C_p = 80.7 cal·mol^{-1}·K^{-1}
337.6 J·mol^{-1}·K^{-1}
Temperature range 102–311 K. Value is unsmoothed experimental datum.
Molecular Weight 130.2296
Wiswesser Line Notation QX5&1&1
Evaluation C

$C_8H_{18}O$ (liq) 31CLI/AND
4–Methyl–2–heptanol
Heat Capacity 298.5 K, C_p = 74.7 cal·mol^{-1}·K^{-1}
312.5 J·mol^{-1}·K^{-1}
Temperature range 102–298 K. Value is unsmoothed experimental datum.
Molecular Weight 130.2296
Wiswesser Line Notation QY1&1Y3&1
Evaluation C

$C_8H_{18}O$ (liq) 31CLI/AND
5–Methyl–2–heptanol
Heat Capacity 298.5 K, C_p = 70.8 cal·mol^{-1}·K^{-1}
296.2 J·mol^{-1}·K^{-1}
Temperature range 102–298 K. Value is unsmoothed experimental datum.
Molecular Weight 130.2296
Wiswesser Line Notation QY1&2Y2&1
Evaluation C

$C_8H_{18}O$ (liq) 31CLI/AND
6–Methyl–2–heptanol
Heat Capacity 298.5 K, C_p = 75.3 cal·mol^{-1}·K^{-1}
315.1 J·mol^{-1}·K^{-1}
Temperature range 102–298 K. Value is unsmoothed experimental datum.
Molecular Weight 130.2296
Wiswesser Line Notation QY1&3Y1&1
Evaluation C

$C_8H_{18}O$ (liq) 31CLI/AND
3–Methyl–2–heptanol
Heat Capacity 298.5 K, C_p = 71.1 cal·mol^{-1}·K^{-1}
297.5 J·mol^{-1}·K^{-1}
Temperature range 102–298 K. Value is unsmoothed experimental datum.
Molecular Weight 130.2296
Wiswesser Line Notation QY1&Y4&1
Evaluation C

$C_8H_{18}O$ (liq) 31CLI/AND
6–Methyl–3–heptanol
Heat Capacity 298.5 K, C_p = 74.2 cal·mol^{-1}·K^{-1}
310.5 J·mol^{-1}·K^{-1}
Temperature range 102–323 K. Value is unsmoothed experimental datum.
Molecular Weight 130.2296
Wiswesser Line Notation QY2&2Y1&1
Evaluation C

$C_8H_{18}O$ (liq) 31CLI/AND
4–Methyl–3–heptanol
Heat Capacity 298.5 K, C_p = 73.9 cal·mol^{-1}·K^{-1}
309.2 J·mol^{-1}·K^{-1}
Temperature range 102–298 K. Value is unsmoothed experimental datum.
Molecular Weight 130.2296
Wiswesser Line Notation QY2&Y3&1
Evaluation C

$C_8H_{18}O$ (liq) 31CLI/AND
2–Methyl–4–heptanol
Heat Capacity 298.5 K, C_p = 79.3 cal·mol^{-1}·K^{-1}
331.8 J·mol^{-1}·K^{-1}
Temperature range 102–298 K. Value is unsmoothed experimental datum.
Molecular Weight 130.2296
Wiswesser Line Notation QY3&1Y1&1
Evaluation C

$C_8H_{18}O$ (liq) 31CLI/AND
4–Octanol
Heat Capacity 298.5 K, C_p = 80.7 cal·mol^{-1}·K^{-1}
337.6 J·mol^{-1}·K^{-1}
Temperature range 102–311 K. Value is unsmoothed experimental datum.
Molecular Weight 130.2296
Wiswesser Line Notation QY4&3
Evaluation C

$C_8H_{18}O$ (liq) 31CLI/AND
2–Methyl–1–heptanol
Heat Capacity 298.5 K, C_p = 74.8 cal·mol^{-1}·K^{-1}
313.0 J·mol^{-1}·K^{-1}
Temperature range 102–311 K. Value is unsmoothed experimental datum.
Molecular Weight 130.2296
Wiswesser Line Notation QY5&1
Evaluation C

$C_8H_{18}O$ (liq) 31CLI/AND
3–Octanol
Heat Capacity 298.5 K, C_p = 80.9 cal·mol^{-1}·K^{-1}
338.5 J·mol^{-1}·K^{-1}
Temperature range 102–298 K. Value is unsmoothed experimental datum.
Molecular Weight 130.2296
Wiswesser Line Notation QY5&2
Evaluation C

$C_8H_{18}O$ (liq) 31CLI/AND
2–Octanol
Heat Capacity 298.5 K, C_p = 78.9 cal·mol^{-1}·K^{-1}
330.1 J·mol^{-1}·K^{-1}
Temperature range 102–298 K. Value is unsmoothed experimental datum.
Molecular Weight 130.2296
Wiswesser Line Notation QY6&1
Evaluation C

$C_8H_{18}O_2$ (liq) 73KUS/SUU
4,7–Dioxadecane; 1,2–Di–n–propoxyethane
Heat Capacity 298.15 K, C_p = 73.85 cal·mol^{-1}·K^{-1}
309.0 J·mol^{-1}·K^{-1}
One temperature
Molecular Weight 146.2290
Wiswesser Line Notation 3O2O3
Evaluation B

Table of Heat Capacities, Entropies, and Phase Transition Properties — Continued

$C_8H_{18}O_3$ (liq) 78ROU/PER 2
3,6,9–Trioxaundecane; Bis(2–ethoxyethyl) ether
Heat Capacity 298.1 K, C_p = 83.05 cal·mol⁻¹·K⁻¹
347.5 J·mol⁻¹·K⁻¹
Temperature range 283–313 K
Molecular Weight 162.2284
Wiswesser Line Notation 2O2O2O2
Evaluation B

$C_8H_{18}O_4$ (liq) 66BEA/CLE
2,5,8,11–Tetraoxadodecane; Triglyme
Heat Capacity 298.15 K, C_p = 88.0 cal·mol⁻¹·K⁻¹
368.2 J·mol⁻¹·K⁻¹
Temperature range 90–350 K
Entropy 298.15 K, S = 117.8 cal·mol⁻¹·K⁻¹
492.9 J·mol⁻¹·K⁻¹
Extrapolation below 90 K, 24.8 cal.mol⁻¹.K⁻¹
Phase Changes
c/liq 229.3 K, ΔH = 5668 cal·mol⁻¹
23715 J·mol⁻¹
ΔS = 24.7 cal·mol⁻¹·K⁻¹
103.4 J·mol⁻¹·K⁻¹
Molecular Weight 178.2278
Wiswesser Line Notation 1O2O2O2O1
Evaluation A(C_p),C(S)

$C_8H_{18}O_5$ (liq) 79STE/TAM
1,11–Dihydroxy–3,6,9–trioxaundecane;
Tetraethylene glycol
Heat Capacity 298 K, C_p = 100.2 cal·mol⁻¹·K⁻¹
419.2 J·mol⁻¹·K⁻¹
Temperature range 273–533 K
Molecular Weight 194.2272
Wiswesser Line Notation Q2O2O2O2Q
Evaluation B

$C_8H_{18}S$ (liq) 61MCC/FIN
5–Thianonane; Di–n–butyl sulfide
Heat Capacity 298.15 K, C_p = 67.96 cal·mol⁻¹·K⁻¹
284.34 J·mol⁻¹·K⁻¹
Temperature range 11–370 K
Entropy 298.15 K, S = 96.82 cal·mol⁻¹·K⁻¹
405.09 J·mol⁻¹·K⁻¹
Phase Changes
c/liq 198.13 K, ΔH = 4643 cal·mol⁻¹
19426 J·mol⁻¹
ΔS = 23.43 cal·mol⁻¹·K⁻¹
98.05 J·mol⁻¹·K⁻¹
Molecular Weight 146.2902
Wiswesser Line Notation 4S4
Evaluation A

$C_8H_{20}BrN$ (c) 74BUR/VER
Tetraethylammonium bromide
Heat Capacity 298 K, C_p = 59.0 cal·mol⁻¹·K⁻¹
246.9 J·mol⁻¹·K⁻¹
Temperature range 273–373 K
Phase Changes
c/liq 447 K, ΔH = 4550 cal·mol⁻¹
20300 J·mol⁻¹
ΔS = 11 cal·mol⁻¹·K⁻¹
45 J·mol⁻¹·K⁻¹
Molecular Weight 210.1567
Wiswesser Line Notation 2K2&2&2 E
Evaluation B

$C_8H_{20}Ge$ (liq) 72MAS/RAB
Tetraethylgermane; Germanium tetraethyl
Heat Capacity 298.15 K, C_p = 74.00 cal·mol⁻¹·K⁻¹
309.6 J·mol⁻¹·K⁻¹
Temperature range 60–300 K
Phase Changes
c/liq 180.3 K, ΔH = 3015 cal·mol⁻¹
12615 J·mol⁻¹
ΔS = 16.7 cal·mol⁻¹·K⁻¹
70.0 J·mol⁻¹·K⁻¹
Molecular Weight 188.8360
Wiswesser Line Notation 2–GE–2&2&2
Evaluation B

$C_8H_{20}IN$ (c) 73JOH/MAR
Tetraethylammonium iodide
Heat Capacity 298.15 K, C_p = 60.8 cal·mol⁻¹·K⁻¹
254.4 J·mol⁻¹·K⁻¹
Temperature range 12–310 K
Entropy 298.15 K, S = 74.4 cal·mol⁻¹·K⁻¹
311.3 J·mol⁻¹·K⁻¹
Molecular Weight 257.1572
Wiswesser Line Notation 2 K2&2&2 &I
Evaluation A

$C_8H_{20}Pb$ (liq) 56SCO/GOO
Tetraethyllead
Heat Capacity 298.15 K, C_p = 74.1 cal·mol⁻¹·K⁻¹
310.0 J·mol⁻¹·K⁻¹
One temperature
Molecular Weight 323.4460
Wiswesser Line Notation 2–PB–2&2&2
Evaluation B

$C_8H_{20}Si$ (liq) 72MAS/RAB
Tetraethylsilane; Silicon tetraethyl
Heat Capacity 298.15 K, C_p = 71.25 cal·mol⁻¹·K⁻¹
298.1 J·mol⁻¹·K⁻¹
Temperature range 60–300 K
Phase Changes
c/liq 190.6 K, ΔH = 3205 cal·mol⁻¹
13410 J·mol⁻¹
ΔS = 16.8 cal·mol⁻¹·K⁻¹
70.4 J·mol⁻¹·K⁻¹
Molecular Weight 144.3315
Wiswesser Line Notation 2–SI–2&2&2
Evaluation B

$C_8H_{20}Sn$ (liq) 72MAS/RAB
Tetraethylstannane; Tin tetraethyl
Heat Capacity 298.15 K, C_p = 72.10 cal·mol⁻¹·K⁻¹
3010.7 J·mol⁻¹·K⁻¹
Temperature range 60–300 K
Phase Changes
c,II/c,I 121.4 K, ΔH = –260 cal·mol⁻¹
–1090 J·mol⁻¹
Metastable transition
c,I/liq 141.9 K, ΔH = 2134 cal·mol⁻¹
8929 J·mol⁻¹
ΔS = 15.0 cal·mol⁻¹·K⁻¹
62.9 J·mol⁻¹·K⁻¹
Molecular Weight 234.9360
Wiswesser Line Notation 2–SN–2&2&2
Evaluation B

Table of Heat Capacities, Entropies, and Phase Transition Properties — Continued

$C_9H_6N_2O_2$ (liq) 62STR/BAR
2,4–Tolylenediisocyanate;
1–Methyl–2,4–diisocyanatobenzene
Heat Capacity 298 K, C_p = 68.8 cal·mol^{-1}·K^{-1}
287.8 J·mol^{-1}·K^{-1}
One temperature
Molecular Weight 174.1586
Wiswesser Line Notation OCNR B1 ENCO
Evaluation D

C_9H_7N (liq) 16BRA
Quinoline
Heat Capacity 283 K, C_p = 45.5 cal·mol^{-1}·K^{-1}
190.4 J·mol^{-1}·K^{-1}
Mean value, 0 to 20°C
Molecular Weight 129.1610
Wiswesser Line Notation T66 BNJ
Evaluation C

C_9H_7N (liq) 34KOL/UDO
Quinoline
Heat Capacity 302.5 K, C_p = 46.1 cal·mol^{-1}·K^{-1}
192.9 J·mol^{-1}·K^{-1}
One temperature
Molecular Weight 129.1610
Wiswesser Line Notation T66 BNJ
Evaluation C

C_9H_7N (liq) 34KOL/UDO 2
Quinoline
Heat Capacity 302.4 K, C_p = 46.1 cal·mol^{-1}·K^{-1}
192.9 J·mol^{-1}·K^{-1}
One temperature
Molecular Weight 129.1610
Wiswesser Line Notation T66 BNJ
Evaluation C

C_9H_7N (liq) 34RAD/JUL
Quinoline
Heat Capacity 290 K, C_p = 39.3 cal·mol^{-1}·K^{-1}
164.4 J·mol^{-1}·K^{-1}
One temperature
Molecular Weight 129.1610
Wiswesser Line Notation T66 BNJ
Evaluation C

C_9H_7N (liq) 36PAR/TOD
Quinoline
Heat Capacity 298.1 K, C_p = 47.61 cal·mol^{-1}·K^{-1}
199.20 J·mol^{-1}·K^{-1}
Temperature range 90–300 K
Entropy 298.1 K, S = 51.9 cal·mol^{-1}·K^{-1}
217.1 J·mol^{-1}·K^{-1}
Extrapolation below 90 K, 13.13 cal·mol^{-1}K^{-1}
Phase Changes
c/liq 258.4 K, ΔH = 2581 cal·mol^{-1}
10799 J·mol^{-1}
ΔS = 9.99 cal·mol^{-1}·K^{-1}
41.79 J·mol^{-1}·K^{-1}
Molecular Weight 129.1610
Wiswesser Line Notation T66 BNJ
Evaluation B(C_p),C(S)

C_9H_7N (liq) 51TSC/KRI
Quinoline
Heat Capacity 298 K, C_p = 48.7 cal·mol^{-1}·K^{-1}
203.8 J·mol^{-1}·K^{-1}
One temperature
Molecular Weight 129.1610
Wiswesser Line Notation T66 BNJ
Evaluation C

$C_9H_7NO_4$ (c) 41SAT/SOG
o–Nitrocinnamic acid
Heat Capacity 323 K, C_p = 57.5 cal·mol^{-1}·K^{-1}
240.6 J·mol^{-1}·K^{-1}
Temperature range 0 to 100°C. Mean value.
Molecular Weight 193.1586
Wiswesser Line Notation WNR B1U1VQ
Evaluation C
Same data in 40SAT/SOG 2.

$C_9H_7NO_4$ (c) 41SAT/SOG
m–Nitrocinnamic acid
Heat Capacity 323 K, C_p = 57.4 cal·mol^{-1}·K^{-1}
240.2 J·mol^{-1}·K^{-1}
Temperature range 0 to 100°C. Mean value.
Molecular Weight 193.1586
Wiswesser Line Notation WNR C1U1VQ
Evaluation C
Same data in 40SAT/SOG 2.

$C_9H_7NO_4$ (c) 41SAT/SOG
p–Nitrocinnamic acid
Heat Capacity 323 K, C_p = 56.9 cal·mol^{-1}·K^{-1}
238.1 J·mol^{-1}·K^{-1}
Temperature range 0 to 100°C. Mean value.
Molecular Weight 193.1586
Wiswesser Line Notation WNR D1U1VQ
Evaluation C
Same data in 40SAT/SOG 2.

$C_9H_7O_5$ (c) 78MAR/CIO 2
Trimellitic anhydride
Heat Capacity 298.15 K, C_p = 59.5 cal·mol^{-1}·K^{-1}
248.9 J·mol^{-1}·K^{-1}
Temperature range 298–540 K
Phase Changes
c/liq 385 K, ΔH = 2501 cal·mol^{-1}
10464 J·mol^{-1}
ΔS = 6.5 cal·mol^{-1}·K^{-1}
27.2 J·mol^{-1}·K^{-1}
Molecular Weight 195.1513
Wiswesser Line Notation T56 BVOVJ GVQ
Evaluation D

C_9H_8 (liq) 59STU/SIN
Indene
Heat Capacity 298.15 K, C_p = 44.68 cal·mol^{-1}·K^{-1}
186.94 J·mol^{-1}·K^{-1}
Temperature range 15–320 K
Entropy 298.15 K, S = 51.47 cal·mol^{-1}·K^{-1}
215.35 J·mol^{-1}·K^{-1}
Phase Changes
c/liq 271.70 K, ΔH = 2438 cal·mol^{-1}
10201 J·mol^{-1}
ΔS = 8.976 cal·mol^{-1}·K^{-1}
37.55 J·mol^{-1}·K^{-1}
Molecular Weight 116.1622
Wiswesser Line Notation L56 BHJ
Evaluation A

Table of Heat Capacities, Entropies, and Phase Transition Properties — Continued

$C_9H_8O_2$ (c) 39SAT/SOG 2
Cinnamic acid
Heat Capacity 323 K, C_p = 47.2 cal·mol^{-1}·K^{-1}
197.5 J·mol^{-1}·K^{-1}
Temperature range 0 to 100°C. Mean value.
Molecular Weight 148.1610
Wiswesser Line Notation QV1U1R
Evaluation C
Same data in 40SAT/SOG.

$C_9H_9Cl_3$ (c) 72LAG
1,2,3–Trichloro–4,5,6–trimethylbenzene
Heat Capacity 298.15 K, C_p = 60.31 cal·mol^{-1}·K^{-1}
252.34 J·mol^{-1}·K^{-1}
Temperature range 20–300 K
Entropy 298.15 K, S = 75.27 cal·mol^{-1}·K^{-1}
314.93 J·mol^{-1}·K^{-1}
Molecular Weight 223.5291
Wiswesser Line Notation GR BG CG D1 E1 F1
Evaluation B
Second order transition between 140 and 270 K.

$C_9H_9NO_2$ (c) 41SAT/SOG 2
m–Aminocinnamic acid
Heat Capacity 323 K, C_p = 54.4 cal·mol^{-1}·K^{-1}
227.6 J·mol^{-1}·K^{-1}
Temperature range 0 to 100°C. Mean value.
Molecular Weight 163.1756
Wiswesser Line Notation ZR C1U1VQ
Evaluation C
Same data as 40SAT/SOG 3.

$C_9H_9NO_3$ (c) 41HUF
Hippuric acid; Benzoylglycine
Heat Capacity 298.4 K, C_p = 51.23 cal·mol^{-1}·K^{-1}
214.35 J·mol^{-1}·K^{-1}
Temperature range 85–298 K. Value is unsmoothed experimental datum.
Entropy 298.1 K, S = 57.2 cal·mol^{-1}·K^{-1}
239.3 J·mol^{-1}·K^{-1}
Extrapolation below 90 K, 18.48 cal·mol^{-1}K^{-1}
Molecular Weight 179.1750
Wiswesser Line Notation QV1MVR
Evaluation A(C_p),C(S)

C_9H_{10} (liq) 59STU/SIN
Indan
Heat Capacity 298.15 K, C_p = 45.47 cal·mol^{-1}·K^{-1}
190.25 J·mol^{-1}·K^{-1}
Temperature range 15–320 K
Entropy 298.15 K, S = 56.01 cal·mol^{-1}·K^{-1}
234.34 J·mol^{-1}·K^{-1}
Phase Changes
c/liq 221.77 K, ΔH = 2055 cal·mol^{-1}
8598 J·mol^{-1}
ΔS = 9.276 cal·mol^{-1}·K^{-1}
38.77 J·mol^{-1}·K^{-1}
Molecular Weight 118.1780
Wiswesser Line Notation L56T&J
Evaluation A

$C_9H_{10}O_2$ (liq) 33KOL/UDO
Benzyl ethanoate; Benzyl acetate
Heat Capacity 292.7 K, C_p = 59.8 cal·mol^{-1}·K^{-1}
250.2 J·mol^{-1}·K^{-1}
One temperature
Molecular Weight 150.1768
Wiswesser Line Notation 1VO1R
Evaluation C

$C_9H_{10}O_2$ (liq) 34KOL/UDO 2
Benzyl ethanoate; Benzyl acetate
Heat Capacity 292.7 K, C_p = 59.8 cal·mol^{-1}·K^{-1}
250.2 J·mol^{-1}·K^{-1}
One temperature
Molecular Weight 150.1768
Wiswesser Line Notation 1VO1R
Evaluation C

$C_9H_{10}O_2$ (liq) 39PHI
Benzyl ethanoate; Benzyl acetate
Heat Capacity 306.0 K, C_p = 36.8 cal·mol^{-1}·K^{-1}
154.0 J·mol^{-1}·K^{-1}
One temperature
Molecular Weight 150.1768
Wiswesser Line Notation 1VO1R
Evaluation C

$C_9H_{10}O_2$ (liq) 79FUC
Benzyl ethanoate; Benzyl acetate
Heat Capacity 298.15 K, C_p = 35.5 cal·mol^{-1}·K^{-1}
148.5 J·mol^{-1}·K^{-1}
One temperature
Molecular Weight 150.1768
Wiswesser Line Notation 1VO1R
Evaluation B

$C_9H_{10}O_2$ (liq) 33KOL/UDO
Ethyl benzoate
Heat Capacity 292.7 K, C_p = 57.8 cal·mol^{-1}·K^{-1}
241.8 J·mol^{-1}·K^{-1}
One temperature
Molecular Weight 150.1768
Wiswesser Line Notation 2OVR
Evaluation C

$C_9H_{10}O_2$ (liq) 34KOL/UDO 2
Ethyl benzoate
Heat Capacity 292.7 K, C_p = 57.8 cal·mol^{-1}·K^{-1}
241.8 J·mol^{-1}·K^{-1}
One temperature
Molecular Weight 150.1768
Wiswesser Line Notation 2OVR
Evaluation C

$C_9H_{10}O_2$ (liq) 36KUR/VOS
Ethyl benzoate
Heat Capacity 290 K, C_p = 67.6 cal·mol^{-1}·K^{-1}
282.8 J·mol^{-1}·K^{-1}
One temperature
Molecular Weight 150.1768
Wiswesser Line Notation 2OVR
Evaluation D

$C_9H_{10}O_2$ (liq) 79FUC
Ethyl benzoate
Heat Capacity 298.15 K, C_p = 58.8 cal·mol^{-1}·K^{-1}
246.0 J·mol^{-1}·K^{-1}
One temperature
Molecular Weight 150.1768
Wiswesser Line Notation 2OVR
Evaluation B

Table of Heat Capacities, Entropies, and Phase Transition Properties — Continued

$C_9H_{10}O_3$ (c) 78GEI/KAR
3-Methyltetrahydrophthalic anhydride
Heat Capacity C_p data not given.
Temperature range 12–360 K. Data deposited VINITI, No. 3381–77, 5 Oct 1977. Includes C_p, S, ΔH fusion, Tm.
Molecular Weight 166.1762
Wiswesser Line Notation TS6 BVOV AUJG F1
Evaluation B(for original data)

$C_9H_{11}NO_2$ (c) 71PRI
Ethyl phenylcarbamate
Heat Capacity 298 K, C_p = 54.3 cal·mol^{-1}·K^{-1}
227.2 J·mol^{-1}·K^{-1}
Temperature range 200–390 K. Complete data deposited VINITI, No. 2713–71, 25 March 1971.
Phase Changes
c/liq 326 K, ΔH = 3889 cal·mol^{-1}
16272 J·mol^{-1}
ΔS = 11.9 cal·mol^{-1}·K^{-1}
49.9 J·mol^{-1}·K^{-1}
Molecular Weight 165.1914
Wiswesser Line Notation 2OVMR
Evaluation B

$C_9H_{11}NO_2$ (c) 39SAT/SOG 2
Ammonium cinnamate
Heat Capacity 323 K, C_p = 60.4 cal·mol^{-1}·K^{-1}
252.7 J·mol^{-1}·K^{-1}
Temperature range 0 to 100°C. Mean value.
Molecular Weight 165.1914
Wiswesser Line Notation QV1U1R &ZH
Evaluation C
Same data in 40SAT/SOG.

$C_9H_{11}NO_2$ (c) 63COL/HUT
Phenylalanine(L)
Heat Capacity 298.15 K, C_p = 48.52 cal·mol^{-1}·K^{-1}
203.01 J·mol^{-1}·K^{-1}
Temperature range 11–305 K
Entropy 298.15 K, S = 51.06 cal·mol^{-1}·K^{-1}
213.64 J·mol^{-1}·K^{-1}
Molecular Weight 165.1914
Wiswesser Line Notation QVYZ1R -L
Evaluation A

$C_9H_{11}NO_2$ (c) 75SPI/WAD
Phenylalanine(L)
Heat Capacity 298.15 K, C_p = 48.5 cal·mol^{-1}·K^{-1}
203.1 J·mol^{-1}·K^{-1}
One temperature
Molecular Weight 165.1914
Wiswesser Line Notation QVYZ1R -L
Evaluation B

$C_9H_{11}NO_3$ (c) 37HUF/ELL
Tyrosine(L)
Heat Capacity 294.6 K, C_p = 51.22 cal·mol^{-1}·K^{-1}
214.30 J·mol^{-1}·K^{-1}
Temperature range 87–295 K. Value is unsmoothed experimental datum.
Entropy 298.15 K, S = 53.0 cal·mol^{-1}·K^{-1}
221.8 J·mol^{-1}·K^{-1}
Extrapolation below 90 K, 15.43 cal·mol^{-1}K^{-1}
Molecular Weight 181.1908
Wiswesser Line Notation QVYZ1R DQ -L
Evaluation B(C_p)C(S)

$C_9H_{11}NO_3$ (c) 63COL/HUT
Tyrosine(L)
Heat Capacity 298.15 K, C_p = 51.73 cal·mol^{-1}·K^{-1}
216.44 J·mol^{-1}·K^{-1}
Entropy 298.15 K, S = 51.15 cal·mol^{-1}·K^{-1}
214.01 J·mol^{-1}·K^{-1}
Molecular Weight 181.1908
Wiswesser Line Notation QVYZ1R DQ -L
Evaluation A

$C_9H_{11}NO_4$ (c) 40CAM/CAM
Salicylic acid–acetamide complex;
Acetamide–salicylic acid complex
Heat Capacity 293 K, C_p = 40.5 cal·mol^{-1}·K^{-1}
169.5 J·mol^{-1}·K^{-1}
One temperature
Molecular Weight 197.1902
Wiswesser Line Notation QVR BQ &ZV1
Evaluation C

C_9H_{12} (liq) 55TAY/JOH
1,2,3–Trimethylbenzene
Heat Capacity 298.15 K, C_p = 51.73 cal·mol^{-1}·K^{-1}
216.44 J·mol^{-1}·K^{-1}
Temperature range 19–301 K
Entropy 298.15 K, S = 64.04 cal·mol^{-1}·K^{-1}
267.94 J·mol^{-1}·K^{-1}
Phase Changes
c,III/c,II 218.70 K, ΔH = 157.4 cal·mol^{-1}
658.6 J·mol^{-1}
ΔS = 0.72 cal·mol^{-1}·K^{-1}
3.01 J·mol^{-1}·K^{-1}
c,II/c,I 230.27 K, ΔH = 319.3 cal·mol^{-1}
1336.0 J·mol^{-1}
ΔS = 1.39 cal·mol^{-1}·K^{-1}
5.80 J·mol^{-1}·K^{-1}
c,I/liq 247.81 K, ΔH = 1955.1 cal·mol^{-1}
8180.1 J·mol^{-1}
ΔS = 7.89 cal·mol^{-1}·K^{-1}
33.01 J·mol^{-1}·K^{-1}
Molecular Weight 120.1938
Wiswesser Line Notation 1R B1 C1
Evaluation A

C_9H_{12} (liq) 31HUF/PAR
1,2,4–Trimethylbenzene; Pseudocumene
Heat Capacity 297.3 K, C_p = 50.7 cal·mol^{-1}·K^{-1}
212.1 J·mol^{-1}·K^{-1}
Temperature range 94–297 K. Value is unsmoothed experimental datum.
Entropy 298.1 K, S = 67.7 cal·mol^{-1}·K^{-1}
283.3 J·mol^{-1}·K^{-1}
Extrapolation below 90 K, 16.68 cal·mol^{-1}K^{-1}.
Phase Changes
c/liq 228.6 K, ΔH = 3023 cal·mol^{-1}
12648 J·mol^{-1}
ΔS = 13.2 cal·mol^{-1}·K^{-1}
55.3 J·mol^{-1}·K^{-1}
Molecular Weight 120.1938
Wiswesser Line Notation 1R B1 D1
Evaluation B(C_p),C(S)

Table of Heat Capacities, Entropies, and Phase Transition Properties — Continued

C_9H_{12} (liq) 47KUR
1,2,4–Trimethylbenzene; Pseudocumene
Heat Capacity 298 K, C_p = 50.9 cal·mol^{-1}·K^{-1}
213.0 J·mol^{-1}·K^{-1}
Temperature range 15 to 168 °C, mean C_p, five temperatures.
Molecular Weight 120.1938
Wiswesser Line Notation 1R B1 D1
Evaluation D

C_9H_{12} (liq) 55HEL/HEI
1,2,4–Trimethylbenzene; Pseudocumene
Heat Capacity 299.8 K, C_p = 50.29 cal·mol^{-1}·K^{-1}
210.41 J·mol^{-1}·K^{-1}
Temperature range 80–220°F
Molecular Weight 120.1938
Wiswesser Line Notation 1R B1 D1
Evaluation B

C_9H_{12} (liq) 57PUT/KIL
1,2,4–Trimethylbenzene; Pseudocumene
Heat Capacity 298.15 K, C_p = 51.38 cal·mol^{-1}·K^{-1}
214.97 J·mol^{-1}·K^{-1}
Temperature range 15–300 K
Entropy 298.15 K, S = 67.73 cal·mol^{-1}·K^{-1}
283.38 J·mol^{-1}·K^{-1}
Phase Changes
c,I/liq 229.33 K, ΔH = 3152.5 cal·mol^{-1}
13190 J·mol^{-1}
ΔS = 13.75 cal·mol^{-1}·K^{-1}
57.53 J·mol^{-1}
Molecular Weight 120.1938
Wiswesser Line Notation 1R B1 D1
Evaluation A

C_9H_{12} (liq) 55TAY/KIL
1,3,5–Trimethylbenzene; Mesitylene
Heat Capacity 298.15 K, C_p = 50.03 cal·mol^{-1}·K^{-1}
209.33 J·mol^{-1}·K^{-1}
Temperature range 20–305 K
Entropy 298.15 K, S = 65.38 cal·mol^{-1}·K^{-1}
273.55 J·mol^{-1}·K^{-1}
Phase Changes
c,I/liq 228.42 K, ΔH = 2274.1 cal·mol^{-1}
9514.8 J·mol^{-1}
ΔS = 9.96 cal·mol^{-1}·K^{-1}
41.65 J·mol^{-1}·K^{-1}
Metastable melting points at 221.46 K and 223.35 K.
Molecular Weight 120.1938
Wiswesser Line Notation 1R C1 E1
Evaluation A

C_9H_{12} (liq) **81REI**
1,3,5–Trimethylbenzene; Mesitylene
Heat Capacity 298 K, C_p = 50.5 cal·mol^{-1}·K^{-1}
211.3 J·mol^{-1}·K^{-1}
Temperature range 292–403 K
Molecular Weight 120.1938
Wiswesser Line Notation 1R C1 E1
Evaluation D

C_9H_{12} (liq) 47KUR
1,3,5–Trimethylbenzene; Mesitylene
Heat Capacity 298 K, C_p = 50.9 cal·mol^{-1}·K^{-1}
213.0 J·mol^{-1}·K^{-1}
Temperature range 15 to 155 °C, mean C_p, five temperatures.
Molecular Weight 120.1938
Wiswesser Line Notation 1R C1 E1
Evaluation D

C_9H_{12} (liq) 55HEL/HEI
1,3,5–Trimethylbenzene; Mesitylene
Heat Capacity 299.8 K, C_p = 48.15 cal·mol^{-1}·K^{-1}
201.46 J·mol^{-1}·K^{-1}
Temperature range 80–220°F
Molecular Weight 120.1938
Wiswesser Line Notation 1R C1 E1
Evaluation B

C_9H_{12} (liq) 68REC
1,3,5–Trimethylbenzene; Mesitylene
Heat Capacity 298 K, C_p = 49.4 cal·mol^{-1}·K^{-1}
206.5 J·mol^{-1}·K^{-1}
Temperature range 24 to 40 °C. Equation only.
Molecular Weight 120.1938
Wiswesser Line Notation 1R C1 E1
Evaluation C

C_9H_{12} (liq) 79WIL/FAR
1,3,5–Trimethylbenzene; Mesitylene
Heat Capacity 298.15 K, C_p = 49.63 cal·mol^{-1}·K^{-1}
207.66 J·mol^{-1}·K^{-1}
One temperature.
Molecular Weight 120.1938
Wiswesser Line Notation 1R C1 E1
Evaluation B

C_9H_{12} (liq) 34KOL/UDO
Isopropylbenzene; Cumene
Heat Capacity 302.0 K, C_p = 47.5 cal·mol^{-1}·K^{-1}
198.7 J·mol^{-1}·K^{-1}
One temperature
Molecular Weight 120.1938
Wiswesser Line Notation 1Y1&R
Evaluation C

C_9H_{12} (liq) 34KOL/UDO 2
Isopropylbenzene; Cumene
Heat Capacity 302.0 K, C_p = 47.5 cal·mol^{-1}·K^{-1}
198.7 J·mol^{-1}·K^{-1}
One temperature
Molecular Weight 120.1938
Wiswesser Line Notation 1Y1&R
Evaluation C

C_9H_{12} (liq) 47KUR
Isopropylbenzene; Cumene
Heat Capacity 298 K, C_p = 50.9 cal·mol^{-1}·K^{-1}
213.0 J·mol^{-1}·K^{-1}
Temperature range 16 to 153 °C, mean C_p, four temperatures.
Molecular Weight 120.1938
Wiswesser Line Notation 1Y1&R
Evaluation D

Table of Heat Capacities, Entropies, and Phase Transition Properties — Continued

C_9H_{12} (liq) 52SCH/SAG
Isopropylbenzene; Cumene
Heat Capacity 299.8 K, C_p = 50.05 cal·mol^{-1}·K^{-1}
209.41 J·mol^{-1}·K^{-1}
Temperature range 300–366 K, (80 to 200°F).
Molecular Weight 120.1938
Wiswesser Line Notation 1Y1&R
Evaluation B

C_9H_{12} (liq) 73KIS/SUG
Isopropylbenzene; Cumene
Heat Capacity 298.15 K, C_p = 51.48 cal·mol^{-1}·K^{-1}
215.40 J·mol^{-1}·K^{-1}
Temperature range 14–314 K. Glass, 14–126 K
Entropy 298.15 K, S = 66.34 cal·mol^{-1}·K^{-1}
277.57 J·mol^{-1}·K^{-1}
Phase Changes
c/liq 177.13 K, ΔH = 1751 cal·mol^{-1}
7326 J·mol^{-1}
ΔS = 9.89 cal·mol^{-1}·K^{-1}
41.36 J·mol^{-1}·K^{-1}
Molecular Weight 120.1938
Wiswesser Line Notation 1Y1&R
Evaluation A

C_9H_{12} (liq) 65MES/TOD
n–Propylbenzene
Heat Capacity 298.15 K, C_p = 51.32 cal·mol^{-1}·K^{-1}
214.72 J·mol^{-1}·K^{-1}
Temperature range 10–370 K
Entropy 298.15 K, S = 68.78 cal·mol^{-1}·K^{-1}
287.78 J·mol^{-1}·K^{-1}
Phase Changes
c,II/liq 171.67 K, ΔH = 2031 cal·mol^{-1}
8498 J·mol^{-1}
ΔS = 11.83 cal·mol^{-1}·K^{-1}
49.50 J·mol^{-1}·K^{-1}
c,I/liq 173.60 K, ΔH = 2215 cal·mol^{-1}
9268 J·mol^{-1}
ΔS = 12.76 cal·mol^{-1}·K^{-1}
53.39 J·mol^{-1}·K^{-1}
Metastable crystals
Molecular Weight 120.1938
Wiswesser Line Notation 3R
Evaluation A

$C_9H_{12}O$ (liq) 75NIC/WAD
3–Phenylpropanol
Heat Capacity 298.15 K, C_p = 67.10 cal·mol^{-1}·K^{-1}
280.74 J·mol^{-1}·K^{-1}
One temperature
Molecular Weight 136.1932
Wiswesser Line Notation Q3R
Evaluation B

$C_9H_{13}N$ (liq) 02LOU
2,N,N–Trimethylaniline
Heat Capacity 370 K, C_p = 67 cal·mol^{-1}·K^{-1}
280 J·mol^{-1}·K^{-1}
Mean value 21 to 184°C
Molecular Weight 135.2084
Wiswesser Line Notation 1N1&R B1
Evaluation D

$C_9H_{13}N$ (liq) 75NIC/WAD
3–Phenylpropylamine
Heat Capacity 298.15 K, C_p = 63.48 cal·mol^{-1}·K^{-1}
265.59 J·mol^{-1}·K^{-1}
One temperature
Molecular Weight 135.2084
Wiswesser Line Notation Z3R
Evaluation B

$C_9H_{14}O_2$ (liq) 79FUC
Glyceryl triacetate
Heat Capacity 298.15 K, C_p = 96.1 cal·mol^{-1}·K^{-1}
402 J·mol^{-1}·K^{-1}
One temperature
Molecular Weight 154.2084
Wiswesser Line Notation 1VOY1OV1Y1OV1
Evaluation B

C_9H_{16} (liq) 70CHA/MCC
cis–Bicyclo[6.1.0]nonane
Heat Capacity 315 K, C_p = 56.2 cal·mol^{-1}·K^{-1}
235.1 J·mol^{-1}·K^{-1}
One temperature
Molecular Weight 124.2254
Wiswesser Line Notation L38TJ -C
Evaluation B

C_9H_{16} (liq) 62GOL/BEL
Hexahydroindan
Heat Capacity 311 K, C_p = 52.5 cal·mol^{-1}·K^{-1}
219.7 J·mol^{-1}·K^{-1}
Temperatures 100, 200, 300°F
Molecular Weight 124.2254
Wiswesser Line Notation L56TJ
Evaluation C

C_9H_{16} (liq) 63GUD/CAM
Hexahydroindan
Heat Capacity 313 K, C_p = 51.9 cal·mol^{-1}·K^{-1}
217.1 J·mol^{-1}·K^{-1}
Temperature range 313–423 K
Molecular Weight 124.2254
Wiswesser Line Notation L56TJ
Evaluation C

C_9H_{16} (liq) 72FIN/MCC
cis–Hexahydroindan
Heat Capacity 298.15 K, C_p = 51.19 cal·mol^{-1}·K^{-1}
214.18 J·mol^{-1}·K^{-1}
Temperature range 10–370 K
Entropy 298.15 K, S = 63.45 cal·mol^{-1}·K^{-1}
265.47 J·mol^{-1}·K^{-1}
Phase Changes
c,III/c,II 182.28 K, ΔH = 1974.8 cal·mol^{-1}
8262.6 J·mol^{-1}
ΔS = 10.83 cal·mol^{-1}·K^{-1}
45.33 J·mol^{-1}·K^{-1}
c,II/c,I 184.9 K, ΔH = 94.4 cal·mol^{-1}
395.0 J·mol^{-1}
ΔS = 0.51 cal·mol^{-1}·K^{-1}
2.14 J·mol^{-1}·K^{-1}
c,I/liq 236.48 K, ΔH = 333.9 cal·mol^{-1}
1397.0 J·mol^{-1}
ΔS = 1.41 cal·mol^{-1}·K^{-1}
5.91 J·mol^{-1}·K^{-1}
Molecular Weight 124.2254
Wiswesser Line Notation L56TJ -C
Evaluation A

Table of Heat Capacities, Entropies, and Phase Transition Properties — Continued

C_9H_{16} (liq) 72FIN/MCC
trans–Hexahydroindan
Heat Capacity 298.15 K, C_p = 50.12 cal·mol⁻¹·K⁻¹ 209.70 J·mol⁻¹·K⁻¹
Temperature range 10–400 K
Entropy 298.15 K, S = 61.87 cal·mol⁻¹·K⁻¹ 258.86 J·mol⁻¹·K⁻¹
Phase Changes
c/liq 213.86 K, ΔH = 2606.4 cal·mol⁻¹ 10905.2 J·mol⁻¹
ΔS = 12.19 cal·mol⁻¹·K⁻¹ 50.99 J·mol⁻¹·K⁻¹
Molecular Weight 124.2254
Wiswesser Line Notation L56TJ -T
Evaluation A

C_9H_{16} (liq) 79PUC/PEA
Allylcyclohexane
Heat Capacity 298.15 K, C_p = 55.8 cal·mol⁻¹·K⁻¹ 233.5 J·mol⁻¹·K⁻¹
One temperature
Molecular Weight 124.2254
Wiswesser Line Notation L6TJ A2U1
Evaluation B

$C_9H_{16}O_2$ (liq) 79FUC
Ethyl cyclohexanecarboxylate
Heat Capacity 298.15 K, C_p = 64.9 cal·mol⁻¹·K⁻¹ 271.5 J·mol⁻¹·K⁻¹
One temperature
Molecular Weight 156.2242
Wiswesser Line Notation L6TJ AVO2
Evaluation B

$C_9H_{17}N_3O_4$ (c) 75DAU/DEL
Tri–L–alanine
Heat Capacity
Temperature range 1–300 K. C_p data given graphically only.
Entropy 273 K, S = 85.1 cal·mol⁻¹ 356.1 J·mol⁻¹·K⁻¹
Molecular Weight 231.2510
Wiswesser Line Notation ZY1&VMY1&VMY1&VQ
Evaluation B

C_9H_{18} (liq) 65MES/TOD 2
n–Butylcyclopentane
Heat Capacity 298.15 K, C_p = 58.64 cal·mol⁻¹·K⁻¹ 245.35 J·mol⁻¹·K⁻¹
Temperature range 12–370 K
Entropy 298.15 K, S = 82.18 cal·mol⁻¹·K⁻¹ 343.84 J·mol⁻¹·K⁻¹
Phase Changes
c/liq 165.18 K, ΔH = 2704 cal·mol⁻¹ 11314 J·mol⁻¹
ΔS = 16.37 cal·mol⁻¹·K⁻¹ 68.49 J·mol⁻¹·K⁻¹
Molecular Weight 126.2412
Wiswesser Line Notation L5TJ A4
Evaluation A

C_9H_{18} (liq) 65FIN/MES
n–Propylcyclohexane
Heat Capacity 298.15 K, C_p = 57.85 cal·mol⁻¹·K⁻¹ 242.04 J·mol⁻¹·K⁻¹
Temperature range 10–380 K
Entropy 298.15 K, S = 74.54 cal·mol⁻¹·K⁻¹ 311.88 J·mol⁻¹·K⁻¹
Phase Changes
c/liq 178.25 K, ΔH = 2479 cal·mol⁻¹ 13072 J·mol⁻¹
ΔS = 13.91 cal·mol⁻¹·K⁻¹ 58.19 J·mol⁻¹·K⁻¹
Molecular Weight 126.2412
Wiswesser Line Notation L6TJ A3
Evaluation A

$C_9H_{18}O$ (liq) 70AND/COU
5–Nonanone; Di–n–butyl ketone
Heat Capacity 298.15 K, C_p = 72.56 cal·mol⁻¹·K⁻¹ 303.6 J·mol⁻¹·K⁻¹
Temperature range 10–320 K
Entropy 298.15 K, S = 95.94 cal·mol⁻¹·K⁻¹ 401.4 J·mol⁻¹·K⁻¹
Phase Changes
c,II/c,I 110 K, ΔH = 89 cal·mol⁻¹ 373 J·mol⁻¹
ΔS = 0.81 cal·mol⁻¹·K⁻¹ 3.39 J·mol⁻¹·K⁻¹
c,I/liq 269.31 K, ΔH = 5960 cal·mol⁻¹ 24930 J·mol⁻¹
ΔS = 22.12 cal·mol⁻¹·K⁻¹ 92.56 J·mol⁻¹·K⁻¹
Molecular Weight 142.2406
Wiswesser Line Notation 4V4
Evaluation A

$C_9H_{18}O$ (liq) 79SAL/PEA
5–Nonanone; Di–n–butyl ketone
Heat Capacity 298.15 K, C_p = 73.2 cal·mol⁻¹·K⁻¹ 306.3 J·mol⁻¹·K⁻¹
One temperature
Molecular Weight 142.2406
Wiswesser Line Notation 4V4
Evaluation B

$C_9H_{18}O_2$ (liq) 24GAR/RAN
Nonanoic acid; Pelargonic acid
Heat Capacity 304 K, C_p = 79.8 cal·mol⁻¹·K⁻¹ 333.9 J·mol⁻¹·K⁻¹
Temperature range -9 to 44C. Mean value 18 to 44C.
Phase Changes
c,II/c,I 268 K, ΔH = 1330 cal·mol⁻¹ 5560 J·mol⁻¹
ΔS = 5.0 cal·mol⁻¹·K⁻¹ 20.7 J·mol⁻¹·K⁻¹
c,I/liq 285.5 K, ΔH = 4850 cal·mol⁻¹ 20290 J·mol⁻¹
ΔS = 17.0 cal·mol⁻¹·K⁻¹ 71.1 J·mol⁻¹·K⁻¹
Molecular Weight 158.2400
Wiswesser Line Notation QV8
Evaluation B

Table of Heat Capacities, Entropies, and Phase Transition Properties — Continued

$C_9H_{18}O_2$ (liq) 82SCH/MIL
Nonanoic acid; Pelargonic acid
Heat Capacity 298.15 K, C_p = 86.609 cal·mol^{-1}·K^{-1}
362.37 J·mol^{-1}·K^{-1}
Temperature range 80–310 K
Phase Changes
c,II/c,I 263.0 K, ΔH = 1948 cal·mol^{-1}
8150 J·mol^{-1}
ΔS = 6.709 cal·mol^{-1}·K^{-1}
28.07 J·mol^{-1}·K^{-1}
c,I/liq 285.53 K, ΔH = 4737.8 cal·mol^{-1}
19823 J·mol^{-1}
ΔS = 16.59 cal·mol^{-1}·K^{-1}
69.42 J·mol^{-1}·K^{-1}
Molecular Weight 158.2400
Wiswesser Line Notation QV8
Evaluation B

C_9H_{20} (liq) 79PUC/PEA
3,3–Diethylpentane
Heat Capacity 298.15 K, C_p = 66.5 cal·mol^{-1}·K^{-1}
278.2 J·mol^{-1}·K^{-1}
One temperature
Molecular Weight 128.2570
Wiswesser Line Notation 2X2&2&2
Evaluation B

C_9H_{20} (liq) 30PAR/HUF
n–Nonane
Heat Capacity 299.1 K, C_p = 67.2 cal·mol^{-1}·K^{-1}
281.2 J·mol^{-1}·K^{-1}
Temperature range 224–299 K. Value is unsmoothed experimental datum.
Molecular Weight 128.2570
Wiswesser Line Notation 9H
Evaluation B

C_9H_{20} (liq) 31HUF/PAR
n–Nonane
Heat Capacity 297.9 K, C_p = 67.1 cal·mol^{-1}·K^{-1}
280.7 J·mol^{-1}·K^{-1}
Temperature range 93–298 K. Value is unsmoothed experimental datum.
Entropy 298.1 K, S = 93.9 cal·mol^{-1}·K^{-1}
392.9 J·mol^{-1}·K^{-1}
Extrapolation below 90 K, 19.86 cal·mol^{-1}K^{-1}.
Phase Changes
c,I/liq 219.2 K, ΔH = 5287 cal·mol^{-1}
22121 J·mol^{-1}
ΔS = 24.1 cal·mol^{-1}·K^{-1}
100.9 J·mol^{-1}·K^{-1}
Includes heat effect due to transition just below melting point.
Molecular Weight 128.2570
Wiswesser Line Notation 9H
Evaluation B(C_p),C(S)

C_9H_{20} (liq) 47OSB/GIN
n–Nonane
Heat Capacity 298.15 K, C_p = 67.88 cal·mol^{-1}·K^{-1}
284.01 J·mol^{-1}·K^{-1}
Temperature range 278–318 K
Molecular Weight 128.2570
Wiswesser Line Notation 9H
Evaluation A

C_9H_{20} (liq) 54FIN/GRO 2
n–Nonane
Heat Capacity 298.15 K, C_p = 67.97 cal·mol^{-1}·K^{-1}
284.39 J·mol^{-1}·K^{-1}
Temperature range 12–320 K
Entropy 298.15 K, S = 94.09 cal·mol^{-1}·K^{-1}
393.67 J·mol^{-1}·K^{-1}
Phase Changes
c,II/c,I 217.2 K, ΔH = 1501 cal·mol^{-1}
6280 J·mol^{-1}
ΔS = 6.91 cal·mol^{-1}·K^{-1}
28.91 J·mol^{-1}·K^{-1}
c,I/liq 219.66 K, ΔH = 3697 cal·mol^{-1}
15468 J·mol^{-1}
ΔS = 16.83 cal·mol^{-1}·K^{-1}
70.42 J·mol^{-1}·K^{-1}
Molecular Weight 128.2570
Wiswesser Line Notation 9H
Evaluation A

C_9H_{20} (liq) 58SWI/ZIE
n–Nonane
Heat Capacity 350 K, C_p = 77.0 cal·mol^{-1}·K^{-1}
322.2 J·mol^{-1}·K^{-1}
Mean value over the temperature range 22 to 129°C.
Molecular Weight 128.2570
Wiswesser Line Notation 9H
Evaluation C

C_9H_{20} (liq) 77MUS
n–Nonane
Heat Capacity Data in document deposited at VINITI, No. 880–77, March 10, 1977.
Molecular Weight 128.2570
Wiswesser Line Notation 9H
Evaluation B(for deposited data)

C_9H_{20} (liq) 79GRO/HAM
n–Nonane
Heat Capacity 298.15 K, C_p = 67.9 cal·mol^{-1}·K^{-1}
284.0 J·mol^{-1}·K^{-1}
One temperature
Molecular Weight 128.2570
Wiswesser Line Notation 9H
Evaluation B

C_9H_{20} (liq) 82WIL/ING
n–Nonane
Heat Capacity 298.15 K, C_p = 67.82 cal·mol^{-1}·K^{-1}
283.8 J·mol^{-1}·K^{-1}
Temperature range 298.15 K, One temperature
Molecular Weight 128.2570
Wiswesser Line Notation 9H
Evaluation A

$C_9H_{21}CaCl_2N_3O_6$ (c,I) 79MAT/MAN
Tris(sarcosine) calcium chloride
Heat Capacity 298.15 K, C_p = 101.65 cal·mol^{-1}·K^{-1}
425.3 J·mol^{-1}·K^{-1}
Temperature range 13–300 K
Entropy 298.15 K, S = 127.68 cal·mol^{-1}·K^{-1}
534.2 J·mol^{-1}·K^{-1}

Table of Heat Capacities, Entropies, and Phase Transition Properties — Continued

Phase Changes
c,II/c,I 130.8 K, ΔH = 107 cal·mol^{-1}
446 J·mol^{-1}
ΔS = 1.0 cal·mol^{-1}·K^{-1}
4.3 J·mol^{-1}·K^{-1}
Second order ferroelectric transition. ΔS is not equal to the isothermal $\Delta H/T$.
Molecular Weight 378.2674
Wiswesser Line Notation OV1M1 3 &.CA.G2
Evaluation A

$C_9H_{21}CaCl_2N_3O_6$ (c) 81LOP/TEL
Tris(sarcosine) calcium chloride
Heat Capacity
Temperature range 50–330 K. Data given graphically.
Phase Changes
c,II/c,I 130.27 K, ΔH = 39.5 cal·mol^{-1}
165.3 J·mol^{-1}
ΔS = 0.32 cal·mol^{-1}·K^{-1}
1.34 J·mol^{-1}·K^{-1}
Second order ferroelectric transition.
Molecular Weight 378.2674
Wiswesser Line Notation QV1M1 3–CA–GG
Evaluation A

$C_{10}F_{16}$ (liq) 81ZHO/KOS
Perfluorobicyclo[4.4.0]dec–1,6–diene
Heat Capacity 298.15 K, C_p = 102.4 cal·mol^{-1}·K^{-1}
428.4 J·mol^{-1}·K^{-1}
Temperature range 6–300 K, 0.41 mole % impurity of sample.
Entropy 298.15 K, S = 117.4 cal·mol^{-1}·K^{-1}
491.2 J·mol^{-1}·K^{-1}
Phase Changes
c,III/c,II 200.0 K, ΔH = 190 cal·mol^{-1}
794 J·mol^{-1}
ΔS = 0.90 cal·mol^{-1}·K^{-1}
3.77 J·mol^{-1}·K^{-1}
c,II/c,I 233.4 K, ΔH = 266 cal·mol^{-1}
1113 J·mol^{-1}
ΔS = 1.14 cal·mol^{-1}·K^{-1}
4.77 J·mol^{-1}·K^{-1}
c,I/liq 264.09 K, ΔH = 2503 cal·mol^{-1}
10473 J·mol^{-1}
ΔS = 9.48 cal·mol^{-1}·K^{-1}
39.66 J·mol^{-1}·K^{-1}
Molecular Weight 424.0844
Wiswesser Line Notation L66 AU FUTJ BF BF CF CF DF DF EF EF GF GF HF HF IF IF JF JF
Evaluation A

$C_{10}F_{18}$ (liq) 81ZHO/KOS 2
cis–Perfluorodecalin
Heat Capacity 298.15 K, C_p = 107.5 cal·mol^{-1}·K^{-1}
449.8 J·mol^{-1}·K^{-1}
Temperature range 6–310 K
Entropy 298.15 K, S = 123.0 cal·mol^{-1}·K^{-1}
514.6 J·mol^{-1}·K^{-1}
Phase Changes
c,II/c,I 232.5 K, ΔH = 1014 cal·mol^{-1}
4243 J·mol^{-1}
ΔS = 4.36 cal·mol^{-1}·K^{-1}
18.24 J·mol^{-1}·K^{-1}
c,I/liq 266.70 K, ΔH = 2462 cal·mol^{-1}
10305 J·mol^{-1}
ΔS = 9.23 cal·mol^{-1}·K^{-1}
38.62 J·mol^{-1}·K^{-1}
Molecular Weight 462.0812
Wiswesser Line Notation L66TTJ AF BF BF CF CF DF DF EF EF FF GF GF HF HF IF IF JF JF –C
Evaluation B

$C_{10}F_{18}$ (liq) 81ZHO/KOS 2
trans–Perfluorodecalin
Heat Capacity 298.15 K, C_p = 106.8 cal·mol^{-1}·K^{-1}
446.8 J·mol^{-1}·K^{-1}
Temperature range 6–310 K
Entropy 298.15 K, S = 123.6 cal·mol^{-1}·K^{-1}
517.1 J·mol^{-1}·K^{-1}
Phase Changes
c/liq 294.61 K, ΔH = 4293 cal·mol^{-1}
17962 J·mol^{-1}
ΔS = 14.6 cal·mol^{-1}·K^{-1}
61.09 J·mol^{-1}·K^{-1}
Molecular Weight 462.0812
Wiswesser Line Notation L66TTJ AF BF BF CF CF DF DF EF EF FF GF GF HF HF IF IF JF JF –T
Evaluation B

$C_{10}H_2N_4$ (c) 76CLA/WOR
1,2,4,5–Tetracyanobenzene
Heat Capacity 298.15 K, C_p = 53.13 cal·mol^{-1}·K^{-1}
222.3 J·mol^{-1}·K^{-1}
Temperature range 5–300 K
Entropy 298.15 K, S = 60.40 cal·mol^{-1}·K^{-1}
252.7 J·mol^{-1}·K^{-1}
Molecular Weight 178.1526
Wiswesser Line Notation NCR BCN DCN ECN
Evaluation A

$C_{10}H_2O_6$ (c) 78DUN/RAH
Pyromellitic dianhydride
Heat Capacity 298.15 K, C_p = 51.55 cal·mol^{-1}·K^{-1}
215.7 J·mol^{-1}·K^{-1}
Temperature range 5–300 K
Entropy 298.15 K, S = 56.67 cal·mol^{-1}·K^{-1}
237.1 J·mol^{-1}·K^{-1}
Molecular Weight 218.1222
Wiswesser Line Notation T C565 DVOV JVOVJ
Evaluation A

$C_{10}H_2O_6$ (c) 78MAR/CIO 2
Pyromellitic dianhydride
Heat Capacity 298.15 K, C_p = 55.3 cal·mol^{-1}·K^{-1}
231.4 J·mol^{-1}·K^{-1}
Temperature range 298–580 K
Phase Changes
c/liq 557.15 K, ΔH = 3783 cal·mol^{-1}
15828 J·mol^{-1}
ΔS = 6.8 cal·mol^{-1}·K^{-1}
28.4 J·mol^{-1}·K^{-1}
Molecular Weight 218.1222
Wiswesser Line Notation T C565 DVOV JVOVJ
Evaluation D

$C_{10}H_6N_2O_4$ (c) 41SAT/SOG 4
1,5–Dinitronaphthalene
Heat Capacity 323 K, C_p = 62.0 cal·mol^{-1}·K^{-1}
259.4 J·mol^{-1}·K^{-1}
Temperature range 0 to 100°C. Mean value.
Molecular Weight 218.1684
Wiswesser Line Notation L66J BNW GNW
Evaluation C
Same data in 40SAT/SOG 5.

Table of Heat Capacities, Entropies, and Phase Transition Properties — Continued

$C_{10}H_6N_2O_4$ (c) 41SAT/SOG 4
1,8–Dinitronaphthalene
Heat Capacity 323 K, C_p = 60.8 cal·mol^{-1}·K^{-1}
254.4 J·mol^{-1}·K^{-1}
Temperature range 0 to 100°C. Mean value.
Molecular Weight 218.1684
Wiswesser Line Notation L66J BNW JNW
Evaluation C
Same data in 40SAT/SOG 5.

$C_{10}H_6OS_2$ (c) 75CUC
Naphthalene–1,8–disulfide–S–oxide
Heat Capacity 298 K, C_p = 123.8 cal·mol^{-1}·K^{-1}
518.0 J·mol^{-1}·K^{-1}
Temperature range 298–473 K
Phase Changes
c/liq 363 K, ΔH = 3200 cal·mol^{-1}
13390 J·mol^{-1}
ΔS = 8.8 cal·mol^{-1}·K^{-1}
36.9 J·mol^{-1}·K^{-1}
Molecular Weight 206.2768
Wiswesser Line Notation T566 1A L CSSJ CO
Evaluation B

$C_{10}H_7Br$ (c) 81CHA/HAG
2–Bromonaphthalene
Phase Changes
c,III/c,I 319 K, ΔH = 1379 cal·mol^{-1}
5770 J·mol^{-1}
ΔS = 4.323 cal·mol^{-1}·K^{-1}
18.09 J·mol^{-1}·K^{-1}
c,I/liq 329 K, ΔH = 3442 cal·mol^{-1}
14400 J·mol^{-1}
ΔS = 10.46 cal·mol^{-1}·K^{-1}
43.77 J·mol^{-1}·K^{-1}
A second order transition occurs between crystalline phases c,I and c,II over the temperature range 275–319 K.
Molecular Weight 207.0693
Wiswesser Line Notation L66J CE
Evaluation A

$C_{10}H_8$ (c) 26AND/LYN
Naphthalene
Heat Capacity 298 K, C_p = 38.1 cal·mol^{-1}·K^{-1}
159.4 J·mol^{-1}·K^{-1}
Temperature range 12 to 300°C
Phase Changes
c/liq 353.1 K, ΔH = 4540 cal·mol^{-1}
19000 J·mol^{-1}
ΔS = 12.9 cal·mol^{-1}·K^{-1}
53.8 J·mol^{-1}·K^{-1}
Molecular Weight 128.1732
Wiswesser Line Notation L66J
Evaluation C

$C_{10}H_8$ (c) 30HUF/PAR
Naphthalene
Heat Capacity 295.1 K, C_p = 39.1 cal·mol^{-1}·K^{-1}
163.6 J·mol^{-1}·K^{-1}
Temperature range 91–295 K
Entropy 298.1 K, S = 39.9 cal·mol^{-1}·K^{-1}
166.9 J·mol^{-1}·K^{-1}
Extrapolation below 90 K, 12.69 cal·mol^{-1}K^{-1}
Molecular Weight 128.1732
Wiswesser Line Notation L66J
Evaluation B(C_p),C(S)

$C_{10}H_8$ (c) 32SPA/THO
Naphthalene
Heat Capacity 303 K, C_p = 40.4 cal·mol^{-1}·K^{-1}
169.0 J·mol^{-1}·K^{-1}
Temperature range 30 to 190°C
Phase Changes
c/liq 353.0 K, ΔH = 4589 cal·mol^{-1}
19200 J·mol^{-1}
ΔS = 13.0 cal·mol^{-1}·K^{-1}
54.4 J·mol^{-1}·K^{-1}
Molecular Weight 128.1732
Wiswesser Line Notation L66J
Evaluation B

$C_{10}H_8$ (c) 33SOU/BRI
Naphthalene
Heat Capacity 294.68 K, C_p = 39.55 cal·mol^{-1}·K^{-1}
165.48 J·mol^{-1}·K^{-1}
Temperature range 15–295 K. Value is unsmoothed experimental datum.
Entropy 298.15 K, S = 38.92 cal·mol^{-1}·K^{-1}
162.84 J·mol^{-1}·K^{-1}
Molecular Weight 128.1732
Wiswesser Line Notation L66J
Evaluation A

$C_{10}H_8$ (c) 34PEA/TAN
Naphthalene
Heat Capacity 297.6 K, C_p = 40.17 cal·mol^{-1}·K^{-1}
168.07 J·mol^{-1}·K^{-1}
Temperature range 94–298 K. Value is unsmoothed experimental datum.
Entropy 298.15 K, S = 39.88 cal·mol^{-1}·K^{-1}
166.86 J·mol^{-1}·K^{-1}
Extrapolation below 90 K, 13.94 cal·mol^{-1}K^{-1}
Molecular Weight 128.1732
Wiswesser Line Notation L66J
Evaluation B(C_p),C(S)

$C_{10}H_8$ (c) 38HIC
Naphthalene
Heat Capacity 301.58 K, C_p = 40.18 cal·mol^{-1}·K^{-1}
168.11 J·mol^{-1}·K^{-1}
Temperature range 58–304 K. Value is unsmoothed experimental datum.
Molecular Weight 128.1732
Wiswesser Line Notation L66J
Evaluation A

$C_{10}H_8$ (c) 41SCH
Naphthalene
Heat Capacity 298.1 K, C_p = 38.6 cal·mol^{-1}·K^{-1}
161.5 J·mol^{-1}·K^{-1}
Temperature range 22 to 200 °C, equations only, in t°C.
$C_p(c) = 0.2595 + 0.001672t$ cal g^{-1}C^{-1} (22 to 80°C);
$C_p(liq) = 0.3360 + 0.0008180t$ cal g^{-1}C^{-1}(80 to 200°C).
Phase Changes
c/liq 353.4 K, ΔH = 4550 cal·mol^{-1}
19040 J·mol^{-1}
ΔS = 12.9 cal·mol^{-1}·K^{-1}
53.9 J·mol^{-1}·K^{-1}
Molecular Weight 128.1732
Wiswesser Line Notation L66J
Evaluation C

Table of Heat Capacities, Entropies, and Phase Transition Properties — Continued

$C_{10}H_8$ (c) 44EIB
Naphthalene
Heat Capacity 298.1 K, C_p = 46.8 cal·mol^{-1}·K^{-1}
195.8 J·mol^{-1}·K^{-1}
Temperature range 30 to 200 °C, equations only in t°C. C_p(c) = 0.365 cal g^{-1}C^{-1} (30 to 80°C); C_p(liq) = 0.329 + 00.000824t cal g^{-1}C^{-1}(80 to 200°C).
Phase Changes
c/liq 353.4 K, ΔH = 4490 cal·mol^{-1}
18790 J·mol^{-1}
ΔS = 12.7 cal·mol^{-1}·K^{-1}
53.2 J·mol^{-1}·K^{-1}
Molecular Weight 128.1732
Wiswesser Line Notation L66J
Evaluation C

$C_{10}H_8$ (c) 50UEB/ORT
Naphthalene
Heat Capacity 298.15 K, C_p = 37.3 cal·mol^{-1}·K^{-1}
156.1 J·mol^{-1}·K^{-1}
Temperature range 293–368 K. Equation only.
Phase Changes
c/liq 353 K, ΔH = 4490 cal·mol^{-1}
18785 J·mol^{-1}
ΔS = 12.7 cal·mol^{-1}·K^{-1}
53.2 J·mol^{-1}·K^{-1}
Molecular Weight 128.1732
Wiswesser Line Notation L66J
Evaluation C

$C_{10}H_8$ (c) 57MCC/FIN
Naphthalene
Heat Capacity 298.15 K, C_p = 39.60 cal·mol^{-1}·K^{-1}
165.69 J·mol^{-1}·K^{-1}
Temperature range 10–370 K
Entropy 298.15 K, S = 40.01 cal·mol^{-1}·K^{-1}
167.40 J·mol^{-1}·K^{-1}
Phase Changes
c/liq 353.43 K, ΔH = 4536 cal·mol^{-1}
18226 J·mol^{-1}
ΔS = 12.32 cal·mol^{-1}·K^{-1}
51.57 J·mol^{-1}·K^{-1}
Molecular Weight 128.1732
Wiswesser Line Notation L66J
Evaluation A

$C_{10}H_8$ (c) 64DAV
Naphthalene
Heat Capacity 330 K, C_p = 51 cal·mol^{-1}·K^{-1}
213 J·mol^{-1}·K^{-1}
Temperature range 298–353 K. Mean value.
Temperature range uncertain.
Phase Changes
c/liq 353 K, ΔH = 4600 cal·mol^{-1}
19250 J·mol^{-1}
ΔS = 13 cal·mol^{-1}·K^{-1}
55 J·mol^{-1}·K^{-1}
Temperature not measured.
Molecular Weight 128.1732
Wiswesser Line Notation L66J
Evaluation D

$C_{10}H_8$ (c) 64RAS/BAS
Naphthalene
Heat Capacity 342 K, C_p = 45.0 cal·mol^{-1}·K^{-1}
188.4 J·mol^{-1}·K^{-1}
Temperatures 342, 384 K
Phase Changes
c/liq 353.5 K, ΔH = 4565 cal·mol^{-1}
19100 J·mol^{-1}
ΔS = 12.9 cal·mol^{-1}·K^{-1}
54.0 J·mol^{-1}·K^{-1}
Molecular Weight 128.1732
Wiswesser Line Notation L66J
Evaluation C

$C_{10}H_8$ (c) 80AND/CON
Naphthalene
Phase Changes
c,I/liq 353.376 K, ΔH = 4552 cal·mol^{-1}
19046 J·mol^{-1}
ΔS = 12.88 cal·mol^{-1}·K^{-1}
53.90 J·mol^{-1}·K^{-1}
Molecular Weight 128.1732
Wiswesser Line Notation L66J
Evaluation A

$C_{10}H_8$ (c) 80RAD/RAD
Naphthalene
Phase Changes
c,I/liq 353.8 K, ΔH = 4541 cal·mol^{-1}
19000 J·mol^{-1}
ΔS = 12.83 cal·mol^{-1}·K^{-1}
53.70 J·mol^{-1}·K^{-1}
Molecular Weight 128.1732
Wiswesser Line Notation L66J
Evaluation A

$C_{10}H_8O$ (c) 26AND/LYN
α–Naphthol; 1–Hydroxynaphthalene
Heat Capacity 298 K, C_p = 39.9 cal·mol^{-1}·K^{-1}
166.9 J·mol^{-1}·K^{-1}
Temperature range 22 to 180°C
Phase Changes
c/liq 368.2 K, ΔH = 5610 cal·mol^{-1}
23470 J·mol^{-1}
ΔS = 15.2 cal·mol^{-1}·K^{-1}
63.6 J·mol^{-1}·K^{-1}
Molecular Weight 144.1726
Wiswesser Line Notation L66J BQ
Evaluation C

$C_{10}H_8O$ (liq) 67PAC
α–Naphthol; 1–Hydroxynaphthalene
Heat Capacity 393 K, C_p = 68 cal·mol^{-1}·K^{-1}
285 J·mol^{-1}·K^{-1}
One temperature
Phase Changes
c/liq 369 K, ΔH = 5550 cal·mol^{-1}
23220 J·mol^{-1}
ΔS = 15.0 cal·mol^{-1}·K^{-1}
62.9 J·mol^{-1}·K^{-1}
Molecular Weight 144.1726
Wiswesser Line Notation L66J BQ
Evaluation C

Table of Heat Capacities, Entropies, and Phase Transition Properties — Continued

$C_{10}H_8O$ (c) 26AND/LYN
β-Naphthol; 2-Hydroxynaphthalene
Heat Capacity 298 K, C_p = 41.3 cal·mol^{-1}·K^{-1}
172.8 J·mol^{-1}·K^{-1}
Temperature range 22 to 205°C
Phase Changes
c/liq 393.6 K, ΔH = 4490 cal·mol^{-1}
18790 J·mol^{-1}
ΔS = 13.4 cal·mol^{-1}·K^{-1}
55.9 J·mol^{-1}·K^{-1}
Molecular Weight 144.1726
Wiswesser Line Notation L66J CQ
Evaluation C

$C_{10}H_9N$ (c) 40CAM/CAM
β-Naphthylamine; 2-Aminonaphthalene
Heat Capacity 293 K, C_p = 30.1 cal·mol^{-1}·K^{-1}
125.9 J·mol^{-1}·K^{-1}
One temperature
Molecular Weight 143.1878
Wiswesser Line Notation L66J CZ
Evaluation C

$C_{10}H_{10}Co$ (c) 78RAB/NIS
Cobaltocene
Heat Capacity 298.15 K, C_p = 47.2 cal·mol^{-1}·K^{-1}
197.5 J·mol^{-1}·K^{-1}
Temperature range 5–300 K
Entropy 298.15 K, S = 56.4 cal·mol^{-1}·K^{-1}
236.0 J·mol^{-1}·K^{-1}
Phase Changes
c,II/c,I 93.5 K, ΔH = 56.9 cal·mol^{-1}
238 J·mol^{-1}
ΔS = 0.65 cal·mol^{-1}·K^{-1}
2.72 J·mol^{-1}·K^{-1}
Lambda transition between 70 to 120 K with a maximum at 93.5 K.
Molecular Weight 189.1223
Wiswesser Line Notation L5ØJ Ø–CO– –ØL5ØJ
Evaluation A

$C_{10}H_{10}Cr$ (c) 78RAB/NIS
Chromocene
Heat Capacity 298.15 K, C_p = 47.6 cal·mol^{-1}·K^{-1}
199.2 J·mol^{-1}·K^{-1}
Temperature range 5–300 K
Entropy 298.15 K, S = 56.4 cal·mol^{-1}·K^{-1}
236.0 J·mol^{-1}·K^{-1}
Phase Changes
c,II/c,I 99.7 K, ΔH = 63.3 cal·mol^{-1}
265 J·mol^{-1}
ΔS = 0.61 cal·mol^{-1}·K^{-1}
2.55 J·mol^{-1}·K^{-1}
Lambda transition between 75 to 140 K with a maximum at 99.7 K.
Molecular Weight 182.1850
Wiswesser Line Notation L5ØJ Ø–CR– –ØL5ØJ
Evaluation A

$C_{10}H_{10}Fe$ (c) 60EDW/KIN
Ferrocene
Heat Capacity
Temperature range 125–200 K. Heat capacity measured and given graphically in region of transition.
Phase Changes
c,II/c,I 169 K, ΔH = 204 cal·mol^{-1}
854 J·mol^{-1}
ΔS = 1.27 cal·mol^{-1}·K^{-1}
5.31 J·mol^{-1}·K^{-1}
Lambda transition at 163.9 K with a secondary transition at 169 K. Data given for overall transition.
Molecular Weight 186.0360
Wiswesser Line Notation L5ØJ Ø–FE– –ØL5ØJ
Evaluation A

$C_{10}H_{10}Fe$ (c) 62EDW/KIN
Ferrocene
Heat Capacity 298.16 K, C_p = 46.82 cal·mol^{-1}·K^{-1}
195.9 J·mol^{-1}·K^{-1}
Temperature range 0–300 K. Debye function used to evaluate heat capacity between 0 and 17 K.
Entropy 298.15 K, S = 51.68 cal·mol^{-1}·K^{-1}
216.2 J·mol^{-1}·K^{-1}
Phase Changes
Maximum transition between 163 and 165 K (163.9 K); secondary transition 171 K (graphical estimate).
Molecular Weight 186.0360
Wiswesser Line Notation L5ØJ Ø–FE– –ØL5ØJ
Evaluation A

$C_{10}H_{10}Fe$ (c) 76AZO/CAL
Ferrocene
Heat Capacity 200 K, C_p = 31.3 cal·mol^{-1}·K^{-1}
131 J·mol^{-1}·K^{-1}
Temperature range 120–200 K
Phase Changes
Maximum peak at 164 K; secondary peak at 169 K; c,III/c,II triclinic/monoclinic transition.
Molecular Weight 186.0360
Wiswesser Line Notation L5ØJ Ø–FE– –ØL5ØJ
Evaluation B

$C_{10}H_{10}Fe$ (c) 81OGA/SOR
Ferrocene
Heat Capacity 298.15 K, C_p = 45.31 cal·mol^{-1}·K^{-1}
189.56 J·mol^{-1}·K^{-1}
Temperature range 13–300 K
Entropy 298.15 K, S = 50.63 cal·mol^{-1}·K^{-1}
211.85 J·mol^{-1}·K^{-1}
Phase Changes
c,III/c,II 163.9 K, ΔH = 215 cal·mol^{-1}
900 J·mol^{-1}
ΔS = 1.31 cal·mol^{-1}·K^{-1}
5.5 J·mol^{-1}·K^{-1}
Lambda transition with a subsidiary C_p maximum at 169 K between metastable LT and undercooled HT phases.
c,II/c,I 242 K, ΔH = 990.7 cal·mol^{-1}
4145 J·mol^{-1}
ΔS = 4.094 cal·mol^{-1}·K^{-1}
17.13 J·mol^{-1}·K^{-1}
Phase transition between stable LT and stable HT phases.
Molecular Weight 186.0360
Wiswesser Line Notation L5ØJ Ø–FE– –ØL5ØJ
Evaluation A

Table of Heat Capacities, Entropies, and Phase Transition Properties — Continued

$C_{10}H_{10}Fe$ (c) 81TOM/CUR
Ferrocene
Heat Capacity 298 K, C_p = 46.0 cal·mol^{-1}·K^{-1}
192.5 J·mol^{-1}·K^{-1}
Temperature range 293–393 K. Equation given.
Phase Changes
c/liq 447.0 K
Molecular Weight 186.0360
Wiswesser Line Notation L5ØJ Ø–FE– –ØL5ØJ
Evaluation B

$C_{10}H_{10}O_4$ (liq) 69RAB/MAR
Dimethyl o–phthalate
Heat Capacity 300 K, C_p = 72.60 cal·mol^{-1}·K^{-1}
303.76 J·mol^{-1}·K^{-1}
Temperature range 80–360 K
Entropy 300 K, S = 87.35 cal·mol^{-1}·K^{-1}
365.47 J·mol^{-1}·K^{-1}
Phase Changes
c/liq 274.18 K, ΔH = 4050 cal·mol^{-1}
16945 J·mol^{-1}
ΔS = 14.8 cal·mol^{-1}·K^{-1}
61.8 J·mol^{-1}·K^{-1}
Molecular Weight 194.1866
Wiswesser Line Notation 1OVR BVO1
Evaluation C
Glass transition temp. 192K; ΔH = 126 cal·mol^{-1}, ΔS = 0.63 cal·mol^{-1}K^{-1}

$C_{10}H_{10}O_4$ (liq) 78MIL
Dimethyl o–phthalate
Heat Capacity 298.15 K, C_p = 73.92 cal·mol^{-1}·K^{-1}
309.3 J·mol^{-1}·K^{-1}
Temperature range 250–370 K. Data graphically and by equation only.
Molecular Weight 194.1866
Wiswesser Line Notation 1OVR BVO1
Evaluation B

$C_{10}H_{10}O_4$ (c) 56SMI/DOL
Dimethyl terephthalate; Dimethyl p–phthalate
Heat Capacity 353 K, C_p = 66.0 cal·mol^{-1}·K^{-1}
276.1 J·mol^{-1}·K^{-1}
Temperature range 80 to 190°C. Equation only.
Phase Changes
c/liq 413.8 K, ΔH = 7670 cal·mol^{-1}
32100 J·mol^{-1}
ΔS = 18.5 cal·mol^{-1}·K^{-1}
77.6 J·mol^{-1}·K^{-1}
Molecular Weight 194.1866
Wiswesser Line Notation 1OVR DVO1
Evaluation B

$C_{10}H_{10}O_4$ (c) 68ELL/CHR
Dimethyl terephthalate; Dimethyl p–phthalate
Heat Capacity 298.15 K, C_p = 62.4 cal·mol^{-1}·K^{-1}
261.1 J·mol^{-1}·K^{-1}
Temperature range 30 to 200°C
Phase Changes
c/liq 413.79 K, ΔH = 7560 cal·mol^{-1}
31631 J·mol^{-1}
ΔS = 18.27 cal·mol^{-1}·K^{-1}
76.44 J·mol^{-1}·K^{-1}
Molecular Weight 194.1866
Wiswesser Line Notation 1OVR DVO1
Evaluation B

$C_{10}H_{10}Mn$ (c) 78RAB/NIS
Manganocene
Heat Capacity 298.15 K, C_p = 49.8 cal·mol^{-1}·K^{-1}
208.4 J·mol^{-1}·K^{-1}
Temperature range 5–300 K
Entropy 298.15 K, S = 60.1 cal·mol^{-1}·K^{-1}
251.5 J·mol^{-1}·K^{-1}
Molecular Weight 185.1270
Wiswesser Line Notation L5ØJ Ø–MN– –ØL5ØJ
Evaluation A

$C_{10}H_{10}Ni$ (c) 78RAB/NIS
Nickelocene
Heat Capacity 298.15 K, C_p = 49.1 cal·mol^{-1}·K^{-1}
205.4 J·mol^{-1}·K^{-1}
Temperature range 5–300 K
Entropy 298.15 K, S = 60.5 cal·mol^{-1}·K^{-1}
253.1 J·mol^{-1}·K^{-1}
Phase Changes
c,II/c,I 100–190 K, ΔH = 43.5 cal·mol^{-1}
182 J·mol^{-1}
ΔS = 0.32 cal·mol^{-1}·K^{-1}
1.34 J·mol^{-1}·K^{-1}
Lambda transition over the temperature range 100 to 190 K.
Molecular Weight 188.8890
Wiswesser Line Notation L5ØJ Ø–NI– –ØL5ØJ
Evaluation A

$C_{10}H_{10}Ni$ (c) 76AZO/CAL
Nickelocene
Heat Capacity 298 K, C_p = 40.9 cal·mol^{-1}·K^{-1}
171 J·mol^{-1}·K^{-1}
Temperature range 130–300 K
Phase Changes
c,II/c,I 170–240 K, ΔS = 1.3 cal·mol^{-1}·K^{-1}
5.3 J·mol^{-1}·K^{-1}
No peak is observed on heat capacity curve, but a deviation from normal variation occurs between 170 to 240 K.
Molecular Weight 188.8890
Wiswesser Line Notation L5ØJ Ø–NI– –ØL5ØJ
Evaluation C

$C_{10}H_{10}V$ (c) 78RAB/NIS
Vanadocene
Heat Capacity 298.15 K, C_p = 48.9 cal·mol^{-1}·K^{-1}
204.6 J·mol^{-1}·K^{-1}
Temperature range 5–300 K
Entropy 298.15 K, S = 57.4 cal·mol^{-1}·K^{-1}
240.2 J·mol^{-1}·K^{-1}
Phase Changes
c,II/c,I 120–200 K, ΔH = 46.8 cal·mol^{-1}
196 J·mol^{-1}
ΔS = 0.29 cal·mol^{-1}·K^{-1}
1.21 J·mol^{-1}·K^{-1}
Lambda transition between 120 to 200 K.
Molecular Weight 181.1305
Wiswesser Line Notation L5ØJ Ø–VA– –ØL5ØJ
Evaluation A

Table of Heat Capacities, Entropies, and Phase Transition Properties — Continued

$C_{10}H_{10}V$ (c) 80CAL/BER
Vanadocene
Heat Capacity 298 K, C_p = 48.6 cal·mol⁻¹·K⁻¹ 203.3 J·mol⁻¹·K⁻¹
Temperature range 100–300 K, C_p between 235–300 K given by equation.
Phase Changes
c,II/c,I 120–235 K, ΔH = 196 cal·mol⁻¹ 820 J·mol⁻¹
ΔS = 1.11 cal·mol⁻¹·K⁻¹ 4.64 J·mol⁻¹·K⁻¹
Lambda transition between 120 to 235 K.
Molecular Weight 181.1305
Wiswesser Line Notation L5ØJ Ø–VA– –ØL5ØJ
Evaluation B

$C_{10}H_{12}$ (liq) 57MCC/FIN
1,2,3,4–Tetrahydronaphthalene
Heat Capacity 298.15 K, C_p = 51.97 cal·mol⁻¹·K⁻¹ 217.44 J·mol⁻¹·K⁻¹
Temperature range 10–320 K
Entropy 298.15 K, S = 60.10 cal·mol⁻¹·K⁻¹ 251.46 J·mol⁻¹·K⁻¹
Phase Changes
c/liq 237.36 K, ΔH = 2975 cal·mol⁻¹ 12447 J·mol⁻¹
ΔS = 12.53 cal·mol⁻¹·K⁻¹ 52.44 J·mol⁻¹·K⁻¹
Molecular Weight 132.2048
Wiswesser Line Notation L66&TJ
Evaluation A

$C_{10}H_{12}O_4$ (c) 65SIL/DAU
2–Monobenzoylglycerol
Heat Capacity 298 K, C_p = 56.5 cal·mol⁻¹·K⁻¹ 236.4 J·mol⁻¹·K⁻¹
One temperature
Molecular Weight 196.2024
Wiswesser Line Notation Q1Y1QOVR
Evaluation B

$C_{10}H_{12}O_4$ (c) 65SIL/DAU
1–Monobenzoylglycerol
Heat Capacity 298 K, C_p = 57.1 cal·mol⁻¹·K⁻¹ 238.9 J·mol⁻¹·K⁻¹
One temperature. β_L form
Molecular Weight 196.2024
Wiswesser Line Notation Q1YQ1OVR
Evaluation B

$C_{10}H_{13}NO_2$ (c) 71PRI
Propyl phenylcarbamate
Heat Capacity 298 K, C_p = 63.0 cal·mol⁻¹·K⁻¹ 263.6 J·mol⁻¹·K⁻¹
Temperature range 200–390 K. Complete data deposited VINITI, No. 2713–71, 25 March 1971.
Phase Changes
c/liq 331 K, ΔH = 5038 cal·mol⁻¹ 21079 J·mol⁻¹
ΔS = 15.2 cal·mol⁻¹·K⁻¹ 63.7 J·mol⁻¹·K⁻¹
Molecular Weight 179.2182
Wiswesser Line Notation 3OVMR
Evaluation B

$C_{10}H_{14}$ (liq) 31HUF/PAR
1,2,3,4–Tetramethylbenzene; Prehnitene
Heat Capacity 291.9 K, C_p = 56.4 cal·mol⁻¹·K⁻¹ 236.0 J·mol⁻¹·K⁻¹
Temperature range 91–292 K. Value is unsmoothed experimental datum.
Entropy 298.1 K, S = 69.5 cal·mol⁻¹·K⁻¹ 290.8 J·mol⁻¹·K⁻¹
Extrapolation below 90 K, 18.24 cal·mol⁻¹K⁻¹
Phase Changes
c/liq 265.4 K, ΔH = 2684 cal·mol⁻¹ 11230 J·mol⁻¹
ΔS = 10.1 cal·mol⁻¹·K⁻¹ 42.3 J·mol⁻¹·K⁻¹
Molecular Weight 134.2206
Wiswesser Line Notation 1R B1 C1 D1
Evaluation B(C_p),C(S)

$C_{10}H_{14}$ (liq) 47KUR
1,2,3,4–Tetramethylbenzene; Prehnitene
Heat Capacity 298 K, C_p = 58.4 cal·mol⁻¹·K⁻¹ 244.3 J·mol⁻¹·K⁻¹
Temperature range 12 to 198 °C, mean C_p, four temperatures.
Molecular Weight 134.2206
Wiswesser Line Notation 1R B1 C1 D1
Evaluation D

$C_{10}H_{14}$ (liq) 31HUF/PAR
1,2,3,5–Tetramethylbenzene; Isodurene
Heat Capacity 297.1 K, C_p = 57.4 cal·mol⁻¹·K⁻¹ 240.2 J·mol⁻¹·K⁻¹
Temperature range 92–297 K. Value is unsmoothed experimental datum.
Entropy 298.1 K, S = 74.1 cal·mol⁻¹·K⁻¹ 310.0 J·mol⁻¹·K⁻¹
Extrapolation below 90 K, 19.25 cal·mol⁻¹K⁻¹
Phase Changes
c,I/liq 248.6 K, ΔH = 3092 cal·mol⁻¹ 12937 J·mol⁻¹
ΔS = 12.4 cal·mol⁻¹·K⁻¹ 52.0 J·mol⁻¹·K⁻¹
Value includes heat effect for solid transition below melting point.
Molecular Weight 134.2206
Wiswesser Line Notation 1R B1 C1 E1
Evaluation B(C_p),C(S)

$C_{10}H_{14}$ (c) 31HUF/PAR
1,2,4,5–Tetramethylbenzene; Durene
Heat Capacity 297.1 K, C_p = 51.4 cal·mol⁻¹·K⁻¹ 215.1 J·mol⁻¹·K⁻¹
Temperature range 92–297.1 K. Value is unsmoothed experimental datum.
Entropy 298.1 K, S = 58.7 cal·mol⁻¹·K⁻¹ 245.6 J·mol⁻¹·K⁻¹
Extrapolation below 90 K, 18.33 cal·mol⁻¹K⁻¹
Molecular Weight 134.2206
Wiswesser Line Notation 1R B1 D1 E1
Evaluation B(C_p),C(S)

$C_{10}H_{14}$ (c) 44EIB
1,2,4,5–Tetramethylbenzene; Durene
Heat Capacity 298.1 K, C_p = 52.6 cal·mol⁻¹·K⁻¹ 220.1 J·mol⁻¹·K⁻¹
Temperature range 25 to 200 °C, equations only in t°C.
C_p(c) = 0.3662 + 0.001033t cal g⁻¹C⁻¹ (25 to 45°C);
C_p(liq) = 0.424 + 0.000589t cal g⁻¹C⁻¹ (79 to 200°C).

Table of Heat Capacities, Entropies, and Phase Transition Properties — Continued

Phase Changes
c/liq 352.4 K, ΔH = 4990 cal·mol^{-1}
20880 J·mol^{-1}
ΔS = 14.2 cal·mol^{-1}·K^{-1}
59.3 J·mol^{-1}·K^{-1}
Molecular Weight 134.2206
Wiswesser Line Notation 1R B1 D1 E1
Evaluation C

$C_{10}H_{14}$ (liq) 47KUR
1,2,4,5–Tetramethylbenzene; Durene
Heat Capacity 353 K, C_p = 65.9 cal·mol^{-1}·K^{-1}
275.7 J·mol^{-1}·K^{-1}
Temperature range 80 to 193 °C, mean C_p, three temperatures.
Molecular Weight 134.2206
Wiswesser Line Notation 1R B1 D1 E1
Evaluation D

$C_{10}H_{14}$ (liq) 30HUF/PAR
tert–Butylbenzene
Heat Capacity 294.3 K, C_p = 56.91 cal·mol^{-1}·K^{-1}
238.11 J·mol^{-1}·K^{-1}
Temperature range 92–294 K. Value is unsmoothed experimental datum.
Entropy 298.1 K, S = 66.6 cal·mol^{-1}·K^{-1}
278.7 J·mol^{-1}·K^{-1}
Extrapolation below 90 K, 16.18 cal·mol^{-1}K^{-1}o
Phase Changes
c/liq 215.0 K, ΔH = 2007 cal·mol^{-1}
8397 J·mol^{-1}
ΔS = 9.33 cal·mol^{-1}·K^{-1}
39.06 J·mol^{-1}·K^{-1}
Molecular Weight 134.2206
Wiswesser Line Notation 1X1&1&R
Evaluation B(C_p),C(S)

$C_{10}H_{14}$ (liq) 31HUF/PAR
p–Cymene; 1–Isopropyl–4–methylbenzene
Heat Capacity 297.1 K, C_p = 56.5 cal·mol^{-1}·K^{-1}
236.4 J·mol^{-1}·K^{-1}
Temperature range 92–297 K. Value is unsmoothed experimental datum.
Entropy 298.1 K, S = 73.3 cal·mol^{-1}·K^{-1}
306.7 J·mol^{-1}·K^{-1}
Extrapolation below 90 K, 19.12 cal·mol^{-1}K^{-1}
Phase Changes
c/liq 204.2 K, ΔH = 2309 cal·mol^{-1}
9661 J·mol^{-1}
ΔS = 11.3 cal·mol^{-1}·K^{-1}
47.3 J·mol^{-1}·K^{-1}
Molecular Weight 134.2206
Wiswesser Line Notation 1Y1&R D1
Evaluation B(C_p),C(S)

$C_{10}H_{14}$ (liq) 47KUR
p–Cymene; 1–Isopropyl–4–methylbenzene
Heat Capacity 298 K, C_p = 57.9 cal·mol^{-1}·K^{-1}
242.3 J·mol^{-1}·K^{-1}
Temperature range 10 to 166 °C, mean C_p, four temperatures.
Molecular Weight 134.2206
Wiswesser Line Notation 1Y1&R D1
Evaluation D

$C_{10}H_{14}$ (liq) 31HUF/PAR
n–Butylbenzene
Heat Capacity 298.2 K, C_p = 57.4 cal·mol^{-1}·K^{-1}
240.2 J·mol^{-1}·K^{-1}
Temperature range 94–298 K. Value is unsmoothed experimental datum.
Entropy 298.1 K, S = 76.8 cal·mol^{-1}·K^{-1}
321.3 J·mol^{-1}·K^{-1}
Extrapolation below 90 K, 18.87 cal·mol^{-1}K^{-1}
Phase Changes
c/liq 184.6 K, ΔH = 2624 cal·mol^{-1}
10979 J·mol^{-1}
ΔS = 14.2 cal·mol^{-1}·K^{-1}
59.5 J·mol^{-1}·K^{-1}
Molecular Weight 134.2206
Wiswesser Line Notation 4R
Evaluation B(C_p),C(S)

$C_{10}H_{14}$ (liq) 65MES/TOD
n–Butylbenzene
Heat Capacity 298.15 K, C_p = 58.16 cal·mol^{-1}·K^{-1}
243.34 J·mol^{-1}·K^{-1}
Temperature range 10–380 K
Entropy 298.15 K, S = 76.77 cal·mol^{-1}·K^{-1}
321.21 J·mol^{-1}·K^{-1}
Phase Changes
c,II/liq 185.14 K, ΔH = 2691 cal·mol^{-1}
11259 J·mol^{-1}
ΔS = 14.53 cal·mol^{-1}·K^{-1}
60.81 J·mol^{-1}·K^{-1}
Metastable crystals
c,I/liq 185.30 K, ΔH = 2682 cal·mol^{-1}
11221 J·mol^{-1}
ΔS = 14.47 cal·mol^{-1}·K^{-1}
60.56 J·mol^{-1}·K^{-1}
Molecular Weight 134.2206
Wiswesser Line Notation 4R
Evaluation A

$C_{10}H_{14}$ (liq) 73GOO 2
n–Butylbenzene
Heat Capacity 298.15 K, C_p = 58 cal·mol^{-1}·K^{-1}
241 J·mol^{-1}·K^{-1}
One temperature
Molecular Weight 134.2206
Wiswesser Line Notation 4R
Evaluation B

$C_{10}H_{14}CuO_4$ (c) 81TEG/FER
Copper acetylacetonate
Heat Capacity 298 K, C_p = 51.0 cal·mol^{-1}·K^{-1}
213.4 J·mol^{-1}·K^{-1}
Temperature range 4.2–450 K
Entropy 298 K, S = 89.1 cal·mol^{-1}·K^{-1}
372.8 J·mol^{-1}·K^{-1}
Molecular Weight 261.7642
Wiswesser Line Notation D6O–CR–O ADJ D1 F1 B–& BD6O–CR–O ADJ D1 F1
Evaluation B

$C_{10}H_{15}N$ (liq) 34KOL/UDO 2
N,N–Diethylaniline
Heat Capacity 302.3 K, C_p = 65.6 cal·mol^{-1}·K^{-1}
274.5 J·mol^{-1}·K^{-1}
One temperature
Molecular Weight 149.2352
Wiswesser Line Notation 2N2&R
Evaluation C

Table of Heat Capacities, Entropies, and Phase Transition Properties — Continued

$C_{10}H_{15}N$ (liq) 34KOL/UDO
N,N–Diethylaniline
Heat Capacity 302.0 K, C_p = 65.6 cal·mol^{-1}·K^{-1}
274.5 J·mol^{-1}·K^{-1}
One temperature
Molecular Weight 149.2352
Wiswesser Line Notation 2N2&R
Evaluation C

$C_{10}H_{15}NO$ (c) 77MEI/BLO
Carvoxime(L)
Heat Capacity 298.15 K, C_p = 61.1 cal·mol^{-1}·K^{-1}
255.7 J·mol^{-1}·K^{-1}
Temperature range 160–385 K
Phase Changes
c/liq 346.5 K, ΔH = 5425 cal·mol^{-1}
22700 J·mol^{-1}
ΔS = 15.7 cal·mol^{-1}·K^{-1}
65.5 J·mol^{-1}·K^{-1}
Molecular Weight 165.2346
Wiswesser Line Notation L6Y BUTJ AUNQ B1 EY1&U1 –L
Evaluation B

$C_{10}H_{15}NO$ (c) 77MEI/BLO
Carvoxime(DL)
Heat Capacity 298.15 K, C_p = 61.0 cal·mol^{-1}·K^{-1}
255.2 J·mol^{-1}·K^{-1}
Temperature range 160–385 K
Phase Changes
c/liq 365.1 K, ΔH = 4068 cal·mol^{-1}
17020 J·mol^{-1}
ΔS = 11.1 cal·mol^{-1}·K^{-1}
46.6 J·mol^{-1}·K^{-1}
Molecular Weight 165.2346
Wiswesser Line Notation L6Y BUTJ AUNQ B1 EY1&U1
Evaluation B

$C_{10}H_{16}$ (liq) 33KOL/UDO
Sabinene
Heat Capacity 297.0 K, C_p = 60.3 cal·mol^{-1}·K^{-1}
252.3 J·mol^{-1}·K^{-1}
One temperature
Molecular Weight 136.2364
Wiswesser Line Notation L35 DYTJ AY1&1 DU1
Evaluation C

$C_{10}H_{16}$ (liq) 34KOL/UDO 2
Sabinene
Heat Capacity 288.3 K, C_p = 60.3 cal·mol^{-1}·K^{-1}
252.9 J·mol^{-1}·K^{-1}
One temperature
Molecular Weight 136.2364
Wiswesser Line Notation L35 BYTJ AY1&1 DU1
Evaluation C

$C_{10}H_{16}$ (c) 71BOY/SAN
Tricyclo[5.2.1.0^{2,6}]decane
Heat Capacity 329 K, C_p = 57.7 cal·mol^{-1}·K^{-1}
241.4 J·mol^{-1}·K^{-1}
Temperature range 329–390 K, 4 temperatures.
Phase Changes
c/liq 352 K, ΔH = 705 cal·mol^{-1}
2950 J·mol^{-1}
ΔS = 2.0 cal·mol^{-1}·K^{-1}
8.4 J·mol^{-1}·K^{-1}
Molecular Weight 136.2364
Wiswesser Line Notation L556/FH 2AF JTJ
Evaluation C

$C_{10}H_{16}$ (c,I) 60CHA/WES
Adamantane; Tricyclo[3.3.1.1^{3,7}]decane
Heat Capacity 298.15 K, C_p = 45.35 cal·mol^{-1}·K^{-1}
189.74 J·mol^{-1}·K^{-1}
Temperature range 5–350 K
Entropy 298.15 K, S = 46.80 cal·mol^{-1}·K^{-1}
195.83 J·mol^{-1}·K^{-1}
Phase Changes
c,II/c,I 208.62 K, ΔH = 807 cal·mol^{-1}
3376 J·mol^{-1}
ΔS = 3.87 cal·mol^{-1}·K^{-1}
16.18 J·mol^{-1}·K^{-1}
Molecular Weight 136.2364
Wiswesser Line Notation L66 B6/B–H/DI A B– C 1B ITJ
Evaluation A

$C_{10}H_{16}$ (c,I) 61WES
Adamantane; Tricyclo[3.3.1.1^{3,7}]decane
Heat Capacity 298.15 K, C_p = 45.35 cal·mol^{-1}·K^{-1}
189.74 J·mol^{-1}·K^{-1}
Temperature range 5–350 K. Only values at 298.15 K given.
Entropy 298.15 K, S = 46.80 cal·mol^{-1}·K^{-1}
195.83 J·mol^{-1}·K^{-1}
Phase Changes
c,II/c,I 208.62 K, ΔH = 807 cal·mol^{-1}
3376 J·mol^{-1}
ΔS = 3.87 cal·mol^{-1}·K^{-1}
16.18 J·mol^{-1}·K^{-1}
Molecular Weight 136.2364
Wiswesser Line Notation L66 B6/B–H/DI A B– C 1B ITJ
Evaluation A
Details reported in other papers.

$C_{10}H_{16}$ (liq) 33KOL/UDO
Limonene
Heat Capacity 293.4 K, C_p = 59.6 cal·mol^{-1}·K^{-1}
249.4 J·mol^{-1}·K^{-1}
One temperature
Molecular Weight 136.2364
Wiswesser Line Notation L6UTJ A1 DY1&U1
Evaluation C

$C_{10}H_{16}$ (liq) 34KOL/UDO 2
Limonene
Heat Capacity 293.3 K, C_p = 59.6 cal·mol^{-1}·K^{-1}
249.4 J·mol^{-1}·K^{-1}
One temperature.
Molecular Weight 136.2364
Wiswesser Line Notation L6UTJ A1 DY1&U1
Evaluation C

$C_{10}H_{16}O$ (c) 31FRA
Camphor
Heat Capacity 298.1 K, C_p = 64.8 cal·mol^{-1}·K^{-1}
271.2 J·mol^{-1}·K^{-1}
Temperature range 307–483 K
Phase Changes
c/liq 451.5 K, ΔH = 1630 cal·mol^{-1}
6820 J·mol^{-1}
ΔS = 3.61 cal·mol^{-1}·K^{-1}
15.10 J·mol^{-1}·K^{-1}
Molecular Weight 152.2358
Wiswesser Line Notation L55 A CVTJ A1 A1 B1
Evaluation C
Synthetic camphor.

Table of Heat Capacities, Entropies, and Phase Transition Properties — Continued

$C_{10}H_{16}O$ (c,II) 35WHI/MOR
Camphor(D)
Heat Capacity 259 K, C_p = 58 cal·mol⁻¹·K⁻¹
243 J·mol⁻¹·K⁻¹
Temperature range 213–259 K
Phase Changes
c,III/c,II 243 K, ΔH = 1860 cal·mol⁻¹
7780 J·mol⁻¹
ΔS = 7.7 cal·mol⁻¹·K⁻¹
32.0 J·mol⁻¹·K⁻¹
Molecular Weight 152.2358
Wiswesser Line Notation L55 A CVTJ A1 A1 B1 -D
Evaluation C

$C_{10}H_{16}O$ (c,I) 53SCH
Camphor(D)
Heat Capacity C_p data given graphically only.
Temperature range 238–247 K
Phase Changes
c,II/c,I 243.9 K, ΔH = 1670 cal·mol⁻¹
6990 J·mol⁻¹
ΔS = 6.8 cal·mol⁻¹·K⁻¹
28.7 J·mol⁻¹·K⁻¹
Molecular Weight 152.2358
Wiswesser Line Notation L55 A CVTJ A1 A1 B1 -D
Evaluation C

$C_{10}H_{16}O$ (c,I) 53SCH
Camphor(DL)
Heat Capacity C_p data given graphically only.
Temperature range -180 to 180 °C
Phase Changes
c,II/c,I 210 K, ΔH = 200 cal·mol⁻¹
840 J·mol⁻¹
ΔS = 1.0 cal·mol⁻¹·K⁻¹
4.0 J·mol⁻¹·K⁻¹
Molecular Weight 152.2358
Wiswesser Line Notation L55 A CVTJ A1 A1 B1
Evaluation C

$C_{10}H_{16}O$ (liq) 33KOL/UDO
Pulegone
Heat Capacity 293.3 K, C_p = 65.7 cal·mol⁻¹·K⁻¹
274.9 J·mol⁻¹·K⁻¹
One temperature
Molecular Weight 152.2358
Wiswesser Line Notation L6VYTJ BUY1&1 E1
Evaluation C

$C_{10}H_{16}O$ (liq) 34KOL/UDO 2
Pulegone
Heat Capacity 293.3 K, C_p = 65.7 cal·mol⁻¹·K⁻¹
274.9 J·mol⁻¹·K⁻¹
One temperature
Molecular Weight 152.2358
Wiswesser Line Notation L6VYTJ BUY1&1 E1
Evaluation C

$C_{10}H_{16}S_4$ (c) 62CHA/WES
1,3,5,7–Tetramethyl–2,4,6,8–tetrathiaadamantane
Heat Capacity 298.15 K, C_p = 70.71 cal·mol⁻¹·K⁻¹
295.85 J·mol⁻¹·K⁻¹
Temperature range 5–350 K
Entropy 298.15 K, S = 71.90 cal·mol⁻¹·K⁻¹
300.83 J·mol⁻¹·K⁻¹
Molecular Weight 264.4764
Wiswesser Line Notation T66 B6/B–H/DI A B– C 1B I AS B–S CS ESTJ B1 D1 F1 H1
Evaluation A

$C_{10}H_{18}$ (liq) 62GOL/BEL
Pinane
Heat Capacity 311 K, C_p = 55.3 cal·mol⁻¹·K⁻¹
231.4 J·mol⁻¹·K⁻¹
Temperatures 100, 200, 300°F
Molecular Weight 138.2522
Wiswesser Line Notation L46 ATJ A1 A1 E1
Evaluation C

$C_{10}H_{18}$ (liq) 63GUD/CAM
Pinane
Heat Capacity 313 K, C_p = 55.4 cal·mol⁻¹·K⁻¹
231.8 J·mol⁻¹·K⁻¹
Temperature range 313–523 K
Molecular Weight 138.2522
Wiswesser Line Notation L46 ATJ A1 A1 E1
Evaluation C

$C_{10}H_{18}$ (liq) 63GUD/CAM
Methylhydroindan
Heat Capacity 313 K, C_p = 62.3 cal·mol⁻¹·K⁻¹
260.7 J·mol⁻¹·K⁻¹
Temperature range 313–423 K
Molecular Weight 138.2522
Wiswesser Line Notation L56TJ X1
Evaluation C

$C_{10}H_{18}$ (liq) 70CHA/MCC
cis–Bicyclo[5.3.0]decane
Heat Capacity 377 K, C_p = 74.4 cal·mol⁻¹·K⁻¹
311.3 J·mol⁻¹·K⁻¹
One temperature
Molecular Weight 138.2522
Wiswesser Line Notation L57TJ –C
Evaluation B

$C_{10}H_{18}$ (liq) 62GOL/BEL
Bicyclopentyl
Heat Capacity 311 K, C_p = 54.9 cal·mol⁻¹·K⁻¹
229.7 J·mol⁻¹·K⁻¹
Temperatures 100, 200, 300°F
Molecular Weight 138.2522
Wiswesser Line Notation L5TJ A– AL5TJ
Evaluation C

$C_{10}H_{18}$ (liq) 76GOO/LEE
Bicyclopentyl
Heat Capacity 298.15 K, C_p = 57.1 cal·mol⁻¹·K⁻¹
238.9 J·mol⁻¹·K⁻¹
One temperature
Molecular Weight 138.2522
Wiswesser Line Notation L5TJ A– AL5TJ
Evaluation B

$C_{10}H_{18}$ (liq) 62GOL/BEL
Decahydronaphthalene; Decalin
Heat Capacity 311 K, C_p = 52.4 cal·mol⁻¹·K⁻¹
219.2 J·mol⁻¹·K⁻¹
Temperatures 100, 200, 300°F
Molecular Weight 138.2522
Wiswesser Line Notation L66TJ
Evaluation D

Table of Heat Capacities, Entropies, and Phase Transition Properties — Continued

$C_{10}H_{18}$ (liq) 53SEY
cis–Decahydronaphthalene; cis–Decalin
Heat Capacity 298 K, C_p = 55.81 cal·mol^{-1}·K^{-1}
233.51 J·mol^{-1}·K^{-1}
Temperature range 293–343 K
Phase Changes
λ-type transition in liquid state at 323 K.
Molecular Weight 138.2522
Wiswesser Line Notation L66TJ -C
Evaluation B

$C_{10}H_{18}$ (liq) 57MCC/FIN
cis–Decahydronaphthalene; cis–Decalin
Heat Capacity 298.15 K, C_p = 55.45 cal·mol^{-1}·K^{-1}
232.00 J·mol^{-1}·K^{-1}
Temperature range 10–350 K
Entropy 298.15 K, S = 63.34 cal·mol^{-1}·K^{-1}
265.01 J·mol^{-1}·K^{-1}
Phase Changes
c/liq 242.78 K, ΔH = 3445 cal·mol^{-1}
14414 J·mol^{-1}
ΔS = 14.19 cal·mol^{-1}·K^{-1}
59.37 J·mol^{-1}·K^{-1}
Molecular Weight 138.2522
Wiswesser Line Notation L66TJ -C
Evaluation A

$C_{10}H_{18}$ (liq) 63GUD/CAM
cis–Decahydronaphthalene; cis–Decalin
Heat Capacity 313 K, C_p = 60.0 cal·mol^{-1}·K^{-1}
251.0 J·mol^{-1}·K^{-1}
Temperature range 313–423 K
Molecular Weight 138.2522
Wiswesser Line Notation L66TJ -C
Evaluation C

$C_{10}H_{18}$ (liq) 53SEY
trans–Decahydronaphthalene; trans–Decalin
Heat Capacity 298 K, C_p = 54.22 cal·mol^{-1}·K^{-1}
226.86 J·mol^{-1}·K^{-1}
Temperature range 293–413 K
Molecular Weight 138.2522
Wiswesser Line Notation L66TJ -T
Evaluation B

$C_{10}H_{18}$ (liq) 57MCC/FIN
trans–Decahydronaphthalene; trans–Decalin
Heat Capacity 298.15 K, C_p = 54.61 cal·mol^{-1}·K^{-1}
228.49 J·mol^{-1}·K^{-1}
Temperature range 10–350 K
Entropy 298.15 K, S = 63.32 cal·mol^{-1}·K^{-1}
264.93 J·mol^{-1}·K^{-1}
Phase Changes
c,II/c,I 216.1 K, ΔH = 510.4 cal·mol^{-1}
2135.5 J·mol^{-1}
ΔS = 2.36 cal·mol^{-1}·K^{-1}
9.88 J·mol^{-1}·K^{-1}
c,I/liq 230.18 K, ΔH = 2268 cal·mol^{-1}
9489 J·mol^{-1}
ΔS = 9.85 cal·mol^{-1}·K^{-1}
41.22 J·mol^{-1}·K^{-1}
Molecular Weight 138.2522
Wiswesser Line Notation L66TJ -T
Evaluation A

$C_{10}H_{18}$ (liq) 63GUD/CAM
trans–Decahydronaphthalene; trans–Decalin
Heat Capacity 313 K, C_p = 59.9 cal·mol^{-1}·K^{-1}
250.6 J·mol^{-1}·K^{-1}
Temperature range 313–423 K
Molecular Weight 138.2522
Wiswesser Line Notation L66TJ -T
Evaluation C

$C_{10}H_{18}O$ (liq) 33KOL/UDO
2,6–Dimethylocta–2,7–dien–6–ol; Coriandrol; Linalool
Heat Capacity 293.1 K, C_p = 89.0 cal·mol^{-1}·K^{-1}
372.4 J·mol^{-1}·K^{-1}
One temperature
Molecular Weight 154.2516
Wiswesser Line Notation 1Y1&U3XQ1&1U1
Evaluation C

$C_{10}H_{18}O$ (liq) 34KOL/UDO 2
2,6–Dimethylocta–2,7–dien–6–ol; Coriandrol; Linalool
Heat Capacity 293.1 K, C_p = 89.0 cal·mol^{-1}·K^{-1}
372.4 J·mol^{-1}·K^{-1}
One temperature
Molecular Weight 154.2516
Wiswesser Line Notation 1Y1&U3XQ1&1U1
Evaluation C

$C_{10}H_{20}$ (liq) 57MCC/FIN 2
1–Decene
Heat Capacity 298.15 K, C_p = 71.78 cal·mol^{-1}·K^{-1}
300.83 J·mol^{-1}·K^{-1}
Temperature range 11–360 K
Entropy 298.15 K, S = 101.58 cal·mol^{-1}·K^{-1}
425.01 J·mol^{-1}·K^{-1}
Phase Changes
c,II/c,I 198.3 K, ΔH = 1900 cal·mol^{-1}
7950 J·mol^{-1}
ΔS = 9.58 cal·mol^{-1}·K^{-1}
40.09 J·mol^{-1}·K^{-1}
c,I/liq 206.89 K, ΔH = 3300 cal·mol^{-1}
13807 J·mol^{-1}
ΔS = 15.95 cal·mol^{-1}·K^{-1}
66.74 J·mol^{-1}·K^{-1}
Molecular Weight 140.2680
Wiswesser Line Notation 9U1
Evaluation A

$C_{10}H_{20}$ (liq) 63GUD/CAM
1,4–Diethylcyclohexane
Heat Capacity 313 K, C_p = 62.4 cal·mol^{-1}·K^{-1}
261.1 J·mol^{-1}·K^{-1}
Temperature range 313–423 K. 70% cis, 30% trans
Molecular Weight 140.2680
Wiswesser Line Notation L6TJ A2 D2
Evaluation C

$C_{10}H_{20}$ (liq) 63GUD/CAM
Diethylcyclohexane
Heat Capacity 313 K, C_p = 62.2 cal·mol^{-1}·K^{-1}
261.9 J·mol^{-1}·K^{-1}
Temperature range 313–424 K
Molecular Weight 140.2680
Wiswesser Line Notation L6TJ A2 X2
Evaluation C

Table of Heat Capacities, Entropies, and Phase Transition Properties — Continued

$C_{10}H_{20}$ (liq) 65FIN/MES
n–Butylcyclohexane
Heat Capacity 298.15 K, C_p = 64.78 cal·mol^{-1}·K^{-1}
271.04 J·mol^{-1}·K^{-1}
Temperature range 10–370 K
Entropy 298.15 K, S = 82.45 cal·mol^{-1}·K^{-1}
344.97 J·mol^{-1}·K^{-1}
Phase Changes
c/liq 198.42 K, ΔH = 3384 cal·mol^{-1}
14159 J·mol^{-1}
ΔS = 17.05 cal·mol^{-1}·K^{-1}
71.36 J·mol^{-1}·K^{-1}
Molecular Weight 140.2680
Wiswesser Line Notation L6TJ A4
Evaluation A

$C_{10}H_{20}$ (liq) 63GUD/CAM
tert–Butylcyclohexane
Heat Capacity 313 K, C_p = 63.3 cal·mol^{-1}·K^{-1}
264.8 J·mol^{-1}·K^{-1}
Temperature range 313–423 K
Molecular Weight 140.2680
Wiswesser Line Notation L6TJ AX1&1&1
Evaluation C

$C_{10}H_{20}FeIN_2S_4$ (c) 80YOS/SOR
Iodo bis–(N,N–diethyldithiocarbamato) Iron(III)
Heat Capacity 296.531 K, C_p = 104.1 cal·mol^{-1}·K^{-1}
435.6 J·mol^{-1}·K^{-1}
Temperature range 0.4–300 K. Data given from 0.4 to 19.8 K via ^{3}He calorimeter, and from 14 to 300 K via an adiabatic calorimeter; unsmoothed experimental data.
Phase Changes
c,III/c,II 1.93 K, ΔH = 3.15 cal·mol^{-1}
13.2 J·mol^{-1}
ΔS = 1.33 cal·mol^{-1}·K^{-1}
5.57 J·mol^{-1}·K^{-1}
Total entropy and enthalpy given for an antiferromagnetic to para–magnetic phase transition with maximum at 1.937 K and a Schottky–type anomalies around 12 K are 11.36 J·mol^{-1}·K^{-1} and 134.5 J·mol^{-1}, respectively, for the temperature range 0.4 ot 40 K.
Molecular Weight 553.9950
Wiswesser Line Notation SUYS&N2&2 2 FE I
Evaluation A

$C_{10}H_{20}O_2$ (c) 24GAR/RAN
Decanoic acid; Capric acid
Heat Capacity 285 K, C_p = 86.3 cal·mol^{-1}·K^{-1}
361.1 J·mol^{-1}·K^{-1}
Temperature range 0 to 65°C. Mean value 0 to 24°C.
Phase Changes
c/liq 304.4 K, ΔH = 6690 cal·mol^{-1}
27990 J·mol^{-1}
ΔS = 22.0 cal·mol^{-1}·K^{-1}
92.0 J·mol^{-1}·K^{-1}
Molecular Weight 172.2668
Wiswesser Line Notation QV9
Evaluation B

$C_{10}H_{20}O_2$ (c) 82SCH/MIL 2
Decanoic acid; Capric acid
Heat Capacity 298.15 K, C_p = 113.67 cal·mol^{-1}·K^{-1}
475.59 J·mol^{-1}·K^{-1}
Temperature range 80–345 K
Phase Changes
c,I/liq 304.55 K, ΔH = 6643.9 cal·mol^{-1}
27798 J·mol^{-1}
ΔS = 21.82 cal·mol^{-1}·K^{-1}
91.28 J·mol^{-1}·K^{-1}
Molecular Weight 172.2668
Wiswesser Line Notation QV9
Evaluation B

$C_{10}H_{22}$ (liq) 30PAR/HUF
n–Decane
Heat Capacity 295.5 K, C_p = 74.0 cal·mol^{-1}·K^{-1}
309.6 J·mol^{-1}·K^{-1}
Temperature range 242–296 K. Value is unsmoothed experimental datum.
Molecular Weight 142.2838
Wiswesser Line Notation 10H
Evaluation B

$C_{10}H_{22}$ (liq) 31HUF/PAR
n–Decane
Heat Capacity 297.7 K, C_p = 74.4 cal·mol^{-1}·K^{-1}
311.3 J·mol^{-1}·K^{-1}
Temperature range 91–298 K. Value is unsmoothed experimental datum.
Entropy 298.1 K, S = 102.5 cal·mol^{-1}·K^{-1}
428.9 J·mol^{-1}·K^{-1}
Extrapolation below 90 K, 22.00 cal·mol^{-1}K^{-1}
Phase Changes
c/liq 243.1 K, ΔH = 6878 cal·mol^{-1}
28778 J·mol^{-1}
ΔS = 28.3 cal·mol^{-1}·K^{-1}
118.4 J·mol^{-1}·K^{-1}
Molecular Weight 142.2838
Wiswesser Line Notation 10H
Evaluation B(C_p),C(S)

$C_{10}H_{22}$ (liq) 47OSB/GIN
n–Decane
Heat Capacity 298.15 K, C_p = 75.04 cal·mol^{-1}·K^{-1}
313.97 J·mol^{-1}·K^{-1}
Temperature range 278–318 K
Molecular Weight 142.2838
Wiswesser Line Notation 10H
Evaluation A

$C_{10}H_{22}$ (liq) 52SCH/SAG
n–Decane
Heat Capacity 299.8 K, C_p = 74.64 cal·mol^{-1}·K^{-1}
312.29 J·mol^{-1}·K^{-1}
Temperature range 80 to 200°F
Molecular Weight 142.2838
Wiswesser Line Notation 10H
Evaluation B

$C_{10}H_{22}$ (liq) 54FIN/GRO 2
n–Decane
Heat Capacity 298.15 K, C_p = 75.16 cal·mol^{-1}·K^{-1}
314.47 J·mol^{-1}·K^{-1}
Temperature range 12–300 K
Entropy 298.15 K, S = 101.79 cal·mol^{-1}·K^{-1}
425.89 J·mol^{-1}·K^{-1}
Phase Changes
c/liq 243.51 K, ΔH = 6863 cal·mol^{-1}
28715 J·mol^{-1}
ΔS = 28.18 cal·mol^{-1}·K^{-1}
117.92 J·mol^{-1}·K^{-1}

Table of Heat Capacities, Entropies, and Phase Transition Properties — Continued

Molecular Weight 142.2838
Wiswesser Line Notation 10H
Evaluation A

$C_{10}H_{22}$ (liq) 75GRI/RAS
n–Decane
Heat Capacity 298 K, C_p = 74.5 cal·mol^{-1}·K^{-1}
311.7 J·mol^{-1}·K^{-1}
Temperature range 300–463 K
Molecular Weight 142.2838
Wiswesser Line Notation 10H
Evaluation B

$C_{10}H_{22}$ (liq) 79GRO/HAM
n–Decane
Heat Capacity 298.15 K, C_p = 74.9 cal·mol^{-1}·K^{-1}
313.3 J·mol^{-1}·K^{-1}
One temperature
Molecular Weight 142.2838
Wiswesser Line Notation 10H
Evaluation B

$C_{10}H_{22}$ (liq) 82WIL/ING
n–Decane
Heat Capacity 298.15 K, C_p = 74.69 cal·mol^{-1}·K^{-1}
312.5 J·mol^{-1}·K^{-1}
Temperature range 298.15 K, One temperature
Molecular Weight 142.2838
Wiswesser Line Notation 10H
Evaluation A

$C_{10}H_{22}$ (liq) 30PAR/HUF
2,7–Dimethyloctane
Heat Capacity 295.0 K, C_p = 72.1 cal·mol^{-1}·K^{-1}
301.7 J·mol^{-1}·K^{-1}
Temperature range 223–295 K. Value is unsmoothed experimental datum.
Molecular Weight 142.2838
Wiswesser Line Notation 1Y1&4Y1&1
Evaluation B

$C_{10}H_{22}$ (liq) 41PAR/WES
5–Methylnonane
Heat Capacity 298.1 K, C_p = 75.15 cal·mol^{-1}·K^{-1}
314.43 J·mol^{-1}·K^{-1}
Temperature range 80–300 K
Entropy 298.1 K, S = 101.3 cal·mol^{-1}·K^{-1}
423.8 J·mol^{-1}·K^{-1}
Extrapolation below 80 K, 19.38 cal·mol^{-1}K^{-1}
Phase Changes
c/liq 186.7 K, ΔH = 3977 cal·mol^{-1}
16640 J·mol^{-1}
ΔS = 21.30 cal·mol^{-1}·K^{-1}
89.13 J·mol^{-1}·K^{-1}
Molecular Weight 142.2838
Wiswesser Line Notation 4Y4&1
Evaluation B(C_p),C(S)

$C_{10}H_{22}$ (liq) 41PAR/WES
4–Methylnonane(DL)
Heat Capacity 298.1 K, C_p = 75.85 cal·mol^{-1}·K^{-1}
317.36 J·mol^{-1}·K^{-1}
Temperature range 80–300 K
Entropy 298.1 K, S = 101.7 cal·mol^{-1}·K^{-1}
425.5 J·mol^{-1}·K^{-1}
Extrapolation below 80 K, 18.88 cal·mol^{-1}K^{-1}
Phase Changes
c/liq 174.7 K, ΔH = 3630 cal·mol^{-1}
15188 J·mol^{-1}
ΔS = 20.78 cal·mol^{-1}·K^{-1}
86.94 J·mol^{-1}·K^{-1}
Molecular Weight 142.2838
Wiswesser Line Notation 5Y3&1 -DL
Evaluation B(C_p),C(S)

$C_{10}H_{22}$ (liq) 41PAR/WES
3–Methylnonane(DL)
Heat Capacity 298.1 K, C_p = 73.85 cal·mol^{-1}·K^{-1}
308.99 J·mol^{-1}·K^{-1}
Temperature range 80–300 K
Entropy 298.1 K, S = 102.1 cal·mol^{-1}·K^{-1}
427.2 J·mol^{-1}·K^{-1}
Extrapolation below 80 K, 18.57 cal·mol^{-1}K^{-1}
Phase Changes
c/liq 188.5 K, ΔH = 4469 cal·mol^{-1}
18698 J·mol^{-1}
ΔS = 23.71 cal·mol^{-1}·K^{-1}
99.20 J·mol^{-1}·K^{-1}
Molecular Weight 142.2838
Wiswesser Line Notation 6Y2&1 -DL
Evaluation B(C_p),C(S)

$C_{10}H_{22}$ (liq) 41PAR/WES
2–Methylnonane
Heat Capacity 298.1 K, C_p = 74.88 cal·mol^{-1}·K^{-1}
313.30 J·mol^{-1}·K^{-1}
Temperature range 80–300 K
Entropy 298.1 K, S = 100.4 cal·mol^{-1}·K^{-1}
420.1 J·mol^{-1}·K^{-1}
Extrapolation below 80 K, 19.46 cal·mol^{-1}K^{-1}
Phase Changes
c/liq 198.8 K, ΔH = 4180 cal·mol^{-1}
17489 J·mol^{-1}
ΔS = 21.03 cal·mol^{-1}·K^{-1}
87.97 J·mol^{-1}·K^{-1}
Molecular Weight 142.2838
Wiswesser Line Notation 7Y1&1
Evaluation B(C_p),C(S)

$C_{10}H_{22}N_2O_{10}$ (c,I) 77CLA/CLE
Urea–trioxane inclusion compound
Heat Capacity 298.15 K, C_p = 95.6 cal·mol^{-1}·K^{-1}
400.0 J·mol^{-1}·K^{-1}
Temperature range 15–300 K
Phase Changes
c,IV/c,III 189.91 K, ΔH = 170.9 cal·mol^{-1}
715.1 J·mol^{-1}
ΔS = 0.90 cal·mol^{-1}·K^{-1}
3.77 J·mol^{-1}·K^{-1}
ΔS given in article does not agree with ΔH and T.
c,III/c,II 200.8 K, ΔH = 151.4 cal·mol^{-1}
633.5 J·mol^{-1}
ΔS = 0.75 cal·mol^{-1}·K^{-1}
3.15 J·mol^{-1}·K^{-1}
ΔS given in article does not agree with ΔH and T.
c,II/c,I 242.51 K, ΔH = 448.9 cal·mol^{-1}
1878 J·mol^{-1}
ΔS = 1.85 cal·mol^{-1}·K^{-1}
7.74 J·mol^{-1}·K^{-1}
ΔS given in article does not agree with ΔH and T.
Molecular Weight 330.2912
Wiswesser Line Notation T6O CO EOTJ 3 &ZVZ
Evaluation A

Table of Heat Capacities, Entropies, and Phase Transition Properties — Continued

$C_{10}H_{22}O$ (liq) 79SVE
1-Decanol; n-Decyl alcohol
Heat Capacity 301 K, C_p = 90.1 cal·mol^{-1}·K^{-1}
377 J·mol^{-1}·K^{-1}
Temperature range 301–461 K
Phase Changes
liq/g 323.15 K, ΔH = 18685 cal·mol^{-1}
78180 J·mol^{-1}
ΔS = 57.82 cal·mol^{-1}·K^{-1}
241.93 J·mol^{-1}·K^{-1}
No pressure measurement.
Molecular Weight 158.2832
Wiswesser Line Notation Q10
Evaluation B

$C_{10}H_{22}O_2$ (liq) 47CON/ELV
6-Methyl-5,7-dioxaundecane; Acetaldehyde dibutyl acetal
Heat Capacity 298 K, C_p = 84.18 cal·mol^{-1}·K^{-1}
352.20 J·mol^{-1}·K^{-1}
Temperature range 298–353 K
Molecular Weight 174.2826
Wiswesser Line Notation 4OY1&O4
Evaluation B

$C_{10}H_{22}O_6$ (liq) 79STE/TAM
1,14-Dihydroxy-3,6,9,12-tetraoxatetradecane; Pentaethylene glycol
Heat Capacity 298 K, C_p = 123.2 cal·mol^{-1}·K^{-1}
515.5 J·mol^{-1}·K^{-1}
Temperature range 273–513 K
Molecular Weight 238.2802
Wiswesser Line Notation Q2O2O2O2O2Q
Evaluation B

$C_{10}H_{22}S$ (liq) 70FIN/MCC
1-Decanethiol; n-Decyl mercaptan
Heat Capacity 298.15 K, C_p = 83.75 cal·mol^{-1}·K^{-1}
350.41 J·mol^{-1}·K^{-1}
Temperature range 10–370 K
Entropy 298.15 K, S = 113.08 cal·mol^{-1}·K^{-1}
476.13 J·mol^{-1}·K^{-1}
Phase Changes
c/liq 247.86 K, ΔH = 7963 cal·mol^{-1}
33317 J·mol^{-1}
ΔS = 32.13 cal·mol^{-1}·K^{-1}
134.42 J·mol^{-1}·K^{-1}
Molecular Weight 174.3438
Wiswesser Line Notation SH10
Evaluation A

$C_{11}H_{10}$ (liq) 57MCC/FIN
1-Methylnaphthalene
Heat Capacity 298.15 K, C_p = 53.63 cal·mol^{-1}·K^{-1}
224.39 J·mol^{-1}·K^{-1}
Temperature range 10–370 K
Entropy 298.15 K, S = 60.90 cal·mol^{-1}·K^{-1}
254.81 J·mol^{-1}·K^{-1}
Phase Changes
c,II/c,I 240.70 K, ΔH = 1190 cal·mol^{-1}
4979 J·mol^{-1}
ΔS = 4.94 cal·mol^{-1}·K^{-1}
20.68 J·mol^{-1}·K^{-1}
c,I/liq 242.70 K, ΔH = 1660 cal·mol^{-1}
6945 J·mol^{-1}
ΔS = 6.84 cal·mol^{-1}·K^{-1}
28.62 J·mol^{-1}·K^{-1}
Molecular Weight 142.2000
Wiswesser Line Notation L66J B1
Evaluation A

$C_{11}H_{10}$ (c) 57MCC/FIN
2-Methylnapththalene
Heat Capacity 298.15 K, C_p = 46.84 cal·mol^{-1}·K^{-1}
195.98 J·mol^{-1}·K^{-1}
Temperature range 10–400 K
Entropy 298.15 K, S = 52.58 cal·mol^{-1}·K^{-1}
219.99 J·mol^{-1}·K^{-1}
Phase Changes
c,II/c,I 288.5 K, ΔH = 1340 cal·mol^{-1}
5606 J·mol^{-1}
ΔS = 4.64 cal·mol^{-1}·K^{-1}
19.43 J·mol^{-1}·K^{-1}
c,I/liq 307.73 K, ΔH = 2898 cal·mol^{-1}
12125 J·mol^{-1}
ΔS = 9.42 cal·mol^{-1}·K^{-1}
39.40 J·mol^{-1}·K^{-1}
Molecular Weight 142.2000
Wiswesser Line Notation L66J C1
Evaluation A

$C_{11}H_{10}$ (liq) 31HUF/PAR
2-Methylnaphthalene
Heat Capacity 310.4 K, C_p = 54.5 cal·mol^{-1}·K^{-1}
228.0 J·mol^{-1}·K^{-1}
Temperature range 94–310 K. Value is unsmoothed experimental datum.
Entropy 298.1 K, S = 48.7 cal·mol^{-1}·K^{-1}
203.8 J·mol^{-1}·K^{-1}
Extrapolation below 90 K, 15.74 cal·mol^{-1}K^{-1}. Value is for crystal. Did not observe transition at 288 K with ΔS = 4.6 cal·mol^{-1}·K^{-1}.
Phase Changes
c/liq 307.2 K, ΔH = 28.60 cal·mol^{-1}
11966 J·mol^{-1}
ΔS = 9.3 cal·mol^{-1}·K^{-1}
39.0 J·mol^{-1}·K^{-1}
Molecular Weight 142.2000
Wiswesser Line Notation L66J C1
Evaluation B(C_p),C(S)

$C_{11}H_{12}N_2O$ (c) 41SAT/SOG 2
Antipyrine
Heat Capacity 323 K, C_p = 64.1 cal·mol^{-1}·K^{-1}
268.2 J·mol^{-1}·K^{-1}
Temperature range 0 to 100°C. Mean value.
Molecular Weight 188.2286
Wiswesser Line Notation T5NNVJ A1 BR& E1
Evaluation C
Same data as 40SAT/SOG 3.

$C_{11}H_{12}N_2O_2$ (c) 63COL/HUT
Tryptophane(L)
Heat Capacity 298.15 K, C_p = 56.92 cal·mol^{-1}·K^{-1}
238.15 J·mol^{-1}·K^{-1}
Temperature range 11–305 K
Entropy 298.15 K, S = 60.00 cal·mol^{-1}·K^{-1}
251.04 J·mol^{-1}·K^{-1}
Molecular Weight 204.2280
Wiswesser Line Notation T56 BMJ D1YZVQ -L
Evaluation A

Table of Heat Capacities, Entropies, and Phase Transition Properties — Continued

$C_{11}H_{12}N_2O_4$ (c) 41HUF
Hippurylglycine
Heat Capacity 296.7 K, C_p = 66.33 cal·mol^{-1}·K^{-1}
277.52 J·mol^{-1}·K^{-1}
Temperature range 85–297 K
Entropy 298.1 K, S = 75.2 cal·mol^{-1}·K^{-1}
314.6 J·mol^{-1}·K^{-1}
Extrapolation below 90 K, 24.14 cal·mol^{-1}K^{-1}
Molecular Weight 236.2268
Wiswesser Line Notation QV1MV1MVR
Evaluation A(C_p),C(S)

$C_{11}H_{12}O_2$ (liq) 58DVO
Benzyl methacrylate
Heat Capacity 296.6 K, C_p = 64.5 cal·mol^{-1}·K^{-1}
269.9 J·mol^{-1}·K^{-1}
One temperature
Molecular Weight 176.2146
Wiswesser Line Notation 1UY1&VO1R
Evaluation C

$C_{11}H_{12}O_2$ (liq) **81REI**
trans–Ethyl cinnamate
Heat Capacity 298 K, C_p = 65.5 cal·mol^{-1}·K^{-1}
274.1 J·mol^{-1}·K^{-1}
Temperature range 289–465 K
Molecular Weight 176.2146
Wiswesser Line Notation 2OV1U1R -T
Evaluation D

$C_{11}H_{13}N_3O$ (c) 41SAT/SOG 2
Aminoantipyrine
Heat Capacity 323 K, C_p = 70.4 cal·mol^{-1}·K^{-1}
294.6 J·mol^{-1}·K^{-1}
Temperature range 0 to 100°C. Mean value.
Molecular Weight 203.2432
Wiswesser Line Notation T5NNVJ A1 BR& DZ E1
Evaluation C
Same data as 40SAT/SOG 3.

$C_{11}H_{14}O_2$ (liq) **81REI**
Ethyl hydrocinnamate
Heat Capacity 298 K, C_p = 68.5 cal·mol^{-1}·K^{-1}
286.6 J·mol^{-1}·K^{-1}
Temperature range 289–457 K
Molecular Weight 178.2304
Wiswesser Line Notation 2OV2R
Evaluation D

$C_{11}H_{15}Cl$ (c) 82GYO/YOS
Chloropentamethylbenzene
Heat Capacity 300 K, C_p = 58.46 cal·mol^{-1}·K^{-1}
244.60 J·mol^{-1}·K^{-1}
Temperature range 3–300 K, C_p for phases II, III, and IV.
Entropy 300 K, S = 74.45 cal·mol^{-1}·K^{-1}
311.48 J·mol^{-1}
Temperature range 300 K, S for phases II, III, and IV.
Phase Changes
c,IV/c,III 84.0 K, ΔH = 222 cal·mol^{-1}
930 J·mol^{-1}
ΔS = 2.65 cal·mol^{-1}·K^{-1}
11.1 J·mol^{-1}·K^{-1}
c,III/c,II 153.5 K, ΔH = 318 cal·mol^{-1}
1330 J·mol^{-1}
ΔS = 2.1 cal·mol^{-1}·K^{-1}
8.8 J·mol^{-1}·K^{-1}
c,II′/c,II 180.4 K, ΔH = 318 cal·mol^{-1}
1330 J·mol^{-1}
ΔS = 1.84 cal·mol^{-1}·K^{-1}
7.7 J·mol^{-1}·K^{-1}
c,III′/c,II 187.2 K, ΔH = 358.5 cal·mol^{-1}
1500 J·mol^{-1}
ΔS = 2.0 cal·mol^{-1}·K^{-1}
8.5 J·mol^{-1}·K^{-1}
Phase changes c,II′/c,II and c,III′/c,II are monotropic transitions.
Molecular Weight 182.6925
Wiswesser Line Notation GR B1 C1 D1 E1 F1
Evaluation A

$C_{11}H_{16}$ (c,I) 31HUF/PAR
Pentamethylbenzene
Heat Capacity 283.8 K, C_p = 60.0 cal·mol^{-1}·K^{-1}
251.0 J·mol^{-1}·K^{-1}
Temperature range 92–304 K. Value is unsmoothed experimental datum.
Entropy 298.1 K, S = 70.3 cal·mol^{-1}·K^{-1}
294.1 J·mol^{-1}·K^{-1}
Extrapolation below 90 K, 24.59 cal·mol^{-1}K^{-1}
Phase Changes
c,II/c,I 296.8 K, ΔH = 473 cal·mol^{-1}
1979 J·mol^{-1}
ΔS = 1.6 cal·mol^{-1}·K^{-1}
6.7 J·mol^{-1}·K^{-1}
Molecular Weight 148.2474
Wiswesser Line Notation 1R B1 C1 D1 E1
Evaluation B(C_p),C(S)

$C_{11}H_{16}$ (c) 44EIB
Pentamethylbenzene
Heat Capacity 298.1 K, C_p = 64.6 cal·mol^{-1}·K^{-1}
270.3 J·mol^{-1}·K^{-1}
Temperature range 25 to 200 °C, equations only, in t°C.
C_p(c) = 0.3914 + 0.001760t cal g^{-1}C^{-1} (25 to 40°C);
C_p(liq) = 0.432 + 0.000425t cal g^{-1}C^{-1} (55 to 200°C).
Phase Changes
c/liq 328.2 K, ΔH = 2550 cal·mol^{-1}
10670 J·mol^{-1}
ΔS = 7.8 cal·mol^{-1}·K^{-1}
32.5 J·mol^{-1}·K^{-1}
Molecular Weight 148.2474
Wiswesser Line Notation 1R B1 C1 D1 E1
Evaluation C

$C_{11}H_{20}$ (liq) 62GOL/BEL
Ethylhydroindan
Heat Capacity 311 K, C_p = 67.3 cal·mol^{-1}·K^{-1}
281.6 J·mol^{-1}·K^{-1}
Temperatures 100, 200, 300°F
Molecular Weight 152.2790
Wiswesser Line Notation L56TJ X2
Evaluation C

$C_{11}H_{20}$ (liq) 63GUD/CAM
Ethylhydroindan
Heat Capacity 313 K, C_p = 66.9 cal·mol^{-1}·K^{-1}
279.9 J·mol^{-1}·K^{-1}
Temperature range 313–423 K
Molecular Weight 152.2790
Wiswesser Line Notation L56TJ X2
Evaluation C

Table of Heat Capacities, Entropies, and Phase Transition Properties — Continued

$C_{11}H_{20}$ (liq) 62GOL/BEL
α–Methyldecalin
Heat Capacity 311 K, C_p = 64.3 cal·mol^{-1}·K^{-1}
269.0 J·mol^{-1}·K^{-1}
Temperatures 100, 200, 300°F
Molecular Weight 152.2790
Wiswesser Line Notation L66TJ B1
Evaluation C

$C_{11}H_{20}$ (liq) 63GUD/CAM
α–Methyldecalin
Heat Capacity 313 K, C_p = 63.7 cal·mol^{-1}·K^{-1}
266.5 J·mol^{-1}·K^{-1}
Temperature range 313–423 K
Molecular Weight 152.2790
Wiswesser Line Notation L66TJ B1
Evaluation C

$C_{11}H_{20}$ (liq) 62GOL/BEL
β–Methyldecalin
Heat Capacity 311 K, C_p = 60.5 cal·mol^{-1}·K^{-1}
253.1 J·mol^{-1}·K^{-1}
Temperatures 100, 200, 300°F
Molecular Weight 152.2790
Wiswesser Line Notation L66TJ C1
Evaluation C

$C_{11}H_{20}$ (liq) 63GUD/CAM
β–Methyldecalin
Heat Capacity 313 K, C_p = 61.8 cal·mol^{-1}·K^{-1}
258.6 J·mol^{-1}·K^{-1}
Temperature range 313–423 K
Molecular Weight 152.2790
Wiswesser Line Notation L66TJ C1
Evaluation C

$C_{11}H_{20}O_2$ (liq) 81LEB/YEV
Undecanolactone
Heat Capacity 298.15 K, C_p = 81.91 cal·mol^{-1}·K^{-1}
342.7 J·mol^{-1}·K^{-1}
Temperature range 13.8–388 K, 0.92 mole % impurity in sample.
Entropy 298.15 K, S = 88.31 cal·mol^{-1}·K^{-1}
369.5 J·mol^{-1}·K^{-1}
Phase Changes
c,II/c,I 250.2 K, ΔH = 803 cal·mol^{-1}
3360 J·mol^{-1}
ΔS = 3.20 cal·mol^{-1}·K^{-1}
13.4 J·mol^{-1}·K^{-1}
c,I/liq 275.33 K, ΔH = 3014 cal·mol^{-1}
12610 J·mol^{-1}
ΔS = 10.9 cal·mol^{-1}·K^{-1}
45.8 J·mol^{-1}·K^{-1}
Molecular Weight 184.2778
Wiswesser Line Notation T–12–VOTJ
Evaluation A

$C_{11}H_{20}O_2$ (liq) 81LEB/YEV 2
Undecanolactone
Heat Capacity 298.15 K, C_p = 81.91 cal·mol^{-1}·K^{-1}
342.7 J·mol^{-1}·K^{-1}
Temperature range 13.8–388 K, 0.92 mole % impurity in sample.
Entropy 298.15 K, S = 88.31 cal·mol^{-1}·K^{-1}
369.5 J·mol^{-1}·K^{-1}
Phase Changes
c,II/c,I 250.2 K, ΔH = 803 cal·mol^{-1}
3360 J·mol^{-1}
ΔS = 3.20 cal·mol^{-1}·K^{-1}
13.4 J·mol^{-1}·K^{-1}
c,I/liq 275.33 K, ΔH = 3014 cal·mol^{-1}
12610 J·mol^{-1}
ΔS = 10.9 cal·mol^{-1}·K^{-1}
45.8 J·mol^{-1}·K^{-1}
Molecular Weight 184.2778
Wiswesser Line Notation T–12–VOTJ
Evaluation A

$C_{11}H_{22}$ (liq) 57MCC/FIN 2
1–Undecene
Heat Capacity 298.15 K, C_p = 78.86 cal·mol^{-1}·K^{-1}
329.95 J·mol^{-1}·K^{-1}
Temperature range 11–360 K
Entropy 298.15 K, S = 109.12 cal·mol^{-1}·K^{-1}
456.56 J·mol^{-1}·K^{-1}
Does not include S_0.
Phase Changes
c,II/c,I 217.3 K, ΔH = 2202 cal·mol^{-1}
9213 J·mol^{-1}
ΔS = 10.13 cal·mol^{-1}·K^{-1}
42.40 J·mol^{-1}·K^{-1}
c,I/liq 223.99 K, ΔH = 4061 cal·mol^{-1}
16991 J·mol^{-1}
ΔS = 18.13 cal·mol^{-1}·K^{-1}
75.86 J·mol^{-1}·K^{-1}
Molecular Weight 154.2948
Wiswesser Line Notation 10U1
Evaluation A

$C_{11}H_{22}O_2$ (liq) 79FUC
Methyl decanoate; Methyl caprate
Heat Capacity 298.15 K, C_p = 91.5 cal·mol^{-1}·K^{-1}
382.8 J·mol^{-1}·K^{-1}
One temperature
Molecular Weight 186.2936
Wiswesser Line Notation 9VO1
Evaluation B

$C_{11}H_{22}O_2$ (c,I) 24GAR/RAN
Undecanoic acid
Heat Capacity 293 K, C_p = 99.2 cal·mol^{-1}·K^{-1}
415.1 J·mol^{-1}·K^{-1}
Temperature range 0 to 66°C. Mean value 14 to 21°C.
Phase Changes
c,II/c,I 290 K, ΔH = 1840 cal·mol^{-1}
7700 J·mol^{-1}
ΔS = 6.3 cal·mol^{-1}·K^{-1}
26.6 J·mol^{-1}·K^{-1}
c,I/liq 301.4 K, ΔH = 6000 cal·mol^{-1}
25100 J·mol^{-1}
ΔS = 19.9 cal·mol^{-1}·K^{-1}
83.3 J·mol^{-1}·K^{-1}
Molecular Weight 186.2936
Wiswesser Line Notation QV10
Evaluation B

$C_{11}H_{22}O_2$ (c) 82SCH/MIL
Undecanoic acid
Heat Capacity 298.15 K, C_p = 183.7 cal·mol^{-1}·K^{-1}
768.8 J·mol^{-1}·K^{-1}
Temperature range 80–330 K.

Table of Heat Capacities, Entropies, and Phase Transition Properties — Continued

Phase Changes
c,II/c,I 290.3 K, ΔH = 1948 cal·mol^{-1}
8150 J·mol^{-1}
ΔS = 6.709 cal·mol^{-1}·K^{-1}
28.07 J·mol^{-1}·K^{-1}
c,I/liq 301.63 K, ΔH = 6209.4 cal·mol^{-1}
25980 J·mol^{-1}
ΔS = 20.59 cal·mol^{-1}·K^{-1}
86.13 J·mol^{-1}·K^{-1}
Molecular Weight 186.2936
Wiswesser Line Notation QV10
Evaluation B

$C_{11}H_{24}$ (liq) 31HUF/PAR
n–Undecane
Heat Capacity 298.0 K, C_p = 81.9 cal·mol^{-1}·K^{-1}
342.7 J·mol^{-1}·K^{-1}
Temperature range 92–298 K. Value is unsmoothed experimental datum.
Entropy 298.1 K, S = 110.9 cal·mol^{-1}·K^{-1}
464.0 J·mol^{-1}·K^{-1}
Extrapolation below 90 K, 23.92 cal·mol^{-1}K^{-1}
Phase Changes
c,II/c,I 236.1 K, ΔH = 1515 cal·mol^{-1}
6339 J·mol^{-1}
ΔS = 6.4 cal·mol^{-1}·K^{-1}
26.8 J·mol^{-1}·K^{-1}
c,I/liq 247.2 K, ΔH = 5333 cal·mol^{-1}
22313 J·mol^{-1}
ΔS = 21.6 cal·mol^{-1}·K^{-1}
90.3 J·mol^{-1}·K^{-1}
Molecular Weight 156.3106
Wiswesser Line Notation 11H
Evaluation B(C_p),C(S)

$C_{11}H_{24}$ (liq) 54FIN/GRO 2
n–Undecane
Heat Capacity 298.15 K, C_p = 82.47 cal·mol^{-1}·K^{-1}
345.05 J·mol^{-1}·K^{-1}
Temperature range 12–320 K
Entropy 298.15 K, S = 109.50 cal·mol^{-1}·K^{-1}
458.15 J·mol^{-1}·K^{-1}
Phase Changes
c,II/c,I 236.6 K, ΔH = 1639 cal·mol^{-1}
6858 J·mol^{-1}
ΔS = 6.93 cal·mol^{-1}·K^{-1}
23.99 J·mol^{-1}·K^{-1}
c,I/liq 247.59 K, ΔH = 5301 cal·mol^{-1}
22179 J·mol^{-1}
ΔS = 21.41 cal·mol^{-1}·K^{-1}
89.58 J·mol^{-1}·K^{-1}
Molecular Weight 156.3106
Wiswesser Line Notation 11H
Evaluation A

$C_{11}H_{24}$ (liq) 71MES/FIN
2–Methyldecane
Heat Capacity 298.15 K, C_p = 81.55 cal·mol^{-1}·K^{-1}
341.21 J·mol^{-1}·K^{-1}
Temperature range 11–390 K
Entropy 298.15 K, S = 108.46 cal·mol^{-1}·K^{-1}
453.80 J·mol^{-1}·K^{-1}
Phase Changes
c/liq 224.31 K, ΔH = 5996 cal·mol^{-1}
25087 J·mol^{-1}
ΔS = 26.73 cal·mol^{-1}·K^{-1}
111.84 J·mol^{-1}·K^{-1}
Molecular Weight 156.3106
Wiswesser Line Notation 8Y1&1
Evaluation A

$C_{11}H_{24}O$ (liq) 82VAS/PET
Undecanol-1; n–Undecyl alcohol
Heat Capacity 303 K, C_p = 100.3 cal·mok^{-1}·K^{-1}
419.8 J·mol^{-1}·K^{-1}
Temperature range 303–508 K
Molecular Weight 172.3100
Wiswesser Line Notation Q11
Evaluation B

$C_{11}H_{24}O$ (liq) 75AND/MAR
2–Oxadodecane; Methyl n–decyl ether
Heat Capacity 298.15 K, C_p = 88.62 cal·mol^{-1}·K^{-1}
370.8 J·mol^{-1}·K^{-1}
Temperature range 12–350 K
Entropy 298.15 K, S = 117.23 cal·mol^{-1}·K^{-1}
490.5 J·mol^{-1}·K^{-1}
Phase Changes
c/liq 243.47 K, ΔH = 7581 cal·mol^{-1}
31720 J·mol^{-1}
ΔS = 31.14 cal·mol^{-1}·K^{-1}
130.31 J·mol^{-1}·K^{-1}
Molecular Weight 172.3100
Wiswesser Line Notation 1O01
Evaluation A

$C_{11}H_{26}N_2O$ (c,I) 65PEM/PAR
Urea–n–decane adduct
Heat Capacity 298.15 K, C_p = 30.23 cal·mol^{-1}·K^{-1}
126.48 J·mol^{-1}·K^{-1}
Temperature range 12–300 K. Value for aduct with 1 mole of urea.
Entropy 298.15 K, S = 33.21 cal·mol^{-1}·K^{-1}
138.95 J·mol^{-1}·K^{-1}
Phase Changes
Transition at 110.9 K with ΔH = 203 cal(mol hydrocarbon)$^{-1}$.
Molecular Weight 202.3392
Wiswesser Line Notation ZVZ &10H
Evaluation B
Sample 78.45 percent urea.

$C_{12}F_{10}$ (c) 71PAU/RAK
Decafluorobiphenyl; Perfluorobiphenyl
Heat Capacity 300.7 K, C_p = 77.3 cal·mol^{-1}·K^{-1}
323.4 J·mol^{-1}·K^{-1}
Temperature range 13–350 K. Complete article deposited at VINITI, No. 2536–71, 2 January 1971.
Entropy 298.15 K, S = 90.97 cal·mol^{-1}·K^{-1}
380.62 J·mol^{-1}·K^{-1}
Phase Changes
Melting point and enthalpy of fusion given only in complete paper. ΔS fusion = 15.26 cal·mol^{-1} K^{-1}.
Molecular Weight 334.1160
Wiswesser Line Notation FR BF CF DF EF FR BF CF DF EF FF
Evaluation B

$C_{12}F_{22}$ (c) 65COX/GUN
Docosafluorobicyclohexyl; Perfluorobicyclohexyl
Heat Capacity 298 K, C_p = 215 cal·mol^{-1}·K^{-1}
899 J·mol^{-1}·K^{-1}
One temperature
Molecular Weight 562.0968
Wiswesser Line Notation L6TJ AF AF BF BF CF CF DF DF EF EF FF F- FL6TJ AF AF BF BF CF CF DF DF EF EF FF
Evaluation B

Table of Heat Capacities, Entropies, and Phase Transition Properties — Continued

$C_{12}H_6F_6$ (liq) 78RIP/WRI
Benzene:hexafluorobenzene complex
Heat Capacity 305 K, C_p = 88.4 cal·mol^{-1}·K^{-1}
370 J·mol^{-1}·K^{-1}
Temperature range 60–320 K
Phase Changes
Transitions observed at 199.0, 247.5, 272.0 K; no heat data given.
c,I/liq 297.2 K, ΔH = 5230 cal·mol^{-1}
21870 J·mol^{-1}
ΔS = 17.6 cal·mol^{-1}·K^{-1}
73.6 J·mol^{-1}·K^{-1}
Molecular Weight 264.1698
Wiswesser Line Notation FR BF CF DF EF FF &R
Evaluation C

$C_{12}H_6N_2O_2$ (c) 62STR/BAR
1,5–Naphthylenediisocyanate;
1,5–Diisocyanatonaphthalene
Heat Capacity 298 K, C_p = 53.4 cal·mol^{-1}·K^{-1}
223.6 J·mol^{-1}·K^{-1}
One temperature
Molecular Weight 210.1916
Wiswesser Line Notation L66J BNCO GNCO
Evaluation D

$C_{12}H_8$ (c) 69SAD/STE
Acenaphthylene
Heat Capacity 298 K, C_p = 39.8 cal·mol^{-1}·K^{-1}
166.4 J·mol^{-1}·K^{-1}
Temperature range 20 to 89 °C, equation only; liquid, 90 to 150 °C, equation only.
Phase Changes
c/liq 362.6 K, ΔH = 1659 cal·mol^{-1}
6940 J·mol^{-1}
ΔS = 4.57 cal·mol^{-1}·K^{-1}
19.14 J·mol^{-1}·K^{-1}
Molecular Weight 152.1952
Wiswesser Line Notation L566 1A LJ
Evaluation C

$C_{12}H_8N_2O_5$ (c) 78MAR/CIO
4,4′–Dinitrodiphenyl ether; Bis(4–nitrophenyl) ether
Heat Capacity 298 K, C_p = 100.8 cal·mol^{-1}·K^{-1}
421.7 J·mol^{-1}·K^{-1}
Temperature range 298–491 K. Values for solid seem odd; minimum at 345 of 66 cal·mol^{-1}K^{-1}, rising to 132 at 410 K.
Phase Changes
c/liq 418.2 K, ΔH = 2460 cal·mol^{-1}
10295 J·mol^{-1}
ΔS = 5.9 cal·mol^{-1}·K^{-1}
24.6 J·mol^{-1}·K^{-1}
Molecular Weight 260.2056
Wiswesser Line Notation WNR DOR DNW
Evaluation D

$C_{12}H_8OS_2$ (c) 75CUC
Diphenylene–2,2′disulfide–S–oxide
Heat Capacity 303 K, C_p = 178.1 cal·mol^{-1}·K^{-1}
745.2 J·mol^{-1}·K^{-1}
Temperature range 303–523 K
Phase Changes
c/liq 407 K, ΔH = 4300 cal·mol^{-1}
17990 J·mol^{-1}
ΔS = 10.6 cal·mol^{-1}·K^{-1}
44.2 J·mol^{-1}·K^{-1}
Molecular Weight 232.3146
Wiswesser Line Notation T B666 HSSJ HO
Evaluation B

$C_{12}H_9Cl$ (c) 74GEI/DZH
2–Chlorobiphenyl
Heat Capacity 298.15 K, C_p = 49.72 cal·mol^{-1}·K^{-1}
208.03 J·mol^{-1}·K^{-1}
Temperature range 12–330 K
Entropy 298.15 K, S = 72.60 cal·mol^{-1}·K^{-1}
303.76 J·mol^{-1}·K^{-1}
Phase Changes
c/liq 304.94 K, ΔH = 3470 cal·mol^{-1}
14518 J·mol^{-1}
ΔS = 11.38 cal·mol^{-1}·K^{-1}
47.61 J·mol^{-1}·K^{-1}
Molecular Weight 188.6561
Wiswesser Line Notation GR BR
Evaluation A

$C_{12}H_9Cl$ (c) 77GEI/KAR
2–Chlorobiphenyl
Heat Capacity 298.15 K, C_p = 50.65 cal·mol^{-1}·K^{-1}
211.92 J·mol^{-1}·K^{-1}
Based on previously published work of authors, not available in detail, on C_p 12–370 K.
Entropy 298.15 K, S = 60.78 cal·mol^{-1}·K^{-1}
254.30 J·mol^{-1}·K^{-1}
Phase Changes
c/liq 304.94 K, ΔH = 3470 cal·mol^{-1}
14518 J·mol^{-1}
ΔS = 11.38 cal·mol^{-1}·K^{-1}
47.61 J·mol^{-1}·K^{-1}
Molecular Weight 188.6561
Wiswesser Line Notation GR BR
Evaluation B

$C_{12}H_9Cl$ (c) 77GEI/KAR
4–Chlorobiphenyl
Heat Capacity 298.15 K, C_p = 58.26 cal·mol^{-1}·K^{-1}
243.76 J·mol^{-1}·K^{-1}
Based on previously published work of authors, not available in detail, on C_p 12–370 K.
Entropy 298.15 K, S = 61.40 cal·mol^{-1}·K^{-1}
256.90 J·mol^{-1}·K^{-1}
Phase Changes
c/liq 348.55 K, ΔH = 3183 cal·mol^{-1}
13318 J·mol^{-1}
ΔS = 9.13 cal·mol^{-1}·K^{-1}
38.21 J·mol^{-1}·K^{-1}
Molecular Weight 188.6561
Wiswesser Line Notation GR DR
Evaluation B

$C_{12}H_9Cl_3Si$ (c) 77GEI/KAR
o–Trichlorosilylbiphenyl
Heat Capacity 298.15 K, C_p = 80.75 cal·mol^{-1}·K^{-1}
337.86 J·mol^{-1}·K^{-1}
Based on previously published work of authors, not available in detail, on C_p 12–370 K.
Entropy 298.15 K, S = 83.27 cal·mol^{-1}·K^{-1}
348.40 J·mol^{-1}·K^{-1}
Phase Changes
c/liq 339.18 K, ΔH = 4952 cal·mol^{-1}
20719 J·mol^{-1}
ΔS = 14.68 cal·mol^{-1}·K^{-1}
61.09 J·mol^{-1}·K^{-1}
Molecular Weight 287.6476
Wiswesser Line Notation G–SI–GGR BR
Evaluation B

Table of Heat Capacities, Entropies, and Phase Transition Properties — Continued

$C_{12}H_9Cl_3Si$ (c) 77GEI/KAR
p–Trichlorosilylbiphenyl
Heat Capacity 298.15 K, C_p = 69.57 cal·mol⁻¹·K⁻¹ 291.08 J·mol⁻¹·K⁻¹
Based on previously published work of authors, not available in detail, on C_p 12–370 K.
Entropy 298.15 K, S = 78.58 cal·mol⁻¹·K⁻¹ 328.78 J·mol⁻¹·K⁻¹
Phase Changes
c/liq 372.90 K, ΔH = 4438 cal·mol⁻¹ 18569 J·mol⁻¹
ΔS = 11.90 cal·mol⁻¹·K⁻¹ 49.80 J·mol⁻¹·K⁻¹
Molecular Weight 287.6476
Wiswesser Line Notation G–SI–GGR DR
Evaluation B

$C_{12}H_{10}$ (c) 44EIB
Acenaphthene
Heat Capacity 298.1 K, C_p = 50.3 cal·mol⁻¹·K⁻¹ 210.5 J·mol⁻¹·K⁻¹
Temperature range 25 to 200 °C, equations only in t°C.
C_p(c) = 0.2756 + 0.001854t cal g⁻¹C⁻¹ (25 to 60°C);
C_p(liq) = 0.409 + 0.000598t cal g⁻¹C⁻¹ (95 to 200°C).
Phase Changes
c/liq 367.8 K, ΔH = 6000 cal·mol⁻¹ 25100 J·mol⁻¹
ΔS = 16.3 cal·mol⁻¹·K⁻¹ 68.2 J·mol⁻¹·K⁻¹
Molecular Weight 154.2110
Wiswesser Line Notation L566 1A LT&&J
Evaluation C

$C_{12}H_{10}$ (c) 69SAD/STE
Acenaphthene
Heat Capacity 298 K, C_p = 44.4 cal·mol⁻¹·K⁻¹ 185.8 J·mol⁻¹·K⁻¹
Temperature range 20 to 93 °C, equation only; liquid, 93 to 200 °C, equation only.
Phase Changes
c/liq 366.4 K, ΔH = 4836 cal·mol⁻¹ 20233 J·mol⁻¹
ΔS = 13.20 cal·mol⁻¹·K⁻¹ 55.22 J·mol⁻¹·K⁻¹
Molecular Weight 154.2110
Wiswesser Line Notation L566 1A LT&&J
Evaluation C

$C_{12}H_{10}$ (c) 77FIN/MES
Acenaphthene
Heat Capacity 298.15 K, C_p = 45.50 cal·mol⁻¹·K⁻¹ 190.37 J·mol⁻¹·K⁻¹
Temperature range 10–440 K
Entropy 298.15 K, S = 45.14 cal·mol⁻¹·K⁻¹ 188.87 J·mol⁻¹·K⁻¹
Phase Changes
c/liq 366.56 K, ΔH = 5129.6 cal·mol⁻¹ 21462.2 J·mol⁻¹
ΔS = 13.99 cal·mol⁻¹·K⁻¹ 58.55 J·mol⁻¹·K⁻¹
Molecular Weight 154.2110
Wiswesser Line Notation L566 1A LT&&J
Evaluation A

$C_{12}H_{10}$ (c) 30HUF/PAR
Biphenyl; Diphenyl
Heat Capacity 294.4 K, C_p = 46.4 cal·mol⁻¹·K⁻¹ 194.1 J·mol⁻¹·K⁻¹
Temperature range 93–295 K. Value is unsmoothed experimental datum.
Entropy 298.1 K, S = 49.2 cal·mol⁻¹·K⁻¹ 205.9 J·mol⁻¹·K⁻¹
Extrapolation below 90 K, 15.63 cal·mol⁻¹K⁻¹
Molecular Weight 154.2110
Wiswesser Line Notation RR
Evaluation B(C_p),C(S)

$C_{12}H_{10}$ (c) 32SPA/THO
Biphenyl; Diphenyl
Heat Capacity 303 K, C_p = 47.3 cal·mol⁻¹·K⁻¹ 197.9 J·mol⁻¹·K⁻¹
Temperature range 30 to 100°C
Phase Changes
c/liq 341.5 K, ΔH = 4457 cal·mol⁻¹ 18648 J·mol⁻¹
ΔS = 13.1 cal·mol⁻¹·K⁻¹ 54.6 J·mol⁻¹·K⁻¹
Molecular Weight 154.2110
Wiswesser Line Notation RR
Evaluation B

$C_{12}H_{10}$ (c) 41SCH
Biphenyl; Diphenyl
Heat Capacity 298.1 K, C_p = 47.3 cal·mol⁻¹·K⁻¹ 197.9 J·mol⁻¹·K⁻¹
Temperature range 20 to 200 °C, equations only, in t°C.
C_p(c) = 0.2745 + 0.001235t cal g⁻¹C⁻¹ (20 to 69°C);
C_p(liq) = 0.3917 + 0.0005206t cal g⁻¹C⁻¹ (69 to 200°C).
Phase Changes
c/liq 342 K, ΔH = 4444 cal·mol⁻¹ 18594 J·mol⁻¹
ΔS = 13.0 cal·mol⁻¹·K⁻¹ 54.4 J·mol⁻¹·K⁻¹
Molecular Weight 154.2110
Wiswesser Line Notation RR
Evaluation C

$C_{12}H_{10}$ (c) 50UEB/ORT
Biphenyl; Diphenyl
Heat Capacity 298.15 K, C_p = 45.6 cal·mol⁻¹·K⁻¹ 190.8 J·mol⁻¹·K⁻¹
Temperature range 293–368 K. Equation only.
Phase Changes
c/liq 343 K, ΔH = 4440 cal·mol⁻¹ 18575 J·mol⁻¹
ΔS = 12.9 cal·mol⁻¹·K⁻¹ 54.2 J·mol⁻¹·K⁻¹
Molecular Weight 154.2110
Wiswesser Line Notation RR
Evaluation C

$C_{12}H_{10}$ (liq) 31FOR/BRU
Biphenyl; Diphenyl
Heat Capacity 350.8 K, C_p = 62.9 cal·mol⁻¹·K⁻¹ 263.2 J·mol⁻¹·K⁻¹
Temperature range 350–620 K. Value is unsmoothed experimental datum.
Molecular Weight 154.2110
Wiswesser Line Notation RR
Evaluation B

Table of Heat Capacities, Entropies, and Phase Transition Properties — Continued

$C_{12}H_{10}$ (liq) 50KUR
Biphenyl; Diphenyl
Heat Capacity 370 K, C_p = 71.7 $cal{\cdot}mol^{-1}{\cdot}K^{-1}$
300.0 $J{\cdot}mol^{-1}{\cdot}K^{-1}$
Temperature range 98 to 255°C. Mp 70.8°C
Molecular Weight 154.2110
Wiswesser Line Notation RR
Evaluation B

$C_{12}H_{10}$ (liq) 56MCE
Biphenyl; Diphenyl
Heat Capacity 422 K, C_p = 72.0 $cal{\cdot}mol^{-1}{\cdot}K^{-1}$
301.2 $J{\cdot}mol^{-1}{\cdot}K^{-1}$
Temperature range 300 to 600°F
Molecular Weight 154.2110
Wiswesser Line Notation RR
Evaluation C
Quoted in 58WAL/BRO

$C_{12}H_{10}$ (liq) 58WAL/BRO
Biphenyl; Diphenyl
Heat Capacity 370 K, C_p = 68.2 $cal{\cdot}mol^{-1}{\cdot}K^{-1}$
285.3 $J{\cdot}mol^{-1}{\cdot}K^{-1}$
Temperature range 200 to 600°F
Molecular Weight 154.2110
Wiswesser Line Notation RR
Evaluation B

$C_{12}H_{10}$ (c) 80ATA/CHI
Biphenyl; Diphenyl
Heat Capacity C_p data given graphically.
Temperature range 3–300 K.
Phase Changes
c,III/c,II 11.0 K, ΔH = 0.07 $cal{\cdot}mol^{-1}$
0.028 $J{\cdot}mol^{-1}$
ΔS = 0.006 $cal{\cdot}mol^{-1}{\cdot}K^{-1}$
0.026 $J{\cdot}mol^{-1}{\cdot}K^{-1}$
c,III/c,II: anomalous region, 7.5 to 14.0 K
c,II/c,I 40.4 K, ΔH = 1.20 $cal{\cdot}mol^{-1}$
5.02 $J{\cdot}mol^{-1}$
ΔS = 0.031 $cal{\cdot}mol^{-1}{\cdot}K^{-1}$
0.129 $J{\cdot}mol^{-1}{\cdot}K^{-1}$
c,II/c,I: anomalous region 30.0 to 47.0 K
Molecular Weight 154.2110
Wiswesser Line Notation RR
Evaluation A

$C_{12}H_{10}Hg$ (c) 31SMI/AND 2
Diphenylmercury; Mercury diphenyl
Heat Capacity 298.5 K, C_p = 53.9 $cal{\cdot}mol^{-1}{\cdot}K^{-1}$
225.5 $J{\cdot}mol^{-1}{\cdot}K^{-1}$
Temperature range 102–346 K. Value is unsmoothed experimental datum.
Molecular Weight 354.8010
Wiswesser Line Notation R-HG-R
Evaluation B

$C_{12}H_{10}O$ (c) 73GEI/DZH
o-Hydroxybiphenyl
Heat Capacity 298.15 K, C_p = 56.19 $cal{\cdot}mol^{-1}{\cdot}K^{-1}$
235.10 $J{\cdot}mol^{-1}{\cdot}K^{-1}$
Temperature range 12–350K; complete C_p data in paper deposited at VINITI, No. 4748-72, 24 Aug, 1972.
Entropy 298.15 K, S = 62.56 $cal{\cdot}mol^{-1}{\cdot}K^{-1}$
261.75 $J{\cdot}mol^{-1}{\cdot}K^{-1}$
Phase Changes
c/liq 330.6 K, ΔH = 3875 $cal{\cdot}mol^{-1}$
16213 $J{\cdot}mol^{-1}$
ΔS = 11.72 $cal{\cdot}mol^{-1}{\cdot}K^{-1}$
49.04 $J{\cdot}mol^{-1}{\cdot}K^{-1}$
Molecular Weight 170.2104
Wiswesser Line Notation QR BR
Evaluation A

$C_{12}H_{10}O$ (c) 77GEI/KAR
o-Hydroxybiphenyl
Heat Capacity 298.15 K, C_p = 54.40 $cal{\cdot}mol^{-1}{\cdot}K^{-1}$
227.61 $J{\cdot}mol^{-1}{\cdot}K^{-1}$
Based on previously published work of authors, not available in detail, on C_p 12–370 K.
Entropy 298.15 K, S = 62.56 $cal{\cdot}mol^{-1}{\cdot}K^{-1}$
261.75 $J{\cdot}mol^{-1}{\cdot}K^{-1}$
Phase Changes
c/liq 330.60 K, ΔH = 3875 $cal{\cdot}mol^{-1}$
16213 $J{\cdot}mol^{-1}$
ΔS = 11.72 $cal{\cdot}mol^{-1}{\cdot}K^{-1}$
49.04 $J{\cdot}mol^{-1}{\cdot}K^{-1}$
Molecular Weight 170.2104
Wiswesser Line Notation QR BR
Evaluation B

$C_{12}H_{10}O$ (c) 31SMI/AND 2
Diphenyl oxide; Diphenyl ether
Heat Capacity 298.5 K, C_p = 51.6 $cal{\cdot}mol^{-1}{\cdot}K^{-1}$
215.9 $J{\cdot}mol^{-1}{\cdot}K^{-1}$
Temperature range 102–298 K
Molecular Weight 170.2104
Wiswesser Line Notation ROR
Evaluation B

$C_{12}H_{10}O$ (c) 51FUR/GIN
Diphenyl oxide; Diphenyl ether
Heat Capacity 298.15 K, C_p = 51.76 $cal{\cdot}mol^{-1}{\cdot}K^{-1}$
216.56 $J{\cdot}mol^{-1}{\cdot}K^{-1}$
Temperature range 18–570 K
Entropy 298.15 K, S = 55.91 $cal{\cdot}mol^{-1}{\cdot}K^{-1}$
233.91 $J{\cdot}mol^{-1}{\cdot}K^{-1}$
Phase Changes
c/liq 300.02 K, ΔH = 4114.7 $cal{\cdot}mol^{-1}$
17216 $J{\cdot}mol^{-1}$
ΔS = 13.71 $cal{\cdot}mol^{-1}{\cdot}K^{-1}$
57.38 $J{\cdot}mol^{-1}{\cdot}K^{-1}$
Molecular Weight 170.2104
Wiswesser Line Notation ROR
Evaluation A

$C_{12}H_{10}O$ (c) 53GIN/FUR
Diphenyl oxide; Diphenyl ether
Heat Capacity 298.15 K, C_p = 51.76 $cal{\cdot}mol^{-1}{\cdot}K^{-1}$
216.56 $J{\cdot}mol^{-1}{\cdot}K^{-1}$
Temperature range 14–570 K
Phase Changes
c/liq 300.03 K, ΔH = 4114 $cal{\cdot}mol^{-1}$
17215 $J{\cdot}mol^{-1}$
ΔS = 13.71 $cal{\cdot}mol^{-1}{\cdot}K^{-1}$
57.38 $J{\cdot}mol^{-1}{\cdot}K^{-1}$
Molecular Weight 170.2104
Wiswesser Line Notation ROR
Evaluation A

Table of Heat Capacities, Entropies, and Phase Transition Properties — Continued

$C_{12}H_{10}OS$ (c) 31SMI/AND 2
Diphenyl sulfoxide
Heat Capacity 298.5 K, C_p = 57.3 cal·mol^{-1}·K^{-1}
239.7 J·mol^{-1}·K^{-1}
Temperature range 102–323 K. Value is unsmoothed experimental datum.
Molecular Weight 202.2704
Wiswesser Line Notation OSR&R
Evaluation B

$C_{12}H_{10}O_2S$ (c) 31SMI/AND 2
Diphenyl sulfone
Heat Capacity 298.5 K, C_p = 58.4 cal·mol^{-1}·K^{-1}
244.3 J·mol^{-1}·K^{-1}
Temperature range 102–346 K
Molecular Weight 218.2698
Wiswesser Line Notation WSR&R
Evaluation B

$C_{12}H_{10}O_4$ (c) 24LAN
Quinhydrone
Heat Capacity 243.4 K, C_p = 54.6 cal·mol^{-1}·K^{-1}
228.4 J·mol^{-1}·K^{-1}
Temperature range 20–244 K. Value is unsmoothed experimental datum.
Molecular Weight 218.2086
Wiswesser Line Notation L6V DVJ &QR DQ
Evaluation B

$C_{12}H_{10}S$ (liq) 31SMI/AND 2
Diphenyl sulfide
Heat Capacity 298.5 K, C_p = 64.8 cal·mol^{-1}·K^{-1}
271.1 J·mol^{-1}·K^{-1}
Temperature range 102–298 K. Value is unsmoothed experimental datum.
Molecular Weight 186.2710
Wiswesser Line Notation RSR
Evaluation B

$C_{12}H_{11}N_3$ (c) 41SAT/SOG 2
p–Aminoazobenzene
Heat Capacity 323 K, C_p = 66.0 cal·mol^{-1}·K^{-1}
276.1 J·mol^{-1}·K^{-1}
Temperature range 0 to 100°C. Mean value.
Molecular Weight 197.2390
Wiswesser Line Notation ZR DNUNR
Evaluation C
Same data as 40SAT/SOG 3.

$C_{12}H_{12}$ (c) 73GOO
1,8–Dimethylnaphthalene
Heat Capacity 298.15 K, C_p = 57.8 cal·mol^{-1}·K^{-1}
241.8 J·mol^{-1}·K^{-1}
One temperature
Molecular Weight 156.2268
Wiswesser Line Notation L66J B1 J1
Evaluation B

$C_{12}H_{12}$ (c) 77FIN/MES
1,8–Dimethylnaphthalene
Heat Capacity 298.15 K, C_p = 58.03 cal·mol^{-1}·K^{-1}
242.80 J·mol^{-1}·K^{-1}
Temperature range 10–440 K
Entropy 298.15 K, S = 53.71 cal·mol^{-1}·K^{-1}
224.72 J·mol^{-1}·K^{-1}
Phase Changes
c/liq 336.33 K, ΔH = 3767.5 cal·mol^{-1}
46.87 J·mol^{-1}
ΔS = 11.20 cal·mol^{-1}·K^{-1}
46.86 J·mol^{-1}·K^{-1}
Molecular Weight 156.2268
Wiswesser Line Notation L66J B1 J1
Evaluation A

$C_{12}H_{12}$ (c) 77FIN/MES
2,6–Dimethylnaphthalene
Heat Capacity 298.15 K, C_p = 48.65 cal·mol^{-1}·K^{-1}
203.55 J·mol^{-1}·K^{-1}
Temperature range 10–440 K
Entropy 298.15 K, S = 54.46 cal·mol^{-1}·K^{-1}
227.86 J·mol^{-1}·K^{-1}
Phase Changes
c/liq 383.32 K, ΔH = 5988.7 cal·mol^{-1}
25056.7 J·mol^{-1}
ΔS = 15.62 cal·mol^{-1}·K^{-1}
65.37 J·mol^{-1}·K^{-1}
Molecular Weight 156.2268
Wiswesser Line Notation L66J C1 H1
Evaluation A

$C_{12}H_{12}$ (c) 73GOO
2,6–Dimethylnaphthalene
Heat Capacity 298.15 K, C_p = 48.4 cal·mol^{-1}·K^{-1}
202.5 J·mol^{-1}·K^{-1}
One temperature
Molecular Weight 156.2268
Wiswesser Line Notation L66J C1 H1
Evaluation B

$C_{12}H_{12}$ (c) 73GOO
2,7–Dimethylnaphthalene
Heat Capacity 298.15 K, C_p = 48.4 cal·mol^{-1}·K^{-1}
202.5 J·mol^{-1}·K^{-1}
One temperature
Molecular Weight 156.2268
Wiswesser Line Notation L66J C1 I1
Evaluation B

$C_{12}H_{12}$ (c) 77FIN/MES
2,7–Dimethylnaphthalene
Heat Capacity 298.15 K, C_p = 48.85 cal·mol^{-1}·K^{-1}
204.39 J·mol^{-1}·K^{-1}
Temperature range 10–440 K
Entropy 298.15 K, S = 54.63 cal·mol^{-1}·K^{-1}
228.57 J·mol^{-1}·K^{-1}
Phase Changes
c/liq 368.81 K, ΔH = 5581.1 cal·mol^{-1}
23351.3 J·mol^{-1}
ΔS = 15.13 cal·mol^{-1}·K^{-1}
63.32 J·mol^{-1}·K^{-1}
Molecular Weight 156.2268
Wiswesser Line Notation L66J C1 I1
Evaluation A

$C_{12}H_{12}FeO$ (c) 81TOM/CUR
Acetylferrocene
Heat Capacity 298 K, C_p = 58.8 cal·mol^{-1}·K^{-1}
246.0 J·mol^{-1}·K^{-1}
Temperature range 293–353°C. Equation given.
Phase Changes
c/liq 358.7 K
Molecular Weight 228.0732
Wiswesser Line Notation L5ØJ Ø–FE– –ØL5ØJ AV1
Evaluation B

Table of Heat Capacities, Entropies, and Phase Transition Properties — Continued

$C_{12}H_{12}N_2O$ (c) 78MAR/CIO
4,4′–Diaminodiphenyl ether; Bis(4–aminophenyl) ether
Heat Capacity 298 K, C_p = 67.1 cal·mol⁻¹·K⁻¹ 280.7 J·mol⁻¹·K⁻¹
Temperature range 298–502 K
Phase Changes
c/liq 465.4 K, ΔH = 1850 cal·mol⁻¹ 7740 J·mol⁻¹
ΔS = 4.0 cal·mol⁻¹·K⁻¹ 16.6 J·mol⁻¹·K⁻¹
Molecular Weight 200.2396
Wiswesser Line Notation ZR DOR DZ
Evaluation D

$C_{12}H_{12}N_4$ (c) 41SAT/SOG 2
2,4–Diaminoazobenzene
Heat Capacity 323 K, C_p = 73.3 cal·mol⁻¹·K⁻¹ 306.7 J·mol⁻¹·K⁻¹
Temperature range 0 to 100°C. Mean value.
Molecular Weight 212.2536
Wiswesser Line Notation ZR CZ DNUNR
Evaluation C
Same data as 40SAT/SOG 3.

$C_{12}H_{14}N_4O$ (c) 73KAR/SAP
4,4′–Dihydrazodiphenyl oxide
Heat Capacity 300 K, C_p = 97.47 cal·mol⁻¹·K⁻¹ 407.81 J·mol⁻¹·K⁻¹
Temperature range 20–300 K
Entropy 300 K, S = 77.11 cal·mol⁻¹·K⁻¹ 322.63 J·mol⁻¹·K⁻¹
Molecular Weight 230.2688
Wiswesser Line Notation ZMR DOR DMZ
Evaluation B

$C_{12}H_{14}N_4O$ (c) 73KAR/SAP
3,3′,4,4′–Tetraaminodiphenyl oxide;
3,3′,4,4′–Tetraaminodiphenyl ether
Heat Capacity 300 K, C_p = 76.54 cal·mol⁻¹·K⁻¹ 320.24 J·mol⁻¹·K⁻¹
Temperature range 20–300 K
Entropy 300 K, S = 70.70 cal·mol⁻¹·K⁻¹ 295.1 J·mol⁻¹·K⁻¹
Molecular Weight 230.2688
Wiswesser Line Notation ZR BZ DOR CZ DZ
Evaluation B

$C_{12}H_{14}O_4$ (liq) 67CHA/HOR
Diethyl o–phthalate
Heat Capacity 298.15 K, C_p = 87.51 cal·mol⁻¹·K⁻¹ 366.15 J·mol⁻¹·K⁻¹
Temperature range 10–360 K. Glass transition temperature about 180 K. Also data for annealed glass and quenched glass 10–170 K.
Entropy 298.15 K, S = 101.60 cal·mol⁻¹·K⁻¹ 425.08 J·mol⁻¹·K⁻¹
Phase Changes
c/liq 269.92 K, ΔH = 4298 cal·mol⁻¹ 17984 J·mol⁻¹
ΔS = 15.92 cal·mol⁻¹·K⁻¹ 66.63 J·mol⁻¹·K⁻¹
Molecular Weight 222.2402
Wiswesser Line Notation 2OVR BVO2
Evaluation A

$C_{12}H_{14}O_4$ (liq) 79FUC
Diethyl o–phthalate
Heat Capacity 298.15 K, C_p = 85.5 cal·mol⁻¹·K⁻¹ 357.7 J·mol⁻¹·K⁻¹
One temperature
Molecular Weight 222.2402
Wiswesser Line Notation 2OVR BVO2
Evaluation B

$C_{12}H_{14}O_4$ (liq) 56SMI/DOL
Diethyl p–phthalate; Diethyl terephthalate
Heat Capacity 320 K, C_p = 91 cal·mol⁻¹·K⁻¹ 381 J·mol⁻¹·K⁻¹
Temperature range 43 to 75°C. Equation only.
Phase Changes
c/liq ΔH = 5900 cal·mol⁻¹ 24600 J·mol⁻¹
Melting temperature not given. Premelting over a 10 K range was observed.
Molecular Weight 222.2402
Wiswesser Line Notation 2OVR DVO2
Evaluation C

$C_{12}H_{18}$ (c,I) 30HUF/PAR
Hexamethylbenzene
Heat Capacity 294.6 K, C_p = 60.9 cal·mol⁻¹·K⁻¹ 254.8 J·mol⁻¹·K⁻¹
Temperature range 85–294 K. Value is unsmoothed experimental datum.
Entropy 298.1 K, S = 74.0 cal·mol⁻¹·K⁻¹ 309.6 J·mol⁻¹·K⁻¹
Extrapolation below 90 K, 19.69 cal·mol⁻¹K⁻¹
Phase Changes
c,III/c,II 108 K, ΔH = 243 cal·mol⁻¹ 1017 J·mol⁻¹
ΔS = 2.3 cal·mol⁻¹·K⁻¹ 9.5 J·mol⁻¹·K⁻¹
c,II/c,I 151 K, ΔH = 37 cal·mol⁻¹ 155 J·mol⁻¹
ΔS = 0.24 cal·mol⁻¹·K⁻¹ 1.0 J·mol⁻¹·K⁻¹
Molecular Weight 162.2742
Wiswesser Line Notation 1R B1 C1 D1 E1 F1
Evaluation B(C_p),C(S)

$C_{12}H_{18}$ (c,II) 32SPA/THO
Hexamethylbenzene
Heat Capacity 303 K, C_p = 61.7 cal·mol⁻¹·K⁻¹ 258.5 J·mol⁻¹·K⁻¹
Temperature range 30 to 200°C
Phase Changes
c,II/c,I 383.7 K, ΔH = 422 cal·mol⁻¹ 1766 J·mol⁻¹
ΔS = 1.1 cal·mol⁻¹·K⁻¹ 4.60 J·mol⁻¹·K⁻¹
c,I/liq 438.7 K, ΔH = 4933 cal·mol⁻¹ 20640 J·mol⁻¹
ΔS = 11.2 cal·mol⁻¹·K⁻¹ 47.0 J·mol⁻¹·K⁻¹
Molecular Weight 162.2742
Wiswesser Line Notation 1R B1 C1 D1 E1 F1
Evaluation B

Table of Heat Capacities, Entropies, and Phase Transition Properties — Continued

$C_{12}H_{18}$ (c,I) 65FRA/AST
Hexamethylbenzene
Heat Capacity 298.15 K, C_p = 58.71 cal·mol⁻¹·K⁻¹
245.64 J·mol⁻¹·K⁻¹
Temperature range 13–340 K
Entropy 298.15 K, S = 73.21 cal·mol⁻¹·K⁻¹
306.31 J·mol⁻¹·K⁻¹
Phase Changes
c,II/c,I 116.48 K, ΔH = 268.7 cal·mol⁻¹
1128.4 J·mol⁻¹
ΔS = 2.41 cal·mol⁻¹·K⁻¹
10.08 J·mol⁻¹·K⁻¹
Entropy obtained as difference of integral of observed C_p over range 115–128 K, and integral of extrapolated C_p data.
Molecular Weight 162.2742
Wiswesser Line Notation 1R B1 C1 D1 E1 F1
Evaluation A

$C_{12}H_{18}$ (liq) 47KUR
Hexamethylbenzene
Heat Capacity 455 K, C_p = 88.6 cal·mol⁻¹·K⁻¹
370.7 J·mol⁻¹·K⁻¹
Temperature range 183 to 256 °C, mean C_p, two temperatures.
Molecular Weight 162.2742
Wiswesser Line Notation 1R B1 C1 D1 E1 F1
Evaluation D

$C_{12}H_{19}N$ (c) 80RAD/RAD
Carbazole
Heat Capacity 298.15 K, C_p = 45.6 cal·mol⁻¹·K⁻¹
190.8 J·mol⁻¹·K⁻¹
Temperature range 180–410 K. Data given graphically. C_p calculated from equation.
Phase Changes
c,I/liq 521.2 K, ΔH = 6501 cal·mol⁻¹
27200 J·mol⁻¹
ΔS = 12.47 cal·mol⁻¹·K⁻¹
52.19 J·mol⁻¹·K⁻¹
Molecular Weight 177.2888
Wiswesser Line Notation T B656 HMJ
Evaluation B

$C_{12}H_{20}$ (liq) 63GUD/CAM
Perhydromethylcyclopentadiene dimer
Heat Capacity 313 K, C_p = 70.0 cal·mol⁻¹·K⁻¹
292.2 J·mol⁻¹·K⁻¹
Temperature range 313–523 K
Molecular Weight 164.2900
Wiswesser Line Notation L B545TJ X1 X1
Evaluation C

$C_{12}H_{20}$ (liq) 62GOL/BEL
Tetracyclo[6.2.1.1$^{3.6}$]dodecane
Heat Capacity 311 K, C_p = 66.9 cal·mol⁻¹·K⁻¹
279.9 J·mol⁻¹·K⁻¹
Temperatures 100, 200, 300°F
Molecular Weight 164.2900
Wiswesser Line Notation L D595 A D- TJ
Evaluation C

$(C_{12}H_{20}O_4)_n$ (c) 58WUN/DOL
Poly(ethylenesebacate)
Heat Capacity 298 K, C_p = 112 cal·mol⁻¹·K⁻¹
469 J·mol⁻¹·K⁻¹
Temperature range 256–408 K Value per repeating monomer units.
Phase Changes
c/liq 342 K
Molecular Weight 228.2876
Wiswesser Line Notation /*VO2OV8*/
Evaluation C

$C_{12}H_{22}$ (liq) 62GOL/BEL
Isopropylhydroindan
Heat Capacity 311 K, C_p = 78.8 cal·mol⁻¹·K⁻¹
329.7 J·mol⁻¹·K⁻¹
Temperatures 100, 200, 300°F
Molecular Weight 166.3058
Wiswesser Line Notation L56TJ XY1&1
Evaluation C

$C_{12}H_{22}$ (liq) 63GUD/CAM
Isopropylhydroindan
Heat Capacity 313 K, C_p = 78.2 cal·mol⁻¹·K⁻¹
327.2 J·mol⁻¹·K⁻¹
Temperature range 313–423 K
Molecular Weight 166.3058
Wiswesser Line Notation L56TJ XY1&1
Evaluation C

$C_{12}H_{22}$ (liq) 62GOL/BEL
α–Ethyldecalin
Heat Capacity 311 K, C_p = 72.5 cal·mol⁻¹·K⁻¹
303.3 J·mol⁻¹·K⁻¹
Temperatures 100, 200, 300°F
Molecular Weight 166.3058
Wiswesser Line Notation L66TJ B2
Evaluation C

$C_{12}H_{22}$ (liq) 63GUD/CAM
α–Ethyldecalin
Heat Capacity 313 K, C_p = 73.0 cal·mol⁻¹·K⁻¹
305.4 J·mol⁻¹·K⁻¹
Temperature range 313–483 K
Molecular Weight 166.3058
Wiswesser Line Notation L66TJ B2
Evaluation C

$C_{12}H_{22}$ (liq) 62GOL/BEL
β-Ethyldecalin
Heat Capacity 311 K, C_p = 69.4 cal·mol⁻¹·K⁻¹
290.4 J·mol⁻¹·K⁻¹
Temperatures 100, 200, 300°F
Molecular Weight 166.3058
Wiswesser Line Notation L66TJ C2
Evaluation C

$C_{12}H_{22}$ (liq) 63GUD/CAM
β-Ethyldecalin
Heat Capacity 313 K, C_p = 69.9 cal·mol⁻¹·K⁻¹
292.5 J·mol⁻¹·K⁻¹
Temperature range 313–483 K
Molecular Weight 166.3058
Wiswesser Line Notation L66TJ C2
Evaluation C

Table of Heat Capacities, Entropies, and Phase Transition Properties — Continued

$C_{12}H_{22}$ (liq) 63GUD/CAM
Dimethyldecalin
Heat Capacity 313 K, C_p = 72.5 cal·mol^{-1}·K^{-1}
303.3 J·mol^{-1}·K^{-1}
Temperature range 313–483 K
Molecular Weight 166.3058
Wiswesser Line Notation L66TJ X1 X1
Evaluation C

$C_{12}H_{22}$ (liq) 63GUD/CAM
Ethyldecalin
Heat Capacity 313 K, C_p = 71.2 cal·mol^{-1}·K^{-1}
318.8 J·mol^{-1}·K^{-1}
Temperature range 313–423 K
Molecular Weight 166.3058
Wiswesser Line Notation L66TJ X2
Evaluation C

$C_{12}H_{22}$ (liq) 62GOL/BEL
Bicyclohexyl
Heat Capacity 311 K, C_p = 71.8 cal·mol^{-1}·K^{-1}
300.4 J·mol^{-1}·K^{-1}
Temperatures 100, 200, 300°F
Molecular Weight 166.3058
Wiswesser Line Notation L6TJ A– AL6TJ
Evaluation C

$C_{12}H_{22}$ (liq) 63GUD/CAM
Bicyclohexyl
Heat Capacity 313 K, C_p = 71.7 cal·mol^{-1}·K^{-1}
300 J·mol^{-1}·K^{-1}
Temperature range 313–483 K
Molecular Weight 166.3058
Wiswesser Line Notation L6TJ A– AL6TJ
Evaluation C

$C_{12}H_{22}O_2$ (liq) 63GUD/CAM
Dimethoxydecalin
Heat Capacity 313 K, C_p = 79.9 cal·mol^{-1}·K^{-1}
334.3 J·mol^{-1}·K^{-1}
Temperature range 313–423 K
Molecular Weight 198.3046
Wiswesser Line Notation L66TJ XO1 XO1
Evaluation D

$C_{12}H_{22}O_{11}$ (c) 33PAR/HUF
Sucrose
Heat Capacity 297.0 K, C_p = 100.98 cal·mol^{-1}·K^{-1}
422.50 J·mol^{-1}·K^{-1}
Temperature range 94–297 K. Value is unsmoothed experimental datum.
Entropy 298.1 K, S = 86.1 cal·mol^{-1}·K^{-1}
360.2 J·mol^{-1}·K^{-1}
Extrapolation below 90 K, 27.06 cal·mol^{-1}K^{-1}
Molecular Weight 342.2992
Wiswesser Line Notation T6OTJ B1Q CQ DQ EQ FO– BT5OTJ B1Q CQ DQ E1Q –A&BD –B&CEF –A&BD –B&CE
Evaluation B(C_p),C(S)

$C_{12}H_{22}O_{11}$ (c) 50AND/HIG
Sucrose
Heat Capacity 298.15 K, C_p = 101.7 cal·mol^{-1}·K^{-1}
425.5 J·mol^{-1}·K^{-1}
Temperature range 298–363 K
Molecular Weight 342.2992
Wiswesser Line Notation T6OTJ B1Q CQ DQ EQ FO– BT5OTJ B1Q CQ DQ E1Q –A&BD –B&CEF –A&BD –B&CE
Evaluation B

$C_{12}H_{22}O_{11}$ (c) 36FUR/STE
β–Lactose
Heat Capacity 298.44 K, C_p = 97.58 cal·mol^{-1}·K^{-1}
408.27 J·mol^{-1}·K^{-1}
Temperature range 83–298 K. Value is unsmoothed experimental datum.
Entropy 298 K, S = 96.4 cal·mol^{-1}·K^{-1}
403.3 J·mol^{-1}·K^{-1}
Extrapolation below 90 K, 26.00 cal·mol^{-1}K^{-1}
Molecular Weight 342.2992
Wiswesser Line Notation T6OTJ BO CO DQ F1Q EO– BT6OTJ CO DO EO F1O –A&CE –B&BDF –A&C –B&BDEF
Evaluation B(C_p),C(S)

$C_{12}H_{22}O_{11}$ (c) 41AND/STE
β–Lactose
Heat Capacity 289.44 K, C_p = 95.55 cal·mol^{-1}·K^{-1}
399.78 J·mol^{-1}·K^{-1}
Temperature range 65–290 K. Value is unsmoothed experimental datum.
Entropy 298.15 K, S = 92.3 cal·mol^{-1}·K^{-1}
386.2 J·mol^{-1}·K^{-1}
Extrapolation below 90 K, 22.6 cal.mol^{-1}K^{-1}
Molecular Weight 342.2992
Wiswesser Line Notation T6OTJ BQ CQ DQ F1Q EO– BT6OTJ CQ DQ EQ F1Q –A&CE –B&BDF –A&C –B&BDEF
Evaluation B(C_p), C(S)

$C_{12}H_{22}O_{11}$ (c) 81KAW/NIS
Lactose
Heat Capacity 300 K, C_p = 99.8 cal·mol^{-1}·K^{-1}
417.6 J·mol^{-1}·K^{-1}
Temperature range 270–325 K, C_p given as 1.22 J·g^{-1}·K^{-1}.
Molecular Weight 342.2992
Wiswesser Line Notation T6OTJ BQ CQ DQ F1Q EO– BT6OTJ CQ DQ EQ F1Q –A&CE –B&BDF –A&C –B&BDEF
Evaluation A

$C_{12}H_{22}O_{11}$ (c) 81KAW/NIS
Maltose
Heat Capacity 300 K, C_p = 103.9 cal·mol^{-1}·K^{-1}
434.7 J·mol^{-1}·K^{-1}
Temperature range 270–325 K, C_p given as 1.2T J·g^{-1}·K^{-1}.
Molecular Weight 342.2992
Wiswesser Line Notation T6OTJ BQ CQ DQ F1Q EO– BT6OTJ CQ DQ EQ F1Q –A&CE –B&DF –A&BCE –B&DF
Evaluation A

$C_{12}H_{22}O_{11}$•H_2O (c) 41AND/STE
α–Lactose monohydrate
Heat Capacity 297.44 K, C_p = 105.06 cal·mol^{-1}·K^{-1}
439.57 J·mol^{-1}·K^{-1}
Temperature range 60–298 K. Value is unsmoothed experimental datum.
Entropy 298.15 K, S = 99.1 cal·mol^{-1}·K^{-1}
414.6 J·mol^{-1}·K^{-1}
Extrapolation below 90 K, 24.2 cal.mol^{-1}K^{-1}
Molecular Weight 360.3144
Wiswesser Line Notation T6OTJ BQ CQ DQ F1Q EO– BT6OTJ CQ DQ EQ F1Q –A&BCE –B&DF –A&C –B&BDEF &QH
Evaluation B(C_p),C(S)

Table of Heat Capacities, Entropies, and Phase Transition Properties — Continued

$C_{12}H_{22}O_{11}$•H_2O (c) 41AND/STE
β–Maltose monohydrate
Heat Capacity 296.27 K, C_p = 107.72 cal·mol^{-1}·K^{-1}
450.70 J·mol^{-1}·K^{-1}
Temperature range 60–298 K. Value is unsmoothed experimental datum.
Entropy 298.15 K, S = 99.8 cal·mol^{-1}·K^{-1}
417.6 J·mol^{-1}·K^{-1}
Extrapolation below 90 K, 24.2 cal·mol^{-1}K^{-1}
Molecular Weight 360.3144
Wiswesser Line Notation T6OTJ BQ CQ DQ F1Q EO– BT6OTJ CQ DQ EQ F1Q –A&CE –B&BDF –A&BCE –B&DF &QH
Evaluation B(C_p), C(S)

$C_{12}H_{24}$ (liq) 57MCC/FIN 2
1–Dodecene
Heat Capacity 298.15 K, C_p = 86.20 cal·mol^{-1}·K^{-1}
360.66 J·mol^{-1}·K^{-1}
Temperature range 11–360 K
Entropy 298.15 K, S = 115.87 cal·mol^{-1}·K^{-1}
484.80 J·mol^{-1}·K^{-1}
Does not include S_0.
Phase Changes
c,II/c,I 212.9 K, ΔH = 1088 cal·mol^{-1}
4552 J·mol^{-1}
ΔS = 5.11 cal·mol^{-1}·K^{-1}
21.38 J·mol^{-1}·K^{-1}
c,I/liq 237.93 K, ΔH = 4758 cal·mol^{-1}
19907 J·mol^{-1}
ΔS = 20.00 cal·mol^{-1}·K^{-1}
83.67 J·mol^{-1}·K^{-1}
Molecular Weight 168.3216
Wiswesser Line Notation 11U1
Evaluation A

$C_{12}H_{24}N_2O_2$ (c) 53WIL/DOL
N,N′–Di–n–propyladipamide
Heat Capacity 393 K, C_p = 111.9 cal·mol^{-1}·K^{-1}
468.1 J·mol^{-1}·K^{-1}
Temperature range 393–583 K
Phase Changes
c/liq 452 K, ΔH = 8630 cal·mol^{-1}
36110 J·mol^{-1}
ΔS = 19.1 cal·mol^{-1}·K^{-1}
79.9 J·mol^{-1}·K^{-1}
Molecular Weight 228.3338
Wiswesser Line Notation 3MV4VM3
Evaluation C

$C_{12}H_{24}O_2$ (c) 85STO/WIL
Dodecanoic acid; Lauric acid
Heat Capacity 279 K, C_p = 91 cal·mol^{-1}·K^{-1}
381 J·mol^{-1}·K^{-1}
Temperature range 0 to 100°C. Mean value 0 to 12°C.
Phase Changes
c/liq 327 K, ΔH = 10740 cal·mol^{-1}
44940 J·mol^{-1}
ΔS = 33 cal·mol^{-1}·K^{-1}
137 J·mol^{-1}·K^{-1}
Molecular Weight 200.3204
Wiswesser Line Notation QV11
Evaluation D

$C_{12}H_{24}O_2$ (c) 24GAR/RAN
Dodecanoic acid; Lauric acid
Heat Capacity 301 K, C_p = 102.5 cal·mol^{-1}·K^{-1}
428.9 J·mol^{-1}·K^{-1}
Temperature range 18 to 78°C. Mean value 19 to 39°C.
Phase Changes
c/liq 316.9 K, ΔH = 8760 cal·mol^{-1}
36650 J·mol^{-1}
ΔS = 27.6 cal·mol^{-1}·K^{-1}
115.7 J·mol^{-1}·K^{-1}
Molecular Weight 200.3204
Wiswesser Line Notation QV11
Evaluation B

$C_{12}H_{24}O_2$ (c) 82SCH/MIL 2
Dodecanoic acid; Lauric acid
Heat Capacity 298.15 K, C_p = 96.625 cal·mol^{-1}·K^{-1}
404.28 J·mol^{-1}·K^{-1}
Temperature range 80–345 K
Phase Changes
c,I/liq 316.98 K, ΔH = 8674.7 cal·mol^{-1}
36295 J·mol^{-1}
ΔS = 27.369 cal·mol^{-1}·K^{-1}
114.51 J·mol^{-1}·K^{-1}
Molecular Weight 200.3204
Wiswesser Line Notation QV11
Evaluation B

$C_{12}H_{26}$ (liq) 31HUF/PAR
n–Dodecane
Heat Capacity 297.7 K, C_p = 88.7 cal·mol^{-1}·K^{-1}
371.1 J·mol^{-1}·K^{-1}
Temperature range 93–298 K. Value is unsmoothed experimental datum.
Entropy 298.1 K, S = 118.1 cal·mol^{-1}·K^{-1}
497.1 J·mol^{-1}·K^{-1}
Extrapolation below 90 K, 25.10 cal·mol^{-1}K^{-1}
Phase Changes
c/liq 263.5 K, ΔH = 8743 cal·mol^{-1}
36581 J·mol^{-1}
ΔS = 33.2 cal·mol^{-1}·K^{-1}
138.8 J·mol^{-1}·K^{-1}
Molecular Weight 170.3374
Wiswesser Line Notation 12H
Evaluation B(C_p),C(S)

$C_{12}H_{26}$ (liq) 54FIN/GRO 2
n–Dodecane
Heat Capacity 298.15 K, C_p = 89.86 cal·mol^{-1}·K^{-1}
375.93 J·mol^{-1}·K^{-1}
Temperature range 12–320 K
Entropy 298.15 K, S = 117.27 cal·mol^{-1}·K^{-1}
490.66 J·mol^{-1}·K^{-1}
Phase Changes
c/liq 263.59 K, ΔH = 8804 cal·mol^{-1}
36836 J·mol^{-1}
ΔS = 33.40 cal·mol^{-1}·K^{-1}
139.75 J·mol^{-1}·K^{-1}
Molecular Weight 170.3374
Wiswesser Line Notation 12H
Evaluation A

Table of Heat Capacities, Entropies, and Phase Transition Properties — Continued

$C_{12}H_{26}O$ (liq) 79SVE
1-Dodecanol; n-Dodecyl alcohol
Heat Capacity 316 K, C_p = 110.4 cal·mol^{-1}·K^{-1}
462 J·mol^{-1}·K^{-1}
Temperature range 316–486 K
Phase Changes
liq/g 343.15 K, ΔH = 20235 cal·mol^{-1}
84670 J·mol^{-1}
ΔS = 58.97 cal·mol^{-1}·K^{-1}
246.74 J·mol^{-1}·K^{-1}
No pressure measurement.
Molecular Weight 186.3368
Wiswesser Line Notation Q12
Evaluation B

$C_{12}H_{28}IN$ (c) 73JOH/MAR
Tetra-n-propylammonium iodide
Heat Capacity 298.15 K, C_p = 83.4 cal·mol^{-1}·K^{-1}
349.4 J·mol^{-1}·K^{-1}
Temperature range 12–310 K
Entropy 298.15 K, S = 103.3 cal·mol^{-1}·K^{-1}
432.2 J·mol^{-1}·K^{-1}
Phase Changes
c,II/c,I 218.3 K, ΔH = 322.6 cal·mol^{-1}
1350 J·mol^{-1}
ΔS = 1.48 cal·mol^{-1}·K^{-1}
6.19 J·mol^{-1}·K^{-1}
Molecular Weight 313.2644
Wiswesser Line Notation 3 K3&3&3 &I
Evaluation A

$C_{12}H_{36}ClCr_3O_{22}$ (c) 71SOR/TAC
Trisaquo hexacetate chromate chloride hexahydrate
Heat Capacity 278.084 K, C_p = 228.97 cal·mol^{-1}·K^{-1}
958.02 J·mol^{-1}·K^{-1}
Temperature range 1.5–280 K. Value is unsmoothed experimental datum.
Phase Changes
c,II/c,I 211.4 K, ΔH = 794 cal·mol^{-1}
3322 J·mol^{-1}
ΔS = 3.29 cal·mol^{-1}·K^{-1}
13.778 J·mol^{-1}·K^{-1}
Two peaks were observed: at 211.4 K and 215.5 K, ΔH and ΔS given for overall transition.
Molecular Weight 723.8442
Wiswesser Line Notation CR3 O& QH 3 &OV1 6 &G &QH 6
Evaluation A

$C_{13}H_8N_8O_{15}$ (c) 24TAY/RIN
Tetryl-picric acid complex
Heat Capacity 293 K, C_p = 138.4 cal·mol^{-1}·K^{-1}
433.9 J·mol^{-1}·K^{-1}
Temperature range 90–352 K
Molecular Weight 516.2508
Wiswesser Line Notation WNN1&R BNW DNW FNW &WNR BQ CNW ENW
Evaluation C

$C_{13}H_{10}$ (c) 44EIB
Fluorene; Diphenylenemethane
Heat Capacity 298.1 K, C_p = 46.3 cal·mol^{-1}·K^{-1}
189.5 J·mol^{-1}·K^{-1}
Temperature range 25 to 200 °C, equations only, in t°C.
C_p(c) = 0.2479 + 0.001233t cal g^{-1}C^{-1} (25 to 70°C);
C_p(liq) = 0.320 + 0.00845t cal g^{-1}C^{-1}(114 to 200°C).
Phase Changes
c/liq 387.0 K, ΔH = 4750 cal·mol^{-1}
19870 J·mol^{-1}
ΔS = 12.3 cal·mol^{-1}·K^{-1}
51.3 J·mol^{-1}·K^{-1}
Molecular Weight 166.2220
Wiswesser Line Notation L B656 HHJ
Evaluation C

$C_{13}H_{10}$ (c,I) 77FIN/MES
Fluorene; Diphenylenemethane
Heat Capacity 298.15 K, C_p = 48.55 cal·mol^{-1}·K^{-1}
203.13 J·mol^{-1}·K^{-1}
Temperature range 10–440 K
Entropy 298.15 K, S = 49.55 cal·mol^{-1}·K^{-1}
207.32 J·mol^{-1}·K^{-1}
Phase Changes
c,II/c,I 288 K
Second order transition
c,I/liq 387.94 K, ΔH = 4679.3 cal·mol^{-1}
19578.2 J·mol^{-1}
ΔS = 12.06 cal·mol^{-1}·K^{-1}
50.47 J·mol^{-1}·K^{-1}
Molecular Weight 166.2220
Wiswesser Line Notation L B656 HHJ
Evaluation A

$C_{13}H_{10}O_3$ (c) 58SIN/HIL
Diphenyl carbonate
Heat Capacity 298.15 K, C_p = 62.89 cal·mol^{-1}·K^{-1}
263.13 J·mol^{-1}·K^{-1}
Temperature range 15–310 K
Entropy 298.15 K, S = 66.54 cal·mol^{-1}·K^{-1}
278.40 J·mol^{-1}·K^{-1}
Molecular Weight 214.2202
Wiswesser Line Notation ROVOR
Evaluation A

$C_{13}H_{11}Cl$ (liq) 31SMI/AND
Diphenylchloromethane
Heat Capacity 298.5 K, C_p = 69.4 cal·mol^{-1}·K^{-1}
290.4 J·mol^{-1}·K^{-1}
Temperature range 102–311 K. Value is unsmoothed experimental datum.
Molecular Weight 202.6829
Wiswesser Line Notation GYR&R
Evaluation C

$C_{13}H_{12}$ (c) 30HUF/PAR
Diphenylmethane
Heat Capacity 282.5 K, C_p = 53.5 cal·mol^{-1}·K^{-1}
223.8 J·mol^{-1}·K^{-1}
Temperature range 89–312 K. Value is unsmoothed experimental datum.
Entropy 298.1 K, S = 57.2 cal·mol^{-1}·K^{-1}
239.3 J·mol^{-1}·K^{-1}
Extrapolation below 90 K, 18.61 cal·mol^{-1}K^{-1}
Phase Changes
c/liq 298.3 K, ΔH = 4438 cal·mol^{-1}
18569 J·mol^{-1}
ΔS = 14.88 cal·mol^{-1}·K^{-1}
62.25 J·mol^{-1}·K^{-1}
Molecular Weight 168.2378
Wiswesser Line Notation R1R
Evaluation B(C_p),C(S)

Table of Heat Capacities, Entropies, and Phase Transition Properties — Continued

$C_{13}H_{12}$ (liq) 31SMI/AND
Diphenylmethane
Heat Capacity 298.5 K, C_p = 55.8 cal·mol⁻¹·K⁻¹ 233.5 J·mol⁻¹·K⁻¹
Temperature range 102–322 K. Value is unsmoothed experimental datum.
Molecular Weight 168.2378
Wiswesser Line Notation R1R
Evaluation C

$C_{13}H_{12}$ (liq) 50KUR
Diphenylmethane
Heat Capacity 300 K, C_p = 66.9 cal·mol⁻¹·K⁻¹ 279.9 J·mol⁻¹·K⁻¹
Temperature range 29 to 254°C.
Phase Changes
c/liq 299.65 K
Molecular Weight 168.2378
Wiswesser Line Notation R1R
Evaluation B

$C_{13}H_{12}$ (liq) 56DUF/EVE
Diphenylmethane
Heat Capacity 303 K, C_p = 63.6 cal·mol⁻¹·K⁻¹ 266.1 J·mol⁻¹·K⁻¹
Temperature range 303–353 K
Molecular Weight 168.2378
Wiswesser Line Notation R1R
Evaluation B

$C_{13}H_{12}O$ (c) 31SMI/AND
Diphenylcarbinol
Heat Capacity 298.5 K, C_p = 56.6 cal·mol⁻¹·K⁻¹ 236.8 J·mol⁻¹·K⁻¹
Temperature range 102–299 K. Value is unsmoothed experimental datum.
Molecular Weight 184.2372
Wiswesser Line Notation QYR&R
Evaluation C

$C_{13}H_{13}N$ (liq) 81REI
Methyldiphenylamine
Heat Capacity 298 K, C_p = 72.0 cal·mol⁻¹·K⁻¹ 301.2 J·mol⁻¹·K⁻¹
Temperature range 293–467 K
Molecular Weight 183.2524
Wiswesser Line Notation 1NR&R
Evaluation D

$C_{13}H_{14}N_2$ (c) 78MAR/CIO
Bis(4–aminophenyl)methane
Heat Capacity 298 K, C_p = 15.9 cal·mol⁻¹·K⁻¹ 66.5 J·mol⁻¹·K⁻¹
Temperature range 298–469 K. Data seem odd; values for solid increase from 15.9 to 144.9 cal·mol⁻¹K⁻¹ at 370 K to 100.7 at 470 K.
Phase Changes
c/liq 363.7 K, ΔH = 2205 cal·mol⁻¹ 9225 J·mol⁻¹
ΔS = 6.1 cal·mol⁻¹·K⁻¹ 25.4 J·mol⁻¹·K⁻¹
Molecular Weight 198.2670
Wiswesser Line Notation ZR D1R DZ
Evaluation D

$C_{13}H_{14}N_2$ (liq) 66ZAL/STR
Bis(4–aminophenyl)methane
Heat Capacity 388 K, C_p = 54.3 cal·mol⁻¹·K⁻¹ 227.2 J·mol⁻¹·K⁻¹
Temperature range 110 to 120 °C, mean value.
Molecular Weight 198.2670
Wiswesser Line Notation ZR D1R DZ
Evaluation D

$C_{13}H_{15}NO$ (c) 40CAM/CAM
p–Toluidine–phenol complex; Phenol–p–toluidine complex
Heat Capacity 293 K, C_p = 51.8 cal·mol⁻¹·K⁻¹ 216.7 J·mol⁻¹·K⁻¹
One temperature
Molecular Weight 201.2676
Wiswesser Line Notation ZR D1 &QR
Evaluation C

$C_{13}H_{16}N_2O_3$ (c) 40CAM/CAM
Phenol–urea complex; Urea–phenol complex
Heat Capacity 293 K, C_p = 59.8 cal·mol⁻¹·K⁻¹ 250.2 J·mol⁻¹·K⁻¹
One temperature
Molecular Weight 248.2810
Wiswesser Line Notation ZVZ &QR 2
Evaluation C

$C_{13}H_{19}NO_2$ (c) 71PRI
Hexyl phenylcarbamate
Heat Capacity 298 K, C_p = 80.5 cal·mol⁻¹·K⁻¹ 336.8 J·mol⁻¹·K⁻¹
Temperature range 200–390 K. Complete data deposited VINITI, No. 2713–71, 25 March 1971.
Phase Changes
c/liq 328 K, ΔH = 7831 cal·mol⁻¹ 32765 J·mol⁻¹
ΔS = 23.9 cal·mol⁻¹·K⁻¹ 99.9 J·mol⁻¹·K⁻¹
Molecular Weight 221.2986
Wiswesser Line Notation 6OVMR
Evaluation B

$C_{13}H_{22}FeN_6S_3$ (c) 81SOR/OGA
Thiourea–ferrocene
Heat Capacity 273.15 K, C_p = 107.04 cal·mol⁻¹·K⁻¹ 447.86 J·mol⁻¹·K⁻¹
Temperature range 13–280 K
Entropy 273.15 K, S = 123.34 cal·mol⁻¹·K⁻¹ 516.05 J·mol⁻¹·K⁻¹
Phase Changes
c,V/c,IV 147.2 K, ΔH = 62.9 cal·mol⁻¹ 263 J·mol⁻¹
ΔS = 0.43 cal·mol⁻¹·K⁻¹ 1.8 J·mol⁻¹·K⁻¹
c,IV/c,III 159.79 K, ΔH = 352 cal·mol⁻¹ 1473 J·mol⁻¹
ΔS = 2.21 cal·mol⁻¹·K⁻¹ 9.24 J·mol⁻¹·K⁻¹
c,III/c,II 171.4 K, ΔH = 3.3 cal·mol⁻¹ 14 J·mol⁻¹
ΔS = 0.019 cal·mol⁻¹·K⁻¹ 0.079 J·mol⁻¹·K⁻¹
c,II/c,I 185.5 K, ΔH = 8.4 cal·mol⁻¹ 35 J·mol⁻¹
ΔS = 0.45 cal·mol⁻¹·K⁻¹ 0.188 J·mol⁻¹·K⁻¹

Table of Heat Capacities, Entropies, and Phase Transition Properties — Continued

c,III/c,I 220 K, ΔH = 18.4 cal·mol^{-1}
77 J·mol^{-1}
ΔS = 0.086 cal·mol^{-1}·K^{-1}
0.36 J·mol^{-1}·K^{-1}

Molecular Weight 414.3840
Wiswesser Line Notation L5ØJ Ø–FE– –ØL5ØJ 4ZYZUS 3
Evaluation A

$C_{13}H_{24}$ (liq) 62GOL/BEL
α-n-Propyldecalin
Heat Capacity 311 K, C_p = 80.2 cal·mol^{-1}·K^{-1}
335.6 J·mol^{-1}·K^{-1}
Temperatures 100, 200, 300°F
Molecular Weight 180.3326
Wiswesser Line Notation L66TJ B3
Evaluation C

$C_{13}H_{24}$ (liq) 63GUD/CAM
α-n-Propyldecalin
Heat Capacity 313 K, C_p = 80.6 cal·mol^{-1}·K^{-1}
337.2 J·mol^{-1}·K^{-1}
Temperature range 313–483 K
Molecular Weight 180.3326
Wiswesser Line Notation L66TJ B3
Evaluation C

$C_{13}H_{24}$ (liq) 62GOL/BEL
α-Isopropyldecalin
Heat Capacity 311 K, C_p = 75.6 cal·mol^{-1}·K^{-1}
316.3 J·mol^{-1}·K^{-1}
Temperatures 100, 200, 300°F
Molecular Weight 180.3326
Wiswesser Line Notation L66TJ BY1&1
Evaluation C

$C_{13}H_{24}$ (liq) 63GUD/CAM
α-Isopropyldecalin
Heat Capacity 313 K, C_p = 76.1 cal·mol^{-1}·K^{-1}
318.4 J·mol^{-1}·K^{-1}
Temperature range 313–483 K
Molecular Weight 180.3326
Wiswesser Line Notation L66TJ BY1&1
Evaluation C

$C_{13}H_{24}$ (liq) 63GUD/CAM
Isopropyldecalin
Heat Capacity 373 K, C_p = 89.3 cal·mol^{-1}·K^{-1}
373.6 J·mol^{-1}·K^{-1}
Temperature range 373–483 K
Molecular Weight 180.3326
Wiswesser Line Notation L66TJ XY1&1
Evaluation C

$C_{13}H_{24}$ (liq) 63GUD/CAM
2-Methylbicyclohexyl
Heat Capacity 313 K, C_p = 79.2 cal·mol^{-1}·K^{-1}
331.4 J·mol^{-1}·K^{-1}
Temperature range 313–483 K
Molecular Weight 180.3326
Wiswesser Line Notation L6TJ A– BL6TJ A1
Evaluation C

$C_{13}H_{24}$ (liq) 62GOL/BEL
Dicyclohexylmethane
Heat Capacity 311 K, C_p = 84.9 cal·mol^{-1}·K^{-1}
355.2 J·mol^{-1}·K^{-1}
Temperatures 100, 200, 300°F
Molecular Weight 180.3326
Wiswesser Line Notation L6TJ A1– AL6TJ
Evaluation C

$C_{13}H_{24}$ (liq) 63GUD/CAM
Dicyclohexylmethane
Heat Capacity 313 K, C_p = 76.5 cal·mol^{-1}·K^{-1}
320.1 J·mol^{-1}·K^{-1}
Temperature range 313–483 K
Molecular Weight 180.3326
Wiswesser Line Notation L6TJ A1– AL6TJ
Evaluation C

$C_{13}H_{24}O_2$ (c) 81LEB/YEV
Tridecanolactone
Heat Capacity 298.15 K, C_p = 95.20 cal·mol^{-1}·K^{-1}
398.3 J·mol^{-1}·K^{-1}
Temperature range 10.3–330 K
Entropy 298.15 K, S = 96.06 cal·mol^{-1}·K^{-1}
401.9 J·mol^{-1}·K^{-1}
Phase Changes
c,II/c,I 290.63 K, ΔH = 4338 cal·mol^{-1}
18150 J·mol^{-1}
ΔS = 14.9 cal·mol^{-1}·K^{-1}
62.3 J·mol^{-1}·K^{-1}
c,I/liq 300.4 K, ΔH = 2165 cal·mol^{-1}
9058 J·mol^{-1}
ΔS = 7.22 cal·mol^{-1}·K^{-1}
30.2 J·mol^{-1}·K^{-1}
Molecular Weight 212.3314
Wiswesser Line Notation T-14–VOTJ
Evaluation A

$C_{13}H_{26}$ (liq) 49PAR/MOO
n-Heptylcyclohexane
Heat Capacity 298.15 K, C_p = 86.8 cal·mol^{-1}·K^{-1}
363.2 J·mol^{-1}·K^{-1}
Temperature range 80–300 K
Entropy 298.15 K, S = 106.8 cal·mol^{-1}·K^{-1}
446.9 J·mol^{-1}·K^{-1}
Extrapolation below 80 K, 21.52 cal·mol^{-1}K^{-1}
Phase Changes
c/liq 232.8 K, ΔH = 5312 cal·mol^{-1}
22225 J·mol^{-1}
ΔS = 22.8 cal·mol^{-1}·K^{-1}
95.5 J·mol^{-1}·K^{-1}
Molecular Weight 182.3484
Wiswesser Line Notation L6TJ A7
Evaluation B(C_p),C(S)

$C_{13}H_{26}O_2$ (c) 82SCH/MIL
Tridecanoic acid
Heat Capacity 298.15 K, C_p = 92.64 cal·mol^{-1}·K^{-1}
387.6 J·mol^{-1}·K^{-1}
Temperature range 80–340 K.
Phase Changes
c,II/c,I 307.1 K, ΔH = 2087 cal·mol^{-1}
8730 J·mol^{-1}
ΔS = 6.790 cal·mol^{-1}·K^{-1}
28.41 J·mol^{-1}·K^{-1}
c,I/liq 315.01 K, ΔH = 8061.4 cal·mol^{-1}
33729 J·mol^{-1}
ΔS = 25.59 cal·mol^{-1}·K^{-1}
107.07 J·mol^{-1}·K^{-1}

Table of Heat Capacities, Entropies, and Phase Transition Properties — Continued

Molecular Weight 214.3472
Wiswesser Line Notation QV12
Evaluation B

$C_{13}H_{26}O_4$ (c) 65SIL/DAU
1–Monocaprin
Heat Capacity 298 K, C_p = 98.0 cal·mol⁻¹·K⁻¹ — 410.0 J·mol⁻¹·K⁻¹
One temperature. β_L form
Molecular Weight 246.3460
Wiswesser Line Notation Q1YQ1OV9
Evaluation B

$C_{13}H_{28}$ (liq) 54FIN/GRO 2
n–Tridecane
Heat Capacity 298.15 K, C_p = 97.25 cal·mol^{-1}·K^{-1} — 406.89 J·mol^{-1}·K^{-1}
Temperature range 12–310 K
Entropy 298.15 K, S = 124.97 cal·mol^{-1}·K^{-1} — 522.87 J·mol^{-1}·K^{-1}
Phase Changes

c,II/c,I	255.0 K,	ΔH =	1831 cal·mol^{-1} — 7661 J·mol^{-1}
		ΔS =	7.18 cal·mol^{-1}·K^{-1} — 30.04 J·mol^{-1}·K^{-1}
c,I/liq	267.79 K,	ΔH =	6812 cal·mol^{-1} — 28501 J·mol^{-1}
		ΔS =	25.44 cal·mol^{-1}·K^{-1} — 106.43 J·mol^{-1}·K^{-1}

Molecular Weight 184.3642
Wiswesser Line Notation 13H
Evaluation A

$C_{13}H_{28}O$ (c) 74MOS/MOU
1–Tridecanol; n–Tridecyl alcohol
Heat Capacity 298.15 K, C_p = 90.3 cal·mol^{-1}·K^{-1} — 378 J·mol^{-1}·K^{-1}
β–form, 276–299 K
Phase Changes

c,β/c,γ	301.6 K,	ΔH =	860 cal·mol^{-1} — 3600 J·mol^{-1}
		ΔS =	2.9 cal·mol^{-1}·K^{-1} — 11.9 J·mol^{-1}·K^{-1}
c,β/c,α	305.8 K,	ΔH =	5280 cal·mol^{-1} — 22100 J·mol^{-1}
		ΔS =	17.3 cal·mol^{-1}·K^{-1} — 72.3 J·mol^{-1}·K^{-1}
c,γ/c,α	306.6 K,	ΔH =	4470 cal·mol^{-1} — 18700 J·mol^{-1}
		ΔS =	14.6 cal·mol^{-1}·K^{-1} — 61.0 J·mol^{-1}·K^{-1}
c,β/liq	304.6 K,	ΔH =	10780 cal·mol^{-1} — 45120 J·mol^{-1}
		ΔS =	35.4 cal·mol^{-1}·K^{-1} — 148.1 J·mol^{-1}·K^{-1}
c,γ/liq	304.9 K,	ΔH =	9895 cal·mol^{-1} — 41400 J·mol^{-1}
		ΔS =	32.5 cal·mol^{-1}·K^{-1} — 135.8 J·mol^{-1}·K^{-1}
c,α/liq	303.5 K,	ΔH =	5570 cal·mol^{-1} — 23000 J·mol^{-1}
		ΔS =	18.4 cal·mol^{-1}·K^{-1} — 76.8 J·mol^{-1}·K^{-1}

Molecular Weight 200.3636
Wiswesser Line Notation Q13
Evaluation B

$C_{13}H_{28}O$ (liq) 74MOS/MOU
1–Tridecanol; n–Tridecyl alcohol
Heat Capacity 305 K, C_p = 113.8 cal·mol^{-1}·K^{-1} — 476 J·mol^{-1}·K^{-1}
Temperature range 305–346 K
Molecular Weight 200.3636
Wiswesser Line Notation Q13
Evaluation B

$C_{13}H_{30}N_2O$ (c,I) 65PEM/PAR
Urea–n–dodecane adduct
Heat Capacity 298.15 K, C_p = 29.88 cal·mol^{-1}·K^{-1} — 125.02 J·mol^{-1}·K^{-1}
Temperature range 12–300 K. Value for aduct with 1 mole of urea.
Entropy 298.15 K, S = 32.97 cal·mol^{-1}·K^{-1} — 137.95 J·mol^{-1}·K^{-1}
Phase Changes
Transition at 123.2 K with ΔH = 265 cal(mol hydrocarbon)$^{-1}$.
Molecular Weight 230.3928
Wiswesser Line Notation ZVZ &12H
Evaluation B
Sample 78.20 percent urea.

$C_{14}H_8O_2$ (c) 17HIL/DUS
Anthraquinone
Heat Capacity 298.15 K, C_p = 57.4 cal·mol^{-1}·K^{-1} — 240.2 J·mol^{-1}·K^{-1}
Temperature range 293–593 K. From heat content data.
Phase Changes

c/liq	555 K,	ΔH =	7785 cal·mol^{-1} — 32570 J·mol^{-1}
		ΔS =	14.0 cal·mol^{-1}·K^{-1} — 58.7 J·mol^{-1}·K^{-1}

Molecular Weight 208.2160
Wiswesser Line Notation L C666 BV IVJ
Evaluation C

$C_{14}H_{10}$ (c) 31HUF/PAR
Phenanthrene
Heat Capacity 297.5 K, C_p = 55.8 cal·mol^{-1}·K^{-1} — 233.5 J·mol^{-1}·K^{-1}
Temperature range 93–304 K. Value is unsmoothed experimental datum.
Entropy 298.1 K, S = 50.6 cal·mol^{-1}·K^{-1} — 211.7 J·mol^{-1}·K^{-1}
Extrapolation below 90 K, 15.58 cal·mol^{-1}K^{-1}
Molecular Weight 178.2330
Wiswesser Line Notation L B666J
Evaluation B(C_p),C(S)

$C_{14}H_{10}$ (c) 41SCH
Phenanthrene
Heat Capacity 298.1 K, C_p = 54.1 cal·mol^{-1}·K^{-1} — 226.4 J·mol^{-1}·K^{-1}
Temperature range 20 to 200 °C, equations only, in t°C. C_p(c) = 0.2440 + 0.002604t–0.0000111t^2 cal g^{-1}C^{-1}(20 to 98°C); C_p(liq) = 0.3328 + 0.0006760t cal g^{-1}C^{-1}(98 to 200°C).
Phase Changes

c/liq	371.7 K,	ΔH =	4096 cal·mol^{-1} — 17138 J·mol^{-1}
		ΔS =	11.0 cal·mol^{-1}·K^{-1} — 46.1 J·mol^{-1}·K^{-1}

Molecular Weight 178.2330
Wiswesser Line Notation L B666J
Evaluation C

Table of Heat Capacities, Entropies, and Phase Transition Properties — Continued

$C_{14}H_{10}$ (c) 44EIB
Phenanthrene
Heat Capacity 298.1 K, C_p = 49.5 cal·mol⁻¹·K⁻¹
207.1 J·mol⁻¹·K⁻¹
Temperature range 25 to 200 °C, equations only in t°C.
C_p(c) = 0.2003 + 0.00306t cal g⁻¹C⁻¹ (98 to 20°C);
C_p(liq) = 0.292 + 0.000923t cal g⁻¹C⁻¹ (98–200°C).
Phase Changes
c/liq 371.4 K, ΔH = 4100 cal·mol⁻¹
17150 J·mol⁻¹
ΔS = 11.0 cal·mol⁻¹·K⁻¹
46.2 J·mol⁻¹·K⁻¹
Molecular Weight 178.2330
Wiswesser Line Notation L B666J
Evaluation C

$C_{14}H_{10}$ (c,II) 50UEB/ORT
Phenanthrene
Heat Capacity 298.15 K, C_p = 32.2 cal·mol⁻¹·K⁻¹
134.7 J·mol⁻¹·K⁻¹
Temperature range 293–368 K. Equation only.
Phase Changes
c,II/c,I 342 K, ΔH = 607 cal·mol⁻¹
2600 J·mol⁻¹
ΔS = 1.8 cal·mol⁻¹·K⁻¹
7.6 J·mol⁻¹·K⁻¹
c,I/liq 373 K, ΔH = 4450 cal·mol⁻¹
18620 J·mol⁻¹
ΔS = 11.9 cal·mol⁻¹·K⁻¹
49.9 J·mol⁻¹·K⁻¹
Molecular Weight 178.2330
Wiswesser Line Notation L B666J
Evaluation C

$C_{14}H_{10}$ (c) 64RAS/BAS
Phenanthrene
Heat Capacity 343 K, C_p = 63.9 cal·mol⁻¹·K⁻¹
267.4 J·mol⁻¹·K⁻¹
Temperature range 343, 404 K
Phase Changes
c/liq 373.2 K, ΔH = 4302 cal·mol⁻¹
18000 J·mol⁻¹
ΔS = 11.5 cal·mol⁻¹·K⁻¹
48.2 J·mol⁻¹·K⁻¹
Molecular Weight 178.2330
Wiswesser Line Notation L B666J
Evaluation B

$C_{14}H_{10}$ (c,II) 77FIN/MES
Phenanthrene
Heat Capacity 298.15 K, C_p = 52.73 cal·mol⁻¹·K⁻¹
220.62 J·mol⁻¹·K⁻¹
Temperature range 10–440 K
Entropy 298.15 K, S = 51.40 cal·mol⁻¹·K⁻¹
215.06 J·mol⁻¹·K⁻¹
Phase Changes
c,III/c,II ~270 K
Second–order glass–type transition.
c,II/c,I 347.5 K, ΔH = 52 cal·mol⁻¹
218 J·mol⁻¹
ΔS = 0.15 cal·mol⁻¹·K⁻¹
0.63 J·mol⁻¹·K⁻¹
Lambda transition.
c,I/liq 372.38 K, ΔH = 3934.7 cal·mol⁻¹
16462.8 J·mol⁻¹
ΔS = 10.57 cal·mol⁻¹·K⁻¹
44.21 J·mol⁻¹·K⁻¹
Molecular Weight 178.2330
Wiswesser Line Notation L B666J
Evaluation A

$C_{14}H_{10}$ (liq) 17HIL/DUS
Anthracene
Heat Capacity 298.15 K, C_p = 53.0 cal·mol⁻¹·K⁻¹
221.8 J·mol⁻¹·K⁻¹
Temperature range 293–593 K. From heat content data.
Phase Changes
c/liq 489.7 K, ΔH = 6900 cal·mol⁻¹
28870 J·mol⁻¹
ΔS = 14.1 cal·mol⁻¹·K⁻¹
59.0 J·mol⁻¹·K⁻¹
Molecular Weight 178.2330
Wiswesser Line Notation L C666J
Evaluation C

$C_{14}H_{10}$ (c) 31HUF/PAR
Anthracene
Heat Capacity 297.2 K, C_p = 49.5 cal·mol⁻¹·K⁻¹
207.1 J·mol⁻¹·K⁻¹
Temperature range 94–297 K. Value is unsmoothed experimental datum.
Entropy 298.1 K, S = 49.6 cal·mol⁻¹·K⁻¹
207.5 J·mol⁻¹·K⁻¹
Extrapolation below 90 K, 14.98 cal·mol⁻¹K⁻¹
Molecular Weight 178.2330
Wiswesser Line Notation L C666J
Evaluation B(C_p),C(S)

$C_{14}H_{10}$ (c) 50UEB/ORT
Anthracene
Heat Capacity 298.15 K, C_p = 50.8 cal·mol⁻¹·K⁻¹
217.5 J·mol⁻¹·K⁻¹
Temperature range 293–368 K. Equation only.
Phase Changes
c/liq 490 K, ΔH = 6890 cal·mol⁻¹
28830 J·mol⁻¹
ΔS = 14.1 cal·mol⁻¹·K⁻¹
58.0 J·mol⁻¹·K⁻¹
Molecular Weight 178.2330
Wiswesser Line Notation L C666J
Evaluation C

$C_{14}H_{10}$ (c) 68GOU/GIR
Anthracene
Heat Capacity 298.15 K, C_p = 50.31 cal·mol⁻¹·K⁻¹
210.50 J·mol⁻¹·K⁻¹
Temperature range 5–520 K. Only 6 points given; summary article.
Entropy 298.15 K, S = 49.51 cal·mol⁻¹·K⁻¹
207.15 J·mol⁻¹·K⁻¹
Molecular Weight 178.2330
Wiswesser Line Notation L C666J
Evaluation A

$C_{14}H_{10}$ (c) 70GOU/GIR
Anthracene
Heat Capacity 298.15 K, C_p = 50.31 cal·mol⁻¹·K⁻¹
210.50 J·mol⁻¹·K⁻¹
Temperature range 5–500 K
Entropy 298.15 K, S = 49.51 cal·mol⁻¹·K⁻¹
207.15 J·mol⁻¹·K⁻¹
Phase Changes
c/liq 488.93 K, ΔH = 7020 cal·mol⁻¹
29372 J·mol⁻¹
ΔS = 14.36 cal·mol⁻¹·K⁻¹
60.07 J·mol⁻¹·K⁻¹
Note that table of smoothed values indicates ΔH_m = 1550 cal·mol⁻¹ and ΔS_m = 3.17 cal·mol⁻¹K⁻¹.
Molecular Weight 178.2330
Wiswesser Line Notation L C666J
Evaluation A

Table of Heat Capacities, Entropies, and Phase Transition Properties — Continued

$C_{14}H_{10}$ (c) 80RAD/RAD
Anthracene
Heat Capacity 298.15 K, C_p = 50.6 cal·mol⁻¹·K⁻¹
211.7 J·mol⁻¹·K⁻¹
Temperature range 180–430 K. Data given graphically. C_p calculated from equation.
Phase Changes
c,I/liq 490.6 K, ΔH = 6931 cal·mol⁻¹
29000 J·mol⁻¹
ΔS = 14.31 cal·mol⁻¹·K⁻¹
59.1 J·mol⁻¹·K⁻¹
Molecular Weight 178.2330
Wiswesser Line Notation L C666J
Evaluation B

$C_{14}H_{10}$ (c) 31SMI/AND
Diphenylethyne; Diphenylacetylene
Heat Capacity 298.5 K, C_p = 54.0 cal·mol⁻¹·K⁻¹
225.9 J·mol⁻¹·K⁻¹
Temperature range 102–323 K. Value is unsmoothed experimental datum.
Molecular Weight 178.2330
Wiswesser Line Notation R1UU1R
Evaluation C

$C_{14}H_{10}$ (liq) 50KUR
Diphenylethyne; Diphenylacetylene
Heat Capacity 340 K, C_p = 71.1 cal·mol⁻¹·K⁻¹
297.5 J·mol⁻¹·K⁻¹
Temperature range 65 to 303°C. Tm 61°C.
Molecular Weight 178.2330
Wiswesser Line Notation R1UU1R
Evaluation B

$C_{14}H_{10}O_2$ (c) 77DWO/FUC
Benzil; Diphenyl diketone
Heat Capacity C_p data given graphicallyonly.
Temperature range 60–100 K.
Phase Changes
c,II/c,I 84.07 K, ΔH = 10.5 cal·mol⁻¹
44.1 J·mol⁻¹
ΔS = 0.12 cal·mol⁻¹·K⁻¹
0.52 J·mol⁻¹·K⁻¹
Molecular Weight 210.2318
Wiswesser Line Notation RVVR
Evaluation B

$C_{14}H_{10}O_2$ (c) 80AND/CON
Benzil; Diphenyl diketone
Phase Changes
c,I/liq 368.022 K, ΔH = 5630 cal·mol⁻¹
23556 J·mol⁻¹
ΔS = 15.30 cal·mol⁻¹·K⁻¹
64.01 J·mol⁻¹·K⁻¹
Molecular Weight 210.2318
Wiswesser Line Notation RVVR
Evaluation A

$C_{14}H_{12}$ (liq) 31SMI/AND
1,1-Diphenylethylene
Heat Capacity 298.5 K, C_p = 71.5 cal·mol⁻¹·K⁻¹
299.2 J·mol⁻¹·K⁻¹
Temperature range 102–299 K. Value is unsmoothed experimental datum.
Molecular Weight 180.2488
Wiswesser Line Notation 1UYR&R
Evaluation C

$C_{14}H_{12}$ (c) 79LEE/HOS
9,10–Dihydrophenanthrene
Heat Capacity 298.15 K, C_p = 58.27 cal·mol⁻¹·K⁻¹
243.8 J·mol⁻¹·K⁻¹
Temperature range 10–350 K
Entropy 298.15 K, S = 54.83 cal·mol⁻¹·K⁻¹
229.4 J·mol⁻¹·K⁻¹
Phase Changes
c/liq 306.52 K, ΔH = 3057 cal·mol⁻¹
12790 J·mol⁻¹
ΔS = 9.97 cal·mol⁻¹·K⁻¹
41.73 J·mol⁻¹·K⁻¹
Molecular Weight 180.2488
Wiswesser Line Notation L B666&T&J
Evaluation A

$C_{14}H_{12}$ (c) 30PAR/HUF 2
Stilbene; 1,2–Diphenylethylene
Heat Capacity 292.8 K, C_p = 54.3 cal·mol⁻¹·K⁻¹
227.2 J·mol⁻¹·K⁻¹
Temperature range 92–293 K. Value is unsmoothed experimental datum.
Entropy 298.15 K, S = 60.0 cal·mol⁻¹·K⁻¹
251.0 J·mol⁻¹·K⁻¹
Extrapolation below 90 K, 19.81 cal·mol⁻¹K⁻¹
Molecular Weight 180.2488
Wiswesser Line Notation R1U1R
Evaluation B(C_p),C(S)

$C_{14}H_{12}$ (c) 31SMI/AND
Stilbene; 1,2–Diphenylethylene
Heat Capacity 298.5 K, C_p = 55.6 cal·mol⁻¹·K⁻¹
232.6 J·mol⁻¹·K⁻¹
Temperature range 102–346 K. Value is unsmoothed experimental datum.
Molecular Weight 180.2488
Wiswesser Line Notation R1U1R
Evaluation C

$C_{14}H_{12}$ (liq) 50KUR
Stilbene; 1,2–Diphenylethylene
Heat Capacity 410 K, C_p = 82.0 cal·mol⁻¹·K⁻¹
343.1 J·mol⁻¹·K⁻¹
Temperature range 135 to 305°C. Mp 124.2°C
Molecular Weight 180.2488
Wiswesser Line Notation R1U1R
Evaluation B

$C_{14}H_{12}O_2$ (c) 80AND/CON
Diphenylacetic acid
Phase Changes
c,I/liq 420.441 K, ΔH = 7474 cal·mol⁻¹
31271 J·mol⁻¹
ΔS = 17.78 cal·mol⁻¹·K⁻¹
74.38 J·mol⁻¹·K⁻¹
Molecular Weight 212.2476
Wiswesser Line Notation QVYR&R
Evaluation A

$(C_{14}H_{12}Si)_n$ (amorp) 77LEB/EVS
Polyvinylenediphenylsilane
Heat Capacity 298 K, C_p = 79.2 cal·mol⁻¹·K⁻¹
331.2 J·mol⁻¹·K⁻¹
Temperature range 13–334 K. Values per repeating unit. Data deposited VINITI, No. 2360–76, 24 June 1976.
Entropy 298 K, S = 71.2 cal·mol⁻¹·K⁻¹
298.0 J·mol⁻¹·K⁻¹
Molecular Weight 208.3343
Wiswesser Line Notation /*1U1-SI-*R&R/
Evaluation B

Table of Heat Capacities, Entropies, and Phase Transition Properties — Continued

$C_{14}H_{14}$ (liq) 31SMI/AND
1,1-Diphenylethane
Heat Capacity 298.5 K, C_p = 70.5 cal·mol^{-1}·K^{-1}
295.0 J·mol^{-1}·K^{-1}
Temperature range 102–299 K. Value is unsmoothed experimental datum.
Molecular Weight 182.2646
Wiswesser Line Notation 1YR&R
Evaluation C

$C_{14}H_{14}$ (c) 30HUF/PAR
1,2-Diphenylethane; Dibenzyl
Heat Capacity 293.6 K, C_p = 60.0 cal·mol^{-1}·K^{-1}
251.0 J·mol^{-1}·K^{-1}
Temperature range 93–294K. Value is unsmoothed experimental datum.
Entropy 298.1 K, S = 64.6 cal·mol^{-1}·K^{-1}
270.3 J·mol^{-1}·K^{-1}
Extrapolation below 90 K, 21.60 cal·mol^{-1}K^{-1}
Molecular Weight 182.2646
Wiswesser Line Notation R2R
Evaluation B(C_p),C(S)

$C_{14}H_{14}$ (c) 31SMI/AND
1,2-Diphenylethane; Dibenzyl
Heat Capacity 298.5 K, C_p = 60.6 cal·mol^{-1}·K^{-1}
253.6 J·mol^{-1}·K^{-1}
Temperature range 102–299 K. Value is unsmoothed experimental datum.
Molecular Weight 182.2646
Wiswesser Line Notation R2R
Evaluation C

$C_{14}H_{14}$ (c) 41SCH
1,2-Diphenylethane; Dibenzyl
Heat Capacity 298.1 K, C_p = 60.2 cal·mol^{-1}·K^{-1}
251.9 J·mol^{-1}·K^{-1}
Temperature range 20 to 200 °C, equations only in t°C.
C_p(c) = 0.2867 + 0.001743t cal g^{-1}C^{-1} (20 to 51°C);
C_p(liq) = 0.3865 + 0.0005986t cal g^{-1}C^{-1}(51 to -200°C).
Phase Changes
c/liq 324.4 K, ΔH = 5500 cal·mol^{-1}
23010 J·mol^{-1}
ΔS = 17.0 cal·mol^{-1}·K^{-1}
70.9 J·mol^{-1}·K^{-1}
Molecular Weight 182.2646
Wiswesser Line Notation R2R
Evaluation C

$C_{14}H_{14}$ (liq) 50KUR
1,2-Diphenylethane; Dibenzyl
Heat Capacity 330 K, C_p = 76.5 cal·mol^{-1}·K^{-1}
320.1 J·mol^{-1}·K^{-1}
Temperature range 54 to 254°C. Tm = 52.8°C
Molecular Weight 182.2646
Wiswesser Line Notation R2R
Evaluation B

$C_{14}H_{14}FeO_2$ (c) 81TOM/CUR
1,1′-Diacetylferrocene
Heat Capacity 298 K, C_p = 69.1 cal·mol^{-1}·K^{-1}
289.1 J·mol^{-1}·K^{-1}
Temperature range 293–393 K. Equation given.
Phase Changes
c/liq 403.7 K
Molecular Weight 270.1104
Wiswesser Line Notation L5ØJ AV1 Ø-FE- - ØL5ØJ AV1
Evaluation B

$C_{14}H_{14}Hg$ (c) 31SMI/AND 2
Di(p-tolyl)mercury; Mercury di(p-tolyl)
Heat Capacity 298.5 K, C_p = 62.7 cal·mol^{-1}·K^{-1}
262.3 J·mol^{-1}·K^{-1}
Temperature range 102–346 K. Value is unsmoothed experimental datum.
Molecular Weight 382.8546
Wiswesser Line Notation 1R D- 2 .HG
Evaluation B

$C_{14}H_{14}N_2O_3$ (c) 67BAR/POR
p-Azoxyanisole; 4,4′-Dimethoxyazoxybenzene
Heat Capacity 368 K, C_p = 90.9 cal·mol^{-1}·K^{-1}
380.3 J·mol^{-1}·K^{-1}
Temperature range 368–423 K
Phase Changes
c,I/liq 390.8 K, ΔH = 7258 cal·mol^{-1}
30367 J·mol^{-1}
ΔS = 18.6 cal·mol^{-1}·K^{-1}
77.7 J·mol^{-1}·K^{-1}
Solid–nematic transition
liq/liq 407.1 K, ΔH = 176.6 cal·mol^{-1}
736 J·mol^{-1}
ΔS = 0.4 cal·mol^{-1}·K^{-1}
1.8 J·mol^{-1}·K^{-1}
Nematic liquid–isotropic liquid transition
Molecular Weight 258.2762
Wiswesser Line Notation 1OR DNUNO&R DO1
Evaluation B

$C_{14}H_{14}N_2O_3$ (liq) 38KRE
p-Azoxyanisole; 4,4′-Dimethoxyazoxybenzene
Heat Capacity 400 K, C_p = 136 cal·mol^{-1}·K^{-1}
569 J·mol^{-1}·K^{-1}
Value a few degrees below anisotropic–isotropic liquid–liquid transition at 409 K.
Molecular Weight 258.2762
Wiswesser Line Notation 1OR DNUNO&R DO1
Evaluation C

$C_{14}H_{18}$ (c) 71BOY/SAN
1,2,3,4,5,6,7,8-Octahydroanthracene
Heat Capacity 327 K, C_p = 78.3 cal·mol^{-1}·K^{-1}
327.6 J·mol^{-1}·K^{-1}
Temperature range 327–390 K, 4 temperatures.
Phase Changes
c/liq 346 K, ΔH = 4280 cal·mol^{-1}
17910 J·mol^{-1}
ΔS = 12.4 cal·mol^{-1}·K^{-1}
51.8 J·mol^{-1}·K^{-1}
Molecular Weight 186.2962
Wiswesser Line Notation L 666 T&TJ
Evaluation C

$C_{14}H_{20}$ (c,III) 78WES/MCK
Diamantane;
Pentacyclo[7.3.1.1^{4,12}.0^{2,7}.0^{6,11}]tetradecane
Heat Capacity 298.15 K, C_p = 53.35 cal·mol^{-1}·K^{-1}
223.22 J·mol^{-1}·K^{-1}
Temperature range 5–350 K. Data between 350 and 540 K taken from other work.
Entropy 298.15 K, S = 47.84 cal·mol^{-1}·K^{-1}
200.16 J·mol^{-1}·K^{-1}
Phase Changes
Second order transition between 26 and 36 K, with ΔH = 2.5 cal·mol^{-1}, ΔS = 0.08 cal·mol^{-1}K^{-1}.
Molecular Weight 188.3120
Wiswesser Line Notation L666 C6 E6 B C- D G 4ACEF MTJ
Evaluation A

Table of Heat Capacities, Entropies, and Phase Transition Properties — Continued

$C_{14}H_{20}$ (c,III) 78SPI/AND
Diamantane;
Pentacyclo[7.3.1.1^{4,12}.0^{2,7}.0^{6,11}]tetradecane
Heat Capacity 295.56 K, C_p = 52.6 cal·mol^{-1}·K^{-1}
220.2 J·mol^{-1}·K^{-1}
Temperature range 295–540 K. Value is unsmoothed experimental datum.
Phase Changes
c,III/c,II 407.22 K, ΔH = 1062 cal·mol^{-1}
4445 J·mol^{-1}
ΔS = 2.61 cal·mol^{-1}·K^{-1}
10.92 J·mol^{-1}·K^{-1}
c,II/c,I 440.43 K, ΔH = 2141 cal·mol^{-1}
8960 J·mol^{-1}
ΔS = 4.86 cal·mol^{-1}·K^{-1}
20.34 J·mol^{-1}·K^{-1}
c,I/liq 517.92 K, ΔH = 2066 cal·mol^{-1}
8646 J·mol^{-1}
ΔS = 3.99 cal·mol^{-1}·K^{-1}
16.69 J·mol^{-1}·K^{-1}
Molecular Weight 188.3120
Wiswesser Line Notation L666 C6 E6 B C– D G 4ACEF MTJ
Evaluation A

$C_{14}H_{24}$ (liq) 63GUD/CAM
Perhydrophenanthrene
Heat Capacity 313 K, C_p = 78.9 cal·mol^{-1}·K^{-1}
330.1 J·mol^{-1}·K^{-1}
Temperature range 313–583 K
Molecular Weight 192.3436
Wiswesser Line Notation L B666TJ
Evaluation C

$C_{14}H_{26}$ (liq) 63GUD/CAM
α–Isobutyldecalin
Heat Capacity 313 K, C_p = 85.7 cal·mol^{-1}·K^{-1}
358.6 J·mol^{-1}·K^{-1}
Temperature range 313–483 K
Molecular Weight 194.3594
Wiswesser Line Notation L66TJ B2Y1&1
Evaluation C

$C_{14}H_{26}$ (liq) 62GOL/BEL
α–n–Butyldecalin
Heat Capacity 311 K, C_p = 85.3 cal·mol^{-1}·K^{-1}
356.9 J·mol^{-1}·K^{-1}
Temperatures 100, 200, 300°F
Molecular Weight 194.3594
Wiswesser Line Notation L66TJ B4
Evaluation C

$C_{14}H_{26}$ (liq) 63GUD/CAM
α–n–Butyldecalin
Heat Capacity 313 K, C_p = 86.5 cal·mol^{-1}·K^{-1}
361.9 J·mol^{-1}·K^{-1}
Temperature range 313–483 K
Molecular Weight 194.3594
Wiswesser Line Notation L66TJ B4
Evaluation C

$C_{14}H_{26}$ (liq) 62GOL/BEL
α–sec–Butyldecalin
Heat Capacity 311 K, C_p = 83.8 cal·mol^{-1}·K^{-1}
350.6 J·mol^{-1}·K^{-1}
Temperatures 100, 200, 300°F
Molecular Weight 194.3594
Wiswesser Line Notation L66TJ BY2&1
Evaluation C

$C_{14}H_{26}$ (liq) 63GUD/CAM
α–sec–Butyldecalin
Heat Capacity 313 K, C_p = 84.2 cal·mol^{-1}·K^{-1}
352.3 J·mol^{-1}·K^{-1}
Temperature range 313–483 K
Molecular Weight 194.3594
Wiswesser Line Notation L66TJ BY2&1
Evaluation C

$C_{14}H_{26}$ (liq) 63GUD/CAM
tert–Butyldecalin
Heat Capacity 313 K, C_p = 84.2 cal·mol^{-1}·K^{-1}
352.3 J·mol^{-1}·K^{-1}
Temperature range 313–483 K
Molecular Weight 194.3594
Wiswesser Line Notation L66TJ XX1&1&1
Evaluation C

$C_{14}H_{26}$ (liq) 63GUD/CAM
2–Methylbicyclohexylmethane
Heat Capacity 313 K, C_p = 86.3 cal·mol^{-1}·K^{-1}
361.1 J·mol^{-1}·K^{-1}
Temperature range 313–483 K
Molecular Weight 194.3594
Wiswesser Line Notation L6TJ A1– BL6TJ A1
Evaluation C

$C_{14}H_{26}$ (liq) 63GUD/CAM
1,2–Dicyclohexylethane
Heat Capacity 313 K, C_p = 87.9 cal·mol^{-1}·K^{-1}
367.8 J·mol^{-1}·K^{-1}
Temperature range 313–483 K
Molecular Weight 194.3594
Wiswesser Line Notation L6TJ A2– AL6TJ
Evaluation C

$C_{14}H_{26}$ (liq) 63GUD/CAM
1,1–Dicyclohexylethane
Heat Capacity 313 K, C_p = 83.2 cal·mol^{-1}·K^{-1}
348.1 J·mol^{-1}·K^{-1}
Temperature range 313–483 K
Molecular Weight 194.3594
Wiswesser Line Notation L6TJ AY1&– AL6TJ
Evaluation C

$C_{14}H_{26}$ (liq) 63GUD/CAM
2–Ethylbicyclohexyl
Heat Capacity 313 K, C_p = 88.2 cal·mol^{-1}·K^{-1}
369.0 J·mol^{-1}·K^{-1}
Temperature range 313–483 K
Molecular Weight 194.3594
Wiswesser Line Notation L6TJA– BL6TJ A2
Evaluation C

$C_{14}H_{26}O_4Pb$ (c,II) 78ADE/SIM
Lead(II) heptanoate; Lead(II) oenanthate
Heat Capacity 345 K, C_p = 184 cal·mol^{-1}·K^{-1}
770 J·mol^{-1}·K^{-1}
Mean value, 341–351 K. Data only graphically for c,IV. Data also for c,I, 363–371 K and liquid, 413–453 K.

Table of Heat Capacities, Entropies, and Phase Transition Properties — Continued

Phase Changes
c,IV/c,III 336.6 K, ΔH = 4090 cal·mol^{-1}
17100 J·mol^{-1}
ΔS = 12.1 cal·mol^{-1}·K^{-1}
51 J·mol^{-1}·K^{-1}
c,III, c,II, c,I, are mesophases.
c,III/c,II 356.8 K, ΔH = 2030 cal·mol^{-1}
8500 J·mol^{-1}
ΔS = 5.7 cal·mol^{-1}·K^{-1}
24 J·mol^{-1}·K^{-1}
c,II/c,I 361.3 K, ΔH = 2320 cal·mol^{-1}
9700 J·mol^{-1}
ΔS = 6.4 cal·mol^{-1}·K^{-1}
27 J·mol^{-1}·K^{-1}
c,I/liq 374.7 K, ΔH = 335 cal·mol^{-1}
1400 J·mol^{-1}
ΔS = 0.9 cal·mol^{-1}·K^{-1}
3.7 J·mol^{-1}·K^{-1}

Molecular Weight 465.5570
Wiswesser Line Notation OV6 2 .PB
Evaluation C

$C_{14}H_{28}O$ (c) 79SUN/SVE
2–Tetradecanone; n–Dodecyl methyl ketone
Heat Capacity 298.15 K, C_p = 99.2 cal·mol^{-1}·K^{-1}
415.2 J·mol^{-1}·K^{-1}
Temperature range 278–338 K. Equations only.
Phase Changes
c/liq 306.7 K, ΔH = 11740 cal·mol^{-1}
49120 J·mol^{-1}
ΔS = 38.3 cal·mol^{-1}·K^{-1}
160.2 J·mol^{-1}·K^{-1}

Molecular Weight 212.3746
Wiswesser Line Notation 12V1
Evaluation B

$C_{14}H_{28}O_2$ (c) 85STO/WIL
Tetradecanoic acid; Myristic acid
Heat Capacity 298 K, C_p = 125 cal·mol^{-1}·K^{-1}
523 J·mol^{-1}·K^{-1}
Temperature range 0 to 100°C
Phase Changes
c/liq 317 K, ΔH = 8670 cal·mol^{-1}
36280 J·mol^{-1}
ΔS = 27 cal·mol^{-1}·K^{-1}
114 J·mol^{-1}·K^{-1}

Molecular Weight 228.3740
Wiswesser Line Notation QV13
Evaluation D

$C_{14}H_{28}O_2$ (c) 82SCH/MIL 2
Tetradecanoic acid; Myristic acid
Heat Capacity 298.15 K, C_p = 103.25 cal·mol^{-1}·K^{-1}
432.01 J·mol^{-1}·K^{-1}
Temperature range 80–345 K
Phase Changes
c,I/liq 327.32 K, ΔH = 10779 cal·mol^{-1}
45100 J·mol^{-1}
ΔS = 32.933 cal·mol^{-1}·K^{-1}
137.79 J·mol^{-1}·K^{-1}

Molecular Weight 228.3740
Wiswesser Line Notation QV13
Evaluation B

$C_{14}H_{30}$ (liq) 34PAR/LIG
n–Tetradecane
Heat Capacity 290.6 K, C_p = 103.8 cal·mol^{-1}·K^{-1}
434.3 J·mol^{-1}·K^{-1}
Temperature range 93–291 K. Value is unsmoothed experimental datum.
Entropy 298.15 K, S = 134.4 cal·mol^{-1}·K^{-1}
562.3 J·mol^{-1}·K^{-1}
Extapolation below 90 K, 28.74 cal·mol^{-1}K^{-1}
Phase Changes
c,II/c,I 194 K, ΔH = 43.6 cal·mol^{-1}
182.4 J·mol^{-1}
ΔS = 0.22 cal·mol^{-1}·K^{-1}
0.94 J·mol^{-1}·K^{-1}
c,I/liq 288.7 K, ΔH = 10580 cal·mol^{-1}
44267 J·mol^{-1}
ΔS = 326.6 cal·mol^{-1}·K^{-1}
153.3 J·mol^{-1}·K^{-1}

Molecular Weight 198.3910
Wiswesser Line Notation 14H
Evaluation B(C_p),C(S)

$C_{14}H_{30}$ (liq) 54FIN/GRO 2
n–Tetradecane
Heat Capacity 298.15 K, C_p = 104.79 cal·mol^{-1}·K^{-1}
438.44 J·mol^{-1}·K^{-1}
Temperature range 12–300 K
Entropy 298.15 K, S = 132.75 cal·mol^{-1}·K^{-1}
555.43 J·mol^{-1}·K^{-1}
Phase Changes
c/liq 279.03 K, ΔH = 10772 cal·mol^{-1}
45070 J·mol^{-1}
ΔS = 38.61 cal·mol^{-1}·K^{-1}
161.52 J·mol^{-1}·K^{-1}

Molecular Weight 198.3910
Wiswesser Line Notation 14H
Evaluation A

$C_{14}H_{30}O$ (c) 74MOS/MOU
1–Tetradecanol; n–Tetradecyl alcohol
Heat Capacity 298.15 K, C_p = 92.7 cal·mol^{-1}·K^{-1}
388 J·mol^{-1}·K^{-1}
β–form, 286–303 K
Phase Changes
c,β/c,γ 306 K, ΔH = 430 cal·mol^{-1}
1800 J·mol^{-1}
ΔS = 1.4 cal·mol^{-1}·K^{-1}
5.9 J·mol^{-1}·K^{-1}
c,β/c,α 311.2 K, ΔH = 5690 cal·mol^{-1}
23800 J·mol^{-1}
ΔS = 18.3 cal·mol^{-1}·K^{-1}
76.5 J·mol^{-1}·K^{-1}
c,γ/c,α 311.6 K, ΔH = 5260 cal·mol^{-1}
22000 J·mol^{-1}
ΔS = 16.9 cal·mol^{-1}·K^{-1}
70.6 J·mol^{-1}·K^{-1}
c,β/liq 311.0 K, ΔH = 11835 cal·mol^{-1}
49510 J·mol^{-1}
ΔS = 38.1 cal·mol^{-1}·K^{-1}
159.2 J·mol^{-1}·K^{-1}
c,γ/liq 311.2 K, ΔH = 11235 cal·mol^{-1}
47000 J·mol^{-1}
ΔS = 36.1 cal·mol^{-1}·K^{-1}
151.0 J·mol^{-1}·K^{-1}
c,α/liq 310.8 K, ΔH = 6000 cal·mol^{-1}
25100 J·mol^{-1}
ΔS = 19.3 cal·mol^{-1}·K^{-1}
80.8 J·mol^{-1}·K^{-1}

Table of Heat Capacities, Entropies, and Phase Transition Properties — Continued

Molecular Weight 214.3904
Wiswesser Line Notation Q14
Evaluation B

$C_{14}H_{30}O$ (liq) 74MOS/MOU
1-Tetradecanol; n-Tetradecyl alcohol
Heat Capacity 312 K, C_p = 120.9 cal·mol^{-1}·K^{-1}
506 J·mol^{-1}·K^{-1}
Temperature range 312-346 K
Molecular Weight 214.3904
Wiswesser Line Notation Q14
Evaluation B

$C_{14}H_{34}$ (liq) 63GUD/CAM
9-Methylperhydrofluorene
Heat Capacity 313 K, C_p = 77.3 cal·mol^{-1}·K^{-1}
323.4 J·mol^{-1}·K^{-1}
Temperature range 313-583 K
Molecular Weight 202.4226
Wiswesser Line Notation L B656TJ H1
Evaluation C

$C_{14}H_{34}Br_2N_2$ (c) 74BUR/VER
1,2-Bis(triethylammonium)ethane dibromide
Heat Capacity 298 K, C_p = 98.4 cal·mol^{-1}·K^{-1}
411.7 J·mol^{-1}·K^{-1}
Temperature range 273-373 K
Molecular Weight 390.2440
Wiswesser Line Notation 2K2&2&2K2&2&2 &E &E
Evaluation B

$C_{15}H_{10}N_2O_2$ (c) 77LEB/EVS
Bis(4-isocyanatophenyl)methane
Heat Capacity 298.15 K, C_p = 73.4 cal·mol^{-1}·K^{-1}
307.0 J·mol^{-1}·K^{-1}
Temperature range 12-355 K. Data deposited VINITI, No 4328-76, 14 December 1976.
Entropy 298.15 K, S = 79.5 cal·mol^{-1}·K^{-1}
332.5 J·mol^{-1}·K^{-1}
Phase Changes
c/liq 313.57 K, ΔH = 6525 cal·mol^{-1}
27300 J·mol^{-1}
ΔS = 20.8 cal·mol^{-1}·K^{-1}
87.1 J·mol^{-1}·K^{-1}
Molecular Weight 250.2562
Wiswesser Line Notation OCNR D1R DNCO
Evaluation A

$C_{15}H_{10}N_2O_2$ (c) 66ZAL/STR
4,4′-Diphenylmethane diisocyanate
Heat Capacity 334 K, C_p = 58.5 cal·mol^{-1}·K^{-1}
244.8 J·mol^{-1}·K^{-1}
Temperature range 46 to 78 °C, mean value.
Molecular Weight 250.2562
Wiswesser Line Notation OCNR D1R DNCO
Evaluation D
State not given; assumed solid.

$C_{15}H_{16}$ (liq) 56MCE
p-Isopropylbiphenyl
Heat Capacity 422 K, C_p = 100.7 cal·mol^{-1}·K^{-1}
421.3 J·mol^{-1}·K^{-1}
Temperature range 300 to 600°F
Molecular Weight 196.2914
Wiswesser Line Notation 1Y1&R DR
Evaluation C
Quoted in 58WAL/BRO

$C_{15}H_{16}$ (liq) 63VAR/KOP
Isopropylbiphenyl
Heat Capacity 295 K, C_p = 82.2 cal·mol^{-1}·K^{-1}
343.9 J·mol^{-1}·K^{-1}
Temperature range 293-670 K. Value is unsmoothed experimental datum.
Molecular Weight 196.2914
Wiswesser Line Notation 1Y1&R XR
Evaluation B
Probably the para isomer.

$C_{15}H_{16}$ (liq) 64VUK/RAS
Isopropylbiphenyl
Heat Capacity 298 K, C_p = 80.9 cal·mol^{-1}·K^{-1}
338.5 J·mol^{-1}·K^{-1}
Temperature range 38 to 212°C
Molecular Weight 196.2914
Wiswesser Line Notation 1Y1&R XR
Evaluation C
Probably the para isomer.

$C_{15}H_{16}N_2O_3$ (liq) 38KRE
p-Azoxyanisoylphenetole
Heat Capacity 420 K, C_p = 142 cal·mol^{-1}·K^{-1}
594 J·mol^{-1}·K^{-1}
Value a few degrees below anisotropic-isotropic liquid-liquid transition at 428 K.
Molecular Weight 272.3030
Wiswesser Line Notation 2OR DNO&UNR DO1
Evaluation C

$C_{15}H_{21}AlO_6$ (c) 81TEG/FER
Aluminum acetylacetonate
Heat Capacity 298 K, C_p = 76.8 cal·mol^{-1}·K^{-1}
321.4 J·mol^{-1}·K^{-1}
Temperature range 4.2-450 K
Entropy 298 K, S = 114.5 cal·mol^{-1}·K^{-1}
479.1 J·mol^{-1}·K^{-1}
Molecular Weight 324.3088
Wiswesser Line Notation D6O-AL-O ADJ D1 F1 B-& BD6O-AL-O ADJ D1 F1 B-& BD6O-AL-O ADJ D1 F1
Evaluation B

$C_{15}H_{26}$ (liq) 63GUD/CAM
1,3-Dicyclopentylcyclopentane
Heat Capacity 313 K, C_p = 86.9 cal·mol^{-1}·K^{-1}
363.6 J·mol^{-1}·K^{-1}
Temperature range 313-583 K
Molecular Weight 206.3704
Wiswesser Line Notation L5TJ A CL5TJ A- AL5TJ
Evaluation C

$C_{15}H_{26}O_6$ (liq) 76PHI/MAT
Tributyrin; Glyceryl tributyrate
Heat Capacity 313 K, C_p = 136 cal·mol^{-1}·K^{-1}
569 J·mol^{-1}·K^{-1}
Temperature range 313-413 K
Molecular Weight 302.3668
Wiswesser Line Notation 3VO1YOV3&1OV3
Evaluation C

Table of Heat Capacities, Entropies, and Phase Transition Properties — Continued

$C_{15}H_{28}$ (liq) 63GUD/CAM
2–Isopropylbicyclohexyl
Heat Capacity 313 K, C_p = 100.0 cal·mol⁻¹·K⁻¹ 418.4 J·mol⁻¹·K⁻¹
Temperature range 313–483 K
Molecular Weight 208.3862
Wiswesser Line Notation L6TJ A– BL6TJ AY1&1
Evaluation C

$C_{15}H_{28}$ (liq) 63GUD/CAM
2–Ethylbicyclohexylmethane
Heat Capacity 313 K, C_p = 91.9 cal·mol⁻¹·K⁻¹ 384.5 J·mol⁻¹·K⁻¹
Temperature range 313–483 K
Molecular Weight 208.3862
Wiswesser Line Notation L6TJ A1– BL6TJ A2
Evaluation C

$C_{15}H_{28}$ (liq) 63GUD/CAM
Cyclohexyl(ethylcyclohexyl)methane
Heat Capacity 313 K, C_p = 96.3 cal·mol⁻¹·K⁻¹ 402.9 J·mol⁻¹·K⁻¹
Temperature range 373–483 K
Molecular Weight 208.3862
Wiswesser Line Notation L6TJ A1– XL6TJ A2
Evaluation C

$C_{15}H_{28}$ (liq) 62GOL/BEL
1,2–Dicyclohexylpropane
Heat Capacity 422 K, C_p = 119.0 cal·mol⁻¹·K⁻¹ 497.9 J·mol⁻¹·K⁻¹
One temperature
Molecular Weight 208.3862
Wiswesser Line Notation L6TJ AY1&1– AL6TJ
Evaluation C

$C_{15}H_{28}$ (liq) 63GUD/CAM
1,2–Dicyclohexylpropane
Heat Capacity 313 K, C_p = 95.2 cal·mol⁻¹·K⁻¹ 398.3 J·mol⁻¹·K⁻¹
Temperature range 313–583 K
Molecular Weight 208.3862
Wiswesser Line Notation L6TJ AY1&1– AL6TJ
Evaluation C

$C_{15}H_{28}$ (liq) 62GOL/BEL
Cyclohexyl(2–ethylcyclohexyl)methane
Heat Capacity 311 K, C_p = 98.4 cal·mol⁻¹·K⁻¹ 411.7 J·mol⁻¹·K⁻¹
Temperatures 100, 200, 300°F
Molecular Weight 208.3862
Wiswesser Line Notation L6TJ B2 A1– AL6TJ
Evaluation C

$C_{15}H_{28}$ (liq) 62GOL/BEL
Isopropylbicyclohexyl
Heat Capacity 311 K, C_p = 99.8 cal·mol⁻¹·K⁻¹ 417.6 J·mol⁻¹·K⁻¹
Temperatures 100, 200, 300°F
Molecular Weight 208.3862
Wiswesser Line Notation L6TJ XY1&1 A– AL6TJ
Evaluation C

$C_{15}H_{28}O_2$ (c) 81LEB/YEV
Pentadecanolactone
Heat Capacity 298.15 K, C_p = 106.2 cal·mol⁻¹·K⁻¹ 444.3 J·mol⁻¹·K⁻¹
Temperature range 13.8–330 K
Entropy 298.15 K, S = 115.4 cal·mol⁻¹·K⁻¹ 482.8 J·mol⁻¹·K⁻¹
Phase Changes
c,II/c,I 282.98 K, ΔH = 6525 cal·mol⁻¹ 27301 J·mol⁻¹
ΔS = 23.0 cal·mol⁻¹·K⁻¹ 96.2 J·mol⁻¹·K⁻¹
c,I/liq 308.5 K, ΔH = 1668 cal·mol⁻¹ 6979 J·mol⁻¹
ΔS = 5.40 cal·mol⁻¹·K⁻¹ 22.6 J·mol⁻¹·K⁻¹
Molecular Weight 240.3850
Wiswesser Line Notation T–16–VOTJ
Evaluation A

$C_{15}H_{30}$ (liq) 65MES/TOD 2
n–Decylcyclopentane
Heat Capacity 298.15 K, C_p = 101.94 cal·mol⁻¹·K⁻¹ 426.52 J·mol⁻¹·K⁻¹
Temperature range 12–370 K
Entropy 298.15 K, S = 128.71 cal·mol⁻¹·K⁻¹ 538.52 J·mol⁻¹·K⁻¹
Phase Changes
c/liq 251.02 K, ΔH = 2917 cal·mol⁻¹ 33125 J·mol⁻¹
ΔS = 31.5 cal·mol⁻¹·K⁻¹ 132.0 J·mol⁻¹·K⁻¹
Molecular Weight 210.4020
Wiswesser Line Notation L5TJ A10
Evaluation A

$C_{15}H_{30}O$ (c) 79SUN/SVE
2–Pentadecanone; Tridecyl methyl ketone
Heat Capacity 298.15 K, C_p = 102.0 cal·mol⁻¹·K⁻¹ 426.8 J·mol⁻¹·K⁻¹
Temperature range 285–343 K. Equations only.
Phase Changes
c/liq 312.2 K, ΔH = 13042 cal·mol⁻¹ 54570 J·mol⁻¹
ΔS = 41.8 cal·mol⁻¹·K⁻¹ 174.8 J·mol⁻¹·K⁻¹
Molecular Weight 226.4014
Wiswesser Line Notation 13V1
Evaluation B

$C_{15}H_{30}O_2$ (c) 82SCH/MIL
Pentadecanoic acid
Heat Capacity 298.15 K, C_p = 105.95 cal·mol⁻¹·K⁻¹ 443.28 J·mol⁻¹·K⁻¹
Temperature range 80–345 K
Phase Changes
c,II/c,I 318.7 K, ΔH = 1941 cal·mol⁻¹ 8123 J·mol⁻¹
ΔS = 6.087 cal·mol⁻¹·K⁻¹ 25.47 J·mol⁻¹·K⁻¹
c,I/liq 325.68 K, ΔH = 9925 cal·mol⁻¹ 41526 J·mol⁻¹
ΔS = 30.473 cal·mol⁻¹·K⁻¹ 127.50 J·mol⁻¹·K⁻¹
Molecular Weight 242.4008
Wiswesser Line Notation QV14
Evaluation B

$C_{15}H_{30}O_2$ (liq) 79FUC
Methyl tetradecanoate; Methyl myristate
Heat Capacity 298.15 K, C_p = 120.8 cal·mol⁻¹·K⁻¹ 505.4 J·mol⁻¹·K⁻¹
One temperature
Molecular Weight 242.4008
Wiswesser Line Notation 13VO1
Evaluation B

Table of Heat Capacities, Entropies, and Phase Transition Properties — Continued

$C_{15}H_{30}O_4$ (c) 65SIL/DAU
2–Monolaurin
Heat Capacity 298 K, C_p = 104.3 cal·mol^{-1}·K^{-1}
436.4 J·mol^{-1}·K^{-1}
One temperature
Molecular Weight 274.3996
Wiswesser Line Notation Q1Y1QOV11
Evaluation B

$C_{15}H_{30}O_4$ (c) 65SIL/DAU
1–Monolaurin
Heat Capacity 298 K, C_p = 107.0 cal·mol^{-1}·K^{-1}
447.7 J·mol^{-1}·K^{-1}
One temperature
Molecular Weight 274.3996
Wiswesser Line Notation Q1YQ1OV11
Evaluation B
β_L form

$C_{15}H_{32}$ (liq) 54FIN/GRO 2
n–Pentadecane
Heat Capacity 298.15 K, C_p = 112.32 cal·mol^{-1}·K^{-1}
469.95 J·mol^{-1}·K^{-1}
Temperature range 12–310 K
Entropy 298.15 K, S = 140.42 cal·mol^{-1}·K^{-1}
587.52 J·mol^{-1}·K^{-1}
Phase Changes
c,II/c,I 270.9 K, ΔH = 2191 cal·mol^{-1}
9167 J·mol^{-1}
ΔS = 8.09 cal·mol^{-1}·K^{-1}
33.84 J·mol^{-1}·K^{-1}
c,I/liq 283.11 K, ΔH = 8268 cal·mol^{-1}
34593 J·mol^{-1}
ΔS = 29.20 cal·mol^{-1}·K^{-1}
122.19 J·mol^{-1}·K^{-1}
Molecular Weight 212.4178
Wiswesser Line Notation 15H
Evaluation A

$C_{15}H_{32}O$ (c) 74MOS/MOU
1–Pentadecanol; n–Pentadecyl alcohol
Heat Capacity 298.15 K, C_p = 95.6 cal·mol^{-1}·K^{-1}
400 J·mol^{-1}·K^{-1}
Temperature range 295–308 K
Phase Changes
c,β/c,α 316.2 K, ΔH = 5650 cal·mol^{-1}
23650 J·mol^{-1}
ΔS = 17.9 cal·mol^{-1}·K^{-1}
74.8 J·mol^{-1}·K^{-1}
c,β/liq 316.6 K, ΔH = 13080 cal·mol^{-1}
54720 J·mol^{-1}
ΔS = 41.3 cal·mol^{-1}·K^{-1}
172.8 J·mol^{-1}·K^{-1}
c,α/liq 316.9 K, ΔH = 7255 cal·mol^{-1}
30350 J·mol^{-1}
ΔS = 22.9 cal·mol^{-1}·K^{-1}
95.8 J·mol^{-1}·K^{-1}
Molecular Weight 228.4172
Wiswesser Line Notation Q15
Evaluation B

$C_{15}H_{32}O$ (liq) 74MOS/MOU
1–Pentadecanol; n–Pentadecyl alcohol
Heat Capacity 318 K, C_p = 127.9 cal·mol^{-1}·K^{-1}
535 J·mol^{-1}·K^{-1}
Temperature range 318–346 K
Molecular Weight 228.4172
Wiswesser Line Notation Q15
Evaluation B

$C_{15}H_{32}O$ (liq) 82VAS/PET
1–Pentadecanol; n–Pentadecyl alcohol
Heat Capacity 358 K, C_p = 128.4 cal·mol^{-1}·K^{-1}
537.3 J·mol^{-1}·K^{-1}
Temperature range 358–608 K
Molecular Weight 228.4172
Wiswesser Line Notation Q15
Evaluation B

$C_{15}H_{36}Br_2N_2$ (c) 74BUR/VER
1,3–Bis(triethylammonium)propane dibromide
Heat Capacity 298 K, C_p = 105.0 cal·mol^{-1}·K^{-1}
439.3 J·mol^{-1}·K^{-1}
Temperature range 273–373 K
Molecular Weight 404.2708
Wiswesser Line Notation 2K2&2&3K2&2&2 &E &E
Evaluation B

$C_{16}H_8N_4$ (c) 80BOE/WES
Naphthalene–tetracyanoethylene adduct
Heat Capacity 298.15 K, C_p = 78.57 cal·mol^{-1}·K^{-1}
328.7 J·mol^{-1}·K^{-1}
Temperature range 5–300 K
Entropy 298.15 K, S = 91.86 cal·mol^{-1}·K^{-1}
384.3 J·mol^{-1}·K^{-1}
Phase Changes
c,III/c,II 160 K, ΔH = 196.6 cal·mol^{-1}
822.6 J·mol^{-1}
ΔS = 1.22 cal·mol^{-1}·K^{-1}
5.10 J·mol^{-1}·K^{-1}
c,II/c,I 240 K, ΔH = 246 cal·mol^{-1}
1029 J·mol^{-1}
ΔS = 1.20 cal·mol^{-1}·K^{-1}
5.02 J·mol^{-1}·K^{-1}
c,III/c,II: region from 150 to 172.5 K;
c,II/c,I: region from 172.5 to 240 K.
Molecular Weight 256.2660
Wiswesser Line Notation L66J &NCYCN&UYCN&CN
Evaluation A

$C_{16}H_{10}$ (c) 71WON/WES
Fluoranthene; Idryl; 1,2–Benzacenaphthene
Heat Capacity 298.15 K, C_p = 55.03 cal·mol^{-1}·K^{-1}
230.25 J·mol^{-1}·K^{-1}
Temperature range 5–427 K
Entropy 298.15 K, S = 55.11 cal·mol^{-1}·K^{-1}
230.58 J·mol^{-1}·K^{-1}
Phase Changes
c/liq 383.36 K, ΔH = 4476 cal·mol^{-1}
18728 J·mol^{-1}
ΔS = 11.68 cal·mol^{-1}·K^{-1}
48.85 J·mol^{-1}·K^{-1}
Molecular Weight 202.2550
Wiswesser Line Notation L C6566 1A PJ
Evaluation A

$C_{16}H_{10}$ (c) 34JAC/PAR
Pyrene; Benzo(d,e,f)phenanthrene
Heat Capacity 291.1 K, C_p = 54.41 cal·mol^{-1}·K^{-1}
227.65 J·mol^{-1}·K^{-1}
Temperature range 94–292 K. Value is unsmoothed experimental datum.
Entropy 298.1 K, S = 51.4 cal·mol^{-1}·K^{-1}
215.1 J·mol^{-1}·K^{-1}
Extrapolation below 90 K, 14.29 cal·mol^{-1}K^{-1}. Hump in C_p curve around 116 K, probably 2nd order transition. ΔH = 24 cal·mol^{-1}
Molecular Weight 202.2550
Wiswesser Line Notation L666 B6 2AB PJ
Evaluation B(C_p),C(S)

Table of Heat Capacities, Entropies, and Phase Transition Properties — Continued

$C_{16}H_{10}$ (c) 71WON/WES
Pyrene; Benzo(d,e,f)phenanthrene
Heat Capacity 298.15 K, C_p = 54.90 cal·mol^{-1}·K^{-1}
229.70 J·mol^{-1}·K^{-1}
Temperature range 5–484 K
Entropy 298.15 K, S = 53.75 cal·mol^{-1}·K^{-1}
224.89 J·mol^{-1}·K^{-1}
Phase Changes
c,II/c,I 120.8 K, ΔH = 69 cal·mol^{-1}
289 J·mol^{-1}
ΔS = 0.55 cal·mol^{-1}·K^{-1}
2.30 J·mol^{-1}·K^{-1}
c,I/liq 423.81 K, ΔH = 4150 cal·mol^{-1}
17364 J·mol^{-1}
ΔS = 9.79 cal·mol^{-1}·K^{-1}
40.97 J·mol^{-1}·K^{-1}
Molecular Weight 202.2550
Wiswesser Line Notation L666 B6 2AB PJ
Evaluation A

$C_{16}H_{11}N_3O_6$ (c) 80RAD/RAD
Naphthalene-1,3,5-trinitrobenzene adduct
Heat Capacity 298.15 K, C_p = 55.8 cal·mol^{-1}·K^{-1}
233.5 J·mol^{-1}·K^{-1}
Temperature range 220–400 K. Data given graphically. C_p calculated from equation.
Phase Changes
c,III/c,II 220 K, ΔH = 406 cal·mol^{-1}
1700 J·mol^{-1}
ΔS = 1.85 cal·mol^{-1}·K^{-1}
7.22 J·mol^{-1}·K^{-1}
c,II/c,I 424.5 K, ΔH = 382 cal·mol^{-1}
1598 J·mol^{-1}
ΔS = .90 cal·mol^{-1}·K^{-1}
3.77 J·mol^{-1}·K^{-1}
c,I/liq 430.8 K, ΔH = 7624 cal·mol^{-1}
31900 J·mol^{-1}
ΔS = 17.70 cal·mol^{-1}·K^{-1}
74.05 J·mol^{-1}·K^{-1}
Molecular Weight 341.2794
Wiswesser Line Notation L66J &WNR CNW ENW
Evaluation B

$C_{16}H_{12}O_2$ (c) 30PAR/HUF 2
Dibenzoylethylene
Heat Capacity 291.9 K, C_p = 68.5 cal·mol^{-1}·K^{-1}
286.6 J·mol^{-1}·K^{-1}
Temperature range 89–292 K. Value is unsmoothed experimental datum.
Entropy 298.15 K, S = 76.3 cal·mol^{-1}·K^{-1}
319.2 J·mol^{-1}·K^{-1}
Extrapolation below 90 K, 24.88 cal·mol^{-1}K^{-1}
Molecular Weight 236.2696
Wiswesser Line Notation RV1U1VR
Evaluation B(C_p),C(S)

$C_{16}H_{14}O_2$ (c,I) 30PAR/HUF 2
1,2–Dibenzoylethane
Heat Capacity 296.0 K, C_p = 69.6 cal·mol^{-1}·K^{-1}
291.2 J·mol^{-1}·K^{-1}
Temperature range 93–296 K. Value is unsmoothed experimental datum.
Entropy 298.15 K, S = 77.6 cal·mol^{-1}·K^{-1}
324.7 J·mol^{-1}·K^{-1}
Extrapolation below 90 K, 24.79 cal·mol^{-1}K^{-1}
Phase Changes
c,II/c,I 187 K, ΔH = 52 cal·mol^{-1}
218 J·mol^{-1}
ΔS = 0.28 cal·mol^{-1}·K^{-1}
1.17 J·mol^{-1}·K^{-1}
Molecular Weight 238.2854
Wiswesser Line Notation RV2VR
Evaluation B(C_p),C(S)

$C_{16}H_{14}O_2$ (c) 32SPA/THO
1,2–Dibenzoylethane
Heat Capacity 303 K, C_p = 72.2 cal·mol^{-1}·K^{-1}
302.1 J·mol^{-1}·K^{-1}
Temperature range 30 to 190°C
Phase Changes
c/liq 418.6 K, ΔH = 9317 cal·mol^{-1}
38982 J·mol^{-1}
ΔS = 22.3 cal·mol^{-1}·K^{-1}
93.1 J·mol^{-1}·K^{-1}
Molecular Weight 238.2854
Wiswesser Line Notation RV2VR
Evaluation B

$C_{16}H_{16}$ (c) 69SHI/MCN
2,2–Metacyclophane
Heat Capacity 300 K, C_p = 57.5 cal·mol^{-1}·K^{-1}
240.6 J·mol^{-1}·K^{-1}
Temperatures 300, 318 K
Molecular Weight 208.3024
Wiswesser Line Notation L E6 B–10–6 A E– &T&J
Evaluation B

$C_{16}H_{16}$ (c) 69SHI/MCN
2,2–Metaparacyclophane
Heat Capacity 300 K, C_p = 62.5 cal·mol^{-1}·K^{-1}
261.5 J·mol^{-1}·K^{-1}
One temperature
Molecular Weight 208.3024
Wiswesser Line Notation L E6 B–11–6 A E– – &T&J
Evaluation B

$C_{16}H_{16}$ (c) 69SHI/MCN
2,2–Paracyclophane
Heat Capacity 300 K, C_p = 59.3 cal·mol^{-1}·K^{-1}
248.1 J·mol^{-1}·K^{-1}
Temperatures 300, 318 K
Molecular Weight 208.3024
Wiswesser Line Notation L E6 B–11–6 A E– E– – &T&J
Evaluation B

$C_{16}H_{16}$ (c) 70AND/WES 2
2,2–Paracyclophane
Heat Capacity 298.15 K, C_p = 60.31 cal·mol^{-1}·K^{-1}
252.34 J·mol^{-1}·K^{-1}
Temperature range 10–350 K
Entropy 298.15 K, S = 63.50 cal·mol^{-1}·K^{-1}
265.68 J·mol^{-1}·K^{-1}
Phase Changes
Second order transition between 30 and 60 K.
Molecular Weight 208.3024
Wiswesser Line Notation L F6 C–12–6 A B F– F– – &T&J
Evaluation A

Table of Heat Capacities, Entropies, and Phase Transition Properties — Continued

$C_{16}H_{16}N_2O_2$ (c) 67BAR/POR
Anisaldazine
Heat Capacity C_p data given graphically only.
Temperature range 313–500 K.
Phase Changes
c/liq 442.0 K, ΔH = 7710 cal·mol^{-1}
29750 J·mol^{-1}
ΔS = 16.1 cal·mol^{-1}·K^{-1}
67.3 J·mol^{-1}·K^{-1}
Solid–nematic transition
liq/liq 453.6 K, ΔH = 158 cal·mol^{-1}
661 J·mol^{-1}
ΔS = 0.3 cal·mol^{-1}·K^{-1}
1.5 J·mol^{-1}·K^{-1}
Nematic–isotropic transition
Molecular Weight 268.3146
Wiswesser Line Notation 1OR D1UNNU1R DO1
Evaluation B

$C_{16}H_{18}N_2O_3$ (liq) 38KRE
p–Azoxyphenetole
Heat Capacity 430 K, C_p = 145 cal·mol^{-1}·K^{-1}
607 J·mol^{-1}·K^{-1}
Value a few degrees below anisotropic–isotropic liquid–liquid transition at 436 K.
Molecular Weight 286.3298
Wiswesser Line Notation 2OR DNUNO&R DO2
Evaluation C

$C_{16}H_{22}O_3Si_3$ (c) 82KUL/DZH
Diphenyltetramethylcyclotrisiloxane
Heat Capacity 298.15 K, C_p = 110.8 cal·mol^{-1}·K^{-1}
463.6 J·mol^{-1}·K^{-1}
Temperature range 4.2–370 K. Data given graphically except for data at 298.15 K.
Entropy 298.15 K, S = 126.2 cal·mol^{-1}·K^{-1}
528.0 J·mol^{-1}·K^{-1}
Phase Changes
c,I/liq 337.98 K, ΔH = 5304 cal·mol^{-1}
22192 J·mol^{-1}
ΔS = 15.7 cal·mol^{-1}·K^{-1}
65.7 J·mol^{-1}·K^{-1}
Molecular Weight 346.6045
Wiswesser Line Notation T6–SI–O–SI–O–SI–OTJ AR AR C1 C1 E1 E1
Evaluation B

$C_{16}H_{22}O_{11}$ (c) 44CLA/STE
α–Glucose pentaacetate (D)
Heat Capacity 298 K, C_p = 117.5 cal·mol^{-1}·K^{-1}
491.6 J·mol^{-1}·K^{-1}
One temperature
Molecular Weight 390.3432
Wiswesser Line Notation T6OTJ BOV1 COV1 DOV1 EOV1 F1OV1 -A&BCE -B&DF
Evaluation C

$C_{16}H_{22}O_{11}$ (c) 44CLA/STE
β–Glucose pentaacetate (D)
Heat Capacity 298 K, C_p = 119.1 cal·mol^{-1}·K^{-1}
498.3 J·mol^{-1}·K^{-1}
One temperature
Molecular Weight 390.3432
Wiswesser Line Notation T6OTJ BOV1 COV1 DOV1 EOV1 F1OV1 -A&CE -B&BDF
Evaluation C

$C_{16}H_{25}NO_2$ (c) 71PRI
Nonyl phenylcarbamate
Heat Capacity 298.15 K, C_p = 112.6 cal·mol^{-1}·K^{-1}
471.1 J·mol^{-1}·K^{-1}
Temperature range 200–390 K. Complete data deposited in VINITI, No. 2713–71, 25 March 1971.
Phase Changes
c/liq 327 K, ΔH = 6705 cal·mol^{-1}
28054 J·mol^{-1}
ΔS = 20.5 cal·mol^{-1}·K^{-1}
85.8 J·mol^{-1}·K^{-1}
Molecular Weight 263.3790
Wiswesser Line Notation 9OVMR
Evaluation B

$C_{16}H_{26}$ (liq) 63GUD/CAM
1–Cyclohexyl–3–methylhydroindan
Heat Capacity 313 K, C_p = 95.7 cal·mol^{-1}·K^{-1}
400.7 J·mol^{-1}·K^{-1}
Temperature range 313–483 K
Molecular Weight 218.3814
Wiswesser Line Notation L56TJ B1 D- AL6TJ
Evaluation C

$C_{16}H_{28}$ (liq) 63GUD/CAM
2–Ethylperhydrophenanthrene
Heat Capacity 313 K, C_p = 96.3 cal·mol^{-1}·K^{-1}
402.9 J·mol^{-1}·K^{-1}
Temperature range 313–583 K
Molecular Weight 220.3972
Wiswesser Line Notation L B666TJ E2
Evaluation C

$C_{16}H_{30}$ (liq) 63GUD/CAM
Cyclohexyl(isopropylcyclohexyl)methane
Heat Capacity 313 K, C_p = 102.5 cal·mol^{-1}·K^{-1}
428.9 J·mol^{-1}·K^{-1}
Temperature range 313–483 K
Molecular Weight 222.4130
Wiswesser Line Notation L6TJ A1- AL6TJ XY1&1
Evaluation C

$C_{16}H_{30}$ (liq) 63GUD/CAM
1,3–Dicyclohexylbutane
Heat Capacity 313 K, C_p = 95.9 cal·mol^{-1}·K^{-1}
359.4 J·mol^{-1}·K^{-1}
Temperature range 313–583 K
Molecular Weight 222.4130
Wiswesser Line Notation L6TJ AY1&2- AL6TJ
Evaluation C

$C_{16}H_{30}HgO_4$ (liq) 78ADE
Mercuric octanoate; Mercuric caprylate
Heat Capacity 410 K, C_p = 151.9 cal·mol^{-1}·K^{-1}
635.5 J·mol^{-1}·K^{-1}
Temperature range 420–425 K
Mean value. Data graphically only for solid.
Phase Changes
c/liq 387.2 K, ΔH = 14700 cal·mol^{-1}
61500 J·mol^{-1}
ΔS = 38.0 cal·mol^{-1}·K^{-1}
158.8 J·mol^{-1}·K^{-1}
Molecular Weight 487.0006
Wiswesser Line Notation OV7 2 .HG
Evaluation C

Table of Heat Capacities, Entropies, and Phase Transition Properties — Continued

$C_{16}H_{31}NaO_2$ (c) 59WIR/DRO
δ–Sodium palmitate
Heat Capacity 298.15 K, C_p = 107.4 cal·mol⁻¹·K⁻¹ 449.4 J·mol⁻¹·K⁻¹
Temperature range 58–298 K.
Entropy 298.15 K, S = 113.4 cal·mol⁻¹·K⁻¹ 474.5 J·mol⁻¹·K⁻¹
Extrapolation below 50 K, 13.19 cal·mol⁻¹K⁻¹
Molecular Weight 278.4095
Wiswesser Line Notation OV15 .NA
Evaluation B
Sample with 0.12 mol H_2O. Correction of 1.2 cal·mol⁻¹K⁻¹ for H_2O gives S = 112.2 cal·mol⁻¹K⁻¹ for anhydrous salt.

$C_{16}H_{31}NaO_2$ (c) 59WIR/DRO
ω–Sodium palmitate
Heat Capacity 298.15 K, C_p = 118.3 cal·mol⁻¹·K⁻¹ 495.0 J·mol⁻¹·K⁻¹
Temperature range 58–298 K.
Entropy 298.15 K, S = 113.8 cal·mol⁻¹·K⁻¹ 476.1 J·mol⁻¹·K⁻¹
Extrapolation below 50 K, 12.85 cal·mol⁻¹K⁻¹
Molecular Weight 278.4095
Wiswesser Line Notation OV15 .NA
Evaluation B
Sample with 0.017 mol H_2O. Correction of 0.2 cal·mol⁻¹K⁻¹ for H_2O gives S = 113.6 cal·mol⁻¹K⁻¹ for anhydrous salt.

$C_{16}H_{31}NaO_2$•0.01H_2O (c) 59WIR/WOO
β–Sodium palmitate
Heat Capacity 298.15 K, C_p = 110.6 cal·mol⁻¹·K⁻¹ 462.8 J·mol⁻¹·K⁻¹
Temperature range 15–300 K
Entropy 298.15 K, S = 113.0 cal·mol⁻¹·K⁻¹ 472.8 J·mol⁻¹·K⁻¹
Molecular Weight 278.5896
Wiswesser Line Notation OV15 .NA &QH 0.01
Evaluation A

$C_{16}H_{31}NaO_2$•0.409H_2O (c) 59WIR/WOO
β–Sodium palmitate
Heat Capacity 298.15 K, C_p = 115.0 cal·mol⁻¹·K⁻¹ 481.2 J·mol⁻¹·K⁻¹
Temperature range 15–300 K
Entropy 298,15 K, S = 119.3 cal·mol⁻¹·K⁻¹ 499.2 J·mol⁻¹·K⁻¹
Molecular Weight 285.7777
Wiswesser Line Notation OV15 .NA &QH 0.409
Evaluation A

$C_{16}H_{31}NaO_2$•0.482H_2O (c) 59WIR/WOO
ϵ–Sodium palmitate
Heat Capacity 298.15 K, C_p = 113.9 cal·mol⁻¹·K⁻¹ 476.6 J·mol⁻¹·K⁻¹
Temperature range 60–300 K
Entropy 298.15 K, S = 119.7 cal·mol⁻¹·K⁻¹ 500.8 J·mol⁻¹·K⁻¹
Extrapolation below 50 K, 13.89 cal·mol⁻¹K⁻¹
Molecular Weight 287.0928
Wiswesser Line Notation OV15 .NA &QH 0.482
Evaluation A

$C_{16}H_{31}NaO_2$•0.715H_2O (c) 59WIR/WOO
ϵ–Sodium palmitate
Heat Capacity 298.15 K, C_p = 119.9 cal·mol⁻¹·K⁻¹ 501.7 J·mol⁻¹·K⁻¹
Temperature range 15–300 K
Entropy 298.15 K, S = 123.9 cal·mol⁻¹·K⁻¹ 518.4 J·mol⁻¹·K⁻¹
Molecular Weight 291.2904
Wiswesser Line Notation OV15 .NA &QH 0.715
Evaluation A

$C_{16}H_{32}$ (liq) 57MCC/FIN 2
1–Hexadecene
Heat Capacity 298.15 K, C_p = 115.89 cal·mol⁻¹·K⁻¹ 488.88 J·mol⁻¹·K⁻¹
Temperature range 11–360 K
Entropy 298.15 K, S = 140.51 cal·mol⁻¹·K⁻¹ 587.90 J·mol⁻¹·K⁻¹
Does not include S_0.
Phase Changes
c/liq 277.51 K, ΔH = 7216 cal·mol⁻¹ 30192 J·mol⁻¹
ΔS = 26.00 cal·mol⁻¹·K⁻¹ 108.80 J·mol⁻¹·K⁻¹
Molecular Weight 224.4288
Wiswesser Line Notation 15U1
Evaluation A

$C_{16}H_{32}$ (liq) 65FIN/MES
n–Decacyclohexane
Heat Capacity 298.15 K, C_p = 108.09 cal·mol⁻¹·K⁻¹ 452.25 J·mol⁻¹·K⁻¹
Temperature range 10–310 K
Entropy 298.15 K, S = 129.10 cal·mol⁻¹·K⁻¹ 540.15 J·mol⁻¹·K⁻¹
Phase Changes
c/liq 271.43 K, ΔH = 9225 cal·mol⁻¹ 38597 J·mol⁻¹
ΔS = 33.99 cal·mol⁻¹·K⁻¹ 142.20 J·mol⁻¹·K⁻¹
Molecular Weight 224.4288
Wiswesser Line Notation L6TJ A10
Evaluation A

$C_{16}H_{32}O_2$ (c) 25PAR/KEL
Hexadecanoic acid; Palmitic acid
Heat Capacity 292.5 K, C_p = 110.5 cal·mol⁻¹·K⁻¹ 462.3 J·mol⁻¹·K⁻¹
Temperature range 88–293 K. Value is unsmoothed experimental datum.
Entropy 298.1 K, S = 129.9 cal·mol⁻¹·K⁻¹ 543.5 J·mol⁻¹·K⁻¹
Extrapolation below 90 K, 51.20 cal·mol⁻¹K⁻¹
Molecular Weight 256.4276
Wiswesser Line Notation QV15
Evaluation B(C_p),C(S)

$C_{16}H_{32}O_2$ (c) 29PAR/KEL
Hexadecanoic acid; Palmitic acid
Entropy 298.1 K, S = 113.7 cal·mol⁻¹·K⁻¹ 475.7 J·mol⁻¹·K⁻¹
Extrapolation below 90 K, 35.0 cal·mol⁻¹K⁻¹.
Revision of previous data.
Molecular Weight 256.4276
Wiswesser Line Notation QV15
Evaluation C

Table of Heat Capacities, Entropies, and Phase Transition Properties — Continued

$C_{16}H_{32}O_2$ (c) 52WAR/SIN
Hexadecanoic acid; Palmitic acid
Heat Capacity 298 K, C_p = 107 cal·mol^{-1}·K^{-1}
448 J·mol^{-1}·K^{-1}
Temperature range 183–365 K. Three temperatures, each for solid and liquid and equations. C form.
Entropy 298.6 K, S = 104.84 cal·mol^{-1}·K^{-1}
438.65 J·mol^{-1}·K^{-1}
Extrapolation below 90 K, 25.51 cal·mol^{-1}K^{-1}
Phase Changes
c/liq 335.73 K, ΔH = 13120 cal·mol^{-1}
54894 J·mol^{-1}
ΔS = 39.1 cal·mol^{-1}·K^{-1}
163.5 J·mol^{-1}·K^{-1}
Molecular Weight 256.4276
Wiswesser Line Notation QV15
Evaluation C

$C_{16}H_{32}O_2$ (c) 56WIR/DRO
Hexadecanoic acid; Palmitic acid
Heat Capacity 298.15 K, C_p = 110.10 cal·mol^{-1}·K^{-1}
460.66 J·mol^{-1}·K^{-1}
Temperature range 15–302 K
Entropy 298.15 K, S = 108.12 cal·mol^{-1}·K^{-1}
452.37 J·mol^{-1}·K^{-1}
Molecular Weight 256.4276
Wiswesser Line Notation QV15
Evaluation A

$C_{16}H_{32}O_2$ (liq) 67PAC
Hexadecanoic acid; Palmitic acid
Heat Capacity 373 K, C_p = 162 cal·mol^{-1}·K^{-1}
678 J·mol^{-1}·K^{-1}
One temperature
Phase Changes
c/liq 336 K, ΔH = 13130 cal·mol^{-1}
54935 J·mol^{-1}
ΔS = 39.1 cal·mol^{-1}·K^{-1}
163.5 J·mol^{-1}·K^{-1}
Molecular Weight 256.4276
Wiswesser Line Notation QV15
Evaluation C

$C_{16}H_{32}O_2$ (c) 82SCH/MIL 2
Hexadecanoic acid; Palmitic acid
Heat Capacity 298.15 K, C_p = 110.75 cal·mol^{-1}·K^{-1}
463.36 J·mol^{-1}·K^{-1}
Temperature range 80–345 K
Phase Changes
c,I/liq 335.66 K, ΔH = 12837 cal·mol^{-1}
53711 J·mol^{-1}
ΔS = 38.246 cal·mol^{-1}·K^{-1}
160.02 J·mol^{-1}·K^{-1}
Molecular Weight 256.4276
Wiswesser Line Notation QV15
Evaluation B

$C_{16}H_{34}$ (liq) 49PAR/MOO
n–Hexadecane; Cetane
Heat Capacity 298.15 K, C_p = 120.5 cal·mol^{-1}·K^{-1}
504.2 J·mol^{-1}·K^{-1}
Temperature range 80–300 K
Entropy 298.15 K, S = 149.8 cal·mol^{-1}·K^{-1}
626.8 J·mol^{-1}·K^{-1}
Extrapolation below 80 K, 27.88 cal·mol^{-1}K^{-1}
Phase Changes
c/liq 291.1 K, ΔH = 12319 cal·mol^{-1}
51543 J·mol^{-1}
ΔS = 42.3 cal·mol^{-1}·K^{-1}
177.1 J·mol^{-1}·K^{-1}
Molecular Weight 226.4446
Wiswesser Line Notation 16H
Evaluation B(C_p),C(S)

$C_{16}H_{34}$ (liq) 54FIN/GRO 2
n–Hexadecane; Cetane
Heat Capacity 298.15 K, C_p = 119.85 cal·mol^{-1}·K^{-1}
501.45 J·mol^{-1}·K^{-1}
Temperature range 12–320 K
Entropy 298.15 K, S = 148.10 cal·mol^{-1}·K^{-1}
586.18 J·mol^{-1}·K^{-1}
Phase Changes
c/liq 291.34 K, ΔH = 12753 cal·mol^{-1}
53359 J·mol^{-1}
ΔS = 43.77 cal·mol^{-1}·K^{-1}
183.15 J·mol^{-1}·K^{-1}
Molecular Weight 226.4446
Wiswesser Line Notation 16H
Evaluation A

$C_{16}H_{34}$ (liq) 62GOL/BEL
n–Hexadecane; Cetane
Heat Capacity 311 K, C_p = 115.9 cal·mol^{-1}·K^{-1}
484.9 J·mol^{-1}·K^{-1}
Temperatures 100, 200, 300°F
Molecular Weight 226.4446
Wiswesser Line Notation 16H
Evaluation C

$C_{16}H_{34}$ (liq) 74DIA/REN
n–Hexadecane; Cetane
Heat Capacity 298.15 K, C_p = 119.89 cal·mol^{-1}·K^{-1}
501.6 J·mol^{-1}·K^{-1}
Temperature range 300–324 K
Molecular Weight 226.4446
Wiswesser Line Notation 16H
Evaluation A

$C_{16}H_{34}$ (liq) 74PET/TER
n–Hexadecane; Cetane
Heat Capacity 297.79 K, C_p = 119 cal·mol^{-1}·K^{-1}
499 J·mol^{-1}·K^{-1}
Temperature range 297–471 K. Value is unsmoothed experimental datum.
Molecular Weight 226.4446
Wiswesser Line Notation 16H
Evaluation B

$C_{16}H_{34}N_2$ (c,II) 74BUR/VER
1,2–Bis(methyldiallylammonium)ethane dibromide
Heat Capacity 298 K, C_p = 111.2 cal·mol^{-1}·K^{-1}
465.3 J·mol^{-1}·K^{-1}
Temperature range 273–373 K
Phase Changes
c,II/c,I 371 K, ΔH = 720 cal·mol^{-1}
3010 J·mol^{-1}
ΔS = 1.9 cal·mol^{-1}·K^{-1}
8.1 J·mol^{-1}·K^{-1}
Temperature range 370–372 K
Molecular Weight 254.4580
Wiswesser Line Notation 1U2K1&2U1&2K1&2U1&2U1 E 2
Evaluation B

Table of Heat Capacities, Entropies, and Phase Transition Properties — Continued

$C_{16}H_{34}N_2$ (c,II) 74BUR/VER
1,2-Bis(methyldiallylammonium)ethane dibromide
Heat Capacity 298 K, C_p = 111.2 cal·mol⁻¹·K⁻¹
465.3 J·mol⁻¹·K⁻¹
Temperature range 273–373 K
Phase Changes
c,II/c,I 371 K, ΔH = 720 cal·mol⁻¹
3010 J·mol⁻¹
ΔS = 1.9 cal·mol⁻¹·K⁻¹
8.1 J·mol⁻¹·K⁻¹
Temperature range 370–372 K
Molecular Weight 254.4580
Wiswesser Line Notation 1U2K1&2U1&2K1&2U1&2U1 E 2
Evaluation B

$C_{16}H_{34}O$ (c) 56PAR/KEN
1-Hexadecanol; n-Cetyl alcohol
Heat Capacity 290 K, C_p = 105.46 cal·mol⁻¹·K⁻¹
441.24 J·mol⁻¹·K⁻¹
Temperature range 80–290 K
Entropy 298.1 K, S = 108.0 cal·mol⁻¹·K⁻¹
451.9 J·mol⁻¹·K⁻¹
Extrapolation below 80 K, 27.56 cal mol⁻¹K⁻¹
Molecular Weight 242.4440
Wiswesser Line Notation Q16
Evaluation B(C_p),C(S)

$C_{16}H_{34}O$ (c) 74MOS/MOU
1-Hexadecanol; n-Cetyl alcohol
Heat Capacity 298.15 K, C_p = 100.9 cal·mol⁻¹·K⁻¹
422 J·mol⁻¹·K⁻¹
Temperature range 293–311 K
Phase Changes
c,γ/c,α 322.2 K, ΔH = 5665 cal·mol⁻¹
23700 J·mol⁻¹
ΔS = 17.5 cal·mol⁻¹·K⁻¹
73.6 J·mol⁻¹·K⁻¹
c,γ/liq 322.2 K, ΔH = 13955 cal·mol⁻¹
58380 J·mol⁻¹
ΔS = 43.3 cal·mol⁻¹·K⁻¹
181.2 J·mol⁻¹·K⁻¹
c,α/liq 322.3 K, ΔH = 8030 cal·mol⁻¹
33600 J·mol⁻¹
ΔS = 24.9 cal·mol⁻¹·K⁻¹
104.3 J·mol⁻¹·K⁻¹
Molecular Weight 242.4440
Wiswesser Line Notation Q16
Evaluation B

$C_{16}H_{34}O$ (liq) 74MOS/MOU
1-Hexadecanol; n-Cetyl alcohol
Heat Capacity 323 K, C_p = 125.2 cal·mol⁻¹·K⁻¹
524 J·mol⁻¹·K⁻¹
Temperature range 323–346 K
Molecular Weight 242.4440
Wiswesser Line Notation Q16
Evaluation B

$C_{16}H_{36}BrN$ (c) 73VIS/SOM
Tetra-n-butylammonium bromide
Heat Capacity 298.15 K, C_p = 98.5 cal·mol⁻¹·K⁻¹
412 J·mol⁻¹·K⁻¹
Temperature range 278–328 K
Molecular Weight 322.3711
Wiswesser Line Notation 4K4&4&4 E
Evaluation B

$C_{16}H_{36}BrN$ (c,III) 74BUR/VER
Tetra-n-butylammonium bromide
Heat Capacity 298 K, C_p = 114.2 cal·mol⁻¹·K⁻¹
447.8 J·mol⁻¹·K⁻¹
Temperature range 273–373 K
Phase Changes
c,IV/c,III 294 K, ΔH = 16 cal·mol⁻¹
67 J·mol⁻¹
ΔS = 0.05 cal·mol⁻¹·K⁻¹
0.2 J·mol⁻¹·K⁻¹
Temperature range 293–295 K. ΔH maximum value.
c,III/c,II 367 K, ΔH = 390 cal·mol⁻¹
1630 J·mol⁻¹
ΔS = 1.1 cal·mol⁻¹·K⁻¹
4.4 J·mol⁻¹·K⁻¹
Temperature range 366.0–369 K.
c,II/c,I 383 K, ΔH = 80 cal·mol⁻¹
335 J·mol⁻¹
ΔS = 0.2 cal·mol⁻¹·K⁻¹
0.9 J·mol⁻¹·K⁻¹
Temperature range 382–383.5 K
c/liq 395 K, ΔH = 3860 cal·mol⁻¹
16150 J·mol⁻¹
ΔS = 9.77 cal·mol⁻¹·K⁻¹
40.89 J·mol⁻¹·K⁻¹
Molecular Weight 322.3711
Wiswesser Line Notation 4K4&4&4 E
Evaluation B

$C_{16}H_{36}O_4Ti$ (liq) 79SAM/GRI
Tetrabutoxytitanium
Heat Capacity 353 K, C_p = 170 cal·mol⁻¹·K⁻¹
711 J·mol⁻¹·K⁻¹
Temperature range 333–453 K
Molecular Weight 340.3580
Wiswesser Line Notation 4O-TI-O4&O4&O4
Evaluation C

$C_{16}H_{38}Br_2N_2$ (c,II) 74BUR/VER
1,4-Bis(triethylammonium)butane dibromide
Heat Capacity 298 K, C_p = 110.5 cal·mol⁻¹·K⁻¹
462.3 J·mol⁻¹·K⁻¹
Temperature range 273–373 K
Phase Changes
c,II/c,I 518 K, ΔH = 12000 cal·mol⁻¹
50200 J·mol⁻¹
ΔS = 23.2 cal·mol⁻¹·K⁻¹
96.9 J·mol⁻¹·K⁻¹
Temperature range 503–530 K
Molecular Weight 418.2976
Wiswesser Line Notation 2K2&2&4K2&2&2 E 2
Evaluation B

$C_{16}H_{40}Cl_4N_2Ni$ (c,I) 79LAN/WES
bis-Tetraethylammonium tetrachloronickelate
Heat Capacity 298.15 K, C_p = 151.6 cal·mol⁻¹·K⁻¹
634.3 J·mol⁻¹·K⁻¹
Temperature range 5–350 K
Entropy 298.15 K, S = 185.33 cal·mol⁻¹·K⁻¹
775.42 J·mol⁻¹·K⁻¹
Phase Changes
c,III/c,I 222.4 K, ΔH = 2087 cal·mol⁻¹
8732 J·mol⁻¹
ΔS = 9.16 cal·mol⁻¹·K⁻¹
38.32 J·mol⁻¹·K⁻¹
Two transitions, c,III/c,II at 222.1 K and c,II/c,I at 222.7 K. ΔH and ΔS are the sum of the two transitions.
Molecular Weight 461.0174
Wiswesser Line Notation 2K2&2&2 2 NI G4
Evaluation A

Table of Heat Capacities, Entropies, and Phase Transition Properties — Continued

$C_{16}H_{40}Cl_4N_2Zn$ (c,I) 79LAN/WES
bis–Tetramethylammonium tetrachlorozincate
Heat Capacity 298.15 K, C_p = 153.0 cal·mol⁻¹·K⁻¹
640.2 J·mol⁻¹·K⁻¹
Temperature range 5–350 K
Entropy 298.15 K, S = 181.52 cal·mol⁻¹·K⁻¹
759.48 J·mol⁻¹·K⁻¹
Phase Changes
c,III/c,I 228.4 K, ΔH = 2284 cal·mol⁻¹
9556 J·mol⁻¹
ΔS = 9.96 cal·mol⁻¹·K⁻¹
41.67 J·mol⁻¹·K⁻¹
Two transitions, c,III/c,II at 227.9 K and c,II/c,I at 228.9 K. ΔH and ΔS are the sum of the two transitions.
Molecular Weight 467.6974
Wiswesser Line Notation 2K2&2&2 2 ZN G4
Evaluation A

$C_{17}H_{12}$ (c) 50UEB/ORT
Benzanthrene
Heat Capacity 298.15 K, C_p = 65.4 cal·mol⁻¹·K⁻¹
273.6 J·mol⁻¹·K⁻¹
Temperature range 293–368 K. Equation only.
Molecular Weight 216.2818
Wiswesser Line Notation L C6666 1A Q IHJ
Evaluation C

$C_{17}H_{14}FeO$ (c) 81TOM/CUR
Benzoylferrocene
Heat Capacity 298 K, C_p = 92.7 cal·mol⁻¹·K⁻¹
387.9 J·mol⁻¹·K⁻¹
Temperature range 293–363 K. Equation given.
Phase Changes
c/liq 384.2 K
Molecular Weight 290.1440
Wiswesser Line Notation L5ØJ Ø–FE– –ØL5ØJ AVR
Evaluation B

$C_{17}H_{22}O_4$ (liq) 69RAB/MAR
Dibutyl o–phthalate
Heat Capacity 300 K, C_p = 114.0 cal·mol⁻¹·K⁻¹
477.0 J·mol⁻¹·K⁻¹
Temperature range 60–360 K
Entropy 300 K, S = 223.0 cal·mol⁻¹·K⁻¹
933.0 J·mol⁻¹·K⁻¹
Molecular Weight 290.3584
Wiswesser Line Notation 4OVR BVO4
Evaluation C
Glass transition at 173.5K; ΔH = 296 cal·mol⁻¹, ΔS = 1.14 cal·mol⁻¹K⁻¹.

$C_{17}H_{30}$ (liq) 62GOL/BEL
Cyclopentylbicyclohexyl
Heat Capacity 311 K, C_p = 102.9 cal·mol⁻¹·K⁻¹
430.5 J·mol⁻¹·K⁻¹
Temperatures 100, 200, 300°F
Molecular Weight 234.4240
Wiswesser Line Notation L6TJ A– AL6TJ X– AL5TJ
Evaluation C

$C_{17}H_{30}$ (liq) 63GUD/CAM
Cyclopentylbicyclohexyl
Heat Capacity 313 K, C_p = 103.6 cal·mol⁻¹·K⁻¹
433.5 J·mol⁻¹·K⁻¹
Temperature range 313–483 K
Molecular Weight 234.4240
Wiswesser Line Notation L6TJ AAL6TJ X– AL5TJ
Evaluation C

$C_{17}H_{32}$ (liq) 63GUD/CAM
Bis(ethylcyclohexyl)methane
Heat Capacity 313 K, C_p = 111.4 cal·mol⁻¹·K⁻¹
466.1 J·mol⁻¹·K⁻¹
Temperature range 313–483 K
Molecular Weight 236.4398
Wiswesser Line Notation L6TJ A2 X1– AL6TJ X2
Evaluation C

$C_{17}H_{32}$ (liq) 63GUD/CAM
1–Cyclohexyl–1–isopropylcyclohexylethane
Heat Capacity 313 K, C_p = 105.0 cal·mol⁻¹·K⁻¹
439.3 J·mol⁻¹·K⁻¹
Temperature range 313–583 K
Molecular Weight 236.4398
Wiswesser Line Notation L6TJ AY1&– AL6TJ XY1&1
Evaluation C

$C_{17}H_{34}O_2$ (c) 82SCH/MIL
Heptadecanoic acid
Heat Capacity 298.15 K, C_p = 113.70 cal·mol⁻¹·K⁻¹
475.74 J·mol⁻¹·K⁻¹
Temperature range 80–350 K
Phase Changes
c,II/c,I 329.2 K, ΔH = 1777 cal·mol⁻¹
7435 J·mol⁻¹
ΔS = 5.40 cal·mol⁻¹·K⁻¹
22.59 J·mol⁻¹·K⁻¹
c,I/liq 334.25 K, ΔH = 12271 cal·mol⁻¹
51342 J·mol⁻¹
ΔS = 36.711 cal·mol⁻¹·K⁻¹
153.60 J·mol⁻¹·K⁻¹
Molecular Weight 270.4544
Wiswesser Line Notation QV16
Evaluation A

$C_{17}H_{34}O_2$ (c) 56WIR/DRO
Methyl hexadecanoate; Methyl palmitate
Heat Capacity 298.15 K, C_p = 113.40 cal·mol⁻¹·K⁻¹
474.47 J·mol⁻¹·K⁻¹
Temperature range 15–297 K
Entropy 298.15 K, S = 118.33 cal·mol⁻¹·K⁻¹
495.09 J·mol⁻¹·K⁻¹
Molecular Weight 270.4544
Wiswesser Line Notation 15VO1
Evaluation A

$C_{17}H_{34}O_4$ (c) 65SIL/DAU
2–Monomyristin
Heat Capacity 298 K, C_p = 121.0 cal·mol⁻¹·K⁻¹
506.3 J·mol⁻¹·K⁻¹
One temperature.
Molecular Weight 302.4532
Wiswesser Line Notation Q1Y1QOV13
Evaluation B

Table of Heat Capacities, Entropies, and Phase Transition Properties — Continued

$C_{17}H_{34}O_4$ (c) 65SIL/DAU
1–Monomyristin
Heat Capacity 298 K, C_p = 124.3 cal·mol^{-1}·K^{-1}
520.1 J·mol^{-1}·K^{-1}
One temperature. β_L form
Molecular Weight 302.4532
Wiswesser Line Notation Q1YQ1OV13
Evaluation B

$C_{17}H_{36}$ (liq) 67MES/GUT
n–Heptadecane
Heat Capacity 298.15 K, C_p = 127.71 cal·mol^{-1}·K^{-1}
534.34 J·mol^{-1}·K^{-1}
Temperature range 12–380 K
Entropy 298.15 K, S = 155.89 cal·mol^{-1}·K^{-1}
652.24 J·mol^{-1}·K^{-1}
Phase Changes
c,II/c,I 284.27 K, ΔH = 2615.2 cal·mol^{-1}
10942 J·mol^{-1}
ΔS = 9.20 cal·mol^{-1}·K^{-1}
38.50 J·mol^{-1}·K^{-1}
c,I/liq 295.14 K, ΔH = 9599.5 cal·mol^{-1}
40164 J·mol^{-1}
ΔS = 32.53 cal·mol^{-1}·K^{-1}
136.08 J·mol^{-1}·K^{-1}
Molecular Weight 240.4714
Wiswesser Line Notation 17H
Evaluation A

$C_{17}H_{38}N_2O$ (c,I) 65PEM/PAR
Urea–n–hexadecane adduct
Heat Capacity 298.15 K, C_p = 30.61 cal·mol^{-1}·K^{-1}
128.07 J·mol^{-1}·K^{-1}
Temperature range 12–300 K. Value for adduct with 1 mole of urea.
Entropy 298.15 K, S = 33.13 cal·mol^{-1}·K^{-1}
138.62 J·mol^{-1}·K^{-1}
Transitions of 135.3 K and 151.8 K with total ΔH = 422 cal·(mol hydrocarbon)$^{-1}$.
Molecular Weight 286.5000
Wiswesser Line Notation ZVZ &16H
Evaluation A
Sample 76.30 percent urea.

$C_{17}H_{40}Br_2N_2$ (c,II) 74BUR/VER
1,5–Bis(triethylammonium)pentane dibromide
Heat Capacity 298 K, C_p = 118.0 cal·mol^{-1}·K^{-1}
493.7 J·mol^{-1}·K^{-1}
Temperature range 273–373 K
Phase Changes
c,II/c,I 465 K, ΔH = 5850 cal·mol^{-1}
24480 J·mol^{-1}
ΔS = 12.6 cal·mol^{-1}·K^{-1}
52.6 J·mol^{-1}·K^{-1}
Temperature range 458–471 K
Molecular Weight 432.3244
Wiswesser Line Notation 2K2&2&5K2&2&2 E 2
Evaluation B

$C_{18}H_{12}$ (c) 80WON/WES
Naphthacene
Heat Capacity 298.15 K, C_p = 56.54 cal·mol^{-1}·K^{-1}
236.6 J·mol^{-1}·K^{-1}
Temperature range 5–350 K
Entropy 298.15 K, S = 51.48 cal·mol^{-1}·K^{-1}
215.4 J·mol^{-1}·K^{-1}
Molecular Weight 228.2928
Wiswesser Line Notation L C6666J
Evaluation A

$C_{18}H_{12}$ (c) 71WON/WES
Triphenylene; 9,10–Benzophenanthrene
Heat Capacity 298.15 K, C_p = 61.95 cal·mol^{-1}·K^{-1}
259.20 J·mol^{-1}·K^{-1}
Temperature range 5–514 K
Entropy 298.15 K, S = 60.87 cal·mol^{-1}·K^{-1}
254.68 J·mol^{-1}·K^{-1}
Phase Changes
c/liq 471.01 K, ΔH = 5914 cal·mol^{-1}
24744 J·mol^{-1}
ΔS = 12.57 cal·mol^{-1}·K^{-1}
52.53 J·mol^{-1}·K^{-1}
Molecular Weight 228.2928
Wiswesser Line Notation L B6 H666J
Evaluation A

$C_{18}H_{14}$ (c) 72CHA/BES
o–Terphenyl
Heat Capacity 298.15 K, C_p = 65.67 cal·mol^{-1}·K^{-1}
274.75 J·mol^{-1}·K^{-1}
Temperature range 2–350 K. Also data for annealed and quenched glass.
Entropy 298.15 K, S = 71.42 cal·mol^{-1}·K^{-1}
298.81 J·mol^{-1}·K^{-1}
Phase Changes
c/liq 329.35 K, ΔH = 4109 cal·mol^{-1}
17.191 J·mol^{-1}
ΔS = 12.48 cal·mol^{-1}·K^{-1}
52.20 J·mol^{-1}·K^{-1}
Molecular Weight 230.3086
Wiswesser Line Notation RR BR
Evaluation A

$C_{18}H_{14}$ (liq) 72CHA/BES
o–Terphenyl
Heat Capacity 298.15 K, C_p = 88.21 cal·mol^{-1}·K^{-1}
369.05 J·mol^{-1}·K^{-1}
Temperature range 250–360 K. Supercooled liquid below Tm 329.35 K.
Entropy 298.15 K, S = 80.57 cal·mol^{-1}·K^{-1}
337.11 J·mol^{-1}·K^{-1}
Molecular Weight 230.3086
Wiswesser Line Notation RR BR
Evaluation A

$C_{18}H_{14}$ (liq) 58WAL/BRO
m–Terphenyl
Heat Capacity 370 K, C_p = 99.7 cal·mol^{-1}·K^{-1}
417.1 J·mol^{-1}·K^{-1}
Temperature range 200 to 600°F
Molecular Weight 230.3086
Wiswesser Line Notation RR CR
Evaluation B

Table of Heat Capacities, Entropies, and Phase Transition Properties — Continued

$C_{18}H_{14}$ (c,I) 79CAI/DWO
p–Terphenyl
Heat Capacity C_p data not given.
Phase Changes
Temperature range 10–300 K
c,II/c,I 193.3 K, ΔH = 23 cal·mol^{-1}
95 J·mol^{-1}
ΔS = 0.12 cal·mol^{-1}·K^{-1}
0.49 J·mol^{-1}·K^{-1}
Obtained T = 191.0 K, ΔS = 0.45 J·mol^{-1}·K^{-1}. from DSC data.
Molecular Weight 230.3086
Wiswesser Line Notation RR DR
Evaluation C

$C_{18}H_{15}As$ (c) 31SMI/AND 2
Triphenylarsine
Heat Capacity 298.5 K, C_p = 76.8 cal·mol^{-1}·K^{-1}
321.3 J·mol^{-1}·K^{-1}
Temperature range 102–311 K. Value is unsmoothed experimental datum.
Molecular Weight 306.2381
Wiswesser Line Notation R–AS–R&R
Evaluation B

$C_{18}H_{15}Bi$ (c) 79STE
Triphenylbismuthine; Bismuth triphenyl
Heat Capacity 298.15 K, C_p = 78.9 cal·mol^{-1}·K^{-1}
330.2 J·mol^{-1}·K^{-1}
One temperature
Molecular Weight 440.2969
Wiswesser Line Notation R–BI–R&R
Evaluation B

$C_{18}H_{15}Bi$ (c) 31SMI/AND 2
Triphenylbismuthine; Bismuth triphenyl
Heat Capacity 298.5 K, C_p = 78.5 cal·mol^{-1}·K^{-1}
328.4 J·mol^{-1}·K^{-1}
Temperature range 102–323 K
Molecular Weight 440.2969
Wiswesser Line Notation R–BI–R&R
Evaluation B

$C_{18}H_{15}N$ (c) 31SMI/AND 2
Triphenylamine
Heat Capacity 298.5 K, C_p = 71.1 cal·mol^{-1}·K^{-1}
297.5 J·mol^{-1}·K^{-1}
Temperature range 102–346 K. Value is unsmoothed experimental datum.
Molecular Weight 245.3232
Wiswesser Line Notation RNR&R
Evaluation B

$C_{18}H_{15}P$ (c) 31SMI/AND 2
Triphenylphosphine
Heat Capacity 298.5 K, C_p = 74.7 cal·mol^{-1}·K^{-1}
312.5 J·mol^{-1}·K^{-1}
Temperature range 102–298 K. Value is unsmoothed experimental datum.
Molecular Weight 262.2903
Wiswesser Line Notation RPR&R
Evaluation B

$C_{18}H_{15}Sb$ (c) 31SMI/AND 2
Triphenylstibine; Antimony triphenyl
Heat Capacity 298.5 K, C_p = 77.8 cal·mol^{-1}·K^{-1}
325.5 J·mol^{-1}·K^{-1}
Temperature range 102–311 K
Molecular Weight 353.0665
Wiswesser Line Notation R–SB–R&R
Evaluation B

$C_{18}H_{18}$ (c) 44EIB
1–Methyl–7–isopropylphenanthrene; Retene
Heat Capacity 298.1 K, C_p = 70.4 cal·mol^{-1}·K^{-1}
294.6 J·mol^{-1}·K^{-1}
Temperature range 25 to 200 °C, equations only, in t°C.
C_p(c) = 0.2620 + 0.001584t cal·g^{-1}C^{-1} (25 to 60°C);
C_p(liq) = 0.364 + 0.000661t cal·g^{-1}C^{-1} (96 to 200°C).
Phase Changes
c/liq 369.0 K, ΔH = 4310 cal·mol^{-1}
18030 J·mol^{-1}
ΔS = 11.7 cal·mol^{-1}·K^{-1}
48.9 J·mol^{-1}·K^{-1}
Molecular Weight 234.3402
Wiswesser Line Notation L B666J EY1&1 K1
Evaluation C

$C_{18}H_{18}N_2O_5$ (c) 32SPA/THO
Ethyl azoxybenzenedicarboxylate
Heat Capacity 303 K, C_p = 107.8 cal·mol^{-1}·K^{-1}
451.0 J·mol^{-1}·K^{-1}
Temperature range 30 to 150°C
Phase Changes
liq/liq 395.7 K, ΔH = 1300 cal·mol^{-1}
5439 J·mol^{-1}
ΔS = 3.3 cal·mol^{-1}·K^{-1}
13.7 J·mol^{-1}·K^{-1}
Liquid crystal–isotropic liquid transition
c/liq 386.9 K, ΔH = 4896 cal·mol^{-1}
20485 J·mol^{-1}
ΔS = 12.7 cal·mol^{-1}·K^{-1}
52.9 J·mol^{-1}·K^{-1}
Crystal–liquid crystal transition
Molecular Weight 342.3506
Wiswesser Line Notation 2OVR DNO&UNR DVO2
Evaluation B
Uncertain isomer; para assumed.

$C_{18}H_{20}$ (c) 69SHI/MCN
3,3–Paracyclophane
Heat Capacity 300 K, C_p = 77.5 cal·mol^{-1}·K^{-1}
324.3 J·mol^{-1}·K^{-1}
Temperatures 300, 318 K
Molecular Weight 236.3560
Wiswesser Line Notation L G6 C–14–6 A B G– G– – &T&J
Evaluation B

$C_{18}H_{21}NO$ (c) 74SHI/MAE
N–(4–Methoxybenzylidene)–p–butylaniline
Heat Capacity 298.15 K, C_p = 113.6 cal·mol^{-1}·K^{-1}
475.3 J·mol^{-1}·K^{-1}
Temperature range 2–330 K. Nematic liquid crystal.

Table of Heat Capacities, Entropies, and Phase Transition Properties — Continued

Phase Changes
c,I/c 294.45 K, ΔH = 3142.6 cal·mol^{-1}
13148.6 J·mol^{-1}
ΔS = 10.67 cal·mol^{-1}·K^{-1}
44.65 J·mol^{-1}·K^{-1}
c = Nematic liquid crystal. c,I = metastable crystal form.
c,III/c 295.65 K, ΔH = 3809.2 cal·mol^{-1}
15937.7 J·mol^{-1}
ΔS = 12.88 cal·mol^{-1}·K^{-1}
53.91 J·mol^{-1}·K^{-1}
c = Nematic liquid crystal. c,III = stable crystal form.
c/liq 320.14 K, ΔH = 67.9 cal·mol^{-1}
284.1 J·mol^{-1}
ΔS = 0.21 cal·mol^{-1}·K^{-1}
0.89 J·mol^{-1}·K^{-1}
c = Nematic liquid crystal; liquid = isotropic liquid.
Molecular Weight 267.3700
Wiswesser Line Notation 4R DNU1R DO1
Evaluation A

$C_{18}H_{21}NO_2$ (c) 73SOR/SEK
N–(2–Hydroxy–4–methoxybenzylidene)–p–butylaniline
Heat Capacity 300 K, C_p = 107.82 cal·mol^{-1}·K^{-1}
451.12 J·mol^{-1}·K^{-1}
Temperature range 12–375 K
Entropy 300 K, S = 100.57 cal·mol^{-1}·K^{-1}
420.77 J·mol^{-1}·K^{-1}
Phase Changes
liq/liq 335.65 K, ΔH = 212.1 cal·mol^{-1}
887.4 J·mol^{-1}
ΔS = 0.63 cal·mol^{-1}·K^{-1}
2.64 J·mol^{-1}·K^{-1}
nematic–isotropic liquid transition
c/liq 314.52 K, ΔH = 5355 cal·mol^{-1}
24405 J·mol^{-1}
ΔS = 17.03 cal·mol^{-1}·K^{-1}
71.24 J·mol^{-1}·K^{-1}
Molecular Weight 283.3694
Wiswesser Line Notation 4R DNU1R BQ DO1
Evaluation A

$C_{18}H_{22}$ (liq) 56MCE
p,p'–Diisopropylbiphenyl
Heat Capacity 422 K, C_p = 124.4 cal·mol^{-1}·K^{-1}
520.5 J·mol^{-1}·K^{-1}
Temperature range 300 to 600°F
Molecular Weight 238.3718
Wiswesser Line Notation 1Y1&R DR DY1&1
Evaluation C
Quoted in 58WAL/BRO.

$C_{18}H_{22}N_4O_6$ (c) 80RAD/RAD
Carbazole–1,3,5–trinitrobenzene adduct
Heat Capacity 298.15 K, C_p = 96.75 cal·mol^{-1}·K^{-1}
404.8 J·mol^{-1}·K^{-1}
Temperature range 180–430 K. Data given graphically. C_p calculated from equation.
Phase Changes
c,I/liq 477.5 K, ΔH = 10445 cal·mol^{-1}
43702 J·mol^{-1}
ΔS = 21.87 cal·mol^{-1}·K^{-1}
91.52 J·mol^{-1}·K^{-1}
Molecular Weight 390.3950
Wiswesser Line Notation T B656 HMJ &WNR CNW ENW
Evaluation B

$C_{18}H_{24}$ (c) 81JEN/OBR
Triamantane
Phase Changes
c,II/c,I 293.65 K, ΔH = 264 cal·mol^{-1}
1106 J·mol^{-1}
ΔS = 0.90 cal·mol^{-1}·K^{-1}
3.77 J·mol^{-1}·K^{-1}
Room temperature specific heat anomaly.
Molecular Weight 240.3876
Wiswesser Line Notation L666 B666 E6/B–L/CN/JO/EP/HQ B– G -11–AAABBCEFFJJ Q JXTJ
Evaluation A

$C_{18}H_{30}$ (liq) 62GOL/BEL
3–Ethylperhydropyrene
Heat Capacity 311 K, C_p = 109.9 cal·mol^{-1}·K^{-1}
459.8 J·mol^{-1}·K^{-1}
Temperatures 100, 200, 300°F
Molecular Weight 246.4350
Wiswesser Line Notation L666 B6 2AB PJ F2
Evaluation C

$C_{18}H_{32}$ (liq) 62GOL/BEL
1–Cyclohexyl–1,3,3–trimethylhydroindan
Heat Capacity 311 K, C_p = 109.3 cal·mol^{-1}·K^{-1}
457.3 J·mol^{-1}·K^{-1}
Temperatures 100, 200, 300°F
Molecular Weight 248.4368
Wiswesser Line Notation L56TJ B1 B1 D1 D– AL6TJ
Evaluation C

$C_{18}H_{32}$ (liq) 63GUD/CAM
1–Cyclohexyl–1,3,3–trimethylhydroindan
Heat Capacity 313 K, C_p = 109.2 cal·mol^{-1}·K^{-1}
456.9 J·mol^{-1}·K^{-1}
Temperature range 313–483 K
Molecular Weight 248.4508
Wiswesser Line Notation L56TJ B1 B1 D1 D– AL6TJ
Evaluation C

$C_{18}H_{32}$ (liq) 62GOL/BEL
o–Tercyclohexyl
Heat Capacity 311 K, C_p = 101.4 cal·mol^{-1}·K^{-1}
424.3 J·mol^{-1}·K^{-1}
Temperatures 100, 200, 300°F
Molecular Weight 248.4508
Wiswesser Line Notation L6TJ AAL6TJ B– AL6TJ
Evaluation C

$C_{18}H_{32}$ (liq) 63GUD/CAM
o–Tercyclohexyl
Heat Capacity 313 K, C_p = 102.1 cal·mol^{-1}·K^{-1}
427.2 J·mol^{-1}·K^{-1}
Temperature range 313–483 K
Molecular Weight 248.4508
Wiswesser Line Notation L6TJ AAL6TJ B– AL6TJ
Evaluation C

$C_{18}H_{32}$ (liq) 62GOL/BEL
m–Tercyclohexyl
Heat Capacity 311 K, C_p = 109.3 cal·mol^{-1}·K^{-1}
457.3 J·mol^{-1}·K^{-1}
Temperatures 100, 200, 300°F
Molecular Weight 248.4508
Wiswesser Line Notation L6TJ AAL6TJ C– AL6TJ
Evaluation C

Table of Heat Capacities, Entropies, and Phase Transition Properties — Continued

$C_{18}H_{32}$ (liq) 63GUD/CAM
m–Tercyclohexyl
Heat Capacity 373 K, C_p = 120.0 cal·mol^{-1}·K^{-1}
502.1 J·mol^{-1}·K^{-1}
Temperature range 373–483 K
Molecular Weight 248.4508
Wiswesser Line Notation L6TJ AAL6TJ C– AL6TJ
Evaluation C

$C_{18}H_{32}$ (liq) 63GUD/CAM
p–Tercyclohexyl
Heat Capacity 423 K, C_p = 134.9 cal·mol^{-1}·K^{-1}
564.4 J·mol^{-1}·K^{-1}
One temperature
Molecular Weight 248.4508
Wiswesser Line Notation L6TJ AAL6TJ D– AL6TJ
Evaluation C

$C_{18}H_{34}$ (liq) 63GUD/CAM
1,1–Bis(dimethylcyclohexyl)ethane
Heat Capacity 313 K, C_p = 110.2 cal·mol^{-1}·K^{-1}
461.1 J·mol^{-1}·K^{-1}
Temperature range 313–483 K
Molecular Weight 250.4666
Wiswesser Line Notation L6TJ A1 X1 XY1&– AL6TJ X1 X1
Evaluation C

$C_{18}H_{34}$ (liq) 63GUD/CAM
1,1–Bis(ethylcyclohexyl)ethane
Heat Capacity 313 K, C_p = 114.2 cal·mol^{-1}·K^{-1}
477.8 J·mol^{-1}·K^{-1}
Temperature range 313–483 K
Molecular Weight 250.4666
Wiswesser Line Notation L6TJ A2 XY1&– AL6TJ X2
Evaluation C

$C_{18}H_{34}O_4$ (liq) 76PHI/MAT
Di–n–butyl sebacate
Heat Capacity 312 K, C_p = 148 cal·mol^{-1}·K^{-1}
619 J·mol^{-1}·K^{-1}
Temperature range 312–412 K
Molecular Weight 314.4642
Wiswesser Line Notation 4OV8VO4
Evaluation C

$C_{18}H_{34}O_4Pb$ (c,II) 78ADE/SIM
Lead(II) nonate; Lead(II) pelargonate
Heat Capacity 310 K, C_p = 221 cal·mol^{-1}·K^{-1}
926 J·mol^{-1}·K^{-1}
Mean value 353–363 K. Data only graphically for c,III.
Data also for c,I, 370–378 K, and liquid, 413–463 K.
Phase Changes
c,III/c,II 348.9 K, ΔH = 7700 cal·mol^{-1}
32200 J·mol^{-1}
ΔS = 22.1 cal·mol^{-1}·K^{-1}
92 J·mol^{-1}·K^{-1}
c,II, and c,I, are mesophases.
c,II/c,I 367.4 K, ΔH = 3920 cal·mol^{-1}
16400 J·mol^{-1}
ΔS = 10.7 cal·mol^{-1}·K^{-1}
45 J·mol^{-1}·K^{-1}
c,I/liq 384.8 K, ΔH = 310 cal·mol^{-1}
1300 J·mol^{-1}
ΔS = 0.8 cal·mol^{-1}·K^{-1}
3.4 J·mol^{-1}·K^{-1}
Molecular Weight 521.6642
Wiswesser Line Notation OV8 2 .PB
Evaluation C

$C_{18}H_{36}$ (liq) 49PAR/MOO
n–Dodecylcyclohexane
Heat Capacity 298.15 K, C_p = 147.1 cal·mol^{-1}·K^{-1}
615.5 J·mol^{-1}·K^{-1}
Extrapolation below 80 K, 28.51 cal mol^{-1}K^{-1}
Phase Changes
Hump in specific heat curve at 229–249 K.
c/liq 258.8 K, ΔH = 10955 cal·mol^{-1}
45836 J·mol^{-1}
ΔS = 38.3 cal·mol^{-1}·K^{-1}
160.4 J·mol^{-1}·K^{-1}
Molecular Weight 252.4824
Wiswesser Line Notation L6TJ A12
Evaluation B(C_p),C(S)

$C_{18}H_{36}$ (liq) 63GUD/CAM
Hexaethylcyclohexane
Heat Capacity 313 K, C_p = 126.7 cal·mol^{-1}·K^{-1}
530.1 J·mol^{-1}·K^{-1}
Temperature range 313–483 K
Molecular Weight 252.4824
Wiswesser Line Notation L6TJ A2 B2 C2 D2 E2 F2
Evaluation C

$C_{18}H_{36}N_2O_2$ (c) 53WIL/DOL
N,N′–Di–n–hexyladipamide
Heat Capacity 373 K, C_p = 154.1 cal·mol^{-1}·K^{-1}
644.7 J·mol^{-1}·K^{-1}
Temperature range 373–583 K
Phase Changes
c/liq 432 K, ΔH = 9750 cal·mol^{-1}
4079.5 J·mol^{-1}
ΔS = 22.6 cal·mol^{-1}·K^{-1}
94.4 J·mol^{-1}·K^{-1}
Molecular Weight 312.4946
Wiswesser Line Notation 6MV4VM6
Evaluation C

$C_{18}H_{36}O_2$ (c) 50SIN/WAR
Octadecanoic acid; Stearic acid
Heat Capacity 298.15 K, C_p = 134.3 cal·mol^{-1}·K^{-1}
561.9 J·mol^{-1}·K^{-1}
Temperature range 154–350 K
Entropy 298.15 K, S = 104.1 cal·mol^{-1}·K^{-1}
435.6 J·mol^{-1}·K^{-1}
Extrapolation below 90 K, 15.4 cal mol^{-1}K^{-1}.
Phase Changes
Transition between B & C forms previously reported as 52.90°C is high. Irreversible and slow change observed as low as 35.2°C.
c/liq 342.65 K, ΔH = 16360 cal·mol^{-1}
68450 J·mol^{-1}
ΔS = 47.7 cal·mol^{-1}·K^{-1}
1998 J·mol^{-1}·K^{-1}
Molecular Weight 284.4812
Wiswesser Line Notation QV17
Evaluation C

Table of Heat Capacities, Entropies, and Phase Transition Properties — Continued

$C_{18}H_{36}O_2$ (c) 82SCH/MIL 2
Octadecanoic acid; Stearic acid
Heat Capacity 298.15 K, C_p = 119.87 cal·mol⁻¹·K⁻¹
501.55 J·mol⁻¹·K⁻¹
Temperature range 80–355 K
Phase Changes
c,I/liq 342.49 K, ΔH = 14629 cal·mol⁻¹
61208 J·mol⁻¹
ΔS = 42.572 cal·mol⁻¹·K⁻¹
178.12 J·mol⁻¹·K⁻¹
Molecular Weight 280.4496
Wiswesser Line Notation QV17
Evaluation B

$C_{18}H_{37}Cl$ (liq) 75STR/SUN
1–Chloroöctadecane; n–Octadecyl chloride
Heat Capacity 301 K, C_p = 145.0 cal·mol⁻¹·K⁻¹
606.8 J·mol⁻¹·K⁻¹
One temperature
Phase Changes
liq/g 298.15 K, ΔH = 23460 cal·mol⁻¹
98160 J·mol⁻¹
ΔS = 78.69 cal·mol⁻¹·K⁻¹
329.22 J·mol⁻¹·K⁻¹
ΔH vaporization from equation using data from 68WAD.
Molecular Weight 288.9433
Wiswesser Line Notation G18
Evaluation B

$C_{18}H_{38}$ (c) 67MES/GUT
n–Octadecane
Heat Capacity 298.15 K, C_p = 116.07 cal·mol⁻¹·K⁻¹
485.64 J·mol⁻¹·K⁻¹
Temperature range 12–380 K
Entropy 298.15 K, S = 114.77 cal·mol⁻¹·K⁻¹
480.20 J·mol⁻¹·K⁻¹
Phase Changes
c/liq 301.33 K, ΔH = 14748 cal·mol⁻¹
61706 J·mol⁻¹
ΔS = 48.94 cal·mol⁻¹·K⁻¹
204.6 J·mol⁻¹·K⁻¹
Molecular Weight 254.4982
Wiswesser Line Notation 18H
Evaluation A

$C_{18}H_{38}$ (liq) 49PAR/MOO
n–Octadecane
Heat Capacity 300 K, C_p = 134.9 cal·mol⁻¹·K⁻¹
564.4 J·mol⁻¹·K⁻¹
Temperature range 80–300 K
Entropy 298.15 K, S = 166.5 cal·mol⁻¹·K⁻¹
696.6 J·mol⁻¹·K⁻¹
Supercooled liquid. Extrapolation below 80 K, 30.52 cal·mol⁻¹K⁻¹.
Phase Changes
Hump in specific heat curve at 228–240 K.
c/liq 301.3 K, ΔH = 14456 cal·mol⁻¹
60484 J·mol⁻¹
ΔS = 48.0 cal·mol⁻¹·K⁻¹
200.7 J·mol⁻¹·K⁻¹
Molecular Weight 254.4982
Wiswesser Line Notation 18H
Evaluation B(C_p),C(S)

$C_{18}H_{42}Br_2N_2$ (c,II) 74BUR/VER
1,6–Bis(triethylammonium)hexane dibromide
Heat Capacity 298 K, C_p = 124.6 cal·mol⁻¹·K⁻¹
521.3 J·mol⁻¹·K⁻¹
Temperature range 273–373 K
Phase Changes
c,II/c,I 495 K, ΔH = 4500 cal·mol⁻¹
18830 J·mol⁻¹
ΔS = 9.1 cal·mol⁻¹·K⁻¹
38.0 J·mol⁻¹·K⁻¹
Temperature range 482–505 K
Molecular Weight 446.3512
Wiswesser Line Notation 2K2&2&6K2&2&2 E 2
Evaluation B

$C_{18}D_{14}$ (c,I) 79CAI/DWO
p–Terphenyl–d_{14}
Phase Changes
c,II/c,I 178.8 K, ΔH = 46 cal·mol⁻¹
193 J·mol⁻¹
ΔS = 0.26 cal·mol⁻¹·K⁻¹
1.08 J·mol⁻¹·K⁻¹
Molecular Weight 244.3954
Wiswesser Line Notation RR DR &1B–F/2BCEF/5B–F/H–2 14
Evaluation C

$C_{18.1}H_{18.6}O_6$ (c) 60PAR/STA
β–Quinol–methane clathrate
Heat Capacity 296.62 K, C_p = 99.66 cal·mol⁻¹·K⁻¹
416.98 J·mol⁻¹·K⁻¹
Temperature range 13–298 K. Value is unsmoothed experimental datum.
Molecular Weight 332.9836
Wiswesser Line Notation QR DQ 3 &1H 0.165
Evaluation A

$C_{18.4}H_{19.8}O_6$ (c) 60PAR/STA
β–Quinol–methane clathrate
Heat Capacity 297.83 K, C_p = 102.39 cal·mol⁻¹·K⁻¹
428.40 J·mol⁻¹·K⁻¹
Temperature range 13–298 K. Value is unsmoothed experimental datum.
Molecular Weight 337.6681
Wiswesser Line Notation QR DQ 3 &1H 0.457
Evaluation A

$C_{18.4}H_{44.6}N_2S$ (c,I) 72COP/GAN
Thiourea–2,2–dimethylbutane adduct;
2,2–Dimethylbutane–thiourea adduct
Heat Capacity 298.15 K, C_p = 36.8 cal·mol⁻¹·K⁻¹
154.0 J·mol⁻¹·K⁻¹
Temperature range 12–300 K. Values for one mole of thiourea.
Entropy 298.15 K, S = 45.7 cal·mol⁻¹·K⁻¹
191.4 J·mol⁻¹·K⁻¹
Does not include possible zero–point entropy.
Phase Changes
c,IV/c,III 69.9 K, ΔH = 215 cal·mol⁻¹
899 J·mol⁻¹
ΔS = 2.99 cal·mol⁻¹·K⁻¹
12.62 J·mol⁻¹·K⁻¹

Table of Heat Capacities, Entropies, and Phase Transition Properties — Continued

c,III/c,II 89.5 K, ΔH = 804 cal·mol^{-1}
3364 J·mol^{-1}
ΔS = 8.96 cal·mol^{-1}·K^{-1}
8.96 J·mol^{-1}·K^{-1}
P = 37.50 kPa
c,II/c,I 169.6 K, ΔH = 166 cal·mol^{-1}
696 J·mol^{-1}
ΔS = 0.98 cal·mol^{-1}·K^{-1}
4.12 J·mol^{-1}·K^{-1}
Molecular Weight 326.0281
Wiswesser Line Notation 2X1&1&1 2.90 &ZYZUS
Evaluation A

$C_{18.7}H_{21}O_6$ (c) 60PAR/STA
β–quinol–methane clathrate
Heat Capacity 287.79 K, C_p = 101.27 cal·mol^{-1}·K^{-1}
423.71 J·mol^{-1}·K^{-1}
Temperature range 13–298 K. Value is unsmoothed experimental datum.
Molecular Weight 342.4488
Wiswesser Line Notation QR DQ 3 &1H 0.755
Evaluation A

$C_{19}H_{15}Cl$ (c) 31SMI/AND
Triphenylchloromethane
Heat Capacity 298.5 K, C_p = 74.5 cal·mol^{-1}·K^{-1}
311.7 J·mol^{-1}·K^{-1}
Temperature range 102–346 K. Value is unsmoothed experimental datum.
Molecular Weight 278.7805
Wiswesser Line Notation GXR&R&R
Evaluation C

$C_{19}H_{16}$ (c) 17HIL/DUS
Triphenylmethane
Heat Capacity 298.15 K, C_p = 62.4 cal·mol^{-1}·K^{-1}
261.1 J·mol^{-1}·K^{-1}
Temperature range 293–418 K. From heat content data.
Phase Changes
c/liq 365.5 K, ΔH = 4350 cal·mol^{-1}
18200 J·mol^{-1}
ΔS = 11.9 cal·mol^{-1}·K^{-1}
49.8 J·mol^{-1}·K^{-1}
Molecular Weight 244.3354
Wiswesser Line Notation RYR&R
Evaluation C

$C_{19}H_{16}$ (c) 30HUF/PAR
Triphenylmethane
Heat Capacity 294.3 K, C_p = 70.4 cal·mol^{-1}·K^{-1}
294.6 J·mol^{-1}·K^{-1}
Temperature range 89–294 K. Value is unsmoothed experimental datum.
Entropy 298.1 K, S = 74.6 cal·mol^{-1}·K^{-1}
312.1 J·mol^{-1}·K^{-1}
Extrapolation below 90 K, 23.84 cal mol^{-1}K^{-1}
Molecular Weight 244.3354
Wiswesser Line Notation RYR&R
Evaluation B(C_p),C(S)

$C_{19}H_{16}$ (c) 31SMI/AND
Triphenylmethane
Heat Capacity 298.15 K, C_p = 70.6 cal·mol^{-1}·K^{-1}
295.4 J·mol^{-1}·K^{-1}
Temperature range 102–346 K. Value is unsmoothed experimental datum.
Molecular Weight 244.3354
Wiswesser Line Notation RYR&R
Evaluation C

$C_{19}H_{16}$ (c) 32SPA/THO
Triphenylmethane
Heat Capacity 303 K, C_p = 73.8 cal·mol^{-1}·K^{-1}
308.8 J·mol^{-1}·K^{-1}
Temperature range 30 to 110°C
Phase Changes
c/liq 365.3 K, ΔH = 5253 cal·mol^{-1}
21979 J·mol^{-1}
ΔS = 14.4 cal·mol^{-1}·K^{-1}
60.2 J·mol^{-1}·K^{-1}
Molecular Weight 244.3354
Wiswesser Line Notation RYR&R
Evaluation B

$C_{19}H_{16}$ (c) 44EIB
Triphenylmethane
Heat Capacity 298.1 K, C_p = 79.7 cal·mol^{-1}·K^{-1}
333.5 J·mol^{-1}·K^{-1}
Temperature range 30–200 °C, equations only, in t°C. C_p(c) = 0.326 cal·g^{-1}C^{-1} (30 to 90°C); C_p(liq) = 0.325 + 0.000889t cal g^{-1}C^{-1} (92 to 200°C).
Phase Changes
c/liq 365.6 K, ΔH = 5000 cal·mol^{-1}
20920 J·mol^{-1}
ΔS = 13.7 cal·mol^{-1}·K^{-1}
57.2 J·mol^{-1}·K^{-1}
Molecular Weight 244.3354
Wiswesser Line Notation RYR&R
Evaluation C

$C_{19}H_{16}$ (c) 79STE
Triphenylmethane
Heat Capacity 298.15 K, C_p = 70.7 cal·mol^{-1}·K^{-1}
295.6 J·mol^{-1}·K^{-1}
One temperature
Molecular Weight 244.3354
Wiswesser Line Notation RYR&R
Evaluation B

$C_{19}H_{16}$ (liq) 50KUR
Triphenylmethane
Heat Capacity 373 K, C_p = 108.5 cal·mol^{-1}·K^{-1}
454.0 J·mol^{-1}·K^{-1}
Temperature range 100 to 343°C
Molecular Weight 244.3354
Wiswesser Line Notation RYR&R
Evaluation B

$C_{19}H_{16}O$ (c) 31SMI/AND
Triphenylcarbinol
Heat Capacity 298.5 K, C_p = 76.2 cal·mol^{-1}·K^{-1}
318.8 J·mol^{-1}·K^{-1}
Temperature range 102–346 K. Value is unsmoothed experimental datum.
Molecular Weight 260.3348
Wiswesser Line Notation QXR&R&R
Evaluation C

$C_{19}H_{21}ClNO$ (c) 77TSU/SOR
p–n–Hexyloxybenzylideneamino–p′–chlorobenzene
Heat Capacity 300 K, C_p = 103 cal·mol^{-1}·K^{-1}
430 J·mol^{-1}·K^{-1}
Temperature range 300–385K. Data graphically only.

Table of Heat Capacities, Entropies, and Phase Transition Properties — Continued

Phase Changes
liq/liq 333.90 K, ΔH = 2950 cal·mol^{-1}
12350 J·mol^{-1}
ΔS = 8.8 cal·mol^{-1}·K^{-1}
37.0 J·mol^{-1}·K^{-1}
Smectic–smectic transition
c/liq 327.70 K, ΔH = 2600 cal·mol^{-1}
10880 J·mol^{-1}
ΔS = 7.9 cal·mol^{-1}·K^{-1}
33.2 J·mol^{-1}·K^{-1}
Crystal–smectic transition
Molecular Weight 314.8340
Wiswesser Line Notation GR DNU1UR DO6
Evaluation C(C_p); B(phase transition)

$C_{19}H_{22}O_2$ (c) 72YOU/HAL
4–Methoxy–4′–butoxy–trans–stilbene
Phase Changes
liq/liq 435 K, ΔH = 181 cal·mol^{-1}
757 J·mol^{-1}
ΔS = 0.4 cal·mol^{-1}·K^{-1}
1.7 J·mol^{-1}·K^{-1}
Nematic–isotropic liquid transition
c/liq 442 K, ΔH = 9480 cal·mol^{-1}
39665 J·mol^{-1}
ΔS = 21.4 cal·mol^{-1}·K^{-1}
89.7 J·mol^{-1}·K^{-1}
Crystal–isotropic liquid transition
Molecular Weight 282.3816
Wiswesser Line Notation 4OR D1U1R DO1 -T
Evaluation B

$C_{19}H_{36}$ (liq) 63GUD/CAM
4–n–Heptylbicyclohexyl
Heat Capacity 313 K, C_p = 115.9 cal·mol^{-1}·K^{-1}
484.9 J·mol^{-1}·K^{-1}
Temperature range 313–483 K
Molecular Weight 264.4934
Wiswesser Line Notation L6TJ A– AL6TJ D7
Evaluation C

$C_{19}H_{36}$ (liq) 63GUD/CAM
Bis(2,4,6–trimethylcyclohexyl)methane
Heat Capacity 373 K, C_p = 141.2 cal·mol^{-1}·K^{-1}
590.8 J·mol^{-1}·K^{-1}
Temperature range 373–483 K
Molecular Weight 264.4934
Wiswesser Line Notation L6TJ A1 C1 E1 B1– AL6TJ B1 D1 F1
Evaluation C

$C_{19}H_{36}$ (liq) 62GOL/BEL
1,1–Dicyclohexylheptane
Heat Capacity 311 K, C_p = 127.0 cal·mol^{-1}·K^{-1}
531.4 J·mol^{-1}·K^{-1}
Temperatures 100, 200, 300°F
Molecular Weight 264.4934
Wiswesser Line Notation L6TJ AY6&– AL6TJ
Evaluation C

$C_{19}H_{36}$ (liq) 63GUD/CAM
1,1–Dicyclohexylheptane
Heat Capacity 313 K, C_p = 127.8 cal·mol^{-1}·K^{-1}
534.7 J·mol^{-1}·K^{-1}
Temperature range 313–583 K
Molecular Weight 264.4934
Wiswesser Line Notation L6TJ AY6&– AL6TJ
Evaluation C

$C_{19}H_{38}O_2$ (c) 82SCH/MIL
Nonadecanoic acid
Heat Capacity 298.15 K, C_p = 125.56 cal·mol^{-1}·K^{-1}
525.35 J·mol^{-1}·K^{-1}
Temperature range 80–355 K
Phase Changes
c,II/c,I 338.0 K, ΔH = 2193 cal·mol^{-1}
9177 J·mol^{-1}
ΔS = 6.90 cal·mol^{-1}·K^{-1}
27.15 J·mol^{-1}·K^{-1}
c,I/liq 341.23 K, ΔH = 13771 cal·mol^{-1}
57618 J·mol^{-1}
ΔS = 40.356 cal·mol^{-1}·K^{-1}
168.85 J·mol^{-1}·K^{-1}
Molecular Weight 298.5080
Wiswesser Line Notation QV18
Evaluation B

$C_{19}H_{38}O_4$ (c) 40CLA/STE
2–Monopalmitin
Heat Capacity 298 K, C_p = 145.4 cal·mol^{-1}·K^{-1}
607.1 J·mol^{-1}·K^{-1}
One temperature
Molecular Weight 330.5068
Wiswesser Line Notation Q1Y1QOV15
Evaluation B

$C_{19}H_{38}O_4$ (c) 65SIL/DAU
2–Monopalmitin
Heat Capacity 298 K, C_p = 133.5 cal·mol^{-1}·K^{-1}
558.6 J·mol^{-1}·K^{-1}
One temperature.
Molecular Weight 330.5068
Wiswesser Line Notation Q1Y1QOV15
Evaluation B

$C_{19}H_{38}O_4$ (c) 40CLA/STE
1–Monopalmitin
Heat Capacity 298 K, C_p = 144.1 cal·mol^{-1}·K^{-1}
602.9 J·mol^{-1}·K^{-1}
One temperature
Molecular Weight 330.5068
Wiswesser Line Notation Q1YQ1OV15
Evaluation B

$C_{19}H_{38}O_4$ (c) 65SIL/DAU
1–Monopalmitin
Heat Capacity 298 K, C_p = 135.5 cal·mol^{-1}·K^{-1}
566.9 J·mol^{-1}·K^{-1}
One temperature. β_L–form
Molecular Weight 330.5068
Wiswesser Line Notation Q1YQ1OV15
Evaluation B

$C_{19}H_{40}$ (liq) 69ATK/LAR
n–Nonadecane
Heat Capacity 353 K, C_p = 153 cal·mol^{-1}·K^{-1}
640 J·mol^{-1}·K^{-1}
Temperature range 353–453 K. Equation only.
Molecular Weight 268.5250
Wiswesser Line Notation 19H
Evaluation C

$C_{19.6}H_{41.2}N_2S$ (c,I) 72COP/GAN
Thiourea–cyclohexane adduct;
Cyclohexane–thiourea adduct
Heat Capacity 298.15 K, C_p = 33.2 cal·mol^{-1}·K^{-1}
139.1 J·mol^{-1}·K^{-1}
Temperature range 14–297 K. Values for one mole of thiourea.

Table of Heat Capacities, Entropies, and Phase Transition Properties — Continued

Entropy 298.15 K, S = 41.3 cal·mol^{-1}·K^{-1}
173.0 J·mol^{-1}·K^{-1}
Does not include possible zero–point entropy.
Phase Changes
c,VI/c,V 128.8 K, ΔH = 846 cal·mol^{-1}
3540 J·mol^{-1}
ΔS = 6.56 cal·mol^{-1}·K^{-1}
27.50 J·mol^{-1}·K^{-1}
Transition 130–150 K, ΔH = 1026 J·mol^{-1}(thiourea), ΔS = 7.25 J·mol^{-1} K^{-1}.
Transition 153–161 K, ΔH = 112 J·mol^{-1}(thiourea), ΔS = 0.78 J·mol^{-1} K^{-1}.
c,III/c,II 170.8 K, ΔH = 105 cal·mol^{-1}
440 J·mol^{-1}
ΔS = 0.62 cal·mol^{-1}·K^{-1}
2.60 J·mol^{-1}·K^{-1}
Transition 210–240 K, ΔH = 260 J·mol^{-1}(thiourea), ΔS = 1.18 J·mol^{-1}·K^{-1}.
Molecular Weight 337.0145
Wiswesser Line Notation ZYZUS &L6TJ 3.10
Evaluation A

$C_{20}H_{10}N_4$ (c) 78BOE/WES
Naphthalene–1,2,4,5–tetracyanobenzene
Heat Capacity 298.15 K, C_p = 88.51 cal·mol^{-1}·K^{-1}
370.3 J·mol^{-1}·K^{-1}
Temperature range 5–300 K
Entropy 298.15 K, S = 100.82 cal·mol^{-1}·K^{-1}
421.8 J·mol^{-1}·K^{-1}
Phase Changes
c,II/c,I 75 K, ΔH = 46 cal·mol^{-1}
192 J·mol^{-1}
ΔS = 0.71 cal·mol^{-1}·K^{-1}
2.97 J·mol^{-1}·K^{-1}
Molecular Weight 306.3258
Wiswesser Line Notation L66J &NCR BCN DCN ECN
Evaluation A

$C_{20}H_{10}O_6$ (c) 80BOE/WES 2
Naphthalene–pyromellitic dianhydride adduct
Heat Capacity 298.15 K, C_p = 93.86 cal·mol^{-1}·K^{-1}
392.7 J·mol^{-1}·K^{-1}
Temperature range 5–300 K
Entropy 298.15 K, S = 98.36 cal·mol^{-1}·K^{-1}
411.5 J·mol^{-1}·K^{-1}
Molecular Weight 346.2954
Wiswesser Line Notation T C565 DVOV JVOVJ &L66J
Evaluation A

$C_{20}H_{12}$ (c) 80WON/WES
Perylene
Heat Capacity 298.15 K, C_p = 65.71 cal·mol^{-1}·K^{-1}
274.9 J·mol^{-1}·K^{-1}
Temperature range 5–575 K
Entropy 298.15 K, S = 63.23 cal·mol^{-1}·K^{-1}
264.6 J·mol^{-1}·K^{-1}
Phase Changes
c,I/liq 550.95 K, ΔH = 7618 cal·mol^{-1}
31874 J·mol^{-1}
ΔS = 13.84 cal·mol^{-1}·K^{-1}
57.91 J·mol^{-1}·K^{-1}
Molecular Weight 252.3148
Wiswesser Line Notation L666 L6 K6 2AL TJ
Evaluation A

$C_{20}H_{13}N_3O_6$ (c) 80RAD/RAD
Anthracene–1,3,5–trinitrobenzene adduct
Heat Capacity 298.15 K, C_p = 101.5 cal·mol^{-1}·K^{-1}
424.7 J·mol^{-1}·K^{-1}
Temperature range 180–420 K. Data given graphically. C_p calculated from equation.
Phase Changes
c,I/liq 438.6 K, ΔH = 9656 cal·mol^{-1}
40401 J·mol^{-1}
ΔS = 22.02 cal·mol^{-1}·K^{-1}
92.11 J·mol^{-1}·K^{-1}
Molecular Weight 391.3392
Wiswesser Line Notation L C666J &WNR CNW ENW/
Evaluation B

$C_{20}H_{14}$ (c) 70AND/WES
Triptycene; 9,10–o–Benzeno–9,10–dihydroanthracene
Heat Capacity 298.15 K, C_p = 67.56 cal·mol^{-1}·K^{-1}
282.67 J·mol^{-1}·K^{-1}
Temperature range 5–550 K
Entropy 298.15 K, S = 65.48 cal·mol^{-1}·K^{-1}
273.97 J·mol^{-1}·K^{-1}
Phase Changes
c/liq 527.18 K, ΔH = 7236 cal·mol^{-1}
30275 J·mol^{-1}
ΔS = 13.73 cal·mol^{-1}·K^{-1}
57.43 J·mol^{-1}·K^{-1}
Molecular Weight 254.3306
Wiswesser Line Notation L6 H66 O66/GT 2AF T GH NHJ
Evaluation A

$C_{20}H_{16}$ (c) 31SMI/AND
Triphenylethylene
Heat Capacity 298.5 K, C_p = 73.9 cal·mol^{-1}·K^{-1}
309.2 J·mol^{-1}·K^{-1}
Temperature range 102–322 K. Value is unsmoothed experimental datum.
Molecular Weight 256.3464
Wiswesser Line Notation RYR&U1R
Evaluation C

$C_{20}H_{16}O_6Si$ (c) 75LEB/MIL
Dianisyldiethynylsilane
Heat Capacity 298.15 K, C_p = 96.2 cal·mol^{-1}·K^{-1}
402.5 J·mol^{-1}·K^{-1}
Temperature range 60–315 K. Data deposited in VINITI, No. 1667–75, 11 June 1975.
Entropy 298.15 K, S = 106.7 cal·mol^{-1}·K^{-1}
446.4 J·mol^{-1}·K^{-1}
Molecular Weight 380.4283
Wiswesser Line Notation 1UU1 2–SI–1R DO1 2
Evaluation B

Table of Heat Capacities, Entropies, and Phase Transition Properties — Continued

$C_{20}H_{18}$ (c) 31SMI/AND
1,1,1–Triphenylethane
Heat Capacity 298.5 K, C_p = 75.7 cal·mol^{-1}·K^{-1}
316.7 J·mol^{-1}·K^{-1}
Temperature range 102–346 K. Value is unsmoothed experimental datum.
Molecular Weight 258.3622
Wiswesser Line Notation 1XR&R&R
Evaluation C

$C_{20}H_{18}$ (c) 31SMI/AND
1,1,2–Triphenylethane
Heat Capacity 298.5 K, C_p = 76.4 cal·mol^{-1}·K^{-1}
319.7 J·mol^{-1}·K^{-1}
Temperature range 102–311 K. Value is unsmoothed experimental datum.
Molecular Weight 258.3622
Wiswesser Line Notation RYR&1R
Evaluation C

$C_{20}H_{24}O_2$ (c) 72YOU/HAL
4–Methoxy–4′–pentoxy–trans–stilbene
Phase Changes
liq/liq 427 K, ΔH = 187 cal·mol^{-1}
782 J·mol^{-1}
ΔS = 0.4 cal·mol^{-1}·K^{-1}
1.8 J·mol^{-1}·K^{-1}
Nematic–isotropic liquid transition
c/liq 435 K, ΔH = 9840 cal·mol^{-1}
41170 J·mol^{-1}
ΔS = 22.6 cal·mol^{-1}·K^{-1}
94.6 J·mol^{-1}·K^{-1}
Crystal–isotropic liquid transition
Molecular Weight 296.4084
Wiswesser Line Notation 5OR D1U1R DO1 –T
Evaluation B

$C_{20}H_{34}$ (liq) 63GUD/CAM
Diethylperhydropyrene
Heat Capacity 313 K, C_p = 122.1 cal·mol^{-1}·K^{-1}
510.9 J·mol^{-1}·K^{-1}
Temperature range 313–583 K
Molecular Weight 274.4886
Wiswesser Line Notation L666 B6 2AB PTJ X2 X2
Evaluation C

$C_{20}H_{40}Br_2N_2$ (c) 74BUR/VER
1,2–Bis(triallylammonium)ethane dibromide
Heat Capacity 298 K, C_p = 124.4 cal·mol^{-1}·K^{-1}
520.5 J·mol^{-1}·K^{-1}
Temperature range 273–373 K
Molecular Weight 468.3574
Wiswesser Line Notation 1U2K2U1&2U1&2K2U1&2U1&2U1 E 2
Evaluation B

$C_{20}H_{34}O_5Si_5$ (liq) 77KUL/DZH
Octamethyldiphenylcyclopentasiloxane
Heat Capacity No C_p data given.
Temperature range 12–300 K.
Data in paper, deposited VINITI, No. 986–77, 14 March, 1977.
Molecular Weight 494.9131
Wiswesser Line Notation T-10-O-SI-O-SI-O-SI-O-SI-O-SITJ A1 B1 C1 D1 E1 X1 X1 X1 XR& XR
Evaluation B(for deposited paper)

$C_{20}H_{36}$ (liq) 63GUD/CAM
Bis(cyclohexylmethyl)cyclohexane
Heat Capacity 373 K, C_p = 144.3 cal·mol^{-1}·K^{-1}
603.8 J·mol^{-1}·K^{-1}
Temperature range 373–583 K
Molecular Weight 276.5044
Wiswesser Line Notation L6TJ A1AL6TJ X1- AL6TJ
Evaluation C

$C_{20}H_{38}HgO_4$ (liq) 78ADE
Mercuric decanoate; Mercuric caprate
Heat Capacity 410 K, C_p = 195.3 cal·mol^{-1}·K^{-1}
817.0 J·mol^{-1}·K^{-1}
Mean value 403–420 K. Data only graphically for solid.
Phase Changes
c,II/c,I 380.8 K, ΔH = 1270 cal·mol^{-1}
5300 J·mol^{-1}
ΔS = 3.3 cal·mol^{-1}·K^{-1}
13.9 J·mol^{-1}·K^{-1}
c,I/liq 389.3 K, ΔH = 16800 cal·mol^{-1}
70200 J·mol^{-1}
ΔS = 43.1 cal·mol^{-1}·K^{-1}
180.3 J·mol^{-1}·K^{-1}
Molecular Weight 543.1078
Wiswesser Line Notation OV9 2 .HG
Evaluation C

$C_{20}H_{40}O_2$ (c) 82SCH/MIL 2
Eicosanoic acid
Heat Capacity 298.15 K, C_p = 130.29 cal·mol^{-1}·K^{-1}
545.14 J·mol^{-1}·K^{-1}
Temperature range 80–355 K
Phase Changes
c,I/liq 348.23 K, ΔH = 16540 cal·mol^{-1}
69204 J·mol^{-1}
ΔS = 47.49 cal·mol^{-1}·K^{-1}
198.7 J·mol^{-1}·K^{-1}
Molecular Weight 312.5348
Wiswesser Line Notation QV19
Evaluation B

$C_{20}H_{42}$ (c) 30PAR/HUF
n–Eicosane
Heat Capacity 279.1 K, C_p = 144.7 cal·mol^{-1}·K^{-1}
602.5 J·mol^{-1}·K^{-1}
Temperature range 94–280 K. Value is unsmoothed experimental datum.
Entropy 298.15 K, S = 133.5 cal·mol^{-1}·K^{-1}
558.6 J·mol^{-1}·K^{-1}
Extrapolation below 90 K, 15.46 cal mol^{-1}K^{-1}
Phase Changes
c/liq 309.7 K, ΔH = 14693 cal·mol^{-1}
61476 J·mol^{-1}
ΔS = 47.44 cal·mol^{-1}·K^{-1}
198.50 J·mol^{-1}·K^{-1}
Molecular Weight 282.5518
Wiswesser Line Notation 20H
Evaluation B(C_p),C(S)

$C_{20}H_{44}Sn$ (liq) 72MAS/RAB
Tetraamylstannane; Tin tetraamyl
Heat Capacity 298.15 K, C_p = 156.0 cal·mol^{-1}·K^{-1}
652.7 J·mol^{-1}·K^{-1}
Temperature range 60–300 K

Table of Heat Capacities, Entropies, and Phase Transition Properties — Continued

Phase Changes
c,II/c,I 181.4 K, ΔH = –2370 cal·mol^{-1}
–9920 J·mol^{-1}
Metastable transition
Molecular Weight 403.2576
Wiswesser Line Notation 5-SN-5&5&5
Evaluation B

$C_{20}H_{46}Br_2N_2$ (c,II) 74BUR/VER
1,8–Bis(triethylammonium)octane dibromide
Heat Capacity 298 K, C_p = 137.3 cal·mol^{-1}·K^{-1}
574.5 J·mol^{-1}·K^{-1}
Temperature range 273–373 K
Phase Changes
c,II/c,I 438 K, ΔH = 2900 cal·mol^{-1}
12130 J·mol^{-1}
ΔS = 6.6 cal·mol^{-1}·K^{-1}
27.7 J·mol^{-1}·K^{-1}
Temperature range 430–446 K
Molecular Weight 474.4048
Wiswesser Line Notation 2K2&2&8K2&2&2 E 2
Evaluation B

$C_{21}H_{14}N_2O_4$ (c) 78MAR/CIO
Bis(4–(N–maleicimido)phenyl)methane
Heat Capacity 298 K, C_p = 28.2 cal·mol^{-1}·K^{-1}
118.0 J·mol^{-1}·K^{-1}
Temperature range 298–571 K. Values for solid seem odd; increase from 28.2 to 82.5 cal·mol^{-1}K^{-1} at 420 K.
Phase Changes
c/liq 430.9 K, ΔH = 4355 cal·mol^{-1}
18220 J·mol^{-1}
ΔS = 10.1 cal·mol^{-1}·K^{-1}
42.3 J·mol^{-1}·K^{-1}
Molecular Weight 358.3526
Wiswesser Line Notation T5VNVJ BR D1R D- BT5VNVJ
Evaluation D

$C_{21}H_{15}N_{11}O_{20}$ (c) 24TAY/RIN
Tetryl–bis(trinitrotoluene) complex
Heat Capacity 293 K, C_p = 228.4 cal·mol^{-1}·K^{-1}
955.6 J·mol^{-1}·K^{-1}
Temperature range 90–333 K
Molecular Weight 741.4112
Wiswesser Line Notation WNN1&R BNW DNW FNW &WNR B CNW ENW 2
Evaluation C

$C_{21}H_{16}$ (c) 77FIN/MES
1,2′–Dinaphthylmethane
Heat Capacity 298.15 K, C_p = 75.17 cal·mol^{-1}·K^{-1}
314.51 J·mol^{-1}·K^{-1}
Temperature range 10–440 K
Entropy 298.15 K, S = 74.27 cal·mol^{-1}·K^{-1}
310.75 J·mol^{-1}·K^{-1}
Phase Changes
c/liq 369.55 K, ΔH = 7303.3 cal·mol^{-1}
30557.0 J·mol^{-1}
ΔS = 19.76 cal·mol^{-1}·K^{-1}
82.69 J·mol^{-1}·K^{-1}
Molecular Weight 268.3574
Wiswesser Line Notation L66J B1- CL66J
Evaluation A

$C_{21}H_{25}NO$ (c) 78JOH/HAY
Octyloxycyanobiphenyl
Heat Capacity C_p given grpahically only.
Temperature range 334–348 K.
Phase Changes
Smectic A–nematic transition at 340 K
Molecular Weight 307.4346
Wiswesser Line Notation NCR DR DO8
Evaluation D

$C_{21}H_{26}O_2$ (c) 72YOU/HAL
4–Methoxy–4′–hexoxy–trans–stilbene
Phase Changes
liq/liq 426 K, ΔH = 189 cal·mol^{-1}
791 J·mol^{-1}
ΔS = 0.4 cal·mol^{-1}·K^{-1}
1.9 J·mol^{-1}·K^{-1}
Nematic–isotropic liquid transition
c/liq 430 K, ΔH = 9890 cal·mol^{-1}
41380 J·mol^{-1}
ΔS = 23.0 cal·mol^{-1}·K^{-1}
96.2 J·mol^{-1}·K^{-1}
Crystal–isotropic liquid transition
Molecular Weight 310.4352
Wiswesser Line Notation 6OR D1U1R DO1 -T
Evaluation B

$C_{21}H_{28}N_7O_{14}P_2 \cdot 3H_2O$ (c) 79YAN/RUP
Nicotinamide adenine dinucleotide trihydrate; NAD
Heat Capacity 298 K, C_p = 185 cal·mol^{-1}·K^{-1}
773 J·mol^{-1}·K^{-1}
One temperature
Molecular Weight 718.4839
Wiswesser Line Notation T56 BN DN FN HNJ IZ D- BT5OTJ CQ DQ E1OPQOPOPQOPO1- BT5OTJ CQ DQ E- AT6NJ CVZ &QH 3
Evaluation C

$C_{21}H_{42}Br_2N_2$ (c) 74BUR/VER
1,3–Bis(triallylammonium)propane dibromide
Heat Capacity 298 K, C_p = 128.6 cal·mol^{-1}·K^{-1}
538.1 J·mol^{-1}·K^{-1}
Temperature range 273–373 K
Molecular Weight 482.3842
Wiswesser Line Notation 1U2K2U1&2U1&3K2U1&2U1&2U1 2 E
Evaluation B

$C_{21}H_{38}$ (liq) 63GUD/CAM
9–(2′–Ethylhexyl)perhydrofluorene
Heat Capacity 313 K, C_p = 145.9 cal·mol^{-1}·K^{-1}
610.4 J·mol^{-1}·K^{-1}
Temperature range 313–583 K
Molecular Weight 290.5312
Wiswesser Line Notation L B656TJ H1Y 4&2
Evaluation C

$C_{21}H_{38}$ (liq) 63GUD/CAM
1,1,3–Tricyclohexylpropane
Heat Capacity 373 K, C_p = 152.5 cal·mol^{-1}·K^{-1}
638.1 J·mol^{-1}·K^{-1}
Temperature range 373–583 K
Molecular Weight 290.5312
Wiswesser Line Notation L6TJ AY2AL6TJ&- AL6TJ
Evaluation C

Table of Heat Capacities, Entropies, and Phase Transition Properties — Continued

$C_{21}H_{38}O_6$ (liq) 76PHI/MAT
Tricaproin; Glyceryl tricaproate
Heat Capacity 313 K, C_p = 180 cal·mol^{-1}·K^{-1}
753 J·mol^{-1}·K^{-1}
Temperature range 313–413 K
Molecular Weight 386.5276
Wiswesser Line Notation 5VO1YOV5&1OV5
Evaluation C

$C_{21}H_{40}$ (liq) 63GUD/CAM
4–n–Nonylbicyclohexyl
Heat Capacity 313 K, C_p = 136.9 cal·mol^{-1}·K^{-1}
578.8 J·mol^{-1}·K^{-1}
Temperature range 313–483 K
Molecular Weight 292.5470
Wiswesser Line Notation L6TJ A -AL6TJ D9
Evaluation C

$C_{21}H_{42}O_4$ (c) 65SIL/DAU
2–Monostearin
Heat Capacity 298 K, C_p = 145.9 cal·mol^{-1}·K^{-1}
610.4 J·mol^{-1}·K^{-1}
One temperature.
Molecular Weight 358.5604
Wiswesser Line Notation Q1Y1QOV17
Evaluation B

$C_{21}H_{42}O_4$ (c,α) 55WAR/VIC
1–Monostearin
Heat Capacity 298.2 K, C_p = 207.0 cal·mol^{-1}·K^{-1}
866.0 J·mol^{-1}·K^{-1}
Temperature range –74 to 114°C. Give experimental points and equations for solid and liquid states. Sub–alpha form, $C_p = 0.4977 + 0.00318t$ cal·g^{-1}C^{-1}, –13 to 40°C; liquid, $C_p = 0.5118 + 0.00182t$ cal g^{-1} C^{-1}, 87–100C.
Phase Changes
c,α/liq 347.2 K, ΔH = 14124 cal·mol^{-1}
59095 J·mol^{-1}
ΔS = 40.7 cal·mol^{-1}·K^{-1}
170.2 J·mol^{-1}·K^{-1}
Molecular Weight 358.5604
Wiswesser Line Notation Q1YQ1OV17
Evaluation C

$C_{21}H_{42}O_4$ (c) 65SIL/DAU
1–Monostearin
Heat Capacity 298 K, C_p = 145.9 cal·mol^{-1}·K^{-1}
610.4 J·mol^{-1}·K^{-1}
One temperature. β_L–form
Molecular Weight 358.5604
Wiswesser Line Notation Q1YQ1OV17
Evaluation B

$C_{21}H_{46}N_2O$ (c,I) 65PEM/PAR
Urea–n–eicosane adduct
Heat Capacity 298.15 K, C_p = 30.32 cal·mol^{-1}·K^{-1}
126.86 J·mol^{-1}·K^{-1}
Temperature range 12–300 K. Value for aduct with 1 mole of urea.
Entropy 298.15 K, S = 33.09 cal·mol^{-1}·K^{-1}
138.45 J·mol^{-1}·K^{-1}
Phase Changes
Transition at 189.3 K with ΔH = 667 cal(mol hydrocarbon)$^{-1}$.
Molecular Weight 342.6072
Wiswesser Line Notation ZVZ &20H
Evaluation A
Sample 75.95 percent urea.

$C_{21.9}H_{45.8}N_2S$ (c,I) 72COP/GAN
Cycloheptane–thiourea adduct;
Thiourea–cycloheptane adduct
Heat Capacity 298.15 K, C_p = 35.4 cal·mol^{-1}·K^{-1}
148.3 J·mol^{-1}·K^{-1}
Temperature range 12–300 K. Values for one mole of thiourea.
Entropy 298.15 K, S = 42.8 cal·mol^{-1}·K^{-1}
179.0 J·mol^{-1}·K^{-1}
Does not include possible zero–point entropy.
Phase Changes
c,V/c,IV 162.4 K, ΔH = 88 cal·mol^{-1}
368 J·mol^{-1}
ΔS = 0.54 cal·mol^{-1}·K^{-1}
2.24 J·mol^{-1}·K^{-1}
c,IV/c,III 241 K, ΔH = 136 cal·mol^{-1}
568 J·mol^{-1}
ΔS = 0.56 cal·mol^{-1}·K^{-1}
2.36 J·mol^{-1}·K^{-1}
c,III/c,II 262 K, ΔH = 22 cal·mol^{-1}
93 J·mol^{-1}
ΔS = 0.09 cal·mol^{-1}·K^{-1}
0.36 J·mol^{-1}·K^{-1}
Shallow hump at 270–290 K, ΔH = 4802 J. mol^{-1}(thiourea), ΔS = 34.50.
Molecular Weight 369.6969
Wiswesser Line Notation ZYZUS &L7TJ 2.99
Evaluation A

$C_{22}H_{20}N_2O_4$ (c) 73HAM/MIT
N,N′–Bis(m–methoxyphenyl)terephthalamide
Heat Capacity 298.15 K, C_p = 109.5 cal·mol^{-1}·K^{-1}
458.1 J·mol^{-1}·K^{-1}
One temperature
Molecular Weight 376.4110
Wiswesser Line Notation 1OR CMVR DVMR CO1
Evaluation C

$C_{22}H_{20}N_2O_4$ (c) 73HAM/MIT
N,N′–Bis(p–methoxyphenyl)terephthalamide
Heat Capacity 298.15 K, C_p = 111.8 cal·mol^{-1}·K^{-1}
467.8 J·mol^{-1}·K^{-1}
One temperature
Molecular Weight 376.4110
Wiswesser Line Notation 1OR DMVR DVMR DO1
Evaluation C

$C_{22}H_{24}O_6$ (c) 74AND/BAC
Di(p–methoxyphenyl)–trans–cyclohexane–1,4–dicarboxylate
Heat Capacity 297.0 K, C_p = 114.4 cal·mol^{-1}·K^{-1}
478.5 J·mol^{-1}·K^{-1}
Temperature range 297–550 K. Value is unsmoothed experimental datum.
Phase Changes
c,II/c,I 411.58 K, ΔH = 1149 cal·mol^{-1}
4806 J·mol^{-1}
ΔS = 2.79 cal·mol^{-1}·K^{-1}
11.69 J·mol^{-1}·K^{-1}
c,I/liq,I 416.16 K, ΔH = 7610 cal·mol^{-1}
31841 J·mol^{-1}
ΔS = 18.29 cal·mol^{-1}·K^{-1}
76.51 J·mol^{-1}·K^{-1}
Fusion of crystal I to nematic liquid crystal
liq,I/liq,II 516.0 K, ΔH = 685 cal·mol^{-1}
2865 J·mol^{-1}
ΔS = 1.4 cal·mol^{-1}·K^{-1}
5.7 J·mol^{-1}·K^{-1}
Transition between nematic liquid crystal and isotropic liquid.

Table of Heat Capacities, Entropies, and Phase Transition Properties — Continued

Molecular Weight 384.4280
Wiswesser Line Notation L6TJ AVOR BO1& DVOR DO1
Evaluation A

$C_{22}H_{28}O_2$ (c) 72YOU/HAL
4–Methoxy–4′–heptoxy–trans–stilbene
Phase Changes
liq/liq 421 K, ΔH = 159 cal·mol⁻¹ 665 J·mol⁻¹
ΔS = 0.4 cal·mol⁻¹·K⁻¹ 1.6 J·mol⁻¹·K⁻¹
Nematic–isotropic liquid transition
c/liq 423 K, ΔH = 10220 cal·mol⁻¹ 42760 J·mol⁻¹
ΔS = 24.2 cal·mol⁻¹·K⁻¹ 101.1 J·mol⁻¹·K⁻¹
Crystal–isotropic liquid transition
Molecular Weight 324.4620
Wiswesser Line Notation 7OR D1U1R DO1 -T
Evaluation B

$C_{22}H_{36}Br_2N_2$ (c,II) 74BUR/VER
1,4–Bis(triallylammonium)butene–2 dibromide
Heat Capacity 298 K, C_p = 139.2 cal·mol⁻¹·K⁻¹ 582.4 J·mol⁻¹·K⁻¹
Temperature range 273–373 K
Phase Changes
c,II/c,I 430 K, ΔH = 2200 cal·mol⁻¹ 9200 J·mol⁻¹
ΔS = 5.1 cal·mol⁻¹·K⁻¹ 21.4 J·mol⁻¹·K⁻¹
Temperature range 425–436 K
Molecular Weight 494.3952
Wiswesser Line Notation 1U2K2U1&2U1&2U2K2U1&2U1&2U1 &E &E
Evaluation B

$C_{22}H_{42}O_4$ (liq) 76PHI/MAT
Di–n–hexyl sebacate
Heat Capacity 315 K, C_p = 175 cal·mol⁻¹·K⁻¹ 732 J·mol⁻¹·K⁻¹
Temperature range 315–414 K
Molecular Weight 370.5714
Wiswesser Line Notation 6OV8VO6
Evaluation C

$C_{22}H_{42}O_4Pb$ (c,II) 78ADE/SIM
Lead(II) undecanoate
Heat Capacity 365 K, C_p = 214 cal·mol⁻¹·K⁻¹ 894 J·mol⁻¹·K⁻¹
Mean value 363–371 K. Data only graphically for c,III. Data also for c,I, and liquid.
Phase Changes
c,III/c,II 360.9 K, ΔH = 12100 cal·mol⁻¹ 50700 J·mol⁻¹
ΔS = 33.5 cal·mol⁻¹·K⁻¹ 140 J·mol⁻¹·K⁻¹
c,II, and c,I, are mesophases.
c,II/c,I 377.0 K, ΔH = 6570 cal·mol⁻¹ 27500 J·mol⁻¹
ΔS = 17.4 cal·mol⁻¹·K⁻¹ 73 J·mol⁻¹·K⁻¹
c,I/liq 383.7 K, ΔH = 260 cal·mol⁻¹ 1100 J·mol⁻¹
ΔS = 0.7 cal·mol⁻¹·K⁻¹ 2.9 J·mol⁻¹·K⁻¹
Molecular Weight 577.7714
Wiswesser Line Notation OV10 2 .PB
Evaluation C

$C_{22}H_{44}N_2O_2$ (c) 53WIL/DOL
N,N′–Di–n–hexylsebacamide
Heat Capacity 333–483 K, C_p = 254.34 cal·mol⁻¹·K⁻¹ 1064.1 J·mol⁻¹·K⁻¹
Phase Changes
c/liq 415 K, ΔH = 12830 cal·mol⁻¹ 53680 J·mol⁻¹
ΔS = 30.9 cal·mol⁻¹·K⁻¹ 129.3 J·mol⁻¹·K⁻¹
Molecular Weight 368.6018
Wiswesser Line Notation 6MV8VM6
Evaluation C

$C_{22}H_{46}$ (c,II) 31GAR/VAN
n–Docosane
Heat Capacity 299 K, C_p = 134.7 cal·mol⁻¹·K⁻¹ 563.6 J·mol⁻¹·K⁻¹
Temperature range 280–347 K. Mean value 17–35 °C, β–form.
Phase Changes
c,II/c,I 313.4 K, ΔH = 2140 cal·mol⁻¹ 8950 J·mol⁻¹
ΔS = 6.8 cal·mol⁻¹·K⁻¹ 28.6 J·mol⁻¹·K⁻¹
β–α transition
c,I/liq 317.0 K, ΔH = 3635 cal·mol⁻¹ 15210 J·mol⁻¹
ΔS = 11.5 cal·mol⁻¹·K⁻¹ 48.0 J·mol⁻¹·K⁻¹
Molecular Weight 310.6054
Wiswesser Line Notation 22H
Evaluation B

$C_{22}H_{46}$ (liq) 69ATK/LAR
n–Docosane
Heat Capacity 353 K, C_p = 177 cal·mol⁻¹·K⁻¹ 739 J·mol⁻¹·K⁻¹
Temperature range 353–453 K. Equation only.
Molecular Weight 310.6054
Wiswesser Line Notation 22H
Evaluation C

$C_{22}H_{48}$ (liq) 58SEL/AST
Bis(cyclopentane)–2,2–dimethylbutane adduct
Heat Capacity 298.15 K, C_p = 35.53 cal·mol⁻¹·K⁻¹ 148.66 J·mol⁻¹·K⁻¹
Temperature range 14–300 K
Entropy 298.15 K, S = 164.82 cal·mol⁻¹·K⁻¹ 689.61 J·mol⁻¹·K⁻¹
Does not include zero point entropy.
Phase Changes
c,II/c,I 83.2 K, ΔH = 986.7 cal·mol⁻¹ 4128.4 J·mol⁻¹
ΔS = 11.86 cal·mol⁻¹·K⁻¹ 49.62 J·mol⁻¹·K⁻¹
c,I/liq 137.71 K, ΔH = 600.8 cal·mol⁻¹ 2513.7 J·mol⁻¹
ΔS = 4.36 cal·mol⁻¹·K⁻¹ 18.25 J·mol⁻¹·K⁻¹
Molecular Weight 312.6212
Wiswesser Line Notation L5TJ 2 &2X1&1&1
Evaluation A

Table of Heat Capacities, Entropies, and Phase Transition Properties — Continued

$C_{22}H_{50}Br_2N_2$ (c,II) 74BUR/VER
1,10–Bis(triethylammonium)decane dibromide
Heat Capacity 298 K, C_p = 153.0 cal·mol^{-1}·K^{-1}
640.2 J·mol^{-1}·K^{-1}
Temperature range 273–373 K
Phase Changes
c,II/c,I 444 K, ΔH = 200 cal·mol^{-1}
840 J·mol^{-1}
ΔS = 0.5 cal·mol^{-1}·K^{-1}
1.9 J·mol^{-1}·K^{-1}
Temperature range 437–452 K
Molecular Weight 502.4584
Wiswesser Line Notation 2K2&2&10K2&2&2 E 2
Evaluation B

$C_{23}H_{30}O_2$ (c) 72YOU/HAL
4–Methoxy–4′–octoxy–trans–stilbene
Phase Changes
liq/liq 419 K, ΔH = 232 cal·mol^{-1}
971 J·mol^{-1}
ΔS = 0.6 cal·mol^{-1}·K^{-1}
2.3 J·mol^{-1}·K^{-1}
Nematic–isotropic liquid transition
c/liq 424 K, ΔH = 10020 cal·mol^{-1}
41925 J·mol^{-1}
ΔS = 23.6 cal·mol^{-1}·K^{-1}
98.9 J·mol^{-1}·K^{-1}
Crystal–isotropic liquid transition
Molecular Weight 338.4888
Wiswesser Line Notation 8OR D1U1R DO1 -T
Evaluation B

$C_{23}H_{44}O_5$ (c,α) 55WAR/VIC
1–Aceto–3–stearin
Heat Capacity 298.2 K, C_p = 224.3 cal·mol^{-1}·K^{-1}
938.5 J·mol^{-1}·K^{-1}
Temperature range –71 to 72°C. Give experimental points and equations for 2 solid and liquid states. Sub–alpha form, C_p = 0.4471 + 0.00133t cal·g^{-1}·C^{-1}, (-73 to -1°C); alpha form, C_p = 0.4513 + 0.00434t cal·g^{-1}·C^{-1}, (– 1 to 27 °C); liquid, C_p = 0.2290 + 0.0068t cal·g^{-1}·C^{-1} (57 to 87°C).
Phase Changes
c,α/liq 319.9 K, ΔH = 9963 cal·mol^{-1}
41685 J·mol^{-1}
ΔS = 31.1 cal·mol^{-1}·K^{-1}
130.3 J·mol^{-1}·K^{-1}
Molecular Weight 400.5976
Wiswesser Line Notation 17VO1YQ1OV1
Evaluation C

$C_{23}H_{48}$ (liq) 69ATK/LAR
n–Tricosane
Heat Capacity 353 K, C_p = 185 cal·mol^{-1}·K^{-1}
772 J·mol^{-1}·K^{-1}
Temperature range 353–453 K. Equation only.
Molecular Weight 324.6322
Wiswesser Line Notation 23H
Evaluation C

$C_{24}H_{12}$ (c) 80WON/WES
Coronene
Heat Capacity 298.15 K, C_p = 74.99 cal·mol^{-1}·K^{-1}
313.8 J·mol^{-1}·K^{-1}
Temperature range 5–350 K
Entropy 298.15 K, S = 67.13 cal·mol^{-1}·K^{-1}
280.9 J·mol^{-1}·K^{-1}
Phase Changes
c,II/c,I 225 K, ΔH = 106 cal·mol^{-1}
444 J·mol^{-1}
ΔS = 0.44 cal·mol^{-1}·K^{-1}
1.84 J·mol^{-1}·K^{-1}
Molecular Weight 300.3588
Wiswesser Line Notation L666 B6 C6 D6 E6 6ABCDEF A&J
Evaluation A

$C_{24}H_{18}$ (liq) 58WAL/BRO
m–Quaterphenyl
Heat Capacity 370 K, C_p = 132.3 cal·mol^{-1}·K^{-1}
553.5 J·mol^{-1}·K^{-1}
Temperature range 200 to 600°F
Molecular Weight 306.4062
Wiswesser Line Notation RR CR CR
Evaluation C

$C_{24}H_{18}$ (c) 36PAR/TOD
1,3,5–Triphenylbenzene
Heat Capacity 298.1 K, C_p = 85.64 cal·mol^{-1}·K^{-1}
358.32 J·mol^{-1}·K^{-1}
Temperature range 90–300 K
Entropy 298.1 K, S = 87.8 cal·mol^{-1}·K^{-1}
367.4 J·mol^{-1}·K^{-1}
Extrapolation below 90 K, 27.40 cal mol^{-1}K^{-1}
Molecular Weight 306.4062
Wiswesser Line Notation RR CR ER
Evaluation B(C_p),C(S)

$C_{24}H_{18}FeO_2$ (c) 81TOM/CUR
1,1′–Dibenzoylferrocene
Heat Capacity 298 K, C_p = 111.4 cal·mol^{-1}·K^{-1}
466.1 J·mol^{-1}·K^{-1}
Temperature range 293–353 K. Equation given.
Phase Changes
c/liq 379.7 K
Molecular Weight 394.2520
Wiswesser Line Notation L5ØJ AVR Ø–FE– – ØL5ØJ AVR
Evaluation B

$C_{24}H_{20}Si$ (c) 31SMI/AND 2
Tetraphenylsilane
Heat Capacity 298.5 K, C_p = 95.6 cal·mol^{-1}·K^{-1}
395.8 J·mol^{-1}·K^{-1}
Temperature range 102–346 K. Value is unsmoothed experimental datum.
Molecular Weight 336.5075
Wiswesser Line Notation R–SI–R&R&R
Evaluation B

$C_{24}H_{20}Sn$ (c) 31SMI/AND 2
Tetraphenylstannane; Tetraphenyl Tin
Heat Capacity 298.5 K, C_p = 101.9 cal·mol^{-1}·K^{-1}
426.3 J·mol^{-1}·K^{-1}
Temperature range 102–346 K. Value is unsmoothed experimental datum.
Molecular Weight 427.1120
Wiswesser Line Notation R–SN–R&R&R
Evaluation B

Table of Heat Capacities, Entropies, and Phase Transition Properties — Continued

$C_{24}H_{22}N_4O$ (c) 73KAR/SAP
4,4′–Dianilino–3,3′–diaminodiphenyl oxide;
4,4′–Dianilino–3,3′–diaminodiphenyl ether
Heat Capacity 300 K, C_p = 109.5 cal·mol⁻¹·K⁻¹
458.1 J·mol⁻¹·K⁻¹
Temperature range 20–300 K
Entropy 300 K, S = 103.2 cal·mol⁻¹·K⁻¹
431.8 J·mol⁻¹·K⁻¹
Molecular Weight 382.4640
Wiswesser Line Notation RMR CZ DOR CZ DMR
Evaluation B

$C_{24}H_{34}$ (liq) 60KAR/STR
1,1–Diphenyldodecane
Heat Capacity 298.15 K, C_p = 141.9 cal·mol⁻¹·K⁻¹
593.7 J·mol⁻¹·K⁻¹
Temperature range 10–300 K
Entropy 298.15 K, S = 163.7 cal·mol⁻¹·K⁻¹
684.9 J·mol⁻¹·K⁻¹
Phase Changes
c,II/c,I 191 K, ΔH = 461 cal·mol⁻¹
1929 J·mol⁻¹
ΔS = 2.4 cal·mol⁻¹·K⁻¹
10.1 J·mol⁻¹·K⁻¹
c,I/liq 281.4 K, ΔH = 9284 cal·mol⁻¹
38844 J·mol⁻¹
ΔS = 33.0 cal·mol⁻¹·K⁻¹
138.0 J·mol⁻¹·K⁻¹
Molecular Weight 322.5326
Wiswesser Line Notation 11YR&R
Evaluation B

$C_{24}H_{38}O_4$ (liq) 69RAB/MAR
Bis(2–ethylhexyl) phthalate
Heat Capacity 300 K, C_p = 160.0 cal·mol⁻¹·K⁻¹
669.4 J·mol⁻¹·K⁻¹
Temperature range 80–360 K
Entropy 300 K, S = 180.5 cal·mol⁻¹·K⁻¹
755.2 J·mol⁻¹·K⁻¹
Phase Changes
c,II/c,I 182.5 K, ΔH = 248 cal·mol⁻¹
1038 J·mol⁻¹
ΔS = 1.39 cal·mol⁻¹·K⁻¹
5.82 J·mol⁻¹·K⁻¹
Glass phase transition
Molecular Weight 390.5618
Wiswesser Line Notation 4Y2&1OVR BVO1&Y4&2
Evaluation C

$C_{24}H_{40}$ (liq) 60KAR/STR
1–Cyclohexyl–1–phenyldodecane
Heat Capacity 298.15 K, C_p = 146.1 cal·mol⁻¹·K⁻¹
611.3 J·mol⁻¹·K⁻¹
Phase Changes
c/liq 275.8 K, ΔH = 8406 cal·mol⁻¹
35171 J·mol⁻¹
ΔS = 30.5 cal·mol⁻¹·K⁻¹
127.5 J·mol⁻¹·K⁻¹
Molecular Weight 328.5800
Wiswesser Line Notation L6TJ AY11&R
Evaluation B

$C_{24}H_{46}$ (c) 60KAR/STR
2,11–Dicyclohexyldodecane
Heat Capacity 298.1 K, C_p = 133.2 cal·mol⁻¹·K⁻¹
557.3 J·mol⁻¹·K⁻¹
Temperature range 13–298 K. Values above 260 K show extensive premelting effects. Value given is corrected for premelting. Experimental value at 298.09 K is 852.27 cal mol⁻¹K⁻¹.
Entropy 298.15 K, S = 130.4 cal·mol⁻¹·K⁻¹
545.6 J·mol⁻¹·K⁻¹
Phase Changes
c/liq 300.58 K, ΔH = 10581 cal·mol⁻¹
44271 J·mol⁻¹
ΔS = 35.2 cal·mol⁻¹·K⁻¹
147.3 J·mol⁻¹·K⁻¹
Molecular Weight 334.6274
Wiswesser Line Notation L6TJ AY1&8Y1&– AL6TJ
Evaluation B

$C_{24}H_{46}$ (c) 60KAR/STR
1,1–Dicyclohexyldodecane
Heat Capacity 298.15 K, C_p = 134.5 cal·mol⁻¹·K⁻¹
562.7 J·mol⁻¹·K⁻¹
Temperature range 10–300 K
Entropy 298.15 K, S = 130.4 cal·mol⁻¹·K⁻¹
545.6 J·mol⁻¹·K⁻¹
Phase Changes
c/liq 300.6 K, ΔH = 10580 cal·mol⁻¹
44267 J·mol⁻¹
ΔS = 35.2 cal·mol⁻¹·K⁻¹
147.3 J·mol⁻¹·K⁻¹
Molecular Weight 334.6274
Wiswesser Line Notation L6TJ AY11&– AL6TJ
Evaluation B

$C_{24}H_{46}HgO_4$ (liq) 78ADE
Mercuric dodecanoate; Mercuric laurate
Heat Capacity 420 K, C_p = 260.3 cal·mol⁻¹·K⁻¹
1089.2 J·mol⁻¹·K⁻¹
Mean value, 413–430 K. Data for solid only graphically.
Phase Changes
c/liq 394.2 K, ΔH = 22660 cal·mol⁻¹
94800 J·mol⁻¹
ΔS = 57.5 cal·mol⁻¹·K⁻¹
240.5 J·mol⁻¹·K⁻¹
Molecular Weight 599.2150
Wiswesser Line Notation OV11 2 .HG
Evaluation C

$C_{24}H_{50}$ (c) 49PAR/MOO
n–Tetracosane
Heat Capacity 298.15 K, C_p = 174.7 cal·mol⁻¹·K⁻¹
723.3 J·mol⁻¹·K⁻¹
Temperature range 80–300 K. Specific heat at 290–300 K rapidly increasing; possible premelting effects. Value may be high.
Entropy 298.15 K, S = 155.6 cal·mol⁻¹·K⁻¹
651.0 J·mol⁻¹·K⁻¹
Extrapolation below 80 K, 38.36 cal mol⁻¹K⁻¹
Phase Changes
Hump in specific heat curve at 250–265 K.
Molecular Weight 338.6590
Wiswesser Line Notation 24H
Evaluation B(C_p),C(S)

Table of Heat Capacities, Entropies, and Phase Transition Properties — Continued

$C_{24}H_{50}$ (liq) 69ATK/LAR
n–Tetracosane
Heat Capacity 353 K, C_p = 192 cal·mol^{-1}·K^{-1}
805 J·mol^{-1}·K^{-1}
Temperature range 353–453 K. Equation only.
Molecular Weight 338.6590
Wiswesser Line Notation 24H
Evaluation C

$C_{24}H_{52}ClNO_4$ (c,IV) 73AND/GOR
Tetra–n–hexylammonium perchlorate
Heat Capacity 298.15 K, C_p = 178 cal·mol^{-1}·K^{-1}
744 J·mol^{-1}·K^{-1}
Temperature range 300–382 K
Phase Changes

c,IV/c,III	333.57 K,	ΔH =	5495 cal·mol^{-1}
			22990 J·mol^{-1}
		ΔS =	16.5 cal·mol^{-1}·K^{-1}
			68.9 J·mol^{-1}·K^{-1}
c,III/c,II	355.91 K,	ΔH =	1396 cal·mol^{-1}
			5839 J·mol^{-1}
		ΔS =	3.91 cal·mol^{-1}·K^{-1}
			16.4 J·mol^{-1}·K^{-1}
c,II/c,I	367.51 K,	ΔH =	635 cal·mol^{-1}
			2658 J·mol^{-1}
		ΔS =	1.74 cal·mol^{-1}·K^{-1}
			7.2 J·mol^{-1}·K^{-1}
c,I/liq	379.18 K,	ΔH =	3908 cal·mol^{-1}
			16350 J·mol^{-1}
		ΔS =	10.3 cal·mol^{-1}·K^{-1}
			43.1 J·mol^{-1}·K^{-1}

Molecular Weight 454.1321
Wiswesser Line Notation 6K6&6&6 G–O4
Evaluation B

$C_{24.8}H_{51.6}N_2S$ (c,I) 72COP/GAN
Thiourea–cyclooctane adduct;
Cyclooctane–thiourea adduct
Heat Capacity 298.15 K, C_p = 39.3 cal·mol^{-1}·K^{-1}
164.3 J·mol^{-1}·K^{-1}
Temperature range 12–300 K. Values for one mole of thiourea.
Entropy 298.15 K, S = 44.7 cal·mol^{-1}·K^{-1}
187.1 J·mol^{-1}·K^{-1}
Does not include possible zero–point entropy.
Phase Changes

c,IV/c,III	187.2 K,	ΔH =	504 cal·mol^{-1}
			2110 J·mol^{-1}
		ΔS =	2.69 cal·mol^{-1}·K^{-1}
			11.26 J·mol^{-1}·K^{-1}
c,III/c,II	240 K,	ΔH =	5118 cal·mol^{-1}
			21414 J·mol^{-1}
		ΔS =	22.56 cal·mol^{-1}·K^{-1}
			94.41 J·mol^{-1}·K^{-1}
c,II/c,I	265 K,	ΔH =	168 cal·mol^{-1}
			703 J·mol^{-1}
		ΔS =	0.63 cal·mol^{-1}·K^{-1}
			2.65 J·mol^{-1}·K^{-1}

Molecular Weight 410.5149
Wiswesser Line Notation ZYZUS &L8TJ 2.98
Evaluation A

$C_{25}H_{20}$ (c) 31SMI/AND
Tetraphenylmethane
Heat Capacity 298.5 K, C_p = 88.0 cal·mol^{-1}·K^{-1}
368.2 J·mol^{-1}·K^{-1}
Temperature range 102–346 K. Value is unsmoothed experimental datum.
Molecular Weight 320.4330
Wiswesser Line Notation RXR&R&R
Evaluation C

$C_{25}H_{20}$ (c) 31SMI/AND 2
Tetraphenylmethane
Heat Capacity 298.5 K, C_p = 88.0 cal·mol^{-1}·K^{-1}
368.2 J·mol^{-1}·K^{-1}
Temperature range 102–346 K. Value is unsmoothed experimental datum.
Molecular Weight 320.4330
Wiswesser Line Notation RXR&R&R
Evaluation C

$C_{25}H_{34}O_2S$ (c) 81CHR/RIC
4–n–Pentylphenyl–4′–n–heptyloxythiobenzoate
Heat Capacity C_p data given graphically only.
Temperature range 90–370 K.
Phase Changes

c,III/c,II	183.53 K,	ΔH =	279 cal·mol^{-1}
			1167 J·mol^{-1}
		ΔS =	1.44 cal·mol^{-1}·K^{-1}
			6.03 J·mol^{-1}·K^{-1}
c,II/c,I	272 K,	ΔH =	54.4 cal·mol^{-1}
			228 J·mol^{-1}
		ΔS =	0.21 cal·mol^{-1}·K^{-1}
			0.88 J·mol^{-1}·K^{-1}
c,I/nematic liq.	325.87 K,	ΔH =	6815 cal·mol^{-1}
			28514 J·mol^{-1}
		ΔS =	21.14 cal·mol^{-1}·K^{-1}
			88.45 J·mol^{-1}·K^{-1}
Nematic liquid/isotropic liquid			
	352.2 K,	ΔH =	610 cal·mol^{-1}
			2552 J·mol^{-1}
		ΔS =	1.28 cal·mol^{-1}·K^{-1}
			5.36 J·mol^{-1}·K^{-1}

Molecular Weight 398.6024
Wiswesser Line Notation 7OR DVSR D5
Evaluation A

$C_{25}H_{41}DO_3$ (c,I) 77IKE/HAT
p–n–Octadecyloxybenzoic acid–d
Heat Capacity 302 K, C_p = 33 cal·mol^{-1}·K^{-1}
137 J·mol^{-1}·K^{-1}
Temperature range 302–420 K. Value is unsmoothed experimental datum for c,I phase.
Phase Changes

c,I/liq	374.2 K,	ΔH =	15703 cal·mol^{-1}
			65700 J·mol^{-1}
		ΔS =	42 cal·mol^{-1}·K^{-1}
			176 J·mol^{-1}·K^{-1}
c,I–smectic transition			
c,II/liq	365.7 K,	ΔH =	8720 cal·mol^{-1}
			36500 J·mol^{-1}
		ΔS =	24 cal·mol^{-1}·K^{-1}
			101 J·mol^{-1}·K^{-1}
c,II–smectic transition			
c,III/liq	340.2 K		
c,III–smectic transition temperature			
liq/liq	402.5 K,	ΔH =	3440 cal·mol^{-1}
			14400 J·mol^{-1}
		ΔS =	8.5 cal·mol^{-1}·K^{-1}
			36 J·mol^{-1}·K^{-1}

Smectic–isotropic liquid transition

Table of Heat Capacities, Entropies, and Phase Transition Properties — Continued

Molecular Weight 391.6112
Wiswesser Line Notation QVR DO18 &1/H-2
Evaluation B

$C_{25}H_{42}O_3$ (c) 76IKE/HAT
p-n-Octadecyloxybenzoic acid
Heat Capacity 320 K, C_p = 35 cal·mol^{-1}·K^{-1}
145 J·mol^{-1}·K^{-1}
Temperature range 320–420 K. Value is unsmoothed experimental datum on c,I phase.
Phase Changes
c,I/liq 379.5 K, ΔH = 16085 cal·mol^{-1}
67300 J·mol^{-1}
ΔS = 42 cal·mol^{-1}·K^{-1}
177 J·mol^{-1}·K^{-1}
c,I-smectic transition
c,II/liq 371.0 K, ΔH = 9297 cal·mol^{-1}
38900 J·mol^{-1}
ΔS = 25 cal·mol^{-1}·K^{-1}
105 J·mol^{-1}·K^{-1}
c,II–smectic transition
c,III/liq 342.5 K
c,III–smectic transition temperature
liq/liq 408.5 K, ΔH = 3179 cal·mol^{-1}
13300 J·mol^{-1}
ΔS = 8 cal·mol^{-1}·K^{-1}
33 J·mol^{-1}·K^{-1}
Smectic–isotropic transition
Molecular Weight 390.6050
Wiswesser Line Notation QVR DO18
Evaluation B

$C_{25}H_{46}$ (liq) 62GOL/BEL
4′-n-Heptyl-m-tercyclohexyl
Heat Capacity 311 K, C_p = 159.8 cal·mol^{-1}·K^{-1}
668.6 J·mol^{-1}·K^{-1}
Temperatures 100, 200, 300°F
Molecular Weight 346.6384
Wiswesser Line Notation L6TJ AAL6TJ C- AL6TJ D7
Evaluation C

$C_{25}H_{46}$ (liq) 63GUD/CAM
4-n-Heptyltercyclohexyl
Heat Capacity 373 K, C_p = 179.9 cal·mol^{-1}·K^{-1}
752.7 J·mol^{-1}·K^{-1}
Temperature range 373–483 K
Molecular Weight 346.6384
Wiswesser Line Notation L6TJ AAL6TJ X- AL6TJ D7
Evaluation C

$C_{25}H_{46}O_6$ (c,α) 55WAR/VIC
1,2-Diaceto-3-stearin
Heat Capacity 298.2 K, C_p = 216.1 cal·mol^{-1}·K^{-1}
904.2 J·mol^{-1}·K^{-1}
Temperature range -69 to 97°C. Give experimental points and equations for 2 solid and liquid forms. Sub-alpha form, $C_p = 0.4315 + 0.00104t$ cal·g^{-1}·C^{-1}, (-73 to -1°C); alpha form, $C_p = 0.4349 + 0.00213t$ cal·g^{-1}·C^{-1}, (-3 to 33°C); liquid $C_p = 0.3790 + 0.00195t$ cal·g^{-1}·C^{-1}, (77 to 97°C).
Phase Changes
c,α/liq 208.3 K, ΔH = 10889 cal·mol^{-1}
45560 J·mol^{-1}
ΔS = 52.3 cal·mol^{-1}·K^{-1}
218.7 J·mol^{-1}·K^{-1}
Molecular Weight 442.6348
Wiswesser Line Notation 17VO1YOV1&1OV1
Evaluation C

$C_{25}H_{52}$ (c) 30PAR/HUF
n-Pentacosane
Heat Capacity 294.5 K, C_p = 183.8 cal·mol^{-1}·K^{-1}
769.0 J·mol^{-1}·K^{-1}
Temperature range 91–295 K. Value is unsmoothed experimental datum.
Entropy 298.15 K, S = 160.4 cal·mol^{-1}·K^{-1}
671.1 J·mol^{-1}·K^{-1}
Extrapolation below 90 K, 49.2 cal mol^{-1}K^{-1}
Molecular Weight 352.6858
Wiswesser Line Notation 25H
Evaluation B(C_p),C(S)

$C_{25}H_{52}$ (liq) 32SPA/THO
n-Pentacosane
Heat Capacity 333 K, C_p = 195.0 cal·mol^{-1}·K^{-1}
815.9 J·mol^{-1}·K^{-1}
Temperature range 60 to 100°C
Phase Changes
c/liq 326.6 K, ΔH = 18975 cal·mol^{-1}
79391 J·mol^{-1}
ΔS = 58.1 cal·mol^{-1}·K^{-1}
243.1 J·mol^{-1}·K^{-1}
Molecular Weight 352.6858
Wiswesser Line Notation 25H
Evaluation B

$C_{26}H_{12}N_4$ (c) 76CLA/WOR
1,2,4,5-Tetracyanobenzene-pyrene complex
Heat Capacity 298.15 K, C_p = 103.25 cal·mol^{-1}·K^{-1}
432.0 J·mol^{-1}·K^{-1}
Temperature range 13–295 K
Entropy 298.15 K, S = 111.40 cal·mol^{-1}·K^{-1}
466.1 J·mol^{-1}·K^{-1}
Phase Changes
Broad transition in 220–250 K. ΔH estimated as 2150 J mol^{-1}K^{-1}, $\Delta S = 9.2$ J·mol^{-1}K^{-1}
Molecular Weight 380.4076
Wiswesser Line Notation L666 B6 2AB PJ &NCR BCN DCN ECN
Evaluation A

$C_{26}H_{12}O_6$ (c) 78DUN/RAH
Pyrene-pyromellitic dianhydride charge transfer complex
Heat Capacity 298.15 K, C_p = 104.2 cal·mol^{-1}·K^{-1}
435.9 J·mol^{-1}·K^{-1}
Temperature range 5–300 K
Entropy 298.15 K, S = 111.0 cal·mol^{-1}·K^{-1}
464.6 J·mol^{-1}·K^{-1}
Molecular Weight 420.3772
Wiswesser Line Notation T C565 DVOV JVOVJ &L666 B6 2AB PJ
Evaluation A

$C_{26}H_{12}O_6$ (c) 80BOE/WES 2
Pyrene-pyromellitic dianhydride charge transfer transfer
Heat Capacity 298.15 K, C_p = 106.7 cal·mol^{-1}·K^{-1}
446.4 J·mol^{-1}·K^{-1}
Temperature range 5–300 K
Entropy 298.15 K, S = 111.1 cal·mol^{-1}·K^{-1}
464.8 J·mol^{-1}·K^{-1}

Table of Heat Capacities, Entropies, and Phase Transition Properties — Continued

Phase Changes
c,II/c,I 155 K, ΔH = 51.0 cal·mol^{-1}
213.4 J·mol^{-1}
ΔS = 0.32 cal·mol^{-1}·K^{-1}
1.34 J·mol^{-1}·K^{-1}

Molecular Weight 420.3772
Wiswesser Line Notation T C565 DVOV JVOVJ &L666 B6 2AB PJ
Evaluation A

$C_{26}H_{20}$ (c) 31SMI/AND
Tetraphenylethylene
Heat Capacity 298.5 K, C_p = 92.5 cal·mol^{-1}·K^{-1}
387.6 J·mol^{-1}·K^{-1}
Temperature range 102–346 K. Value is unsmoothed experimental datum.
Molecular Weight 332.4440
Wiswesser Line Notation RYR&UYR&R
Evaluation C

$C_{26}H_{22}$ (c) 31SMI/AND
1,1,1,2–Tetraphenylethane
Heat Capacity 298.5 K, C_p = 94.5 cal·mol^{-1}·K^{-1}
395.4 J·mol^{-1}·K^{-1}
Temperature range 102–346 K. Value is unsmoothed experimental datum.
Molecular Weight 334.4598
Wiswesser Line Notation RXR&R&1R
Evaluation C

$C_{26}H_{22}$ (c) 31SMI/AND
1,1,2,2–Tetraphenylethane
Heat Capacity 298.5 K, C_p = 94.8 cal·mol^{-1}·K^{-1}
399.6 J·mol^{-1}·K^{-1}
Temperature range 102–346 K. Value is unsmoothed experimental datum.
Molecular Weight 334.4598
Wiswesser Line Notation RYR&YR&R
Evaluation C

$C_{26}H_{26}O_3Si_3$ (c) 82KUL/DZH
Dimethyltetraphenylcyclotrisiloxane
Heat Capacity 298.15 K, C_p = 136.6 cal·mol^{-1}·K^{-1}
571.5 J·mol^{-1}·K^{-1}
Temperature range 4.7–300 K. Data given graphically except for data at 298.15 K.
Entropy 298.15 K, S = 150.6 cal·mol^{-1}·K^{-1}
630.1 J·mol^{-1}·K^{-1}
Phase Changes
c,I/liq 361.06 K, ΔH = 6740 cal·mol^{-1}
28200 J·mol^{-1}
ΔS = 18.7 cal·mol^{-1}·K^{-1}
78.2 J·mol^{-1}·K^{-1}
Molecular Weight 470.7461
Wiswesser Line Notation T6-SI-O-SI-O-SI-OTJ A1 A1 CR CR ER ER
Evaluation B

$C_{26}H_{38}N_2O_3$ (c,I) 79RAC/NGU
4,4′-Bis(n-heptyloxy)azoxybenzene
Heat Capacity 300 K, C_p = 115 cal·mol^{-1}·K^{-1}
480 J·mol^{-1}·K^{-1}
Temperature range 90–420K. Data graphically only.
Phase Changes
liq/liq 368.2 K, ΔH = 130 cal·mol^{-1}
545 J·mol^{-1}
ΔS = 0.4 cal·mol^{-1}·K^{-1}
1.5 J·mol^{-1}·K^{-1}
Smectic–nematic transition
liq/liq 397.25 K, ΔH = 198 cal·mol^{-1}
830 J·mol^{-1}
ΔS = 0.5 cal·mol^{-1}·K^{-1}
2.1 J·mol^{-1}·K^{-1}
Nematic–isotropic liquid transition
c,I/liq 347.75 K, ΔH = 6560 cal·mol^{-1}
27450 J·mol^{-1}
ΔS = 18.9 cal·mol^{-1}·K^{-1}
78.9 J·mol^{-1}·K^{-1}
Melting of stable crystal phase
Molecular Weight 426.5978
Wiswesser Line Notation 7OR DNUNO&R DO7
Evaluation B

$C_{26}H_{50}O_4$ (liq) 76PHI/MAT
Di-n-octyl sebacate
Heat Capacity 318 K, C_p = 203 cal·mol^{-1}·K^{-1}
849 J·mol^{-1}·K^{-1}
Temperature range 318–393 K
Molecular Weight 426.6786
Wiswesser Line Notation 8OV8VO8
Evaluation C

$C_{26}H_{50}O_4Pb$ (c,II) 78ADE/SIM
Lead(II) tridecanoate
Heat Capacity 375 K, C_p = 294 cal·mol^{-1}·K^{-1}
1230 J·mol^{-1}·K^{-1}
Mean value 371–377 K. Data only graphically for c,III. Also data for liquid.
Phase Changes
c,III/c,II 368.7 K, ΔH = 13960 cal·mol^{-1}
58400 J·mol^{-1}
ΔS = 37.9 cal·mol^{-1}·K^{-1}
158 J·mol^{-1}·K^{-1}
c,II, and c,I, are mesophases.
c,II/c,I 381.5 K, ΔH = 9270 cal·mol^{-1}
38800 J·mol^{-1}
ΔS = 24.3 cal·mol^{-1}·K^{-1}
102 J·mol^{-1}·K^{-1}
Molecular Weight 633.8786
Wiswesser Line Notation OV12 2 .PB
Evaluation C

$C_{26}H_{54}$ (c,II) 31GAR/VAN
n–Hexacosane
Heat Capacity 304 K, C_p = 162.0 cal·mol^{-1}·K^{-1}
677.8 J·mol^{-1}·K^{-1}
Temperature range 295–358 K. Mean value 22–40 °C, β–form.
Phase Changes
c,II/c,I 323.3 K, ΔH = 3070 cal·mol^{-1}
12840 J·mol^{-1}
ΔS = 9.5 cal·mol^{-1}·K^{-1}
39.7 J·mol^{-1}·K^{-1}
β–α transition
c,I/liq 329.3 K, ΔH = 5150 cal·mol^{-1}
21550 J·mol^{-1}
ΔS = 15.6 cal·mol^{-1}·K^{-1}
65.4 J·mol^{-1}·K^{-1}
Molecular Weight 366.7126
Wiswesser Line Notation 26H
Evaluation B

Table of Heat Capacities, Entropies, and Phase Transition Properties — Continued

$C_{26}H_{54}$ (c,II) 76AND/MAR
n–Hexacosane
Heat Capacity 298.15 K, C_p = 158.03 cal·mol⁻¹·K⁻¹ → $C_p = 158.03$ cal·mol^{-1}·K^{-1}
661.2 J·mol^{-1}·K^{-1}
Temperature range 10–360 K
Entropy 298.15 K, S = 159.42 cal·mol^{-1}·K^{-1}
667.0 J·mol^{-1}·K^{-1}
Phase Changes
c,II/c,I 325.5 K, ΔH = 7990 cal·mol^{-1}
33420 J·mol^{-1}
ΔS = 24.5 cal·mol^{-1}·K^{-1}
102.7 J·mol^{-1}·K^{-1}
c,I/liq 329.25 K, ΔH = 14510 cal·mol^{-1}
60700 J·mol^{-1}
ΔS = 44.1 cal·mol^{-1}·K^{-1}
184.4 J·mol^{-1}·K^{-1}
Molecular Weight 366.7126
Wiswesser Line Notation 26H
Evaluation A

$C_{26}H_{54}$ (liq) 69ATK/LAR
n–Hexacosane
Heat Capacity 353 K, C_p = 208 cal·mol^{-1}·K^{-1}
870 J·mol^{-1}·K^{-1}
Temperature range 353–453 K. Equation only.
Molecular Weight 366.7126
Wiswesser Line Notation 26H
Evaluation C

$C_{27}H_{38}O_2$ (c) 72YOU/HAL
4–Methoxy–4′–dodecoxy–trans–stilbene
Phase Changes
liq/liq 409 K, ΔH = 5230 cal·mol^{-1}
21880 J·mol^{-1}
ΔS = 12.8 cal·mol^{-1}·K^{-1}
53.5 J·mol^{-1}·K^{-1}
Smectic–isotropic liquid transition
c/liq 415 K, ΔH = 14020 cal·mol^{-1}
58660 J·mol^{-1}
ΔS = 33.8 cal·mol^{-1}·K^{-1}
141.3 J·mol^{-1}·K^{-1}
Crystal–isotropic liquid transition
Molecular Weight 394.5960
Wiswesser Line Notation 12OR D1U1R DO1 –T
Evaluation B

$C_{27}H_{48}$ (liq) 49PAR/MOO
11–Phenyleicosane
Heat Capacity 300 K, C_p = 183.0 cal·mol^{-1}·K^{-1}
765.7 J·mol^{-1}·K^{-1}
Temperature range 80–300 K
Entropy 298.15 K, S = 207.4 cal·mol^{-1}·K^{-1}
867.8 J·mol^{-1}·K^{-1}
Extrapolation below 80 K, 41.12 cal mol^{-1}K^{-1}
Phase Changes
c/liq 294.3 K, ΔH = 15481 cal·mol^{-1}
64772 J·mol^{-1}
ΔS = 52.6 cal·mol^{-1}·K^{-1}
220.1 J·mol^{-1}·K^{-1}
Molecular Weight 372.6762
Wiswesser Line Notation 10Y10&R
Evaluation B(C_p),C(S)

$C_{27}H_{50}$ (liq) 63GUD/CAM
4–n–Nonyltercyclohexyl
Heat Capacity 373 K, C_p = 196.1 cal·mol^{-1}·K^{-1}
820.5 J·mol^{-1}·K^{-1}
Temperature range 373–483 K
Molecular Weight 374.6920
Wiswesser Line Notation L6TJ AAL6TJ X– AL6TJ X9
Evaluation C

$C_{27}H_{50}O_6$ (liq) 76PHI/MAT
Trioctanoin; Glyceryl trioctanoate
Heat Capacity 338 K, C_p = 220 cal·mol^{-1}·K^{-1}
920 J·mol^{-1}·K^{-1}
Temperature range 338–413 K
Molecular Weight 470.6884
Wiswesser Line Notation 7VO1YOV7&1OV7
Evaluation C

$C_{27}H_{54}$ (liq) 49PAR/MOO
11–Cyclohexyleicosane
Heat Capacity 298.15 K, C_p = 188.2 cal·mol^{-1}·K^{-1}
787.4 J·mol^{-1}·K^{-1}
Temperature range 80–300 K
Entropy 298.15 K, S = 211.3 cal·mol^{-1}·K^{-1}
844.1 J·mol^{-1}·K^{-1}
Extrapolation below 80 K, 41.80 cal mol^{-1}K^{-1}
Phase Changes
c/liq 269.9 K, ΔH = 11638 cal·mol^{-1}
48693 J·mol^{-1}
ΔS = 43.1 cal·mol^{-1}·K^{-1}
180.4 J·mol^{-1}·K^{-1}
Molecular Weight 378.7236
Wiswesser Line Notation L6TJ AY10&10
Evaluation B(C_p),C(S)

$C_{27}H_{56}$ (c,β) 38VER
n–Heptacosane
Heat Capacity 313 K, C_p = 198 cal·mol^{-1}·K^{-1}
828 J·mol^{-1}·K^{-1}
One temperature. Also data for α form, 267 cal·mol^{-1}K^{-1} at 328 K, and liquid, 198 cal mol^{-1}K^{-1} at 338 K.
Molecular Weight 380.7394
Wiswesser Line Notation 27H
Evaluation C

$C_{28}H_{32}N_2O_2$ (c) 79KOB/KAM
Terephthal–bis–n–butylaniline
Heat Capacity C_p data given graphically only.
Temperature range 88–300 K.
Phase Changes
c,X/c,IX 99 K, ΔH = 67 cal·mol^{-1}
280 J·mol^{-1}
ΔS = 0.68 cal·mol^{-1}·K^{-1}
2.83 J·mol^{-1}·K^{-1}
c,IX/c,VIII 140 K, ΔH = 300 cal·mol^{-1}
1255 J·mol^{-1}
ΔS = 2.1 cal·mol^{-1}·K^{-1}
9.0 J·mol^{-1}·K^{-1}
Molecular Weight 428.5730
Wiswesser Line Notation 4NR&VR DVN4&R
Evaluation C

Table of Heat Capacities, Entropies, and Phase Transition Properties — Continued

$C_{28}H_{32}O_4Si_4$ (c) 76KUL/DZH
1,1,3,3–Tetramethyl–5,5,7,7–tetraphenylcyclotetrasiloxane
Heat Capacity C_p not given.
Temperature range 12–370 K. Data deposited in VINITI, No. 1191–76, 13 April 1976.
Entropy 298.15 K, S = 193.9 cal·mol^{-1}·K^{-1}
811.3 J·mol^{-1}·K^{-1}
Phase Changes
c,III/c,II 186.5 K, ΔH = 58 cal·mol^{-1}
243 J·mol^{-1}
ΔS = .31 cal·mol^{-1}·K^{-1}
1.3 J·mol^{-1}·K^{-1}
c,II/c,I 271.5 K, ΔH = 250 cal·mol^{-1}
1046 J·mol^{-1}
ΔS = 0.92 cal·mol^{-1}·K^{-1}
3.85 J·mol^{-1}·K^{-1}
c,I/liq 346.21 K, ΔH = 6465 cal·mol^{-1}
27050 J·mol^{-1}
ΔS = 18.67 cal·mol^{-1}·K^{-1}
78.13 J·mol^{-1}·K^{-1}
Molecular Weight 544.9004
Wiswesser Line Notation T8–SI–O–SI–O–SI–O–SI–OTJ A1 A1 C1 C1 ER ER GR GR
Evaluation B

$C_{28}H_{54}HgO_4$ (liq) 78ADE
Mercuric tetradecanoate; Mercuric myristate
Heat Capacity 415 K, C_p = 271.4 cal·mol^{-1}·K^{-1}
1135.5 J·mol^{-1}·K^{-1}
Mean value, 413–421 K. Data only graphically for solid.
Phase Changes
c,II/c,I 382.4 K, ΔH = 13840 cal·mol^{-1}
57900 J·mol^{-1}
ΔS = 36.2 cal·mol^{-1}·K^{-1}
151.4 J·mol^{-1}·K^{-1}
c,I probably a smectic phase.
c,I/liq 387.0 K, ΔH = 9560 cal·mol^{-1}
40000 J·mol^{-1}
ΔS = 24.7 cal·mol^{-1}·K^{-1}
103.4 J·mol^{-1}·K^{-1}
Molecular Weight 655.3222
Wiswesser Line Notation OV13 2 .HG
Evaluation C

$C_{28}H_{56}Ni_4O_{16}$ (c) 78SOR/YOS
Tetrakis[μ_3–methoxo–2,4–pentanedionato(methanol) nickel(II)]
Heat Capacity 284.91 K, C_p = 243.4 cal·mol^{-1}·K^{-1}
1018.5 J·mol^{-1}·K^{-1}
Temperature range 0.4–285 K. Value is unsmoothed experimental datum.
Molecular Weight 883.5408
Wiswesser Line Notation D6O–NI–O ADTJ BO1 BO1 D1 F1 4
Evaluation B

$C_{28}H_{58}$ (liq) 69ATK/LAR
n–Octacosane
Heat Capacity 353 K, C_p = 224 cal·mol^{-1}·K^{-1}
937 J·mol^{-1}·K^{-1}
Temperature range 353–453 K. Equation only.
Molecular Weight 394.7662
Wiswesser Line Notation 28H
Evaluation C

$C_{29}H_{41}O_2$ (c,I) 69KOS/SUG
Galvinoxyl radical;
2,6–Di–tert–butyl–4–(3,5–di–tert–butyl–4–oxocyclohexa-2,5–dienylidene methyl)phenoxyl
Heat Capacity 298.15 K, C_p = 156.38 cal·mol^{-1}·K^{-1}
654.28 J·mol^{-1}·K^{-1}
Temperature range 12–300 K
Entropy 298.15 K, S = 160.16 cal·mol^{-1}·K^{-1}
670.10 J·mol^{-1}·K^{-1}
Phase Changes
c,II/c,I 81.5 K, ΔH = 359.6 cal·mol^{-1}
1504.6 J·mol^{-1}
ΔS = 4.46 cal·mol^{-1}·K^{-1}
18.67 J·mol^{-1}·K^{-1}
Magnetic transition temperature
Molecular Weight 421.6417
Wiswesser Line Notation L6V DYJ BX1&1&1 DU1R DO CX1&1&1 EX1&1&1& FX1&1&1
Evaluation A

$C_{29}H_{42}O_2$ (c) 69KOS/SUG
2,6–Di–tert–butyl–4–(3,5–di–tert–butyl–4–oxocyclo–hexa–2,5–dienylidene methyl)phenol
Heat Capacity 298.15 K, C_p = 156.28 cal·mol^{-1}·K^{-1}
654.28 J·mol^{-1}·K^{-1}
Temperature range 12–300 K
Entropy 298.15 K, S = 158.28 cal·mol^{-1}·K^{-1}
662.24 J·mol^{-1}·K^{-1}
Molecular Weight 422.6496
Wiswesser Line Notation L6V DYJ BX1&1&1 DU1R DQ CX1&1&1 EX1&1&1& FX1&1&1
Evaluation A

$C_{30}H_{58}O_4$ (liq) 76PHI/MAT
Di–n–decyl sebacate
Heat Capacity 368 K, C_p = 240 cal·mol^{-1}·K^{-1}
1004 J·mol^{-1}·K^{-1}
Temperature range 368–240 K
Molecular Weight 482.7858
Wiswesser Line Notation 10OV8VO10
Evaluation C

$C_{30}H_{58}O_4Pb$ (c,II) 78ADE/SIM
Lead(II) pentadecanoate
Heat Capacity 380 K, C_p = 342 cal·mol^{-1}·K^{-1}
1430 J·mol^{-1}·K^{-1}
Mean value, 376–380 K. Data only graphically for c,III.
Phase Changes
c,III/c,II 374.7 K, ΔH = 15320 cal·mol^{-1}
64100 J·mol^{-1}
ΔS = 40.9 cal·mol^{-1}·K^{-1}
171 J·mol^{-1}·K^{-1}
c,II, and c,I, are mesophases.
c,II/c,I 384.6 K, ΔH = 11500 cal·mol^{-1}
48100 J·mol^{-1}
ΔS = 29.9 cal·mol^{-1}·K^{-1}
125 J·mol^{-1}·K^{-1}
Molecular Weight 689.9858
Wiswesser Line Notation OV14 2 .PB
Evaluation C

Table of Heat Capacities, Entropies, and Phase Transition Properties — Continued

$C_{30}H_{62}$ (c,II) 31GAR/VAN
n–Triacontane
Heat Capacity 301 K, C_p = 193.3 cal·mol^{-1}·K^{-1}
808.8 J·mol^{-1}·K^{-1}
Temperature range 289–373 K. Mean value 16–38 °C, β–form.
Phase Changes
c,II/c,I 332.2 K, ΔH = 3690 cal·mol^{-1}
15440 J·mol^{-1}
ΔS = 11.1 cal·mol^{-1}·K^{-1}
46.5 J·mol^{-1}·K^{-1}
β–α transition
c,I/liq 338.7 K, ΔH = 6960 cal·mol^{-1}
29120 J·mol^{-1}
ΔS = 20.5 cal·mol^{-1}·K^{-1}
86.0 J·mol^{-1}·K^{-1}
Molecular Weight 422.8198
Wiswesser Line Notation 30H
Evaluation B

$C_{31}H_{64}$ (liq) 45FIS/NAY
11–n–Decylheneicosane
Heat Capacity 300 K, C_p = 227.0 cal·mol^{-1}·K^{-1}
949.8 J·mol^{-1}·K^{-1}
Temperature range 12–296 K (Penn State), 80–297 K (Stanford); at 300 K, C_p = 227.0 cal·mol^{-1}·K^{-1} (Penn State), 230.2 cal·mol^{-1}·K^{-1} (Stanford).
Entropy 298.15 K, S = 259.6 cal·mol^{-1}·K^{-1}
1086.2 J·mol^{-1}·K^{-1}
Penn State entropy above. From Stanford data, with extrapolation of 52.1 cal mol^{-1}K^{-1}, S = 262.5 cal mol^{-1}K^{-1}.
Phase Changes
c/liq 282.34 K, ΔH = 17019 cal·mol^{-1}
71207 J·mol^{-1}
ΔS = 60.28 cal·mol^{-1}·K^{-1}
252.20 J·mol^{-1}·K^{-1}
Stanford data give ΔH = 16980, Tm 282.2 K.
Molecular Weight 436.8466
Wiswesser Line Notation 10Y10&10
Evaluation B
Impurity from melting data, 3.4–3.5 mol%

$C_{31}H_{64}$ (c,β) 38VER
n–Unatriacontane
Heat Capacity 323 K, C_p = 218 cal·mol^{-1}·K^{-1}
912 J·mol^{-1}·K^{-1}
One temperature. Also data for α form, 424 cal·mol^{-1}K^{-1} at 338 K, and liquid 262 cal mol^{-1}K^{-1} at 348 K.
Molecular Weight 436.8466
Wiswesser Line Notation 31H
Evaluation C

$C_{32}H_{26}$ (c) 31SMI/AND
Pentaphenylethane
Heat Capacity 298.5 K, C_p = 113.2 cal·mol^{-1}·K^{-1}
473.6 J·mol^{-1}·K^{-1}
Temperature range 102–346 K. Value is unsmoothed experimental datum.
Molecular Weight 410.5574
Wiswesser Line Notation RYR&XR&R&R
Evaluation C

$C_{32}H_{64}HgO_4$ (liq) 78ADE
Mercuric hexadecanoate; Mercuric palmitate
Heat Capacity 410 K, C_p = 290.9 cal·mol^{-1}·K^{-1}
1217.3 J·mol^{-1}·K^{-1}
Mean value, 396–421 K. Data only graphically for solid.
Phase Changes
c,II/c,I 383.4 K, ΔH = 11830 cal·mol^{-1}
49500 J·mol^{-1}
ΔS = 30.9 cal·mol^{-1}·K^{-1}
129.1 J·mol^{-1}·K^{-1}
c,I probably a smectic phase.
c,I/liq 390.3 K, ΔH = 14220 cal·mol^{-1}
59500 J·mol^{-1}
ΔS = 36.4 cal·mol^{-1}·K^{-1}
152.4 J·mol^{-1}·K^{-1}
Molecular Weight 713.4452
Wiswesser Line Notation OV15 2 .HG
Evaluation C

$C_{32}H_{66}$ (c,α) 38VER
n–Dotriacontane
Heat Capacity 338 K, C_p = 253 cal·mol^{-1}·K^{-1}
1059 J·mol^{-1}·K^{-1}
One temperature
Molecular Weight 450.8734
Wiswesser Line Notation 32H
Evaluation C

$C_{32}H_{66}$ (c,β) 38VER
n–Dotriacontane
Heat Capacity 333 K, C_p = 248 cal·mol^{-1}·K^{-1}
1038 J·mol^{-1}·K^{-1}
One temperature
Molecular Weight 450.8734
Wiswesser Line Notation 32H
Evaluation C

$C_{32}H_{66}$ (c) 49PAR/MOO
n–Dotriacontane
Heat Capacity 298.15 K, C_p = 209.7 cal·mol^{-1}·K^{-1}
877.4 J·mol^{-1}·K^{-1}
Temperature range 80–300 K
Entropy 298.15 K, S = 203.5 cal·mol^{-1}·K^{-1}
851.4 J·mol^{-1}·K^{-1}
Extrapolation below 80 K, 51.33 cal mol^{-1}K^{-1}
Molecular Weight 450.8734
Wiswesser Line Notation 32H
Evaluation B(C_p),C(S)

$C_{32}H_{66}$ (liq) 38VER
n–Dotriacontane
Heat Capacity 348 K, C_p = 324 cal·mol^{-1}·K^{-1}
1356 J·mol^{-1}·K^{-1}
One temperature
Molecular Weight 450.8734
Wiswesser Line Notation 32H
Evaluation C

$C_{33}H_{62}O_6$ (liq) 76PHI/MAT
Tridecanoin; Glyceryl tridecanoate
Heat Capacity 313 K, C_p = 265 cal·mol^{-1}·K^{-1}
1109 J·mol^{-1}·K^{-1}
Temperature range 313–388 K
Molecular Weight 554.8492
Wiswesser Line Notation 9VO1YOV9&1OV9
Evaluation C

Table of Heat Capacities, Entropies, and Phase Transition Properties — Continued

$C_{33}H_{68}$ (c) 30PAR/HUF
n–Tritriacontane
Heat Capacity 294.4 K, C_p = 215.3 cal·mol⁻¹·K⁻¹
900.8 J·mol⁻¹·K⁻¹
Temperature range 94–294 K. Value is unsmoothed experimental datum.
Entropy 298.15 K, S = 209.8 cal·mol⁻¹·K⁻¹
877.8 J·mol⁻¹·K⁻¹
Extrapolation below 90 K, 63.8 cal mol⁻¹K⁻¹
Molecular Weight 464.9002
Wiswesser Line Notation 33H
Evaluation B(C_p),C(S)

$C_{33}H_{68}$ (liq) 32SPA/THO
n–Tritriacontane
Heat Capacity 353 K, C_p = 265.9 cal·mol⁻¹·K⁻¹
1112.5 J·mol⁻¹·K⁻¹
Temperature range 80 to 110°C
Phase Changes
c/liq 344.2 K, ΔH = 25105 cal·mol⁻¹
105039 J·mol⁻¹
ΔS = 72.9 cal·mol⁻¹·K⁻¹
305.2 J·mol⁻¹·K⁻¹
Molecular Weight 464.9002
Wiswesser Line Notation 33H
Evaluation B

$C_{34}H_{64}O_4$ (liq) 76PHI/MAT
Di–n–dodecyl sebacate
Heat Capacity 368 K, C_p = 267 cal·mol⁻¹·K⁻¹
1117 J·mol⁻¹·K⁻¹
Temperature range 368–408 K
Molecular Weight 536.8772
Wiswesser Line Notation 12OV8VO12
Evaluation C

$C_{34}H_{66}O_4Pb$ (c,III) 78ADE/SIM
Lead(II) heptadecanoate
Heat Capacity Data only graphically for c,III.
Phase Changes
c,III/c,II 378.7 K, ΔH = 16250 cal·mol⁻¹
68000 J·mol⁻¹
ΔS = 42.9 cal·mol⁻¹·K⁻¹
180 J·mol⁻¹·K⁻¹
c,II, and c,I, are mesophases.
c,II/c,I 387.4 K, ΔH = 13170 cal·mol⁻¹
55100 J·mol⁻¹
ΔS = 34.0 cal·mol⁻¹·K⁻¹
142 J·mol⁻¹·K⁻¹
Molecular Weight 746.0930
Wiswesser Line Notation OV16 2 .PB
Evaluation C

$C_{34}H_{70}$ (c,II) 31GAR/VAN
n–Tetratriacontane
Heat Capacity 303 K, C_p = 212.1 cal·mol⁻¹·K⁻¹
887.4 J·mol⁻¹·K⁻¹
Temperature range 293–373 K. Mean value 20–40 °C β–form.
Phase Changes
c,II/c,I 341.1 K, ΔH = 5500 cal·mol⁻¹
23010 J·mol⁻¹
ΔS = 16.1 cal·mol⁻¹·K⁻¹
67.5 J·mol⁻¹·K⁻¹
β–α transition
c,I/liq 345.6 K, ΔH = 9150 cal·mol⁻¹
38280 J·mol⁻¹
ΔS = 26.5 cal·mol⁻¹·K⁻¹
110.8 J·mol⁻¹·K⁻¹
Molecular Weight 478.9270
Wiswesser Line Notation 34H
Evaluation B

$C_{34}H_{70}$ (liq) 69ATK/LAR
n–Tetratriacontane
Heat Capacity 353 K, C_p = 275 cal·mol⁻¹·K⁻¹
1149 J·mol⁻¹·K⁻¹
Temperature range 353–453 K. Equation only.
Molecular Weight 478.9270
Wiswesser Line Notation 34H
Evaluation C

$C_{35}H_{72}$ (c,II) 31GAR/VAN
n–Pentatriacontane
Heat Capacity 302 K, C_p = 218.9 cal·mol⁻¹·K⁻¹
915.9 J·mol⁻¹·K⁻¹
Temperature range 290–373 K. Mean value, 17–41 °C, β–form.
Phase Changes
c,II/c,I 344.7 K, ΔH = 4840 cal·mol⁻¹
20250 J·mol⁻¹
ΔS = 14.0 cal·mol⁻¹·K⁻¹
58.7 J·mol⁻¹·K⁻¹
β–α transition
c,I/liq 347.2 K, ΔH = 10180 cal·mol⁻¹
42590 J·mol⁻¹
ΔS = 29.3 cal·mol⁻¹·K⁻¹
122.7 J·mol⁻¹·K⁻¹
Molecular Weight 492.9538
Wiswesser Line Notation 35H
Evaluation B

$C_{36}H_{24}$ (c) 67MAG
1,3,5–Tri–2–naphthylbenzene
Heat Capacity 300 K, C_p = 115 cal·mol⁻¹·K⁻¹
481 J·mol⁻¹·K⁻¹
Temperature range 200–560 K. Estimated, data graphically only. Also data for glass.
Phase Changes
c/liq 472 K, ΔH = 10140 cal·mol⁻¹
42425 J·mol⁻¹
ΔS = 21.5 cal·mol⁻¹·K⁻¹
89.9 J·mol⁻¹·K⁻¹
Molecular Weight 456.5856
Wiswesser Line Notation L66J C– R CCL66J& E– CL66J
Evaluation C

$C_{36}H_{30}O_3Si_3$ (c) 82KUL/DZH
Hexaphenylcyclotrisiloxane
Heat Capacity 298.15 K, C_p = 163.4 cal·mol⁻¹·K⁻¹
683.7 J·mol⁻¹·K⁻¹
Temperature range 4.7–300 K. Data given graphically except for data at 298.15 K.
Entropy 298.15 K, S = 175.8 cal·mol⁻¹·K⁻¹
735.5 J·mol⁻¹·K⁻¹
Molecular Weight 594.8877
Wiswesser Line Notation T6–SI–O–SI–O–SI–OTJ AR AR CR CR ER ER
Evaluation B

$C_{36}H_{70}HgO_4$ (liq) 78ADE
Mercuric octadecanoate; Mercuric stearate
Heat Capacity 410 K, C_p = 329.2 cal·mol⁻¹·K⁻¹
1377.2 J·mol⁻¹·K⁻¹
Mean value, 391–433 K. Data only graphically for solid.

Table of Heat Capacities, Entropies, and Phase Transition Properties — Continued

Phase Changes
c,II/c,I 355.2 K, ΔH = 1050 cal·mol^{-1}
4400 J·mol^{-1}
ΔS = 3.0 cal·mol^{-1}·K^{-1}
12.4 J·mol^{-1}·K^{-1}
c,I/liq 393.2 K, ΔH = 27840 cal·mol^{-1}
116500 J·mol^{-1}
ΔS = 70.8 cal·mol^{-1}·K^{-1}
296.3 J·mol^{-1}·K^{-1}
Molecular Weight 767.5366
Wiswesser Line Notation OV17 2 .HG
Evaluation C

$C_{36}H_{74}$ (liq) 69ATK/LAR
n–Hexatriacontane
Heat Capacity 353 K, C_p = 288 cal·mol^{-1}·K^{-1}
1206 J·mol^{-1}·K^{-1}
Temperature range 353–453 K. Equation only.
Molecular Weight 506.9806
Wiswesser Line Notation 36H
Evaluation C

$C_{37}H_{54}Si$ (liq) 60BAR/BOL
Tribenzyl–n–hexyldecylsilane
Heat Capacity 313 K, C_p = 216 cal·mol^{-1}·K^{-1}
904 J·mol^{-1}·K^{-1}
Temperature range 40 to 240°C
Molecular Weight 526.9191
Wiswesser Line Notation 16–S1–1R&1R&1R
Evaluation B

$C_{38}H_{70}O_8$ (liq) 76PHI/MAT
Dihexyl hexamethylenesebacate
Heat Capacity 328 K, C_p = 309 cal·mol^{-1}·K^{-1}
1293 J·mol^{-1}·K^{-1}
Temperature range 328–408 K
Molecular Weight 654.9662
Wiswesser Line Notation 6OV8VO6OV8VO6
Evaluation C

$C_{38}H_{72}O_4$ (liq) 76PHI/MAT
Di–n–tetradecyl sebacate
Heat Capacity 343 K, C_p = 298 cal·mol^{-1}·K^{-1}
1247 J·mol^{-1}·K^{-1}
Temperature range 343–433 K
Molecular Weight 592.9844
Wiswesser Line Notation 14OV8VO14
Evaluation C

$C_{38}H_{74}O_4Pb$ (c,II) 78ADE/SIM
Lead(II) nonadecanoate
Heat Capacity Data only graphically for c,III.
Phase Changes
c,III/c,II 383.8 K, ΔH = 18140 cal·mol^{-1}
75900 J·mol^{-1}
ΔS = 47.3 cal·mol^{-1}·K^{-1}
198 J·mol^{-1}·K^{-1}
c,II, and c,I, are mesophases.
c,II/c,I 389.1 K, ΔH = 15370 cal·mol^{-1}
64300 J·mol^{-1}
ΔS = 39.5 cal·mol^{-1}·K^{-1}
165 J·mol^{-1}·K^{-1}
Molecular Weight 802.2002
Wiswesser Line Notation OV18 2 .PB
Evaluation C

$C_{39}H_{74}O_6$ (liq) 76PHI/MAT
Trilaurin; Glyceryl trilaurate
Heat Capacity 323 K, C_p = 312 cal·mol^{-1}·K^{-1}
1305 J·mol^{-1}·K^{-1}
Temperature range 323–398 K
Molecular Weight 639.0100
Wiswesser Line Notation 11VO1YOV11&1OV11
Evaluation C

$C_{39}H_{74}O_6$ (liq) 47CHA/SIN
Trilaurin; Glyceryl trilaurate
Heat Capacity 330.7 K, C_p = 324.0 cal·mol^{-1}·K^{-1}
1355.6 J·mol^{-1}·K^{-1}
Temperature range 90–370 K. Value is unsmoothed experimental datum.
Entropy 298.15 K, S = 265.1 cal·mol^{-1}·K^{-1}
1071.5 J·mol^{-1}·K^{-1}
For c,β: Extrapolation below 90 K, 81.0 cal mol^{-1}K^{-1}.
Phase Changes
c,β/liq 319.5 K, ΔH = 29520 cal·mol^{-1}
123510 J·mol^{-1}
ΔS = 92.4 cal·mol^{-1}·K^{-1}
386.6 J·mol^{-1}·K^{-1}
Molecular Weight 639.0100
Wiswesser Line Notation 11VO1YOV11&1OV11
Evaluation C

$C_{41}H_{72}O_2$ (c) 67BAR/POR
Cholesteryl myristate
Heat Capacity C_p data given graphically only.
Temperature range 270–270 K
Phase Changes
liq/liq 352.9 K, ΔH = 310 cal·mol^{-1}
1300 J·mol^{-1}
ΔS = 0.9 cal·mol^{-1}·K^{-1}
3.7 J·mol^{-1}·K^{-1}
Smectic–cholesteric transition
liq/liq 358.7 K, ΔH = 245 cal·mol^{-1}
1025 J·mol^{-1}
ΔS = 0.7 cal·mol^{-1}·K^{-1}
2.9 J·mol^{-1}·K^{-1}
Cholesteric–isotropic transition
c/liq 346.8 K, ΔH = 11160 cal·mol^{-1}
46690 J·mol^{-1}
ΔS = 32.2 cal·mol^{-1}·K^{-1}
134.6 J·mol^{-1}·K^{-1}
Solid–smectic transition
Molecular Weight 597.0186
Wiswesser Line Notation L E5 B666 LUTJ A1 E1FY1&3Y1&1 OOV13
Evaluation B

$C_{42}H_{66}O_{12}$ (c) 80SOR/TSU
Benzene–hexa–n–hexanoate
Heat Capacity 298.15 K, C_p = 309.5 cal·mol^{-1}·K^{-1}
1294.9 J·mol^{-1}·K^{-1}
Temperature range 13–393 K
Entropy 298.15 K, S = 329.8 cal·mol^{-1}·K^{-1}
1380.0 J·mol^{-1}·K^{-1}

Table of Heat Capacities, Entropies, and Phase Transition Properties — Continued

Phase Changes
c,IV/c,III 251.58 K, ΔH = 6134 cal·mol^{-1}
25665 J·mol^{-1}
ΔS = 24.54 cal·mol^{-1}·K^{-1}
102.7 J·mol^{-1}·K^{-1}
Transition has a large amount of first-order character.
c,III/c,II 291.46 K, ΔH = 2944 cal·$^{-1}$
12272 J·mol^{-1}
ΔS = 11.16 cal·mol^{-1}·K^{-1}
46.69 J·mol^{-1}·K^{-1}
Anomalous transition
c,II/c,I 348.27 K, ΔH = 3886 cal·mol^{-1}
16259 J·mol^{-1}
ΔS = 11.16 cal·mol^{-1}·K^{-1}
46.69 J·mol^{-1}·K^{-1}
c,I/liq 368.74 K, ΔH = 8007 cal·mol^{-1}
33501 J·mol^{-1}
ΔS = 21.72 cal·mol^{-1}·K^{-1}
90.88 J·mol^{-1}·K^{-1}
Molecular Weight 762.9762
Wiswesser Line Notation 5VOR BOV5 COV5 DOV5 EOV5 FOV5
Evaluation A

$C_{42}H_{82}O_4$ (liq) 76PHI/MAT
Di-n-hexadecyl sebacate
Heat Capacity 353 K, C_p = 349 cal·mol^{-1}·K^{-1}
1460 J·mol^{-1}·K^{-1}
Temperature range 353–384 K
Molecular Weight 651.1074
Wiswesser Line Notation 16OV8VO16
Evaluation C

$C_{42}H_{86}$ (liq) 69ATK/LAR
n-Dotetracontane
Heat Capacity 353 K, C_p = 341 cal·mol^{-1}·K^{-1}
1425 J·mol^{-1}·K^{-1}
Temperature range 353–453 K. Equation only.
Molecular Weight 591.1414
Wiswesser Line Notation 42H
Evaluation C

$C_{43}H_{26}AsN_8$ (c) 77KOS/SOR 2
Methyltriphenylarsonium bis[7,7,8,8-tetracyanoquinodimethanide]
Heat Capacity 300 K, C_p = 210.26 cal·mol^{-1}·K^{-1}
879.74 J·mol^{-1}·K^{-1}
Temperature range 12–350 K
Entropy 300 K, S = 234.72 cal·mol^{-1}·K^{-1}
982.05 J·mol^{-1}·K^{-1}
Molecular Weight 729.6536
Wiswesser Line Notation L6Y DYJ AYCN&CN DYCN&CN 2 &1-AS-R&R&R
Evaluation A

$C_{43}H_{26}N_8P$ (c,II) 77KOS/SOR 2
Methyltriphenylphosphonium bis(7,7,8,8-tetracyanoquinodomethanide)
Heat Capacity 300 K, C_p = 205.37 cal·mol^{-1}·K^{-1}
859.26 J·mol^{-1}·K^{-1}
Temperature range 12–350 K
Entropy 300 K, S = 220.53 cal·mol^{-1}·K^{-1}
922.69 J·mol^{-1}·K^{-1}

Phase Changes
c,II/c,I 315.65 K, ΔH = 485 cal·mol^{-1}
2030 J·mol^{-1}
ΔS = 1.54 cal·mol^{-1}·K^{-1}
6.43 J·mol^{-1}·K^{-1}
Molecular Weight 685.7058
Wiswesser Line Notation L6Y DYJ AYCN&CN DYCN&CN 2 & 1PR&R&R
Evaluation A

$C_{45}H_{86}O_6$ (liq) 47CHA/SIN
Trimyristin; Glyceryl trimyristate
Heat Capacity 331.5 K, C_p = 371.7 cal·mol^{-1}·K^{-1}
1555.2 J·mol^{-1}·K^{-1}
Temperature range 89–365 K. Value is unsmoothed experimental datum. Data for c,α, 192–247 K.
Entropy 298.15 K, S = 297.8 cal·mol^{-1}·K^{-1}
1246.0 J·mol^{-1}·K^{-1}
For c,β. Extrapolation below 90 K, 91.1 cal mol^{-1}K^{-1}.
Phase Changes
c,α/liq 305.5 K, ΔH = 25020 cal·mol^{-1}
104685 J·mol^{-1}
ΔS = 81.9 cal·mol^{-1}·K^{-1}
342.7 J·mol^{-1}·K^{-1}
c,β/liq 330.2 K, ΔH = 36375 cal·mol^{-1}
152195 J·mol^{-1}
ΔS = 110.2 cal·mol^{-1}·K^{-1}
460.9 J·mol^{-1}·K^{-1}
Molecular Weight 723.1708
Wiswesser Line Notation 13VO1YOV13&1OV13
Evaluation C

$C_{45}H_{86}O_6$ (liq) 76PHI/MAT
Trimyristin; Glyceryl trimyristate
Heat Capacity 333 K, C_p = 354 cal·mol^{-1}·K^{-1}
1481 J·mol^{-1}·K^{-1}
Temperature range 333–433 K
Molecular Weight 723.1708
Wiswesser Line Notation 13VO1YOV13&1OV13
Evaluation C

$C_{46}H_{47}N_7O_2$ (c) 40CAM/CAM
p-Nitrosodimethylaniline-β-naphthylamine complex; β-Naphthylamine-p-nitrosodimethylaniline complex
Heat Capacity 293 K, C_p = 205.0 cal·mol^{-1}·K^{-1}
857.7 J·mol^{-1}·K^{-1}
One temperature
Molecular Weight 729.9230
Wiswesser Line Notation ONR DN1&1 2 &L66J CZ 3
Evaluation C

$C_{46}H_{90}O_4$ (liq) 76PHI/MAT
Di-n-octadecyl sebacate
Heat Capacity 353 K, C_p = 354 cal·mol^{-1}·K^{-1}
1481 J·mol^{-1}·K^{-1}
Temperature range 353–354 K
Molecular Weight 707.2146
Wiswesser Line Notation 18OV8VO18
Evaluation C

$C_{48}H_{78}O_{12}$ (c) 81SOR/SUG
Benzene-hexa-n-heptanoate
Heat Capacity 298.15 K, C_p = 359.7 cal·mol^{-1}·K^{-1}
1505.0 J·mol^{-1}·K^{-1}
Temperature range 13–393 K
Entropy 298.15 K, S = 365.7 cal·mol^{-1}·K^{-1}
1530.1 J·mol^{-1}·K^{-1}

Table of Heat Capacities, Entropies, and Phase Transition Properties — Continued

Phase Changes
c,IV/c,III 129 K, ΔH = 268 cal·mol^{-1}
1120 J·mol^{-1}
ΔS = 2.02 cal·mol^{-1}·K^{-1}
8.46 J·mol^{-1}·K^{-1}
anomalous or diffuse first–order transition
c,III/c,II 222.80 K
anomalous transition
c,II/c,I 230.81 K
first–order transition
c,III/c,I ΔH = 2750 cal·mol^{-1}
11500 J·mol^{-1}
ΔS = 12.06 cal·mol^{-1}·K^{-1}
50.44 J·mol^{-1}·K^{-1}
c,I/mesophase 353.79 K, ΔH = 7698 cal·mol^{-1}
32210 J·mol^{-1}
ΔS = 21.77 cal·mol^{-1}·K^{-1}
91.08 J·mol^{-1}·K^{-1}
mesophase/isotropic liquid
359.28 K, ΔH = 5148 cal·mol^{-1}
21540 J·mol^{-1}
ΔS = 14.32 cal·mol^{-1}·K^{-1}
59.93 J·mol^{-1}·K^{-1}
triple point is 353.81 K.
Molecular Weight 847.1370
Wiswesser Line Notation 6VOR BOV6 COV6 DOV6 EOV6 FOV6
Evaluation A

$C_{48}H_{98}$ (liq) 69ATK/LAR
n–Octatetracontane
Heat Capacity 353 K, C_p = 381 cal·mol^{-1}·K^{-1}
1595 J·mol^{-1}·K^{-1}
Temperature range 353–453 K. Equation only.
Molecular Weight 675.3022
Wiswesser Line Notation 48H
Evaluation C

$C_{51}H_{98}O_6$ (liq) 47CHA/SIN
Tripalmitin; Glyceryl tripalmitate
Heat Capacity 338.8 K, C_p = 419.0 cal·mol^{-1}·K^{-1}
1753.1 J·mol^{-1}·K^{-1}
Temperature range 87–369 K. Value is unsmoothed experimental datum. Data for c,α, 195–252 K.
Entropy 298.15 K, S = 331.6 cal·mol^{-1}·K^{-1}
1387.4 J·mol^{-1}·K^{-1}
For c,β: extrapolation below 90 K, 99.2 cal mol^{-1}K^{-1}.
Phase Changes
c,α/liq 317.9 K, ΔH = 30195 cal·mol^{-1}
126335 J·mol^{-1}
ΔS = 95.0 cal·mol^{-1}·K^{-1}
397.4 J·mol^{-1}·K^{-1}
c,β/liq 338.9 K, ΔH = 42870 cal·mol^{-1}
179370 J·mol^{-1}
ΔS = 126.5 cal·mol^{-1}·K^{-1}
529.3 J·mol^{-1}·K^{-1}
Molecular Weight 807.3316
Wiswesser Line Notation 15VO1YOV15&1OV15
Evaluation C

$C_{51}H_{98}O_6$ (liq) 76PHI/MAT
Tripalmitin; Glyceryl tripalmitate
Heat Capacity 343 K, C_p = 398 cal·mol^{-1}·K^{-1}
1665 J·mol^{-1}·K^{-1}
Temperature range 343–418 K
Molecular Weight 807.3316
Wiswesser Line Notation 15VO1YOV15&1OV15
Evaluation C

$C_{54}H_{90}O_{12}$ (c) 82SOR/YOS
Benzene–hexa–n–octanoate
Heat Capacity 298.15 K, C_p = 509.4 cal·mol^{-1}·K^{-1}
2131.3 J·mol^{-1}·K^{-1}
Temperature range 13–393 K
Entropy 298.15 K, S = 361.9 cal·mol^{-1}·K^{-1}
1514.2 J·mol^{-1}·K^{-1}
Phase Changes
c,II/c,I 301.89 K, ΔH = 11702 cal·mol^{-1}
48960 J·mol^{-1}
ΔS = 39.20 cal·mol^{-1}·K^{-1}
164.01 J·mol^{-1}·K^{-1}
c,I/mesophase 355.10 K, ΔH = 11011 cal·mol^{-1}
46070 J·mol^{-1}
ΔS = 31.03 cal·mol^{-1}·K^{-1}
129.81 J·mol^{-1}·K^{-1}
mesophase/isotropic liquid
357.09 K, ΔH = 4594 cal·mol^{-1}
19220 J·mol^{-1}
ΔS = 12.85 cal·mol^{-1}·K^{-1}
53.77 J·mol^{-1} K^{-1}
Molecular Weight 931.2978
Wiswesser Line Notation 7OVR BOV7 COV7 DOV7 EOV7 FOV7
Evaluation A

$C_{54}H_{98}O_{12}$ (liq) 76PHI/MAT
Dihexyl bis(hexamethylenesebacate)
Heat Capacity 333 K, C_p = 437 cal·mol^{-1}·K^{-1}
1828 J·mol^{-1}·K^{-1}
Temperature range 333–433 K
Molecular Weight 939.3610
Wiswesser Line Notation 6OV8VO6OV8VO6OV8VO6
Evaluation C

$C_{57}H_{110}O_6$ (liq) 47CHA/SIN
Tristearin; Glyceryl tristearate
Heat Capacity 346.5 K, C_p = 470.6 cal·mol^{-1}·K^{-1}
1969.0 J·mol^{-1}·K^{-1}
Temperature range 96–372 K. Value is unsmoothed experimental datum. Data for c,α, 192–226 K.
Entropy 298.15 K, S = 366.8 cal·mol^{-1}·K^{-1}
1534.7 J·mol^{-1}·K^{-1}
For c,β: extrapolation below 90 K, 110.6 cal mol^{-1}K^{-1}.
Phase Changes
c,α/liq 327.2 K, ΔH = 34675 cal·mol^{-1}
145080 J·mol^{-1}
ΔS = 106.0 cal·mol^{-1}·K^{-1}
443.4 J·mol^{-1}·K^{-1}
c,β/liq 345.7 K, ΔH = 48580 cal·mol^{-1}
203260 J·mol^{-1}
ΔS = 140.5 cal·mol^{-1}·K^{-1}
588.0 J·mol^{-1}·K^{-1}
Molecular Weight 891.4924
Wiswesser Line Notation 17VO1YOV17&1OV17
Evaluation C

$C_{57}H_{110}O_6$ (liq) 76PHI/MAT
Tristearin; Glyceryl tristearate
Heat Capacity 353 K, C_p = 472 cal·mol^{-1}·K^{-1}
1975 J·mol^{-1}·K^{-1}
Temperature range 353–453 K
Molecular Weight 891.4924
Wiswesser Line Notation 17VO1YOV17&1OV17
Evaluation C

Table of Heat Capacities, Entropies, and Phase Transition Properties — Continued

$C_{508}H_{752}N_{130}O_{150}S_{12}Zn$ (c) 69HUT/COL

Bovine zinc insulin, anhydrous

Heat Capacity 298.15 K, C_p = 29.96 cal·100g^{-1}·K^{-1}
125.4 J·100g^{-1}·K^{-1}

Temperature range 10–310 K

Entropy 298.15 K, S = 31.44 cal·100g^{-1}·K^{-1}
131.5 J·100g^{-1}·K^{-1}

Anhydrous protein from beef pancreas and consists of a sequence of 96 amino acids.

Molecular Weight 11,530.4098

Evaluation A

$C_{508}H_{752}N_{130}O_{150}S_{12}Zn$ (c) 69HUT/COL

Bovine zinc insulin, hydrated

Heat Capacity 298.15 K, C_p = 31.55 cal·100g^{-1}·K^{-1}
132.0 J·100g^{-1}·K^{-1}

Temperature range 10–310 K

Entropy 298.15 K, S = 32.53 cal·100g^{-1}·K^{-1}
136.1 J·100g^{-1}·K^{-1}

Anhydrous protein from beef pancreas and consists of a sequence of 96 amino acids. Hydrated bovine zinc insulin contains 4.0% water and would require the addition of 26.7 moles of H_2O to the empirical formula.

Molecular Weight 11530.4098

Evaluation A

$C_{1077}H_{1736}N_{304}O_{343}S_{12}$ (c) 69HUT/COL

Bovine chymotrypsinogen A, anhydrous

Heat Capacity 298.15 K, C_p = 30.90 cal·100g^{-1}·K^{-1}
129.3 J·100g^{-1}·K^{-1}

Temperature range 10–310 K

Entropy 298.15 K, S = 32.27 cal·100g^{-1}·K^{-1}
135.0 J·100g^{-1}·K^{-1}

Empirical formula and molecular weight calculated from compositional data on chymotrypsinogen in 69HUT/COL which shows a sequence of 245 amino acids.

Molecular Weight 24816.1124

Evaluation A

$C_{1077}H_{1736}N_{304}O_{343}S_{12}$ (c) 69HUT/COL

Bovine chymotrypsinogen A, hydrated

Heat Capacity 298.15 K, C_p = 38.34 cal·100g^{-1}·K^{-1}
160.4 J·100g^{-1}·K^{-1}

Temperature range 10–310 K

Entropy 298.15 K, S = 36.35 cal·100g^{-1}·K^{-1}
152.1 J·100g^{-1}·K^{-1}

Empirical formula and molecular weight calculated from compositional data on chymotrypsinogen in 69HUT/COL which shows a sequence of 245 amino acids.
Hydrated bovine chymotrypsinogen A contains 10.7% water and would require the addition of 165 moles of H_2O to the empirical formula.

Molecular Weight 24816.1124

Evaluation A

9. Compound Name—Formula Index

Benzothiazole	C_7H_5NS
Benzothiophene	C_8H_6S
Benzotrifluoride	$C_7H_5F_3$
Benzoylferrocene	$C_{17}H_{14}FeO$
Benzoylglycine	$C_9H_9NO_3$
Benzyl acetate	$C_9H_{10}O_2$
Benzyl alcohol	C_7H_8O
Benzylamine	C_7H_9N
Benzyl chloride	C_7H_7Cl
Benzyl ethanoate	$C_9H_{10}O_2$
Benzyl methacrylate	$C_{11}H_{12}O_2$
Bibenzyl	$C_{14}H_{14}$
1-Bicyclobutyl cyanide	C_5H_5N
cis-Bicyclo[5.3.0]decane	$C_{10}H_{18}$
cis-Bicyclo[4.1.0]heptane	C_7H_{12}
Bicyclohexyl	$C_{12}H_{22}$
1-Bicyclo[3.1.0]hexyl cyanide	C_7H_9N
cis-Bicyclo[6.1.0]nonane	C_9H_{16}
Bicyclo[2.2.2]octane	C_8H_{14}
cis-Bicyclo[3.3.0]octane	C_8H_{14}
trans-Bicyclo[3.3.0]octane	C_8H_{14}
cis-Bicyclo[4.2.0]octane	C_8H_{14}
Bicyclo[2.2.2]octene-2	C_8H_{12}
1-Bicyclo[2.1.0]pentyl cyanide	C_6H_7N
Biphenyl	$C_{12}H_{10}$
Bis(4-aminophenyl) ether	$C_{12}H_{12}N_2O$
Bis(4-aminophenyl)methane	$C_{13}H_{14}N_2$
2,2-Bis(bromomethyl)-1,3-dibromopropane	$C_5H_8Br_4$
2,2-Bis(chloromethyl)-1,3-dichloropropane	$C_5H_8Cl_4$
3,3-Bis-(chloromethyl)oxacyclobutane	$C_5H_8Cl_2O$
3,3-Bis-(chloromethyl)polyoxacyclobutane	$(C_5H_8Cl_2O)_n$
Bis(cyclohexylmethyl)cyclohexane	$C_{20}H_{36}$
Bis(cyclopentane)-2,2-dimethylbutane adduct	$C_{22}H_{48}$
1,2-Bis(difluoramino)propane	$C_3H_6F_4N_2$
1,1-Bis(dimethylcyclohexyl)ethane	$C_{18}H_{34}$
Bis(2-ethoxyethyl) ether	$C_8H_{18}O_3$
1,1-Bis(ethylcyclohexyl)ethane	$C_{18}H_{34}$
Bis(ethylcyclohexyl)methane	$C_{17}H_{32}$
Bis(2-ethylhexyl) phthalate	$C_{24}H_{38}O_4$
2,2-Bis(fluoromethyl)-1,3-difluoropropane	$C_5H_8F_4$
4,4'-Bis(n-heptyloxy)azoxybenzene	$C_{26}H_{38}N_2O_3$
2,2-Bis(hydroxymethyl)-1,3-dihydroxypropane	$C_5H_{12}O_4$
2,2-Bis(iodomethyl)-1,3-diiodopropane	$C_5H_8I_4$
Bis(4-isocyanatophenyl)methane	$C_{15}H_{10}N_2O_2$
Bis(4-(N-maleicimido)phenyl)methane	$C_{20}H_{12}N_2O_4$
N,N'-Bis(m-methoxyphenyl)terephthalamide	$C_{22}H_{20}N_2O_4$
N,N'-Bis(p-methoxyphenyl)terephthalamide	$C_{22}H_{20}N_2O_4$
1,2-Bis(methyldiallylammonium)ethane dibromide	$C_{16}H_{34}N_2$
Bis(4-nitrophenyl) ether	$C_{12}H_8N_2O_5$
Bis-tetraethylammonium tetrachloronickelate	$C_{16}H_{40}Cl_4N_2Ni$
Bis-(tetrafluoropropyl)carbonate	$C_7H_6F_8O_3$
Bis-tetramethylammonium tetrachlorozincate	$C_{16}H_{40}Cl_4N_2Zn$
1,4-Bis(triallylammonium)butene-2 dibromide	$C_{22}H_{42}Br_2N_2$
1,2-Bis(triallylammonium)ethane dibromide	$C_{20}H_{40}Br_2N_2$
1,3-Bis(triallylammonium)propane dibromide	$C_{21}H_{42}Br_2N_2$
1,4-Bis(triethylammonium)butane dibromide	$C_{16}H_{38}Br_2N_2$
1,10-Bis(triethylammonium)decane dibromide	$C_{22}H_{50}Br_2N_2$
1,2-Bis(triethylammonium)ethane dibromide	$C_{14}H_{34}Br_2N_2$
1,6-Bis(triethylammonium)hexane dibromide	$C_{18}H_{42}Br_2N_2$
1,8-Bis(triethylammonium)octane dibromide	$C_{20}H_{46}Br_2N_2$
1,5-Bis(triethylammonium)pentane dibromide	$C_{17}H_{40}Br_2N_2$
1,3-Bis(triethylammonium)propane dibromide	$C_{15}H_{36}Br_2N_2$
Bis(2,4,6-trimethylcyclohexyl)methane	$C_{19}H_{36}$
Bis(trinitrotoluene)-tetryl complex	$C_7H_{15}N_{11}O_{20}$
Bismuth triphenyl	$C_{18}H_{15}Bi$
Biuret	$C_2H_5N_3O_2$
Bromobenzene	C_6H_5Br
1-Bromobutane	C_4H_9Br
1-Bromo-2-chloroethane	C_2H_4BrCl
Bromoethane	C_2H_5Br
Bromoethene	C_2H_3Br
Bromoform	$CHBr_3$
1-Bromohexane	$C_6H_{13}Br$
Bromomethane	CH_3Br
1-Bromo-3-methylbutane	$C_5H_{11}Br$
1-Bromo-2-methylpropane	C_4H_9Br
2-Bromo-2-methylpropane	C_4H_9Br
2-Bromonaphthalene	$C_{10}H_7Br$
1-Bromopentane	$C_5H_{11}Br$
4-Bromophenol	C_6H_5BrO
1-Bromopropane	C_3H_7Br
2-Bromopropane	C_3H_7Br
Bromotrichloromethane	$CBrCl_3$
1,2-Butadiene	C_4H_6
1,3-Butadiene	C_4H_6
Butanal	C_4H_8O
n-Butane	C_4H_{10}
1,4-Butanediamide	$C_4H_8N_2O_2$
1,4-Butanedinitrile	$C_4H_4N_2$
1,4-Butanedioic acid	$C_4H_6O_4$
1,3-Butanediol	$C_4H_{10}O_2$
1,4-Butanediol	$C_4H_{10}O_2$
2,3-Butanediol	$C_4H_{10}O_2$
1,2,3,4-Butanetetrol	$C_4H_{10}O_4$
1-Butanethiol	$C_4H_{10}S$
2-Butanethiol	$C_4H_{10}S$
Butanoic acid	$C_4H_8O_2$
1-Butanol	$C_4H_{10}O$
2-Butanol	$C_4H_{10}O$
Butanone	C_4H_8O
Butanonitrile	C_4H_7N
Butanoyl chloride	C_4H_7ClO
1-Butene	C_4H_8
cis-2-Butene	C_4H_8
trans-2-Butene	C_4H_8
cis-2-Butenedioic acid	$C_4H_4O_4$
trans-2-Butenedioic acid	$C_4H_4O_4$
2-n-Butoxyethanol	$C_6H_{14}O_2$
1-n-Butoxy-2-methoxyethane	$C_7H_{16}O_2$
n-Butryaldehyde	C_4H_8O
N-n-Butylacetamide	$C_6H_{13}NO$
N-tert-Butylacetamide	$C_6H_{13}NO$
n-Butyl acetate	$C_6H_{12}O_2$
tert-Butyl acetate	$C_6H_{12}O_2$
n-Butyl alcohol	$C_4H_{10}O$
sec-Butyl alcohol	$C_4H_{10}O$
tert-Butyl alcohol	$C_4H_{10}O$
n-Butylamine	$C_4H_{11}N$
tert-Butylamine	$C_4H_{11}N$
n-Butylbenzene	$C_{10}H_{14}$
tert-Butylbenzene	$C_{10}H_{14}$
n-Butyl bromide	C_4H_9Br
tert-Butyl bromide	C_4H_9Br
Butylchloral	$C_4H_5Cl_3O$
tert-Butyl chloride	C_4H_9Cl
tert-Butyl cyanide	C_5H_9N
n-Butylcyclohexane	$C_{10}H_{20}$
tert-Butylcyclohexane	$C_{10}H_{20}$
n-Butylcyclopentane	C_9H_{18}
α-n-Butyldecalin	$C_{14}H_{26}$
α-sec-Butyldecalin	$C_{14}H_{26}$
tert-Butyldecalin	$C_{14}H_{26}$
N-n-Butylethanamide	$C_6H_{13}NO$
n-Butyl ethanoate	$C_6H_{12}O_2$
tert-Butylethylene	C_6H_{12}
tert-Butyl ethyl ether	$C_6H_{14}O$
n-Butyl mercaptan	$C_4H_{10}S$
sec-Butyl mercaptan	$C_4H_{10}S$
tert-Butyl mercaptan	$C_4H_{10}S$
Butyl methacrylate	$C_8H_{14}O_2$
tert-Butyl methyl ether	$C_5H_{12}O$
Butyl 2-methylpropenoate	$C_8H_{14}O_2$

Butyl methyl sulfide	$C_5H_{12}S$
1–Butyne	C_4H_6
2–Butyne	C_4H_6
n–Butyric acid	$C_4H_8O_2$
γ–Butyrolactone	$C_4H_6O_2$
Butyryl chloride	C_4H_7ClO
C	
Cadmium dimethyl	C_2H_6Cd
Caffeine	$C_8H_{10}N_4O_2$
Camphor	$C_{10}H_{16}O$
Camphor(D)	$C_{10}H_{16}O$
Camphor(DL)	$C_{10}H_{16}O$
n–Caproic acid	$C_6H_{12}O_2$
ϵ–Caprolactam	$C_6H_{11}NO$
Capryl alcohol	$C_8H_{18}O$
Caprylene	C_8H_{16}
Caprylic acid	$C_8H_{16}O_2$
Carbamide	CH_4N_2O
Carbazole	$C_{12}H_9N$
Carbazole–1,3,5–trinitrobenzene adduct	$C_{18}H_{22}N_4O_6$
Carbon, baked	C
Carbon, diamond	C
Carbon, glassy	C
Carbon, graphite	C
Carbon, irradiated graphite	C
Carbon, natural graphite	C
Carbon, pyrolytic graphite	C
Carbon diselenide	CSe_2
Carbon disulfide	CS_2
Carbon tetrabromide	CBr_4
Carbon tetrachloride	CCl_4
Carbon tetrafluoride	CF_4
Carbonyl chloride	CCl_2O
Carbonyl sulfide	COS
Carvoxime(DL)	$C_{10}H_{15}NO$
Carvoxime(L)	$C_{10}H_{15}NO$
Catechol	$C_6H_6O_2$
Cesium butyrate	$C_4H_7CsO_2$
Cesium propionate	$C_3H_5CsO_2$
Cetane	$C_{16}H_{34}$
n–Cetyl alcohol	$C_{16}H_{34}O$
Chloral	C_2HCl_3O
Chloranil	$C_6Cl_4O_2$
p–Chloranil potassium	$C_6Cl_4KO_2$
Chloroacetic acid	$C_2H_3ClO_2$
m–Chloroaniline	C_6H_6ClN
p–Chloroaniline	C_6H_6ClN
Chlorobenzene	C_6H_5Cl
2–Chlorobenzoic acid	$C_7H_5ClO_2$
3–Chlorobenzoic acid	$C_7H_5ClO_2$
4–Chlorobenzoic acid	$C_7H_5ClO_2$
2–Chlorobiphenyl	$C_{12}H_9Cl$
4–Chlorobiphenyl	$C_{12}H_9Cl$
cis–3–Chloro–2–butenoic acid	$C_4H_5ClO_2$
trans–3–Chloro–2–butenoic acid	$C_4H_5ClO_2$
β–Chlorocrotonic acid	$C_4H_5ClO_2$
Chlorodifluoromethane	$CHClF_2$
4–Chloro–1,3–dioxolan–2–one	$C_4H_5ClO_3$
Chloroethane	C_2H_5Cl
Chloroform	$CHCl_3$
β–Chloroisocrotonic acid	$C_4H_5ClO_2$
Chloromethane	CH_3Cl
1–Chloro–3–methylbutane	$C_5H_{11}Cl$
4–Chloromethyl–1,3–dioxolan–2–one	$C_5H_7ClO_3$
1–Chloro–2–methylpropane	C_4H_9Cl
2–Chloro–2–methylpropane	C_4H_9Cl
2–Chloro–2–nitropropane	$C_3H_6ClNO_2$
1–Chloroöctadecane	$C_{18}H_{37}Cl$
Chloropentafluorobenzene	C_6ClF_5
3–Chloro–1,1,1,3,3–pentafluoropropane	$C_3H_2ClF_5$
Chloropentamethylbenzene	$C_{11}H_{15}Cl$
o–Chlorophenol	C_6H_5ClO
m–Chlorophenylisocyanate	C_7H_4ClNO
p–Chlorophenylisocyanate	C_7H_4ClNO
1–Chloropropane	C_3H_7Cl
3–Chloropropene–1	C_3H_5Cl
α–Chlorotoluene	C_7H_7Cl
Chlorotrifluoroethene	C_2ClF_3
Chlorotrifluoroethylene	C_2ClF_3
Cholesteryl myristate	$C_{41}H_{72}O_2$
Chromocene	$C_{10}H_{10}Cr$
Cinnamic acid	$C_9H_8O_2$
Citric acid monohydrate	$C_6H_8O_7 \cdot H_2O$
Citrulline(DL)	$C_6H_{13}N_3O_3$
Cobaltocene	$C_{10}H_{10}Co$
Copper acetylacetonate	$C_{10}H_{14}CuO_4$
Coriandrol	$C_{10}H_{18}O$
Coronene	$C_{24}H_{12}$
Creatine	$C_4H_9N_3O_2$
Creatine hydrate	$C_4H_9N_3O_2 \cdot H_2O$
Creatinine	$C_4H_7N_3O$
m–Cresol	C_7H_8O
o–Cresol	C_7H_8O
p–Cresol	C_7H_8O
Cumene	C_9H_{12}
1–Cyanobicyclobutane	C_5H_5N
2–Cyanobicyclo[2,2,1]heptane(endo)	$C_8H_{11}N$
2–Cyanobicyclo[2,2,1]heptane(exo)	$C_8H_{11}N$
1–Cyanobicyclo[3.1.0]hexane	C_7H_9N
1–Cyanobicyclo[2.1.0]pentane	C_6H_7N
Cyanocyclobutane	C_5H_7N
Cyanocyclohexane	$C_7H_{11}N$
Cyanocyclopentane	C_6H_9N
Cyanocyclopropane	C_4H_5N
Cyanoethane	C_3H_5N
Cyanogen	C_2N_2
Cyanomethane	C_2H_3N
1–Cyano–3–methylenecyclobutane	C_6H_7N
2–Cyano–2–methylpropane	C_5H_9N
2–Cyanopropane	C_4H_7N
Cyclobutane	C_4H_8
Cyclobutyl cyanide	C_5H_7N
Cycloheptane	C_7H_{14}
Cycloheptane–thiourea adduct	$C_{21.9}H_{45.8}N_2S$
Cycloheptanol	$C_7H_{14}O$
Cycloheptatriene	C_7H_8
1,3–Cyclohexadiene	C_6H_8
1,4–Cyclohexadiene	C_6H_8
Cyclohexane	C_6H_{12}
Cyclohexane–d_{12}	C_6D_{12}
Cyclohexanethiol	$C_6H_{12}S$
Cyclohexane–thiourea adduct	$C_{19.6}H_{41.2}N_2S$
Cyclohexanol	$C_6H_{12}O$
Cyclohexanone	$C_6H_{10}O$
Cyclohexene	C_6H_{10}
Cyclohexyl alcohol	$C_6H_{12}O$
Cyclohexyl cyanide	$C_7H_{11}N$
11–Cyclohexyleicosane	$C_{27}H_{54}$
Cyclohexyl(ethylcyclohexyl)methane	$C_{15}H_{28}$
Cyclohexyl(2–ethylcyclohexyl)methane	$C_{15}H_{28}$
1–Cyclohexyl–1–isopropylcyclohexylethane	$C_{17}H_{32}$
Cyclohexyl(isopropylcyclohexyl)methane	$C_{16}H_{30}$
Cyclohexyl mercaptan	$C_6H_{12}S$
1–Cyclohexyl–3–methylhydroindan	$C_{16}H_{26}$
1–Cyclohexyl–1–phenyldodecane	$C_{24}H_{40}$
1–Cyclohexyl–1,3,3–trimethylhydroindan	$C_{18}H_{32}$
Cycloocta–1,5–diene	C_8H_{12}
Cyclooctane	C_8H_{16}
Cyclooctane–thiourea adduct	$C_{24.8}H_{51.6}N_2S$
Cyclopentane	C_5H_{10}
Cyclopentanethiol	$C_5H_{10}S$
Cyclopentanol	$C_5H_{10}O$
Cyclopentene	C_5H_8

Cyclopentylamine ... $C_5H_{11}N$
Cyclopentylbicyclohexyl ... $C_{17}H_{30}$
Cyclopentyl cyanide ... C_6H_9N
Cyclopentyl mercaptan ... $C_5H_{10}S$
Cyclopentyl methyl sulfide ... $C_6H_{12}S$
Cyclopentyl-1-thiaethane ... $C_6H_{12}S$
Cyclopropane ... C_3H_6
Cyclopropyl amine ... C_3H_7N
Cyclopropyl cyanide ... C_4H_5N
Cycloöctane ... C_8H_{16}
Cycloöctatetraene ... C_8H_8
p-Cymene ... $C_{10}H_{14}$
Cysteine(L) ... $C_3H_7NO_2S$
Cystine(L) ... $C_6H_{12}N_2O_4S_2$
Cytosine ... $C_4H_5N_3O$

D

Decafluorobiphenyl ... $C_{12}F_{10}$
Decahydroonaphthalene ... $C_{10}H_{18}$
cis-Decahydronaphthalene ... $C_{10}H_{18}$
trans-Decahydronaphthalene ... $C_{10}H_{18}$
Decalin ... $C_{10}H_{18}$
cis-Decalin ... $C_{10}H_{18}$
trans-Decalin ... $C_{10}H_{18}$
n-Decane ... $C_{10}H_{22}$
1-Decanethiol ... $C_{10}H_{22}S$
Decanoic acid ... $C_{10}H_{20}O_2$
1-Decanol ... $C_{10}H_{22}O$
1-Decene ... $C_{10}H_{20}$
1-Decene-urea adduct ... $C_{2.3}H_{6.7}N_2O$
n-Decyl alcohol ... $C_{10}H_{22}O$
n-Decylcyclohexane ... $C_{16}H_{32}$
n-Decylcyclopentane ... $C_{15}H_{30}$
11-n-Decylheneicosane ... $C_{31}H_{64}$
n-Decyl mercaptan ... $C_{10}H_{22}S$
Dextrose ... $C_6H_{12}O_6$
1,2-Diaceto-3-stearin ... $C_{25}H_{46}O_6$
1,1′-Diacetylferrocene ... $C_{14}H_{14}FeO_2$
Diallyl ... C_6H_{10}
Diamantane ... $C_{14}H_{20}$
2,4-Diaminoazobenzene ... $C_{12}H_{12}N_4$
4,4′-Diaminodiphenyl ether ... $C_{12}H_{12}N_2O$
4,4′-Diaminodiphenylmethane ... $C_{13}H_{14}N_2$
1,2-Diaminoethane ... $C_2H_8N_2$
1,2-Diamino-2-methylpropane ... $C_4H_{12}N_2$
1,2-Diaminopropane ... $C_3H_{10}N_2$
Diamond ... C
4,4′-Dianilino-3,3′-diaminodiphenyl oxide ... $C_{24}H_{22}N_4O$
Dianisyldiethynylsilane ... $C_{20}H_{16}O_6Si$
1,4-Diazabicyclo[2,2,2]octane ... $C_6H_{12}N_2$
Dibenzoylethane ... $C_{16}H_{14}O_2$
1,2-Dibenzoylethane ... $C_{16}H_{14}O_2$
Dibenzoylethylene ... $C_{16}H_{12}O_2$
1,1′-Dibenzoylferrocene ... $C_{24}H_{18}FeO_2$
Dibenzyl ... $C_{14}H_{14}$
Dibromoacetic acid ... $C_2H_2Br_2O_2$
1,4-Dibromobenzene ... $C_6H_4Br_2$
1,2-Dibromoethane-d_1 ... $C_2H_3DBr_2$
1,2-Dibromoethane-1,1-d_2 ... $C_2H_2D_2Br_2$
1,2-Dibromoethane-1,2-d_2 ... $C_2H_2D_2Br_2$
1,2-Dibromoethane-d_3 ... $C_2HD_3Br_2$
1,2-Dibromoethane-d_4 ... $C_2D_4Br_2$
1,2-Dibromoethane ... $C_2H_4Br_2$
Dibromoethanoic acid ... $C_2H_2Br_2O_2$
Dibromomethane ... CH_2Br_2
2,4-Dibromophenol ... $C_6H_4Br_2O$
1,2-Dibromopropane ... $C_3H_6Br_2$
1,3-Dibromopropane ... $C_3H_6Br_2$
1,2-Dibromotetrafluoroethane ... $C_2Br_2F_4$
Di-n-butyl ketone ... $C_9H_{18}O$
Dibutyl phthalate ... $C_{17}H_{22}O_4$
Di-n-butyl sebacate ... $C_{18}H_{34}O_4$
Dibutyl sulfide ... $C_8H_{18}S$
Di-tert-butyldiazene N-oxide ... $C_8H_{18}N_2O$
2,6-Di-tert-butyl-4-(3,5-di-tert-butyl-4-oxocyclohexa-2,5-dienylidene methyl)phenol ... $C_{29}H_{42}O_2$
2,6-Di-tert-butyl-4-(3,5-di-tert-butyl-4-oxocyclohexa-2,5-dienylidene methyl)phenoxyl ... $C_{29}H_{41}O_2$
Dichloroacetic acid ... $C_2H_2Cl_2O_2$
1,4-Dichlorobenzene ... $C_6H_4Cl_2$
Dichlorodifluoromethane ... CCl_2F_2
4,5-Dichloro-1,3-dioxolan-2-one ... $C_4H_4Cl_2O_3$
1,1-Dichloroethane ... $C_2H_4Cl_2$
1,2-Dichloroethane ... $C_2H_4Cl_2$
Dichloroethanoic acid ... $C_2H_2Cl_2O_2$
1,1-Dichloroethene ... $C_2H_2Cl_2$
1,2-Dichloroethene ... $C_2H_2Cl_2$
cis-1,2-Dichloroethene ... $C_2H_2Cl_2$
trans-1,2-Dichloroethene ... $C_2H_2Cl_2$
1,2-Dichloroethylene ... $C_2H_2Cl_2$
Dichlorofluoromethane ... $CHCl_2F$
Dichloromethane ... CH_2Cl_2
1,2-Dichloropropane ... $C_3H_6Cl_2$
2,2-Dichloropropane ... $C_3H_6Cl_2$
1,2-Dichloro-1,1,2,2-tetrafluoroethane ... $C_2Cl_2F_4$
Dicyandiamide ... $C_2H_4N_4$
1,3-Dicyanopropane ... $C_5H_6N_2$
1,3-Dicyclohexylbutane ... $C_{16}H_{30}$
1,1-Dicyclohexyldodecane ... $C_{24}H_{46}$
2,11-Dicyclohexyldodecane ... $C_{24}H_{46}$
1,1-Dicyclohexylethane ... $C_{14}H_{26}$
1,2-Dicyclohexylethane ... $C_{14}H_{26}$
1,1-Dicyclohexylheptane ... $C_{19}H_{36}$
Dicyclohexylmethane ... $C_{13}H_{24}$
1,2-Dicyclohexylpropane ... $C_{15}H_{28}$
1,3-Dicyclopentylcyclopentane ... $C_{15}H_{26}$
Di-n-decyl sebacate ... $C_{30}H_{58}O_4$
Di-n-dodecyl sebacate ... $C_{34}H_{64}O_4$
1,1-Diethoxyethane ... $C_6H_{14}O_2$
1,2-Diethoxyethane ... $C_6H_{14}O_2$
Diethylamine ... $C_4H_{11}N$
N,N-Diethylaniline ... $C_{10}H_{15}N$
Diethyl carbonate ... $C_5H_{10}O_3$
Diethylcyclohexane ... $C_{10}H_{20}$
1,4-Diethylcyclohexane ... $C_{10}H_{20}$
Diethyl disulfide ... $C_4H_{10}S_2$
Diethylene glycol ... $C_4H_{10}O_3$
Diethyleneimide oxide ... C_4H_9NO
Diethyl ethanedioate ... $C_6H_{10}O_4$
Diethyl ether ... $C_4H_{10}O$
Diethyl ketone ... $C_5H_{10}O$
Diethyl malonate ... $C_7H_{12}O_4$
Diethyl oxalate ... $C_6H_{10}O_4$
3,3-Diethylpentane ... C_9H_{20}
Diethylperhydropyrene ... $C_{20}H_{34}$
Diethyl o-phthalate ... $C_{12}H_{14}O_4$
Diethyl p-phthalate ... $C_{12}H_{14}O_4$
Diethyl succinate ... $C_8H_{14}O_4$
Diethyl sulfide ... $C_4H_{10}S$
Diethyl terephthalate ... $C_{12}H_{14}O_4$
1,2-Difluorobenzene ... $C_6H_4F_2$
1,3-Difluorobenzene ... $C_6H_4F_2$
1,1-Difluoro-1-chloroethane ... $C_2H_3ClF_2$
Diglyme ... $C_6H_{14}O_3$
Di-n-hexadecyl sebacate ... $C_{42}H_{82}O_4$
N,N′-Di-n-hexyladipamide ... $C_{18}H_{36}N_2O_2$
N,N′-Di-n-hexylsebacamide ... $C_{22}H_{44}N_2O_2$
Dihexyl bis(hexamethylenesebacate) ... $C_{54}H_{98}O_{12}$
Dihexyl hexamethylenesebacate ... $C_{38}H_{70}O_8$
Di-n-hexyl sebacate ... $C_{22}H_{42}O_4$
4,4′-Dihydrazodiphenyl oxide ... $C_{12}H_{14}N_4O$
9,10-Dihydrophenanthrene ... $C_{14}H_{12}$
Dihydrosulfide carbonsulfide ... CH_2S_3

1,2–Dihydroxybenzene	$C_6H_6O_2$
1,3–Dihydroxybenzene	$C_6H_6O_2$
1,4–Dihydroxybenzene	$C_6H_6O_2$
1,3–Dihydroxybutane	$C_4H_{10}O_2$
1,4–Dihydroxybutane	$C_4H_{10}O_2$
2,3–Dihydroxybutane	$C_4H_{10}O_2$
1,8–Dihydroxy–3,6–dioxaoctane	$C_6H_{14}O_4$
1,2–Dihydroxyethane	$C_2H_6O_2$
1,2–Dihydroxyethane–d$_2$	$C_2H_4D_2O_2$
1,5–Dihydroxy–3–oxapentane	$C_4H_{10}O_3$
1,2–Dihydroxypropane	$C_3H_8O_2$
1,14–Dihydroxy–3,6,9,12–tetraoxatetradecane	$C_{10}H_{22}O_6$
2,5–Dihydroxytoluene	$C_7H_8O_2$
1,11–Dihydroxy–3,6,9–trioxaundecane	$C_8H_{18}O_5$
1,4–Diiodobenzene	$C_6H_4I_2$
Diiodomethane	CH_2I_2
Diisobutylene	C_8H_{16}
1,4–Diisocyanatobenzene	$C_8H_4N_2O_2$
1,6–Diisocyanatohexane	$C_8H_{12}N_2O_2$
1,5–Diisocyanatonaphthalene	$C_{12}H_6N_2O_2$
4,4'–(Diisocynanatophenyl)methane	$C_{15}H_{10}N_2O_2$
p,p'–Diisopropylbiphenyl	$C_{18}H_{22}$
Diisopropyl ketone	$C_7H_{14}O$
Diisopropyl sulfide	$C_6H_{14}S$
2,5–Diketopiperazine	$C_4H_6N_2O_2$
Dimethanolurea	$C_3H_8N_2O_3$
4,4'–Dimethoxyazoxybenzene	$C_{14}H_{14}N_2O_3$
Dimethoxydecalin	$C_{12}H_{22}O_2$
1,2–Dimethoxyethane	$C_4H_{10}O_2$
Dimethoxymethane	$C_3H_8O_2$
Di(p–methoxyphenyl)–trans–cyclohexane–1,4–dicarboxylate	$C_{22}H_{24}O_6$
Dimethylamine	C_2H_7N
β–Dimethylaminopropionitrile	$C_5H_{10}N_2$
N,N–Dimethylaniline	$C_8H_{11}N$
1,3–Dimethylbenzene	C_8H_{10}
1,4–Dimethylbenzene	C_8H_{10}
2,2–Dimethylbutane	C_6H_{14}
2,2–Dimethylbutane–thiourea adduct	$C_{18.4}H_{44.6}N_2S$
2,3–Dimethylbutane	C_6H_{14}
3,3–Dimethyl–2–butanone	$C_6H_{12}O$
3,3–Dimethyl–1–butene	C_6H_{12}
2,3–Dimethyl–2–butene	C_6H_{12}
2,4–Dimethyl–2–butene	C_6H_{12}
Dimethyl cadmium	C_2H_6Cd
1,1–Dimethylcyclohexane	C_8H_{16}
1–cis–2–Dimethylcyclohexane	C_8H_{16}
1–cis–3–Dimethylcyclohexane	C_8H_{16}
1–cis–4–Dimethylcyclohexane	C_8H_{16}
1–trans–2–Dimethylcyclohexane	C_8H_{16}
1–trans–3–Dimethylcyclohexane	C_8H_{16}
1–trans–4–Dimethylcyclohexane	C_8H_{16}
1,1–Dimethylcyclopentane	C_7H_{14}
1,2–Dimethylcyclopentane	C_7H_{14}
1–cis–2–Dimethylcyclopentane	C_7H_{14}
1–trans–3–Dimethylcyclopentane	C_7H_{14}
Dimethyldecalin	$C_{12}H_{22}$
Dimethyl disulfide	$C_2H_6S_2$
Dimethyl ether	C_2H_6O
N–(1,1–Dimethylethyl)ethanamide	$C_6H_{13}NO$
1,1,–Dimethylethyl ethanoate	$C_6H_{12}O_2$
N,N–Dimethylformamide	C_3H_7NO
Dimethyl fumarate	$C_6H_8O_4$
2,5–Dimethylhexane	C_8H_{18}
3,3–Dimethylhexane	C_8H_{18}
N,N–Dimethylhydrazine	$C_2H_8N_2$
Dimethyl ketone	C_3H_6O
Dimethyl maleate	$C_6H_8O_4$
Dimethylmalonitrile	$C_5H_6N_2$
N,N–Dimethylmethanamide	C_3H_7NO
1,8–Dimethylnaphthalene	$C_{12}H_{12}$
2,6–Dimethylnaphthalene	$C_{12}H_{12}$
2,7–Dimethylnaphthalene	$C_{12}H_{12}$
2,6–Dimethylocta–2,7–dien–6–ol	$C_{10}H_{18}O$
2,7–Dimethyloctane	$C_{10}H_{22}$
3,3–Dimethyl–2–oxabutane	$C_5H_{12}O$
4,4–Dimethyl–3–oxahexane	$C_7H_{16}O$
2,4–Dimethyl–3–oxapentane	$C_6H_{14}O$
3,3–Dimethyl–2–oxapentane	$C_6H_{14}O$
4,4–Dimethyl–3–oxapentane	$C_6H_{14}O$
2,2–Dimethylpentane	C_7H_{16}
2,3–Dimethylpentane	C_7H_{16}
2,4–Dimethylpentane	C_7H_{16}
3,3–Dimethylpentane	C_7H_{16}
2,4–Dimethyl–3–pentanone	$C_7H_{14}O$
Dimethyl o–phthalate	$C_{10}H_{10}O_4$
Dimethyl p–phthalate	$C_{10}H_{10}O_4$
2,N–Dimethylpropanamide	$C_5H_{11}NO$
2,2–Dimethylpropane	C_5H_{12}
N,N–Dimethyl–1,3–propanediamine	$C_5H_{14}N_2$
2,2–Dimethyl–1–propanol	$C_5H_{12}O$
2,2–Dimethylpropanoic acid	$C_5H_{10}O_2$
2,2–Dimethylpropionitrile	C_5H_9N
Dimethylsulfide	C_2H_6S
Dimethyl sulfone	$C_2H_6O_2S$
Dimethyl sulfoxide	C_2H_6OS
Dimethyl terephthalate	$C_{10}H_{10}O_4$
Dimethyltetraphenylcyclotrisiloxane	$C_{26}H_{26}O_3Si_3$
3,3–Dimethyl–2–thiabutane	$C_5H_{12}S$
2,4–Dimethyl–3–thiapentane	$C_6H_{14}S$
2,5–Dimethylthiophene	C_6H_8S
1,2'–Dinaphthylmethane	$C_{21}H_{16}$
1,2–Dinitrobenzene	$C_6H_4N_2O_4$
1,3–Dinitrobenzene	$C_6H_4N_2O_4$
1,4–Dinitrobenzene	$C_6H_4N_2O_4$
4,4'–Dinitrodiphenyl ether	$C_{12}H_8N_2O_5$
1,5–Dinitronaphthalene	$C_{10}H_6N_2O_4$
1,8–Dinitronaphthalene	$C_{10}H_6N_2O_4$
2,2–Dinitropropane	$C_3H_6N_2O_4$
2,4–Dinitrotoluene	$C_7H_6N_2O_4$
Di–n–octadecyl sebacate	$C_{46}H_{90}O_4$
Di–n–octyl sebacate	$C_{26}H_{504}$
4,7–Dioxadecane	$C_8H_{18}O_2$
2,5–Dioxaheptane	$C_5H_{12}O_2$
2,5–Dioxahexane	$C_4H_{10}O_2$
1,3–Dioxane	$C_4H_8O_2$
1,4–Dioxane	$C_4H_8O_2$
1,4–Dioxane–2,5–dione	$C_4H_4O_4$
2,5–Dioxanonane	$C_7H_{16}O_2$
2,5–Dioxaoctane	$C_6H_{14}O_2$
3,6–Dioxaoctane	$C_6H_{14}O_2$
2,4–Dioxapentane	$C_3H_8O_2$
1,3–Dioxepane	$C_5H_{10}O_2$
1,3–Dioxolane	$C_3H_6O_2$
Diphenyl	$C_{12}H_{10}$
Diphenylacetic acid	$C_{14}H_{12}O_2$
Diphenylacetylene	$C_{14}H_{10}$
Diphenylcarbinol	$C_{13}H_{12}O$
Diphenyl carbonate	$C_{13}H_{10}O_3$
Diphenylchloromethane	$C_{13}H_{11}Cl$
Diphenyl diketone	$C_{14}H_{10}O_2$
1,1–Diphenyldodecane	$C_{24}H_{34}$
Diphenylene–2,2'disulfide–S–oxide	$C_{12}H_8OS_2$
Diphenylenemethane	$C_{13}H_{10}$
1,1–Diphenylethane	$C_{14}H_{14}$
1,2–Diphenylethane	$C_{14}H_{14}$
Diphenyl ether	$C_{12}H_{10}O$
1,1–Diphenylethylene	$C_{14}H_{12}$
1,2–Diphenylethylene	$C_{14}H_{12}$
Diphenylmercury	$C_{12}H_{10}Hg$
Diphenylmethane	$C_{13}H_{12}$
4,4'–Diphenylmethanediisocyanate	$C_{15}H_{10}N_2O_2$

Diphenyl oxide	$C_{12}H_{10}O$
Diphenyl sulfide	$C_{12}H_{10}S$
Diphenyl sulfone	$C_{12}H_{10}O_2S$
Diphenyl sulfoxide	$C_{12}H_{10}OS$
Diphenyltetramethylcyclotrisiloxane	$C_{16}H_{22}O_3Si_3$
1,2-Di-n-propoxyethane	$C_8H_{18}O_2$
N,N'-Di-n-propyladipamide	$C_{12}H_{24}N_2O_2$
Dipropyl disulfide	$C_6H_{14}S_2$
Dipropyl sulfide	$C_6H_{14}S$
Di-n-tetradecyl sebacate	$C_{38}H_{72}O_4$
4,5-Dithia-1,8-octanedioic acid	$C_6H_{10}O_4S_2$
2,3-Dithiabutane	$C_2H_6S_2$
3,4-Dithiahexane	$C_4H_{10}S_2$
4,5-Dithiaoctane	$C_6H_{14}S_2$
β,β'-Dithiodilactic acid	$C_6H_{10}O_4S_2$
Di(p-tolyl)mercury	$C_{14}H_{14}Hg$
Docosafluorobicyclohexyl	$C_{12}F_{22}$
n-Docosane	$C_{22}H_{46}$
n-Dodecane	$C_{12}H_{26}$
Dodecanoic acid	$C_{12}H_{24}O_2$
1-Dodecanol	$C_{12}H_{26}O$
1-Dodecene	$C_{12}H_{24}$
1-Dodecene-urea adduct	$C_{2.2}H_{6.5}N_2O$
n-Dodecyl alcohol	$C_{12}H_{26}O$
n-Dodecylcyclohexane	$C_{18}H_{36}$
Dodecyl methyl ketone	$C_{14}H_{28}O$
n-Dotetracontane	$C_{42}H_{86}$
n-Dotriacontane	$C_{32}H_{66}$
Dulcitol	$C_6H_{14}O_6$
Durene	$C_{10}H_{14}$

E

n-Eicosane	$C_{20}H_{42}$
Eicosanoic acid	$C_{20}H_{40}O_2$
1-Eicosene-urea adduct	$C_{2.4}H_{6.8}N_2O$
Enanthal	$C_7H_{14}O$
Erythritol	$C_4H_{10}O_4$
Ethanal	C_2H_4O
Ethanamide	C_2H_5NO
Ethane	C_2H_6
Ethanedioic acid	$C_2H_2O_4$
1,2-Ethanediol	$C_2H_6O_2$
1,2-Ethanediol-d_2	$C_2H_4D_2O_2$
Ethanethiol	C_2H_6S
Ethanoic acid	$C_2H_4O_2$
Ethanoic anhydride	$C_4H_6O_3$
Ethanol	C_2H_6O
Ethanol-d_1	C_2H_5DO
Ethenyl ethanoate	$C_5H_8O_2$
2-Ethoxyethanol	$C_4H_{10}O_2$
1-Ethoxy-2-methoxyethane	$C_5H_{12}O_2$
N-Ethylacetamide	C_4H_9NO
Ethyl acetate	$C_4H_8O_2$
Ethyl acetoacetate	$C_6H_{10}O_3$
Ethyl alcohol	C_2H_6O
Ethyl alcohol-d_1	C_2H_5DO
Ethyl azoxybenzenedicarboxylate	$C_{18}H_{18}N_2O_5$
Ethylbenzene	C_8H_{10}
Ethyl benzoate	$C_9H_{10}O_2$
2-Ethylbicyclohexyl	$C_{14}H_{26}$
2-Ethylbicyclohexylmethane	$C_{15}H_{28}$
Ethyl bromide	C_2H_5Br
Ethyl butanoate	$C_6H_{12}O_2$
Ethyl butyrate	$C_6H_{12}O_2$
Ethyl chloride	C_2H_5Cl
Ethyl 2-chloropropanoate	$C_5H_9ClO_2$
Ethyl α-chloropropionate	$C_5H_9ClO_2$
trans-Ethyl cinnamate	$C_{11}H_{12}O_2$
Ethyl cyanide	C_3H_5N
Ethylcyclohexane	C_8H_{16}
Ethyl cyclohexanecarboxylate	$C_9H_{16}O_2$
Ethylcyclopentane	C_7H_{14}
1-Ethylcyclopentene	C_7H_{12}
Ethyl cyclopropanecarboxylate	$C_6H_{10}O_2$
Ethyldecalin	$C_{12}H_{22}$
α-Ethyldecalin	$C_{12}H_{22}$
β-Ethyldecalin	$C_{12}H_{22}$
Ethyl dichloroacetate	$C_4H_6Cl_2O_2$
Ethylene bromide	$C_2H_4Br_2$
Ethyl 2,3-dichloropropanoate	$C_5H_8Cl_2O_2$
Ethyl α,β-dichloropropionate	$C_5H_8Cl_2O_2$
Ethylene carbonate	$C_3H_4O_3$
Ethylenediamine	$C_2H_8N_2$
Ethylene dibromide	$C_2H_4Br_2$
Ethylene dichloride	$C_2H_4Cl_2$
Ethylenedinitramine	$C_2H_6N_4O_4$
Ethylene glycol	$C_2H_6O_2$
Ethylene glycol-d_2	$C_2H_4D_2O_2$
Ethylene oxide	C_2H_4O
N-Ethylethanamide	C_4H_9NO
Ethyl ethanoate	$C_4H_8O_2$
Ethyl formate	$C_3H_6O_2$
9-(2'-Ethylhexyl)perhydrofluorene	$C_{21}H_{38}$
Ethyl hydrocinnamate	$C_{11}H_{14}O_2$
Ethylhydroindan	$C_{11}H_{20}$
Ethylidene chloride	$C_2H_4Cl_2$
Ethylidenecyclohexane	C_8H_{14}
Ethylidenecyclopentane	C_7H_{12}
Ethyl iodide	C_2H_5I
Ethyl isothiocyanate	C_3H_5NS
Ethyl mercaptan	C_2H_6S
Ethyl methanoate	$C_3H_6O_2$
Ethyl methyl sulfide	C_3H_8S
Ethyl nitrate	$C_2H_5NO_3$
3-Ethylpentane	C_7H_{16}
2-Ethylperhydrophenanthrene	$C_{16}H_{28}$
3-Ethylperhydropyrene	$C_{18}H_{30}$
4-Ethylphenol	$C_8H_{10}O$
Ethyl phenylcarbamate	$C_9H_{11}NO_2$
Ethyl n-propyl ether	$C_5H_{12}O$
Ethyl propyl ketone	$C_6H_{12}O$
Ethyltetryl	$C_8H_7N_5O_8$
Ethyl trichloroacetate	$C_4H_5Cl_3O_2$

F

Ferrocene	$C_{10}H_{10}Fe$
Fluoranthene	$C_{16}H_{10}$
Fluorene	$C_{13}H_{10}$
Fluorobenzene	C_6H_5F
Fluoroform	CHF_3
4-Fluorotoluene	C_7H_7F
Fluorotrichloromethane	CCl_3F
Formaldehyde,dimethylacetal	$C_3H_8O_2$
Formamide	CH_3NO
Formic acid	CH_2O_2
Freon 11	CCl_3F
Freon 12	$CHClF_2$
Freon 14	CF_4
Freon 21	$CHCl_2F$
Freon 22	$CHClF_2$
Freon 23	CHF_3
Freon 113	$C_2Cl_3F_3$
Freon 114	$C_2Cl_2F_4$
Freon 143	$C_2H_3F_3$
Fructose	$C_6H_{12}O_6$
Furan	C_4H_4O
Furfural	$C_5H_4O_2$
α-Furfural	$C_5H_4O_2$
Furfuraldehyde	$C_5H_4O_2$
α-Furfuryl alcohol	$C_5H_6O_2$
Fumaric acid	$C_4H_4O_4$

G

Galactitol	$C_6H_{14}O_6$
Galactose	$C_6H_{12}O_6$

α-Galactose(D)	$C_6H_{12}O_6$
Gallium triphenyl	$C_6H_{15}Ga$
Galvinoxyl radical	$C_{29}H_{41}O_2$
Germanium tetraethyl	$C_8H_{20}Ge$
Glassy carbon	C
Glucose	$C_6H_{12}O_6$
Glucose(D)	$C_6H_{12}O_6$
α-Glucose(D)	$C_6H_{12}O_6$
α-Glucose pentaacetate (D)	$C_{16}H_{22}O_{11}$
β-Glucose pentaacetate (D)	$C_{16}H_{22}O_{11}$
Glutamic acid(D)	$C_5H_9NO_4$
Glutamic acid(L)	$C_5H_9NO_4$
Glutamic acid hydrochloride	$C_5H_{10}ClNO_4$
Glutamine(L)	$C_5H_{10}N_2O_3$
Glutaronitrile	$C_5H_6N_2$
Glycerol	$C_3H_8O_3$
Glycerol-d$_3$	$C_3H_5D_3O_3$
Glyceryl triacetate	$C_9H_{14}O_2$
Glyceryl tributyrate	$C_{15}H_{26}O_6$
Glyceryl tricaproate	$C_{21}H_{38}O_6$
Glyceryl tridecanoate	$C_{33}H_{62}O_6$
Glyceryl trilaurate	$C_{39}H_{74}O_6$
Glyceryl trimyristate	$C_{45}H_{86}O_6$
Glyceryl trioctanoate	$C_{27}H_{50}O_6$
Glyceryl tripalmitate	$C_{51}H_{98}O_6$
Glyceryl tristearate	$C_{57}H_{110}O_6$
Glycine	$C_2H_5NO_2$
Glycol	$C_2H_6O_2$
Glycol-d$_2$	$C_2H_4D_2O_2$
Glycolide	$C_4H_4O_4$
Glycylglycine	$C_4H_8N_2O_3$
Grain alcohol	C_2H_6O
Graphite	C
Graphite, Acheson	C
Graphite, Acheson, irradiated	C
Graphite, natural Taiguinski	C
Graphite, pyrolytic	C
Guanidine carbonate	$C_2H_7N_3O_3$
Guanine	$C_5H_5N_5O$

H

Heptacosane	$C_{27}H_{56}$
n-Heptadecane	$C_{17}H_{36}$
Heptadecanoic acid	$C_{17}H_{34}O_2$
n-Heptaldehyde	$C_7H_{14}O$
Heptanal	$C_7H_{14}O$
n-Heptane	C_7H_{16}
1-Heptanethiol	$C_7H_{16}S$
Heptanoic acid	$C_7H_{14}O_2$
1-Heptanol	$C_7H_{16}O$
4-Heptanol	$C_7H_{16}O$
1-Heptene	C_7H_{14}
n-Heptyl alcohol	$C_7H_{16}O$
4-n-Heptylbicyclohexyl	$C_{19}H_{36}$
n-Heptylcyclohexane	$C_{13}H_{26}$
n-Heptyl mercaptan	$C_7H_{16}S$
4-n-Heptyltercyclohexyl	$C_{25}H_{46}$
4′-n-Heptyl-m-tercyclohexyl	$C_{25}H_{46}$
Hexachlorobenzene	C_6Cl_6
Hexachloroethane	C_2Cl_6
n-Hexacosane	$C_{26}H_{54}$
Hexadecafluoro-3-butyltetrahydrofuran	$C_8F_{16}O$
Hexadecafluorodimethylcyclohexane	C_8F_{16}
Hexadecafluoroheptane	C_7F_{16}
Hexadecane	$C_{16}H_{34}$
n-Hexadecane	$C_{16}H_{34}$
Hexadecanoic acid	$C_{16}H_{32}O_2$
1-Hexadecanol	$C_{16}H_{34}O$
1-Hexadecene	$C_{16}H_{32}$
1-Hexadecene-urea adduct	$C_{2.3}H_{6.7}N_2O$
n-Hexadecyl alcohol	$C_{16}H_{34}O$
1,5-Hexadiene	C_6H_{10}
Hexaethylcyclohexane	$C_{18}H_{36}$
Hexafluoroacetone	C_3F_6O
Hexafluorobenzene	C_6F_6
Hexafluoropropanone	C_3F_6O
Hexahydroindan	C_9H_{16}
cis-Hexahydroindan	C_9H_{16}
trans-Hexahydroindan	C_9H_{16}
Hexamethylbenzene	$C_{12}H_{18}$
Hexamethylcyclotrisiloxane	$C_6H_{18}O_3Si_3$
Hexamethyldisilane	$C_6H_{18}Si_2$
Hexamethyldisiloxane	$C_6H_{18}OSi_2$
Hexamethylenediisocyanate	$C_8H_{12}N_2O_2$
Hexamethyleneimine	$C_6H_{13}N$
Hexamethylenetetramine	$C_6H_{12}N_4$
n-Hexane	C_6H_{14}
1-Hexanethiol	$C_6H_{14}S$
Hexanitroethane	$C_2N_6O_{12}$
Hexanoic acid	$C_6H_{12}O_2$
1-Hexanol	$C_6H_{14}O$
3-Hexanol	$C_6H_{14}O$
2-Hexanone	$C_6H_{12}O$
3-Hexanone	$C_6H_{12}O$
Hexaphenylcyclotrisiloxane	$C_{36}H_{30}O_3Si_3$
n-Hexatriacontane	$C_{36}H_{74}$
1-Hexene	C_6H_{12}
Hexogen	$C_3H_6N_6O_6$
n-Hexyl alcohol	$C_6H_{14}O$
n-Hexylamine	$C_6H_{15}N$
n-Hexyl bromide	$C_6H_{13}Br$
n-Hexyl mercaptan	$C_6H_{14}S$
n-Hexyl methyl ketone	$C_8H_{16}O$
p-n-Hexyloxybenzylideneamino-p′-chlorobenzene	$C_{19}H_{21}ClNO$
Hexyl phenylcarbamate	$C_{13}H_{19}NO_2$
Hippuric acid	$C_9H_9NO_3$
Hippurylglycine	$C_{11}H_{12}N_2O_4$
Histidine hydrochloride(L)	$C_6H_{10}ClN_3O_2$
Hydrogen cyanide	CHN
Hydroquinone	$C_6H_6O_2$
o-Hydroxyacetanilide	$C_8H_9NO_2$
2-Hydroxybenzaldehyde	$C_7H_6O_3$
Hydroxybenzene	C_6H_6O
m-Hydroxybenzoic acid	$C_7H_6O_3$
o-Hydroxybenzoic acid	$C_7H_6O_3$
p-Hydroxybenzoic acid	$C_7H_6O_3$
o-Hydroxybiphenyl	$C_{12}H_{10}O$
N-(2-Hydroxy-4-methoxybenzylidene)-p-butylaniline	$C_{18}H_{21}NO_2$
1-Hydroxynaphthalene	$C_{10}H_8O$
2-Hydroxynaphthalene	$C_{10}H_8O$
2-Hydroxypropanoic acid(D)	$C_3H_6O_3$
2-Hydroxypropanoic acid (DL)	$C_3H_6O_3$
2-Hydroxypropanoic acid(L)	$C_3H_6O_3$
Hypoxanthine	$C_5H_4N_4O$

I

Idryl	$C_{16}H_{10}$
2-Imino-4-thiazolidone	$C_3H_6N_2S$
Indan	C_9H_{10}
Indene	C_9H_8
Iodobenzene	C_6H_5I
Iodobis-(N,N-diethyldithiocarbamato)Iron(III)	$C_{10}H_{20}FeIN_2S_4$
Iodoethane	C_2H_5I
Iodomethane	CH_3I
1-Iodo-3-methylbutane	$C_5H_{11}I$
1-Iodo-2-methylpropane	C_4H_9I
1-Iodopropane	C_3H_7I
Isoamyl alcohol	$C_5H_{12}O$
Isoamyl bromide	$C_5H_{11}Br$
Isoamyl chloride	$C_5H_{11}Cl$
Isoamyl iodide	$C_5H_{11}I$
Isoamyl mercaptan	$C_5H_{12}S$
Isobutane	C_4H_{10}
Isobutene	C_4H_8

Isobutyl acetate	$C_6H_{12}O_2$
Isobutyl alcohol	$C_4H_{10}O$
Isobutylamine	$C_4H_{11}N$
Isobutyl bromide	C_4H_9Br
Isobutyl chloride	C_4H_9Cl
α-Isobutyldecalin	$C_{14}H_{26}$
Isobutyl formate	$C_5H_{10}O_2$
Isobutyl iodide	C_4H_9I
Isobutyl mercaptan	$C_4H_{10}S$
Isobutyric acid	$C_4H_8O_2$
Isobutyryl chloride	C_4H_7ClO
Isodurene	$C_{10}H_{14}$
Isoleucine(L)	$C_6H_{13}NO_2$
Isoöctane	C_8H_{18}
Isopentane	C_5H_{12}
Isophthalic acid	$C_8H_6O_4$
Isoprene	C_5H_8
2-Isopropoxyethanol	$C_5H_{12}O_2$
N-Isopropylacetamide	$C_5H_{11}NO$
Isopropyl acetate	$C_5H_{10}O_2$
Isopropyl alcohol	C_3H_8O
Isopropylamine	C_3H_9N
Isopropylbenzene	C_9H_{12}
Isopropylbicyclohexyl	$C_{15}H_{28}$
2-Isopropylbicyclohexyl	$C_{15}H_{28}$
Isopropylbiphenyl	$C_{15}H_{16}$
p-Isopropylbiphenyl	$C_{15}H_{16}$
Isopropyl bromide	C_3H_7Br
Isopropyl cyanide	C_4H_7N
Isopropyldecalin	$C_{13}H_{24}$
α-Isopropyldecalin	$C_{13}H_{24}$
Isopropyl ether	$C_6H_{14}O$
Isopropylhydroindan	$C_{12}H_{22}$
Isopropyl mercaptan	C_3H_8S
1-Isopropyl-4-methylbenzene	$C_{10}H_{14}$
Isopropyl methyl ketone	$C_5H_{10}O$
Isopropyl methyl sulfide	$C_4H_{10}S$
Isovaleric acid	$C_5H_{10}O_2$

J,K,L

Lactic acid(D)	$C_3H_6O_3$
Lactic acid (DL)	$C_3H_6O_3$
Lactic acid(L)	$C_3H_6O_3$
Lactose	$C_{12}H_{22}O_{11}$
β-Lactose	$C_{12}H_{22}O_{11}$
α-Lactose monohydrate	$C_{12}H_{22}O_{11}\cdot H_2O$
Latex	$(C_5H_8)_n$
Lauric acid	$C_{12}H_{24}O_2$
Lead(II) heptadecanoate	$C_{34}H_{66}O_4Pb$
Lead(II) heptanoate	$C_{14}H_{26}O_4Pb$
Lead(II) nonadecanoate	$C_{38}H_{74}O_4Pb$
Lead(II) nonate	$C_{18}H_{34}O_4Pb$
Lead(II) oenanthate	$C_{14}H_{26}O_4Pb$
Lead(II) oxalate	C_2O_4Pb
Lead(II) pelargonate	$C_{18}H_{34}O_4Pb$
Lead(II) pentadecanoate	$C_{30}H_{58}O_4Pb$
Lead(II) tridecanoate	$C_{26}H_{50}O_4Pb$
Lead(II) undecanoate	$C_{22}H_{42}O_4Pb$
Leucine(DL)	$C_6H_{13}NO_2$
Leucine(L)	$C_6H_{13}NO_2$
Leucylglycine(DL)	$C_8H_{16}N_2O_3$
Limonene	$C_{10}H_{16}$
Linalool	$C_{10}H_{18}O$
Lithium acetate	$C_2H_3LiO_2$
Lithium formate	$CHLiO_2$
Lithium propionate	$C_3H_5LiO_2$

M

Maleic acid	$C_4H_4O_4$
Maleic anhydride	$C_4H_2O_3$
Malononitrile	$C_3H_2N_2$
Maltose	$C_{12}H_{22}O_{11}$
β-Maltose hydrate	$C_{12}H_{22}O_{11}\cdot H_2O$
Mandelic acid	$C_8H_8O_3$
Manganocene	$C_{10}H_{10}Mn$
Mannitol	$C_6H_{14}O_6$
Mannitol(D)	$C_6H_{14}O_6$
Mannose	$C_6H_{12}O_6$
Marlex 50 polymer	$(CH_2)_n$
Melamine	$C_3H_6N_6$
Mercuric caprate	$C_{20}H_{38}HgO_4$
Mercuric caprylate	$C_{16}H_{30}HgO_4$
Mercuric decanoate	$C_{20}H_{38}HgO_4$
Mercuric dodecanoate	$C_{24}H_{46}HgO_4$
Mercuric hexadecanoate	$C_{32}H_{64}HgO_4$
Mercuric laurate	$C_{24}H_{46}HgO_4$
Mercuric myristate	$C_{28}H_{54}HgO_4$
Mercuric octadecanoate	$C_{36}H_{70}HgO_4$
Mercuric octanoate	$C_{16}H_{30}HgO_4$
Mercuric palmitate	$C_{32}H_{64}HgO_4$
Mercuric stearate	$C_{36}H_{70}HgO_4$
Mercuric tetradecanoate	$C_{28}H_{54}HgO_4$
Mercury di(p-tolyl)	$C_{14}H_{14}Hg$
Mercury diphenyl	$C_{12}H_{10}Hg$
Mesitylene	C_9H_{12}
Mesityl oxide	$C_6H_{10}O$
2,2-Metacyclophane	$C_{16}H_{16}$
2,2-Metaparacyclophane	$C_{16}H_{16}$
Methanamide	CH_3NO
Methanethiol	CH_4S
Methanoic acid	CH_2O_2
Methanol	CH_4O
Methanol-d_1	CH_3DO
Methionine	$C_5H_{11}NO_2S$
Methoxybenzene	C_7H_8O
4-Methoxybenzoic acid	$C_8H_8O_3$
N-(p-Methoxybenzylidene)-p-n-butylaniline	$C_{18}H_{21}NO$
4-Methoxy-4′-butoxy-trans-stilbene	$C_{19}H_{22}O_2$
4-Methoxy-4′-dodecoxy-trans-stilbene	$C_{27}H_{38}O_2$
2-Methoxyethanol	$C_3H_8O_2$
4-Methoxy-4′-heptoxy-trans-stilbene	$C_{22}H_{28}O_2$
4-Methoxy-4′-hexoxy-trans-stilbene	$C_{21}H_{26}O_2$
Methoxymethane	C_2H_6O
4-Methoxy-4′-octoxy-trans-stilbene	$C_{23}H_{30}O_2$
4-Methoxy-4′-pentoxy-trans-stilbene	$C_{20}H_{24}O_2$
Methyl acetate	$C_3H_6O_2$
Methyl acrylate	$C_4H_6O_2$
Methylal	$C_3H_8O_2$
Methyl alcohol	CH_4O
Methyl alcohol-d_1	CH_3DO
Methylamine	CH_5N
2-Methyl-1-aminopropane	$C_4H_{11}N$
2-Methyl-2-aminopropane	$C_4H_{11}N$
Methyl ammonium alum	$CH_6AlNO_8S_2\cdot 12H_2O$
Methylammonium chloride	CH_6ClN
2-Methylaniline	C_7H_9N
3-Methylaniline	C_7H_9N
4-Methylaniline	C_7H_9N
N-Methylaniline	C_7H_9N
Methylbenzene	C_7H_8
Methyl benzoate	$C_8H_8O_2$
2-Methylbenzoic acid	$C_8H_8O_2$
3-Methylbenzoic acid	$C_8H_8O_2$
4-Methylbenzoic acid	$C_8H_8O_2$
2-Methylbicyclo[2,2,1]heptane(endo)	C_8H_{14}
2-Methylbicyclo[2,2,1]heptane(exo)	C_8H_{14}
Methyl bicyclobutane-1-carboxylate	$C_6H_8O_2$
2-Methylbicyclohexyl	$C_{13}H_{24}$
2-Methylbicyclohexylmethane	$C_{14}H_{26}$
Methyl bromide	CH_3Br
2-Methyl-1,3-butadiene	C_5H_8
3-Methyl-1,2-butadiene	C_5H_8
N-Methylbutanamide	$C_5H_{11}NO$
2-Methylbutane	C_5H_{12}

3-Methyl-1-butanethiol	$C_5H_{12}S$
3-Methyl-2-butanethiol	$C_5H_{12}S$
Methyl butanoate	$C_5H_{10}O_2$
3-Methylbutanoic acid	$C_5H_{10}O_2$
3-Methyl-1-butanol	$C_5H_{12}O$
3-Methylbutanone	$C_5H_{10}O$
2-Methyl-1-butene	C_5H_{10}
2-Methyl-2-butene	C_5H_{10}
Methyl butyrate	$C_5H_{10}O_2$
Methyl caprate	$C_{11}H_{22}O_2$
Methyl chloride	CH_3Cl
Methyl chloroform	$C_2H_3Cl_3$
Methyl cyanide	C_2H_3N
Methyl cyclobutanecarboxylate	$C_6H_{10}O_2$
Methylcyclohexane	C_7H_{14}
2-Methylcyclohexanol	$C_7H_{14}O$
3-Methylcyclohexanol	$C_7H_{14}O$
4-Methylcyclohexanol	$C_7H_{14}O$
2-Methylcyclohexanone	$C_7H_{12}O$
3-Metgylcyclohexanone	$C_7H_{12}O$
4-Methylcyclohexanone	$C_7H_{12}O$
Methylcyclopentane	C_6H_{12}
1-Methylcyclopentene	C_6H_{10}
3-Methylcyclopentene	C_6H_{10}
2-Methylcyclothiapentane	$C_5H_{10}S$
3-Methylcyclothiapentane	$C_5H_{10}S$
α-Methyldecalin	$C_{11}H_{20}$
β-Methyldecalin	$C_{11}H_{20}$
2-Methyldecane	$C_{11}H_{24}$
Methyl decanoate	$C_{11}H_{22}O_2$
1-Methyl-2,4-diisocyanatobenzene	$C_9H_6N_2O_2$
Methyl-2,2-dimethylpropanoate	$C_6H_{12}O_2$
4-Methyl-3,5-dioxaheptane	$C_6H_{14}O_2$
6-Methyl-5,7-dioxaundecane	$C_{10}H_{22}O_2$
4-Methyl-1,3-dioxolan-2-one	$C_5H_{10}O_3$
Methyldiphenylamine	$C_{13}H_{13}N$
Methyl enanthoate	$C_8H_{16}O_2$
Methylene bromide	CH_2Br_2
Methylene chloride	CH_2Cl_2
Methylenecyclobutane	C_5H_8
3-Methylenecyclobutyl cyanide	C_6H_7N
Methylenecyclohexane	C_7H_{12}
Methylene iodide	CH_2I_2
Methyl ethanoate	$C_3H_6O_2$
N-(1-Methylethyl)ethanamide	$C_5H_{11}NO$
1-Methylethyl ethanoate	$C_5H_{10}O_2$
Methyl ethyl ketone	C_4H_8O
Methyl ethyl ketoxime	C_4H_9NO
N-Methylformamide	C_2H_5NO
Methyl formate	$C_2H_4O_2$
2-Methylfuran	C_5H_6O
N-Methylglycine	$C_3H_7NO_2$
2-Methylheptane	C_8H_{18}
3-Methylheptane	C_8H_{18}
4-Methylheptane	C_8H_{18}
Methyl heptanoate	$C_8H_{16}O_2$
5-Methyl-1-heptanol	$C_8H_{18}O$
2-Methyl-2-heptanol	$C_8H_{18}O$
3-Methyl-2-heptanol	$C_8H_{18}O$
4-Methyl-2-heptanol	$C_8H_{18}O$
5-Methyl-2-heptanol	$C_8H_{18}O$
6-Methyl-2-heptanol	$C_8H_{18}O$
4-Methyl-3-heptanol	$C_8H_{18}O$
6-Methyl-3-heptanol	$C_8H_{18}O$
2-Methyl-4-heptanol	$C_8H_{18}O$
4-Methyl-4-heptanol	$C_8H_{18}O$
Methyl hexadecanoate	$C_{17}H_{34}O_2$
2-Methylhexane	C_7H_{16}
3-Methylhexane	C_7H_{16}
Methyl hexyl ketone	$C_8H_{16}O$
Methylhydrazine	CH_6N_2
Methylhydroindan	$C_{10}H_{18}$
Methyl iodide	CH_3I
1-Methyl-4-isopropylbenzene	$C_{10}H_{14}$
Methyl isopropyl ether	$C_4H_{10}O$
1-Methyl-7-isopropylphenanthrene	$C_{18}H_{18}$
Methyl mercaptan	CH_4S
Methyl methacrylate	$C_5H_8O_2$
N-Methylmethanamide	C_2H_5NO
Methyl methanoate	$C_2H_4O_2$
Methyl 2-methylpropenoate	$C_5H_8O_2$
Methyl myristate	$C_{15}H_{30}O_2$
Methyl n-butyl ether	$C_5H_{12}O$
Methyl n-butyl ketone	$C_6H_{12}O$
Methyl n-decyl ether	$C_{11}H_{24}O$
Methyl n-propyl ether	$C_4H_{10}O$
Methyl n-propyl sulfide	$C_4H_{10}S$
1-Methylnaphthalene	$C_{11}H_{10}$
2-Methylnaphthalene	$C_{11}H_{10}$
Methyl nitrate	CH_3NO_3
2-Methylnonane	$C_{10}H_{22}$
3-Methylnonane(DL)	$C_{10}H_{22}$
4-Methylnonane(DL)	$C_{10}H_{22}$
5-Methylnonane	$C_{10}H_{22}$
Methyl oenanthoate	$C_8H_{16}O_2$
5-Methyl-3-oxahex-1-ene	$C_6H_{12}O$
4-Methyl-3-oxa-1-pentanol	$C_5H_{12}O_2$
2-Methyloxirane	C_3H_6O
Methyl palmitate	$C_{17}H_{34}O_2$
N-Methylpentanamide	$C_6H_{13}NO$
2-Methylpentane	C_6H_{14}
2-Methylpentane	C_6H_{14}
3-Methylpentane	C_6H_{14}
Methyl pentanoate	$C_6H_{12}O_2$
4-Methylpenten-3-one-2	$C_6H_{10}O$
9-Methylperhydrofluorene	$C_{14}H_{34}$
Methyl phenylcarbamate	$C_8H_9NO_2$
Methyl phenyl ether	C_7H_8O
Methyl phenyl ketone	C_8H_8O
Methylphosphonyl chlorofluoride	CH_3ClFOP
Methylphosphonyl dichloride	CH_3Cl_2OP
Methylphosphonyl difluoride	CH_3F_2OP
N-Methylpiperidine	$C_6H_{13}N$
2-Methylpiperidine	$C_6H_{13}N$
4-Methylpiperidine	$C_6H_{13}N$
Methyl pivalate	$C_6H_{12}O_2$
N-Methylpropanamide	C_4H_9NO
2-Methylpropane	C_4H_{10}
2-Methyl-1,2-propanediamine	$C_4H_{12}N_2$
2-Methyl-1-propanethiol	$C_4H_{10}S$
2-Methyl-2-propanethiol	$C_4H_{10}S$
Methyl propanoate	$C_4H_8O_2$
2-Methylpropanoic acid	$C_4H_8O_2$
2-Methyl-1-propanol	$C_4H_{10}O$
2-Methyl-2-propanol	$C_4H_{10}O$
2-Methylpropanoyl chloride	C_4H_7ClO
2-Methylpropene	C_4H_8
Methyl propenoate	$C_4H_6O_2$
Methyl propionate	$C_4H_8O_2$
2-Methylpropionitrile	C_4H_7N
2-Methylpropyl ethanoate	$C_6H_{12}O_2$
Methyl propyl ketone	$C_5H_{10}O$
2-Methylpropyl methanoate	$C_5H_{10}O_2$
2-Methylpyridine	C_6H_7N
3-Methylpyridine	C_6H_7N
1-Methylpyrrolidine	$C_5H_{11}N$
1-Methyl-2-pyrrolidone	C_5H_9NO
Methyl salicylate	$C_8H_8O_3$
2-Methylsuccinic acid	$C_5H_8O_4$
Methyl tert-butyl ether	$C_5H_{12}O$
Methyl tert-butyl ketone	$C_6H_{12}O$
Methyl tert-butyl sulfide	$C_5H_{12}S$

Methyl tetradecanoate	$C_{15}H_{30}O_2$
3-Methyltetrahydrophthalic anhydride	$C_9H_{10}O_3$
N-Methyl-N,2,4,6-tetranitroaniline	$C_7H_5N_5O_8$
Methyltetryl	$C_8H_7N_5O_8$
3-Methyl-2-thiabutane	$C_4H_{10}S$
2-Methylthiazole	C_4H_5NS
2-Methylthiolane	$C_5H_{10}S$
3-Methylthiolane	$C_5H_{10}S$
2-Methylthiophene	C_5H_6S
Methyltriphenylarsonium bis(7,7,8,8-tetracyanoquinodimethanide)	$C_{43}H_{26}AsN_8$
Methyltriphenylphosphonium bis(7,7,8,8-tetracyanoquinodomethanide	$C_{43}H_{26}N_8P$
N-Methylvaleramide	$C_6H_{13}NO$
Methyl valerate	$C_6H_{12}O_2$
1-Monobenzoylglycerol	$C_{10}H_{12}O_4$
2-Monobenzoylglycerol	$C_{10}H_{12}O_4$
1-Monocaprin	$C_{13}H_{26}O_4$
1-Monolaurin	$C_{15}H_{30}O_4$
2-Monolaurin	$C_{15}H_{30}O_4$
1-Monomyristin	$C_{17}H_{34}O_4$
2-Monomyristin	$C_{17}H_{34}O_4$
1-Monopalmitin	$C_{19}H_{38}O_4$
2-Monopalmitin	$C_{19}H_{38}O_4$
1-Monostearin	$C_{21}H_{42}O_4$
2-Monostearin	$C_{21}H_{42}O_4$
Morpholine	C_4H_9NO
Myristic acid	$C_{14}H_{28}O_2$

N

NAD	$C_{21}H_{28}N_7O_{14}P_2\cdot3H_2O$
Naphthacene	$C_{18}H_{12}$
Naphthalene	$C_{10}H_8$
1,5-Naphthylenediisocyanate	$C_{12}H_6N_2O_2$
Naphthalene-1,8-disulfide-S-oxide	$C_{10}H_6OS_2$
Naphthalene-pyromellitic dianhydride adduct	$C_{20}H_{10}O_6$
Naphthalene-1,2,4,5-tetracyanobenzene adduct	$C_{20}H_{10}N_4$
Naphthalene-tetracyanoethylene adduct	$C_{16}H_8N_4$
Naphthalene-1,3,5-trinitrobenzene adduct	$C_{16}H_{11}N_3O_6$
α-Naphthol	$C_{10}H_8O$
β-Naphthol	$C_{10}H_8O$
β-Naphthylamine	$C_{10}H_9N$
β-Naphthylamine-p-nitrosodimethylaniline complex	$C_{46}H_{47}N_7O_2$
Neopentane	C_5H_{12}
Nickelocene	$C_{10}H_{10}Ni$
Nicotinamide adenine dinucleotide trihydrate	$C_{21}H_{28}N_7O_{14}P_2\cdot3H_2O$
p-Nitroacetanilide	$C_8H_9N_2O_3$
5-Nitro-2-aminotoluene	$C_7H_8N_2O_2$
3-Nitro-4-aminotoluene	$C_7H_8N_2O_2$
2-Nitroaniline	$C_6H_6N_2O_2$
3-Nitroaniline	$C_6H_6N_2O_2$
4-Nitroaniline	$C_6H_6N_2O_2$
Nitrobenzene	$C_6H_5NO_2$
2-Nitrobenzoic acid	$C_7H_5NO_4$
3-Nitrobenzoic acid	$C_7H_5NO_4$
4-Nitrobenzoic acid	$C_7H_5NO_4$
4-Nitro-1-chlorobenzene	$C_6H_4ClNO_2$
m-Nitrocinnamic acid	$C_9H_7NO_4$
o-Nitrocinnamic acid	$C_9H_7NO_4$
p-Nitrocinnamic acid	$C_9H_7NO_4$
Nitroethane	$C_2H_5NO_2$
Nitroguanidine	$CH_4N_4O_2$
Nitromethane	CH_3NO_2
p-Nitrosodimethylaniline	$C_8H_{10}N_2O$
p-Nitrosodimethylaniline-β-naphthylamine complex	$C_{46}H_{47}N_7O_2$
2-Nitrotoluene	$C_7H_7NO_2$
3-Nitrotoluene	$C_7H_7NO_2$
4-Nitrotoluene	$C_7H_7NO_2$
n-Nonadecane	$C_{19}H_{40}$
Nonadecanoic acid	$C_{19}H_{38}O_2$
n-Nonane	C_9H_{20}
Nonanoic acid	$C_9H_{18}O_2$
5-Nonanone	$C_9H_{18}O$
4-n-Nonylbicyclohexyl	$C_{21}H_{40}$
Nonyl phenylcarbamate	$C_{16}H_{25}NO_2$
4-n-Nonyltercyclohexyl	$C_{27}H_{50}$
Norbornadiene	C_7H_8
Norbornene	C_7H_{10}
Nortricyclene	C_7H_{10}

O

n-Octacosane	$C_{28}H_{58}$
n-Octadecane	$C_{18}H_{38}$
1-Octadecene-urea adduct	$C_{2.4}H_{6.8}N_2O$
n-Octadecyl chloride	$C_{18}H_{37}Cl$
p-n-Octadecyloxybenzoic acid	$C_{25}H_{42}O_3$
p-n-Octadecyloxybenzoic acid-d	$C_{25}H_{41}DO_3$
Octafluorocyclobutane	C_4F_8
Octafluoropropane	C_3F_8
Octafluorotoluene	C_7F_8
1,2,3,4,5,6,7,8-Octahydroanthracene	$C_{14}H_{18}$
Octahydroazocine	$C_7H_{15}N$
Octamethyldiphenylcyclopentasiloxane	$C_{20}H_{34}O_5Si_5$
n-Octane	C_8H_{18}
Octanoic acid	$C_8H_{16}O_2$
1-Octanol	$C_8H_{18}O$
2-Octanol	$C_8H_{18}O$
3-Octanol	$C_8H_{18}O$
4-Octanol	$C_8H_{18}O$
2-Octanone	$C_8H_{16}O$
n-Octatetracontane	$C_{48}H_{98}$
1-Octene	C_8H_{16}
2-Octene	C_8H_{16}
Octogen(α)	$C_4H_8N_8O_8$
Octogen(β)	$C_4H_8N_8O_8$
Octogen(γ)	$C_4H_8N_8O_8$
Octogen(δ)	$C_4H_8N_8O_8$
n-Octyl alcohol	$C_8H_{18}O$
Octyloxycyanobiphenyl	$C_{21}H_{25}NO$
Oenanthal	$C_7H_{14}O$
Ornithine(DL)	$C_5H_{12}N_2O_2$
Ornithine dihydrochloride	$C_5H_{14}Cl_2N_2O_2$
Orthoformic acid	CH_4O_3
3-Oxabicyclo[3.2.2]nonane	$C_8H_{14}O$
3-Oxa-1-butanol	$C_3H_8O_2$
2-Oxa-3,3-dimethylbutane	$C_5H_{12}O$
2-Oxadodecane	$C_{11}H_{24}O$
2-Oxaheptane	$C_6H_{14}O$
3-Oxa-1-heptanol	$C_6H_{14}O_2$
3-Oxahept-1-ene	$C_6H_{12}O$
2-Oxahexane	$C_5H_{12}O$
3-Oxahexane	$C_5H_{12}O$
3-Oxa-1-hexanol	$C_5H_{12}O_2$
2-Oxa-3-methylbutane	$C_4H_{10}O$
3-Oxa-1-pentanol	$C_4H_{10}O_2$
Oxalic acid	$C_2H_2O_4$
Oxamide	$C_2H_4N_2O_2$
Oxane	$C_5H_{10}O$
2-Oxapentane	$C_4H_{10}O$
3-Oxapentane	$C_4H_{10}O$
2-Oxapropane	C_2H_6O
Oxetane	C_3H_6O
Oxirane	C_2H_4O
Oxolane	C_4H_8O

P

Palmitic acid	$C_{16}H_{32}O_2$
2,2-Paracyclophane	$C_{16}H_{16}$
3,3-Paracyclophane	$C_{18}H_{20}$
Paraldehyde	$C_6H_{12}O_3$
Pelargonic acid	$C_9H_{18}O_2$
Pentachloroethane	C_2HCl_5
Pentachlorophenol	C_6HCl_5O
n-Pentacosane	$C_{25}H_{52}$
Pentacycloformaldehyde	$C_5H_{10}O_5$

Pentacyclo[7.3.1.1^{4,12}.0^{2,7}.0^{6,11}]tetradecane	$C_{14}H_{20}$
n-Pentadecane	$C_{15}H_{32}$
Pentadecanoic acid	$C_{15}H_{30}O_2$
1-Pentadecanol	$C_{15}H_{32}O$
Pentadecanolactone	$C_{15}H_{28}O_2$
2-Pentadecanone	$C_{15}H_{30}O$
n-Pentadecyl alcohol	$C_{15}H_{32}O$
1,2-Pentadiene	C_5H_8
1-cis-3-Pentadiene	C_5H_8
1-trans-3-Pentadiene	C_5H_8
1,4-Pentadiene	C_5H_8
2,3-Pentadiene	C_5H_8
Pentaerythritol	$C_5H_{12}O_4$
Pentaerythrityl tetrabromide	$C_5H_8Br_4$
Pentaerythrityl tetrachloride	$C_5H_8Cl_4$
Pentaerythrityl tetrafluoride	$C_5H_8F_4$
Pentaerythrityl tetraiodide	$C_5H_8I_4$
Pentaethylene glycol	$C_{10}H_{22}O_6$
Pentafluoroaniline	$C_6H_2F_5N$
Pentafluorobenzene	C_6HF_5
Pentafluorochlorobenzene	C_6ClF_5
Pentafluorochloroethane	C_2ClF_5
Pentafluoronitrobenzene	$C_6F_5NO_2$
Pentafluorophenol	C_6HF_5O
2,3,4,5,6-Pentafluorotoluene	$C_7H_3F_5$
Pentamethylbenzene	$C_{11}H_{16}$
n-Pentanal	$C_5H_{10}O$
n-Pentane	C_5H_{12}
1,5-Pentanediol	$C_5H_{12}O_2$
1-Pentanethiol	$C_5H_{12}S$
Pentanoic acid	$C_5H_{10}O_2$
1-Pentanol	$C_5H_{12}O$
3-Pentanol	$C_5H_{12}O$
2-Pentanone	$C_5H_{10}O$
3-Pentanone	$C_5H_{10}O$
Pentanoyl chloride	C_5H_9ClO
Pentaphenylethane	$C_{32}H_{26}$
Pentatriacontane	$C_{35}H_{72}$
1-Pentene	C_5H_{10}
2-Pentene	C_5H_{10}
cis-2-Pentene	C_5H_{10}
trans-2-Pentene	C_5H_{10}
Penton	$(C_5H_8Cl_2O)_n$
Pentoxan	$C_5H_{10}O_5$
n-Pentyl alcohol	$C_5H_{12}O$
n-Pentylamine	$C_5H_{13}N$
n-Pentylammonium chloride	$C_5H_{14}ClN$
n-Pentyl bromide	$C_5H_{11}Br$
Pentyl ethanoate	$C_7H_{14}O_2$
4-n-Pentylphenyl-4′-n-heptyloxythiobenzoate	$C_{25}H_{34}O_2S$
Perchlorobenzene	C_6Cl_6
Perchlorophenol	C_6HCl_5O
Perfluoro-3-butyltetrahydrofuran	$C_8F_{16}O$
Perfluorobenzene	C_6F_6
Perfluorobicyclo[4,4,0]dec-1,6-diene	$C_{10}F_{16}$
Perfluorobicyclohexyl	$C_{12}F_{22}$
Perfluorobiphenyl	$C_{12}F_{10}$
cis-Perfluorodecalin	$C_{10}F_{18}$
trans-Perfluorodecalin	$C_{10}F_{18}$
Perfluorodimethylcyclohexane	C_8F_{16}
Perfluoroheptane	C_7F_{16}
N-perfluorohexane	C_6F_{14}
Perfluoromethylcyclohexane	C_7F_{14}
Perfluoromethyldiethylamine	$C_5F_{13}N$
N-perfluoroöctane	C_8F_{18}
Perfluoropiperidine	$C_5F_{11}N$
Perfluoropropane	C_3F_8
Perfluorotriethylamine	$C_6F_{15}N$
Perfluorotoluene	C_7F_8
Perhydroazepine	$C_6H_{13}N$
Perhydromethylcyclopentadiene dimer	$C_{12}H_{20}$
Perhydrophenanthrene	$C_{14}H_{24}$
Perylene	$C_{20}H_{12}$
Phenanthrene	$C_{14}H_{10}$
Phenol	C_6H_6O
Phenol-p-toluidine complex	$C_{13}H_{15}NO$
Phenol-urea complex	$C_{13}H_{16}N_2O_3$
2-Phenoxyethanol	$C_8H_{10}O_2$
Phenylacetylene	C_8H_6
Phenylalanine(L)	$C_9H_{11}NO_2$
Phenylchloromethane	C_7H_7Cl
11-Phenyleicosane	$C_{27}H_{48}$
1,4-Phenylenediisocyanate	$C_8H_4N_2O_2$
2-Phenylethanol	$C_8H_{10}O$
2-Phenylethylamine	$C_8H_{11}N$
Phenylethylene	C_8H_8
α-Phenylglycine(D)	$C_8H_9NO_2$
N-phenylglycine	$C_8H_9NO_2$
Phenyl isothiocyanate	C_7H_5NS
Phenyl mercaptan	C_6H_6S
Phenylmethylamine	C_7H_9N
Phenyl methyl ether	C_7H_8O
Phenyl methyl sulfide	C_7H_8S
3-Phenylpropanol	$C_9H_{12}O$
3-Phenylpropylamine	$C_9H_{13}N$
Phenyl-1-thiaethane	C_7H_8S
Phenyltrichlorosilane	$C_6H_5Cl_3Si$
Phosgene	CCl_2O
Phthalic acid	$C_8H_6O_4$
m-Phthalic acid	$C_8H_6O_4$
o-Phthalic acid	$C_8H_6O_4$
p-Phthalic acid	$C_8H_6O_4$
Phthalic anhydride	$C_8H_4O_3$
Phthalyl dichloride	$C_8H_4Cl_2O_2$
α-Picoline	C_6H_7N
β-Picoline	C_6H_7N
Picric acid	$C_6H_3N_3O_7$
Pinane	$C_{10}H_{18}$
Piperidine	$C_5H_{11}N$
Pivalic acid	$C_5H_{10}O_2$
Poly-L-alanine	$(C_3H_5NO)_n$
cis-1,4-Polybutadiene	$(C_4H_6)_n$
trans-1,4-Polybutadiene	$(C_4H_6)_n$
Poly(1-butene)	$(C_4H_8)_n$
Polychlorotrifluoroethylene	$(C_2ClF_3)_n$
Polyethylene	$(CH_2)_n$
Polyethylene, branched	$(CH_2)_n$
Polyethylene, branched, annealed	$(CH_2)_n$
Polyethylene, branched, DYNH CT-1660	$(CH_2)_n$
Polyethylene, linear	$(CH_2)_n$
Polyethylene, linear high density	$(CH_2)_n$
Polyethylene, linear high molecular weight	$(CH_2)_n$
Polyethylene, linear, Marlex 50	$(CH_2)_n$
Polyethylene, linear, Rigidex 50	$(CH_2)_n$
Polyethylene, linear, W. N. C. 18	$(CH_2)_n$
Poly(ethylenesebacate)	$(C_{12}H_{20}O_4)_n$
Polyglycine	$(C_2H_3NO)_n$
Poly(methyl methacrylate)	$(C_5H_8O_2)_n$
Poly(4-methyl-1-pentene)	$(C_6H_{12})_n$
Polyoctadiene	$(C_8H_{12})_n$
Polyoxacyclobutane	$(C_3H_6O)_n$
Polyoxymethylene	$(CH_2O)_n$
Polypropylene	$(C_3H_6)_n$
Polypropylene, syndiotactic	$(C_3H_6)_n$
Polystyrene	$(C_8H_8)_n$
Polystyrene, atactic	$(C_8H_8)_n$
Polystyrene, isotactic	$(C_8H_8)_n$
Polystyrene, isotactic, annealed	$(C_8H_8)_n$
Polytetrafluoroethylene, annealed	$(C_2F_4)_n$
Polytetrafluoroethylene, drawn	$(C_2F_4)_n$
Polytetrafluoroethylene, molded	$(C_2F_4)_n$
Polytetrafluoroethylene, powder	$(C_2F_4)_n$

Polytetrafluoroethylene, quenched	$(C_2F_4)_n$
Polytrifluorochloroethylene	$(C_2ClF_3)_n$
Polytrifluorovinyl chloride	$(C_2ClF_3)_n$
Polyvinyl chloride	$(C_2H_3Cl)_n$
Polyvinylenediphenylsilane	$(C_{14}H_{12}Si)_n$
Potassium acetate	$C_2H_3KO_2$
Potassium butyrate	$C_4H_7KO_2$
Potassium propionate	$C_3H_5KO_2$
Potassium thiocyanate	CKNS
Prehnitene	$C_{10}H_{14}$
Proline(L)	$C_5H_9NO_2$
Propaldehyde	C_3H_6O
Propanal	C_3H_6O
Propane	C_3H_8
1,2-Propanediamine	$C_3H_{10}N_2$
1,2-Propanediol	$C_3H_8O_2$
1-Propanethiol	C_3H_8S
2-Propanethiol	C_3H_8S
1,2,3-Propanetriol	$C_3H_8O_3$
1,2,3-Propanetriol-d_3	$C_3H_5D_3O_3$
Propanoic acid	$C_3H_6O_2$
1-Propanol	C_3H_8O
2-Propanol	C_3H_8O
Propanone	C_3H_6O
Propanoyl chloride	C_3H_5ClO
Propene	C_3H_6
Propenenitrile	C_3H_3N
3-Propen-1-ol	C_3H_6O
Propionic acid	$C_3H_6O_2$
Propionitrile	C_3H_5N
Propionyl chloride	C_3H_5ClO
2-n-Propoxyethanol	$C_5H_{12}O_2$
1-n-Propoxy-2-methoxyethane	$C_6H_{15}O_2$
N-(n-Propyl)acetamide	$C_5H_{11}NO$
n-Propyl acetate	$C_5H_{10}O_2$
n-Propyl alcohol	C_3H_8O
n-Propylamine	C_3H_9N
n-Propylbenzene	C_9H_{12}
n-Propyl bromide	C_3H_7Br
n-Propyl chloride	C_3H_7Cl
n-Propyl cyanide	C_4H_7N
n-Propylcyclohexane	C_9H_{18}
α-n-Propyldecalin	$C_{13}H_{24}$
Propylene	C_3H_6
Propylene carbonate	$C_4H_6O_3$
Propylene glycol	$C_3H_8O_2$
Propylene oxide	C_3H_6O
N-(n-Propyl)ethanamide	$C_5H_{11}NO$
Propyl ethanoate	$C_5H_{10}O_2$
n-Propyl ether	$C_6H_{14}O$
Propyl ethyl sulfide	$C_5H_{12}S$
n-Propyl iodide	C_3H_7I
n-Propyl mercaptan	C_3H_8S
n-Propyl methyl ketone	$C_5H_{10}O$
Propyl phenylcarbamate	$C_{10}H_{13}NO_2$
Pseudocumene	C_9H_{12}
Pulegone	$C_{10}H_{16}O$
Pyrazine	$C_4H_4N_2$
Pyrene	$C_{16}H_{10}$
Pyrene-pyromellitic dianhydride adduct	$C_{26}H_{12}O_6$
Pyrene-pyromellitic dianhydride charge transfer complex	$C_{26}H_{12}O_6$
Pyridine	C_5H_5N
Pyrocatechol	$C_6H_6O_2$
Pyrollidine	C_4H_9N
Pyromellitic dianhydride	$C_{10}H_2O_6$
Pyrotartaric acid	$C_5H_8O_4$
Pyrrole	C_4H_5N
Pyrrolidine	C_4H_9N
Pyrrolidine-2-carboxylic acid(L)	$C_5H_9NO_2$

Q

Quadricyclane	C_7H_8
m-Quaterphenyl	$C_{24}H_{18}$
Quinhydrone	$C_{12}H_{10}O_4$
Quinol	$C_6H_6O_2$
Quinoline	C_9H_7N
β-Quinol-methane clathrate	$C_{18.1}H_{18.6}O_6$
Quinone	$C_6H_4O_2$
Quinuclidine	$C_7H_{13}N$

R

Resorcinol	$C_6H_6O_2$
Retene	$C_{18}H_{18}$
Rubber	$(C_5H_8)_n$
Rubidium butyrate	$C_4H_7O_2Rb$
Rubidium formate	CHO_2Rb
Rubidium propionate	$C_3H_5O_2Rb$

S

Sabinene	$C_{10}H_{16}$
Salicylaldehyde	$C_7H_6O_2$
Salicylic acid	$C_7H_6O_3$
Salicylic acid-acetamide complex	$C_9H_{11}NO_4$
Sarcosine	$C_3H_7NO_2$
Semicarbazide hydrochloride	CH_6ClN_3O
Serine(DL)	$C_3H_7NO_3$
Serine(L)	$C_3H_7NO_3$
Silicon tetraethyl	$C_8H_{20}Si$
Silicon tetramethyl	$C_4H_{12}Si$
Sodium acetate	$C_2H_3NaO_2$
Sodium acetate trihydrate	$C_2H_3NaO_2 \cdot 3H_2O$
Sodium formate	$CHNaO_2$
Sodium methoxide	CH_3NaO
Sodium p-nitrophenoxide dihydrate	$C_6H_4NNaO_3 \cdot 2H_2O$
Sodium oxalate	$C_2Na_2O_4$
β-Sodium palmitate	$C_{16}H_{31}NaO_2 \cdot 0.01H_2O$
β-Sodium palmitate	$C_{16}H_{31}NaO_2 \cdot 0.409H_2O$
δ-Sodium palmitate	$C_{16}H_{31}NaO_2$
ε-Sodium palmitate	$C_{16}H_{31}NaO_2 \cdot 0.482H_2O$
ε-Sodium palmitate	$C_{16}H_{31}NaO_2 \cdot 0.715H_2O$
ω-Sodium palmitate	$C_{16}H_{31}NaO_2$
Sodium potassium tartrate tetrahydrate	$C_4H_4KNaO_6 \cdot 4H_2O$
Sodium propionate	$C_3H_5NaO_2$
Sorbose(L)	$C_6H_{12}O_6$
Spiropentane	C_5H_8
Squaric acid	$C_4H_2O_4$
Stearic acid	$C_{18}H_{36}O_2$
Stilbene	$C_{14}H_{12}$
Styrene	C_8H_8
Succinamide	$C_4H_8N_2O_2$
Succinic acid	$C_4H_6O_4$
Succinimide	$C_4H_5NO_2$
Succinonitrile	$C_4H_4N_2$
Sucrose	$C_{12}H_{22}O_{11}$

T

THAM	$C_4H_{11}NO_3$
γ-TNT	$C_7H_5N_3O_6$
TRIS	$C_4H_{11}NO_3$
Tartaric acid	$C_4H_6O_6$
Taurine	$C_2H_7NO_3S$
Teflon, annealed	$(C_2F_4)_n$
Teflon, drawn	$(C_2F_4)_n$
Teflon, molded	$(C_2F_4)_n$
Teflon, powder	$(C_2F_4)_n$
Teflon, quenched	$(C_2F_4)_n$
m-Tercyclohexyl	$C_{18}H_{32}$
o-Tercyclohexyl	$C_{18}H_{32}$
p-Tercyclohexyl	$C_{18}H_{32}$
Terephthal-bis-n-butylaniline	$C_{28}H_{32}N_2O_2$
m-Terphenyl	$C_{18}H_{14}$
o-Terphenyl	$C_{18}H_{14}$
p-Terphenyl	$C_{18}H_{14}$
p-Terphenyl-d_{14}	$C_{18}D_{14}$

Terephthalic acid	$C_8H_6O_4$
3,3′,4,4′-Tetraaminodiphenyl ether	$C_{12}H_{14}N_4O$
3,3′,4,4′-Tetraaminodiphenyl oxide	$C_{12}H_{14}N_4O$
Tetraamylstannane	$C_{20}H_{44}Sn$
1,3,5,7-Tetraazatricyclo[3.3.1.1^{3,7}]decane	$C_6H_{12}N_4$
1,1,2,2-Tetrabromoethane	$C_2H_2Br_4$
Tetrabromomethane	CBr_4
Tetrabutoxytitanium	$C_{16}H_{36}O_4Ti$
Tetra-n-butylammonium bromide	$C_{16}H_{36}BrN$
1,2,4,5-Tetrachlorobenzene	$C_6H_2Cl_4$
Tetrachloro-p-benzoquinone	$C_6Cl_4O_2$
Tetrachlorobis-(methylammonium) cadmium II	$C_2H_{12}CdCl_4N_2$
Tetrachlorobis-(methylammonium) manganese II	$C_2H_{12}Cl_4MnN_2$
Tetrachlorobis-(2-propeneammonium) cadmium II	$C_6H_{16}CdCl_4N_2$
Tetrachlorobis-(n-propylammonium) cadmium II	$C_6H_{20}CdCl_4N_2$
Tetrachlorobis-(n-propylammonium)manganese II	$C_6H_{20}Cl_4MnN_2$
1,1,2,2-Tetrachlorodifluoroethane	$C_2Cl_4F_2$
1,1,2,2-Tetrachloro-1,2-difluoroethane	$C_2Cl_4F_2$
1,1,2,2-Tetrachloroethane	$C_2H_2Cl_4$
Tetrachloroethene	C_2Cl_4
Tetrachloroethylene	C_2Cl_4
Tetrachloromethane	CCl_4
1,1,1,3-Tetrachloropropane	$C_3H_4Cl_4$
n-Tetracosane	$C_{24}H_{50}$
1,2,4,5-Tetracyanobenzene	$C_{10}H_2N_4$
1,2,4,5-Tetracyanobenzene-pyrene complex	$C_{26}H_{12}N_4$
Tetracyclo[6,2,1,1^{3,6}]dodecane	$C_{12}H_{20}$
n-Tetradecane	$C_{14}H_{30}$
Tetradecanoic acid	$C_{14}H_{28}O_2$
1-Tetradecanol	$C_{14}H_{30}O$
2-Tetradecanone	$C_{14}H_{28}O$
n-Tetradecyl alcohol	$C_{14}H_{30}O$
Tetraethylammonium bromide	$C_8H_{20}BrN$
Tetraethylammonium iodide	$C_8H_{20}IN$
Tetraethylene glycol	$C_8H_{18}O_5$
Tetraethylgermane	$C_8H_{20}Ge$
Tetraethyllead	$C_8H_{20}Pb$
Tetraethylsilane	$C_8H_{20}Si$
Tetraethylstannane	$C_8H_{20}Sn$
1,1,2,2-Tetrafluoro-1,2-dichloroethane	$C_2Cl_2F_4$
1,2,3,4-Tetrafluorobenzene	$C_6H_2F_4$
1,2,3,5-Tetrafluorobenzene	$C_6H_2F_4$
1,2,4,5-Tetrafluorobenzene	$C_6H_2F_4$
Tetrafluoroethene	C_2F_4
Tetrafluoroethylene	C_2F_4
Tetrafluoromethane	CF_4
Tetra-n-hexylammonium perchlorate	$C_{24}H_{52}ClNO_4$
1,2,3,4-Tetrahydroxybutane	$C_4H_{10}O_4$
Tetrahydrofuran	C_4H_8O
α-Tetrahydrofurfuryl alcohol	$C_5H_{10}O_2$
1,2,3,4-Tetrahydronaphthalene	$C_{10}H_{12}$
Tetrahydrophthalic anhydride	$C_8H_8O_3$
Tetrahydropyran	$C_5H_{10}O$
1,2,3,4-Tetrahydroxybutane	$C_4H_{10}O_4$
Tetrakis(methylthia)methane	$C_5H_{12}S_4$
Tetrakis[μ_3-methoxo-2,4-pentanedionato (methanol)nickel(II)]	$C_{28}H_{56}Ni_4O_{16}$
Tetramethylammonium bromide	$C_4H_{12}BrN$
Tetramethylammonium chloride	$C_4H_{12}ClN$
Tetramethylammonium hydrogen dichloride	$C_4H_{13}Cl_2N$
Tetramethylammonium iodide	$C_4H_{12}IN$
1,2,3,4-Tetramethylbenzene	$C_{10}H_{14}$
1,2,3,5-Tetramethylbenzene	$C_{10}H_{14}$
1,2,4,5-Tetramethylbenzene	$C_{10}H_{14}$
2,2,3,3-Tetramethylbutane	C_8H_{18}
Tetramethyldisilacyclobutane	$C_6H_{16}Si_2$
Tetramethyldisiletan	$C_6H_{16}Si_2$
Tetramethylethylene	C_6H_{12}
1,3,5,7-Tetramethyl-2,4,6,8,9,10-hexathiaadamantane	$C_8H_{12}S_6$
Tetramethyl lead	$C_4H_{12}Pb$
Tetramethylsilane	$C_4H_{12}Si$
1,1,3,3-Tetramethyl-5,5,7,7-tetraphenyl-cyclotrisiloxane	$C_{28}H_{32}O_4Si_4$
1,3,5,7-Tetramethyl-2,4,6,8-tetrathiaadamantane	$C_{10}H_{16}S_4$
2,4,6,N-Tetranitro-ethylaniline	$C_8H_7N_5O_8$
2,4,6,N-Tetranitro-N-methylaniline	$C_7H_5N_5O_8$
2,4,6,N-Tetranitro-N-methyltoluidine	$C_8H_7N_5O_8$
1,3,5,7-Tetranitro-1,3,5,7-tetraazacyclooctane(α)	$C_4H_8N_8O_8$
1,3,5,7-Tetranitro-1,3,5,7-tetraazacyclooctane(β)	$C_4H_8N_8O_8$
1,3,5,7-Tetranitro-1,3,5,7-tetraazacyclooctane(γ)	$C_4H_8N_8O_8$
1,3,5,7-Tetranitro-1,3,5,7-tetraazacyclooctane(δ)	$C_4H_8N_8O_8$
2,5,8,11-Tetraoxadodecane	$C_8H_{18}O_4$
1,1,1,2-Tetraphenylethane	$C_{26}H_{22}$
1,1,2,2-Tetraphenylethane	$C_{26}H_{22}$
Tetraphenylethylene	$C_{26}H_{20}$
Tetraphenylmethane	$C_{25}H_{20}$
Tetraphenylsilane	$C_{24}H_{20}Si$
Tetraphenylstannane	$C_{24}H_{20}Sn$
Tetraphenyltin	$C_{24}H_{20}Sn$
Tetrapropylammonium iodide	$C_{12}H_{28}IN$
n-Tetratriacontane	$C_{34}H_{70}$
Tetroxan	$C_4H_8O_4$
Tetryl	$C_7H_5N_5O_8$
Tetryl-bis(trinitrotoluene) complex	$C_{21}H_{15}N_{11}O_{20}$
Tetryl-picric acid complex	$C_{13}H_8N_8O_{15}$
Thiabutane	C_3H_8S
Thiacyclobutane	C_3H_6S
Thiacyclohexane	$C_5H_{10}S$
Thiacyclopentane	C_4H_8S
4-Thiaheptane	$C_6H_{14}S$
2-Thiahexane	$C_5H_{12}S$
3-Thiahexane	$C_5H_{12}S$
5-Thianonane	$C_8H_{18}S$
2-Thiapentane	$C_4H_{10}S$
3-Thiapentane	$C_4H_{10}S$
2-Thiapropane	C_2H_6S
Thiazole	C_3H_3NS
β-Thiolactic acid	$C_3H_6O_2S$
3-Thiolpropanoic acid	$C_3H_6O_2S$
Thiophene	C_4H_4S
Thiophenol	C_6H_6S
Thiourea-cycloheptane adduct	$C_{21.9}H_{45.8}N_2S$
Thiourea-cyclohexane adduct	$C_{19.6}H_{41.2}N_2S$
Thiourea-cyclooctane adduct	$C_{24.8}H_{51.6}N_2S$
Thiourea-2,2-dimethylbutane adduct	$C_{18.4}H_{44.6}N_2S$
Thiourea-ferrocene adduct	$C_{13}H_{22}FeN_6S_3$
Thymine	$C_5H_6N_2O_2$
Tin tetraamyl	$C_{20}H_{44}Sn$
Tin tetraethyl	$C_8H_{20}Sn$
Toluene	C_7H_8
m-Toluic acid	$C_8H_8O_2$
o-Toluic acid	$C_8H_8O_2$
p-Toluic acid	$C_8H_8O_2$
m-Toluidine	C_7H_9N
o-Toluidine	C_7H_9N
p-Toluidine	C_7H_9N
p-Toluidine-phenol complex	$C_{13}H_{15}NO$
2,4-Tolylenediisocyanate	$C_9H_6N_2O_2$
Triacontane	$C_{30}H_{62}$
Tri-L-alanine	$C_9H_{17}N_3O_4$
Triamantane	$C_{18}H_{24}$
s-Triazine	$C_3H_3N_3$
Tribenzyl-n-hexyldecylsilane	$C_{37}H_{54}Si$
Tribromomethane	$CHBr_3$
1,2,3-Tribromopropane	$C_3H_5Br_3$
Tributyrin	$C_{15}H_{26}O_6$
Tricaproin	$C_{21}H_{38}O_6$
α,α,α-Trichloroacetaldehyde	C_2HCl_3O
1,2,4-Trichlorobenzene	$C_6H_3Cl_3$
2,2,3-Trichlorobutanal	$C_4H_5Cl_3O$
2,2,2-Trichloroethanal	C_2HCl_3O

1,1,1-Trichloroethane	$C_2H_3Cl_3$
1,1,2-Trichloroethane	$C_2H_3Cl_3$
Trichloroethene	C_2HCl_3
Trichloroethylene	C_2HCl_3
Trichlorofluoromethane	CCl_3F
Trichloromethane	$CHCl_3$
1,2,3-Trichloropropane	$C_3H_5Cl_3$
o-Trichlorosilylbiphenyl	$C_{12}H_9Cl_3Si$
p-Trichlorosilylbiphenyl	$C_{12}H_9Cl_3Si$
1,3,5-Trichloro-2,4,6-trifluorobenzene	$C_6Cl_3F_3$
1,1,2-Trichloro-1,2,2-trifluoroethane	$C_2Cl_3F_3$
1,1,1-Trichloro-3,3,3-trifluoropropane	$C_3H_2Cl_3F_3$
1,2,3-Trichloro-4,5,6-trimethylbenzene	$C_9H_9Cl_3$
n-Tricosane	$C_{23}H_{48}$
Tricyclo[3.3.1.1^{3,7}]decane	$C_{10}H_{16}$
Tricyclo[5.2.1.0^{2,6}]decane	$C_{10}H_{16}$
Tricyclo[2.2.1.0^{2,6}]heptane	C_7H_{10}
1,1,3-Tricyclohexylpropane	$C_{21}H_{38}$
n-Tridecane	$C_{13}H_{28}$
Tridecanoic acid	$C_{13}H_{26}O_2$
Tridecanoin	$C_{33}H_{62}O_6$
1-Tridecanol	$C_{13}H_{28}O$
Tridecanolactone	$C_{13}H_{24}O_2$
n-Tridecyl alcohol	$C_{13}H_{28}O$
Tridecyl methyl ketone	$C_{15}H_{30}O$
Triethylamine	$C_6H_{15}N$
Triethylamineborane	$C_6H_{18}BN$
Triethylantimony	$C_6H_{15}Sb$
Triethylarsine	$C_6H_{15}As$
Triethylborane	$C_6H_{15}B$
Triethylenediamine	$C_6H_{12}N_2$
Triethylene glycol	$C_6H_{14}O_4$
Triethylgallium	$C_6H_{15}Ga$
Triethylindium	$C_6H_{15}In$
Triethylstibine	$C_6H_{15}Sb$
Trifluoroacetonitrile	C_2F_3N
Trifluoroacetyl fluoride	C_2F_4O
Trifluorochloroethene	C_2ClF_3
Trifluorochloroethylene	C_2ClF_3
1,1,1-Trifluoro-3-chloropropane	$C_3H_4ClF_3$
1,1,1-Trifluoro-3,3-dichloropropane	$C_3H_3Cl_2F_3$
1,1,1-Trifluoroethane	$C_2H_3F_3$
Trifluoromethane	CHF_3
Trifluoromethyl cyanide	C_2F_3N
α,α,α,-Trifluorotoluene	$C_7H_5F_3$
1,1,2-Trifluoro-1,2,2-trichloroethane	$C_2Cl_3F_3$
Triglycine sulfate	$C_6H_{13}N_3O_8S$
Triglyme	$C_8H_{18}O_4$
1,2,3-Trihydroxypropane	$C_3H_8O_3$
1,2,3-Trihydroxypropane-d$_3$	$C_3H_5D_3O_3$
Trilaurin	$C_{39}H_{74}O_6$
Trimellitic anhydride	$C_9H_7O_5$
Trimethylacetonitrile	C_5H_9N
Trimethylaluminum	C_3H_9Al
Trimethylamine	C_3H_9N
Trimethylamineborane	$C_3H_{12}BN$
2,N,N-Trimethylaniline	$C_9H_{13}N$
1,2,3-Trimethylbenzene	C_9H_{12}
1,2,4-Trimethylbenzene	C_9H_{12}
1,3,5-Trimethylbenzene	C_9H_{12}
Trimethylborane	C_3H_9B
2,2,3-Trimethylbutane	C_7H_{16}
Trimethylene oxide	C_3H_6O
Trimethylgallium	C_3H_9Ga
Trimethylhydrazine	$C_3H_{10}N_2$
2,2,4-Trimethylpentane	C_8H_{18}
2,3,3-Trimethylpentane	C_8H_{18}
2,3,4-Trimethylpentane	C_8H_{18}
2,4,4-Trimethyl-1-pentene	C_8H_{16}
2,4,4-Trimethyl-2-pentene	C_8H_{16}
2,2,N-Trimethylpropanamide	$C_6H_{13}NO$
Trimethylsulfonylmethane	$C_4H_{10}O_6S_3$
2,4,6-Trimethyl-1,3,5-trioxacyclohexane	$C_6H_{12}O_3$
2,4,6-Trimethyl-1,3,5-trioxane	$C_6H_{12}O_3$
Trimyristin	$C_{45}H_{86}O_6$
1,3,5-Trinaphthylbenzene	$C_{36}H_{24}$
1,3,5-Tri-2-naphthylbenzene	$C_{36}H_{24}$
1,3,5-Trinitrobenzene	$C_6H_3N_3O_6$
2,4,6-Trinitro-N-(methylnitro)-m-toluidine	$C_8H_7N_5O_8$
2,4,6-Trinitrophenol	$C_6H_3N_3O_7$
2,4,6-Trinitrophenylethyl nitramine	$C_8H_7N_5O_8$
2,4,6-Trinitrophenylmethyl nitramine	$C_7H_5N_5O_8$
2,4,5-Trinitrotoluene	$C_7H_5N_3O_6$
2,4,6-Trinitrotoluene	$C_7H_5N_3O_6$
1,3,5-Trinitro-1,3,5-triazacyclohexane	$C_3H_6N_6O_6$
Trioctanoin	$C_{27}H_{50}O_6$
2,5,8-Trioxanonane	$C_6H_{14}O_3$
3,6,9-Trioxaundecane	$C_8H_{18}O_3$
Tripalmitin	$C_{51}H_{98}O_6$
Triphenylamine	$C_{18}H_{15}N$
Triphenylarsine	$C_{18}H_{15}As$
1,3,5-Triphenylbenzene	$C_{24}H_{18}$
Triphenylbismuthine	$C_{18}H_{15}Bi$
Triphenylcarbinol	$C_{19}H_{16}O$
Triphenylchloromethane	$C_{19}H_{15}Cl$
Triphenylene	$C_{18}H_{12}$
1,1,1-Triphenylethane	$C_{20}H_{18}$
1,1,2-Triphenylethane	$C_{20}H_{18}$
Triphenylethylene	$C_{20}H_{16}$
Triphenylmethane	$C_{19}H_{16}$
Triphenylphosphine	$C_{18}H_{15}P$
Triphenylstibine	$C_{18}H_{15}Sb$
Triptycene	$C_{20}H_{14}$
Trishydrate hexacetate chromate chloride hexahydrate	$C_{12}H_{36}ClCrO_{22}$
Tris(hydroxymethyl)aminomethane	$C_4H_{11}NO_3$
Tris(sarcosine)calcium chloride	$C_9H_{21}CaCl_2N_3O_6$
Tristearin	$C_{57}H_{110}O_6$
n-Tritriacontane	$C_{33}H_{68}$
Tryptophane(L)	$C_{11}H_{12}N_2O_2$
Tyrosine(L)	$C_9H_{11}NO_3$

U

n-Unatriacontane	$C_{31}H_{64}$
n-Undecane	$C_{11}H_{24}$
n-Undecane-urea adduct	$C_{2.2}H_{6.6}N_2O$
Undecanoic acid	$C_{11}H_{22}O_2$
Undecanol-1	$C_{11}H_{24}O$
Undecanolactone	$C_{11}H_{20}O_2$
1-Undecene	$C_{11}H_{22}$
Undecyl alcohol	$C_{11}H_{24}O$
Uracil	$C_4H_4N_2O_2$
Urea	CH_4N_2O
Urea-n-decane adduct	$C_{11}H_{26}N_2O$
Urea-1-decene adduct	$C_{2.3}H_{6.7}N_2O$
Urea-n-dodecane adduct	$C_{13}H_{30}N_2O$
Urea-1-dodecene adduct	$C_{2.2}H_{6.5}N_2O$
Urea-n-eicosane adduct	$C_{21}H_{46}N_2O$
Urea-1-eicosene adduct	$C_{2.4}H_{6.8}N_2O$
Urea-n-hexadecane adduct	$C_{17}H_{38}N_2O$
Urea-1-hexadecene adduct	$C_{2.3}H_{6.7}N_2O$
Urea-1-octadecene adduct	$C_{2.4}H_{6.8}N_2O$
Urea-phenol complex	$C_{13}H_{16}N_2O_3$
Urea-trioxane inclusion compound	$C_{10}H_{22}N_2O_{10}$
Urea-n-undecane adduct	$C_{2.2}H_{6.6}N_2O$
Uric acid	$C_5H_4N_4O_3$

V

Valeral	$C_5H_{10}O$
Valeraldehyde	$C_5H_{10}O$
n-Valeric acid	$C_5H_{10}O_2$
Valeryl chloride	C_5H_9ClO
Valine(L)	$C_5H_{11}NO_2$
Vanadocene	$C_{10}H_{10}V$

X,Y,Z

10. Bibliography

81REI von Reis, M. A., Die specifische Wärme flüssiger organischer Verbindungen und ihre Beziehung zu deren Moleculargewicht, *Ann. Physik* [3] **13**, 447–464 (1881).

84WER Werner, E., Recherches sur les phenols bromes, *Ann. chim. phys.* [6] **3**, 567–574 (1884).

85STO/WIL Stohmann, F., and Wilsing, H., Ueber die specifische Wärme und die Schmelzwärme der Myristinsäure und der Laurinsäure, *J. Prakt. Chem.* **140**, 80–93 (1885).

02LOU Louguinine, W., Etude de la chaleur latente de vaporisation de l'aniline, de l'orthotoluidine, de quelques-uns de leurs derives ainsi que d'autres substances de la chimie organique, *Ann. Chim. Phys.* [7] **27**, 105–144 (1902).

07WAL Walden, P., Über organische Lösungs- und Ionisierungsmittel. VI. Teil. Lösungswärmen, *Z. physik. Chem.* **58**, 479–496 (1907).

16BRA Bramley, A., The study of binary mixtures. Part IV. Heats of reaction and specific heats, *J. Chem. Soc. (London)* **109**, 496–515 (1916).

16RIC/SHI Richards, T. W., and Shipley, J. W., The Compressibility of Certain Typical Hydrocarbons, Alcohols, and Ketones, *J. Am. Chem. Soc.* **38**, 989–999 (1916).

17HIL/DUS Hildebrand, J. H., Duschak, A. D., Foster, A. H., and Beebe, C. W., The specific heats and heats of fusion of triphenylmethane, anthraquinone and anthracene, *J. Am. Chem. Soc.* **39**, 2293–2297 (1917).

17MAT/KRA Mathews, J. H., Krause, E. L., and Bohnson, V. L., A contribution to the thermal chemistry of pyridine, *J. Am. Chem. Soc.* **39**, 398–413 (1917).

19DEJ Dejardin, G., Pressions maxima des vapeurs du benzene et du cyclohexane aux temperatures moyennes et calcul de leurs chaleurs specifiques principales, *Ann. phys.* [9] **11**, 253–291 (1919).

20GIB/LAT Gibson, G. E., Latimer, W. M., and Parks, G. S., Entropy changes at low temperatures. I. Formic acid and urea. A test of the third law of thermodynamics, *J. Am. Chem. Soc.* **42**, 1533–1542 (1920).

22LAT Latimer, W. M., The distribution of thermal energy in the tetrachlorides of carbon, silicon, titanium and tin, *J. Am. Chem. Soc.* **44**, 90–97 (1922).

22SIM Simon, F., Untersuchungen über die spezifische Wärme bei tiefen Temperaturen, *Ann. Physik* [4] **68**, 241–280 (1922).

23GIB/GIA Gibson, G. E., and Giauque, W. F., The third law of thermodynamics. Evidence from the specific heats of glycerol that the entropy of a glass exceeds that of a crystal at the absolute zero, *J. Am. Chem. Soc.* **45**, 93–104 (1923).

24GAR/RAN Garner, W. E., and Randall, F. C., Alternation in the heats of crystallisation of the normal monobasic fatty acids. Part I., *J. Chem. Soc.* **125**, 881–896 (1924).

24HER/BLO Herz, W., and Bloch, W., Physikalisch-chemische Untersuchungen an Verbindungen der Cyklohexanreihe, *Z. physik. Chem.* **110**, 23–39 (1924).

24KEY/BEA Keyes, F. G., and Beattie, J. A., A calorimeter for measuring specific heats and heats of vaporization of liquids. The specific heat and heat of vaporization of liquid ethyl ether at 0° and 12°, *J. Am. Chem. Soc.* **46**, 1753–1760 (1924).

24LAN Lange, F., Untersuchungen über die spezifische Wärme bei tiefen Temperaturen, *Z. physik. Chem.* **110**, 343–362 (1924).

24SHO Shorthose, D. N., *G. B. DSIR, Food Invest. Board, Spec. Rept.* No. **19**, 16pp. (1924).

24TAY/RIN Taylor, C. A., and Rinkenbach, W. H., The specific heats of trinitrotoluene, tetryl, picric acid and their molecular complexes, *J. Am. Chem. Soc.* **46**, 1504–1510 (1924).

24WIL/DAN Willams, J. W., and Daniels, F., The specific heats of certain organic liquids at elevated temperatures, *J. Am. Chem. Soc.* **46**, 903–917 (1924).

25MAA/WAL Maass, O., and Waldbauer, L. J., The specific heats and latent heats of fusion of ice and of several organic compounds, *J. Am. Chem. Soc.* **47**, 1–9 (1925).

25PAR Parks, G. S., Thermal data on organic compounds. I. The heat capacities and free energies of methyl, ethyl and normal-butyl alcohols, *J. Am. Chem. Soc.* **47**, 338–345 (1925).

25PAR/KEL Parks, G. S., and Kelley, K. K., Thermal data on organic compounds. II. The heat capacities of five organic compounds. The entropies and free energies of some homologous series of aliphatic compounds, *J. Am. Chem. Soc.* **47**, 2089–2097 (1925).

25WIL Williams, J. W., A study of the physical properties of nitromethane, *J. Am. Chem. Soc.* **47**, 2644–2652 (1925).

25WIL/DAN Williams, J. W., and Daniels, F., The specific heats of binary mixtures, *J. Am. Chem. Soc.* **47**, 1490–1503 (1925).

26AND Andrews, D. H., The specific heats of some isomers of the type ortho, meta and para C_6H_4XY from 110 to 340K, *J. Am. Chem. Soc.* **48**, 1287–1298 (1926).

26AND/LYN Andrews, D. H., Lynn, G., and Johnston, J., The heat capacities and heat of crystallization of some isomeric aromatic compounds, *J. Am. Chem. Soc.* **48**, 1274–1287 (1926).

26PAR/AND Parks, G. S., and Anderson, C. T., Thermal data on organic compounds. III. The heat capacities, entropies and free energies of tertiary butyl alcohol, mannitol, erythritol and normal butyric acid, *J. Am. Chem. Soc.* **48**, 1506–1512 (1926).

26PAR/HUF Parks, G. S., and Huffman, H. M., Thermal data on organic compounds. IV. The heat capacities, entropies and free energies of normal propyl alcohol, ethyl ether and dulcitol, *J. Am. Chem. Soc.* **48**, 2788–2793 (1926).

27BEN/WEN Bennewitz, K., and Wendroth, H., Untersuchungen im kritischen Gebiet. II. Bestimmung der wahren spezifischen Wärme C_p des flüssigen Äthyläthers ober- und unterhalb der kritischen Temperatur, *Z. physik. Chem.* **125**, 111–134 (1927).

27PAR/HUF Parks, G. S., and Huffman, H. M., Studies on glass. I. The transition between the glassy and liquid states in the case of some simple organic compounds, *J. Phys. Chem.* **31**, 1842–1855 (1927).

28AND/HAW Andrews, D. H., and Haworth, E., An application of the rule of Dulong and Petit to molecules, *J. Am. Chem. Soc.* **50**, 2998–3002 (1928).

28CLU/HAR Clusius, K., and Harteck, P., Über die spezifischen Wärmen einiger fester Körper bei tiefen Temperaturen, *Z. physik. Chem.* **134**, 243–263 (1928).

28LAN Lang, H. R., On the measurement of the variation of the specific heat of aniline with temperature, using the continuous flow electric method, *Proc. Roy. Soc. (London) A* **118**, 138–156 (1928).

28PAR/KEL Parks, G. S., and Kelley, K. K., The application of the third law of thermodynamics to some organic reactions, *J. Phys. Chem.* **32**, 734–750 (1928).

28SKA/SAX Skau, E. L., and Saxton, B., The freezing point-solubility relations of geometrical isomers. I. The β-chlorocrotonic acids, *J. Am. Chem. Soc.* **50**, 2693–2701 (1928).

29HER/LOR Herz, W., and Lorentz, E., Physikalisch–chemische Untersuchungen an Dioxan, *Z. physik. Chem. A* **140**, 406–422 (1929).

29KEL Kelley, K. K., Cyclohexanol and the third law of thermodynamics, *J. Am. Chem. Soc.* **51**, 1400–1406 (1929).

29KEL 2 Kelley, K. K., The heat capacities of ethyl and hexyl alcohols from 16°K to 298°K and the corresponding entropies and free energies, *J. Am. Chem. Soc.* **51**, 779–786 (1929).

29KEL 3 Kelley, K. K., The heat capacities of isopropyl alcohol and acetone from 16 to 298K, and the corresponding enthalpies and free energies, *J. Am. Chem. Soc.* **51**, 1145–1150 (1929).

29KEL 4 Kelley, K. K., The heat capacity of toluene from 14K to 298K. The entropy and the free energy of formation, *J. Am. Chem. Soc.* **51**, 2738–2741 (1929).

29KEL 5 Kelley, K. K., The heat capacity of methyl alcohol from 16K to 298K and the corresponding entropy and free energy, *J. Am. Chem. Soc.* **51**, 180–187 (1929).

29MIT/HAR Mitsukuri, S., and Hara, K., Specific heats of acetone, methyl–, ethyl–, and n–propyl–alcohols at low temperatures, *Bull. Chem. Soc. Japan* **4**, 77–81 (1929).

29PAR/KEL Parks, G. S., Kelley, K. K., and Huffman, H. M., Thermal data on organic compounds. V. A revision of the entropies and free energies of nineteen organic compounds, *J. Am. Chem. Soc.* **51**, 1969–1973 (1929).

30HUF/PAR Huffman, H. M., Parks, G. S., and Daniels, A. C., Thermal data on organic compounds. VII. The heat capacities, entropies and free energies of twelve aromatic hydrocarbons, *J. Am. Chem. Soc.* **52**, 1547–1558 (1930).

30HUF/PAR 2 Huffman, H., M., Parks, G. S., and Thomas, S. B., Thermal data on organic compounds. VIII. The heat capacities, entropies and free energies of the isomeric heptanes, *J. Am. Chem. Soc.* **52**, 3241–3251 (1930).

30PAR/HUF Parks, G. S., Huffman, H. M., and Thomas, S. B., Thermal data on organic compounds. VI. The heat capacities, entropies and free energies of some saturated, non–benzenoid hydrocarbons, *J. Am. Chem. Soc.* **52**, 1032–1041 (1930).

30PAR/HUF 2 Parks, G. S., and Huffman, H. M., Thermal data on organic compounds. IX. A study of the effect of unsaturation on the heat capacities, entropies and free energies of some hydrocarbons and other compounds, *J. Am. Chem. Soc.* **52**, 4381–4391 (1930).

30WAS Wassermann, A., Die zwischenmolekularen Bindungsfestigkeiten der Fumar– und Maleinsäure and ihrer Dimethylester, *Z. Physik. Chem. A* **146**, 418–445 (1930).

30WIE/HUB Wiebe, R., Hubbard, K. H., and Brevoort, M. J., The heat capacity of saturated liquid ethane from the boiling point to the critical temperature and heat fusion of the solid, *J. Am. Chem. Soc.* **52**, 611–622 (1930).

31BLA/LET Blacet, F. E., Leighton, P. A., and Bartlett, E. P., The specific heats of five pure organic liquids and of ethyl alcohol–water mixtures, *J. Phys. Chem.* **35**, 1935–1943 (1931).

31CLI/AND Cline, J. K., and Andrews, D. H., Thermal energy studies. III. The octanols, *J. Am. Chem. Soc.* **53**, 3668–3673 (1931).

31DEE Deese, R. F., Jr., Thermal energy studies. IV. Comparison of continuous and discontinuous methods of measuring heat capacities. Heat capacities of some aliphatic bromides, *J. Am. Chem. Soc.* **53**, 3673–3683 (1931).

31FIO/GIN Fiock, E. F., Ginnings, D. C., and Holton, W. B., Calorimetric determinations of thermal properties of methyl alcohol, ethyl alcohol, and benzene, *J. Research NBS* **6**, 881–800 (1931).

31FOR/BRU Forrest, H. O., Brugmann, E. W., and Cummings, L. W. T., The specific heat of diphenyl, *Ind. Eng. Chem.* **23**, 37–39 (1931).

31FRA Frandsen, M., Cryoscopic constant, heat of fusion, and heat capacity of camphor, *Bur. Stand. J. Res.* **7**, 477–483 (1931).

31GAR/VAN Garner, W. E., Van Bibber, K., and King, A. M., The melting points and heats of crystallisation of the normal long–chain hydrocarbons, *J. Chem. Soc. (London)* **1931**, 1533–1541.

31BUF/FLE Buffington, R. M., and Fleischer, J., Thermodynamic properties of dichlorodifluoromethane, a new refrigerant. IV Specific heat of liquid and vapor and latent heat of vaporization, *Ind. Eng. Chem.* **23**, 1290–1292 (1931).

31HUF/PAR Huffman, H. M., Parks, G. S., and Barmore, M., Thermal data on organic compounds. X. Further studies on the heat capacities, entropies and free energies of hydrocarbons, *J. Am. Chem. Soc.* **53**, 3876–3888 (1931).

31SMI/AND Smith, R. H., and Andrews, D. H., Thermal energy studies. I. Phenyl derivatives of methane, ethane and some related compounds, *J. Am. Chem. Soc.* **53**, 3644–3660 (1931).

31SMI/AND 2 Smith, R. H., and Andrews, D. H., Thermal energy studies. II. Phenyl derivatives of metals, *J. Am. Chem. Soc.* **53**, 3661–3667 (1931).

31SWI/RYB Swietoslawski, W., Rybicka, S., and Solodkowska, W., Sur un microcalorimetre adiabatique, adapte aux mesures de la chaleur specifique de substances solides et liquides, *Rocz. Chem.* **11**, 65–77 (1931).

31SWI/RYB 2 Swietoslawski, W., Rybicka, M., and Solodkowska, W., Sur un microcalorimetre adiabatique, adapte aux mesures de la chaleur specifique des substances solides et liquides, *Bull. Int. Acad. Pol. Sci. Lett. Cl. Sci. Math Nat. Ser A.* **1931**, 322–335

32HUF/BOR Huffman, H. M., and Borsook, H., Thermal data. I. The heat capacities, entropies and free energies of seven organic compounds containing nitrogen, *J. Am. Chem. Soc.* **54**, 4297–4301 (1932).

32NEI/KUR Neiman, M. B., and Kurlyankin, I. A., Thermodynamic studies of solutions. II. Study of the thermodynamics of aqueous solutions of ethylene glycol at different temperatures, *Zhur. Obshch. Khim.* **2**, 318–321 (1932).

32NEU Neumann, M. B., Die Untersuchung der Wärmekapazität vom binären System $CH_3COOH + H_2O$ bei verschiedenen Temperaturen, *Z. physik. Chem.* **A158**, 258–264 (1932).

32RIC/WAL Richards, W. T., and Wallace, J. H., Jr., The specific heats of five organic liquids from their adiabatic temperature–pressure coefficients, *J. Am. Chem. Soc.* **54**, 2705–2713 (1932).

32SPA/THO Spaght, M. E, Thomas, S. B., and Parks, G. S., Some heat capacity data on organic compounds obtained with a radiation calorimeter, *J. Phys. Chem.* **36**, 882–888 (1932).

32STU Sturley, K. R., Fresh data on the latent heats and heat conductivities of some aquacrystalline compounds, *J. Soc. Chem. Ind. Trans. Comm.* **51**, 271T–273T (1932).

32TRE Trew, V. C. G., Physical properties of mixtures of acetone and bromoform, *Trans. Faraday Soc.* **28**, 509–514 (1932).

33FER/MIL Ferguson, A., and Miller, J. T., A method for the determination of the specific heats of liquids, and a determination of the specific heats of aniline and benzene over the approximate range 20°C to 50°C, *Proc. Phys. Soc. London* **45**, 194–207 (1933).

33KOL/UDO de Kolossowsky, N. A., and Udowenko, W. W., Mesure des chaleurs specifique moleculaires de quelques liquides, *Compt. rend.* **197**, 519–520 (1933).

33PAR/HUF Parks, G. S., Huffman, H. M., and Barmore, M., Thermal data on organic compounds. XI. The heat capacities, entropies and free energies of ten compounds containing oxygen or nitrogen, *J. Am. Chem. Soc.* **55**, 2733–2740 (1933).

33ROT/MEY Roth, W. A., and Meyer, I., Einige physikalisch-chemische Konstanten des Dioxans, *Z. Elektrochem.* **39**, 35–37 (1933).

33SOU/BRI Southard, J. C., and Brickwedde, F. G., Low temperature specific heats. I. An improved calorimeter for use from 14 to 300K. The heat capacity and entropy of naphthalene, *J. Am. Chem. Soc.* **55**, 4378–4384 (1933).

33SOU/MIL Southard, C., Milner, R. T., and Hendricks, S. B., Low temperature specific heats: III. Molecular rotation in crystalline primary normal amyl ammonium chloride, *J. Chem. Phys.* **1**, 95–102 (1933).

33TRE/WAT Trew, V. C. G., and Watkins, G. M. C., Some physical properties of mixtures of certain organic liquids, *Trans. Faraday Soc.* **29**, 1310–1318 (1933).

34JAC/PAR Jacobs, C. J., and Parks, G. S., Thermal data on organic compounds. XIV. Some heat capacity, entropy and free energy data for cyclic substances, *J. Am. Chem. Soc.* **56**, 1513–1517 (1934).

34KOL/UDO de Kolossowsky, N., and Udowenko, W. W., Determination des chaleurs specifiques des liquides, *Compt. rend.* **198**, 1394–1395 (1934).

34KOL/UDO 2 Kolosovskii, N. A., and Udovenko, V. V., Specific heat of liquids. II., *Zhur. Obshchei Khim.* **4**, 1027–1033 (1934).

34MEH Mehl, W., Physikalische Eigenschaften des Vinylbromids, *Z. physik. Chem.A* **169**, 312–313 (1934).

34MEH 2 Mehl, W., Die spezifische Wärme einiger flüssiger Kältemittel, *Z. ges. Kalte-Ind.* **41**, 152–153 (1934).

34PAR/LIG Parks, G. S., and Light, D. W., Thermal data on organic compounds. XIII. The heat capacities and entropies of n-tetadecane and the hydroxybenzoic acids. The relative free energies of some benzenoid position isomers, *J. Am. Chem. Soc.* **56**, 1511–1513 (1934).

34PAR/THO Parks, G. S., and Thomas, S. B., The heat capacities of crystalline, glassy and undercooled liquid glucose, *J. Am. Chem. Soc.* **56**, 1423 (1934).

34PAR/TOD Parks, G. S., and Todd, S. S., Some heat capacity data for liquid nitrobenzene, no indication of allotropy, *J. Chem. Phys.* **2**, 440–441 (1934).

34PEA/TAN Pearce, J. N., and Tanner, W. B., The heat capacity and the free energy of formation of naphthalene, *Proc. Iowa Acad. Sci.* **41**, 123–126 (1934).

34RAD/JUL Radulescu, D., and Jula, O., Beiträge zur Bestimmung der Abstufung der Polarität des Aminstickstoffes in den organischen Verbindungen, *Z. physik. Chem. B* **26**, 390–393 (1934).

35AOY/KAN Aoyama, S., and Kanda, E., Studies on the heat capacities at low temperature. Report I. Heat capacities of some organic substances at low temperature, *Sci. Rept. Tohoku Imp. Univ.* [1] **24**, 107–115 (1935).

35BEK/MAT Bekkedahl, N., and Matheson, H., Heat capacity, entropy, and free energy of rubber hydrocarbon, *J. Research NBS* **15**, 503–515 (1935).

35HUF/ELL Huffman, H. M., and Ellis, E. L., Thermal data. III. The heat capacities, entropies and free energies of four organic compounds containing sulfur, *J. Am. Chem. Soc.* **57**, 46–48 (1935).

35MIL Miller, P., The free energy of furfural and some of its derivatives, *Iowa State Coll. J. Sci.* **10**, 91–93 (1935).

35STI/HUF Stiehler, R. D., and Huffman, H. M., Thermal data. V. The heat capacities, entropies and free energies of adenine, hypoxanthine, guanine, xanthine, uric acid, allantoin and alloxan, *J. Am. Chem. Soc.* **57**, 1741–1743 (1935).

35WHI/MOR White, A. H., and Morgan, S. O., Transitions in camphor and chemically related compounds. II. Vibration of atomic groups, *J. Am. Chem. Soc.* **57**, 2078–2086 (1935).

36AST/MES Aston, J. G., and Messerly, G. H., Heat capacities and entropies of organic compounds. II. Thermal and vapor pressure data for tetramethylmethane from 13.22°K to the boiling point. The entropy from its Raman spectrum, *J. Am. Chem. Soc.* **58**, 2354–2361 (1936).

36ERN/WAT Ernst, R. C., Watkins, C. H., and Ruwe, H. H., The physical properties of the ternary system ethyl alcohol-glycerin-water, *J. Phys. Chem.* **40**, 627–635 (1936).

36EVA/EDL Evans, T. W., and Edlund, K. R., Tertiary alkyl ethers preparation and properties, *Ind. Eng. Chem.* **28**, 1186–1188 (1936).

36FUR/STE Furtsch, E. F., and Stegeman, G., The heat capacity and entropy of beta-lactose, *J. Am. Chem. Soc.* **58**, 881–882 (1936).

36KHO/KAL Khokhlovkin, M. A., and Kalacheva, A. V., The specific heat of 2,3-butanediol, *Sintet. Kauchuk* **5**, No.1, 25–27 (1936).

36KUR/VOS Kurnakov, N. S., and Voskresenskaya, N. K., Calorimetry of liquid binary systems, *Izv. Akad. Nauk SSSR, Otdel. Mat. i Estestv. Nauk. Ser. Khim*, **1936**, 439–461.

36PAR/THO Parks, G. S., Thomas, S. B., and Light, D. W., XII. Some new heat capacity data for organic glasses. The entropy and free energy of dl-lactic acid, *J. Chem. Phys.* **4**, 64–69 (1936).

36PAR/TOD Parks, G. S., Todd, S. S., and Moore, W. A., Thermal data on organic compounds. XVI. Some heat capacity, entropy and free energy data for typical benzene derivatives and heterocyclic compounds, *J. Am. Chem. Soc.* **58**, 398–401 (1936).

36PAR/TOD 2 Parks, G. S., Todd, S. S., and Shomate, C. H., Thermal data on organic compounds. XVII. Some heat capacity, entropy and free energy data for five higher olefins, *J. Am. Chem. Soc.* **58**, 2505–2508 (1936).

36PEA/BAK Pearce, J. N., and Bakke, H. M., The heat capacity and the free energy of formation of pyridine, *Proc. Iowa Acad. Sci.* **43**, 171–174 (1936).

36TOD/PAR Todd, S. S., and Parks, G. S., Thermal data on organic compounds. XV. Some heat capacity, entropy and free energy data for the isomeric butenes, *J. Am. Chem. Soc.* **58**, 134–137 (1936).

37AHL/BLA Ahlberg, J. E., Blanchard, E. R., and Lundberg, W. O., The heat capacities of benzene, methyl alcohol and glycerol at very low temperatures, *J. Chem. Phys.* **5**, 537–551 (1937).

37AST/SIL Aston, J. G., Siller, C. W., and Messerly, G. H., Heat capacities and entropies of organic compounds. III. Methylamine from 12K to the boiling point. Heat of vaporization and vapor pressure. The entropy from molecular data, *J. Am. Chem. Soc.* **59**, 1743–1751 (1937).

37BEK/WOO Bekkedahl, N., and Wood, L. A., Entropy of isoprene from heat-capacity measurements, *J. Res. NBS* **19**, 551–558 (1937).

37BRO/MAN Brown, O. L. I., and Manov, G. G., The heat capacity of carbon disulfide from 15 to 300°K. The entropy and heat of fusion of carbon disulfide, *J. Am. Chem. Soc.* **59**, 500–502 (1937).

37CHE/CHE Cherbov, S. I., and Chernyak, E. L., Measurement of the specific heat of powdered aluminum oxide and sodium oxalate, *Zh. Prikl. Khim.* **10**, 1220–1222 (1937).

37HUF/ELL Huffman, H. M., and Ellis, E. L., Thermal data. VIII. The heat capacities, entropies and free energies of some amino acids, *J. Am. Chem. Soc.* **59**, 2150–2152 (1937).

37KEM/GIA Kemp, J. D., and Giauque, W. F., Carbonyl sulfide. The heat capacity, vapor pressure, and heats of fusion and vaporization. The third law of thermodynamics and orientation equilibrium in the solid, *J. Am. Chem. Soc.* **59**, 79–84 (1937).

37PAR/SHO Parks, G. S. Shomate, C. H. Kennedy, W. D. and Crawford, B. L.,Jr., The entropies of n-butane and isobutane, with some heat capacity data for isobutane, *J. Chem. Phys.* **5**, 359–363 (1937).

37PER Perlick, A., Kalorimetrische Messungen an Schwefeldioxyd, Methylenchlorid, Difluormonochloraethan und Tetrafluordichloraethan, *Z. ges. Kalte-Ind.* **44**, 201–206 (1937).

37PER 2 Perlick, A., Calorimetric investigations on dichloromethane, difluoromonochloroethane, and tetrafluorodichloroethane, *Bull. Int. Inst. Refrig. Annex* **18**, No. 1, 1–9 (1937).

37STU Stull, D. R., A semi-micro calorimeter for measuring heat capacities at low temperatures, *J. Am. Chem. Soc.* **59**, 2726–2733 (1937).

37VOL Vold, R. D., A calorimetric test of the solubility equation for regular solutions, *J. Am. Chem. Soc.* **59**, 1515–1521 (1937).

37WIT/KEM Witt, R. K., and Kemp, J. D., The heat capacity of ethane from 15°K to the boiling point. The heat of fusion and the heat of vaporization, *J. Am. Chem. Soc.* **59**, 273–276 (1937).

38EGA/KEM Egan, C. J., and Kemp, J. D., Methyl bromide. The heat capacity, vapor pressure, heats of transition, fusion and vaporization. Entropy and density of the gas, *J. Am. Chem. Soc.* **60**, 2097–2101 (1938).

38HIC Hicks, J. F. G., Jr., A low temperature calorimeter. The heat capacity and entropy of thallium from 14 to 300°K. Low temperature studies. No. 3, *J. Am. Chem. Soc.* **60**, 1000–1004 (1938).

38KEM/EGA Kemp, J. D., and Egan, C. J., Hindered rotation of the methyl groups in propane. The heat capacity, vapor pressure, heats of fusion and vaporization of propane. Entropy and density of the gas, *J. Am. Chem. Soc.* **60**, 1521–1525 (1938).

38KEN/SHO Kennedy, W. D., Shomate, C. H., and Parks, G. S., Thermal data on organic compounds. XVIII. The heat capacity and entropy of t-butylethylene, *J. Am. Chem. Soc.* **60**, 1507–1509 (1938).

38KRE Kreutzer, C., Kalorimetrische Messungen beim Übergang von der anisotropen zur isotropen flüssigen Phase, *Ann. Physik.* [5] **33**, 192–209 (1938).

38MAZ Mazur, J., Über die spezifische Wärme des Äthyläthers, *Acta Phys. Pol* **7**, 318–326 (1938–1939).

38PIT Pitzer, K. S., The heat capacity of diamond from 70 to 300K, *J. Chem. Phys.* **6**, 68–70 (1938).

38RIE Riedel, L., Bestimmung der thermischen Eigenschaften von Trifluortrichloräthan, *Z. ges. Kalte.-Ind.* **45**, 221–225 (1938).

38VER Verdonk, G., Calorimetric analysis of paraffins – Contribution to the general subject of "calorimetric analysis" by J. Straub and R. N. M. A. Malataux, *Chem. Weekblad* **35**, 741–743 (1938).

39AST/EID Aston, J. G., Eidinoff, M. L., and Forster, W. S., The heat capacity and entropy, heats of fusion and vaporization and the vapor pressure of dimethylamine, *J. Am. Chem. Soc.* **61**, 1539–1543 (1939).

39BYK Bykov, V. T., Heat of mixing of liquids, *Zhur. Fiz. Khim.* **13**, 1013–1019 (1939).

39FRE/HIL Frederick, K. J., and Hildebrand, J. H., Specific heats and heats of fusion and transition of carbon tetrabromide, *J. Am. Chem. Soc.* **61**, 1555–1558 (1939).

39GIA/RUE Giauque, W. F., and Ruehrwein, R. A., The entropy of hydrogen cyanide. Heat capacity, heat of vaporization and vapor pressure. Hydrogen bond polymerization of the gas in chains of indefinite length, *J. Am. Chem. Soc.* **61**, 2626–2633 (1939).

39MAZ Mazur, J., Über die spezifische Wärme des Nitrobenzols, *Acta Phys. Pol.* **7**, 290–304 (1939).

39MAZ 2 Mazur, J., Über die spezifische Wärme des Äthyläthers, *Acta Phys. Pol.* **7**, 318–326 (1939).

39MAZ 3 Mazur, J., Über die spezifische Wärme des Äthyläthers, des Nitrobenzols und des Schwefelkohlenstoffs, *Z. Physik.* **113**, 710–720 (1939).

39PHI Phillip, N. M., Adiabatic and isothermal compressibilities of liquids, *Proc. Indian Acad. Sci. A* **9**, 109–120 (1939).

39RAI Railing, W. E., The specific heat of some ethylene halides, *J. Am. Chem. Soc.* **61**, 3349–3353 (1939).

39RIE Riedel, L., Determination of the thermal properties of trifluorotrichloroethane, *Bull. Int. Inst. Refrig. Annex* **20**, No. 3, 1–10 (1939).

39RUE/GIA Ruehrwein, R. A., and Giauque, W. F., The entropy of cyanogen. Heat capacity and vapor pressure of solid and liquid. Heats of fusion and vaporization, *J. Am. Chem. Soc.* **61**, 2940–2944 (1939).

39SAT/SOG Satoh, S., and Sogabe, T., The specific heats of some solid aliphatic acids and their ammonium salts and the atomic heat of nitrogen, *Sci. Pap. Inst. Phys. Chem. Res. (Tokyo)* **36**, 97–105 (1939).

39SAT/SOG 2 Satoh, S., and Sogabe, T., The specific heats of some solid aromatic acids and their ammonium salts and the atomic heat of nitrogen, *Sci. Pap. Inst. Phys. Chem. Res. (Tokyo)* **36**, 449–457 (1939).

40AST/KEN Aston, J. G., Kennedy, R. M., and Schumann, S. C., The heat capacity and entropy, heats of fusion and vaporization and the vapor pressure of isobutane, *J. Am. Chem. Soc.* **62**, 2059–2063 (1940).

40AST/MES Aston, J. G., and Messerly, G. H., The heat capacity and entropy, heats of fusion and vaporization and the vapor pressure of n-butane, *J. Am. Chem. Soc.* **62**, 1917–1933 (1940).

40AWB/GRI Awberg, J. H., and Griffiths, E., The specific heat of liquid methyl chloride, *Proc. Phys. Soc. (London)* **52**, 770–776 (1940).

40BEN/MCH Benning, A. F., McHarness, R. C., Markwood, W. H., Jr., and Smith, W. J., Thermodynamic properties of fluorochloromethanes and -ethanes. Heat capacity of the liquid and vapor of three fluorochloromethanes and trifluorotrichloroethane, *Ind. and Eng. Chem.* **32**, 976–980 (1940).

40BUR Burlew, J. S., Measurement of the heat capacity of a small volume of liquid by the piezo-thermometric method. III. Heat capacity of benzene and of toluene from 8°C. to the boiling point, *J. Am. Chem. Soc.* **62**, 696–700 (1940).

40CAM/CAM Campbell, A. N., and Campbell, A. J. R., The heats of solution, heats of formation, specific heats and equilibrium diagrams of certain molecular compounds, *J. Am. Chem. Soc.* **62**, 291–297 (1940).

40CLA/STE Clarke, T. H., and Stegeman, G., α- and β-Monopalmitin, heats of combustion and specific heats at 25C°, *J. Am. Chem. Soc.* **62**, 1815–1817 (1940).

40COU/PIT Coulter, L. V., Pitzer, K. S., and Latimer, W. M., The entropies of large ions. The heat capacity, entropy and heat of solution of potassium chloropatinate, tetramethylammonium iodide and uranyl nitate hexahydrate, *J. Am. Chem. Soc.* **62**, 2845–2851 (1940).

40HUF/ELL Huffman, H. M., Ellis, E. L., and Borsook, H., Thermal data XI. The heat capacities and entropies of guanidine carbonate, glutamic acid hydrochloride, ornithine dihydrochloride, d-lactic acid and l-lactic acid, *J. Am. Chem. Soc.* **62**, 297–299 (1940).

40HUF/FOX Huffman, H. M., and Fox, S. W., Thermal data. XIII. The heat capacities and entropies of creatine hydrate, dl-citrulline, dl-ornithine, l-proline and taurine, *J. Am. Chem. Soc.* **62**, 3464–3465 (1940).

40MES/AST Messerly, G. H., and Aston, J. G., The heat capacity and entropy, heats of fusion and vaporization and the vapor pressure of methyl chloride, *J. Am. Chem. Soc.* **62**, 887–890 (1940).

40MES/AST 2 Messerly, J. G., and Aston, J. G., The heat capacity and entropy, heats of fusion and vaporization and the vapor pressure of methyl chloride, *J. Am. Chem. Soc.* **62**, 3529 (1940).

40MES/KEN Messerly, G. H., and Kennedy, R. M., The heat capacity and entropy, heats of fusion and vaporization and the vapor pressure of n-pentane, *J. Am. Chem. Soc.*

62, 2988–2991 (1940).

40PIT Pitzer, K. S., The thermodynamics of n–heptane and 2,2,4–trimethylpentane, including heat capacities, heats of fusion and vaporization and entropies, *J. Am. Chem. Soc.* **62**, 1224–1227 (1940).

40PIT 2 Pitzer, K. S., The heat capacities, heats of transition and fusion, and entropies of ethylene dichloride and ethylene dibromide, *J. Am. Chem. Soc.* **62**, 331–335 (1940).

40RIE Riedel, L., Bestimmung der spezifischen Wärme von Äthychlorid und Methylenchlorid im flüssigen Zustand, *Z. ges. Kalte–Ind.* **47**, 87 (1940).

40SAT/SOG Satoh, S., and Sogabe, T., The specific heat of some solid aromatic acids and their ammonium salts and the atomic heat of nitrogen, *Rikagaku Kenkyusho Iho* **19**, 576–583 (1940).

40SAT/SOG 2 Satoh, S., and Sogabe, T., The heat capacities of some organic compounds containing nitrogen and the atomic heat of nitrogen(1), *Rikagaku Kenkyusho Iho* **19**, 1192–1197 (1940).

40SAT/SOG 3 Satoh, S., and Sogabe, T., The heat capacities of some organic compounds containing nitrogen and the atomic heat of nitrogen (2), *Rikagoku Kenkyusho Ito* **19**, 1244–1250 (1940).

40SAT/SOG 4 Satoh, S., and Sogabe, T., The heat capacities of some organic compounds containing nitrogen and the atomic heat of nitrogen (3), *Rikagaku Kenkyusho Ito* **19**, 1311–1317 (1940).

40SAT/SOG 5 Satoh, K., and Sogabe, T., The heat capacities of some organic compounds containing nitrogen and the atomic heat of nitrogen (4), *Rikagoku Kenkyusho Ito* **19**, 1348–1352 (1940).

41AND/STE Anderson, A. G., and Stegeman, G., The heat capacities and entropies of three disaccharides, *J. Am. Chem. Soc.* **63**, 2119–2121 (1941).

41AST/KEN Aston, J. G., Kennedy, R. M., and Messerly, G. H., The heat capacity and entropy, heats of fusion and vaporization and the vapor pressure of silicon tetramethyl, *J. Am. Chem. Soc.* **63**, 2343–2348 (1941).

41HUF Huffman, H. M., Thermal data. XIV. The heat capacities and entropies of some compounds having the peptide bond, *J. Am. Chem. Soc.* **63**, 688–689 (1941).

41JAC/STE Jack, G. W., and Stegeman, G., The heat capacities and entropies of two monosaccharides, *J. Am. Chem. Soc.* **63**, 2121–2123 (1941).

41KEN/SAG Kennedy, R. M., Sagenkahn, M., and Aston, J. G., The heat capacity and entropy, heats of fusion and vaporization, and the vapor pressure of dimethyl ether. The density of gaseous dimethyl ether, *J. Am. Chem. Soc.* **63**, 2267–2272 (1941).

41NEL/NEW Nelson, E. W., and Newton, R. F., The heat capacity of glucose glass, *J. Am. Chem. Soc.* **63**, 2178–2182 (1941).

41OSB/GAR Osborne, D. W., Garner, C. S., Doescher, R. N., and Yost, D. M., The heat capacity, entropy, heats of fusion and vaporization and vapor pressure of fluorotrichloromethane, *J. Am. Chem. Soc.* **63**, 3496–3499 (1941).

41PAR/WES Parks, G. S., West, T. J., and Moore, G. E., Thermal data on organic compounds. XXI. Some heat capacity, entropy and free energy data for the four methylnonanes, *J. Am. Chem. Soc.* **63**, 1133–1135 (1941).

41PIT/SCO Pitzer, K. S., and Scott, D. W., The thermodynamics of branched–chain paraffins. The heat capacity, heat of fusion and vaporization, and entropy of 2,3,4–trimethylpentane, *J. Am. Chem. Soc.* **63**, 2419–2422 (1941).

41RIE Riedel, L., Determination of the specific heat of liquid ethyl chloride and liquid methylene chloride, *Bull. Int. Inst., Refrig. Annex* **22**, No 4, 1–3 (1941).

41RIE 2 Riedel, L., Bestimmung der thermischen und kalorischen Eigenschaften von Difluoromonochloräthan, *Z. ges. Kalte–Ind.* **48**, 105–107 (1941).

41SAT/SOG Satoh, S., and Sogabe, T., The heat capacities of some organic compounds containing nitrogen and the atomic heat of nitrogen. (1), *Sci., Pap. Inst. Phys. Chem. Res. (Tokyo)* **38**, 197–203 (1941).

41SAT/SOG 2 Satoh, S., and Sogabe, T., The heat capacities of some organic compounds containing nitrogen and the atomic heat of nitrogen (2), *Sci. Pap. Inst. Phys. Chem. Res. (Tokyo)* **38**, 231–237 (1941).

41SAT/SOG 3 Satoh, S., and Sogabe, T., The heat capacities of some organic compounds containing nitrogen and the atomic heat of nitrogen. (3), *Sci. Pap. Inst. Phys. Chem. Res. (Tokyo)* **38**, 238–245 (1941).

41SAT/SOG 4 Satoh, S. and Sogabe, T., The heat capacities of some organic compounds containing nitrogen and the atomic heat of nitrogen (4), *Bull. Inst. Phys. Chem. Research (Tokyo)* **38**, 246–251 (1941).

41SCH Schmidt, W. R., Thesis, Washington University (St. Louis) (1941).

41STO/FIS Stout, J. W., and Fisher, L. H., The entropy of formic acid. The heat capacity from 15 to 300K. Heats of fusion and vaporization, *J. Chem. Phys.* **9**, 163–168 (1941).

41YOS/OSB Yost, D. M., Osborne, D. W., and Garner, C. S., The heat capacity, entropy, and heats of transition, fusion, and vaporization of dimethylacetylene. Free rotation in the dimethylacetylene molecule, *J. Am. Chem. Soc.* **63**, 3492–3496 (1941).

41ZHD Zhdanov, A. K., Specific heats of some liquids and azeotropic mixtures, *Zhur. Obshch. Khim.* **11**, 471–482 (1941).

42OSB/DOE Osborne, D. W., Doescher, R. N., and Yost, D. M., The heat capacity, heats of fusion and vaporization, vapor pressure and entropy of dimethyl sulfide, *J. Am. Chem. Soc.* **64**, 169–172 (1942).

42RIE Riedel, L., Determination of the thermal and caloric properties of difluoromonochloroethane, *Bull. Int. Inst. Refrig. Annex* **23**, No. 2, 1–5 (1942).

42RUS/OSB Russell, H., Jr, Osborne, D. W., and Yost, D. M., The heat capacity, entropy, heats of fusion, transition, and vaporization and vapor pressures of methyl mercaptan, *J. Am. Chem. Soc.* **64**, 165–169 (1942).

42SCH/AST Schumann, S. C., Aston, J. G., and Sagenkahn, M., The heat capacity and entropy, heats of fusion and vaporization and the vapor pressures of isopentane, *J. Am. Chem. Soc.* **64**, 1039–1043 (1942).

42TRI/ENG Trimble, H. M., Engle, C. J., Brown, R. A., and Schmuck, R. F., The specific heats of morpholine and its aqueous solutions, *J. Am. Chem. Soc.* **64**, 679–681 (1942).

42ZIE/AND Ziegler, W. T., and Andrews, D. H., The heat capacity of benzene–d_6, *J. Am. Chem. Soc.* **64**, 2482–2485 (1942).

43AST/FIN Aston, J. G., Fink, H. L., and Schumann, S. C., The heat capacity and entropy, heats of transition, fusion and vaporization and the vapor pressures of cyclopentane. Evidence for a non–planar structure, *J. Am. Chem. Soc.* **65**, 341–346 (1943),

43AST/SZA Aston, J. G., Szasz, G. J., and Fink, H. L., The heat capacity and entropy, heats of transition, fusion and vaporization and the vapor pressures of cyclohexane. The vibrational frequencies of alicyclic ring systems, *J. Am. Chem. Soc.* **65**, 1135–1139 (1943).

43BAC/PER Backer, H. J., and Perdok, W. G., Transformations, "ordre–desordre" dans les reseaux cristallins de molecules organiques. 1. Properties physiques des molecules du type CX_4, specialement de $C(SCH_3)_4$, *Rec. Trav. Chim.* **62**, 533–549 (1943).

43GUT/HUF Guthrie, G. B., Jr., and Huffman, H. M., Thermal data. XVI. The heat capacity and entropy of isopentane. The absence of a reported anomaly, *J. Am. Chem. Soc.* 65, 1139–1143 (1943).

43PIT/SCO Pitzer, K. S., and Scott, D. W., The thermodynamics and molecular structure of benzene and its methyl derivatives, *J. Am. Chem. Soc.* **65**, 803–829 (1943).

43RUE/HUF Ruehrwein, R. A., and Huffman, H. M., Thermal data. XVII. The heat capacity, entropy, and free energy of formation of cyclohexane. A new method of heat transfer in low temperature calorimetry, *J. Am. Chem. Soc.* **65**, 1620–1625 (1943).

44AST/SAG Aston, J. G., Sagenkahn, M. L., Szasz, G. J., Moessen, G. W., and Zuhr, H. F., The heat capacity and entropy, heats of fusion and vaporization and the vapor pressure of trimethylamine. The entropy from spectroscopic and molecular data, *J. Am. Chem. Soc.* **66**, 1171–1177 (1944).

44CLA/STE Clarke, T. H., and Stegeman, G., The heat of acetylation of α- and β-D-glucose from the heats of combustion of the pentaacetates, *J. Am. Chem. Soc.* **66**, 457–459 (1944).

44EIB Eibert, J., Thesis, Washington University (St. Louis) (1944).

44GUT/SPI Guthrie, G. B., Jr., Spitzer, R. W., and Huffman, H. M., Thermal data. XVIII. The heat capacity, heat of fusion, entropy and free energy of ethylbenzene, *J. Am. Chem. Soc.* **66**, 2120–2121 (1944).

44HIC/HOO Hicks, J. F. G., Hooley, J. G., and Stephenson, C. C., The heat capacity of carbon tetrachloride from 15 to 300K. The heats of transition and of fusion. The entropy from thermal measurements compared with the entropy from molecular data, *J. Am. Chem. Soc.* **66**, 1064–1067 (1944).

44RUB/LEV Rubin, T. R., Levedahl, B. H.,g and Yost, D. M., The heat capacity, heat of transition, vaporization, vapor pressure and entropy of 1,1,1-trichloroethane, *J. Am. Chem. Soc.* **66**, 279–282 (1944).

44RUS/GOL Russell, H., Jr., Golding, D. R. V., and Yost, D. M., The heat capacity, heats of transition, fusion and vaporization, vapor pressure and entropy of 1,1,1-trifluoroethane, *J. Am. Chem. Soc.* **66**, 16–20 (1944).

44SCO/FER Scott, R. B., Ferguson, W. J., and Brickwedde, F. G., Thermodynamic properties of cis-2-butene from 15 to 1500K, *J. Research NBS* **33**, 1–20 (1944).

45DAV/WIE Davis, H. S., and Wiedeman, O. F., Physical properties of acrylonitrile, *Ind. Eng. Chem.* **37**, 482–485 (1945).

45FIS/NAY Fischl, F. B., Naylor, B. F., Ziemer, C. W., Parks, G. S., and Aston, J. G., The heat capacity, heat of fusion and entropy of 11-n-Decylheneicosane, *J. Am. Chem. Soc.* **67**, 2075–2079 (1945).

45GUT/PIT Guttman, L., and Pitzer, K. S., trans-2-Butene. The heat capacity, heats of fusion and vaporization, and vapor pressure. The entropy and barrier to internal rotation, *J. Am. Chem. Soc.* **67**, 324–327 (1945).

45SCO/BRI Scott, R. B., and Brickwedde, F. G., Thermodynamic properties of solid and liquid ethylbenzene from 0 to 300K, *J. Research NBS* **35**, 501–512 (1945).

45SCO/MEY Scott, R. B., Meyers, C. H., Rands, R. D., Jr., Brickwedde, F. G., and Bekkedahl, N., Thermodynamic properties of 1,3-butadiene in the solid, liquid, and vapor states, *J. Research NBS* **35**, 39–85 (1945).

45ZHD Zhdanov, A. K., On the thermal capacity of some pure liquids and azeotropic mixtures, *Zhur. Obshch. Khim.* **15**, 895–902 (1945).

46AST/FIN Aston, J. G., Fink, H. L., Bestul, A. B., Pace, E. L., and Szasz, G. J., The heat capacity and entropy, heats of fusion and vaporization and the vapor pressure of butene-1. The zero point entropy of the glass. The entropy of the gas from molecular data, *J. Am. Chem. Soc.* **68**, 52–57 (1946).

46AST/ZIE Aston, J. G., and Ziemer, C. W., Thermodynamic properties of the three crystalline forms of methylammonium chloride, *J. Am. Chem. Soc.* **68**, 1405–1413 (1946).

46DOU/HUF Douslin, D. R., and Huffman, H. M., Low-temperature thermal data on the five isomeric hexanes, *J. Am. Chem. Soc.* **68**, 1704–1708 (1946).

46DOU/HUF 2 Douslin, D. R., and Huffman, H. M., The heat capacities, heats of transition, heats of fusion and entropies of cyclopentane, methylcyclopentane and methylcyclohexane, *J. Am. Chem. Soc.* **68**, 173–176 (1946).

46PIT/GUT Pitzer, K. S., Guttman, L., and Westrum, E. F., Jr., The heat capacity, heats of fusion and vaporization, vapor pressure, entropy vibration frequencies and barrier to internal rotation of styrene, *J. Am. Chem. Soc.* **68**, 2209–2212 (1946).

46RUE/HUF Ruehrwein, R. A., and Huffman, H. M., Thermal data. XIX. The heat capacity, entropy and free energy of urea, *J. Am. Chem. Soc.* **68**, 1759–1761 (1946).

46RUE/POW Ruehrwein, R. A., and Powell, T. M., The heat capacity, vapor pressure, heats of fusion and vaporization of cyclopropane. Entropy and density of the gas, *J. Am. Chem. Soc.* **68**, 1063–1068 (1946).

47AST/SZA Aston, J. G., and Szasz, G. J., The thermodynamics of butadiene-1,2 from calorimetric and spectroscopic data, *J. Am. Chem. Soc.* **69**, 3108–3114 (1947).

47CHA/SIN Charbonnet, G. H., and Singleton, W. S., Thermal properties of fats and oils. VI. Heat capacity, heats of fusion and transition, and entropy of trilaurin, trimyristin, tripalmitin, and tristearin, *J. Am, Oil Chem. Soc.* **24**, 140–142 (1947).

47CON/ELV Connor, A. Z., Elving, P. J., and Steingiser, S., Specific heat of acetaldyhyde and acetaldehyde dibutyl acetal, *J. Am. Chem. Soc.* **69**, 1532 (1947).

47COR/GIN Corruccini, R. J., and Ginnings, D. C., The enthalpy, entropy and specific heat of liquid p-xylene from 0 to 300°. The heat of fusion, *J. Am. Chem. Soc.* **69**, 2291–2294 (1947).

47JON/GIA Jones, W. M., and Giauque, W. F., The entropy of nitromethane. Heat capacity of solid and liquid. Vapor pressure, heats of fusion and vaporization, *J. Am. Chem. Soc.* **69**, 983–987 (1947).

47KUR Kurbatov, V. Ya., Specific heat of liquids. 1. Specific heat of benzenoid hydrocarbons, *Zhur. Obsh. Khim.* **17**, 1999–2003 (1947).

47OSB/GIN Osborne, N. S., and Ginnings, D. C., Measurements of heat of vaporization and heat capacity of a number of hydrocarbons, *J. Research NBS* **39**, 453–477 (1947).

47SCH/ZOS Schildknecht, C. E., Zoss, A. O., and McKinley, C., Vinyl alkyl ethers, *Ind. Eng. Chem.* **39**, 180–186 (1947).

47SZA/MOR Szasz, G. J., Morrison, J. A., Pace, E. L., and Aston, J. G., Thermal properties of cyclopentane and its use as a standard substance in low temperature thermal measurements, *J. Chem. Phys.* **15**, 562–564 (1947).

47TOD/OLI Todd, S. S., Oliver, G. D., and Huffman, H. M., The heat capacities, heats of fusion and entropies of the six pentenes, *J. Am. Chem. Soc.* **69**, 1519–1525 (1947).

48EUC Eucken, A., Assoziation in Flüssigkeiten, *Z. Elektrochem.* **52**, 255–269 (1948).

48GIA/JON Giauque, W. F., and Jones, W. M., Carbonyl chloride. Entropy. Heat capacity. Vapor pressure. Heats of fusion and vaporization. Comments on solid sulfur dioxide structure, *J. Am. Chem. Soc.* **70**, 120–124 (1948).

48GIN/COR Ginnings, D. C., and Corruccini, R. J., Liquid isopropyl alcohol. Enthalpy, entropy, and specific heat from 0° to 200°C, *Ind. Eng. Chem.* **40**, 1990–1991 (1948).

48GOR/GIA Gordon, J., and Giauque, W. F., The entropy of ethyl chloride. Heat capacity from 13 to 287K. Vapor pressure. Heats of fusion and vaporization, *J. Am. Chem. Soc.* **70**, 1506–1510 (1948).

48HUF/EAT Huffman, H. M., Eaton, M., and Oliver, G. D., The heat capacities, heats of transition, heats of fusion and entropies of cyclopentene and cyclohexene, *J. Am. Chem. Soc.* **70**, 2911–2914 (1948).

48KUR Kurbatov, V. Ya., Heat capacity of liquids. 2. Heat capacity and its relation to temperature for halogenated acyclic hydrocarbons, *Zhur. Obshch. Khim.* **18**, 372–389 (1948).

48OLI/EAT Oliver, G. D., Eaton, M., and Huffman, H. M., The heat capacity, heat of fusion and entropy of benzene, *J. Am. Chem. Soc.* **70**, 1502–1505 (1948).

48PAC/AST Pace, E. L., and Aston, J. G., The thermodynamics of hexafluoroethane from calorimetric and spectroscopic data, *J. Am. Chem. Soc.* **70**, 566–570 (1948).

49GIA/GOR Giauque, W. F., and Gordon, J., The entropy of ethylene oxide. Heat capacity from 14 to 285K. Vapor pressure. Heats of fusion and vaporization, *J. Am. Chem. Soc.* **71**, 2176–2181 (1949).

49HUF/TOD Huffman, H. M., Todd, S. S., and Oliver, G. D., Low temperature thermal data on eight C_8H_{16} alkylcyclohexanes, *J. Am. Chem. Soc.* **71**, 584–592 (1949).

49PAR/MOO Parks, G. S., Moore, G. E., Renquist, M. L., Naylor, B. F., McClaine, L. A., Fujii, P. S., and Hatton, J. A., Thermal data on organic compounds. XXV. Some heat capacity, entropy and free energy data for nine hydrocarbons of high molecular weight, *J. Am. Chem. Soc.* **71**, 3386–3389 (1949).

49SCO/GRO Scott, D. W., Gross, M. E., Oliver, G. D., and Huffman, H. M., Cycloöctatetraene: low–temperature heat capacity, heat of fusion, heat of vaporization, vapor pressure and entropy, *J. Am. Chem. Soc.* 71, 1634–1636 (1949).

49STA/GUP Staveley, L. A. K., and Gupta, A. K., A semi–micro low–temperature calorimeter, and a comparison of some thermodynamic properties of methyl alcohol and methyl deuteroxide, *Trans. Faraday Soc.* **45**, 50–61 (1949).

49WAD/KNO Waddington, G., Knowlton, J. W., Scott, D. W., Olikver, G. D., Todd, S. S., Hubbard, W. N., Smith, J. C., and Huffman, H. M., Thermodynamic properties of thiophene, *J. Am. Chem. Soc.* **71**, 797 (1949).

49WUY/JUN Wuyts, J., and Jungers, J. C., Les chaleurs molaires des bibromures de deuteroethylene, *Bull. Soc. Chim. Belg.* **58**, 80–86 (1949).

50AND/HIG Anderson, G. L., Jr., Higbie, H., and Stegeman, G., The heat capacity of sucrose from 25 to 90°C, *J. Am. Chem. Soc.* **72**, 3798–3799 (1950).

50AST/MAS Aston, J. G., Mastrangelo, S. V. R., and Moessen, G. W., The thermodynamics of 1–butyne from calorimetric and spectroscopic data, *J. Am. Chem. Soc.* **72**, 5287–5291 (1950).

50CRO/SMY Crowe, R. W., and Smyth, C. P., Heat capacities, dielectric constants and molecular rotational freedom in solid trichloroethanes and disubstituted propanes, *J. Am. Chem. Soc.* **72**, 4009–4015 (1950).

50HOU/MAS Hough, E. W., Mason, D. M., and Sage, B. H., Heat capacities of several organic liquids, *J. Am. Chem. Soc.* **72**, 5775–5777 (1950).

50KUR Kurbatov, V. Ya., Specific heat of liquids. III. Specific heat of hydrocarbons with several noncondensed rings, *Zhur. Obshch. Khim.* **20**, 1139–1144 (1950).

50KUS/CRO Kushner, L. M., Crowe, R. W., and Smyth, C. P., The heat capacities and dielectric constants of some alkyl halides in the solid state, *J. Am. Chem. Soc.* **72**, 1091–1098 (1950).

50NIT/SEK Nitta, I., Seki, S., and Momotani, M., On the phase transition in pentaerythritol (I), *Proc. Japan Acad.* **26**, No. 9, 25–29 (1950).

50SCO/FIN Scott, D. W., Finke, H. L., Gross, M. E., Guthrie, G. B., and Huffman, H. M., 2,3–Dithiabutane: low temperature heat capacity, heat of fusion, heat of vaporization, vapor pressure, entropy and thermodynamic functions, *J. Am. Chem. Soc.* **72**, 2424–2430 (1950).

50SCO/FIN 2 Scott, D. W., Finke, H. L., Hubbard, W. N., McCullough, J. P., Gross, M. E., Williamson, K. D., Waddington, G., and Huffman, H. M., Spiropentane: heat capacity, heats of fusion and vaporization, vapor pressure, entropy and thermodynamic functions, *J. Am. Chem. Soc.* **72**, 4664–4668 (1950).

50SIN/WAR Singleton, W. S., Ward, T. L., and Dollear, F. G., Physical properties of fatty acids. I. Some dilatometric and thermal properties of stearic acid in two polymorphic forms, *J. Am. Oil. Chem. Soc.* **27**, 143–146 (1950).

50UEB/ORT Ueberreiter, K., and Orthmann, H.–J., Specifische Wärme, spezifisches Volumen, Temperatur– und Wärme–leittähigkeit einiger disubstituierter Benzole und polycyclischer Systeme, *Z. Natursforsch.* **5a**, 101–108 (1950).

50URA/SID Urazovskii, S. S., and Sidorov, I. A., Concerning singularities in the temperature dependence of the specific heat of liquids and the nature of phase transitions of the second kind, *Doklady Akad. Nauk SSSR* **70**, 859–862 (1950).

51AST/FIN Aston, J. G., Fink, H. L., Janz, G. J., and Russell, K. E., The heat capacity, heats of fusion and vaporization, vapor pressures, entropy and thermodynamic functions of methylhydrazine, *J. Am. Chem. Soc.* **73**, 1939–1943 (1951).

51CON/SAG Connolly, T. J., Sage, B. H., and Lacey, W. N., Isobaric heat capacities at bubble point. n–Hexane, methylcyclopentane, and n–octane, *Ind. Eng. Chem.* **43**, 946–950 (1951).

51DOU/BAL Douglas, T. B., Ball, A. F., and Torgesen, J. L., Heat capacity of crystalline dextrose between 25 and 95°, *J. Am. Chem. Soc.* **73**, 1360–1361 (1951).

51EUC/EIG Eucken, A., and Eigen, M., Untersuchung der Assoziationsstruktur in schwerem Wasser und n–Propanol mit Hilfe thermisch–kalorischer Eigenschaften, insbesondere Messungen der spezifischen Wärmen, *Z. Elektrochem.* **55**, 343–354 (1951).

51FUR/GIN Furukawa, G. T., Ginnings, E. C., McCoskey, R. E., and Nelson, R. A., Calorimetric properties of diphenyl ether from 0 to 570K, *J. Research NBS* **46**, 195–206 (1951).

51FUR/MCC Furukawa, G. T., McCoskey, R. E., and King, G. J., Calorimetric properties of benzoic acid from 0 to 410K, *J. Research NBS* **47**, 256–261 (1951).

51OLI/GRI Oliver, G. D., Grisard, J. W., and Cunningham, C. W., Thermodynamic properties and P–V–T relations of chlorotrifluoroethylene, *J. Am. Chem. Soc.* **73**, 5719–5722 (1951).

51SCO/FIN Scott, D. W., Finke, H. L., McCullough, J. P.,Gross, M. E., Williamson, K. D., Waddington, G., and Huffman, H. M., Thermodynamic properties and rotational isomerism of 2–thiabutane, *J. Am. Chem. Soc.* **75**, 261–265 (1951).

51SIE/CRU Sieg, L., Crützen, J. L., and Jost, W., Zur Thermodynamik von Mischphasen IX. Über das Verdampfungsgleichgewicht Benzol–1–2–Dichloraethan, *Z. physik. Chem.* **198**, 263–269 (1951).

51SUG Suga, H., On the orientational order–disorder transition in crystalline aniline hydrobromide, *Bull. Chem. Soc. Japan* **34**, 426–433 (1961).

51TSC/KRI Tschamler, H., and Krischai, H., Chinolin–m–Kresol, ein stark negatives System, *Monatsh. Chem.* **82**, 259–270 (1951).

52ERD/JAG Erdös, E., Jäger, L., and Pouchly, J., Specificka tepla methylesteru a butylesteru kyseliny methakrylove, *Chem. Listy* **46**, 770 (1952).

52FIN/SCO Finke, H. L., Scott, D. W., Gross, M. E., Waddington, G., and Huffman, H. M., The entropy and vapor pressure of 1–pentanethiol, *J. Am. Chem. Soc.* **74**, 2804–2806 (1952).

52FUR/MCC Furukawa, G. T., McCoskey, R. E., and King, G. J., Calorimetric properties of polytetrafluoroethylene

(Teflon) from 0 to 365°K, *J. Res. NBS* **49**, 273–278 (1952).

52GUT/SCO Guthrie, G. B., Jr., Scott, D. W., Hubbard, W. N., Katz, C., McCullough, J. P., Gross, M. E., Williamson, K. D., and Waddington, G., Thermodynamic properties of furan, *J. Am. Chem. Soc.* **74**, 4662–4669 (1952).

52HOF Hoffman, J. D., The specific heat and degree of crystallinity of polychlorotrifluoroethylene, *J. Am. Chem. Soc.* **74**, 1696–1700 (1952).

52HUB/FIN Hubbard, W. N., Finke, H. L., Scott, D. W., McCullough, J. P., Katz, C., Gross, M. E., Messerly, J. F., Pennington, R. E., and Waddington, G., Thiacyclopentane: heat capacity, heats of fusion and vaporization, vapor pressure, entropy, heat of formation and thermodynamic functions, *J. Am. Chem. Soc.* **74**, 6025–6030 (1952).

52MCC/SCO McCullough, J. P., Scott, D. W., Finke, H. L., Gross, M. E., Williamson, K. D., Pennington, R. E., Waddington, G., and Huffman, H. M., Ethanethiol (ethyl mercaptan): thermodynamic properties in the solid, liquid and vapor states. Thermodynamic functions to 1000K, *J. Am. Chem. Soc.* **74**, 2801–2804 (1952).

52SCH/SAG Schlinger, W. G., and Sage, B. H., Isobaric heat capacities at bubble point. cis–2–Butene, isopropylbenzene, and n–decane, *Ind. Eng. Chem.* **44**, 2454–2456 (1952).

52SCO/DOU Scott, D. W., Douslin, D. R., Gross, M. E., Oliver, G. D., and Huffman, H. M., 2,2,3,3–Tetramethylbutane: Heat capacity, heats of transition, fusion and sublimation, vapor pressure, entropy and thermodynamic functions, *J. Am. Chem. Soc.* **74**, 883–887 (1952).

52SCO/FIN Scott, D. W., Finke, H. L., McCullough, J. P., Gross, M. E., Pennington, R. E., and Waddington, G., 3,4–Dithiahexane: heat capacity, heats of fusion and vaporization, vapor pressure, entropy, and thermodynamic functions, **74**, 2478–2483 (1952).

52SCO/FIN 2 Scott, D. W., Finke, H. L., Hubbard, W. N., McCullough, J. P., Oliver, G. D., Gross, M. E., Katz, C., Williamson, K. D., Waddington, G., and Huffman, H. M., 3–Thiapentane: heat capacity, heats of fusion and vaporization, vapor pressure, entropy, heat of formation and thermodynamic functions, *J. Am. Chem. Soc.* **74**, 4656–4662 (1952).

52SPE/TAM Spengler, H. T., and Tamplin, W. S., Procedures in ice calorimetry, *Anal. Chem.* **24**, 941–944 (1952).

52STE/BER Stephenson, C. C., and Berets, D. J., The heat capacities and entropies of melamine and dicyandiamide, *J. Am. Chem. Soc.* **74**, 882–883 (1952).

52WAR/SIN Ward, T. L., and Singleton, W. S., Physical properties of fatty acids. II. Some dilatometric and thermal properties of palmitic acid, *J. Phys. Chem.* **56**, 696–698 (1952).

53AST/WOO Aston, J. G., Wood, J. L., and Zolki, T. P., The thermodynamic properties and configuration of unsymmetrical dimethylhydrazine, *J. Am. Chem. Soc.* **75**, 6202–6204 (1953).

53DES/TYL DeSorbo, W., and Tyler, W. W., The specific heat of graphite from 13 to 300°K, *J. Chem. Phys.* **21**, 1660–1663 (1953).

53FUR/MCC Furukawa, G. T., McCoskey, R. E., and Reilly, M. L., Heat capacity, heats of fusion and vaporization, and vapor pressure of tetrafluoroethylene, *J. Res. NBS* **51**, 69–72 (1953).

53GIN/FUR Ginnings, D. C., and Furukawa, G. T., Heat capacity standards for the range 14 to 1200°K, *J. Am. Chem. Soc.* **75**, 522–527 (1953).

53GIN/FUR 2 Ginnings, D. C., and Furukawa,G. T., Heat capacity standards for the range 14 to 1200°K, *J. Am. Chem. Soc.* **75**, 6359 (1953).

53GRA/SMI Gray, P., and Smith, P. L., The thermodynamic properties of methyl nitrate, *J. Chem. Soc.* **1953**, 2380–2385.

53GRO/OLI Gross, M. E., Oliver, G. D., and Huffman, H. M., Low–temperature thermal data for some C_7H_{14} alkylcyclopentanes, *J. Am. Chem. Soc.* **75**, 2801–2804 (1953).

53MCC/SCO McCullough, J. P., Scott, D. W., Finke, H. L., Hubbard, W. N., Gross, M. E., Katz, C., Pennington, R. E., Messerly, J. F., and Waddington, G., The thermodynamic properties of 2–methyl–2–propanethiol from 0 to 1000K, *J. Am. Chem. Soc.* **75**, 1818–1824 (1953).

53MCC/SUN McCullough, J. P., Sunner, S., Finke, H. L., Hubbard, W. N., Gross, M. E., Pennington, R. E., Messerly, J. F., Good, W. D., and Waddington, G., The chemical thermodynamic properties of 3–methylthiophene from 0 to 1000K, *J. Am. Chem. Soc.* **75**, 5075–5081 (1953).

53RAT/GWI Rathjens, G. W., Jr., and Gwinn, W. D., Heat capacities and entropy of cyclobutane, *J. Am. Chem. Soc.* **75**, 5629 (1953).

53SCH Schäfer, K., Fenomenos de orden y desorden en solidos, *An. R. Soc. Esp. Fis. Quim. Ser. B* **49**, 161–174 (1953).

53SCO/FIN Scott, D. W., Finke, H. L., Hubbard, W. N., McCullough, J. P., Katz, C., Gross, M. E., Messerly, J. F., Pennington, R. E., and Waddington, G., Thiacyclobutane: heat capacity, heats of transition, fusion and vaporization, vapor pressure, entropy, heat of formation and thermodynamic functions, *J. Am. Chem. Soc.* **75**, 2795–2800 (1953).

53SEY Seyer, W. F., The heat capacity of cis– and trans–decahydronaphthalene and the possible existence of a λ–region for the cis form at 50.1–50.5, *J. Am. Chem. Soc.* **75**, 616–621 (1953).

53WIL/DOL Wilhoit, R. C., and Dole, M., Specific heat of synthetic high polymers. II. Polyhexamethylene adipamide and sebacamide, *J. Phys. Chem.* **57**, 14–21 (1953).

54DOU/FUR Douglas, T. B., Furukawa, G. T., McCoskey, R. E., and Ball, A. F., Calorimetric properties of normal heptane from 0 to 520K, *J. Research NBS* **53**, 139–153 (1954).

54FIN/GRO Finke, H. L., Gross, M. E., Messerly, J. F., and Waddington, G., Benzothiophene: heat capacity, heat of transition, heat of fusion and entropy. An order–disorder transition, *J. Am. Chem. Soc.* **76**, 854–857 (1954).

54FIN/GRO 2 Finke, H. L., Gross, M. E., Waddington, G., and Huffman, H. M., Low–temperature thermal data for the nine normal paraffin hydrocarbons from octane to hexadecane, *J. Am. Chem. Soc.* **76**, 333–341 (1954).

54FUR/MCC Furukawa, G. T., McCoskey, R. E., and Reilly, M. L., Heat capacity, heats of transitions, fusion, and vaporization, and vapor pressure of octafluorocyclobutane, *J. Res. NBS* **52**, 11–16 (1954).

54FUR/PAR Furukawa, T., and Park, R. P., Heat capacity, heats of fusion, transition, and vaporization, and vapor pressure of trimethylborane, $B(CH_3)_3$, *NBS Report* **3649**, 27p (1954).

54GRA/SMI Gray, P., and Smith, P. L., Low–temperature calorimetry and the thermodynamic properties of ethyl nitrate, *J. Chem. Soc.* **1954**, 769–773.

54MCC/FIN McCullough, J. P., Finke, H. L., Hubbard, W. N., Good, W. D., Pennington, R. E., Messerly, J. F., and Waddington, G., The chemical thermodynamic properties of thiacyclohexane from 0 to 100K, *J. Am. Chem. Soc.* **76**, 2661–2669 (1954).

54MCC/FIN 2 McCullough, J. P., Finke, H. L., Scott, D. W., Gross, M. E., Messerly, J. F., Pennington, R. E., and Waddington, G., 2–Propanethiol: experimental thermodynamic studies from 12 to 500K. The chemical thermodynamic properties from 0 to 1000K, *J. Am. Chem. Soc.* **78**, 4796–4802 (1954).

55AST/WIL Aston, J. G., Wills, P. E., and Zolki, T. P., The heat capacities from 10.9°K., heats of transition, fusion and vaporization, vapor pressures and entropy of pentafluorochloroethane, the barrier hindering internal rotation, *J. Am. Chem. Soc.* **77**, 3939–3941 (1955).

55AST/ZOL Aston, J. G., Zolki, T. P., and Wood, J. L., The thermodynamic properties and configuration of trimethylhydrazine identification and correction for a major impurity, *J. Am. Chem. Soc.* **77**, 281–284 (1955).

55DES DeSorbo, W., Low temperature heat capacity of Ceylon graphite, *J. Am. Chem. Soc.* **77**, 4713–4715 (1955).

55FUR Furukawa, G. T., Heat capacity, heats of fusion and vaporization, and vapor pressure of triethylborane, $(C_2H_5)_3B$, *NBS Report No. 3712*, **1955**, 1–16.

55HEL/HEI Helfrey, P. F., Heiser, D. A., and Sage, B. H., Isobaric heat capacities at bubble point. Two trimethylbenzenes and n–heptane, *Ind. Eng. Chem.* **44**, 2385–2388 (1955).

55MAR/DOL Marx, P., and Dole, M., Specific heat of synthetic high polymers. V. A study of the order–disorder transition in polytetrafluoroethylene, *J. Am. Chem. Soc.* **77**, 4771–4774 (1955).

55MCC/FIN McCullough, J. P., Finke, H. L., Messerly, J. F., Pennington, R. E., Hossenlopp, I. A., and Waddington, G., 3–Methyl–2–thiabutane: calorimetric studies from 12 to 500K., the chemical thermodynamic properties from 0 to 1000K, *J. Am. Chem. Soc.* **77**, 6119–6125 (1955).

55RUI Ruiter, L. H., Some thermodynamic properties of the system benzene 1,2–dichloroethane. Part II. The excess molar heat capacities, *Rec. Trav. Chim.* **74**, 1467–1481 (1955).

55SCO/FIN Scott, D. W., Finke, H. L., McCullough, J. P., Gross, M. E., Messerly, J. F., Pennington, R. E., and Waddington, G., 2,3–Dimethyl–2–butene: Thermodynamic properties in the solid, liquid and vapor states, *J. Am. Chem. Soc.* **77**, 4993–4998 (1955).

55STA/TUP Staveley, L. A. K., Tupman, W. I., and Hart, K. R., Some thermodynamic properties of the systems benzene + ethylene dichloride, benzene + carbon tetrachloride, acetone + chloroform, and acetone + carbon disulphide, *Trans. Faraday Soc.* **51**, 323–342 (1955).

55STR Strelkev, I. I., Specific heat and entropy of sodium acetate, *Ukr. Khim. Zh. (Russ. Ed.)* **21**, 551–553 (1955).

55TAY/JOH Taylor, R. D., Johnson, B. H., and Kilpatrick, J. E., Entropy, heat capacity, and heats of transition of 1,2,3–trimethylbenzene, *J. Chem. Phys.* **23**, 1225–1231 (1955).

55TAY/KIL Taylor, R. D., and Kilpatrick, J. E., Entropy, heat capacity, and heats of transition of 1,3,5–trimethylbenzene, *J. Chem. Phys.* **23**, 1232–1235 (1955).

55WAR/VIC Ward, T. L., Vicknair, E. J., Singleton, W. S., and Feuge, R. O., Some thermal properties of 1–monostearin, 1–aceto–3–stearin and 1,2–diaceto–3–stearin, *J. Phys. Chem.* **59**, 4–7 (1955).

56DAV/STA Davies, T., and Staveley, L. A. K., Heat capacity and entropy of trimethylsulphonylmethane, *J. Chem. Soc.* **1956**, 2563

56DUF/EVE Duff, G. M., and Everett, D. H., The heat capacity of the system benzene + diphenylmethane, *J. Chem. Soc. Faraday. Trans. II* **52**, 753–763 (1956).

56FIN/SCO Finke, H. L., Scott, D. W., Gross, M. E., Messerly, J. F., and Waddington, G., Cycloheptane, cycloöctane and 1,3,5–cyclo–oD heptatriene. Low temperature thermal properties, vapor pressure and derived chemical thermodynamic properties, *J. Am. Chem. Soc.* **78**, 5469–5476 (1956).

56LI Li, J. C. M., The thermodynamic properties of cadmium dimethyl: heat capacities from 14 to 291K., heats of transition, fusion and vaporization, vapor pressure up to 296K and the entropy of ideal gas, *J. Am. Chem. Soc.* **78**, 1081–1083 (1956).

56LI/PIT Li, J. C. M., and Pitzer, K. S., The thermodynamic properties of 1,1–dichloroethane: heat capacities from 14 to 294K., heats of fusion and vaporization, vapor pressure and entropy of the ideal gas. The barrier to internal rotation, *J. Am. Chem. Soc.* **78**, 1077–1080 (1956).

56MAR/STA Marshall, J. G., Staveley, L. A. K. and Hart, K. R., A thermodynamic investigation of the transitions in carbon tetrabromide and ammonium chloride, *Trans. Faraday Soc.* **52**, 19–31 (1956).

56MCE McEwen, M., Preliminary engineering study of organic nuclear reactor coolant–moderators, *Monsanto Chemical Co.* (March 31, 1956).

56PAR/KEN Parks, G. S., Kennedy, W. D., Gates, R. R., Mosley, J. R., Moore, G. E., and Renquist, M. L., Thermal data on organic compounds. XXVI. Some heat capacity, entropy and free energy data for seven compounds containing oxygen, *J. Am. Chem. Soc.* **78**, 56–59 (1956).

56PEN/FIN Pennington, R. E., Finke, H. L., Hubbard, W. N., Messerly J. F., Frow, F. R., Hossenlopp, I. A., and Waddington, G., The chemical thermodynamic properties of 2–methylthiophene, *J. Am. Chem. Soc.* **78**, 2055–2060 (1956).

56PEN/SCO Pennington, R. W., Scott, D. W., Finke, H. L., McCullough, J. P., Messerly, J. F., Hossenlopp, I. A., and Waddington, G., The chemical thermodynamic properties and rotational tautomerism of 1–propanethiol, *J. Am. Chem. Soc.* **78**, 3266–3272 (1956).

56POP/KOL Popov, M. M., and Kolesov, V. P., Determination of the true specific heat of solid substances at low temperatures, *Zhur. Obshch. Khim.* **26**, 2385–2393 (1956).

56SCO/GOO Scott, D. W., Good, W. D., and Waddington, G., Tetraethyllead: heat of formation by rotating–bomb calorimetry, *J. Phys. Chem.* **60**, 1090–1095 (1956).

56SCO/MCC Scott, D. W., MCCullough, J. P., Good, W. D., Messerly, J. F., Pennington, R. E., Kincheloe, T. C., Hossenlopp, I. A., Douslin, D. R., and Waddington, G., Fluorobenzene: Thermodynamic properties in the solid, liquid and vapor states, A revised vibrational assignment, *J. Am. Chem. Soc.* **78**, 5457=5463 (1956).

56SCO/MCC 2 Scott, D. W., McCullough, J. P., Hubbard, W. N., Messerly, J. F., Hossenlopp, I. A., Frow, F. R., and Waddington, G., Benzenethiol: thermodynamic properties in the solid, liquid and vapor states, internal rotation of the thiol group, *J. Am. Chem. Soc.* **78**, 5463–5468 (1956).

56SMI/DOL Smith, C. W., and Dole, M., Specific heat of synthetic high polymers. VII. Polyethylene terephthalate, *J. Polym. Sci.* **20**, 37–56 (1956).

56WIR/DRO Wirth, H. E., Droege, J. W., and Wood, J. H., Low temperature heat capacity of palmitic acid and methyl palmitate, *J. Phys. Chem.* **60**, 917–919 (1956).

57CRU/JOS Crützen, J. L., Jost, W., and Sieg, L., Gleichgewichtsmessungen im System Anilin–N–Methylanilin, N–N–Dimethylanilin, 1,2–Äthandiol, *Z. Elektrochem.* **61**, 230–246 (1957).

57DES/TYL DeSorbo, W., and Tyler, W. W., Effect of irradiation on the low–temperature specific heat of graphite, *J. Chem. Phys.* **26**, 244–247 (1957).

57GRE/WES Grenier, G., and Westrum, E. F., Jr., The low temperature heat capacity and thermodynamic properties of sodium methoxide, *J. Am. Chem. Soc.* **79**, 1802–1804 (1957).

57HAR/MOE Harrison, D., and Moelwyn–Hughes, E. A., The heat capacities of certain liquids, *Proc. Roy. Soc. (London)* **239A**, 230–246 (1957).

57MCC/DOU McCullough, J. P., Douslin, D. R., Messerly, J. F., Hossenlopp, I. A., Kincheloe, T. C., and Waddington, G., Pyridine: experimental and calculated chemical thermodynamic properties between 0 and 1500K., a revised vibrational assignment, *J. Am. Chem. Soc.* **79**, 4289–4295 (1957).

57MCC/FIN McCullough, J. P., Finke, H. L., Messerly, J. F., Kincheloe, T. C., and Waddington, G., The low-temperature thermodynamic properties of naphthalene, 1-methylnaphthalene, 2-methylnaphthalene, 1,2,3,4-tetrahydronaphthalene, trans-decahydronaphthalene and cis-decahydronaphthalene, *J. Phys. Chem.* **61**, 1105–1116 (1957).

57MCC/FIN 2 McCullough, J. P., Finke, H. L., Gross, M. E., Messerly, J. F., and Waddington, G., Low temperature calorimetric studies of seven 1-olefins: effect of orientational disorder in the solid state, *J. Phys. Chem.* **61**, 289–301 (1957).

57NEI/WHI Neilson, E. F., and White, D., The heat capacity, heat of fusion, heat of transition and heat of vaporization of chlorodifluoromethane between 16°K and the boiling point, *J. Am. Chem. Soc.* **79**, 5618–5621 (1957).

57PUT/KIL Putnam, W. E., and Kilpatrick, J. E., Entropy, heat capacity, and heats of transition of 1,2,4-trimethylbenzene, *J. Chem. Phys.* **27**, 1075–1080 (1957).

57SCO/FIN Scott, D. W., Finke, H. L., McCullough, J. P., Messerly, J. F., Pennington, R. E., Hossenlopp, I. A., and Waddington, G., 1-butanethiol and 2-thiapentane. Experimental thermodynamic studies between 12 and 500'K., thermodynamic functions by a refined method of increments, *J. Am. Chem. Soc.* **79**, 1062–1068 (1957).

57YAR/KAY Yarrington, R. M., and Kay, W. B., The liquid specific heats of some fluorocarbon compounds, *J. Phys. Chem.* **61**, 1259–1260 (1957).

58BIL/NOL Billings, J. J., and Nolle, A. W., Investigation by nuclear magnetic resonance of the effects of pressure on the thermal transition and rotator phase of 2,2-dinitropropane, *J. Chem. Phys.* **29**, 214–220 (1958).

58DES/MOR Desnoyers, J. E., and Morrison, J. A., The heat capacity of diamond between 12.8° and 277°K, *Phil. Mag.* [8] **3**, 42–48 (1958).

58DVO Dvorak, J., Priprava benzylmethakrylatu, *Chem. Prumysl* **8**, 287–291 (1958).

58HIL/KRA Hildenbrand, D. L., Kramer, W. R., and Stull, D. R., The heat capacities of hexachlorobenzene and pentachlorophenol from 15 to 300K, *J. Phys. Chem.* **62**, 958–959 (1958).

58HUB/DOU Hubbard, W. N., Douslin, D. R., McCullough, J. P., Scott, D. W., Todd, S. S., Messerly, J. F., Hossenlopp, I. A., George, A., and Waddington, G., 2,3-Dithiabutane, 3,4-dithiahexane and 4,5-dithiaoctane: chemical thermodynamic properties from 0 to 1000K, *J. Am. Chem. Soc.* **80**, 3547–3554 (1958).

58LUT/PAN Lutskii, A. E., and Panova, A. N., Specific heat of liquid nitrobenzene, *Zhur. Fiz. Khim.* **32**, 2183–2185 (1958).

58MCC/FIN McCullough, J. P., Finke, H. L., Scott, D. W., Pennington, R. E., Gross, M. E., Messerly, J. F., and Waddington, G., 2-Butanethiol: chemical thermodynamic properties between 0 and 1000K., rotational conformations, *J. Am. Chem. Soc.* **80**, 4786–4793 (1958).

58PEP Peppel, W. J., Preparation and properties of the alkylene carbonates, *Ind. Eng. Chem.* **50**, 767–770 (1958).

58SCO/MCC Scott, D. W., McCullough, J. P., Messerly, J. F., Pennington, R. E., Hossenlopp, I. A., Finke, H. L., and Waddington, G., 2-Methyl-1-propanethiol: chemical thermodynamic properties and rotational isomerism, *J. Am. Chem. Soc.* **80**, 55–59 (1958).

58SEL/AST Selby, R. N., and Aston, J. G., The entropy of the compound formed between cyclopentane and 2,2-dimethylbutane at 0°K, *J. Am. Chem. Soc.* **80**, 5070–5075 (1958).

58SIN/HIL Sinke, G. C., Hildenbrand, D. L., McDonald, R. A., Kramer, W. R., and Stull, D. R., The heat, entropy and free energy of formation of diphenyl carbonate, *J. Phys. Chem.* **62**, 1461–1462 (1958).

58SOC/TRA Sochava, I. V., and Trapeznikova, O. N., Internal rotation and specific heat of some polymers at low temperatures, *Vestn. Leningrad. Univ.* **13**, No. 16, *Ser. Fiz. Khim.* No. 3, 65–72 (1958).

58SWI/ZIE Swietoslawski, W., and Zielenkiewicz, A., Mean specific heat of some ternary azeotropes, *Bull. Acad. Pol. Sci. Ser. Sci. Chim.* **6**, 365–366 (1958).

58SWI/ZIE 2 Swietoslawski, W., and Zielenkiewicz, A., Mean specific heats of binary positive azeotropes, *Bull. Acad. Pol. Sci. Ser. Sci. Chim.* **6**, 367–369 (1958).

58WAL/BRO Walker, B. E., Jr., Brooks, M. S., Ewing, C. T., and Miller, R. R., Specific heat of biphenyl and other polyphenyls. Correlation of specific heat data for phenyl type compounds, *J. Chem. Eng. Data* **3**, 280–282 (1958).

58WUN/DOL Wunderlich, B., and Dole, M., Specific heat of synthetic high polymers. IX. Poly(ethylene sebacate), *J. Polymer. Sci.* **32**, 125–130 (1958).

59BEN/THO Bengough, W. I., and Thomson, R. A. M., A dilatometric method of measuring the heats of radiation-induced reactions. Part 2. The heat of photochemical addition of bromotrichloromethane to vinyl acetate, *Trans. Faraday Soc.* **55**, 268–271 (1959).

59DAI/IVI Dainton, F. S., Ivin, K. J., and Walmsley, D. A. G., The equilibrium between gaseous formaldehyde and solid polyoxymethylene, *Trans. Faraday Soc.* **55**, 61–64 (1959).

59GOO/SCO Good, W. D., Scott, D. W., Lacina, J. L., and McCullough, J. P., Tetramethyllead: heat of formation by rotating-bomb calorimetry, *J. Phys. Chem.* **63**, 1139–1142 (1959).

59HIL/MCD Hildenbrand, D. L., McDonald, R. A., Kramer, W. R., and Stull, D. R., Thermodynamic and spectroscopic study of vinylidene chloride. I. Thermodynamic properties of the solid, liquid, and ideal gas, *J. Chem. Phys.* **30**, 930–934 (1959).

59HIL/SIN Hildenbrand, D. L., Sinke, G. C., McDonald, R. A., Kramer, W. R., and Stull, D. R., Thermodynamic and spectroscopic study of pyrrolidine. I. Thermodynamic properties in the solid, liquid, and vapor states, *J. Chem. Phys.* **31**, 650–654 (1959).

59HUT/BAI Hutchinson, E., and Bailey, L. G., A thermodynamic study of colloidal electrolyte solutions. II. Heat capacities of solubilized systems, experimental, *Z. Physik. Chem. [N.F.]* **21**, 30–37 (1959).

59KAR/STR Karasharli, K. A., and Strelkov, P. G., Thermodynamic properties of 1,1-dicyclohexyldodecane, 14–298K, *Azerb. Khim. Zhur.* No. **4**, 87–92 (1959).

59MCC/DOU McCullough, J. P., Douslin, D. R., Hubbard, W. N., Todd, S. S., Messerly, J. F., Hossenlopp, I. A., Frow, F. R., Dawson, J. P., and Waddington, G., Pyrrolidine, chemical thermodynamic properties between 0 and 1500K., effect of pseudorotation, and an unusual thermal anomaly in the liquid state, *J. Am. Chem. Soc.* **81**, 5884–5890 (1959).

59PAU/KOL Paukov, I. E., Kolesov, V. P., and Skuratov, S. M., Determination of isobaric-isothermal potential at standard conditions for the polymerization reaction of ϵ-caprolactam, *Doklady Akad. Nauk SSSR* **126**, 325–326 (1959).

59SCO/DOU Scott, D. W., Douslin, D. R., Messerly, J. F., Todd, S. S., Hossenlopp, I. A., Kincheloe, T. C., and McCullough, J. P., Benzotrifluoride: chemical thermodynamic properties and internal rotation, *J. Am. Chem. Soc.* **81**, 1015–1017 (1959).

59SCO/DOU Scott, D. W., Douslin, D. R., Messerly, J. F., Todd, S. S., Hossenlopp, I. A., Kincheloe, T. C., and

McCullough, J. P., Benzotrifluoride: chemical thermodynamic properties and internal rotation, *J. Am. Chem. Soc.* **81**, 1015–1020 (1959).

59STU/SIN Stull, D. R., Sinke, G. C., McDonald, R. A., Hatton, W. E., and Hildenbrand, D. L., Thermodynamic properties of indane and indene, *Symp. Thermodynam. Fitzens-Wattens Tiral* No. **48**, 9p (1959).

59SUG/SEK Suga, H., and Seki, S., On the phase transition of hexamethyldisilane, *Bull. Chem. Soc. Japan* **32**, 1088–1093 (1959).

59WES Westrum, E. F., Jr., Ordering transitions in simmetrical molecules, *Symp. Thermodynam. Fitzens-Watens Tiral.* No. **36**, 11p (1959).

59WIR/DRO Wirth, H. E., Droege, J. W., and Wood, J. H., Heat capacities and thermodynamic functions for δ- and ω-sodium palmitate, *J. Phys. Chem.* **63**, 152–153 (1959).

59WIR/WOO Wirth, H. E., Wood, J. H., and Droege, J. W., Low temperature heat capacities and thermodynamic functions of hydrous sodium palmitate, *J. Phys. Chem.* **63**, 149–152 (1959).

60BAR/BOL Barger, J. W., Bolze, C. C., Hughes, R. L., and Medved, T. M., Rapid determination of heat capacity of liquids over extended temperature ranges, *Lubr. End.* **10**, 105–107 (1960).

60CHA/WES Chang, S-S., and Westrum, E. F., Jr., Heat capacities and thermodynamic properties of globular molecules. I. Adamantane and hexamethylenetetramine, *J. Phys. Chem.* **64**, 1547–1551 (1960).

60CHA/WES 2 Chang, S-S., and Westrum, E. F., Jr., Heat capacities and thermodynamic properties of globular molecules. II. Triethylenediamine, *J. Phys. Chem.* **64**, 1551–1552 (1960).

60EDW/KIN Edwards, J. W., Kington, G. L., and Mason, R., The thermodynamic properties of ferrocene. Part 1. The low-temperature transition in ferrocene crystals, *Trans. Faraday Soc.* **56**, 660–669 (1960).

60GIA/OTT Giauque, W. F., and Ott, J. B., The three melting points and heats of fusion of phosgene. Entropy of solids I and II, and atomic exchange disorder in solid II, *J. Am. Chem. Soc.* **82**, 2689–2695 (1960).

60HUT/COL Hutchens, J. O., Cole, A. G., and Stout, J. W., Heat capacities from 11 to 305K. and entropies of l-alanine and glycine, *J. Am. Chem. Soc.* **82**, 4813–4815 (1960).

60KAP/STR Kapustinskii, A. F., Strelkov, I. I., Ganenko, V. E., Alapina, A. V., Stakhanova, M. S., and Selivanova, N. M., Low-temperature heat capacity, entropy, heat of formation, and free energy of formation of lead oxalate, *Russ. J. Phys. Chem.* **34**, 517–521 (1960).

60KAR/STR Karasharli, K. A., and Strelkov, P. G., Thermodynamic properties of dicyclohexyldodecane, *Zhur. Fiz. Khim.* **34**, 693–695 (1960).

60KAR/STR 2 Karasharli, K. A., and Strelkov, P. G., Thermodynamic properties of 1,1-diphenyldodecane, 14–298K, *Dokl. Akad. Nauk Azerb. SSR* **16**, 249–257 (1960).

60KAR/STR 3 Karasharli, K. A., and Strelkov, P. G., Thermodynamic properties of 1-phenyl-1-cyclohexyldodecane, 14–198K, *Dokl. Akad. Nauk Azerb. SSR* **16**, 341–344 (1960).

60KEN/LIN Kenttämaa, J., and Lindberg, J. J., Volumes and heats of mixing of dimethyl sulphoxide-water solutions, *Suom. Kemistilehti B* **33**, 32–35 (1960).

60PAR/STA Parsonage, N. G., and Staveley, L. A. K., Thermodynamic properties of clathrates: II. The heat capacity and entropy of methane in the methane quinol clathrates, *Mol. Phys.* **3**, 59–66 (1960).

60SWI/ZIE Swietoslawski, W., and Zielenkiewicz, A., Mean specific heat in homologous series of binary and ternary positive azeotropes, *Bull. Acad. Pol. Sci. Ser. Sci. Chim.* **8**, 651–653 (1960).

60WES/CHA Westrum, E. F., Jr., Chang, S-S., and Levitin, N. E., The heat capacity and thermodynamic properties of sodium formate from 5 to 350K, . *J. Phys. Chem.* **64**, 1553–1554 (1960).

61BER/SCO Berg, W. T., Scott, D. W., Hubbard, W. N., Todd, S. S., Messerly, J. F., Hossenlopp, I. A., Osborn, A., Douslin, D. R., and McCullough, J. P., The chemical thermodynamic properties of cyclopentanethiol, *J. Phys. Chem.* **65**, 1425–1430 (1961).

61GLA/TIM Glasgow, A. R., and Timmermans, J., Some physical properties of monochloro-, dichloro-, and monobromoacetic acids at 1 atmosphere, *Bull. Soc. Chim. Belg.* **70**, 599–622 (1961).

61GRE Green, J. H. S., Thermodynamic properties of organic oxygen compounds. Part. 5. Ethyl alcohol, *Trans. Faraday Soc.* **57**, 2132–2137 (1961).

61HUF/GRO Huffman, H. M., Gross, M. E., Scott, D. W., and McCullough, I. P., Low temperature thermodynamic properties of six isomeric heptanes, *J. Phys. Chem.* **65**, 495–503 (1961).

61MCC/FIN McCullough, J. P., Finke, H. L., Hubbard, W. N., Todd, S. S., Messerly, J. F., Douslin, D. R., and Waddington, G., Thermodynamic properties of four linear thiaalkanes, *J. Phys. Chem.* **65**, 784–791 (1961).

61PAC/BOB Pace, E. L., and Bobka, R. J., Thermodynamic properties of trifluoroacetonitrile from 12°K to its boiling point, *J. Chem. Phys.* **35**, 454–457 (1961).

61SCO/MES Scott, D. W., Messerly, J. F., Todd, S. S., Guthrie, G. B., Hossenlopp, I. A., Moore, R. T., Osborn, A., Berg, W. T., and McCullough, J. P., Hexamethyldisiloxane: chemical thermodynamic properties and internal rotation about the siloxane linkage, *J. Phys. Chem.* **65**, 1320–1326 (1961).

61WES Westrum, E. F., Jr., The thermophysical properties of three globular molecules, *J. Phys. Chem. Solids* **18**, 83–85 (1961).

62CHA/WES Chang, S-S., and Westrum, E. F., Jr., Heat capacities and thermodynamic properties of globular molecules. III. Two methyl-substituted polythiaadamantanes, *J. Phys. Chem.* **66**, 524–527 (1962).

62CHA/WES 2 Chang, S-S., and Westrum, E. F., Jr., Heat capacities and thermodynamic properties of two tetramethylammonium halides, *J. Chem. Phys.* **36**, 2420–2423 (1962).

62CHA/WES 3 Chang, S-S., and Westrum, E. F., Jr., Thermal properties based upon heat capacity and decomposition pressure of tetramethylammonium hydrogen dichloride. Nature of the hydrogen bond in the $[Cl-H-Cl]^-$ ion, *J. Chem. Phys.* **36**, 2571–2577 (1962).

62DAI/EVA Dainton, F. S., Evans, D. M., Hoare, F. E., and Melia, T. P., Thermodynamic functions of linear high polymers. Part I. Polyoxymethylene, *Polymer* **3**, 263–270 (1962).

62DAI/EVA 2 Dainton, F. S., Evans, D. M., Hoare, F. E., and Melia, T. P., Thermodynamic functions of linear high polymers. Part II. Thermodynamic functions of penton, *Polymer*, **3**, 271–276 (1962).

62DAI/EVA 3 Dainton, F. S., Evans, D. M., Hoare, F. E., and Melia, T. P., Thermodynamic functions of linear high polymers. Part III. Polyethylene, *Polymer* **3**, 277–285 (1962).

62DAI/EVA 4 Dainton, F. S., Evans, D. M., Hoare, F. E., and Melia, T. P., Thermodynamic functions of linear high polymers. Part IV. Stereospecific poly-α-olefines, *Polymer* **3**, 286–296 (1962).

62DAI/EVA 5 Dainton, F. S., Evans, D. M., Hoare, F. E., and Melia, T. P., Thermodynamic functions of linear high polymers. Part V. cis- and trans-1,4-Polybutadiene, *Polymer* **3**, 297–309 (1962).

62EDW/KIN Edwards, J. W., and Kington, G. L., Thermodynamic properties of ferrocene. Part 3. Thermodynamic functions from 0 to 300°K, *Trans.*

Faraday Soc. **58**, 1334–1340 (1962).
62EVA/HOA Evans, D. M., Hoare, F. E., and Melia, T. P., Heat capacity, enthalpy and entropy of citric acid monohydrate, *Trans. Faraday Soc.* **58**, 1511–1514 (1962).
62GOL/BEL Gollis, M. H., Belenyessy, L. I., Gudzinowicz, B. J., Koch, S. D., Smith, J. O., and Wineman, R. J., Evalutaions of pure hydrocarbons as Jet Fuels, *J. Chem. Eng. Data* **7**, 311–316 (1962).
62KAT Katayama, T., Heats of mixing, liquid heat capacities and enthalpy, concentration charts for methanol–water and isopropanol–water systems, *Kagaku Kogaku* **26**, 361–372 (1962).
62KOL/SER Kolesov, V. P., Seregin, E. A., and Skuratov, S. M., Adiabatic calorimeter of small volume for the determination of true heat capacity over the range 12–340K, *Zhur. Fiz. Khim.* **36**, 647–651 (1962).
62LOW/MOE Low, D. I. R., and Moelwyn–Hughes, E. A., The heat capacities of acetone, methyl iodide and mixtures thereof in the liquid state, *Proc. Roy Soc. (London)* **A267**, 384–394 (1962).
62OME Omel'chenko, F. S., Specific heat of furfural, *Izv. Vyssh. Ucheb. Zaved. Pishch. Tekhnol.* No, **2**, 151–152 (1962).
62PAY/WES Payne, D. H., and Westrum, E. F., Jr., Heat capacities and thermodynamic properties of globular molecules. IV. Pentaerythrityl chloride, bromide, and iodide from 6 to 300°K, *J. Phys. Chem.* **66**, 748–751 (1962).
62RAB/NIK Rabinovich, I. B., and Nikolaev, P. N., Isotopic effect in the specific heat of some deutero compounds, *Dokl. Akad. Nauk SSSR* **142**, 1335–1338 (1962).
62SCO/DOU Scott, D. W., Douslin, D. R., Finke, H. L., Hubbard, W. N., Messerly, J. F., Hossenlopp, I. A., and McCullough, J. P., 2–Methyl–2–Butanethiol: Chemical thermodynamic properties and rotational isomerism, *J. Phys. Chem.* **66**, 1334–1341 (1962).
62SCO/GOO Scott, D. W., Good, W. D., Todd, S. S., Messerly, J. F., Berg, W. T., Hossenlopp, I. A., Lacina, J. L., Osborn, A., and McCullough, J. P., 3,3–Dimethyl–2–thiabutane: chemical thermodynamic properties and barriers to internal rotation, *J. Chem. Phys.* **26**, 406–412 (1962). *J. Chem. Phys.* **36**, 406–412 (1962).
62SCO/GUT Scott, D. W., Guthrie, G. B., Messerly, J. F., Todd, S. S., Berg, W. T., Hossenlopp, I. A., and McCullough, J. P, Toluene: thermodynamic properties, molecular vibrations, and internal rotation, *J. Phys. Chem.* **66**, 911–914 (1962).
62SCO/MES Scott, D. W., Messerly, J. F., Todd, S. S., Hossenlopp, I. A., Douslin, D. R., and McCullough, J. P., 4–Fluortoluene: chemical thermodynamic properties, vibrational assignment, and internal rotation, *J. Chem. Physa.* **37**, 867–873 (1962).
62STR/BAR Strepikov, Yu. A., Baranov, Yu. I., and Burmistova, O. A., Determination of the heat of combustion and specific heat of some mono– and diisocyanates, *Izv. Vyssh. Uchebn. Khim. Khim. Tekhnol.* **5**, 387–390 (1962).
62VAL/BRO Valentine, R. H., Brodale, G. E., and Giauque, W. F., Trifluoromethane: entropy, low temperature heat capacity, heats of fusion and vaporization, and vapor pressure, *J. Phys. Chem.* **66**, 392–395 (1962).
62VIC Victor, A. C., Heat capacity of diamond at high temperatures, *J. Chem. Phys.* **36**, 1903–1911 (1962).
62WEB/KIL Weber, L. A., and Kilpatrick, J. E., Entropy and related thermodynamic properties of propionitrile, *J. Chem. Phys.* **36**, 829–834 (1962).
63AND/COU Andon, R. J. L., Counsell, J. F., Herington, E. F. G., and Martin, J. F., Thermodynamic properties of organic oxygen compounds, *Trans. Faraday Soc.* **59**, 830–835 (1963).
63AND/COU 2 Andon, R. J. L., Counsell, J. F., and Martin, J. F., Thermodynamic properties of organic oxygen compounds. Part II. The thermodynamic properties from 10 to 330K of isopropyl alcohol, *Trans. Faraday Soc.* **59**, 1555–1558 (1963).
63BAR/WES Barber, C. M., and Westrum, E. F., Jr., Heat capacities and thermodynamic properties of globular molecules. V. 3–Azabicyclo–[3,3,2]nonane from 5 to 350K, *J. Phys. Chem.* **67**, 2373–2376 (1963).
63COL/HUT Cole, A. G., Hutchens, J. O., and Stout, J. W., Heat capacities from 11 to 305K and entropies of L–phenylalanine, L–proline, L–tryptophane, and L–tyrosine. Some free energies of formation, *J. Phys. Chem.* **67**, 1852–1855 (1963).
63COL/HUT 2 Cole, A. G., Hutchens, J. O., and Stout, J. W., Heat capacities from 11 to 305K. and entropies of L–arginine,HCl, L–histidine. HCl, and L–lysine.HCl, *J. Phys. Chem.* **67**, 2245–2247 (1963).
63GAT/KRE Gattow, G., and Krebs, B., Das Kohlenstoffsulfid–di–(hydrogensulfid) $SC(SH)_2$ und das System H_2S–CS_2 2. Thermochemie des $SC(SH)_2$, *Z. anorg. allgem. Chem.* **322**, 113–128 (1963).
63GOO/TOD Good, W. D., Todd, S. S., Messerly, J. F., Lacina, J. L., Dawson, J. P., Scott, D. W., and McCullough, J. P., Perfluoropiperidine: entropy, heat of formation, and vapor pressure, N–F bond energy, and solid–state transitions, *J. Phys. Chem.* **67**, 1306–1311 (1963).
63GUD/CAM Gudinowicz, B. J., Campbell, R. H., and Adams, J. S., Specific heat measurements of complex saturated hydrocarbons, *J. Chem. Eng. Data* **8**, 201–214 (1963).
63HUT/COL Hutchens, J. O., Cole, A. G., and Stout, J. W., Heat capacities from 11 to 305K., entropies, and free energies of formation of L–valine, L–isoleucine, and L–laucine, *J. Phys. Chem.* **67**, 1128–1130 (1963).
63HUT/COL 2 Hutchens, J. O., Cole, A. G., Robie, R. A.,, and Stout, J. W., Heat capacities from 11 to 305K, entropies and free energies of formation of L–asparagine monohydrate, L–aspartic acid, L–glutamic acid, and L–glutamine, *J. Biol. Chem.* **238**, 2407–2412 (1963).
63MCC/MES McCullough, J. P., Messerly, J. F., Moore, R. T., and Todd, S. S., Trimethylaluminum: thermodynamic functions in the solid and liquid states, 0–380°K., vapor pressure, heat of vaporization, and entropy in the ideal gas state, *J. Phys. Chem.* **67**, 677–679 (1963).
63OET Oetting, F. L., The heat capacity and entropy of 2–methyl–2–propanol from 15 to 330°K, *J. Phys. Chem.* **67**, 2757–2761 (1963).
63SCO/GOO Scott, D. W., Good, W. D., Guthrie, G. B., Todd, S. S., Hossenlopp, I. A., Osborn, A. G., and McCullough, J. P., Chemical thermodynamic properties and internal rotation of methylpyridines. II. 3–Methylpyridine, *J. Phys. Chem.* **67**, 685–689 (1963).
63SCO/HUB Scott, D. W., Hubbard, W. N., Messerly, J. F., Todd, S. S., Hossenlopp, I. A., Good, W. D., Douslin, D. R., and McCullough, J. P., Chemical thermodynamic properties and internal rotation of methylpyridines. I. 2–Methylpyridine, *J. Phys. Chem.* **67**, 680–685 (1963).
63SCO/MES Scott, D. W., Messerly, J. F., Todd, S. S., Hossenlopp, I. A., Osborn, A., and McCullough, J. P., 1,2–Difluorobenzene: Chemical thermodynamic properties and vibrational assignment, *J. Chem. Phys.* **38**, 532–539 (1963).
63TRO/WES Trowbridge, J. C., and Westrum, E. F., Jr., Heat capacities and thermodynamic properties of globular molecules. VII. Transition and fusion of triethylenediamine, *J. Phys. Chem.* **67**, 2381–2385 (1963).
63VAR/KOP Vargaftik, N. B., Kopylov, N. I., Lapushkin, S. A., Pyatibratov, S. N., and Sokolov, S. N., Specific heat of some monoisopropylbiphenyls, *Izv. Vyssh. Ucheb. Zaved. Neft Gaz* **6**, No.3, 75–78 (1963).
63WUL/WES Wulff, C. A., and Westrum, E. F., Jr., Heat capacities and thermodynamic properties of globular molecules. VI. Succinonitrile, *J. Phys. Chem.* **67**,

2376–2381 (1963).

64DAV David, D. J., Determination of specific heat and heat of fusion by differential thermal analysis. Study of theory and operating parameters, *Anal. Chem.* **36**, 2162–2166 (1964).

64FUR/REI Furukawa, G. T., Reilly, M. L., Piccirelli, J. H., and Tenenbaum, M., Thermodynamic properties of some methylphosphonyl dihalides from 15 to 335 K, *J. Res. NBS A* **68**, 367–379 (1964).

64HUT/COL Hutchens, J. O., Cole, A. G., and Stout, J. W., Heat capacities and entropies of L–cystine and L–methionine. The transition of L–methionine near 305.5K, *J. Biolog. Chem.* **239**, 591–595 (1964).

64HUT/COL 2 Hutchens, J. O., Cole, A. G., and Stout, J. W., Heat capacities from 11 to 305K, entropies, enthalpy, and free energy of formation of L–serine, *J. Biol. Chem.* **239**, 4194–4195 (1964).

64MCE/KIL McEachern, D. M., Jr., and Kilpatrick, J. E., Entropy and related thermodynamic properties of dimethoxymethane, *J. Chem. Phys.* **41**, 3127–3131 (1964).

64MOE/THO Moelwyn–Hughes, E. A., and Thorpe, P. L., The physical and thermodynamic properties of some associated solutions. II. Heat capacities and compressibilities, *Proc. Roy Soc. (London)* **278A**, 574–587 (1964).

64OET Oetting, F. L., Low–temperature heat capacity and related thermodynamic functions of propylene oxide, *J. Chem. Phys.* **41**, 149–153 (1964).

64RAS/BAS Rastogi, R. P., and Bassi, P. S., Mechanism of eutectic crystallization, *J. Phys. Chem.* **68**, 2398–2406 (1964).

64SER/GOR Seregin, E. A., Goroshko, N. N., Kolesov, V. P., Belikova, N. A., Skuratov, S. M., and Plate, A. F., Low temperature heat capacity and thermodynamic functions for endo– and exo–2–methylbicyclo[2,2,1]heptane, *Dokl. Akad. Nauk SSR* **159**, 1381–1384 (1964).

64SIN/OET Sinke, G. C., and Oetting, F. L., The chemical thermodynamic properties of methyl ethyl ketone, *J. Phys. Chem.* **68**, 1354–1358 (1964).

64TRO/WES Trowbridge, J. C., and Westrum, E. F., Jr., Heat capacities and thermodynamic properties of globular molecules. X. Fusion of pentaerythrityl fluoride, *J. Phys. Chem.* **68**, 255–258 (1964).

64VUK/RAS Vukalovich, M. P., Rasskazov, D. S., Popov, V. N., and Babikov, Yu. M., Thermophysical properties of monoisopropylbiphenyl, *Teploenergetika* **11**, No.6, 56–58 (1964).

64WUL/WES Wulff, C. A., and Westrum, E. F., Jr., Heat capacities and thermodynamic properties of globular molecules. XI. Melting of 3–azabicyclo[3.2.2]nonane, *J. Phys. Chem.* **68**, 430–431 (1964).

65ABU/DOL Abu–Isa, I., and Dole, M., Specific heat of synthetic high polymers. XII. Atactic and isotactic polystyrene, *J. Phys. Chem.* **69**, 2668–2675 (1965).

65CAR/WES Carlson, H. G., and Westrum, E. F., Jr., 2,5–Dimethylthiophene. Heat capacities and thermodynamic properties from 5 to 300K and fusion of stable and metastable phases, *J. Phys. Chem.* **69**, 1524–1530 (1965).

65CAR/WES 2 Carlson, H. G., and Westrum, E. F., Jr., 2–Methylfuran–heat capacity and thermodynamic properties from 5 to 310K, *J. Chem. Eng. Data* **10**, 134–135 (1965).

65CLE/WON Clever, H. L., Wong, W–K. and Westrum, E. F., Jr., Heat capacities and thermodynamic properties of globular molecules. XIII. Transition and fusion of pentaerythrityl chloride and bromide, transition of pentaerythrityl iodide, *J. Phys. Chem.* **69**, 1209–1213 (1965).

65CLE/WUL Clever, H. L., Wulff, C. A., and Westrum, E. F., Jr., Glutaronitrile. Calorimetrically determined thermal properties from 5 to 350K and statistical gaseous entropy, *J. Phys. Chem.* **69**, 1983–1988 (1965).

65COU/GRE Counsell, J. F., Green, J. H. S., Hales, J. L., and Martin, J. F., Thermodynamic properties of fluorine compounds. Part 2. Physical and thermodynamic properties of hexafluorobenzene, *Trans. Faraday Soc.* 61, 212–218 (1965).

65COU/HAL Counsell, J. F., Hales, J. L., and Martin, J. F., Thermodynamic properties of organic oxygen compounds. Part 16. Butyl alcohol, *Trans. Faraday Soc.* **61**, 1869–1875 (1965).

65COX/GUN Cox, J. D., Gundry, H. A., and Head, A. J., Thermodynamic properties of fluorine compounds. Part 4. Enthalpy of formation of carbon tetrafluoride, *J. Chem. Soc. Faraday Trans. II.* **61**, 1594–1600 (1965).

65DOU/HAR Douglas, T. B., and Harman, A. W., Relative enthalpy of polytetrafluoroethylene from 0 to 440°C, *J. Res. NBS A* **69**, 149–157 (1965).

65EGA/WAK Egan, E. P. Wakefield, Z. T. and Farr, T. D., Low temperature heat capacity and entropy of oxamide, 10–310K, *J. Chem. Eng. Data* **10**, 138–140 (1965).

65FIN/MES Finke, H. L., Messerly, J. F., and Todd, S. S., Thermodynamic properties of n–propyl–,n–butyl–, and n–decyl–substituted cyclohexane from 10 to 370K, *J. Phys. Chem.* **69**, 2094–2100 (1965).

65FRA/AST Frankosky, M., and Aston, J. G., The heat capacity and entropy of hexamethylbenzene from 13 to 340K. An estimate of the internal rotation barrier, *J. Phys. Chem.* **69**, 3126–3132 (1965).

65GUM/KOS Gumbatov, D. O., and Kostrykov, V. N., Thermodynamic studies at low temperatures. The specific heat, entropy, enthalpy, and internal rotation potential of $C_6H_5SiCl_3$, *Zhur. Fiz. Khim.* **39**, 116–122 (1965).

65KAR/BAI Karasz, F. E., Bair, H. E., and O'Reilly, J. M., Thermal properties of atactic and isotactic polystyrene, *J. Phys. Chem.* **69**, 2657–2667 (1965).

65KAR/STR Karasharli, K. A., and Strelkov, P. G., Thermodynamic properties of 1,1–dicyclohexyldodecane,1,1–diphenyldodecane,1–phenyl–1–cyclohexyldodecane at low temperatures, *Doklady Akad. Nauk SSSR* **131**, 568–569 (1965).

65MCD McDonald, R. A., Heat content and heat capacity of an extruded graphite from 341° to 1723°K, *J. Chem. Eng. Data* **10**, 243 (1965).

65MCD/KIL McDougall, L. A., and Kilpatrick, J. E., Entropy and related thermodynamic properties of n–valeric acid, *J. Chem. Phys.* **42**, 2307–2310 (1965).

65MES/TOD Messerly, J. F., Todd, S. S., and Finke, H. L., Low–temperature thermodynamic properties of n–propyl– and n–butylbenzene, *J. Phys. Chem.* **69**, 4304–4311 (1965).

65MES/TOD 2 Messerly, J. F., Todd, S. S., and Finke, H. L., Low–temperature thermodynamic properties of n–propyl–, n–butyl–, and n–decyl–substituted cyclopentanes, *J. Phys. Chem.* **69**, 353–359 (1965).

65OET Oetting, F. L., Absolute entropies of the methyl alkyl ketones at 298.15K, *J. Chem. Eng. Data* **10**, 122–125 (1965).

65PEM/PAR Pemberton, R. C., and Parsonage, N. G., Thermodynamic properties of urea + hydrocarbon adductsPart 1. Heat capacities of the adducts of n–$C_{10}H_{22}$, n–$C_{12}H_{26}$, n–$C_{16}H_{34}$, and n.$C_{20}H_{42}$ from 12 to 300K, *Trans. Faraday Soc.* **61**, 2112–2121 (1965).

65PUT/MCE Putnam, W. E., McEachern, D. M., Jr., and Kilpatrick, J. E., Entropy and related thermodynamic properties of acetonitrile (methyl cyanide), *J. Chem. Phys.* **42**, 749–755 (1965).

65SIL/DAU Silbert, L. S., Daubert, B. F., and Mason, L. S., The heats of combustion, formation, and isomerizaion of isomeric monoglycerides, *J. Phys. Chem.* **69**, 2887–2894

(1965).
65SOM/COO Somsen, G., and Coops, J., Enthalpies of solvation of alkali halides in formamide. I. The enthalpies of solution of alkali halides in formamide, *Rec. Trav. chim.* **84**, 985–1002 (1965).
65SUG/SEK Suga, H., and Seki, S., An automatic adiabatic calorimeter for low temperatures. The heat capacity of standard benzoic acid, *Bull. Chem. Soc. Japan* **38**, 1000–1006 (1965).
65TUN/MIS Tungusov, V. P., and Mishchenko, K. P., Specific heat of pure ethylene glycol and solution of NaI and KI in ethylene glycol at 25°C, *Zhur. Fiz. Khim.* **39**, 2968–2972 (1965).
65WUN Wunderlich, B., Specific heat of polyethylene single crystals, *J. Phys. Chem.* **69**, 2078–2081 (1965).
65ZAL/KOC Zalikin, A. A., Kochetkov, V. L., and Strepikheev, Yu. A., Some physical and physico-chemical constants of m- and p-chloroanilines, m- and p-chlorophenylisocyanates, *Khim. Prom.* **41**, 338 (1965).
66BEA/CLE Beaumont, R. H., Clegg, B., Gee, G., Herbert, J. B. M., Marks, D. J., Roberts, R. C., and Sims, D., Heat capacities of propylene oxide and of some polymers of ethylene and propylene oxides, *Polymer* **7**, 401–416 (1966).
66DWO/GUI Dworkin, A., and Guillamin, M., Chaleur specifique du chlorure de tertiobutyle a basse temperature, *J. chim. Phys. physicochim. Biol.* **63**, 53–58 (1966).
66GAT/DRA Gattow, G., and Dräger, M., Das Kohlenstoffidselenid 2. Thermochemie des CSe_2, *Z. anorg. allgem. Chem.* **343**, 232–243 (1966).
66GEL Geller, B. E., Some physicochemical properties of dimethylformamide, *Zhur. Fiz. Khim.* **40**, 1956–1958 (1966).
66LIU/ZIE Liu, K. F., and Ziegler, W. T., Heat capacity from 80° to 300°K., melting point and heat of fusion of nitroethane, *J. Chem. Eng. Data* **11**, 187–189 (1966).
66NIK/RAB Nikolaev, P. N., Rabinovich, I. B., Gal'perin, V. A., and Tsvetkov, V. G., Isotopic effect on the specific heat and compressibility of deuterocyclohexane, *Zhur. Fiz. Khim.* **40**, 1091–1097 (1966).
66SAS/YOK Sasaki, K., and Yokotake, T., Thermodynamic properties of the products in SO_3–NH_3 reaction. II. Specific heats of sulfamide and ammonium sulfamate, *Tokyo Kogyo Shikenshi Hokohu* **61**, 309–314 (1966).
66ZAL/STR Zalikin, and A. A., Strepikheev, Yu. A., Some physico-chemical constants of 4,4′-diaminodiphenylmethane and 4,4′-diphenylmethanediisocyanate, *Zhur. Priklad. Khim.* **39**, 2607 (1966).
67AND/COU Andon, R. J. L., Counsell, J. F., Lees, E. B., Martin, J. F., and Mash, C. J., Thermodynamic properties of organic oxygen compounds. Part 17. Low-temperature heat capacity and entropy of the cresols, *Trans. Faraday Soc.* **63**, 1115–1121 (1967).
67BAR/POR Barral, E. M., II, Porter, R. S., and Johnson, J. F., Specific heats of nematic, smectic, and cholesteric liquid crystals by differential scanning calorimetry, *J. Phys. Chem.* **71**, 895–900 (1967).
67CHA/HOR Chang, S. S., Horman, J. A., and Bestul, A. B., Heat capacities and related thermal data for diethyl phthalate crystal, glass, and liquid to 360K, *J. Research NBS* **71A**, 293–305 (1967).
67MAG Magill, J. H., Physical properties of aromatic hydrocarbons. III. A test of the Adam–Gibbs relaxation model for glass formers based on the heat-capacity data of 1,3,5-tri-α-naphthylbenzene, *J. Chem. Phys.* **47**, 2802–2807 (1967).
67MEL/TYS Melia, T. P., and Tyson, A., Thermodynamics of addition polymerisation. Part 2. The heat capacity, entropy and enthalpy of isotatic poly(4-methyl-1-pentene), *Makromol. Chem.* **109**, 87–95(1967).
67MES/GUT Messerly, J. F., Guthrie, G. B., Todd, S. S., and Finke, H. L., Low-temperature thermal data for n-Pentane, n-Heptadecane, and n-Octadecane, *J. Chem. Eng. Data* **12**, 338–346 (1967).
67MES/TOD Messerly, J. F., Todd, S. S., and Guthrie, G. B., Jr., Low-temperature thermal properties of cyclohexanethiol and 2,4-dimethyl-3-thiapentane, *J. Chem. Eng. Data* **12**, 426–429 (1967).
67NIK/RAB Nikolaev, P. N., Rabinovich, I. B., and Lebedev, B. V., Specific heat of H- and D-ethyl alcohol in the interval 80–250K, *Zhur. Fiz.* Khim. **41**, 1294–1299 (1967).
67NIK/RAB 2 Nikolaev, P. N., and Rabinovich, I. B., Heat capacity of ethylene glycol and ethylene deuterioglycol in the temperature range 80–300K, *Zhur. Fiz. Khim.* **41**, 2191–2194 (1967).
67PAC Pacor, P., Applicability of the Du Pont 900 DTA apparatus in quantitative differential thermal analysis, *Anal. Chim. Acta* **37**, 200–208 (1967).
67PAC/PLA Pace, E. L., and Plaush, A. C., Thermodynamic properties of octafluoropropane from 14K to its normal boiling point. An estimate of the barrier to internal rotation from the entropy and heat capacity of the gas, *J. Chem. Phys.* **47**, 38–43 (1967).
67PLA/PAC Plaush, A. C., and Pace, E. L., Thermodynamic properties of hexafluoroacetone from 12K to its normal boiling point. An estimate of the barrier to internal rotation from the entropy of the gas, *J. Chem. Phys.* **47**, 44–48 (1967).
67RAS/GAN Rastorguev, Yu. L., and Ganiev, Yu. A., Study of the heat capacity of selected solvents, *Izv. Vyssh. Uchebn. Zaved. Neft Gaz.* **10**, No. 1, 79–82 (1967).
67RIB/WES Ribner, A., and Westrum, E. F., Jr., Dimethylmalononitrile. Low-temperature heat capacity, vapor pressure, density, and chemical thermodynamics of the crystalline, liquid, and gaseous phases, *J. Phys. Chem.* **71**, 1208–1215 (1967).
67SCO/BER Scott, D. W., Berg, W. T., Hossenlopp, I. A., Hubbard, W. N., Messerly, J. F., Todd, S. S., Douslin, D. R., McCullough, J. P., and Waddington, G., Pyrrole: chemical thermodynamic properties, *J. Phys. Chem.* **71**, 2263–2270 (1967).
67SMI/GOO Smith, N. K., and Good, W. D., Enthalpy of formation of triethylamineborane, *J. Chem. Eng. Data* **12**, 570–572 (1967).
67SMI/GOO 2 Smith, N. K., and Good, W. D., Enthalpies of combustion and formation of propylamine, isopropylamine, and tert-butylamine, *J. Chem. Eng. Data* **12**, 572–574 (1967).
67WES/RIB Westrum, E. F., Jr., and Ribner, A., Trimethylacetonitrile. Low-temperature heat capacity, vapor pressure, and chemical thermodynamics of the crystalline, liquid, and gaseous phases, *J. Phys. Chem.* **71**, 1216–1224 (1967).
68ADA/SUG Adachi, K., Suga, H., and Seki, S., Phase changes in crystalline and glassy-crystalline cyclohexanol, *Bull. Chem. Soc. Japan* **41**, 1073–1087 (1968).
68AGU/TEL Aguliar, J., and Tello, M. J., Anomalias del calor especifico del sulfato de triglicina, *An. Fis.* **64**, 19–23 (1968).
68AND/COU Andon, R. J. L., Counsell, J. F., and Martin, J. F., Thermodynamic properties of organic oxygen compounds. Part XX. The low-temperature heat capacity and entropy of C_4 and C_5 ketones, *J. Chem. Soc. A* **1968**, 1894–1897.
68AND/COU 2 Andon, R. J. L., Counsell, J. F., Hales, J. L., Lees, E. B., and Martin, J. F., Thermnodynamic properties of fluorine compounds. Part VII. Heat capacity and entropy of pentafluorochlorobenzene and pentafluorophenol, *J. Chem. Soc. A* **1968**, 2357–2361
68ASH/STE Ashworth, T., and Steeple, H., The continuous

heating method of calorimetry and its application to the study of low temperature specific heat anomalies in methyl-ammonium alum, *Cryogenics* **8**, 225–234 (1968).

68CAR/WES Carlson, H. G., and Westrum, E. F., Jr., Thermal study of the glass-type transformation of 2-methylthiophene, *J. Chem. Eng. Data* **13**, 273–274 (1968).

68CHA/BES Chang, S. S., and Bestul, A. B., Heat capacities for atactic polystyrene of narrow molecular weight distribution to 360K, *J. Polymer Sci. Part A-2* **6**, 849–860 (1968).

68COU/HAL Counsell, J. F., Hales, J. L., and Martin, J. F., Thermodynamic properties of fluorine compounds. Part VI. The heat capacity and entropy of pentafluorobenzene, *J. Chem. Soc. A* **1968**, 2042–2044.

68COU/HAL 2 Counsell, J. F., Hales, J. L., Lees, E. B., and Martin, J. F., Thermodynamic properties of fluorine compounds. Part VIII. The heat capacity and entropy of 2,3,4,5,6-pentafluorotoluene, *J. Chem. Soc. A* **1968**, 2994–2996

68COU/LEE Counsell, J. F., Lees, E. B., and Martin, J. F., Thermodynamic properties of organic oxygen compounds. Part XIX. Low-temperature heat capacity and entropy of propan-1-ol, 2-methyl-propan-1-ol, and pentan-1-ol, *J. Chem. Soc. C* **1968**, 1819–1823

68ELL/CHR Elliott, J. H., and Chris, M. D., Some thermodynamic properties of high purity dimethyl terephthalate, *J. Chem. Eng. Data* **13**, 475–479 (1968).

68GEE/MEL Gee, D. R., and Melia, T. P., Thermodynamics of addition polymerisation Part 4. The heat capacity, entropy, enthalpy, and free energy of syndiotactic polypropylene, *Makroknol. Chem.* **116**, 122–129 (1968).

68GOU/GIR Goursot, P., Girdhar, H. L., and Westrum, E. F., Jr., Mesure de la capacite calorifique de l'anthracene de 5 a 520K, *Compt. rend. C* **266**, 949–950 (1968).

68GOU/WES Goursot, P., and Westrum, E. F., Jr., Heat capacity and thermodynamic properties of 2-methylthiazole from 5 to 350K, *J. Chem. Eng. Data* **13**, 468–471 (1968).

68GOU/WES 2 Goursot, P., and Westrum, E. F., Jr., Heat capacity and thermodynamic functions of thiazole from 5 to 340K, *J. Chem. Eng. Data* **13**, 471–475 (1968).

68GOU/WES 3 Goursot, P., and Westrum, E. F., Jr., Mesure de la capacite calorifique du methyl-2 thiazole de 5 a 300K, *Compt. rend. C* **266**, 667–668 (1968).

68REC Recko, W. M., Excess heat capacity of the binary systems formed by n-propyl alcohol with benzene, mesitylene and cyclohexane, *Bull. Acad. Pol. Sci. Ser. Sci. Chim.* **16**, 549–552 (1968).

68WAD Wadsö, I., Heats of vaporization of organic compounds II., Chlorides, bromides, and iodides, *Acta Chem. Scand.* **22**, 2438–2444 (1968).

68WES/WUL Westrum, E. F., Jr., and Wulff, C. A., Thermodynamic properties and third-law cycle for malononitrile, *J. Chem. Eng. Data* **13**, 239–242 (1968).

69ATK/LAR Atkinson, C. M. L., Larkin, J. A., and Richardson, M. J., Enthalpy changes in molten n-alkanes and polyethylene, *J. Chem. Thermodynam.* **1**, 435–445 (1969).

69BER/WES Berman, H. A., and West, E. D., Heat capacity of liquid nitromethane from 35 to 200^{0}C, *J. Chem. Eng. Data* **14**, 107–109 (1969).

69CLE/MEL Clegg, G. A., and Melia, T. P., Thermodynamics of polymerisation of heterocyclic compounds. Part III. The heat capacity, entropy, enthalpy, and free energy of tetroxan, *Makromol. Chem.* **123**, 184–193 (1969).

69CLE/MEL 2 Clegg, G. A., and Melia, T. P., Thermodynamics of polymerisation of heterocyclic compounds, Part IV. The heat capacity, entropy, enthalpy and free energy of paraldehyde, *Makromol. Chem.* **123**, 194–202 (1969).

69CLE/MEL 3 Clegg, G. A., and Melia, T. P., Heat capacity measurements on pentoxan, *Makromol. Chem.* **130**, 258–260 (1969).

69COP/PAR Cope, A. F. G., and Parsonage, N. G., Thermodynamic prperties of urea + hydrocarbon adducts from 12 to 300 K. The heat capacities and entropies of the adducts of the n-paraffin $C_{11}H_{24}$ and the 1-olefins $C_{10}H_{20}$, $C_{16}H_{32}$, and $C_{20}H_{40}$, *J. Chem. Thermodynam.* **1**, 99–110 (1969).

69ENO/SHI Enokida, H., Shinoda, T., and Mashiko, Y., Thermodynamic properties of neopentane from 4K to the melting point and comparison with spectroscopic data, *Bull. Chem. Soc. Japan* **42**, 84–91 (1969).

69GOU/WES Goursot, P., and Westrum, E. F., Jr., Heat capacity and thermodynamic properties of benzothiazole from 5° to 320°K, *J. Chem. Eng. Data* **14**, 1–3 (1969).

69HUT/COL Hutchens, J. O., Cole, A. G., and Stout, J. W., Heat capacities from 11 to 305° K and entropies of hydrated and anhydrous bovine zinc insulin and bovine chymotripsinogen A, *J. Biol. Chem.* **244** 26–32 (1969).

69HUT/COL 2 Hutchens, J. O., Cole, A. G., and Stout, J. W., Heat capacities from 11 to 305 K, entropies, and free energy of formation of glycylglycine, *J. Biol. Chem.* **244**, 33–35 (1969).

69JUS Justice, B. H., Low temperature thermodynamic properties of aluminum trichloride, *J. Chem. Eng. Data* **14**, 4–5 (1969).

69KOS/SUG Kosaki, A., Suga, H., Seki, S., Mukai, K., and Deguchi, Y., Thermodynamic properties of galvinoxyl radical and its phenol derivative, mechanism of the phase transition, *Bull. Chem. Soc. Japan* **42**, 1525–1530 (1969).

69PAU/GLU Paukov, I. E., and Glukhikh, L. K., The true heat capacity in the range 13–350°K and the absolute entropy and enthalpy of 1,3,5-trichloro-2,4,6-trifluorobenzene, *Zhur. Fiz. Khim.* **43**, 226–228 (1969).

69PAU/GLU 2 Paukov, I. E., and Glukhikh, L. K., The true heat capacity over the range 13–300K and the entropy and enthalpy of chloropentafluorobenzene, *Zhur. Fiz. Khim.* **43**, 1350–1352 (1969).

69PAU/LAV Paukov, I. E., and Lavrent'eva, M. N., Thermodynamic properties of pentafluorobenzene over the range 12–300K, *Zhur. Fiz. Khim.* **43**, 2938–2941 (1969).

69PAU/LAV 2 Paukov, I. E., Lavrent'eva, M. N., and Anisimov, M. P., Heat capacity, phase transitions, entropy, and enthalpy of pentafluorophenol at low temperatures, *Zhur. Fiz. Khim.* **43**, 785–787 (1969).

69PAU/LAV 3 Paukov, I. E., Lavrent'eva, M. N., and Anisimov, M. P., Heat capacity at low temperatures, phase transitions, entropy and enthalpy of pentafluoroaniline, *Zhur. Fiz. Khim.* **43**, 2941–2943 (1969).

69RAB/MAR Rabinovich, I. E., Martynenko, L. Ya., and Maslova, V. A., Specific heat and physical properties of some dialkyl phthalates, *Tr. Khim. Khim. Tekhnol.* **1969**, No. 2, 10–14.

69SAD/STE Sadowska, K. W., Stepniewska, G. B., and Recko, W. M., Specific heat and enthalpy of fusion of acenaphthene and acenaphthylene, *Przem. Chem.* **48**, 282–285 (1969).

69SHI/MCN Shieh, C-F., McNally, D., and Boyd, R. H., The heats of combustion and strain energies of some cyclophanes, *Tetrahedron* **25**, 3653–3665 (1969).

69SMI/PAC Smith, J. H., and Pace, E. L., An estimate of the liberational barrier in the solid perfluoroalkanes from heat capacity data, *J. Phys. Chem.* **73**, 2368–2372 (1969).

69SMI/PAC 2 Smith, J. H., and Pace, E. L., The thermodynamic properties of carbon tetrafluoride from 12°K to its boiling point. The significance of the parameter ν, *J. Phys. Chem.* **73**, 4232–4236 (1969).

69SOU/GOU Soulie, M. A., Goursot, P., Peneloux, A., and Metzger, J., Properties thermochimiques du methyl-2 thiazole, *J. chim. Phys. physicochim. Biol.* **66**, 607–610 {1969}

69SOU/GOU 2 Soulie, M. A., Goursot, P., Peneloux, A., and

Metzger, J., Properties thermochimiques du thiazole, *J. chim. Phys. physicochim. Biol.* **66**, 603–606 (1969).

69WIL/SCH Wilhelm, E., Schano, R., Becker, G., Findenegg, G. H., and Kohler, F., Molar heat capacity at constant volume. Binary mixtures of 1,2-dichloroethane and 1,2-dibromoethane with cyclohexane, *Trans. Faraday Soc.* **65**, 1443–1455 (1969).

70AND/COU Andon, R. J. L., Counsell, J. F., Lees, E. B., and Martin, J. F., Thermodynamic properties of organic oxygen compounds. Part XXIII. Low-temperature heat capacity and entropy of C_6, C_7, and C_9 ketones, *J. Chem. Soc. A* **1970**, 833–837.

70AND/WES Andrews, J. T. S., and Westrum, E. F., Jr., The heat capacity and thermodynamic functions of crystalline and liquid triptycene, *J. Chem. Thermodynam.* **2**, 245–253 (1970).

70AND/WES 2 Andrews, J. T. S., and Westrum, E. F., Jr., Heat capacity and thermodynamic properties of [2.2]paracyclophane. The mechanism of the 50°K transition, *J. Phys. Chem.* **74**, 2170– 2171 (1970).

70CHA/MCC Chang, S., McNally, D., Shary-Tehrany, S., Hickey, M. J., and Boyd, R. H., The heats of combustion and strain energies of bicyclo[n.m.0]alkanes, *J. Am. Chem. Soc.* **92**, 3109–3118 (1970).

70CLE/WES Clever, H. L., and Westrum, E. F., Jr., Dimethylsulfoxide and dimethylsulfone. Heat capacities, enthalpies of fusion, and thermodynamic properties, *J. Phys. Chem.* **74**, 1309–1317 (1970).

70FIN/MCC Finke, H. L., McCullough, J. P., Messerly, J. F., Guthrie, G. B., and Douslin, D. R., Chemical thermodynamic properties for 1-alkanethiols, *J. Chem. Thermodynam.* **2**, 27–41 (1970).

70FIN/TOD Finke, H. L., Todd, S. S., and Messerly, J. F., Trimethylamineborane and triethylamineborane: low-temperature thermodynamic properties, *J. Chem. Thermodynam.* **2**, 129–138 (1970).

70GOU/GIR Goursot, P., Girdhar, H. L., and Westrum, E. F., Jr., Thermodynamics of polynculear aromatic molecules. III. Heat capacities and enthalpies of fusion of anthracene, *J. Phys. Chem.* **74**, 2538–2541 (1970).

70KOL/SER Kolesov, V. P., Seregin, E. A., Belikova, N. A., Skuratov, S. M., and Plate, A. F., Specific heat in the interval 12–350K and thermodynamic functions of the endo- and exo-isomers of 2-cyanobicyclo[2,2,1]heptane, *Termodin. Termokhim. Konstanty* **1970**, 164–172.

70KRI/LIC Krien, G., Licht, H. H., and Trimborn, F., A phase transition of hexanitroethane (HNE), *Explosivstoffe* **9**, 203–207 (1970).

70LUT/VOL Lutcov, A. I., Volga, V. I., and Dymov, B. K., Thermal conductivity, electric resistivity and specific heat of dense graphites, *Carbon* **8**, 753–760 (1970).

70MES/FIN Messerly, J. F., and Finke, H. L., Hexafluorobenzene and 1,3-difluorobenzene. Low-temperature calorimetric studies and chemical thermodynamic properties, *J. Chem. Thermodyn.* **2**, 867–880 (1970).

70MES/TOD Messerly, J. F., Todd, S. S., and Guthrie, G. B., Chemical thermodynamic properties of the pentadienes, *J. Chem. Eng. Data* **15**, 227–232 (1970).

70PAZ/PAZ Paz Andrade, M. I., Paz, J. M., and Recacho, E., Contribucion a la microcalorimetria de los calores especificos de solidos y liquidos, *An. Quim.* **66**, 961–967 (1970).

70REE/SEE Reese, B. O., Seely, L. B., Shaw, R. P., and Tegg, D., Coefficient of thermal expansion of, and sound speed through, nitromethane and four bis(disfluoramino)alkanes, *Chem. Eng. Data* **15**, 140–142 (1970).

70TAK/WFS Takahashi, Y., and Westrum, E. F., Jr., Glassy carbon. Low-temperature thermodynamic properties, *J. Chem. Thermodyn.* **2**, 847–854 (1970).

70VAN/WES Vanderzee, C. E., Westrum, E. F., Jr., Heat capacities and thermodynamic properties of ammonium and potassium thiocyanates from 5 to 340K, *J. Chem. Thermodynam.* **2**, 417–429 (1970).

70VAN/WES 2 Vanderzee, C. E., and Westrum, E. F., Jr., Succinic acid. Heat capacities and thermodynamic properties from 5 to 328K. An efficient drying procedure, *J. Chem. Thermodynam.* **2**, 681–687 (1970).

70WES/WON Westrum, E. F., Jr., Wong, W-K., and Morawetz, E., Thermodynamics of globular molecules. XVIII. Heat capacities and transitional behavior of 1-azabicyclo[2.2.2]octane and 3-oxabicyclo[3.2.2]nonane. Sublimation behavior of five globular molecules, *J. Phys. Chem.* **74** 2542–2547 (1970).

70WON/WES Wong, W-K., and Westrum, E. F., Jr., Thermodynamics of globular molecules. XVII. Heat capacities and transition behavior of bicyclo[2.2.2]octane and bicyclo[2.2.2]octene, *J. Phys. Chem.* **74**, 1303–1308 (1970).

70YOS/SAK Yoshida, S., Sakiyama, M., and Seki, S., Thermodynamic studies of solid polyethers. I. Poly(oxacylobutane),$[-(CH_2)_3O-]_n$ and its hydrate, *Polymer Journal* **1**, 573–581 (1970).

71ADA/SUG Adachi, K., Suga, H., and Seki, S., Calorimetric study of the glassy state. VI. Phase changes in crystalline and glassy-crystalline 2,3-dimethylbutane, *Bull. Chem. Soc. Japan* **44**, 78–89 (1971).

71AND/CON Andon, R. J., Connett, J. E., Counsell, J. F., Lees, E. B., and Martin, J. F., Thermodynamic properties of organic oxygen compounds. Part. XXVII. (±)-Butan-2-ol and (+)-Butan-2-ol, *J. Chem. Soc A* **1971**, 661–664.

71ATA/CHI Atake, T., and Chihara, H., Heat capacity of solid carbon tetrachloride from 3 to 50 K, *J. Chem. Thermodynam.* **3**, 51–60 (1971).

71BOY/SAN Boyd, R. H., Sanwal, S. N., Shary-Tehrany, S., and McNally, D., The thermochemistry, thermodynamic functions, and molecular structures of some cyclic hydrocarbons, *J. Phys. Chem.* **75**, 1265–1271 (1971).

71CAR/WES Carlson, H. G., and Westrum, E. F., Jr., Methanol: heat capacity, enthalpies of transition amd melting, and thermodynamic properties from 5-300K, *J. Chem. Phys.* **54**, 1464–1471 (1971).

71CHI/THO Chick, M. C. and Thorpe, B. W., Polymorphism in 2,4,5-trinitrotoluene, *Australian J. Chem.* **24**, 191–195 (1971).

71COU/LEE Counsell, J. F., Lee, D. A., and Martin, J. F., Thermodynamic properties of organic oxygen compounds. Part XXVI. Diethyl ether, *J. Chem. Soc. A* **1971**, 313–316.

71DES/BHA Deshpande, D. D., and Bhatagadde, L. G., Heat capacities at constant volume, free volumes, and rotational freedom in some liquids, *Aust. J. Chem.* **24**, 1817–1822 (1971).

71HAL/BAL Hall, H. K., Jr., and Baldt, J. H., Thermochemistry of strained-ring bridgehead nitriles and esters, *J. Am. Chem. Soc.* **93**, 140–145 (1971).

71HYD/SUB Hyder Khan, V., and Subrahmanyam, S. V., Excess thermodynamic functions of the systems: benzene + p-xylene and benzene + p-dioxan, *Trans. Faraday Soc.* **67**, 2282–2291 (1971).

71KOL/VOR Kolesov, V. P., and Vorob'ev, V. N., Heat capacity at 12–300K and thermodynamic functions of 1,1,1-trichloro-3,3,3-trifluoropropane, *Zhur. Fiz. Khim.* **45**, 1293–1294 (1971).

71KON/WAD Konicek, J., and Wadsö, I., Thermochemical properties of some carboxylic acids, amines and N-substituted amides in aqueous solution, *Acta Chem. Scand.* **25**, 1541–1551 (1971).

71MES/FIN Messerly, J. F., and Finke, H. L., Low-temperature thermal properties of 2-methylheptane and 2-methyldecane: the thermodynamic properties of the

2-methylalkanes, *J. Chem. Thermodynam.* **3**, 675–687 (1971).

71PAU Paukov, I. E., Low-temperature heat capacity, phase transition, absolute entropy, and enthalpy of pentafluorotoluene, *Zhur. Fiz. Khim.* **45**, 1297–1298 (1971).

71PAU 2 Paukov, I. E., Specific heat from low temperatures, absolute entropy and enthalpy of pentafluoronitrobenzene, *Zhur. Fiz. Khim.* **45**, 1293–1294 (1971).

71PAU/RAK Paukov, I. E., and Rakhmenkulov, F. S., Low-temperature heat capacity and standard thermodynamic functions of decafluorobiphenyl, *Zhur. Fiz. Khim.* **45**, 1296–1297 (1971).

71PRI Privalko, V. P., Heat capacities of certain alkyl phenylcarbamates, *Zhur. Fiz. Khim.* **45**, 1585 (1971).

71SOR/TAC Sorai, M., Tachiki, M., Suga, H., and Seki, S., Magnetic and thermal properties of crystals including isolated clusters. I. Heat capacity and infrared spectrum of $[Cr_3O(CH_3COO)_6(H_2O)_3]Cl{\cdot}6H_2O$ crystal between 1.5 and 280K, *J. Phys. Soc. Japan* **30**, 750–759 (1971).

71VAS/PET Vasil'ev, I. A., Petrov, V. M., Ignat'ev, V. M., and Vvedenskii, A. A., Thermodynamic functions of a series of aliphatic amines. III. Entropy of n-propylamine, *Zhur. Fiz. Khim.* **45**, 1316 (1971).

71WON/WES Wong, W-K., and Westrum, E. F., Jr., Thermodynamics of polynuclear aromatic molecules. I. Heat capacities and enthalpies of fusion of pyrene, fluoranthene, and triphenylene, *J. Chem. Thermodynam.* **3**, 105–124 (1971).

72ADA/SUG Adachi, K., Suga, H., and Seki, S., Calorimetric study of the glassy state. VII. Phase changes between the crystalline phases of cycloheptanol with various degrees of stability, *Bull. Chem. Soc. Japan* **45**, 1960–1972 (1972).

72ARE/MIL Arentsen, J. G., and Van Miltenburg, J. C., Carbon tetrachloride. Determination of the enthalpy of transition from metastable face-centered cubic carbon tetrachloride to the stable rhombohedral modification, *J. Chem. Thermodynam.* **4**, 789–791 (1972).

72ARV/WES Arvidsson, K., and Westrum, E. F., Jr., Tris(hydroxymethane.yl)aminomethane. Heat capacities and thermodynamic properties from 5 to 350K, *J. Chem. Thermodynam.* **4**, 449–453 (1972).

72CHA/BES Chang, S. S., and Bestul, A. B., Heat capacity and thermodynamic properties of o-terphenyl crystal, glass, and liquid, *J. Chem.D Phys.* **56**, 503–516 (1972).

72COP/GAN Cope, A. F. G., Gannon, D. J., and Parsonage, N. G., Thermodynamic properties of thiourea + hydrocarbon adducts from 12 to 300K. The heat capacities and entropies of the adducts of c-C_6H_{12}, c-C_7H_{14}, c-C_8H_{16}, and $(CH_3)_3C.C_2H_5$, *J. Chem. Thermodynam.* **4**, 829–842 (1972).

72FIN/MCC Finke, H. L., McCullough, J. P., Messerly, J. F., Osborn, A., and Douslin, D. R., cis- and trans-Hexahydroindan. Chemical thermodynamic properties and isomerizaton equilibrium, *J. Chem. Thermodynam.* **4**, 477–494 (1972).

72FIN/MES Finke, H. L., Messerly, J. F., and Todd, S. S., Thermodynamic properties of acrylonitrile, 1-aminopropane, 2-aminopropane, and 2-methyl-2-aminopropane, *J. Chem. Thermodynam.* **4**, 359–374 (1972).

72GAN/PAR Gannon, D., J., and Parsonage, N. G., Thermodynamic properties of urea + hydrocarbon adducts from 12 to 300K. The heat capacities and entropies of the adducts of the 1-alkenes $C_{12}H_{24}$, $C_{14}H_{28}$, and $C_{18}H_{36}$, *J. Chem. Thermodynam.* **4**, 745–753 (1972)

72KAW/OTA Kawaizumi, F., Otake, T., Nomura, H., and Miyahara, Y., Heat capacities of aqueous solutions of ethylene glycol, propylene glycol and 1,3-butanediol, *Nippon Kagaku. Kaishi* **1972**, 1773–1776.

72KOL/VOR Kolesov, V. P., and Vorob'ev, V. N., Specific heat in the interval 12–300K and phase changes for some chlorinated 1,1,1-trifluoropropanes, *Doklady Akad. Nauk SSSR* **203**, 116–119 (1972).

72LAG Lagarrigue, M., Etude de la chaleur specifique des derives chlores de l'hexamethylbenzene entre 30 et 300 K. I. 1,2,3-trichloro-4,5,6-trimethylbenzene, *Mol. Cryst. Liquid Cryst.* **17**, 237–249 (1972).

72MAS/FAM Maslova, V. A., and Faminskaya, L. A., Specific heat and phase changes of triethylarsine and triethylgallium, *Trudy Khim. Khim. Tekhnol.* No. **2**, 51–53 (1972).

72MAS/RAB Maslova, V. A., Rabinovich, I. B., Nistoalov, V. P., and Faminskaya, L. A., Specific heat and phase transitions of some alkyl compounds of silicon, germanium, and tin, *Tr. Khim. Khim. Tekhnol.* No. **2**, 44–50 (1972).

72MIL Van Miltenburg, J. C., Construction of an adiabatic calorimeter. Thermodynamic properties of standard n-heptane from 155 to 270K and of 2,2-dichloropropane from 135 to 270K, *J. Chem. Thermodynam.* **4**, 773–782 (1972).

72PAC/HOD Pace, E. L., and Hodge, P. E., Thermodynamic properties of Trifluoroacetyl fluoride from 14 K to its boiling temperature, *J. Chem. Thermodynam.* **4**, 441–447 (1972).

72SHE/BEL Sheindlin, A. E., Belevich, I. S., and Kozhevnikov, I. G., Enthalpy and specific heat of Graphite in the temperature range 273-3650°K, *Teplofiz. Vys. Temp.* **10**, 997–1001 (1972).

72YOU/HAL Young, W. R., Haller, I., and Aviram, A., Mesomorphism in the 4,4'-dialkoxy-trans-stilbenes, *Mol. Cryst. Liquid Cryst.* **15**, 311–317 (1972).

73AKH/EKS Akhundov, T. S., and Eksaev, R. A., Experimental measurements of the isobaric heat capacity of liquid toluene in the temperature range 30–300°C and to 250 bar pressure, *Izv. Vyssh. Uchebn. Zaved. Neft Gaz* **16**, No. 2, 68–72 (1973).

73ALV/BIL Alvarez, J., and Biltonen, R., Nucleic acid-solvent interactions: temperature dependence of the heat of solution of thymine in water and ethanol, *Biopolymers* **12**, 1815–1828 (1973).

73AND/COU Andon, R. J. L., Counsell, J. F., Lee, D. A., and Martin, J. F., Thermodynamic properties of aliphatic halogen compounds. Part 2. Heat capacity of 1,1,1-trichloroethane, *J. Chem. Soc. Faraday Trans. I* **69**, 1721–1726 (1973).

73AND/GOR Andrews, J. T. S., and Gordon, J. E., Adiabatic calorimetry of organic salts. Tetra-n-hexylammonium perchlorate, *J. Chem. Soc. Faraday Trans. I* **69**, 546–554 (1973).

73AND/MAR Andon, R. J. L., and Martin, J. F., Thermodynamic properties of fluorine compounds. Part II. Low-temperature heat capacities of the three tetrafluorobenzenes, *J. Chem. Soc. Faraday Trans. I* **69**, 761–770 (1973).

73AND/MAR 2 Andon, R. J. L., and Martin, J. F., Thermodynamic properties of fluorine compounds. Part 12. Low-temperature heat capacity and entropy of 1,3,5-trichloro-2,4,6-trifluorobenzene, *J. Chem. Soc. Faraday Trans I* **69**, 871–875 (1973).

73CHA/BES Chang, S. S., and Bestul, A. B., Heat capacities of polyethylene from 2 to 360 K. I. Standard samples of linear and branched polyethylene whole polymer, *J. Research NBS* **77A**, 395–405 (1973).

73CHI/MAS Chihara, H., and Masukane, K., Heat capacity of solid tetrachloro-p-benzoquinone (chloranil) between 11 and 300K. Phase transition at 92K, *J. Chem. Phys.* **59**, 5397–5403 (1973).

73DZH/GUS Dzhafarov, O. I., Guseinov, Z. A., Karasharli, K. A., and Zeinalov, T. D., Heat capacity of tetramethyldisiletan at low temperatures, *Zhur. Fiz. Khim.* **47**, 2448 (1973).

73FIN/MES Finke, H. L., and Messerly, J. F., 3-Methylpentane and 3-methylheptane: low-temperature thermodynamic properties, *J. Chem. Thermodynam.* **5**, 247–257 (1973).

73GEI/DZH Geidarov, Kh. I., Dzhafarov, O. I., Karasharli, K. A., and Kostryukov, V. N., Thermodynamics of o-hydroxybiphenyl at low temperatures, *Zhur. Fiz. Khim.* **47**, 275 (1973).

73GOO Good, W. D., The enthalpies of combustion and formation of 1,8-dimethylnaphthalene, 2,3-dimethylnaphthalene, 2,6-dimethylnaphthalene, and 2,7-dimethylnaphthalene, *J. Chem. Thermodynam.* **5**, 715–720 (1973).

73GOO 2 Good, W. D., The enthalpies of combustion and formation of n-butylbenzene, the diethylbenzenes, the methyl-n-propylbenzenes, and the methyl-isopropylbenzenes, *J. Chem. Thermodynam.* **5**, 707–714 (1973).

73HAL/SMI Hall, H. K., Jr., Smith, C. D., and Baldt, J. H., Enthalpies of formation of nortricyclene, norbornene, norbornadiene, and quadricyclane, *J. Am. Chem. Soc.* **95**, 3197–3201 (1973).

73HAM/AYE Hamilton, W. S., and Ayers, D. A., Heat of combustion of 3-amino-5-methylisoxazole, *J. Chem. Eng. Data* **18**, 366–367 (1973).

73HAM/MIT Hamilton, W. S., Mitchell, G. M., and Ayers, D. A., Heat of combustion of N,N′-bis(m-methoxyphenyl)terephthalamide and N,N′-bis(p-methoxyphenyl)terephthalamide, *J. Chem. Eng. Data* **18**, 364–366 (1973).

73JOH/MAR Johnson, D. A., and Martin, J. F., Thermodynamic properties of tetra-alkylammonium ions in aqueous solution, *J. Chem. Soc. Dalton* **1973**, 1585–1590.

73KAR/SAP Karyakin, N. V., Sapozhnikov, V. N., Mochilov, A. N., Aron, B. M., Chernova, V. I., and Rabinovich, I. B., Low temperature heat capacity and thermodynamic properties of some amines, *Tr. Khim. Khim. Tekhnol.* No. **1**, 53–54 (1973).

73KIS/SUG Kishimoto, K., Suga, H., and Syuzo, S., Calorimeric study of the glassy state. VIII. Heat capacity and relaxational phenomena of isopropylbenzene, *Bull. Chem. Soc. Japan* **46**, 3020–3031 (1973).

73KRI/LIC Krien, G., Licht, H. H., and Zierath, J., Thermochemische Untersuchungen an Nitraminen, *Thermochim. Acta* **6**, 465–472 (1973).

73KUS/SUU Kusano, K., Suurkuusk, J., and Wadsö, Thermochemistry of solutions of biochemical model compounds. 2. Alkoxyethanols and 1,2-dialkoxyethanes in water, *J. Chem. Thermodynam.* **5**, 757–767 (1973).

73MAS/NOV Maslova, V. A., Novoselova, N. V., Moseeva, E. M., Berezhnaya, N. D., and Rabinovich, I. B., Specific heat and phase changes of triethylindium, triethyl antimony, and trimethyl gallium, *Trudy. Khim. Khim. Tekhnol.* No. **2**, 51–52 (1973).

73SHI/ENO Shinoda, T., Enokida, H., Maeda, Y., Tomita, H., and Mashiko, Y., Heat Capacity of tetramethylsilane in the range from 2 to 26 K and premelting range, *Bull. Chem. Soc. Japan* **46**, 48–52 (1973).

73SOR/SEK Sorai, M., and Seki, S., Heat capacity of N-(o-hydroxy-p-methoxybenzylidene)-p-butylaniline: a glassy nematic liquid crystal, *Mol. Cryst. Liquid Cryst.* **23**, 299–327 (1973).

73SUB/RAS Subrahmanyam, S. V., and Rajagopal, E., Excess thermodynamic functions of the systems isooctane + carbon tetrachloride and isoctane + cyclohexane, *Z. physik. Chem. [NF]* **85**, 256–268 (1973).

73VAS/KOR Vasil'ev, I. A., and Korkhov, A. D., The heat capacities and enthalpies of fusion and the thermodynamic properties of solid and liquid ethylene carbonate, *Zhur. Fiz. Khim.* **47**, 2710 (1973).

73VIS/SOM de Visser, C., and Somsen, G., Enthalpies of solution and heat capacities of tetra-n-butylammonium bromide in several solvents from 278–328 K, *J. Chem. Thermodynam.* **5**, 147–152 (1973).

74AND/BAC Andrews, J. T. S., and Bacon, W. E., Adiabatic calorimetry of liquid crystals I. Di-(p-methoxyphenyl)-trans-cyclohexane-1,4-dicarboxylate, *J. Chem. Thermodynam.* **6**, 515–523 (1974).

74AND/COU Andon, R. J. L., Counsell, J. F., Lee, D. A., and Martin, J. F., Thermodynamic properties of organic oxygen compounds. Part 36. Heat capacity of isopropyl ether, *J. Chem. Soc. Faraday Trans. I* **70**, 1914–1917 (1974).

74AND/MAR Andon, R. J. L., and Martin, J. F., Thermodynamic properties of fluorine compounds. Part 14. Low-temperature heat capacity and entropy of octafluorotoluene, *J. Chem. Soc. Faraday Trans. I* **70**, 605–608 (1974).

74BUR/VER Burns, I. A., and Verrall, R. E., Thermodynamics of tetraalkyl- and bis-tetraalkylammonium bromides. II. Heat capacities of solid state from 273 to 373 K, *Thermochim. Acta* **9**, 277–287 (1974).

74CHA Chang, S. S., Heat capacities of polyethylene from 2 to 360 K. II. Two high density linear polyethylene samples and thermodynamic properties of crystalline linear polyethylene, *J. Research NBS A* **78**, 387–400 (1974).

74DIA/REN Diaz Pena, M. D., and Renuncio, J. A. R., Construccion de un calorimetro adiabatico. Capacidad calorifica de mezclas n-hexano + n-Hexdecano, *An. Quim.* **70**, 113–120 (1974).

74GEI/DZH Geidarov, Kh. I., Dzhafarov, O. I., Karasharli, K. A., and Kostryukov, V. N., The heat capacity of 2-chlorobiphenyl in the range 12–327 K, *Zhur. Fiz. Khim.* **48**, 1147–1149 (1974).

74KOL/VOR Kolesov, V. P., Vorob'ev, V. N., Sarzhina, E. A., Pentin, Yu. A., and Timoshenkova, Yu. D., Heat capacity from 12 to 300 K, phase transitions, and thermodynamic functions of 1,1,1-trifluoro-3-chloropropane and 1,1,1,3-tetrachloropropane, *J. Chem. Thermodynam.* **6**, 613–628 (1974).

74MES/FIN Messerly, J. F., Finke, H. L., and Todd, S. S., Low-temperature thermal studies on six organo-sulfur compounds, *J. Chem. Thermodynam.* **6**, 635–657 (1974).

74MOS/MOU Mosselman, C., Mourik, J., and Dekker, H., Enthalpies of phase change and heat capacities of some long-chain alcohols. Adiabatic semi-microcalorimeter for studies of polymorphism, *J. Chem. Thermodynam.* **6**, 477–487 (1974).

74PET/TER Petit, J. C. and TerMinassian, L., Measurements of $(\partial V/\partial T)_p$, $(\partial V/\partial p)_T$, and $(\partial H/\partial T)_p$ by flux calorimetry, *J. Chem. Thermodynam.* **6**, 1139–1152 (1974).

74SHI/MAE Shinoda, T., Maeda, Y., and Enokido, H., Thermodynamic properties of N-(p-methoxybenzylidene)-p-n-butyl-aniline(MBBA) from 2 K to its isotropic-liquid phase, *J. Chem. Thermodynam.* **6**, 921–934 (1974).

74VIS/SOM de Visser, C., and Somsen, G., Molar heat capacities of binary mixtures of water and some amides at 298.15K, *Z. physik. Chem. [NF]* **92**, 159–162 (1974).

74VOR/KOL Vorob'ev, V. N., Kolesov, V. P., Sarzhina, E. A., Kuramshina, G. M., and Pentin, Yu. A., The heat capacity, thermodynamic functions and conformational isomerism of 3-chloro-1,1,1,3,3-pentafluoropropane in the range 12–300K, *Zhur. Fiz. Khim.* **48**, 239 (1974).

75AMB/CON Ambrose, D., Connett, J. E., Green, J. H. S., Hales, J. L., Head, A. J., and Martin, J. F., Thermodynamic properties of organic oxygen compounds. 42. Physical and thermodynamic properties of benzaldehyde, *J. Chem. Thermodynam.* **7**, 1143–1157 (1975).

75AND/COU Andon, R. J. L., Counsell, J. F., Lee, D. A., and

Martin, J. F., Thermodynamic properties of organic oxygen compounds. 39. Heat capacity of n-propyl ether, *J. Chem. Thermodynam.* **7**, 587–592 (1975).

75AND/MAR Andon, R. J. L., and Martin, J. F., Thermodynamic properties of organic oxygen compounds. 40. Heat capacity and entropy of six ethers, *J. Chem. Thermodynam.* **7**, 593–606 (1975).

75CHA/WES Chang, S-S., Westrum, E. F., Jr., and Carlson, H. G., Heat capacities of polyethylene III. One linear and one branched sample from 5 to 350 K, *J. Research NBS* **79A**, 437–441 (1975).

75CUC Cuccuru, A., Thermal properties of some cyclic disulfide-S-oxides:Naphthalene-1,8-disulfide-S-oxide and diphenylene-2,2′-disulfide-S-oxide, *Thermochim. Acta* **13**, 96–99 (1975).

75DAU/DEL Daurel, M., Delhaes, P., and Dupart, E., Variations thermiques, entre 1 et 300K, de la chaleur specifique de la L-alanine, tri(L-alanine) et de la poly(L-alanine), *Biopolymers* **14**, 801–823 (1975).

75FER/SAN Ferloni, P., Sanesi, M., and Franzosini, P., Phase transitions in the alkali C_1–n.C_4 alkanoates, Q *Z. Naturforsch.* **30a**, 1447–1454 (1975).

75GRI/RAS Grigor'ev, B. A., Rastorguev, Yu. L., and Yanin, G. S., Experimental determination of the isobaric specific heat of n-alkanes, *Iz. Vyssh. Uchebn. Zaved. Neft Gaz* **18**, No. 10, 63–66 (1975).

75GRO/BEN Grolier, J-P., E., Benson, G. C., and Picker, P., Simultaneous measurements of heat capacities and densities or organic liquid mixtures-systems containing ketones, *J. Chem. Eng. Data* **20**, 243–246 (1975).

75HOL/ZIE Holzhauer, J. K., and Ziegler, W. T., Temperature dependence of excess thermodynamic properties of n-heptane-toluene, methylcyclohexane-toluene, and n-heptane-methylcyclohexane systems, *J. Phys. Chem.* **79**, 590–603 (1975).

75LEB/MIL Lebedev, B. V., Milov, V. I., Sladkov, A. M., and Luneva, L. K., Thermodynamic properties of dianisyldiethylnylsilane and polydianisyldiethynylsilane, *Zhur. Fiz. Khim.* **49**, 2457–2458 (1975).

75MAS/SCO Masi, J. F., and Scott, R. B., Some thermodynamic properties of bromobenzene from 0 to 1500K, *J. Research NBS* **79A**, 619–628 (1975).

75MES/FIN Messerly, J. F., Finke, H. L., Osborn, A. G., and Douslin, D. R., Low-temperature calorimetric and vapor-pressure studies on alkanediamines, *J. Chem. Thermodynam.* **7**, 1029–1046 (1975).

75NIC/WAD Nichols, N., and Wadsö, I., Thermochemistry of solutions of biochemical model compounds. 3. Some benzene derivatives in aqueous solution, *J. Chem. Thermodynam.* **7**, 329–336 (1975).

75RAK/GUT Rakhmenkulov, S. S., Gutov, S. A., and Paukov, I. E., The heat capacity of hexachloroethane in the temperature range 13.7-360 K and the temperatures and enthalpies of its phase transitions, *Zhur. Fiz. Khim.* **49**, 2722 (1975).

75SPI/WAD Spink, C. H., and Wadsö, I., Thermochemistry of solutions of biochemical model compounds. 4. The partial molar heat capacities of some amino acids in aqueous solution, *J. Chem. Thermodynam.* **7**, 561–572 (1975).

75STR/SUN Stridh, G., and Sunner, S., Enthalpies of formation of some 1-chloroalkanes and the CH_2-increment in the 1-chloroalkane series, *J. Chem. Thermodynam.* **7**, 161–168 (1975).

76AND/MAR Andon, R. J. L., and Martin, J. F., Thermodynamic properties of hexacosane, *J. Chem. Thermodynam.* **8**, 1159–1166 (1976).

76ARV/FAL Arvidsson, K., Falk, B., and Sunner, S., A small sample low temperature adiabatic heat capacity calorimeter with an automatic data acquisition system, *Chem. Scr.* **10**, 193–200 (1976).

76AZO/CAL Azokpota, C., Calvarin, G., and Pommier, C., Heat capacity of molecular compounds with order-disorder transition: nickelocene, $Ni(C_5H_5)_2$, and ferrocene, $Fe(C_5H_5)_2$, *J. Chem. Thermodynam.* **8**, 283–287 (1976).

76CHA Chang, S-S., Heat capacities of polyethylene IV. High molecular weight linear polyethylene, *J. Research NBS A* **80**, 51–57 (1976).

76CLA/WOR Clayton, P. R., Worswick, R. D., and Staveley, L. A. K., The heat capacity of 1,2,4,5-tetracyanobenzene and of its 1:1 charge-transfer complex with pyrene from 10 K to 300 K, *Mol. Cryst. Liq. Cryst.* **36**, 153–163 (1976).

76CON/GIN Conte, G., Gianni, P., Matteoli, E., and Mengheri, M., Capacita termiche molari di alcuni composti organici mono- e bifunzionali nel liquido puro e in soluzione acquosa a 25C, *Chim. Ind.(Milan)* **58**, 225 (1976).

76DWO/FIG Dworkin, A., Figuiere, P., Ghelfenstei, M., and Szwarc, H., Heat capacities, enthalpies of transition, and thermodynamic properties of the three solid phases of p-dichlorobenzene from 20 to 330 K, *J. Chem. Thermodynam.* **8**, 835–844 (1976).

76FOR/BEN Fortier, J-L., Benson, G. C., and Picker, P., Heat capacities of some organic liquids determined with the picker flow calorimeter, *J. Chem. Thermodynam.* **8**, 289–299 (1976).

76GEI/WOL Geipel, G., and Wolf, G., Die molare Wärmekapazität von Cyclohexadien(1,4) und Cyclohexadien(1,3) im Temperaturgebiet von 10...300K, *Z. phys. Chem.* **257**, 587–593 (1976).

76GOO/LEE Good, W. D., and Lee, S. H., The enthalpies of formation of selected naphthalenes, diphenylmethanes, and bicyclic hydrocarbons, *J. Chem. Thermodynam.* **8**, 643–650 (1976).

76IKE/HAT Ikeda, M. and Hatakeyama, T., Thermal studies on the phase transition of p-n-Octadecyloxybenzoic acid (Part I), *Mol. Cryst. Liq. Cryst.* **33**, 201–212 (1976).

76KUL/DZH Kuliev, A. M., Dzhafarov, O. I., Karasharli, K. A., and Kuznetsova, A. G., The true heat capacity of 1,1,3,3-tetramethyl-5,5,7,7-tetraphenylcyclotetrasiloxane in the range 12-370K, *Zhur. Fiz. Khim.* **50**, 1903 (1976).

76LEB/LIT Lebedev, B. V., Lityagov, V. Ya., and Korshak K, Yu. V., Thermodynamic properties of polyoctadiene in the range 0-326 K, *Zhur. Fiz. Khim.* **50**, 2439 (1976).

76MAS/PET Masood, A. K. M., Pethrick, R. A., and Swinton, F. L., Physicochemical studies of super-cooled liquids. Cyclic carbonates and α-β-unsaturated aldehydes, *J.Chem. Soc. Faraday Trans. I*, **72**, 20–28 (1976).

76PHI/MAT Phillips, J. C., and Mattamal, M. M., Correlation of liquid heat capacities for carboxylic esters, *J. Chem. Eng. Data* **21**, 228–232 (1976).

76ROD Roder, H. M., Measurements of the specific heats, C_σ, and C_v, of dense gaseous and liquid ethane, *J. Research NBS* **80A**, 739–759 (1976).

76ROD 2 Roder, H. M., The heats of transition of solid ethane, *J. Chem. Phys.* **65**, 1371–1373 (1976).

77CHA Chang, S-S., Heat capacity and thermodynamic properties of poly(vinyl chloride), *J. Research NBS* **82**, 9–18 (1977).

77CLA/CLE Claude, R., Clement, R., and Dworkin, A., Heat capacity of urea-trioxane inclusion compound, *J. Chem. Thermodynam.* **9**, 1199–1204 (1977).

77DWO/FUC Dworkin, A., and Fuchs, A. H., Heat capacity of benzil near its phase transition, *J. Chem. Phys.* **67**, 1789–1790 (1977).

77FIN/MES Finke, H. L., Messerly, J. F., Lee, S. H., Osborn, A. G., and Douslin, D. R., Comprehensive thermodynamic studies of seven aromatic hydrocarbons, *J. Chem. Thermodynam.* **9**, 937–956 (1977).

77GEI/KAR Geidarov, Kh. I., Karasharli, K. A., and Dzhafarov, O. I., Analysis of the results of the thermodynamic investigation of diphenyl derivatives, *Azerb. Khim. Zh.* No.5, 112–115 (1977).

77HAI/SUG Haida, O., Suga, H., and Seki, S., Calorimetric study of the glassy state XII. Plural glass–transition phenomena of ethanol, *J. Chem. Thermodynam.* **9**, 1133–1148 (1977).

77HAI/SUG 2 Haida, O., Suga, H., and Seki, S., Calorimetric study of the glassy state. XI. Plural glass transition phenomena of cyclohexene, *Bull. Chem. Soc. Japan* **50**, 802–809 (1977).

77HAR/ATA Harada, M., Atake, T., and Chihara, H., Thermodynamic properties of polymorphic phases of tetramethylsilane, *J. Chem. Thermodynam.* **9**, 523–534 (1977).

77IKE/HAT Ikeda, M., and Hatakeyama, T., Thermal properties of p–n–octadecyloxybenzoic acid (Part II), *Mol. Cryst. Liq. Cryst.* **39**, 109–121 (1977).

77KOS/SOR Kosaki, A., Sorai, M., Suga, H., and Seki, S., Heat capacity of potassium p–chloranil anion radical salt between 13 and 330K, *Bull. Chem. Soc. Japan* **50**, 810–816 (1977).

77KOS/SOR 2 Kosaki, A., Sorai, M., Suga, H., and Seki, S., Thermodynamic properties and phase transitions of methyltriphenylphosphonium and methyltriphenylarsonium bis(7,7,8,8–tetracyanoquinodimethanide)s and their solid solutions, *Bull. Chem. Soc. Japan* **50**, 817–824 (1977).

77KUL/DZH Kuliev, A. M., Dzhafarov, O. I., Karasharli, K. A., and Kuznetsova, A. G., True heat capacity of octamethyldiphenylcyclopentasiloxane at 12–300K, *Zhur. Fiz. Khim.* **51**, 1564 (1977).

77KUL/DZH 2 Kuliev, A. M., Dzhafarov, O. I., Karasharli, K. A., and Kuznetsova, A. G., Study of the true heat capacity of hexamethylcyclotrisiloxane in the interval 12–350K, *Zhur. Fiz. Khim.* **51**, 1564 (1977).

77LEB/EVS Lebedev, B. V., Evstropov, A. A., Milov, V. I., Sladkov, A. M., and Vasneva, N. A., Thermodynamic properties of polyvinylenediphenylsilane in the range 13–334K, *Zhur. Fiz. Khim.* **51**, 275 (1977).

77LEB/EVS 2 Lebedev, B. V., Evstropov, A. A., Sadikov, G. B., and Larina, L. F., Thermodynamics of di(4–isocyanatophenyl)methane, *Zhur. Fiz. Khim.* **51**, 1285 (1977).

77MEI/BLO Meijer, E. L. Blok, J. G. Kroon, J. and Oonk, H. A. J., The carvoxime system. IV. Heat capacities and enthalpies of melting of dl–carvoxime, l–carvoxime and standard n–heptane, *Thermochim. Acta* **20**, 325–334 (1977).

77MUS Mustafaev, R. A., Investigation of the isobaric heat capacity of n–nonane at high temperatures and pressures, *High Temp.* **15**, 683 (1977).

77SAB/LAF Sabbah, R., and Laffitte, M., The enthalpy of formation of sarcosine in solid state, *J. Chem. Thermodynam.* **9**, 1107–1108 (1977).

77TSU/SOR Tsuji, K., Soral, M., Suga, H., and Seki, S., A new phase in p–n–hexyloxybenzylideneamino–p′–chlorobenzene, *Mol. Cryst. Liq. Cryst.* **41**, 81–87 (1977).

77VIS/PER de Visser, C., Perron, G., and Desnoyers, J. E., Volumes and heat capacities of ternary aqueous systems at 25°C. Mixtures of urea, tert–butyl alcohol, N,N–Dimethylformamide, and water, *J. Am. Chem. Soc.* **99**, 5894–5900 (1977).

77VOR/PRI Vorob'ev, A. F., Privalova, N. M., and Rekharskii, M. V., Hermetically sealed calorimeter with automatic adiabatic enclosure for determining the specific heats of liquids, *Zhur. Fiz. Khim.* **51**, 1843–1846 (1977).

78ADE Adeosun, S. O., A differential thermal analysis study of phase transitions in some mercury(II) carboxylates, *J. Thermal Anal.* **14**, 235–243 (1978).

78ADE/SIM Adeosun, S. O., and Sime, S. J., The properties of molten carboxylates: Part 7. Odd–even variations in melting and mesophase formation in the lead(II) carboxylates, *J. Thermal Anal.* **27**, 319–327 (1978).

78BOE/WES Boerio–Goates, J., Westrum, E. F., Jr., and Fyfe, C. A., Thermodynamic studies of orientational disorder in π–π molecular compounds, *Mol. Cryst. Liq. Cryst.* **48**, 209–218 (1978).

78DUN/RAH Dunn, A. G., Rahman, A., and Staveley, L. A. K., The heat capacity of pyromellitic dianhydride and of its 1–1 charge–transfer complex with pyrene, *J. Chem. Thermodynam.* **10**, 787–796 (1978).

78EVS/BEL Evstropov, A. A., Belov, V. I., Lebedev, B. V., Kiparisova, E. G., and Bykova, T. A., Thermodynamic properties of glycolide in the temperature range 0–400 K, *Zhur. Fiz. Khim.* **52**, 242 (1978).

78GEI/KAR Geidarov, Kh. I., Karasharli, K. A., Dzhafarov, O. I., and Guseinov, Z. A., Study of the true specific heat of 3–methyltetrahydrophthalic anhydride in the interval 12–360K, *Zhur. Fiz. Khim.* **52**, 805–806 (1978).

78GEI/KAR 2 Geidarov, Kh. I., Karasharli, K. A., and Dzhafarov, O. I., Study of the true heat capacity of tetrahydrophthalic anhydride in the range 12–390K, *Zhur. Fiz. Khim.* **52**, 804 (1978).

78GOD/RAC Godlewska, M., and Rachwalska, M., An adiabatic calorimetric study of the phase situation in 2,2–dinitropropane, *Phys. Status Solidi A* **47**, 661–665 (1978).

78GOO Goodwin, R. D., Specific heats of saturated and compressed liquid propane, *J. Res. NBS* **83**, 449–458 (1978).

78JOH/HAY Johnson, D. L., Hayes, C. F., deHoff, R. J., and Schantz, C. A., Specific heat near the nematic–smectic–A transition of octyloxycyanobiphenyl, *Phys. Rev. B.* **18**, 4902–4912 (1978).

78KIL Kilday, M. V., Enthalpies of solution of nucleic acid bases. 1. Adenine in water, *J. Res. NBS* **83**, 347–369 (1978).

78KIL 2 Kilday, M. V., Enthalpies of solution of the nucleic acid bases. 2. Thymine in water, *J. Res. NBS* **83**, 529–537 (1978).

78KIL 3 Kilday, M. V., Enthalpies of solution of the nucleic acid bases. 3. Cytosine in water, *J. Res. NBS* **83**, 539–546 (1978).

78KIL 4 Kilday, M. V., Enthalpies of solution of the nucleic acid bases. 4. Uracil in water, *J. Res. NBS* **83**, 547–554 (1978).

78KIS/SUG Kishimoto, K., Suga, H., and Seki, S., Calorimetric study of the glassy state. XIV. Calorimetric study on unusual glass transition phenomena in $CFCl_2$–$CFCl_2$, *Bull. Chem. Soc. Japan* **51**, 1691–1696 (1978).

78KOS/KOL Kosarukina, E. A., Kolesov, V. P., Vorob'ev, V. N., and Vogel, L., The heat capacity of 1,1,2,2–tetrachloro–1,2–difluoroethane in the temperature range 12–310 K, *Zhur. Fiz. Khim.* **52**, 509 (1978).

78LEB/RAB Lebedev, B. V., Rabinovich, I. B., Milov, V. I., and Lityagov, V. Ya., Thermodynamic properties of tetrahydrofuran from 8 to 322 K, *J. Chem. Thermodynam.* **10**, 321–329 (1978).

78MAR/CIO Marchidan, D. I., and Ciopec, M., Relative enthalpies and related thermodynamic functions of some organic compounds by drop calorimetry, *J. Therm. Anal.* **14**, 131–150 (1978).

78MAR/CIO 2 Marchidan, D. I., and Ciopec, M., Thermodynamic properties of maleic, trimellitic and pryomellitic anhydrides, *Rev. Roum. Chim.* **23**, 19–29 (1978).

78MIL Mills, K. C., The heat capacity of dimethyl phthalate, 250–370 K, *Thermochim. Acta* **23**, 390–392 (1978).

78RAB/NIS Rabinovich, I. B., Nistratov, V. P., Sheiman, M. S., and Burchalova, G. V., Heat capacities of dicyclopentadienyl compounds of vanadium, chromium, manganese, cobalt, and nickel, *J. Chem. Thermodynam.* **10**, 523–536 (1978).

78RIP/WRI Ripmeester, J. A., Wright, D. A., Fyfe, C. A., and Boyd, R. K., Molecular motion and phase transitions in solid hexafluorobenzene+benzene complex by nuclear magnetic resonance and heat capacity measurements, *J. Chem. Soc. Faraday Trans. II* **74**, 1164–1178 (1978).

78ROU/PER Roux, G., Perron, G., and Desnoyers, J. E., Model systems for hydrophobic interactions: volumes and heat capacities of n–alkoxyethanols in water, *J. Solution Chem.* **7**, 639–654 (1978).

78ROU/PER 2 Roux, G., Perron, G., and Desnoyers, J. E., The heat capacities and volumes of some low molecular weight amides, ketones, esters, and ethers in water over the whole solubility range, *Can. J. Chem.* **56**, 2808–2814 (1978).

78SAB/LAF Sabbah, R., and Laffitte, M., Enthalpy of formation of L–serine in the solid state, *Thermochim. Acta* **23**, 192–195 (1978).

78SOR/YOS Sorai, M., Yoshikawa, M., Arai, N., Suga, H., and Seki, S., Heat capacity of $[Ni(OCH_3)(acac)(CH_3OH)]_4$ from 0.4 to 285 K: spin interaction and tunnel–splitting of internal rotation of methyl group, *J. Phys. Chem. Solids* **39**, 413–425 (1978).

78SPI/AND Spinella, G. M., Andrews, J. T. S., and Bacon, W. E., Diamantane: heat capacity (300 to 540 K) and thermodynamics of transition and fusion, *J. Chem. Thermodynam.* **10**, 1023–1032 (1978).

78TAT/MAT Tatsumi, M., Matsuo, T., Suga, H., and Seki, S., Phase transitions of Rochelle salt as studied by high–resolution heat capacity measurements, *J. Phys. Chem. Solids* **39**, 427–434 (1978).

78WES/MCK Westrum, E. F., Jr., McKervey, M. A., Andrews, J. T. S., Fort, R. C., Jr., and Clark, T., Heat capacity and thermodynamic properties of diamantane from 5 to 540 K, *J. Chem. Thermodynam.* **10**, 959–965 (1978).

79BAR/HEL Barth, E., Helwig, J., Maier, H. D., Müser, H. E., and Petersson, J., Specific heat of the two dimensional antiferroelectric squaric acid, *Z. Phys. B* **34**, 393–397 (1979).

79BOT/CAM Bothe, H., and Cammenga, H. K., Phase transitions and thermodynamic properties of anhydrous caffeine, *J. Therm. Anal.* **16**, 267–275 (1979).

79BOY/COM Boyd, R. K., Comper, J., and Ferguson, G., Entropy changes and structural implications for crystalline phases of pyrazine, *Can. J. Chem.* **57**, 3056–3060 (1979).

79BRI/MIL Briels, W. J., and van Miltenburg, J. C., s–Triazine: heat capacities between 130 and 380 K, the solid state transition, and the enthalpy of melting, *J. Chem. Phys.* **70**, 1064–1066 (1979).

79BRO/ZIE Brown, G. N., Jr., and Ziegler, W. T., Temperature dependence of excess thermodynamic properties of ethanol + n–heptane and 2–propanol + n–heptane solutions, *J. Chem. Eng. Data* **24**, 319–330 (1979).

79CAI/DWO Cailleau, H., and Dworkin, A., Calorimetric study of the phase transition of para–terphenyl, *Mol. Cryst. Liq. Cryst.* **50**, 217–222 (1979).

79FUC Fuchs, R., Heat capacities of some liquid aliphatic, alicyclic, and aromatic esters at 298.15 K, *J. Chem. Thermodynam.* **11**, 959–961 (1979).

79GRO/HAM Grolier, J–P. E., Hamedi, M. H., Wilhelm, E., and Kehiaian, H. V., Excess heat capacities of binary mixtures of carbon tetrachloride with n–alkanes at 298.15 K, *Thermochim. Acta* **31**, 79–84 (1979).

79KIY/D'A Kiyohara, O., D'Arcy, P. J., and Benson, G. C., Ultrasonic velocities, compressibilities, and heat capacities of water + tetrahydrofuran mixtures at 298.15K, *Can. J. Chem.* **57**, 1006–1010 (1979).

79KOB/KAM Kobayashi, S., Kamiishi, Y., and Iwayanagi, S., A new crystal phase of terephthal–bis–n–butylaniline, *Mol. Cryst. Liq. Cryst.* **51**, 63–66 (1979).

79LAN/WES Landee, C. P., and Westrum, E. F., Jr., Heat capacities of bis– tetraethylammonium tetrachloronickelate and tetrachlorozincate I. Structural transitions and thermophysical results, *J. Chem. Thermodynam.* **11**, 247–260 (1979).

79LEB/LIT Lebedev, B. V., Lityagov, V. Ya., Krentsina, T. I., and Milov, V. I., Thermodynamic properties of tetrahydrofuran in the range 8–322K, *Zhur. Fiz. Khim.* **53**, 264–265 (1979).

79LEE/HOS Lee–Bechtold, S. H., Hossenlopp, I. A., Scott, D. W., Osborn, A. G., and Good, W. D., A comprehensive thermodynamic study of 9,10–dihydrophenanthrene, *J. Chem. Thermodynam.* **11**, 469–482 (1979).

79MAT/MAN Matsuo, T., Mansson, M., and Sunner, S., Thermodynamic properties and the ferroelectric phase transition of the trisarcosine calcium chloride complex, *Acta Chem. Scand. A* **33**, 781–787 (1979).

79OTT/GOA Ott, J. B., Goates, J. R., and Grigg, R. B., Excess volumes, enthalpies, and Gibbs free energies for mixtures of benzene + p–xylene, *J. Chem. Thermodynam.* **11**, 1167–1173 (1979).

79PUC/PEA Fuchs, R., and Peacock, L. A., Heats of vaporization and gaseous heats of formation of some five– and six–membered ring alkenes, *Can. J. Chem.* **57**, 2302–2304 (1979).

79RAC/NGU Rachwalska, M., and Nguyen Xuan Phuc, An adiabatic calorimetric study of phase situation in HOAB (4,4′–di–n–heptyloxyazoxybenzene), *Acta Phys. Pol. A* **55**, 95–105 (1979).

79RIC/SAV Richardson, M. J., and Savill, N. G., Free energy changes by differential scanning calorimetry: application to a low energy but metastable form of p–nitrotoluene, *Thermochim. Acta* **30**, 327–337 (1979).

79SAL/PEA Saluja, P. P. S., Peacock, L. A., and Fuchs, R., Enthalpies of interaction of aliphatic ketones with polar and nonpolar solvents, *J. Am. Chem. Soc.* **101**, 1958–1962 (1979).

79SAM/GRI Samsonov, D. P., Grinberg, E. E., and Efremov, A. A., Determination of the dissociation energy of the tetrabutoxytitanium dimer, *Zhur. Prikl. Khim.* **52**, 1909–1911 (1979).

79SCH/OFF Schaake, R. C. F., Offringa, J. C. A., van der Berg, G. J. K., and van Miltenburg, J. C., Phase transitions in solids, studied by adiabatic calorimetry. I. Design and test of an automatic adiabatic calorimeter, *J. Royal Netherlands Chem. Soc.* **98**, 408–412 (1979).

79STE Steele, W. V., The standard enthalpies of formation of the triphenyl compounds of the group V elements. 2. Triphenylbismuth and the Ph–Bi mean bond–dissociation energy, *J. Chem. Thermodynam.* **11**, 187–192 (1979).

79STE/TAM Stephens, M. A., and Tamplin, W. S., Saturated liquid specific heats of ethylene glycol homologues, *J. Chem. Eng. Data* **24**, 81–82 (1979).

79SUN/SVE Sunner, S., Svensson, Ch., and Zelepuga, A. S., Enthalpies of vaporization at 298.15K for some 2–alkanones and methyl alkanoates, *J. Chem. Thermodynam.* **11**, 491–495 (1979).

79SVE Svensson, Ch., Enthalpies of vaporization of 1–decanol and 1–dodecanol and their influence on the CH_2–increment for the enthalpies of formation, *J. Chem. Thermodynam.* **11**, 593–596 (1979).

79VIS/SOM de Visser, C., and Somsen, G., Thermochemical behavior of mixtures of N,N–dimethylformamide with dimethylsulfoxide, acetonitrile, and N–methylformamide: volumes and heat capacities, *J. Solution Chem.* **8**, 593–600 (1979).

79WIL/FAR Wilhelm, E., Faradjzadeh, A., and Grolier, J.–P. E., Molar excess heat capacities and excess volumes of 1,2–dichloroethane + cyclooctane, + mesitylene, and + tetrachloromethane, *J. Chem. Thermodynam.* **11**, 979–984 (1979).

79WIL/GRO Wilhelm, E., Grolier, J. –P. E., and Karbalai Ghassemi, M. H., Molar heat capacity of binary liquid mixtures: 1,2–dichloroethane + cyclohexane and 1,2–dichloroethane + methylcyclohexane, *Thermochim. Acta* **28**, 59–69 (1979).

79YAN/RUP Yang, P–H., and Rupley, J. A., Heat capacity of nicotinamide adenine dinucleotide in aqueous solution and in the crystal, *J. Chem. Thermodynam.* **11**, 301–302 (1979).

79ZHO/KOS Zhogin, D. Yu., Kosarukina, E. A., Kolesov, V. P., Baburina, I. I., Pentin, Yu. A., Izmest'ev, I. V. and Danilov, A. V., Heat capacity in the range 9–300 K, phase transitions, thermodynamic functions, and conformations of perfluorotriethylamine, *Zhur. Obshch. Khim.*, **49**, 629–637 (1979).

80AND/CON Andon, R. J. L., and Connett, J. E., Calibrants for thermal analysis. Measurement of their enthalpies of fusion by adiabatic calorimetry, *Thermochim. Acta* **42**, 241–247 (1980).

80ATA/CHI Ataki, T., and Chihara, H., Heat capacity anomalies due to successive phase transitions in 1,1′–biphenyl, *Solid State Commun.* **35**, 131–134 (1980).

80BOE/WES Boerio–Goates, J., and Westrum, E. F., Jr., Thermodynamic studies of orientational disorder in π–molecular compounds. II. Heat capacity of naphthalene–tetracyanoethylene, *Mol. Cryst. Liq. Cryst.* **60**, 237–248 (1980).

80BOE/WES 2 Boerio–Goates, J., and Westrum, E. F., Jr., Thermodynamic studies of orientational disorder in π–molecular compounds. III. Heat capacities of naphthalene–pyromellitic dianhydride and pyrene–pyromellitic dianhydride, *Mol. Cryst. Liq. Cryst.* **60**, 249–266 (1980).

80CAL/BER Calvarin, G., Berar, J. F., Chhor, K., and Pommier, C., Etude de la transition de phase ordre–desordre du vanadocene $V(C_5H_5)_2$, *Mol. Cryst. Liq. Cryst.* **59**, 149–158 (1980).

80LEB/DOB Lebedeva, N. D., Dobychin, S. L., Stepanov, V. B., Katin, Yu. A., and Kozlova, N. M., Heat capacity, thermodynamic functions, and vapor pressure of bis–(tetrafluoropropyl)–carbonate, *Zhur. Priklad. Khim.* **53**, 803–806 (1980).

80RAD/RAD Radomska, M., and Radomski, R., Calorimetric studies of binary systems of 1,3,5–trinitrobenzene with naphthalene, anthracene, and carbazole. I. Phase transitions and heat capacities of the pure components and charge–transfer complexes, *Thermochim. Acta* **40**, 405–414 (1980).

80RAM/CER Ramos, S., del Cerro, J., and Zamora, M., Specific heat of triglycine sulfate at several applied electric fields near the critical temperature, *Phys. Status Solidi A* **61**, 307–313 (1980).

80SAB/SKO Sabbath, R., and Skoulika, S., Thermodynamique de composes azotes. Partie VI. Etude thermochimique de la N–phenylglycine et de la D–α–phenylglycine, *Thermochim. Acta* **36**, 179–187 (1980).

80SOR/TSU Sorai, M., Tsuji, K., Suga, H., and Seki, S., Studies on disc–like molecules. I. Heat capacity of benzene–hexa–n–hexanoate from 13 to 393 K, *Mol. Cryst. Liq. Cryst.* **80**, 33–58 (1980).

80TAY/GRO Taylor, R. E., and Groot, H., Thermophysical properties of POCO graphite, *High Temp.–High Press.* **12**, 147–160 (1980).

80YOS/SOR Yoshikawa, M., Sorai, M., Suga, H., and Seki, S., Heat capacity and phase transition of a five–coordinated antiferromagnet, iodobis (N,N–diethyldithiocarbamato)iron (III), in the temperature range from 0.4 to 300K, *J. Phys. Chem. Solids* **41**, 1295–1302 (1980).

80ZHO/KOS Zhogin, D. Yu., Kosarukina, E. A., and Kolesov, V. P., Heat capacity, phase transformations), and thermodynamic functions of perfluoro(N,methyldiethylamine, *Zhur. Obshchei. Khim.* **50**, 249–255 (1980).

81BYS Bystrom, K., Enthalpies of combustion, vaporization, and formation for di–n–propyldiazene N–oxide and di–t–butyldiazene N–oxide, *J. Chem. Thermodynam.* **13**, 139–145 (1981).

81CHA/HAG Chanh, N. B., Haget, Y., Meresse, A., Louer, D., and Shirley, R., Destructive and nondestructive phase transitions of the molecular crystal 2–bromonaphthalene, *J. Phys. Chem. Solids* **42**, 217–225 (1981).

81CHR/RIC Chrusciel, J., Richter, L., and Rachwalska, M., Study of the phase situation in 4–n–pentylphenyl–4′–n–heptyloxythiobenzoate ($\overline{7}$S5), *Mol. Cryst. Liq. Cryst.* **75**, 155–167 (1981).

81FIN/KUM Finegold, L., and Kumar, P. K., Specific heat of polyglycine I and II in the temperature interval 150–375K, *Thermochim. Acta* **48**, 51–59 (1981).

81FIN/MES Finke, H. L., Messerly, J. F., and Lee–Bechtold, S. H., Thermodynamic properties of cyclopropylamine, cyclopentylamine, and methylenecyclobutane, *J. Chem. Thermodynam.* **13**, 345–355 (1981).

81JEN/OBR Jenkins, T. E., and O'Brien, P., A room temperature specific heat anomaly in triamantane ($C_{18}H_{24}$), *Phys. Status Solidi A* **67**, K161–K162 (1981).

81KAW/NIS Kawaizumi, F., Nishio, N., Nomura, H., and Miyahara, Y., Heat–capacity measurements of aqueous solutions of mono–, di–, and tri–saccharides using an isoperibol twin calorimeter, *J. Chem. Thermodynam.* **13**, 89–98 (1981).

81KIL Kilday, M. V., Enthalpies of solution of the nucleic acid bases. 6. Guanine in aqueous HCl and NaOH, and guanine hydrochlorides in aqueous HCl, *J. Res. NBS* **86**, 367–381 (1981).

81KOL/KOS Kolesov, V. P., Kosarukina, E. A., Zhogin, D. Yu., Poloznikova, M. E., and Pentin, Yu. A., Heat capacities, phase transitions, and thermodynamic functions of 1,1,2,2–tetrafluoro–1,2–dichloroethane and 1,1,2–trifluoro–1,2,2–trichloroethane, *J. Chem. Thermodynam.* **13**, 115–129 (1981).

81LEB/KUL Lebedev, B. V., Kulagina, T. G., and Kiparisova, E. G., Thermodynamics of 2,5–diketopiperazine in the 0–330K range, *Zhur. Obshch. Khim.* **51**, 199–203 (1981).

81LEB/YEV Lebedev, B. V., Yevstropov, A. A., and Kiparisova, Y. G., Thermodynamic properties of undecanolactone over the range 0–330K, *Termodin. Org. Soedin.* **1981**, 9–15.

81LEB/YEV 2 Lebedev, B. V., Yevstropov, A. A., and Kiparisova, Y. E., Thermodynamics of undecanolactone, tridecanolactone, and pentadecanolactone from 0 to 340 K, *J. Chem. Thermodynam.* **13**, 1185–1204 (1981).

81LOP/TEL Lopez–Echarri, A., and Tello, M. J., Calorimetric study of the ferroelectric trissarcosine calcium chloride in the range 50–330 K, *J. Phys. D (Appl. Phys.)* **14**, 71–77 (1981).

81OGA/SOR Ogasahara, K., Sorai, M., and Suga, H., Thermodynamic properties of ferrocene crystal, *Mol. Cryst. Liq. Cryst.* **71**, 189–211 (1981).

81RAH/CLA Rahman, A., Clayton, P. R., and Staveley, L. A. K., The heat capacity of the layer compound methylammonium cadmium tetrachloride, $(CH_3NH_3)_2CdCl_4$, from 2.3 to 301K, *J. Chem. Thermodynam.* **13**, 735–744 (1981).

81SHE/KAM Sheiman, M. S., Kamelova, G. P., and Burchalova, G. V., Heat capacity and thermodynamic functions of dimethanolurea, *Termodin. Org. Soedin* **1981**, 36–37.

81SOR/OGA Sorai, M., Ogasahara, K., and Suga, H., Heat capacity and phase transitions of thiourea–ferrocene channel inclusion compound, *Mol. Cryst. Liq. Cryst.* **73**, 231–254 (1981).

81SOR/SUG Sorai, M., and Suga, H., Studies on mesogenic disc–like molecules II. Heat Capacity of benzene–hexa–n–heptanoate from 13 to 393K, *Mol. Cryst. Liq. Cryst.* **73**, 47–69 (1981).

81TEG/FER Teghil, R., Ferro, D., Bencivenni, L., and Pelino, M., A thermodynamic study of the sublimation processes of aluminum and copper acetylacetonates, *Thermochim. Acta* **44**, 213–222 (1981).

81TOM/CUR Tomassetti, M., Curini, R., D'Ascenzo, G., and Ortaggi, G., Heat capacities of ferrocene, acetyl–, 1,1'-diacetyl–, benzoyl–, and 1,1'-dibenzolferrocene by DSC, *Thermochim. Acta* **48**, 333–341 (1981).

81WHI/GRA White, M. A., Granville, N. W., Davies, N. J., and Staveley, L. A. K., The heat capacity of the layer compounds $(CH_3CH_2CH_2NH_3)_2MnCl_4$ and $(CH_3CH_2CH_2NH_3)_2CdCl_4$ from 10K to 300K, *J. Phys. Chem. Solids* **42**, 953–965 (1981).

81ZHO/KOS Zhogin, D. Yu., Kosarukina, E. A., Kolesov, V. P., Prokudin, I. P., and Mel'nichenko, B. A., The heat capacity in the temperature range 6–300 K, phase transformations, and thermodynamic functions of perfluorobicyclo[4,4,0]dec–1,6–ene, *Zhur. Fiz. Khim.* **55**, 98–103 (1981).

81ZHO/KOS 2 Zhogin, D. Yu., Kosarukina, E. A., and Kolesov, V. P., The heat capacities in the range 6–310K, phase transformations, and thermodynamic functions of perfluorodecalins, *Zhur. Fiz. Khim.* **55**, 1955–1960 (1981).

82CAM/REY Campos–Vallette, M., Rey–Lafon, M., and Lagnier, R., Vibrational and calorimetric study of phase transitions in n–C_6F_{14} and n–C_8F_{18} crystals, *Chem. Phys. Lett.* **89**, 189–192 (1982).

82DZH/KAR Dzhafarov, O. I., Karasharli, K. A., and Kuliev, A. M., Study of the real heat capacity of β–dimethylaminopropionitrile in the range 12–300K, *Azerb. Khim. Zhur.*, **No. 2** 92–95 (1982).

82DZH/KAR 2 Dzhafarov, O. I., Karasharli, K. A., and Kuliev, A. M., Study of the real heat capacity of N,N,–dimethyl–1,3–propanediamine in the range 12–300K, *Azerb. Khim. Zhur.*, **No. 3** 111–113 (1982).

82GOR/GRI Gorbunova, N. I., Grigoriev, V. A., Simonov, V. M., and Shipova, V. A., Heat capacity of liquid benzene and hexafluorobenzene at atmospheric pressure, *Int. J. Thermophysics* **3**, 1–15 (1982).

82GRO/ING Grolier, J. -P. E., Inglese, A., and Wilhelm, E., Excess volumes and excess heat capacities of tetrachloroethene + cyclohexane, + methylcyclohexane, + benzene, and + toluene at 298.15 K, *J. Chem. Thermodynam.* **14**, 523–529 (1982).

82GYO/YOS Gyoten, H., Yoshimoto, Y., Atake, T., and Chihara, H., Thermodynamic studies of phase transitions in chloropentamethylbenzene, *J. Chem. Phys.*. **77**, 5097–5107 (1982).

82KOS/ZHO Kosarukina, E.A., Zhogin, D. Yu., Kolesov, V. P., Kuramshina, G. M., Pentin, Yu. A., Izmest'ev, I. V., and Danilov, A. V. Heat capacity in the range 8–300 K, thermodynamic functions, and conformational isomerism of 1,2–dibromotetrafluoroethane, *Zhur. Fiz. Khim.* **56**, 1892–1896 (1982).

82KUL/DZH Kuliev, A. M., Dzhafarov, O. I., Karasharli, K. A., and Kostryukov, V. N., Heat capacity and thermodynamic functions of cyclic methylphenylsiloxanes, *Zhur. Fiz. Khim.* **56**, 310–313 (1982).

82LEB/NOV Lebedev, R. S., Novikov, V. V., and Sidorov, A. A., Temperature dependence of the heat capacity of 2–imino–4–thiazolidone, *Zhur. Fiz. Khim.* **56**, 993–994 (1982).

82LUF/REE Luff, B. B., and Reed, R. B., Low–temperature heat capacity and entropy of oxalic acid and of biuret, *J. Chem. Eng. Data* **27**, 290–292 (1982).

82MAR/AND Martin, J. F., and Andon, R. J. L., Thermodynamic properties of organic oxygen compounds. Part LII. Molar heat capacity of ethanoic, propanoic, and butanoic acids, *J. Chem. Thermodynam.* **14**, 679–688 (1982).

82MOR/MAT Moriya, K., Matsuo, T., and Suga, H., Low temperature adiabatic calorimeter with a built–in cryo–refigerator, *J. Chem. Thermodynam.* **14**, 1143–1148 (1982).

82SCH/MIL Schaake, R. C. F., van Miltenburg, J. C., and De Kruif, C. G., Thermodynamic properties of the normal alkanoic acids. I. Molar heat capacities of seven odd–numbered normal alkanoic acids, *J. Chem. Thermodynam.* **14**, 763–769 (1982).

82SCH/MIL 2 Schaake, R. C. F., van Miltenburg, J. C., and De Kruif, C. G., Thermodynamic properties of the normal alkanoic acids. II. Molar heat of seven even–numbered normal alkanoic acids, *J. Chem. Thermodynam.* **14**, 771–778 (1982).

82SOR/YOS Sorai, M., Yoshioka, H., and Suga, H., Studies on mesogenic disc–like molecules III. Heat capacity of benzene–hexa–n–octanoate from 13 to 393K, *Mol. Cryst. Liq. Cryst.* **84**, 39–54 (1982).

82TAN Tanaka, R., Determination of excess heat capacities of (benzene + tetrachloromethane and + cyclohexane) between 293.15 and 303.15 K by use of a Picker flow calorimeter, *J. Chem. Thermodynam.* **14**, 259–268 (1982).

82VAS/PET Vasil'ev, I. A., Petrov, V. M., and Treibsho, E. I., Study of the specific heat of some liquid higher alcohols, *Zhur. Priklad. Khim.* **55**, 2116–2117 (1982).

82WIL/ING Wilhelm, E., Inglese, A., Quint, J. R., and Grolier, J.-P. E., Molar excess volumes and excess heat capacities of (1,2,4–trichlorobenzene + an n–alkane), *J. Chem. Thermodynam.* **14**, 303–308 (1982).

82WHI/GRA White, M. A., Granville, N. W., and Staveley, L. A. K., The heat capacity of the layer compound tetrachlorobis(methylammonium) manganese–II $(CH_3NH_3)_2MnCl_4$, from 10 to 300K, *J. Phys. Chem. Solids* **43**, 341–349 (1982).

82WHI/STA White, M. A., and Staveley, L. A. K., The heat capacity of $(CH_2{=}CHCH_2NH_3)_2CdCl_4$. The influence of the carbon–carbon double bond on the structural phase transitions, *J. Chem. Thermodynam.* **14**, 859–864 (1982).

Journal of Physical and Chemical Reference Data Cumulative Listing of Reprints and Supplements

Reprints from Volume 1

1. Gaseous Diffusion Coefficients, *T.R. Marrero and E.A. Mason,* Vol. 1, No. 1, pp. 1–118 (1972) $7.00
2. Selected Values of Critical Supersaturation for Nucleation of Liquids from the Vapor, *G.M. Pound,* Vol. 1, No. 1, pp. 119–134 (1972) $3.00
3. Selected Values of Evaporation and Condensation Coefficients for Simple Substances, *G.M. Pound,* Vol. 1, No. 1, pp. 135–146 (1972) $3.00
4. Atlas of the Observed Absorption Spectrum of Carbon Monoxide between 1060 and 1900 Å, *S.G. Tilford and J.D. Simmons,* Vol. 1, No. 1, pp. 147–188 (1972) $4.50
5. Tables of Molecular Vibrational Frequencies, Part 5, *T. Shimanouchi,* Vol. 1, No. 1, pp. 189–216 (1972) (superseded by No.103) $4.00
6. Selected Values of Heats of Combustion and Heats of Formation of Organic Compounds Containing the Elements C, H, N, O, P, and S, *Eugene S. Domalski,* Vol. 1, No. 2, pp. 221–278 (1972) $5.00
7. Thermal Conductivity of the Elements, *C.Y. Ho, R.W. Powell, and P.E. Liley,* Vol. 1, No. 2, pp. 279–422 (1972) $7.50
8. The Spectrum of Molecular Oxygen, *Paul H. Krupenie,* Vol. 1, No. 2, pp. 423–534 (1972) $6.50
9. A Critical Review of the Gas-Phase Reaction Kinetics of the Hydroxyl Radical, *Wm. E. Wilson, Jr.,* Vol. 1, No. 2, pp. 535–574 (1972) $4.50
10. Molten Salts: Volume 3, Nitrates, Nitrites, and Mixtures, Electrical Conductance, Density, Viscosity, and Surface Tension Data, *G.J. Janz, Ursula Krebs, H.F. Siegenthaler, and R.P.T. Tomkins,* Vol. 1, No. 3, pp. 581–746 (1972) $8.50
11. High Temperature Properties and Decomposition of Inorganic Salts—Part 3. Nitrates and Nitrites, *Kurt H. Stern,* Vol. 1, No. 3, pp. 747–772 (1972) $4.00
12. High-Pressure Calibration: A Critical Review, *D.L. Decker, W.A. Bassett, L. Merrill, H.T. Hall, and J.D. Barnett,* Vol. 1, No. 3, pp. 773–836 (1972) $5.00
13. The Surface Tension of Pure Liquid Compounds, *Joseph J. Jasper,* Vol. 1, No. 4, pp. 841–1009 (1972) $8.50
14. Microwave Spectra of Molecules of Astrophysical Interest, I. Formaldehyde, Formamide, and Thioformaldehyde, *Donald R. Johnson, Frank J. Lovas, and William H. Kirchhoff,* Vol. 1, No. 4, pp. 1011–1046 (1972) $4.50
15. Osmotic Coefficients and Mean Activity Coefficients of Uni-univalent Electrolytes in Water at 25° C, *Walter J. Hamer and Yung-Chi Wu,* Vol. 1, No. 4, pp. 1047-1099 (1972) $5.00
16. The Viscosity and Thermal Conductivity Coefficients of Gaseous and Liquid Fluorine, *H.J.M. Hanley and R. Prydz,* Vol. 1, No. 4, pp. 1101-1113 (1972) $3.00

Reprints from Volume 2

17. Microwave Spectra of Molecules of Astrophysical Interest, II. Methylenimine, *William H. Kirchhoff, Donald R. Johnson, and Frank J. Lovas,* Vol. 2, No. 1, pp. 1-10 (1973) $3.00
18. Analysis of Specific Heat Data in the Critical Region of Magnetic Solids, *F.J. Cook,* Vol. 2, No. 1, pp. 11-24 (1973) $3.00
19. Evaluated Chemical Kinetic Rate Constants for Various Gas Phase Reactions, *Keith Schofield,* Vol. 2, No. 1, pp. 25–84 (1973) $5.00
20. Atomic Transition Probabilities for Forbidden Lines of the Iron Group Elements. (A Critical Data Compilation for Selected Lines), *M.W. Smith and W.L. Wiese,* Vol. 2, No. 1, pp. 85–120 (1973) $4.50
21. Tables of Molecular Vibrational Frequencies, Part 6, *T. Shimanouchi,* Vol. 2, No. 1, pp. 121–162 (1973) (superseded by No. 103) $4.50
22. Compilation of Energy Band Gaps in Elemental and Binary Compound Semiconductors and Insulators, *W.H. Strehlow and E.L. Cook,* Vol. 2, No. 1, pp. 163–200 (1973) $4.50
23. Microwave Spectra of Molecules of Astrophysical Interest, III. Methanol, *R.M. Lees, F.J. Lovas, W.H. Kirchhoff, and D.R. Johnson,* Vol. 2, No. 2, pp. 205–214 (1973) $3.00
24. Microwave Spectra of Molecules of Astrophysical Interest, IV. Hydrogen Sulfide, *Paul Helminger, Frank C. De Lucia, and William H. Kirchhoff,* Vol. 2, No. 2, pp. 215–224 (1973) $3.00
25. Tables of Molecular Vibrational Frequencies, Part 7, *T. Shimanouchi,* Vol. 2, No. 2, pp. 225–256 (1973) (superseded by No. 103) $4.00

Journal of Physical and Chemical Reference Data
Reprint and Supplement Orders

To: American Chemical Society
Distribution Office
1155 Sixteenth Street, N.W.
Washington, D.C. 20036

Name: ______________________

Title: ______________________

Organization: ______________________

Address: ______________________

City: ____________ State: ____________

Country: ____________ Zip: ____________

I am a member of ______________________
(ACS or AIP Member or Affiliated Society)

ORDERS FOR REPRINTS AND SUPPLEMENTS MUST BE PREPAID.
*Foreign orders for Reprints, add $2.50 for each reprint for postage and handling. Foreign orders for Supplements and Reprint Packages, add $4.00 for each Supplement and Reprint Package for postage and handling. California residents add 6% state use tax. Make checks payable to the American Chemical Society.

*Effective January 1985, supplement prices outside the U.S. and Canada will be increased by approximately 20%.

BULK RATES: Subtract 20% from the listed price for orders of 50 or more of any one item.

Please ship the following reprints and supplements:

Reprint No./Package ______, ______ copies, $________

Reprint No./Package ______, ______ copies, $________

Reprint No./Package ______, ______ copies, $________

Reprint No./Package ______, ______ copies, $________

Reprint No./Package ______, ______ copies, $________

Suppl. 1 Vol. 2	☐ Hardcover ☐ Softcover	________	copies $________
Suppl. 1 Vol. 3	☐ Hardcover ☐ Softcover	________	copies $________
Suppl. 1 Vol. 6	☐ Hardcover ☐ Softcover	________	copies $________
Suppl. 1 Vol. 10	☐ Hardcover only	________	copies $________
Suppl. 1 Vol. 11	☐ Hardcover only	________	copies $________
Suppl. 2 Vol. 11	☐ Hardcover only	________	copies $________

Total Enclosed $ ____________

(Continuation of Cumulative Listing of Reprints)

26. Energy Levels of Neutral Helium (^{4}He I), *W.C. Martin,* Vol. 2, No. 2, pp. 257–266 (1973) $3.00

27. Survey of Photochemical and Rate Data for Twenty-eight Reactions of Interest in Atmospheric Chemistry, *R.F. Hampson, Editor, W. Braun, R.L. Brown, D. Garvin, J.T. Herron, R.E. Huie, M.J. Kurylo, A.H. Laufer, J.D. McKinley, H. Okabe, M.D. Scheer, W. Tsang, and D.H. Stedman,* Vol. 2, No. 2, pp. 267–312 (1973) $4.50

28. Compilation of the Static Dielectric Constant of Inorganic Solids, *K.F. Young and H.P.R. Frederikse,* Vol. 2, No. 2, pp. 313–410 (1973) $6.50

29. Soft X-Ray Emission Spectra of Metallic Solids: Critical Review of Selected Systems, *A.J. McAlister, R.C. Dobbyn, J.R. Cuthill, and M.L. Williams,* Vol. 2, No. 2, pp. 411–426 (1973) $3.00

30. Ideal Gas Thermodynamic Properties of Ethane and Propane, *J. Chao, R.C. Wilhoit, and B.J. Zwolinski,* Vol. 2, No. 2, pp. 427–438 (1973) $3.00

31. An Analysis of Coexistence Curve Data for Several Binary Liquid Mixtures Near Their Critical Points, *A. Stein and G.F. Allen,* Vol. 2, No. 3, pp. 443–466 (1973) $4.00

32. Rate Constants for the Reactions of Atomic Oxygen (O ^{3}P) with Organic Compounds in the Gas Phase, *John T. Herron and Robert E. Huie,* Vol. 2, No. 3, pp. 467–518 (1973) $5.00

33. First Spectra of Neon, Argon, and Xenon 136 in the 1.2-4.0 μm Region, *Curtis J. Humphreys,* Vol. 2, No. 3, pp. 519–530 (1973) $3.00

34. Elastic Properties of Metals and Alloys, I. Iron, Nickel, and Iron-Nickel Alloys, *H.M. Ledbetter and R.P. Reed,* Vol. 2, No. 3, pp. 531–618 (1973) $6.00

35. The Viscosity and Thermal Conductivity Coefficients of Dilute Argon, Krypton, and Xenon, *H.J.M. Hanley,* Vol. 2, No. 3, pp. 619–642 (1973) $4.00

36. Diffusion in Copper and Copper Alloys, Part I. Volume and Surface Self-Diffusion in Copper, *Daniel B. Butrymowicz, John R. Manning, and Michael E. Read,* Vol. 2, No. 3, pp. 643–656 (1973) $3.00

37. The 1973 Least-Squares Adjustment of the Fundamental Constants, *E. Richard Cohen and B.N. Taylor,* Vol. 2, No. 4, pp. 663–734 (1973) $5.50

38. The Viscosity and Thermal Conductivity Coefficients of Dilute Nitrogen and Oxygen, *H.J.M. Hanley and James F. Ely,* Vol. 2, No. 4, pp. 735–756 (1973) $4.00

39. Thermodynamic Properties of Nitrogen Including Liquid and Vapor Phases from 63 K to 2000 K with Pressures to 10,000 Bar, *Richard T. Jacobsen and Richard B. Stewart,* Vol. 2, No. 4, pp. 757–922 (1973) $8.50

40. Thermodynamic Properties of Helium 4 from 2 to 1500 K at Pressures to 10^8 Pa, *Robert T. McCarty,* Vol. 2, No. 4, pp. 923–1042 (1973) $7.00

Reprints from Volume 3

41. Molten Salts: Volume 4, Part 1, Fluorides and Mixtures, Electrical Conductance, Density, Viscosity, and Surface Tension Data, *G.J. Janz, G.L. Gardner, Ursula Krebs, and R.P.T. Tomkins,* Vol. 3, No. 1, pp. 1–115 (1974) $7.00

42. Ideal Gas Thermodynamic Properties of Eight Chloro- and Fluoromethanes, *A.S. Rodgers, J. Chao, R. C. Wilhoit, and B.J. Zwolinski,* Vol. 3, No. 1, pp. 117–140 (1974) $4.00

43. Ideal Gas Thermodynamic Properties of Six Chloroethanes, *J. Chao, A.S. Rodgers, R.C. Wilhoit, and B.J. Zwolinski,* Vol. 3, No. 1, pp. 141–162 (1974) $4.00

44. Critical Analysis of Heat-Capacity Data and Evaluation of Thermodynamic Properties of Ruthenium, Rhodium, Palladium, Iridium, and Platinum from 0 to 300 K. A Survey of the Literature Data on Osmium, *George T. Furukawa, Martin L. Reilly, and John S. Gallagher,* Vol. 3, No. 1, pp. 163–209 (1974) $4.50

45. Microwave Spectra of Molecules of Astrophysical Interest, V. Water Vapor, *Frank C. De Lucia, Paul Helminger, and William H. Kirchhoff,* Vol. 3, No. 1, pp. 211–219 (1974) $3.00

46. Microwave Spectra of Molecules of Astrophysical Interest, VI. Carbonyl Sulfide and Hydrogen Cyanide, *Arthur G. Maki,* Vol. 3, No. 1, pp. 221–244 (1974) $4.00

47. Microwave Spectra of Molecules of Astrophysical Interest, VII. Carbon Monoxide, Carbon Monosulfide, and Silicon Monoxide, *Frank J. Lovas and Paul H. Krupenie,* Vol. 3, No. 1, pp. 245–257 (1974) $3.00

48. Microwave Spectra of Molecules of Astrophysical Interest, VIII. Sulfur Monoxide, *Eberhard Tiemann,* Vol. 3, No. 1, pp. 259–268 (1974) $3.00

49. Tables of Molecular Vibrational Frequencies, Part 8, *T. Shimanouchi,* Vol. 3, No. 1, pp. 269–308 (1974) (superseded by No. 103) $4.50

50. JANAF Thermochemical Tables, 1974 Supplement, *M.W. Chase, J.L. Curnutt, A.T. Hu, H. Prophet, A.N. Syverud, and L.C. Walker,* Vol. 3, No. 2, pp. 311–480 (1974) $8.50

51. High Temperature Properties and Decomposition of Inorganic Salts, Part 4. Oxy-Salts of the Halogens, *Kurt H. Stern,* Vol. 3, No. 2, pp. 481–526 (1974) $4.50

52. Diffusion in Copper and Copper Alloys, Part II. Copper-Silver and Copper-Gold Systems, *Daniel B. Butrymowicz, John R. Manning, and Michael E. Read,* Vol. 3, No. 2, pp. 527–602 (1974) $5.50

53. Microwave Spectral Tables I. Diatomic Molecules, *Frank J. Lovas and Eberhard Tiemann,* Vol. 3, No. 3, pp. 609–770 (1974) $8.50

54. Ground Levels and Ionization Potentials for Lanthanide and Actinide Atoms and Ions, *W.C. Martin, Lucy Hagan, Joseph Reader, and Jack Sugar,* Vol. 3, No. 3, pp. 771–780 (1974) $3.00

55. Behavior of the Elements at High Pressures, *John Francis Cannon,* Vol. 3, No. 3, pp. 781–824 (1974) $4.50

56. Reference Wavelengths from Atomic Spectra in the Range 15 Å to 25000 Å, *Victor Kaufman and Bengt Edlén,* Vol. 3, No. 4, pp. 825–895 (1974) $5.50

57. Elastic Properties of Metals and Alloys. II. Copper, *H.M. Ledbetter and E.R. Naimon,* Vol. 3, No. 4, pp.897–935 (1974) $4.50

58. A Critical Review of H-Atom Transfer in the Liquid Phase: Chlorine Atom, Alkyl, Trichloromethyl, Alkoxy, and Alkylperoxy Radicals, *D.G. Hendry, T. Mill, L. Piszkiewicz, J.A. Howard, and H.K. Eigenmann,* Vol. 3, No. 4, pp. 937–978 (1974) $4.50

59. The Viscosity and Thermal Conductivity Coefficients for Dense Gaseous and Liquid Argon, Krypton, Xenon, Nitrogen, and Oxygen, *H.J.M. Hanley, R.D. McCarty, and W.M. Haynes,* Vol. 3, No. 4, pp. 979–1017 (1974) $4.50

Reprints from Volume 4

60. JANAF Thermochemical Tables, 1975 Supplement, *M.W. Chase, J.L. Curnutt, H. Prophet, R.A. McDonald, and A.N. Syverud,* Vol. 4, No. 1, pp. 1–175 (1975) $8.50

61. Diffusion in Copper and Copper Alloys, Part III. Diffusion in Systems Involving Elements of the Groups IA, IIA, IIIB, IVB, VB, VIB, and VIIB, *Daniel B. Butrymowicz, John R. Manning, and Michael E. Read,* Vol. 4, No. 1, pp. 177–249 (1975) $6.00

62. Ideal Gas Thermodynamic Properties of Ethylene and Propylene, *Jing Chao and Bruno J. Zwolinski,* Vol. 4, No. 1, pp. 251–261 (1975) $3.00

63. Atomic Transition Probabilities for Scandium and Titanium (A Critical Data Compilation of Allowed Lines), *W.L. Wiese and J.R. Fuhr,* Vol. 4, No. 2, pp. 263–352 (1975) $6.00

(Continuation of Cumulative Listing of Reprints)

64. Energy Levels of Iron, Fe I through Fe XXVI, *Joseph Reader and Jack Sugar,* Vol. 4, No. 2, pp. 353–440 (1975) $6.00

65. Ideal Gas Thermodynamic Properties of Six Fluoroethanes, *S.S. Chen, A.S. Rodgers, J. Chao, R.C. Wilhoit, and B.J. Zwolinski,* Vol. 4, No. 2, pp. 441–456 (1975) $3.00

66. Ideal Gas Thermodynamic Properties of the Eight Bromo- and Iodomethanes, *S.A. Kudchadker and A.P. Kudchadker,* Vol. 4, No. 2, pp. 457–470 (1975) $3.00

67. Atomic Form Factors, Incoherent Scattering Functions, and Photon Scattering Cross Sections, *J.H. Hubbell, Wm.J. Veigele, E.A. Briggs, R.T. Brown, D.T. Cromer, and R.J. Howerton,* Vol. 4, No. 3, pp. 471–538 (1975) $5.50

68. Binding Energies in Atomic Negative Ions, *H. Hotop and W.C. Lineberger,* Vol. 4, No. 3, pp. 539–576 (1975) $4.50

69. A Survey of Electron Swarm Data, *J. Dutton,* Vol. 4, No. 3, pp. 577–856 (1975) $12.00

70. Ideal Gas Thermodynamic Properties and Isomerization of *n*-Butane and Isobutane, *S.S. Chen, R.C. Wilhoit, and B.J. Zwolinski,* Vol. 4, No. 4, pp. 859–869 (1975) $3.00

71. Molten Salts: Volume 4, Part 2, Chlorides and Mixtures, Electrical Conductance, Density, Viscosity, and Surface Tension Data, *G.J. Janz, R.P.T. Tomkins, C.B. Allen, J.R. Downey, Jr., G.L. Gardner, U. Krebs, and S.K. Singer,* Vol. 4, No. 4, pp. 871–1178 (1975) $13.00

72. Property Index to Volumes 1–4 (1972–1975), Vol. 4, No. 4, pp. 1179–1192 (1975) $3.00

Reprints from Volume 5

73. Scaled Equation of State Parameters for Gases in the Critical Region, *J.M.H. Levelt Sengers, W.L. Greer, and J.V. Sengers,* Vol. 5, No. 1, pp. 1–51 (1976) $5.00

74. Microwave Spectra of Molecules of Astrophysical Interest, IX. Acetaldehyde, *A. Bauder, F.J. Lovas, and D.R. Johnson,* Vol. 5, No. 1, pp. 53–77 (1976) $4.00

75. Microwave Spectra of Molecules of Astrophysical Interest, X. Isocyanic Acid, *G. Winnewisser, W.H. Hocking, and M.C.L. Gerry,* Vol. 5, No. 1, pp. 79–101 (1976) $4.00

76. Diffusion in Copper and Copper Alloys, Part IV. Diffusion in Systems Involving Elements of Group VIII, *Daniel B. Butrymowicz, John R. Manning, and Michael E. Read,* Vol. 5, No. 1, pp. 103–200 (1976) $6.50

77. A Critical Review of the Stark Widths and Shifts of Spectral Lines from Non-Hydrogenic Atoms, *N. Konjevic and J.R. Roberts,* Vol. 5, No. 2, pp. 209–257 (1976) $5.00

78. Experimental Stark Widths and Shifts for Non-Hydrogenic Spectral Lines of Ionized Atoms (A Critical Review and Tabulation of Selected Data), *N. Konjevic and W.L. Wiese,* Vol. 5, No. 2, pp. 259–308 (1976) $5.00

79. Atlas of the Absorption Spectrum of Nitric Oxide (NO) between 1420 and 1250 Å, *E. Miescher and F. Alberti,* Vol. 5, No. 2, pp. 309–317 (1976) $3.00

80. Ideal Gas Thermodynamic Properties of Propanone and 2-Butanone, *Jing Chao and Bruno J. Zwolinski,* Vol. 5, No. 2, pp. 319–328 (1976) $3.00

81. Refractive Index of Alkali Halides and Its Wavelength and Temperature Derivatives, *H.H. Li,* Vol. 5, No. 2, pp. 329–528 (1976) $9.50

82. Tables of Critically Evaluated Oscillator Strengths for the Lithium Isoelectronic Sequence, *G.A. Martin and W.L. Wiese,* Vol. 5, No. 3, pp. 537–570 (1976) $4.50

83. Ideal Gas Thermodynamic Properties of Six Chlorofluoromethanes, *S.S. Chen, R. C. Wilhoit, and B.J. Zwolinski,* Vol. 5, No. 3, pp. 571–580 (1976) $3.00

84. Survey of Superconductive Materials and Critical Evaluation of Selected Properties, *B.W. Roberts,* Vol. 5, No. 3, pp. 581–821 (1976) $12.50

85. Nuclear Spins and Moments, *Gladys H. Fuller,* Vol. 5, No. 4, pp. 835–1092 (1976) $11.50

86. Nuclear Moments and Moment Ratios as Determined by Mössbauer Spectroscopy, *J.G. Stevens and B.D. Dunlap,* Vol. 5, No. 4, pp. 1093–1121 (1976) $4.00

87. Rate Coefficients for Ion-Molecule Reactions, I. Ions Containing C and H, *L. Wayne Sieck and Sharon G. Lias,* Vol. 5, No. 4, pp. 1123–1146 (1976) $4.00

88. Microwave Spectra of Molecules of Astrophysical Interest, XI. Silicon Sulfide, *Eberhard Tiemann,* Vol. 5, No. 4, pp. 1147–1156 (1976) $3.00

89. Property Index and Author Index to Volumes 1–5 (1972–1976), Vol. 5, No. 4, pp. 1161–1183 $4.00

Reprints from Volume 6

90. Diffusion in Copper and Copper Alloys, Part V. Diffusion in Systems Involving Elements of Group VA, *Daniel B. Butrymowicz, John R. Manning, and Michael E. Read,* Vol. 6, No. 1, pp. 1–50 (1977) $5.00

91. The Calculated Thermodynamic Properties of Superfluid Helium-4, *James S. Brooks and Russell J. Donnelly,* Vol. 6, No. 1, pp. 51–104 (1977) $5.00

92. Thermodynamic Properties of Normal and Deuterated Methanols, *S.S. Chen, R.C. Wilhoit, and B.J. Zwolinski,* Vol. 6, No. 1, pp. 105–112 (1977) $3.00

93. The Spectrum of Molecular Nitrogen, *Alf Lofthus and Paul H. Krupenie,* Vol. 6, No. 1, pp. 113–307 (1977) $9.50

94. Energy Levels of Chromium, Cr I through Cr XXIV, *Jack Sugar and Charles Corliss,* Vol. 6, No. 2, pp. 317–383 (1977) $5.50

95. The Activity and Osmotic Coefficients of Aqueous Calcium Chloride at 298.15 K, *Bert R. Staples and Ralph L. Nuttall,* Vol. 6, No. 2, pp. 385–407 (1977) $4.00

96. Molten Salts: Volume 4, Part 3, Bromides and Mixtures; Iodides and Mixtures–Electrical Conductance, Density, Viscosity, and Surface Tension Data, *G.J. Janz, R.P.T. Tomkins, C.B. Allen, J.R. Downey, Jr., and S.K. Singer,* Vol. 6, No. 2, pp. 409–596 (1977) $9.00

97. The Viscosity and Thermal Conductivity Coefficients for Dense Gaseous and Liquid Methane, *H.J.M. Hanley, W.M. Haynes, and R.D. McCarty,* Vol. 6, No. 2, pp. 597–609 (1977) $3.00

98. Phase Diagrams and Thermodynamic Properties of Ternary Copper-Silver Systems, *Y. Austin Chang, Daniel Goldberg, and Joachim P. Neumann,* Vol. 6, No. 3, pp. 621–673 (1977) $5.00

99. Crystal Data Space-Group Tables, *Alan D. Mighell, Helen M. Ondik, and Bettijoyce Breen Molino,* Vol. 6, No. 3, pp. 675–829 (1977) $8.00

100. Energy Levels of One-Electron Atoms, *Glen W. Erickson,* Vol. 6, No. 3, pp. 831–869 (1977) $4.50

101. Rate Constants for Reactions of ClO_x of Atmospheric Interest, *R.T. Watson,* Vol. 6, No. 3, pp. 871–917 (1977) $4.50

102. NMR Spectral Data: A Compilation of Aromatic Proton Chemical Shifts in Mono- and Di-Substituted Benzenes, *B.L. Shapiro and L.E. Mohrmann,* Vol. 6, No. 3, pp. 919–991 (1977) $5.50

103. Tables of Molecular Vibrational Frequencies. Consolidated Volume II. *T. Shimanouchi,* Vol. 6, No. 3, pp. 993–1102 (1977) (supersedes Nos. 5, 21, 25, 49) $6.50

104. Effects of Isotopic Composition, Temperature, Pressure, and Dissolved Gases on the Density of Liquid Water, *George S. Kell,* Vol. 6, No. 4, pp. 1109–1131 (1977) $4.00

105. Viscosity of Water Substance–New International Formulation and Its Background, *A. Nagashima,* Vol. 6, No. 4, pp. 1133–1166 (1977) $4.50

106.A Correlation of the Existing Viscosity and Thermal Conductivity Data of Gaseous and Liquid Ethane, *H.J.M. Hanley, K.E. Gubbins, and S. Murad,* Vol. 6, No. 4, pp. 1167–1180 (1977) $3.00

107.Elastic Properties of Zinc: A Compilation and a Review, *H.M. Ledbetter,* Vol. 6, No. 4, pp. 1181–1203 (1977) $4.00

108.Behavior of the AB-Type Compounds at High Pressures and High Temperatures, *Leo Merrill,* Vol. 6, No. 4, pp. 1205–1252 (1977) $4.50

109.Energy Levels of Manganese, Mn I through Mn XXV, *Charles Corliss and Jack Sugar,* Vol. 6, No. 4, pp. 1253–1329 (1977) $5.50

Reprints from Volume 7

110.Tables of Atomic Spectral Lines for the 10 000 Å to 40 000 Å Region, *Michael Outred,* Vol. 7, No. 1, pp. 1–262 (1978) $11.50

111.Evaluated Activity and Osmotic Coefficients for Aqueous Solutions: The Alkaline Earth Metal Halides, *R.N. Goldberg and R.L. Nuttall,* Vol. 7, No. 1, pp. 263–310 (1978) $4.50

112.Microwave Spectra of Molecules of Astrophysical Interest XII. Hydroxyl Radical, *Robert A. Beaudet and Robert L. Poynter,* Vol. 7, No. 1, pp. 311–362 (1978) $5.00

113.Ideal Gas Thermodynamic Properties of Methanoic and Ethanoic Acids, *Jing Chao and Bruno J. Zwolinski,* Vol. 7, No. 1, pp. 363–377 (1978) $3.00

114.Critical Review of Hydrolysis of Organic Compounds in Water Under Environmental Conditions, *W. Mabey and T. Mill,* Vol. 7, No. 2, pp. 383–415 (1978) $4.50

115.Ideal Gas Thermodynamic Properties of Phenol and Creosols, *S.A. Kudchadker, A.P. Kudchadker, R.C. Wilhoit, and B.J. Zwolinski,* Vol. 7, No. 2, pp. 417–423 (1978) $3.00

116.Densities of Liquid $CH_{4-a}X_a$ (X = Br,I) and $CH_{4-(a+b+c+d)}F_aCl_bBr_cI_d$ Halomethanes, *A.P. Kudchadker, S.A.Kudchadker, P.R. Patnaik, and P.P. Mishra,* Vol. 7, No. 2, pp. 425–439 (1978) $3.00

117.Microwave Spectra of Molecules of Astrophysical Interest XIII. Cyanoacetylene, *W.J. Lafferty and F.J. Lovas,* Vol. 7, No. 2, pp. 441–493 (1978) $5.00

118.Atomic Transition Probabilities for Vanadium, Chromium, and Manganese (A Critical Data Compilation of Allowed Lines), *S.M. Younger, J.R. Fuhr, G.A. Martin, and W.L. Wiese,* Vol. 7, No. 2, pp. 495–629 (1978) $7.50

119.Thermodynamic Properties of Ammonia, *Lester Haar and John S. Gallagher,* Vol. 7, No. 3, pp. 635–792 (1978) $8.00

120.JANAF Thermochemical Tables, 1978 Supplement, *M.W. Chase, Jr., J.L. Curnutt, R.A. McDonald, and A.N. Syverud,* Vol. 7, No. 3, pp. 793–940 (1978) $8.00

121.Viscosity of Liquid Water in the Range −8 °C to 150 °C, *Joseph Kestin, Mordechai Sokolov, and William A. Wakeham,* Vol. 7, No. 3, pp. 941–948 (1978) $3.00

122.The Molar Volume (Density) of Solid Oxygen in Equilibrium with Vapor, *H.M. Roder,* Vol. 7, No. 3, pp. 949–957 (1978) $3.00

123.Thermal Conductivity of Ten Selected Binary Alloy Systems, *C.Y. Ho, M.W. Ackerman, K.Y. Wu, S.G. Oh, and T.N. Havill,* Vol. 7, No. 3, pp. 959–1177 (1978) $10.00

124.Semi-Empirical Extrapolation and Estimation of Rate Constants for Abstraction of H from Methane by H, O, HO, and O_2, *Robert Shaw,* Vol. 7, No. 3, pp. 1179–1190 (1978) $3.00

125.Energy Levels of Vanadium, V I through V XXIII, *Jack Sugar and Charles Corliss,* Vol. 7, No. 3, pp. 1191–1262 (1978) $5.50

126.Recommended Atomic Electron Binding Energies, 1*s* to $6p_{3/2}$, for the Heavy Elements, $Z = 84$ to 103, *F.T. Porter and M.S. Freedman,* Vol. 7, No. 4, pp. 1267–1284 (1978) $4.00

127.Ideal Gas Thermodynamic Properties of $CH_{4-(a+b+c+d)}F_aCl_bBr_cI_d$ Halomethanes, *Shanti A. Kudchadker and Arvind P. Kudchadker,* Vol. 7, No. 4, pp. 1285–1307 (1978) $4.00

128.Critical Review of Vibrational Data and Force Field Constants for Polyethylene, *John Barnes and Bruno Fanconi,* Vol. 7, No. 4, pp. 1309–1321 (1978) $3.00

129.Tables of Molecular Vibrational Frequencies, Part 9, *Takehiko Shimanouchi, Hiroatsu Matsuura, Yoshiki Ogawa, and Issei Harada,* Vol. 7, No. 4, pp. 1323–1443 (1978) $7.00

130.Microwave Spectral Tables. II. Triatomic Molecules, *Frank J. Lovas,* Vol. 7, No. 4, pp. 1445–1750 (1978) $13.00

Reprints from Volume 8

131.Energy Levels of Titanium, Ti I through Ti XXII, *Charles Corliss and Jack Sugar,* Vol. 8, No. 1, pp. 1–62 (1979) $5.00

132.The Spectrum and Energy Levels of the Neutral Atom of Boron (B I), *G.A. Odintzova and A.R. Striganov,* Vol. 8, No. 1, pp. 63–67 (1979) $3.00

133.Relativistic Atomic Form Factors and Photon Coherent Scattering Cross Sections, *J.H. Hubbell and I. Øverbø,* Vol. 8, No. 1, pp. 69–105 (1979) $4.50

134.Microwave Spectra of Molecules of Astrophysical Interest. XIV. Vinyl Cyanide (Acrylonitrile), *M.C.L. Gerry, K. Yamada, and G. Winnewisser,* Vol. 8, No. 1, pp. 107–123 (1979) $4.00

135.Molten Salts: Volume 4, Part 4, Mixed Halide Melts. Electrical Conductance, Density, Viscosity, and Surface Tension Data, *G.J. Janz, R.P.T. Tomkins, and C.B. Allen,* Vol. 8, No. 1, pp. 125–302 (1979) $9.00

136.Atomic Radiative and Radiationless Yields for *K* and *L* Shells, *M.O. Krause,* Vol. 8, No. 2, pp. 307–327 (1979) $4.00

137.Natural Widths of Atomic *K* and *L* Levels, $K\alpha$ X-ray Lines and Several *KLL* Auger Lines, *M.O. Krause and J.H. Oliver,* Vol. 8, No. 2, pp. 329–338 (1979) $3.00

138.Electrical Resistivity of Alkali Elements, *T.C. Chi,* Vol. 8, No. 2, pp. 339–438 (1979) $6.50

139.Electrical Resistivity of Alkaline Earth Elements, *T.C. Chi,* Vol. 8, No. 2, pp. 439–497 (1979) $5.00

140.Vapor Pressures and Boiling Points of Selected Halomethanes, *A.P. Kudchadker, S.A. Kudchadker, R.P. Shukla, and P.R. Patnaik,* Vol. 8, No. 2, pp. 499–517 (1979) $4.00

141.Ideal Gas Thermodynamic Properties of Selected Bromoethanes and Iodoethane, *S.A. Kudchadker and A.P. Kudchadker,* Vol. 8, No. 2, pp. 519–526 (1979) $3.00

142.Thermodynamic Properties of Normal and Deuterated Naphthalenes, *S.S. Chen, S.A. Kudchadker, and R.C. Wilhoit,* Vol. 8, No. 2, pp. 527–535 (1979) $3.00

143.Microwave Spectra of Molecules of Astrophysical Interest. XV. Propyne, *A. Bauer, D. Boucher, J. Burie, J. Demaison, and A. Dubrulle,* Vol. 8, No. 2, pp. 537–558 (1979) $4.00

144.A Correlation of the Viscosity and Thermal Conductivity Data of Gaseous and Liquid Propane, *P.M. Holland, H.J.M. Hanley, K.E. Gubbins, and J.M. Haile,* Vol. 8, No. 2, pp. 559–575 (1979) $4.00

145.Microwave Spectra of Molecules of Astrophysical Interest. XVI. Methyl Formate, *A. Bauder,* Vol. 8, No. 3, pp. 583–618 (1979) $4.50

(Continuation of Cumulative Listing of Reprints)

146. Molecular Structures of Gas-Phase Polyatomic Molecules Determined by Spectroscopic Methods, *Marlin D. Harmony, Victor W. Laurie, Robert L. Kuczkowski, R.H. Schwendeman, D.A. Ramsay, Frank J. Lovas, Walter J. Lafferty, and Arthur G. Maki,* Vol. 8, No. 3, pp. 619–721 (1979) $6.50

147. Critically Evaluated Rate Constants for Gaseous Reactions of Several Electronically Excited Species, *Keith Schofield,* Vol. 8, No. 3, pp. 723–798 (1979) $5.50

148. A Review, Evaluation, and Correlation of the Phase Equilibria, Heat of Mixing, and Change in Volume on Mixing for Liquid Mixtures of Methane + Ethane, *M.J. Hiza, R.C. Miller, and A.J. Kidnay,* Vol. 8, No. 3, pp. 799–816 (1979) $4.00

149. Energy Levels of Aluminum, Al I through Al XIII, *W.C. Martin and Romuald Zalubas,* Vol. 8, No. 3, pp. 817–864 (1979) $4.50

150. Energy Levels of Calcium, Ca I through Ca XX, *Jack Sugar and Charles Corliss,* Vol. 8, No. 3, pp. 865–916 (1979) $5.00

151. Evaluated Activity and Osmotic Coefficients for Aqueous Solutions: Iron Chloride and the Bi-univalent Compounds of Nickel and Cobalt, *R.N. Goldberg, R.L. Nutall, and B.R. Staples,* Vol. 8, No. 4, pp. 923–1003 (1979) $6.00

152. Evaluated Activity and Osmotic Coefficients for Aqueous Solutions: Bi-univalent Compounds of Lead, Copper, Manganese, and Uranium, *R.N. Goldberg,* Vol. 8, No. 4, pp. 1005–1050 (1979) $4.50

153. Microwave Spectra of Molecules of Astrophysical Interest. XVII. Dimethyl Ether, *F.J. Lovas, H. Lutz, and H. Dreizler,* Vol. 8, No. 4, pp. 1051–1107 (1979) $5.00

154. Energy Levels of Potassium, K I through K XIX, *Charles Corliss and Jack Sugar,* Vol. 8, No. 4, pp. 1109–1145 (1979) $4.50

155. Electrical Resistivity of Copper, Gold, Palladium, and Silver, *R.A. Matula,* Vol. 8, No. 4, pp. 1147–1298 (1979) $8.00

Reprints from Volume 9

156. Energy Levels of Magnesium, Mg I through Mg XII, *W.C. Martin and Romuald Zalubas,* Vol 9, No. 1, pp. 1–58(1980) $6.00

157. Microwave Spectra of Molecules of Astrophysical Interest. XVIII. Formic Acid, *Edmond Willemot, Didier Dangoisse, Nicole Monnanteuil, and Jean Bellet,* Vol. 9, No. 1, pp. 59–160 (1980) $7.50

158. Refractive Index of Alkaline Earth Halides and Its Wavelength and Temperature Derivatives, *H.H. Li,* Vol. 9, No. 1, pp. 161–289 (1980). $8.50

159. Evaluated Kinetic and Photochemical Data for Atmospheric Chemistry, *D.L. Baulch, R.A. Cox, R.F. Hampson, Jr., J.A. Kerr, J. Troe, and R.L. Watson,* Vol. 9, No. 2, pp. 295–471 (1980) $10.00

160. Energy Levels of Scandium, Sc I through Sc XXII, *Jack Sugar and Charles Corliss,* Vol. 9, No. 2, pp. 473–511 (1980) $5.50

161. A Compilation of Kinetic Parameters for the Thermal Degradation of *n*-Alkane Molecules, *D.L. Allara and Robert Shaw,* Vol. 9, No. 3, pp. 523–559 (1980) $5.50

162. Refractive Index of Silicon and Germanium and Its Wavelength and Temperature Derivatives, *H.H. Li,* Vol. 9, No. 3, pp. 561–658 (1980) $7.50

163. Microwave Spectra of Molecules of Astrophysical Interest XIX. *Methyl Cyanide, D. Boucher, J. Burie, A. Bauer, A. Dubrulle, and J. Demaison,* Vol. 9, No. 3, pp. 659–719 (1980). $6.00

164. A Review, Evaluation, and Correlation of the Phase Equilibria, Heat of Mixing, and Change in Volume on Mixing for Liquid Mixtures of Methane + Propane, *R.C. Miller, A.J. Kidnay, and M.J. Hiza,* Vol. 9, No. 3, pp. 721–734 (1980) $4.00

165. Saturation States of Heavy Water, *P.G. Hill and R.D. Chris MacMillan,* Vol. 9, No. 3, pp. 735–749(1980) $4.00

166. The Solubility of Some Sparingly Soluble Lead Salts: An Evaluation of the Solubility in Water and Aqueous Electrolyte Solution, *H. Lawrence Clever and Francis J. Johnston,* Vol. 9, No. 3, pp. 751–784 (1980) $5.50

167. Molten Salts Data as Reference Standards for Density, Surface Tension, Viscosity, and Electrical Conductance: KNO_3 and NaCl, *George J. Janz,* Vol. 9, No. 4, pp. 791–829 (1980) $5.50

168. Molten Salts: Volume 5, Part 1, Additional Single and Multi-Component Salt Systems. Electrical Conductance, Density, Viscosity, and Surface Tension Data, *G.J. Janz and R.P. Tomkins,* Vol. 9, No. 4, pp. 831–1021 (1980) $10.50

169. Pair, Triplet, and Total Atomic Cross Sections (and Mass Attenuation Coefficients) for 1 MeV–100 GeV Photons in Elements $Z=1$ to 100, *J.H. Hubbell, H.A. Gimm, and I. Øverbø,* Vol. 9, No. 4, pp. 1023–1147 (1980) $8.00

170. Tables of Molecular Vibrational Frequencies, Part 10, *Takehiko Shimanouchi, Hiroatsu Matsuura, Yoshiki Ogawa, and Issei Harada,* Vol. 9, No. 4, pp. 1149–1254 (1980) $7.50

171. An Improved Representative Equation for the Dynamic Viscosity of Water Substance, *J.T.R. Watson, R.S. Basu, and J.V. Sengers,* Vol. 9, No. 4, pp. 1255–1290 (1980) $5.50

172. Static Dielectric Constant of Water and Steam, *M. Uematsu and E. U. Franck,* Vol. 9, No. 4, pp. 1291–1306 (1980) $4.00

173. Compilation and Evaluation of Solubility Data in the Mercury (I) Chloride-Water System, *Y. Marcus,* Vol. 9, No. 4, pp. 1307–1329 (1980) $5.00

Reprints from Volume 10

174. Evaluated Activity and Osmotic Coefficients for Aqueous Solutions: Bi-Univalent Compounds of Zinc, Cadmium, and Ethylene Bis(Trimethylammonium) Chloride and Iodide, *R. N. Goldberg,* Vol. 10, No. 1, pp. 1–55 (1981) $6.00

175. Tables of the Dynamic and Kinematic Viscosity of Aqueous KCl Solutions in the Temperature Range 25–150 °C and the Pressure Range 0.1–35 MPa, *Joseph Kestin, H. Ezzat Khalifa, and Robert J. Correia,* Vol. 10, No. 1, pp. 57–70 (1981) $4.00

176. Tables of the Dynamic and Kinematic Viscosity of Aqueous NaCl Solutions in the Temperature Range 20–150 °C and the Pressure Range 0.1–35 MPa, *Joseph Kestin, H. Ezzat Khalifa, and Robert J. Correia,* Vol. 10, No. 1, pp. 71–87 (1981) $5.00

177. Heat Capacity and Other Thermodynamic Properties of Linear Macromolecules. I. Selenium, *Umesh Gaur, Hua-Cheng Shu, Aspy Mehta, and Bernhard Wunderlich,* Vol. 10, No. 1, pp. 89–117 (1981) $5.00

178. Heat Capacity and Other Thermodynamic Properties of Linear Macromolecules. II. Polyethylene, *Umesh Gaur and Bernhard Wunderlich,* Vol. 10, No. 1, pp. 119–152 (1981) $5.50

179. Energy Levels of Sodium, Na I through Na XI, *W. C. Martin and Romuald Zalubas,* Vol. 10, No. 1, pp. 153–195 (1981) $5.50

180. Energy Levels of Nickel, Ni I through Ni XXVIII, *Charles Corliss and Jack Sugar,* Vol. 10, No. 1, pp. 197–289 (1981) $7.00

(Continuation of Cumulative Listing of Reprints)

181. Ion Product of Water Substance, 0–1000 °C, 1–10,000 bars New International Formulation and Its Background, *William L. Marshall and E. U. Franck,* Vol. 10, No. 2, pp. 295–304 (1981) $4.00

182. Atomic Transition Probabilities for Iron, Cobalt, and Nickel (A Critical Data Compilation of Allowed Lines), *J. R. Fuhr, G. A. Martin, W. L. Wiese, and S. M. Younger,* Vol. 10, No. 2, pp. 305–565 (1981) $12.50

183. Thermodynamic Tabulations for Selected Phases in the System $CaO\text{-}Al_2O_3SiO_2\text{-}H_2O$ at 101.325 kPa (1 atm) between 273.15 and 1800 K, *John L. Haas, Jr., Gilpin R. Robinson, Jr., and Bruce S. Hemingway,* Vol. 10, No. 3, pp. 575–669 (1981) $7.00

184. Evaluated Activity and Osmotic Coefficients for Aqueous Solutions: Thirty-Six Uni-Bivalent Electrolytes, R. N. Goldberg, Vol. 10, No. 3, pp. 671–764 (1981) $7.00

185. Activity and Osmotic Coefficients of Aqueous Alkali Metal Nitrites, *Bert R. Staples,* Vol. 10, No. 3, pp. 765–778 (1981) $4.00

186. Activity and Osmotic Coefficients of Aqueous Sulfuric Acid at 298.15 K, *Bert R. Staples,* Vol. 10, No. 3, pp. 779–798 (1981) $5.00

187. Rate Constants for the Decay and Reactions of the Lowest Electronically Excited Singlet State of Molecular Oxygen in Solution, *Francis Wilkinson and James G. Brummer,* Vol. 10, No. 4, pp. 809–999 (1981) $10.00

188. Heat Capacity and Other Thermodynamic Properties of Linear Macromolecules. III. Polyoxides, *Umesh Gaur and Bernhard Wunderlich,* Vol. 10, No. 4, pp. 1001–1049 (1981) $5.50

189. Heat Capacity and Other Thermodynamic Properties of Linear Macromolecules. IV. Polypropylene, *Umesh Gaur and Bernhard Wunderlich,* Vol. 10, No. 4, pp. 1051–1064 (1981) $4.00

190. Tables of N_2O Absorption Lines for the Calibration of Tunable Infrared Lasers from 522 cm^{-1} to 657 cm^{-1} and from 1115 cm^{-1} to 1340 cm^{-1}, *W. B. Olson, A. G. Maki, and W. J. Lafferty,* Vol. 10, No. 4, pp. 1065–1084 (1981) $5.00

191. Microwave Spectra of Molecules of Astrophysical Interest. XX. Methane, *I. Ozier, M. C. L. Gerry, and A. G. Robiette,* Vol. 10, No. 4, pp. 1085–1095 (1981) $4.00

192. Energy Levels of Cobalt, Co I through Co XXVII,, *Jack Sugar and Charles Corliss,* Vol. 10, No. 4, pp. 1097–1174 (1981) $6.50

193. A Critical Review of Henry's Law Constants for Chemicals of Environmental Interest, *Donald Mackay and Wan Ying Shiu,* Vol. 10, No. 4, pp. 1175–1199 (1981) $5.00

194. Property, Materials, and Author Indexes to the Journal of Physical and Chemical Reference Data, Vol. 1–10, pp. 1205–1225 (1972–1981) $5.00

Reprints from Volume 11

195. A Fundamental Equation of State for Heavy Water, *P. G. Hill, R. D. Chris MacMillan, and V. Lee,* Vol. 11, No. 1, pp. 1–14 (1982) $5.00

196. Volumetric Properties of Aqueous Sodium Chloride Solutions, *P. S. Z. Rogers and Kenneth S. Pitzer,* Vol. 11, No. 1, pp. 15–81 (1982) $9.00

197. Ideal Gas Thermodynamic Properties of CH_3, CD_3, CD_4, C_2D_2, C_2D_4, C_2D_6, C_2H_6, $CH_3N_2CH_3$, and $CD_3N_2CD_3$, *Krishna M. Pamidimukkala, David Rogers, and Gordon B. Skinner,* Vol. 11, No. 1, pp. 83–99 (1982) $6.00

198. Peak Absorption Coefficients of Microwave Absorption Lines of Carbonyl Sulphide, *Z. Kisiel and D. J. Millen,* Vol. 11, No. 1, pp. 99–116 (1982) $6.00

199. Vibrational Contributions to Molecular Dipole Polarizabilities, *David M. Bishop and Lap M. Cheung,* Vol. 11, No. 1, pp. 119–133 (1982) $5.00

200. Energy Levels of Iron, Fe I through Fe XXVI, *Charles Corliss and Jack Sugar,* Vol. 11, No. 1, pp. 135–241 (1982) $11.00

201. Microwave Spectra of Molecules of Astrophysical Interest. XXI. Ethanol(C_2H_5OH) and Propionitrile (C_2H_5CN), *Frank J. Lovas,* Vol. 11, No. 2, pp. 251–312 (1982) $8.00

202. Heat Capacity and Other Thermodynamic Properties of Linear Macromolecules, V. Polystyrene, *Umesh Gaur and Bernhard Wunderlich,* Vol. 11, No. 2, pp. 313–325 (1982) $5.00

203. Evaluated Kinetic and Photochemical Data for Atmospheric Chemistry: Supplement 1, CODATA Task Group on Chemical Kinetics, *D. L. Baulch, R. A. Cox, P. J. Crutzen, R. F. Hampson, Jr., J. A. Kerr (Chairman), J. Troe, and R. T. Watson,* Vol. 11, No. 2, pp. 327–496 (1982) $15.00

204. Molten Salts Data: Diffusion Coefficients in Single and Multi-Component Salt Systems, *G. J. Janz and N. P. Bansal,* Vol. 11, No. 3, pp. 505–693 (1982) $16.00

205. JANAF Thermochemical Tables, 1982 Supplement, *M. W. Chase, Jr., J. L. Curnutt, J. R. Downey, Jr., R. A. McDonald, A. N. Syverud, and E. A. Valenzuela,* Vol. 11, No. 3, pp. 695–940 (1982) $20.00

206. Critical Evaluation of Vapor-Liquid Equilibrium, Heat of Mixing, and Volume Change of Mixing Data. General Procedures, *Buford D. Smith, Ol Muthu, Ashok Dewan, and Matthew Gierlach,* Vol. 11, No. 3, pp. 941–951 (1982) $5.00

207. Rate Coefficients for Vibrational Energy Transfer Involving the Hydrogen Halides, *Stephen R. Leone,* Vol. 11, No. 3, pp. 953–996 (1982) $7.00

208. Behavior of the AB_2-Type Compounds at High Pressures and High Temperatures, *Leo Merrill,* Vol. 11, No. 4, pp. 1005–1064 (1982) $8.00

209. Heat Capacity and Other Thermodynamic Properties of Linear Macromolecules. VI. Acrylic Polymers, *Umesh Gaur, Suk-fai Lau, Brent B. Wunderlich, and Bernhard Wunderlich,* Vol. 11, No. 4, pp. 1065–1089 (1982) $6.00

210. Molecular Form Factors and Photon Coherent Scattering Cross Sections of Water, *L. R. M. Morin,* Vol. 11, No. 4, pp. 1091–1098 (1982) $5.00

211. Evaluation of Binary *PTxy* Vapor–Liquid Equilibrium Data for C_6 Hydrocarbons. Benzene + Cyclohexane, *Buford D. Smith, Ol Muthu, Ashok Dewan, and Matthew Gierlach,* Vol. 11, No. 4, pp. 1099–1126 (1982) $6.00

212. Evaluation of Binary Excess Enthalpy Data for C_6 Hydrocarbons. Benzene + Cyclohexane, *Buford D. Smith, Ol Muthu, Ashok Dewan, and Matthew Gierlach,* Vol. 11, No. 4, pp. 1127–1149 (1982) $6.00

213. Evaluation of Binary Excess Volume Data for C_6 Hydrocarbons. Benzene + Cyclohexane, *Buford D. Smith, Ol Muthu, Ashok Dewan, and Matthew Gierlach,* Vol. 11, No. 4, pp. 1151–1169 (1982) $6.00

Reprints from Volume 12

214. Thermodynamic Properties of Steam in the Critical Region, *J. M. H. Levelt Sengers, B. Kamgar-Parsi, F. W. Balfour, and J. V. Sengers,* Vol. 12, No. 1, pp. 1–28 (1983) $6.00

215. Heat Capacity and Other Thermodynamic Properties of Linear Macromolecules. VII. Other Carbon Backbone Polymers, *Umesh Gaur, Brent B. Wunderlich, and Bernhard Wunderlich,* Vol. 12, No. 1, pp. 29–63 (1983) $7.00

(Continuation of Cumulative Listing of Reprints)

216.Heat Capacity and Other Thermodynamic Properties of Linear Macromolecules. VIII. Polyesters and Polyamides, *Umesh Gaur, Suk-fai Lau, Brent B. Wunderlich, and Bernhard Wunderlich,* Vol. 12, No. 1, pp. 65–89 (1983) $6.00

217.Heat Capacity and Other Thermodynamic Properties of Linear Macromolecules. IX. Final Group of Aromatic and Inorganic Polymers, *Umesh Gaur, Suk-fai Lau, and Bernhard Wunderlich,* Vol. 12, No. 1, pp. 91–108 (1983) $6.00

218.An Annotated Compilation and Appraisal of Electron Swarm Data in Electronegative Gases, *J. W. Gallagher, E. C. Beaty, J. Dutton, and L. C. Pitchford,* Vol. 12, No. 1, pp. 109–152 (1983) $7.00

219.The Solubility of Oxygen and Ozone in Liquids, *Rubin Battino, Timothy R. Rettich, and Toshihiro Tominaga,* Vol. 12, No. 2, pp. 163–178 (1983) $5.00

220.Recommended Values for the Thermal Expansivity of Silicon from 0 to 1000 K, *C. A. Swenson,* Vol. 12, No. 2, pp. 179–182 (1983) $5.00

221.Electrical Resistivity of Ten Selected Binary Alloy Systems, *C. Y. Ho, M. W. Ackerman, K. Y. Wu, T. N. Havill, R. H. Bogaard, R. A. Matula, S. G. Oh, and H. M. James,* Vol. 12, No. 2, pp. 183–322 (1983) $13.00

222.Energy Levels of Silicon, Si I through Si XIV, *W. C. Martin and Romuald Zalubas,* Vol. 12, No. 2, pp. 323–380 (1983) $8.00

223.Evaluation of Binary *PTxy* Vapor–Liquid Equilibrium Data for C_6 Hydrocarbons. Benzene + Hexane, *Buford D. Smith, Ol Muthu, and Ashok Dewan,* Vol. 12, No. 2, pp. 381–387 (1983) $5.00

224.Evaluation of Binary Excess Enthalpy Data for C_6 Hydrocarbons. Benzene + Hexane, *Buford D. Smith, Ol Muthu, and Ashok Dewan,* Vol. 12, No. 2, pp, 389–393 (1983) $5.00

225.Evaluation of Binary Excess Volume Data for C_6 Hydrocarbons. Benzene + Hexane, *Buford D. Smith, Ol Muthu, and Ashok Dewan,* Vol. 12, No. 2, pp. 395–401 (1983) $5.00

226.Atlas of the High-Temperature Water Vapor Spectrum in the 3000 to 4000 cm^{-1} Region, *A. S. Pine, M. J. Coulombe, C. Camy-Peyret, and J-M. Flaud,* Vol. 12, No. 3, pp. 413–465 (1983) $8.00

227.Small-Angle Rayleigh Scattering of Photons at High Energies: Tabulations of Relativistic HFS Modified Atomic Form Factors, *D. Schaupp, M. Schumacher, F. Smend, P. Rullhusen, and J. H. Hubbell,* Vol. 12, No. 3, pp. 467–512 (1983) $7.00

228.Thermodynamic Properties of D_2O in the Critical Region, *B. Kamgar-Parsi, J. M. H. Levelt Sengers, and J. V. Sengers,* Vol. 12, No. 3, pp. 513–529 (1983) $6.00

229.Chemical Kinetic Data Sheets for High-Temperature Chemical Reactions, *N. Cohen and K. R. Westberg,* Vol. 12, No. 3, pp. 531–590 (1983) $8.00

230.Molten Salts: Volume 5, Part 2. Additional Single and Multi-Component Salt Systems. Electrical Conductance, Density, Viscosity and Surface Tension Data, *G. J. Janz and R. P. T. Tomkins,* Vol. 12, No. 3, pp. 591–815 (1983) $19.00

231.International Tables of the Surface Tension of Water, *N. B. Vargaftik, B. N. Volkov, and L. D. Voljak,* Vol. 12, No. 3, pp. 817–820 (1983) $5.00

232.Evaluated Theoretical Cross Section Data for Charge Exchange of Multiply Charged Ions with Atoms. I. Hydrogen Atom-Fully Stripped Ion Systems, *R. K. Janev, B. H. Bransden, and J. W. Gallagher,* Vol. 12, No. 4, pp. 829–872 (1983) $7.00

233.Evaluated Theoretical Cross Section Data for Charge Exchange of Multiply Charged Ions with Atoms. II. Hydrogen Atom-Partially Stripped Ion Systems, *J. W. Gallagher, B. H. Bransden, and R. K. Janev,* Vol. 12, No. 4, pp. 873–890 (1983) $6.00

234.Recommended Data on the Electron Impact Ionization of Light Atoms and Ions, *K. L. Bell, H. B. Gilbody, J. G. Hughes, A. E. Kingston, and F. J. Smith,* Vol. 12, No. 4, pp. 891–916 (1983) $6.00

235.A Correlation of the Viscosity and Thermal Conductivity Data of Gaseous and Liquid Ethylene, *P. M. Holland, B. E. Eaton, and H. J. M. Hanley,* Vol. 12, No. 4, pp. 917–932 (1983) $5.00

236.Transport Properties of Liquid and Gaseous D_2O over a Wide Range of Temperature and Pressure, *N. Matsunaga and A. Nagashima,* Vol. 12, No. 4, pp. 933–966 (1983) $7.00

237.Thermochemical Data for Gaseous Monoxides, *J. B. Pedley and E. M. Marshall,* Vol. 12, No. 4, pp. 967–1031 (1983) $9.00

238.Vapor Pressure of Coal Chemicals, *J. Chao, C. T. Lin, and T. H. Chung,* Vol. 12, No. 4, pp. 1033–1063 (1983) $6.00

Special Reprints Packages

These special reprints packages offer selected articles in specific subject areas from the JOURNAL OF PHYSICAL AND CHEMICAL REFERENCE DATA, and they are offered at a better rate than when purchased individually. You will have available a complete library of literature for your specific requirements at a fraction of the cost of purchasing back issues of the journal.

Look over the reprints packages available—they are listed by subject area. In the Cumulative Listing of Reprints you will find the titles corresponding to the reprint numbers. You are sure to find building your information bank in this manner to be thorough and economical.

Package A1 (4 Parts) MOLECULAR VIBRATIONAL FREQUENCIES. Consisting of Reprint Nos. 103, 129, 170, NSRDS 39
If purchased individually $26.10
Special package price **$22.00**

Package A2 (4 Parts) JANAF THERMOCHEMICAL TABLES. Consisting of Reprint Nos. 50, 60, 120, NSRDS 37
If purchased individually $40.60
Special package price **$32.00**

Package A3 (19 Parts) ATOMIC ENERGY LEVELS. Consisting of Reprint Nos. 26, 54, 64, 68, 94, 100, 109, 125, 126, 131, 132, 149, 150, 154, 156, 160, 179, 180, 192
If purchased individually $94.00
Special package price **$66.00**

Package A4 (6 Parts) ATOMIC SPECTRA. Consisting of Reprint Nos. 33, 56, 77, 78, 110, 132
If purchased individually $33.00
Special package price **$27.00**

Package A5 (5 Parts) ATOMIC TRANSITION PROBABILITIES. Consisting of Reprint Nos. 20, 63, 82, 118, 182
If purchased individually $35.00
Special package price **$28.00**

Package A6 (20 Parts) MOLECULES IN SPACE. Consisting of Reprint Nos. 14, 17, 23, 24, 45, 46, 47, 48, 74, 75, 88, 112, 117, 134, 143, 145, 153, 157, 163, 191
If purchased individually $82.50
Special package price **$58.00**

Package A7 (9 Parts) CHEMICAL DATA RELEVANT TO AIR POLLUTION. Consisting of Reprint Nos. 1, 9, 19, 27, 32, 87, 101, 147, 159
If purchased individually $50.00
Special package price **$40.00**

Package A8 (7 Parts) MOLECULAR SPECTRA. Consisting of Reprint Nos. 4, 8, 53, 79, 93, 130, 146
If purchased individually $51.50
Special package price **$41.00**

Package A9 (9 Parts) THERMODYNAMIC PROPERTIES OF ELECTROLYTE SOLUTIONS. Consisting of Reprint Nos. 15, 95, 111, 151, 152, 174, 184, 185, 186
If purchased individually $46.00
Special package price **$37.00**

Package B1 (12 Parts) IDEAL GAS THERMODYNAMIC PROPERTIES. Consisting of Reprint Nos. 30, 42, 43, 62, 65, 66, 70, 80, 83, 113, 115, 141
If purchased individually $38.00
Special package price **$31.00**

Package B2 (3 Parts) ELASTIC PROPERTIES OF METALS AND ALLOYS. Consisting of Reprint Nos. 34, 57, 107
If purchased individually $14.50
Special package price **$12.00**

Package B3 (3 Parts) RESISTIVITY. Consisting of Reprint Nos. 138, 139, 155
If purchased individually $19.50
Special package price **$16.00**

Package B4 (7 Parts) MOLTEN SALTS. Consisting of Reprint Nos. 10, 41, 71, 96, 135, 167, 168
If purchased individually $62.50
Special package price **$44.00**

Package B5 (3 Parts) REFRACTIVE INDEX. Consisting of Reprint Nos. 81, 158, 162
If purchased individually $25.50
Special package price **$21.00**

Package B6 (11 Parts) VISCOSITY AND THERMAL CONDUCTIVITY OF FLUIDS. Consisting of Reprint Nos. 16, 35, 38, 59, 97, 106, 121, 144, 171, 175, 176
If purchased individually $43.00
Special package price **$35.00**

Supplements to JPCRD

When the topic demands it, and the quality of the data justifies it, the JOURNAL OF PHYSICAL AND CHEMICAL REFERENCE DATA issues a special Supplement. Each Supplement is a monograph—collected tables of highly significant physical or chemical property data in one complete volume.

Listed below are the special Supplements to JPCRD that have been published. Each is a valuable resource for the physical chemist and chemical physicist. Use the form on the last page to order the Supplements important to you.

THE NBS TABLES OF CHEMICAL THERMODYNAMIC PROPERTIES. *Selected Values for Inorganic and C_1 and C_2 Organic Substances in SI Units* by D.D. Wagman, W.H. Evans, V.B. Parker, R.H. Schumm, I. Halow, S.M. Bailey, K.L. Churney, and R.L. Nuttall. (Supplement No. 2 to Volume 11) 1982, 394 pages. *Just Published!*
Hardbound only $40.00

THERMOPHYSICAL PROPERTIES OF FLUIDS. 1. ARGON, ETHYLENE, PARAHYDROGEN, NITROGEN, NITROGEN TRIFLUORIDE, AND OXYGEN by B.A. Younglove. (Supplement No. 1 to Volume 11) 1982, 368 pages. *Just Published!*
Hardbound only $40.00

EVALUATED KINETIC DATA FOR HIGH TEMPERATURE REACTIONS: VOLUME 4, HOMOGENEOUS GAS PHASE REACTIONS OF HALOGEN- AND CYANIDE-CONTAINING SPECIES by D.L. Baulch, J. Duxbury, S.J. Grant and D.C. Montague. (Supplement No. 1 to Volume 10) 1981, 721 pages.
Hardbound only $80.00

ENERGETICS OF GASEOUS IONS by H.M. Rosenstock, K. Draxl, B.W. Steiner, and J.T. Herron. (Supplement No. 1 to Volume 6) 1977, 783 pages.
Hardbound $70.00
Softbound $65.00

THERMAL CONDUCTIVITY OF THE ELEMENTS: A COMPREHENSIVE REVIEW by C.Y. Ho, R.W. Powell, and P.E. Liley. (Supplement No. 1 to Volume 3) 1974, 796 pages.
Hardbound $60.00
Softbound $55.00

PHYSICAL AND THERMODYNAMIC PROPERTIES OF ALIPHATIC ALCOHOLS by R.C. Wilhoit and B.J. Zwolinski. (Supplement No. 1 to Volume 2) 1973, 420 pages.
Hardbound $33.00
Softbound $30.00